IC Design Insights

From Selected Presentations at CICC 2017

EDITORS

Ali Sheikholeslami
University of Toronto, Canada

Jan Van der Spiegel
University of Pennsylvania, USA

Yanjie Wang
Intel Corporation, USA

SSCS

Tutorials in Solid-State Circuits

For a list of other books in this series, visit www.riverpublishers.com

Series Editor

Jan Van der Spiegel

University of Pennsylvania, USA

Published, sold and distributed by:
River Publishers
Niels Jernes Vej 10
9220 Aalborg Ø Denmark

River Publishers
Lange Geer 44
2611 PW Delft
The Netherlands

Tel.: +45369953197 www.riverpublishers.com

Availability: March 2019

ISBN: Print: 978-87-7022-049-1
E-book: 978-87-7022-048-4

Published by River Publishers.

Library of Congress Cataloging-in-Publication Data: March 2019
Editors: Ali Sheikholeslami, University of Toronto, Canada
Jan Van der Spiegel, University of Pennsylvania, USA
Yanjie Wang, Intel Corporation, USA
Title: IC Design Insights
Sub-title: From Selected Presentations at CICC 2017

Table of contents

Introduction

This book contains a selection of tutorial and invited presentations that were given at the IEEE CICC 2017 in Austin, TX. The selection of the talks was made to provide a comprehensive coverage of key topics, including Circuits Techniques for mm-wave front-ends, RF and mm-wave receivers and frequency synthesis, data and DC-DC converters, and techniques for IoT security. The book is part of an educational initiative of the IEEE Solid-State Circuits Society to offer its members state of the art educational material.

We would like to express our thanks to the organizer of the CICC 2017 tutorials sessions and invited talks, Prof. Ramesh Harjani of the University of Minnesota for his help in identifying the topics and speakers. Also, we like to thank the many world-experts for contributing to this book and sharing their expertise and insight with the readers.

The book is organized into five parts. Part I, which include three chapters (Chapters 1-3), deals with millimeter-wave (mm-wave) transmitter circuits including mm-wave front-end circuits. Prof. Hossein Hashemi (Univ. of S. California) explains the design of mm-wave power amplifiers in the first chapter of this book. Chapter 2, presented by Dr. Tolga Dinc (Texas Instruments) and Prof. Harish Krishnaswamy (Columbia Univ.) describes the details of millimeter-wave full-duplex wireless systems. The final chapter of this part is presented by Prof. John Long (Univ. of Waterloo) where he explains the design of on-chip transformers and their applications to RF and mm-wave front-ends.

Part II of this book presents four chapters (Chapters 4-7) on RF and mm-wave receiver circuits and frequency synthesis. Dr. Amr Fahim (Inphi Corporation) describes the requirements, design and optimizations of local Oscillator (LO) distribution networks for mm-wave frequencies as well as the applications ranging from mm-wave phase-array to high speed optical communication networks. Prof. Foster Dai (Auburn University) mainly focuses on the techniques breaking the trade-off between the high power and low phase noise by both circuits and architectural innovations for area efficient sub-sampling PLL synthesizer. Chapter 6 is presented by Profs. Eric Klumperink and Bram Nauta (both with the University of Twente) where they review recent developments in the field of N-path filters and mixers on selectivity, linearity, and blocker tolerance for multi-band software defined radio receivers. Chapter 7 is presented by Prof. Hua Wang (Georgia Tech.) who explains the operation principle, as well as pros and cons, of Doherty architecture in low-GHz mixed-signal and mm-wave 5G multi-band analog PAs.

The next part consists of four chapters (Chapters 8-11) on Data Converters. Dr. Hui Pan (Broadcom, US) gives a unified framework of data converters with circuit examples of recent implementations, and a discussion of the recent trends of simple ZX-based quantizers combined with signal processing and digital error corrections. The next tutorial by Dr. Lukas Kull and Dr. Danny Luu (while the chapter research was done at IBM Research, Zurich, both authors are currently at Cisco Systems, Zurich, Switzerland) describes a full measurement set-up for measuring high-speed ADCs using needle probes. They explain how to reduce nonlinearity and signal-dependent errors. They also show an area- and power-efficient implementation of a full-rate, on-chip capture memory. The third chapter in this part is presented by Dr. Khiem Nguyen (Analog Devices, US) and focuses on practical design aspects of 3-level dynamic element matching techniques for oversampling Sigma-Delta converters. The final chapter in this part, Chapter 11, is by Dr. Aaron Buchwald (Beechwood Analog) who gives a lively description of time-interleaved ADCs and shares his insight into the journey of designing these ADCs.

Part IV of the book focuses on DC-DC converters and voltage regulation in Chapters 12-13. DC-DC power converter designs are explained in Chapter 12 by Prof. Ramesh Harjani (University of Minnesota). Prof. Hanh Phuc Le (University of Colorado) describes in Chapter 13 the circuits design techniques for integrated switched capacitor DC-DC converter in order to improve power density, efficiency, transient responses and I/O voltage range.

The final part of this book, Part V, includes two chapters (Chapters 14-15) on security techniques for IoT. The first chapter by Drs. Sanu Mathew, Sudhir Sathpathy, Vikram Suresh, and Ram. Krishnamurthy (Intel, US) gives an overview of low-voltage circuits for secure IoT applications. The second chapter describes the challenges and trade-offs for ensuring security of IoT applications, as well as current industrial practices.

This book would not have been possible without the help and dedication of Abira Altvater and support of Michael Kelly of the SSCS office. Also, we like to express our gratitude to Mark de Jongh of River Publishing for his help, patience and encouragement in bringing the book to reality.

Ali Sheikholeslami
Jan Van der Spiegel
Yanjie Wang

PART I

MILLIMETER-WAVE TRANSMITTER CIRCUITS

CHAPTER 01

Millimeter-Wave Power Amplifiers and Transmitters

Hossein Hashemi

University of Southern California, USA

SSCS

Millimeter-wave frequencies enable applications such as high-speed wireless communications, such as those envisioned in the upcoming 5G standards, high-resolution radar, such as those used in advanced driver assistance systems (ADAS), and imaging, such as those used in airport 3D scanners. Power amplifiers determine the overall energy consumption and performance of transceivers. Millimeter-wave operation offers unique challenges and opportunities for power amplifiers. This chapter covers a summary of selected design and implementation issues related to mm-wave power amplifiers with case studies. It touches on active and passive devices, different amplifier operation classes, power combining schemes, and transmitter architectures. This chapter can serve as a basic source of information for those interested to study and contribute to mm-wave power amplifiers and transmitters.

1

Outline

- **Introduction**
- Technology
- Linear Amplifiers
- Switching Amplifiers
- Stacked-Transistor Amplifiers
- Power Combining
- Transmitters
- Conclusions

This presentation covers the basic concepts along with design examples related to mm-wave power amplifiers and transmitters.

2

Millimeter Wave Systems

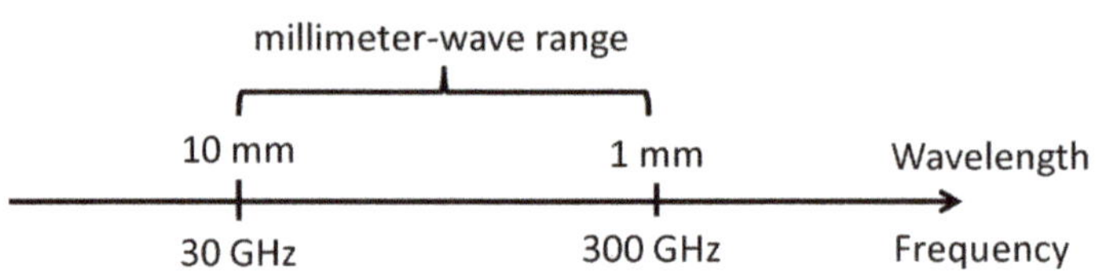

- More available spectrum → higher communication capacity
- Shorter wavelength → higher propagation attenuation through objects → denser frequency reuse
- Shorter wavelength → narrower antenna pattern beam-width → higher imaging resolution + higher communication spatial selectivity

Millimeter-wave corresponds to the range of frequencies where the wavelength of electromagnetic wave is between 1 mm and 10 mm.

Compared to radiofrequencies below 30 GHz, there is more available frequency spectrum (300 GHz – 30 GHz = 270 GHz) that can be used. Therefore, in principle, communication networks with higher capacity can be realized in mm-wave frequencies.

The shorter wavelength at mm-wave frequencies results in higher propagation attenuation for most material. This limits the applicability of mm-waves for conventional broadband systems (radio and television) because these electromagnetic waves get heavily attenuated as they enter buildings. On the other hand, this limited propagation distance in buildings can be leveraged to realize wireless networks with a dense frequency reuse.

The other advantage of shorter wavelength is that, for a given typical antenna size, the antenna pattern will be narrower. The narrow antenna pattern (also known as beam pattern) enhances the resolution of mm-wave imaging systems and offers spatial selectivity for mm-wave wireless communication systems.

3 Generic Transceiver

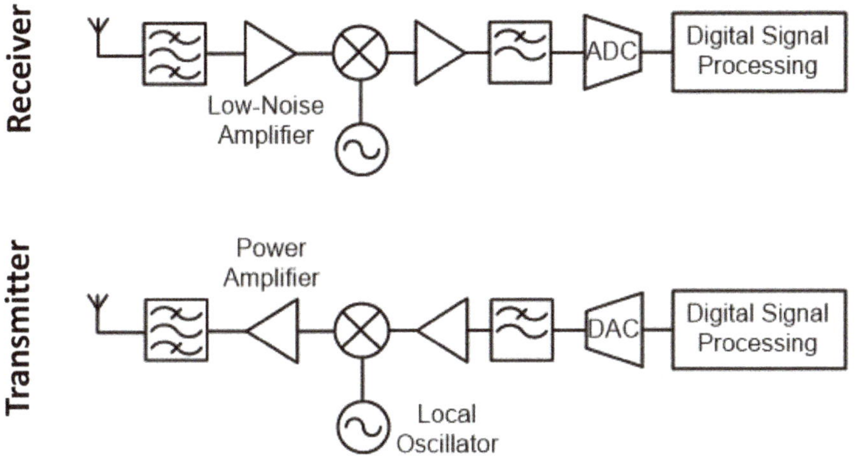

We focus on mm-wave transceivers, and specifically on mm-wave power amplifiers.

4 Example mm-Wave Transmitter Power Limits

	Freq. [GHz]	FCC Power Limits
Licensed Communications	27.5 – 28.35 37 – 40	Base Station: 75 dBm/100 MHz Mobile Station: 43 dBm EIRP Transportable Station: 55 dBm EIRP
Unlicensed Communications	57 – 64	P_{AVG} < 40 dBm(1) P_{PEAK} < 43 dBm(1)
Car Radar	76 – 77	$EIRP_{AVG}$ < 50 dBm $EIRP_{PEAK}$ < 55 dBm
E-Band P2P Communication	71 – 76 81 – 86	P_{PEAK} < 35 dBm $EIRP_{PEAK}$ < 85 dBm(2)

(1) For transmitters located outdoor, P_{AVG} (P_{PEAK}) < 82 (85) dBm minus 2 dB for every dB that the antenna gain is less than 51 dBi.
(2) Minimum antenna gain is 43 dBi. Maximum EIRP is reduced by 2 dB for every dB that the antenna gain is less than 50 dBi.

The transmitted power of communication and imaging systems determines the operation distance. The maximum allowable transmit power levels is typically regulated. In the U.S., the Federal Communication Commission (FCC) has limited the maximum allowable transmit power levels at selected mm-wave frequency bands to the levels shown in this table. It should be noted that the intended application and corresponding regulation for some of the designated mm-wave frequency bands is directional communication schemes.

5 Historical Perspective on Silicon PAs

2000 – 2010

- Standards: 2G, 3G, WiFi
- Frequency: 0.4 – 6 GHz
- Output power: < 33 dBm
- Silicon transistors: f_{max} ~ 70 GHz, BV ~ 3 V
- Passives: Q_L ~ 5, Q_C ~ 1000
- Waveforms:
  - complex modulations with high peak to average power ratio (PAPR)
  - Stringent requirement for out-of-band emissions

2010 – 2020

- Standards: 5G
- Frequency: 30 GHz, 60 GHz
- Output power: 43 dBm EIRP
- Silicon transistors: f_{max} ~ 300 GHz, BV ~ 1 V
- Passives: Q_L ~ 20, Q_C ~ 50
- Waveforms:
  - Complex modulations with high peak to average power ratio (PAPR)
  - Less stringent requirement for out-of-band emissions

Advancements in silicon radiofrequency power amplifiers was dominated by the interest in commercial wireless communication applications for cellular and connectivity standards with a typical range of 10 m to 1000 m. These radiofrequency wireless standards have defined the waveforms to achieve high spectral efficiency (larger data rate for a given bandwidth) resulting in waveforms with high peak to average power ratios. A major challenge in monolithic realization of radiofrequency power amplifiers is the low quality factor of inductors that are needed for impedance transformation at the output of power amplifiers. A second challenge is realization of Watt-level power amplifiers that maintain high power efficiency and spectral purity under large peak-to-average power ratios.

The maximum range envisioned in many commercial mm-wave applications is less compared with those of radiofrequency systems. Furthermore, most mm-wave systems are directional as opposed to omnidirectional or broadcast that is common in radiofrequencies; hence, the power amplifier output can be focused only towards the desired directions. Consequently, the output power requirement for individual mm-wave power amplifiers is typically less when compared to traditional radiofrequency power amplifiers. This is fortunately consistent with the lower possible voltage swings across high-speed low-breakdown-voltage transistors that are needed for mm-wave power amplifier realizations.

6 RF vs. mm-Wave Power Amplifiers

There are differences challenges and opportunities in the design of radiofrequency versus mm-wave power amplifiers.

The first difference is dictated by the device limitations. As the frequency increases and gets close to the transistor's maximum unity gain power frequency (fmax), the transistor gain reduces. This will deteriorate the power gain, and hence the power added efficiency, of power amplifiers. The quality factor of capacitors reduce at higher frequencies; therefore, the capacitive loss of impedance transformation networks and power combiners increases with frequency. Finally, the layout parasitics can have a major effect on the performance of mm-wave power amplifiers.

On the other hand, shorter wavelength of mm-wave frequencies enable compact realization of power amplifiers and antennas leading to efficient spatial power-combining schemes. On-chip transmission line combiners with relatively low loss can also be realized in a compact form factor for mm-wave implementations.

6 RF vs. mm-Wave Power Amplifiers

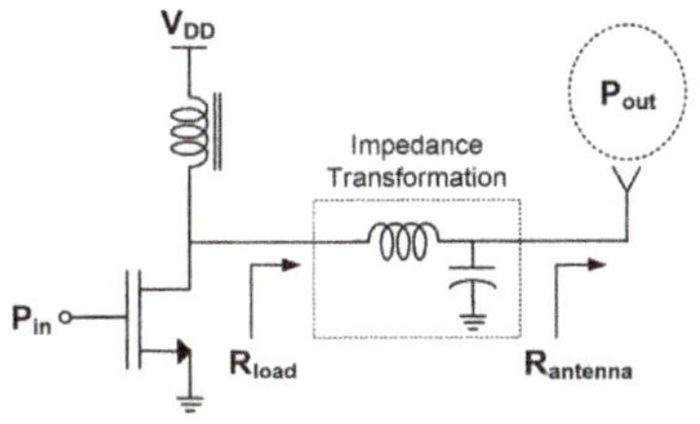

Detrimental effects of transistor loss (limited f_{max}), capacitor loss (limited Q_C), and layout parasitics on the PA performance (gain, efficiency, and output power) increase with frequency.

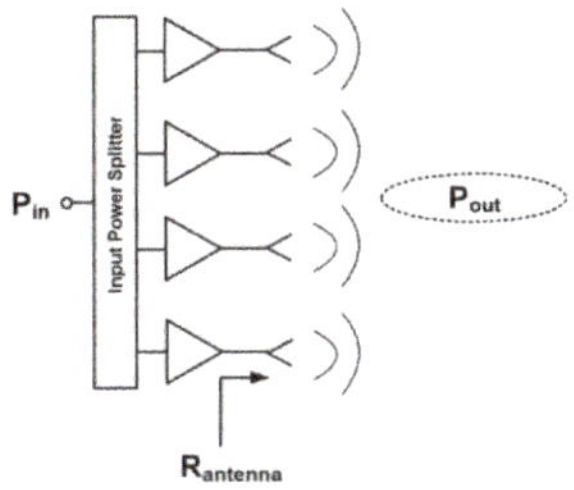

Shorter wavelength of mm-wave frequencies enables using compact on-chip and spatial power combining schemes.

7 Outline

- Introduction
- **Technology**
- Linear Amplifiers
- Switching Amplifiers
- Stacked-Transistor Amplifiers
- Power Combining
- Transmitters
- Conclusions

In this section, we will briefly cover important components such as transistors, capacitors, inductors, and transmission lines along with their characteristics as pertain to designing mm-wave integrated circuits.

8 Speed x Breakdown Voltage in Semiconductors

There is a tradeoff, aka Johnson's FOM, in the charge carrier velocity and electric breakdown fields of semiconductor materials. This sets the ultimate achievable (speed x breakdown voltage) of transistors.

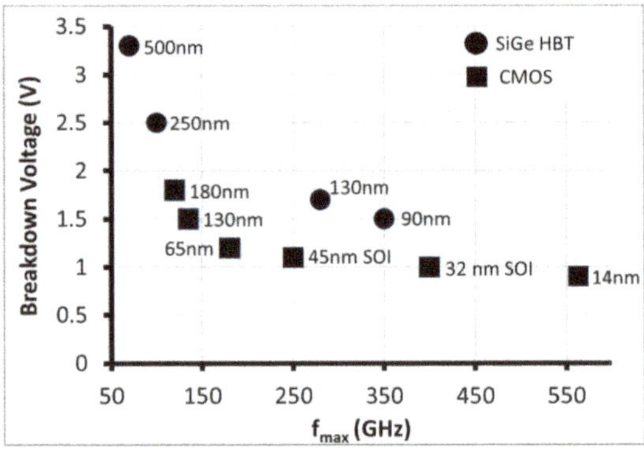

A common approach to increase the transistor speed is to reduce its dimensions (aka scaling). Geometrical scaling leads to an increase electric field (for a fixed voltage) which, to the first order, translates to higher speed. This will, in turn, bring the transistor closer to the breakdown. There is hence a trade-off between the speed and maximum breakdown voltage of any transistor that only depends on the material properties. The maximum unity-power-gain frequency (f_{max}) versus the breakdown voltage of selected silicon transistors are plotted to show this tradeoff.

9 Transistors

Larger transistors, and more transistors in parallel, must be used to support higher output powers. The associated interconnect parasitics often reduce small-signal f_{max}, and large-signal PA performance metrics (power gain, output power, and efficiency).

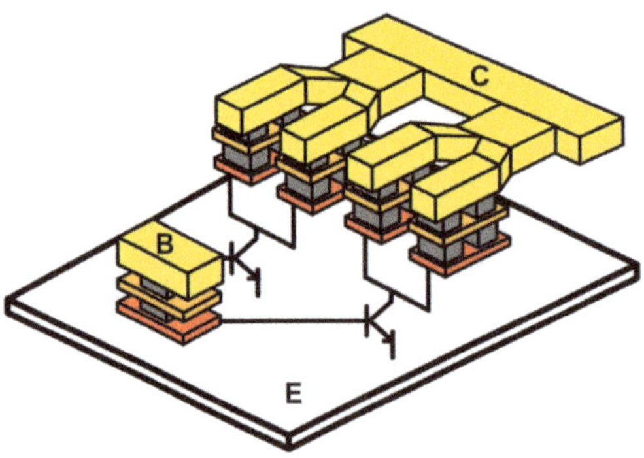

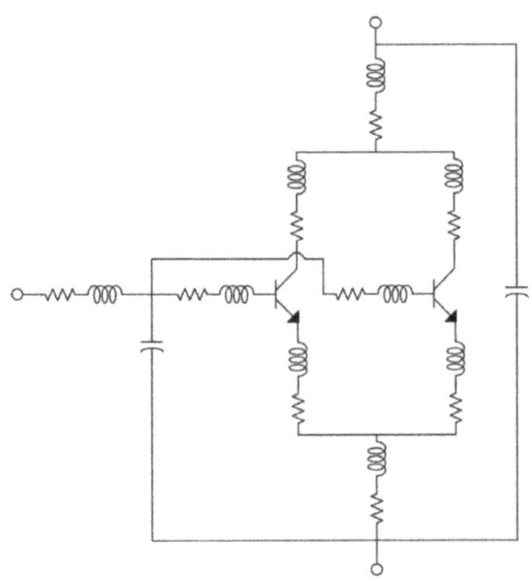

9 Transistors

In a power amplifier, the output power is typically an increasing function of transistor size. The intuitive reasoning is that larger current is needed to deliver larger power, and larger transistors are needed to generate larger currents (for the same voltage). Typically, the large geometries required in power amplifiers are created using multiple parallel transistors. The corresponding interconnects parasitics may significantly degrade the transistor performance at higher frequencies. Specifically, power gain, output power, and efficiency can be reduced.

At mm-wave frequencies, the layout and modelling of such large transistor geometries is quite important to achieve efficient power amplifiers. Unfortunately, common RC extraction tools may be insufficient to capture all the important parasitics; especially, the parasitic inductances, often ignored in commercial RC extractors, can have a significant effect on the power amplifier performance. Electromagnetic simulations may be used to accurately model the layout of large power transistors. It is possible, although not necessary, to build a lumped-component equivalent model of the transistor parasitics to gain intuition about the effect of layout on the circuit performance.

10 Metal-Insulator-Metal (MIM) Capacitors

The quality factor of MIM capacitors drops with frequency due to the resistance and inductance of the MIM metal plates as well the interconnects (including vias).

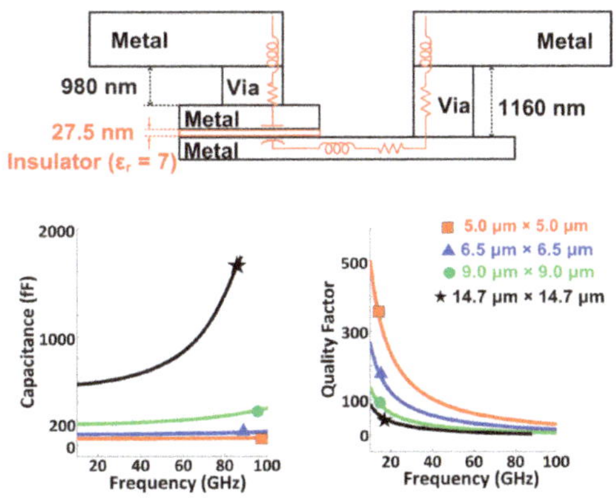

Metal insulator metal (MIM) or metal oxide metal (MOM) capacitors are commonly used at millimeter-wave frequencies. The loss of these capacitors is often dominated by the Ohmic loss of metal plates and associated interconnects (including vias), and modelled with a series resistance R_S. In such cases, the capacitor quality factor drops with frequency as $Q_C=1/(wR_SC)$. This figure shows the electromagnetic simulated values for the capacitance and quality factor of MIM capacitors, with the generic geometry shown on the top, versus frequency. It is seen that the quality factor, for typical capacitor values of 50-200 fF, varies from 100 to less than 10 in the mm-wave frequency range.

The self-resonant frequency (SRF) of these capacitances depends on the inductance of the metal plates and associated interconnects (vias). Electromagnetic simulations are often needed to accurately model capacitors for mm-wave circuits. Equivalent lumped-component circuits may be created to gain intuition about the effect of layout parasitics on the circuit.

11 Spiral Inductors

Objective: Given a desired inductor value, come up with a spiral inductor geometry that leads to the highest quality factor (Q) at the desired frequency range within a constrained footprint.

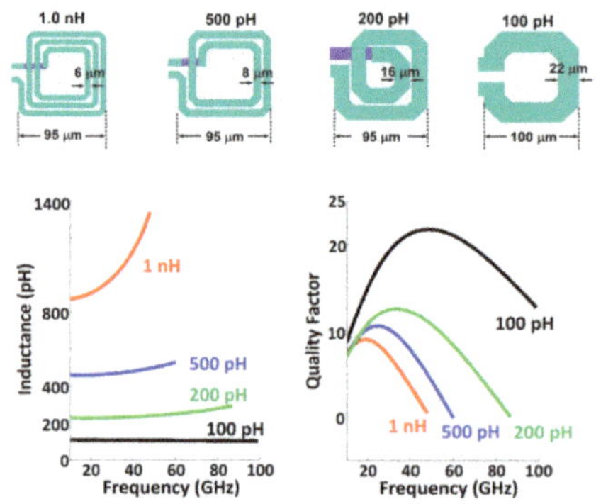

Spiral is an area-efficient geometry to realize on-chip inductors. The exact geometry, such as the total size, metal width, number of turns, and turn spacing, determine the inductance value, quality factor, and self-resonant frequency. At low frequencies, the inductor loss is dominated by the Ohmic loss of metal lines and can be modelled with a series resistance. Therefore, in this regime, the quality factor increases with frequency. At higher frequencies, the parasitic capacitances from the inductor traces to the substrate or between the inductor turns reduce the effective inductance. Therefore, in this regime, as the frequency increases, the inductor quality factor drops (and reaches zero at the self-resonance frequency). In summary, for a given spiral geometry, the quality factor has a peak at a frequency.

Given a constraint on the area, a common objective is to come up with a spiral geometry that gives the desired inductance value with the highest quality factor at the desired frequency. This figure shows representative designs of inductors for operation in the 30-50 GHz frequency range for such spiral geometries given a 100 um x 100 um footprint. In general, the maximum quality factor of on-chip inductors at mm-waves can range from 10 to over 25.

12 Single-Ended Transmission Lines

At mm-wave frequencies, transmission lines may be used for routing the signal, impedance transformation networks, and power dividers/combiners.

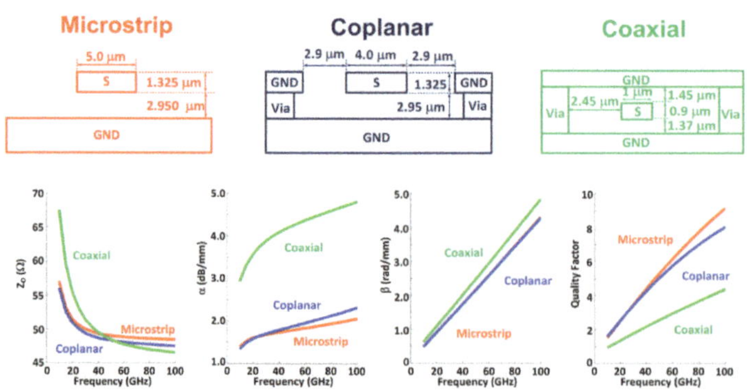

12 Single-Ended Transmission Lines

On-chip transmission lines may be used for impedance transformation networks, power combining, or routing mm-wave signals across the chip. The availability of several metal layers in commercial silicon processes enables creating various types of transmission lines.

The characteristic impedance and propagation constant of these transmission lines depends on the geometry such as the spacing between signal and ground lines, and the width of signal line. With the typical dimensions of on-chip structures, the characteristic impedance of on-chip transmission lines is around 30 Ω – 70 Ω.

The geometries and simulated features of three types of on-chip single-ended transmission lines, all designed to offer a similar characteristic impedance of around 50 Ω, are shown. The coaxial structure has the worst loss and quality factor because of the narrower signal line needed to achieve the same characteristic impedance in presence of more surrounding ground (narrower signal line increases the inductance to compensate for the larger capacitance).

13 Differential Transmission Lines

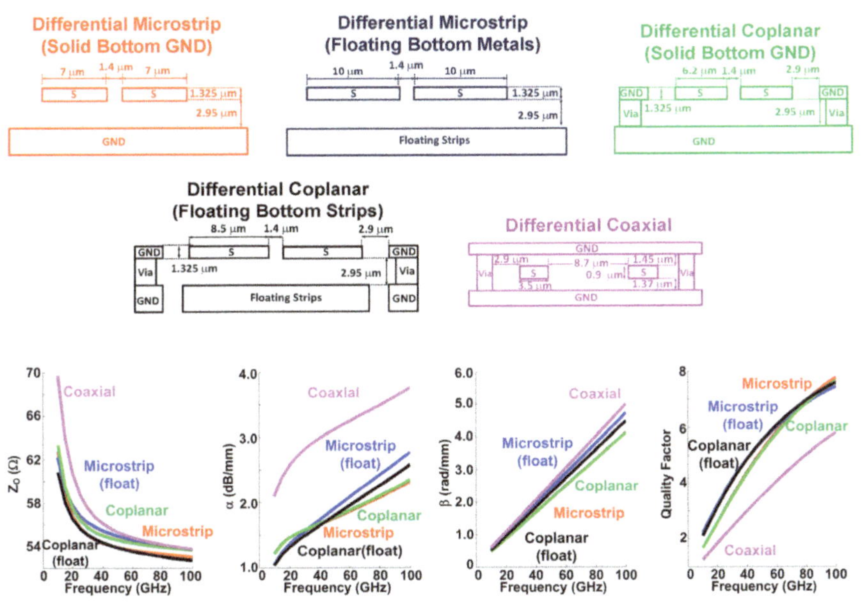

Differential circuits are commonly-used in silicon integrated circuits due to their immunity to common-mode noise and parasitics. Various types of on-chip differential transmission lines can be realized.

The characteristic impedance and propagation constant of these transmission lines depends on the geometry such as the spacing between differential signal and ground lines, and the width of signal line.

The geometries and simulated features of five types of on-chip differential transmission lines, all designed to offer a similar characteristic impedance of around 55 Ω, are shown. The coaxial structure has the worst loss and quality factor because of the narrower signal line needed to achieve the same characteristic impedance in presence of more surrounding ground (narrower signal line increases the inductance to compensate for the larger capacitance).

14

We will now cover the design of monolithic linear power amplifiers at mm-wave frequencies. It should be noted that, while many schematics and case studies show hetero-structure bipolar transistors (HBT) in the amplifier core, most concepts are applicable to amplifiers that utilize field effect transistors (FET) as well (e.g., CMOS).

Outline

- Introduction
- Technology
- **Linear Amplifiers**
- Switching Amplifiers
- Stacked-Transistor Amplifiers
- Power Combining
- Transmitters
- Conclusions

15

Classical Class-A RF Power Amplifiers

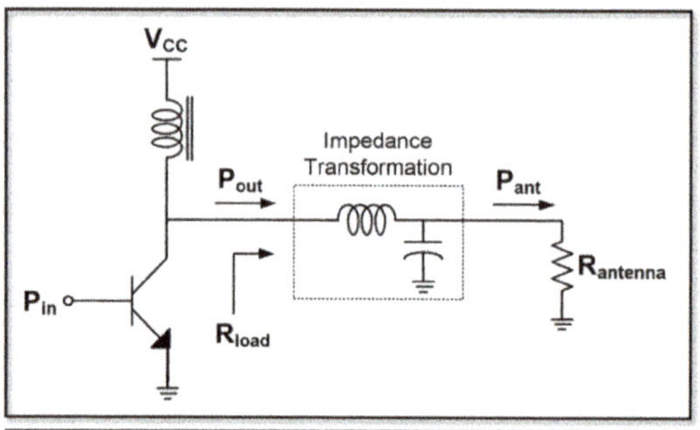

$$R_{load,opt} = \frac{V_{CC} - V_{knee}}{I_{DC}}$$

$$V_{CC} \geq \frac{BV + V_{knee}}{2}$$

$$Q_{xformer} = \sqrt{\frac{R_{antenna}}{R_{load}}}$$

$$P_{out,max} = \frac{(BV - V_{knee})^2}{8R_{load}}$$

$$\eta_{max} = \frac{1}{2}\frac{BV - V_{knee}}{BV + V_{knee}} \times \frac{1}{1 + \frac{Q_{xformer}}{Q_{passives}}}$$

$$PAE_{max} \cong \eta_{max}$$

15 Classical Class-A RF Power Amplifiers

In a Class-A amplifier, the transistor is turned ON and conducts current in the entire time. In an ideal Class-A amplifier, the transistor is always in the forward active region (BJT or HBT) or the saturation region (CMOS FET).

Design and performance expressions for radiofrequency Class-A amplifiers are shown where V_{CC} is the supply voltage, V_{knee} is the knee voltage (lowest collector-emitter voltage for which HBT is in the forward active region or the lowest drain-source voltage for which CMOS FET is in the saturation region), I_{DC} is the DC current through transistor, $R_{load,opt}$ is the optimum load impedance seen by the transistor to result in highest achievable efficiency (e_{max}), BV is the collector (drain) breakdown voltage, $Q_{xformer}$ is the effective quality factor of the impedance transformation network (shown as a LC circuit in this schematic) needed to convert the antenna impedance ($R_{antenna}$) to the desired load impedance for the transistor (R_{load}), $Q_{passive}$ is the quality factor of the components used in the impedance transformation network).

The highest achievable power efficiency, when $BV \gg V_{knee}$ and $Q_{passive} \gg Q_{transformer}$, is 50%. In radio-frequencies, the power added efficiency (PAE) of Class-A amplifier is close to the collector (drain) efficiency given that the power gain is sufficiently high.

16 Class-A PA within the *Safe Operating Region*

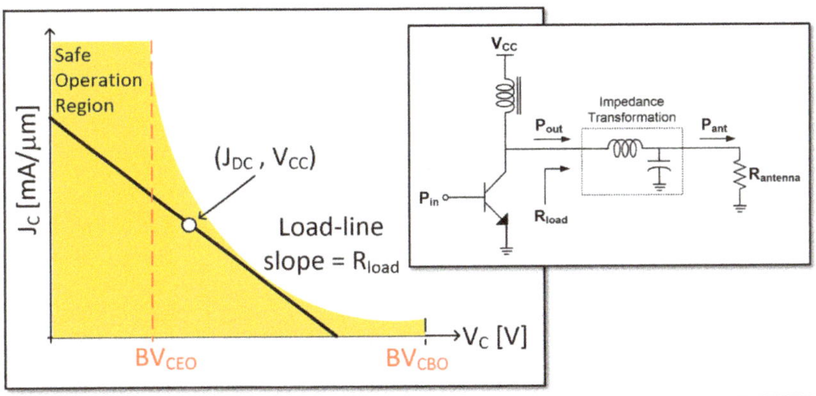

- ❑ Transistor's breakdown voltage depends on its current density.
- ❑ The load line trajectory must remain within the transistor's safe operation region.

It is important to note that the breakdown voltage should not be considered a fixed constant value in designing power amplifiers. In fact, the breakdown voltage of bipolar transistors depends on transistor's collector current density. The "safe operation region" is a combination of collector current densities (J_C) and collector voltages (V_C) that the transistor can operate without breakdown. In a Class-A amplifier, the transistor's dynamic operation follows a straight line on the J_C-V_C coordinate with the slope given by R_L. This so called "load line" must reside within the "safe operation region". It can be seen that, with a proper design, the maximum collector voltage swing of a Class-A amplifier can be above BV_{CEO}.

17 mm-Wave versus RF Class-A PAs

As frequency increases, transistor's maximum power gain reduces. Therefore, the maximum power gain and power-added efficiency (PAE) of Class-A PAs reduce with frequency.

$$G_{p,max} = \frac{G_o}{\left(\frac{\omega}{\omega_{max}}\right)^2}$$

$$PAE_{max} = \eta_{max} \times \left(1 - \frac{1}{G_{P,max}(\omega)}\right)$$

The maximum power gain of transistor decreases with frequency due to the parasitic components such as the base (gate) resistance, the base-emitter (gate-source) capacitance, and the collector-base (drain-gate) capacitance. The unity-power-gain frequency is given by w_{max}. The drop of maximum power gain, $G_{p,max}$, versus frequency is nearly inverse quadratic as shown here. Therefore, in a Class-A amplifier, the maximum achievable power added efficiency drops with frequency due to the loss of maximum power gain. In mm-wave frequencies, the difference between collector (drain) efficiency and the power added efficiency (PAE) can hence be quite significant.

18 mm-Wave Linear PAs (ψ= Conduction Angle)

In the power amplifier terminology, the amplifier is still considered to be operating in the "linear" regime as long as the transistor acts as a current source that changes almost linearly with the input voltage for some part of the waveform (and it can be completely turned off during the other parts of the waveform). Conduction angle (ψ) is defined as the duration in which the transistor conducts current (is not turned off).

Class A: $\psi = 2\pi$, transistor conducts current 100% of the time

Class B: $\psi = \pi$, transistor conducts current 50% of the times

Class AB: $\pi < \psi < 2\pi$, transistor is conducts current between 50% and 100% of the time

Class C: $0 < \psi < \pi$, transistor conducts current between 0 and 50% of the time

The expressions for the achievable performance of linear amplifiers can then be represented using the expressions shown here. In general, for a fixed transistor breakdown voltage, lower conduction angle corresponds to higher collector (drain) efficiency at the expense of lower output power and power gain. It should be reminded again that the breakdown voltage of a transistor is not necessarily a fixed constant value. In fact, it will be shown that amplifiers with smaller conduction angle can generate higher output power as they can sustain a larger voltage swing across the transistor within the "safe operating region".

In all the aforementioned amplifier classes, there is a second-order (LC) filter at the collector to ensure that the collector voltage remains sinusoidal. Other passive impedances can be used at the collector to shape the collector voltage differently. For instance, in Class-F amplifiers, the 50% duty-cycle amplifier (similar to Class-B) is terminated to a passive network to create a square-wave voltage at the collector. The square-wave voltage will have less (ideally zero) overlap with the half-sinusoid current;

18 mm-Wave Linear PAs (ψ= Conduction Angle)

$$R_{load,opt} = \frac{V_{CC} - V_{knee}}{I_{DC}} \pi \frac{1 - \cos\psi/2}{\psi - \sin\psi}$$

$$P_{out,max}(\psi) = \frac{1}{\pi}\frac{\psi - \sin\psi}{1 - \cos\psi/2} \times \frac{(BV - V_{knee})^2}{8R_{load}} \times \zeta(\psi)$$

$$\eta_{max}(\psi) = \frac{\psi - \sin\psi}{2\sin\psi/2 - \psi\cos\psi/2} \times \frac{1}{2}\frac{BV - V_{knee}}{BV + V_{knee}} \times \frac{1}{1 + \frac{Q_{xformer}}{Q_{passives}}}$$

$$G_{p,max}(\psi) = \pi\frac{1 - \cos\psi/2}{2\sin\psi/2 - \psi\cos\psi/2} \times \frac{G_o}{1 + \left(\frac{\omega}{\omega_{max}}\right)^2}$$

$$PAE_{max}(\psi) = \eta_{max}(\psi) \times \left(1 - \frac{1}{G_{P,max}(\omega)}\right)$$

therefore, the efficiency of an ideal Class-F amplifier can reach 100%. In practice, the loss of collector passive impedance network, and the inability to ensure zero overlap between collector current and voltage waveforms, reduces the efficiency.

It should be noted that, for any amplifier class, the inverse of the class can also be realized. For instance, an ideal Class-F^{-1} amplifier will have a square-wave collector current and a half-sinusoidal collector voltage. A Class-F^{-1} amplifier can be realized when the transistor operates as an ON-OFF switch (discussed later).

19 Ex: 45-GHz SiGeHBT Linear PAs (R_{load}= 35Ω)

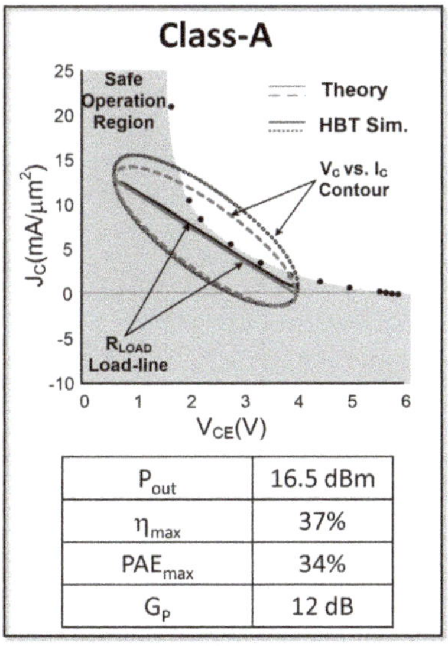

P_{out}	16.5 dBm
η_{max}	37%
PAE_{max}	34%
G_p	12 dB

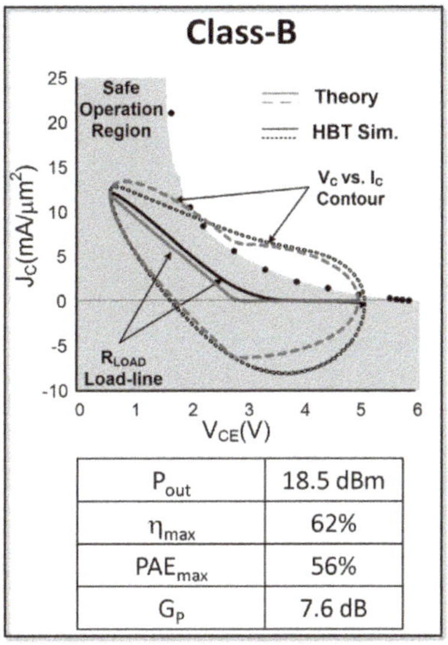

P_{out}	18.5 dBm
η_{max}	62%
PAE_{max}	56%
G_p	7.6 dB

19 Ex: 45-GHz SiGeHBT Linear PAs (R_{load}= 35Ω)

Examples of Class A and Class B amplifier designs at 45 GHz using a 130nm SiGe HBT process driving a load impedance of R_L = 35 Ω are shown here. The J_C-V_C contours represent the current and voltage of the transistor including its collector capacitance. The straight lines correspond to the current and voltage of the transistor without its collector capacitance. It should be noted that at mm-wave frequencies, the collector capacitance is a significant contributor to the overall capacitance at the collector and as such conducts significant current. In a proper design, the collector capacitance will resonate with an inductance at the collector so that the transistor only sees a real impedance value R_L.

The J_C-V_C of a transistor in a Class B amplifier configuration follows two straight lines corresponding to the two regions: one with a slope of R_L when the transistor operates in the forward active region, and the other with a zero slope representing the duration when the transistor is turned OFF and does not conduct any current. This load line enables increasing the maximum collector voltage swing in the transistor, when compared with a transistor in a Class A amplifier, as shown. Consequently, the Class B design can have a higher collector efficiency, higher PAE, and higher output power compared with the Class A design. The power gain of the Class B amplifier is lower though – this is because larger input voltage swing is required, given the 50% transistor duty cycle, to generate a similar amount of current. Therefore, the input power (quadratic in voltage) in the Class B amplifier design is higher as compared with the Class A amplifier design.

20 Outline

- Introduction
- Technology
- Linear Amplifiers
- **Switching Amplifiers**
- Stacked-Transistor Amplifiers
- Power Combining
- Transmitters
- Conclusions

We will now cover the design of monolithic switching power amplifiers at mm-wave frequencies. Once again, it should be noted that, while many schematics and case studies show hetero-structure bipolar transistors (HBT) in the amplifier core, most concepts are applicable to amplifiers that utilize field effect transistors (FET) as well (e.g., CMOS).

21 Classical Class-E RF Power Amplifiers

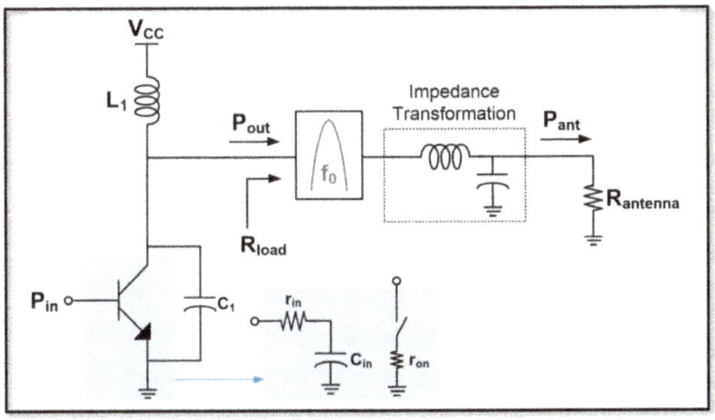

$$I_{DC} = \frac{1}{\pi}\frac{P_{out}}{V_{CC}}$$

$$V_{CC} \geq BV/3.56$$

$$C_1 = \frac{P_{out}}{\pi\omega V_{CC}^2}$$

$$P_{out} \cong \frac{V_{CC}^2}{1.73R_{load}} \qquad P_{in} \cong \frac{1}{2}\left[\frac{\omega r_{in}C_{in}}{1+(\omega r_{in}C_{in})^2}\right]\omega C_{in}V_{in}^2$$

$$\eta_{max} = [1 - 7.43(r_{ON}C_1)\omega] \times \frac{1}{1+\frac{Q_{xformer}}{Q_{passives}}}$$

$$PAE_{max} \cong \eta_{max} \times \left(1 - \frac{1}{G_P}\right)$$

In the power amplifier terminology, in an ideal linear amplifier, the transistor operates as either a current source whose value is proportional to the input voltage or as a open circuit. In practice, the transistor includes parasitic components such as the collector (or drain) capacitance, and the collector (or drain) current relationship with the base (or gate) voltage is not necessarily linear.

In contrast, in an ideal switching power amplifier, the transistor operates as a switch that is either ON (short circuit) or OFF (open circuit). In practice, the ON transistor includes a resistance, r_{ON}, and the OFF transistor includes a collector capacitance C_{OFF}. In many practical cases, the collector capacitance of the ON transistor can be ignored given the relatively small value of r_{ON}. Given that the power consumption of an ideal switch (r_{ON} = 0) is zero, the efficiency of an ideal switching amplifier can reach 100%.

In the classic Class-E amplifier, the transistor switch is turned ON for half the cycle and OFF in the other half. During the ON cycle, the collector voltage is close to zero as the collector current increases from zero to a peak value. During the OFF cycle, the collector current is zero as the collector voltage is similar to half sinusoid. The collector capacitance (C_1) and the collector inductance (L_1) are selected to ensure that the collector current and its derivative are smooth at the OFF to ON transition. A filter, placed between the collector and the load, ensures that only the fundamental component of the signal reaches the load. The C_1 value is proportional to the output power, P_{out}, and inversely proportional to the RF frequency. Given that the collector efficiency improves for lower r_{ON}, larger transistor sizes, to the extent allowed by the C_1 capacitance budget, should be picked.

Design equations and best achievable performance of Class-E amplifier is shown here.

22 Class-E PA within the *Safe Operating Region*

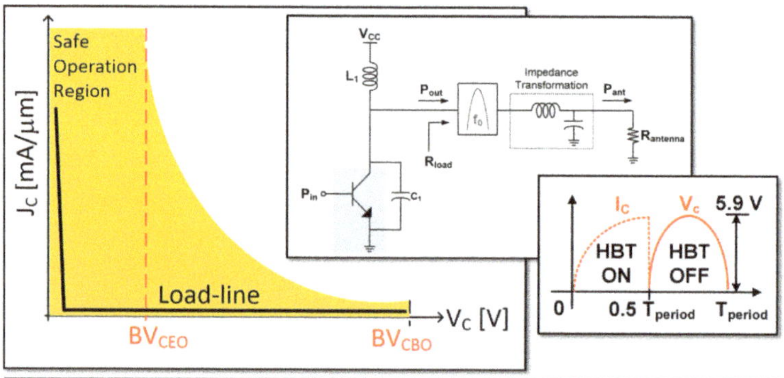

- ❑ Given the non-overlapping nature of I_C and V_{CE} in Class-E amplifiers, the collector node can swing all way up to BV_{CBO}.
- ❑ Power efficiency and output power of Class-E amplifiers can be higher compared with linear amplifiers.

The non-overlapping collector current (I_C) and collector voltage (V_C) of Class-E amplifiers enables a large voltage swing in the "safe operating region" leading to a larger output power. This leads to amplifier designs with higher output power, in addition to higher efficiency, when compared with linear amplifier.

23 mm-Wave versus RF Class-E PAs

- ❑ C_1 entirely consists of transistor's collector capacitance (C_{CS}).
- ❑ (P_{out}, BV, ω) → C_1 → transistor size → P_{in} , η → PAE

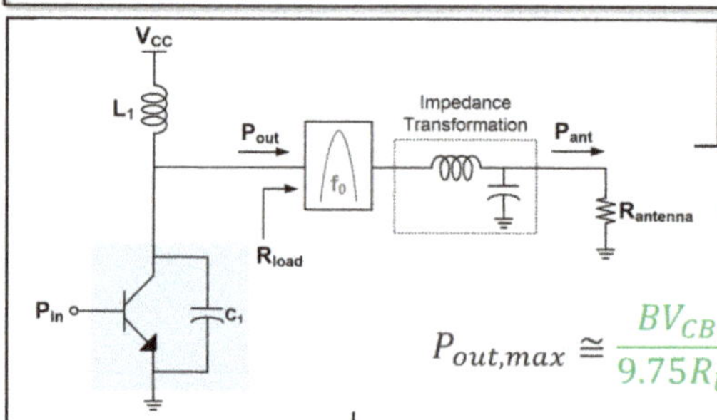

$$G_{p,max} \propto \frac{1}{\omega} \times \frac{BV_{CBO}{}^2}{r_{in}C_{in}}$$

$$P_{out,max} \cong \frac{BV_{CBO}{}^2}{9.75R_{load}} \times \frac{1}{1+2\frac{Q_{xformer}}{Q_{passive}}}$$

$$\eta_{max} \cong \frac{1}{1+4.6\times\omega\times r_{ON}C_{CS}+\frac{2.39}{Q_{passive}}}$$

At higher operation frequencies, the required Class-E C_1 capacitance will be entirely set by the transistor's collector capacitance. In fact, the desired output power, transistor breakdown voltage, and frequency set the transistor value through determining the C_1 value. Once the transistor size is fixed, the input power, collector efficiency, and the power added efficiency can be calculated.

The maximum achievable efficiency of mm-wave Class-E amplifiers improves with technology scaling as scaled transistors offer smaller $r_{ON}C_{CS}$ product. Naturally, efficiency improves with higher quality factor of passive components that are used for impedance transformation.

24

SSCS

Class-E PA Performance vs. Frequency

This slides shows simulated achievable performance of optimally-designed Class-E amplifiers, versus frequency, for a 130nm SiGe HBT process technology. As expected, for a given output power requirement, power gain, maximum efficiency, and maximum power-added efficiency decrease with frequency.

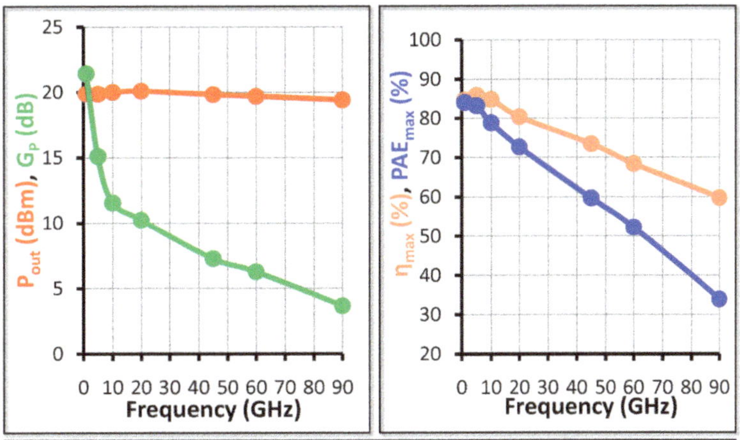

Simulated results in a 130nm SiGe HBT BiCMOS technology.

25

SSCS

Ex: 45-GHz SiGeHBT Class-E PA (R_{load}= 35Ω)

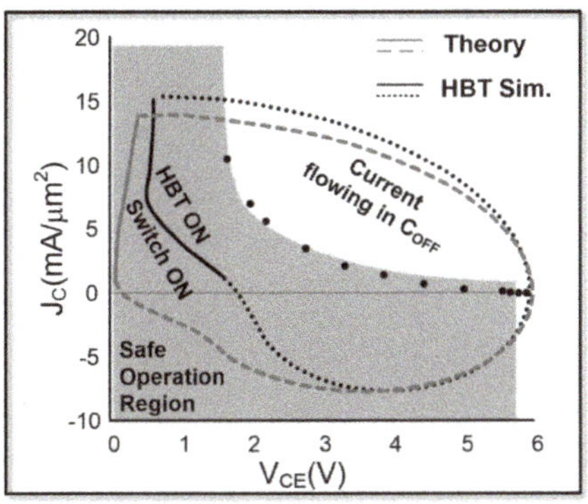

	Class A	Class B	**Class E**
P_{out}	16.5 dBm	18.5 dBm	**20 dBm**
η_{max}	37%	62%	**71%**
PAE_{max}	34%	56%	**60%**
G_P	12 dB	7.6 dB	**7.9 dB**

In a Class-E amplifier, the collector current is zero (transistor OFF) when the collector voltage is high, and the collector voltage is low when the transistor is ON. Therefore, as it can be seen from the plot of collector current density versus collector voltage of a bipolar transistor in a Class-E amplifier, a larger maximum voltage collector swing can be sustained within the safe operation region when compared with linear amplifiers (Class A and Class B). This larger maximum voltage swing results in higher output power that is delivered to the load.

Therefore, for a given load impedance, a Class-E amplifier offers higher output power at a higher power efficiency when compared with Class A and Class B amplifiers. The power gain of Class-E amplifier is smaller compared with Class-A amplifier because a larger input voltage swing is needed to result in a proper transistor switching action.

26

45-GHz Class-E SiGeHBT PA

The schematic, chip microphotograph, and performance summary table of a 45-GHz Class-E amplifier, realized in a 130nm SiGe HBT BiCMOS process, is shown.

The high collector current density can cause impact ionization and base current reversal. This can create instability. To improve the stability, at the input of each stage, resistors and series resonators at half the desired frequency (half-harmonic traps) are placed.

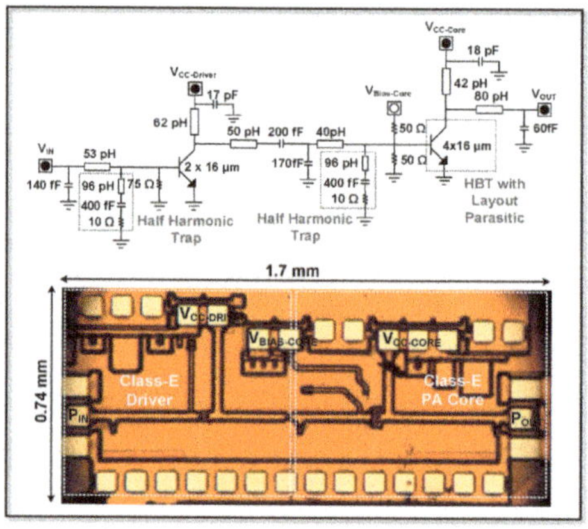

Metric	Measured Performance
Frequency	45 GHz
$P_{out,max}$	20.2 dBm
$G_{p,max}$	10.5 dB
η_{max}	34.5 %
PAE_{max}	31.5 %
P_{-1dB} BW	> 4 GHz
Process	130 nm SiGe BiCMOS

K. Datta, J. Roderick, and H. Hashemi, "20 dBm Q-band Class-E SiGe power amplifier with 31% peak PAE," in *Proceedings of the IEEE CICC*, 2012.

27

88-GHz Class-E SiGeHBT PA

The schematic, chip microphotograph, and performance summary table of a 88-GHz amplifier, realized in a 90nm SiGe HBT BiCMOS process, is shown. Due to a lower power gain of transistors at this frequency, five amplifying stages are used. The first three stages operate in Class-AB primarily for higher power gain, and the final two stages operate in Class-E with higher supply voltage (since Class-E designs can sustain a higher voltage swing without breakdown).

Similar to the previous design, half-harmonic traps and resistors are placed at the input of each stage to eliminate instability.

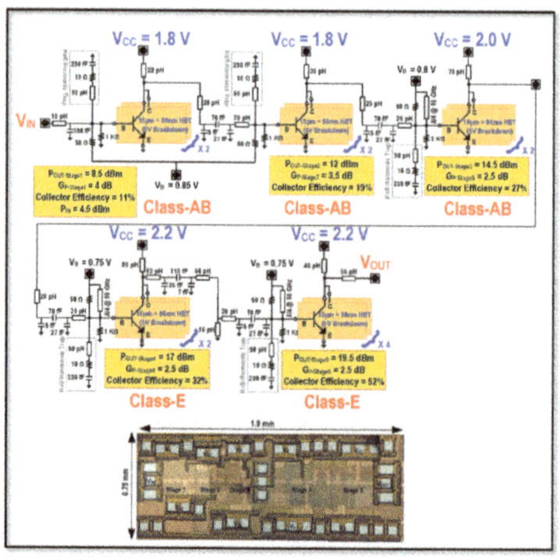

Metric	Measured Performance
Frequency	88 GHz
$P_{out,max}$	19.5 dBm
$G_{p,max}$	15 dB
η_{max}	16.2 %
PAE_{max}	15.9 %
P_{-1dB} BW	> 10 GHz
Process	90 nm SiGe BiCMOS

K. Datta and H. Hashemi, "75 – 100 GHz switching power amplifiers using high-breakdown, high-f_{max} multi-port stacked transistor topologies," in *IEEE RFIC Symposium* Digest, 2016.

28 Design Space for Switching Power Amplifiers

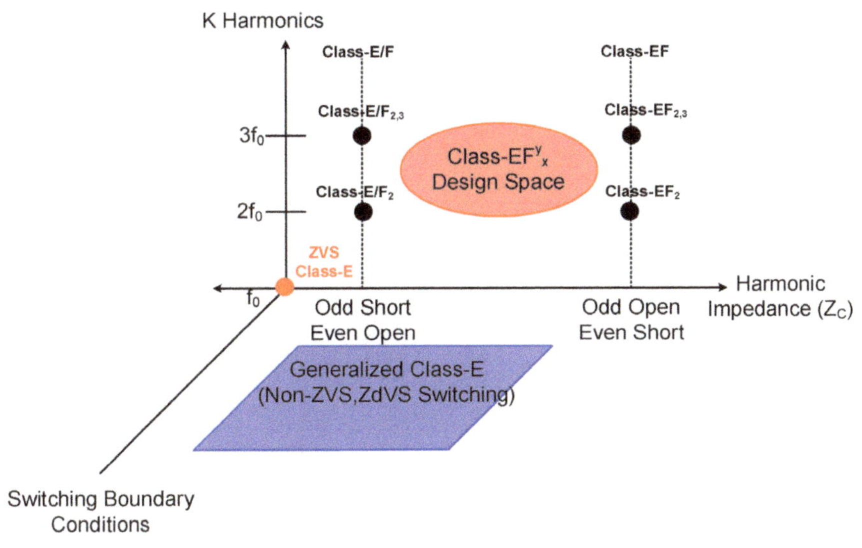

Class-E is only one example of a switching amplifier design where the collector capacitance and inductance are selected to satisfy a specific zero voltage switching (ZVS) and zero derivative-voltage switching (ZVdS) conditions.

In the general case, for a transistor that acts as an ON/OFF switch (ON: short circuit, OFF: open circuit) or for a transistor that acts as a switched current source (ON: current source, OFF: open) other collector impedances can be envisioned to shape the collector voltage and current differently.

For instance, as discussed before, in a Class-F^{-1} amplifier, the collector current is square-wave and the collector voltage is half-sinusoidal. Class-F^{-1} amplifier may be realized with a switching transistor terminated to a passive impedance network that shapes the current and voltage waveforms. Specifically, in order to sustain a square-wave current (which has only odd-harmonic components), the passive impedance network at the collector has to be short circuit at all the odd harmonics (except for the fundamental) and open circuit at all the even harmonics.

More generally, the passive impedance network at the collector of a switching transistor can represent different values at the fundamental and harmonics. For instance, in a Class-EF^{2} amplifier (aka Class-E/F_2), the passive impedance network at the collector is an open circuit only at the 2nd-harmonic (hence F^{2}) while it looks similar to the desired impedance of a Class-E amplifier at the fundamental (capacitive + series RL) and all other harmonics (capacitive). Likewise, in a Class-EF^{3} amplifier (aka Class-E/F_3), the passive impedance network at the collector is a short circuit only at the 3rd-harmonic (hence F^{3}) while it looks similar to the desired impedance of a Class-E amplifier at the fundamental (capacitive + series RL) and all other harmonics (capacitive). In a Class-$EF^{2,3}$ amplifier (aka Class-E/$F_{2,3}$), the passive impedance network at the collector is open circuit at the 2nd-harmonic (hence F^{2}) and short circuit at the 3rd-harmonic (hence F^{3}) while it looks similar to the desired impedance of a Class-E amplifier at the fundamental (capacitive + series RL) and all other harmonics (capacitive).

It is emphasized that, in general, the impedance presented at the collector at the harmonics need not be either open or short circuit. Therefore, the Class EF^{n} is only a subset of switching amplifiers that can be created.

29 Generalized Switching Power Amplifiers

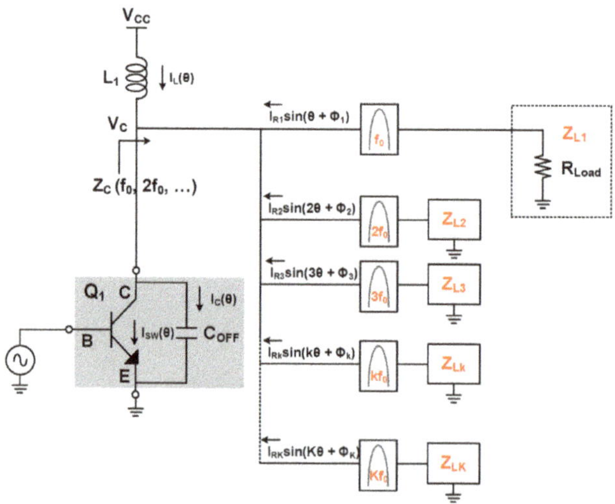

A conceptual schematic of a generic switching amplifier where the collector impedance at each harmonic is set independently is shown here.

30 Harmonic Control in Some Switching PAs

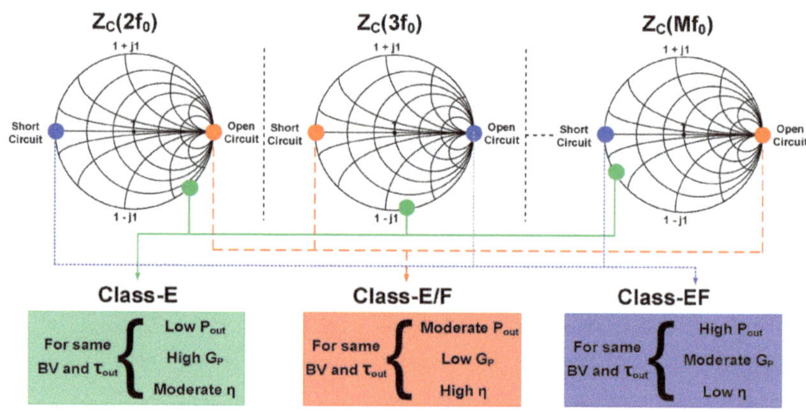

The desired impedances presented at the collector at different harmonics for ideal Class-E, Class-E/F, and Class-EF amplifiers are represented on the Smith chart.

The output power of a Class-EF amplifier can be higher compared with the other two amplifiers due to the square-wave collector voltage that results in a larger fundamental voltage. On the other hand, given the switch ON resistance (or the transistor knee voltage), the square-wave collector voltage cannot reach zero. Therefore, the overlap of the non-zero (but low) square-wave voltage with the current results in undesired transistor power consumption and efficiency degradation.

On the other hand, the square-wave collector current does reach zero when the switch is turned OFF. Therefore, Class-E/F can achieve a higher efficiency compared with the other two amplifiers.

31 Survey of mm-Wave Switching PAs

	JSSC 2016		RFIC 2017	IMS 2016	JSSC 2014	RFIC 2012	RFIC 2015	RFIC 2016
Freq. [GHz]	24	38	26 – 34	40.5	45	60	60	88
$P_{out,max}$ [dBm]	18	16.5	15	18	20.2	9	17.9[(1)]	19.5
PAE_{max}	50%	38%	42% - 46%	43%	31.5%	27%	20.5%[(1)]	15.9%
$G_{P,max}$ [dB]	21	16	10	18	10.5	8.8	21.5[(1)]	15
Topology	Class-F^{-1}		Class-F	Class-F^{-1}	Class-E	Class-E	Class-EF^{-2}	Class-E
Technology	130nm SiGe		65nm CMOS	130nm SiGe	130nm SiGe	32nm SOI	40nm LP CMOS	90nm SiGe

[(1)] 4-Way power combined

Millimeter-wave switching power amplifiers (Class E, Class F^{-1}, Class F, Class EF^{-2}, ...) using SiGe HBT and CMOS transistors have been demonstrated.

A number of switching amplifiers operating at mm-wave frequencies and implemented in commercial silicon processes have been reported. This table summarizes the performance of a few of such implementations.

32 Outline

- Introduction
- Technology
- Linear Amplifiers
- Switching Amplifiers
- **Stacked-Transistor Amplifiers**
- Power Combining
- Transmitters
- Conclusions

In order to increase the output power, the voltage swing across the load impedance must be increased. As discussed before, the maximum voltage swing across the transistor is limited by the technology. Transistors may be stacked to allow for larger voltage swing across the load without exceeding the maximum voltage swing limit across each transistor.

33 Transistor Stacking

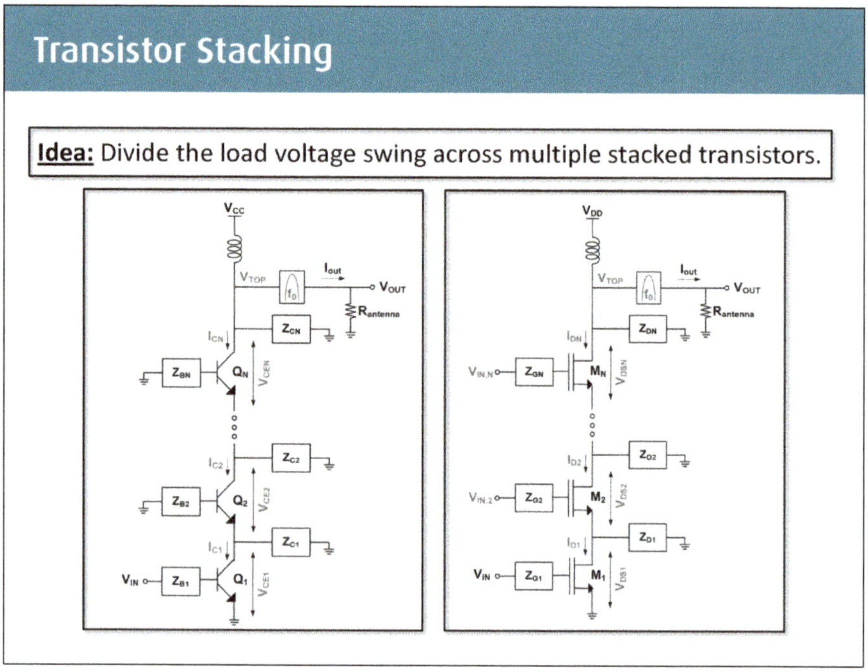

In principle, the voltage swing across the load may be divided across multiple transistors. Impedances connected at or between the intermediate points can ensure that the load voltage swing is equally split across different transistors. The input can be provided to either one or more transistors in the stacked configuration. The transistors can operate in any amplifier class. In fact, as will be shown later, the passive impedances may be chosen so that each of the different transistors in the stacked configuration operates in a different class.

34 Special Case: Class-E Transistor Stacking

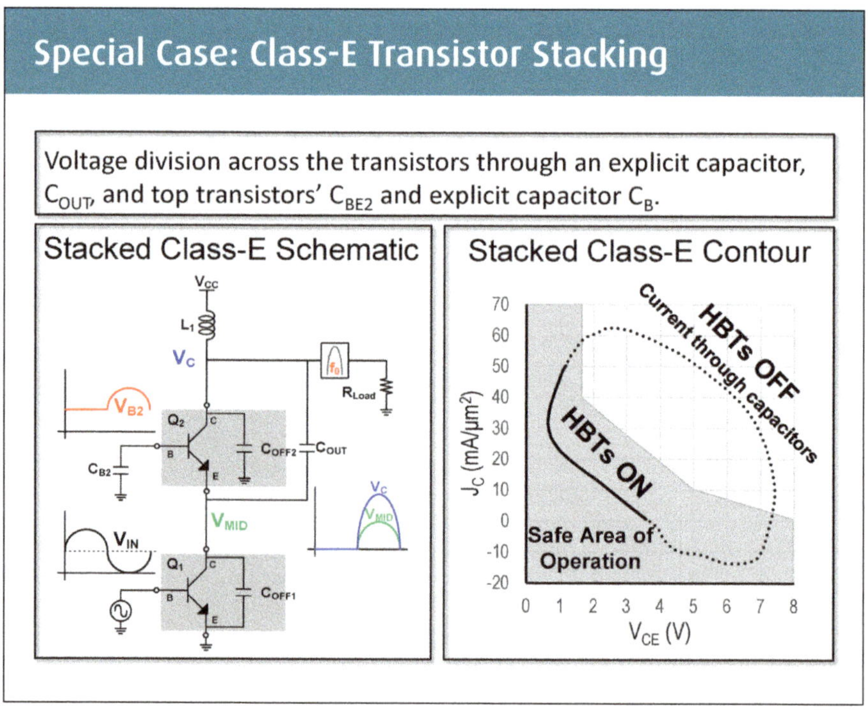

As one special case, stacked transistors may be used in a Class-E amplifier configuration. In one design approach, the input is provided only to the bottommost transistor. Capacitors placed between the collector of two transistors (C_{out}) and at the base of top transistor (C_{B2}) in addition to the transistors' intrinsic capacitances (C_{BE2}, C_{OFF2}, and C_{OFF1}) enable (1) synchronous operation of the transistors, and (2) near-equal collector-emitter voltage swings for the two transistors.

35 Stacked-Transistor Class-E Operation

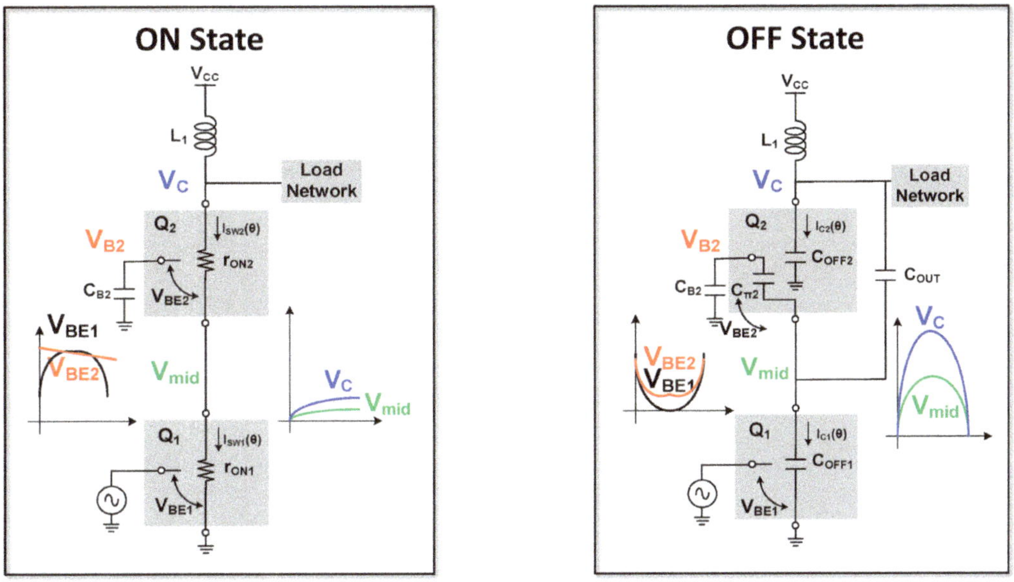

The operation of the two-transistor-stack in a Class-E amplifier design can be intuitively understood as the following.

When the input voltage is high, the bottom transistor turns ON and enters saturation. The collector voltage of the bottom transistor, (shown as V_{mid}) will drop to near zero. This results in large base-emitter voltage in the top transistor (V_{BE2}); therefore, the top transistor also turns ON and enters saturation. In this mode, both transistors may be modelled with an ON resistance (r_{ON1} and r_{ON2}).

When the input voltage is low, the bottom transistor turns OFF and can be modelled with a collector capacitance C_{OFF1}. In the absence of any capacitors, the current of top transistor Q_2 has to follow the current of bottom transistor Q_1. In other words, when Q_1 turns off, the load current reduces (ideally zero) resulting in an increase of output voltage V_C. The capacitive divider network consisting of C_{OUT} and C_{BE2}, C_{B2}, and C_{OFF1} will make the V_{mid} follow the V_C waveform. In a proper design, V_{mid} should be exactly half of V_C to ensure equal voltage stress across Q_1 and Q_2. Furthermore, the base-emitter voltage of Q_2, V_{BE2}, will also be created through the capacitive voltage divider formed by C_{B2} and C_{BE2} from V_{mid}. Therefore, V_{BE2} will decrease as V_C and V_{mid} rise reinforcing synchronous operation of Q_1 and Q_2.

36 Two-Stacked mm-Wave Class-E PA

The expressions for maximum output power and maximum collector efficiency of the aforementioned two-transistor-stack configuration in a Class-E amplifier may be desired.

In theory, a two-stacked-transistor configuration can sustain twice the voltage swing when compared with a single transistor resulting in four-times increase of output power. However, in practice, the voltage swing at the collector of the two transistors cannot be set to be exactly equal. Therefore, the maximum achievable output power will be less.

Compared to the standard Class-E design, the maximum achievable efficiency is lower because of higher total loss due to the two series ON resistances (R_{ON1} + R_{ON2}) as opposed to one ON resistance of a single transistor.

36 Two-Stacked mm-Wave Class-E PA

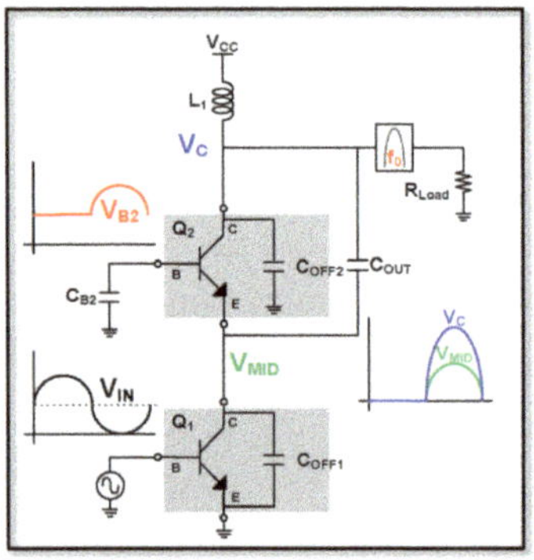

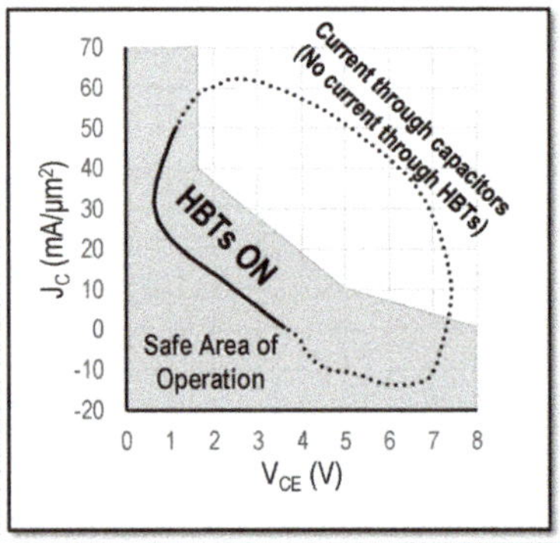

$$P_{out,max} \cong 4\left[\frac{1}{1+\frac{C_{B2}}{C_{OUT}+C_{B2}}}\right]^2 \times \frac{BV_{CBO}{}^2}{9.75R_{load}} \times \frac{1}{1+2\frac{Q_{xformer}}{Q_{passive}}}$$

$$\eta_{max} \cong \frac{1}{1+2\times 4.6\times\omega\times r_{ON}C_{CS}+\frac{2.39}{Q_{passive}}}$$

37 Two-Stacked Class-E PA Performance vs. Freq.

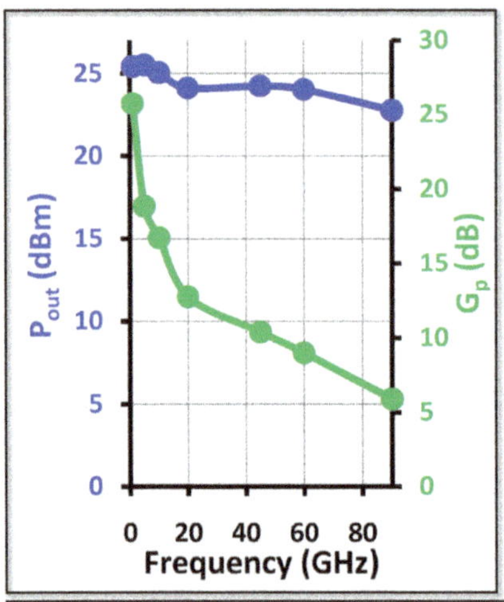

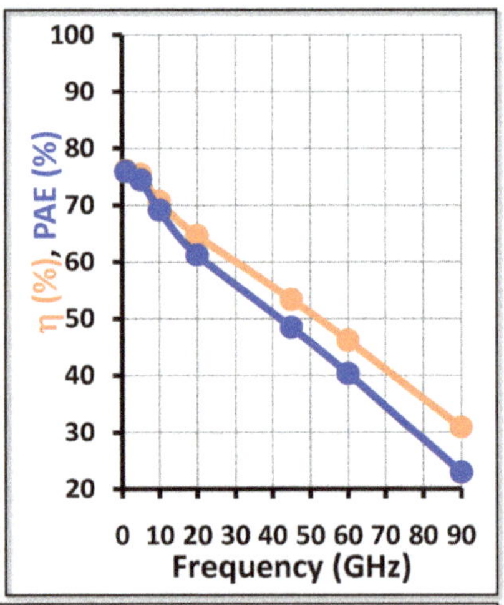

Simulated results in a 130nm SiGe HBT BiCMOS technology.

Similar to the standard Class-E amplifier, at a given output power level, the power gain, maximum efficiency, and maximum PAE of a two-transistor-stacked Class-E amplifier drop with frequency. Simulated performance of optimally-designed such two-transistor-stacked Class-E amplifiers using 130nm SiGe HBTs are shown here.

38 N-Stacked mm-Wave Class-E PA

$$\eta_{max} \cong \frac{1}{1 + N \times 4.6 \times \omega \times \underbrace{r_{ON} C_{OFF}}_{\text{transistor}} \times \underbrace{\left(1 + \frac{C_P}{C_{OFF}}\right)}_{\text{parasitics}} + \underbrace{\frac{1.87}{Q_{cap}}}_{\text{caps}}}$$

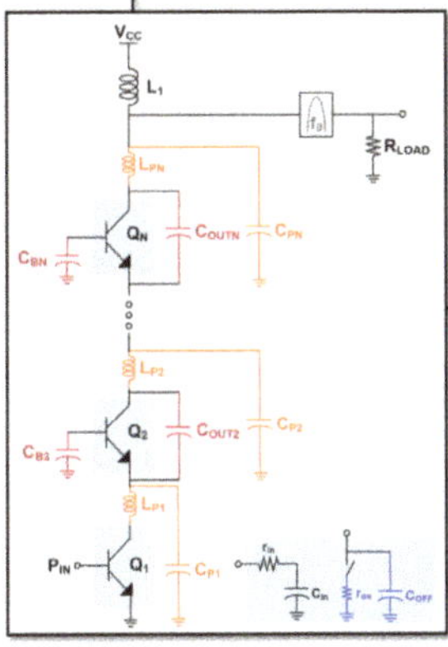

$$P_{out,L_p} \cong \frac{P_{out}}{\left(1 + N \times \frac{L_P}{L_1}\right)^2}$$

- ❑ Layout parasitics degrade the performance of stacked-transistor PAs.
- ❑ Stacked transistors can be laid out to minimize the undesired parasitics and absorb them in the design.

Similarly, an N-transistor-stacked structure in a Class-E configuration leveraging capacitances for voltage division can be created. The performance of stacked-transistor designs degrade with parasitics, especially the collector capacitance and inductance of each transistor. In addition, more parasitic components can stem from the layout of stacked transistors. As it will be shown later, stacked transistors can be laid out in a way to absorb some of these parasitics into the desired capacitors of the amplifier design.

39 41-GHz 2-Stacked Class-E SiGeHBT PA

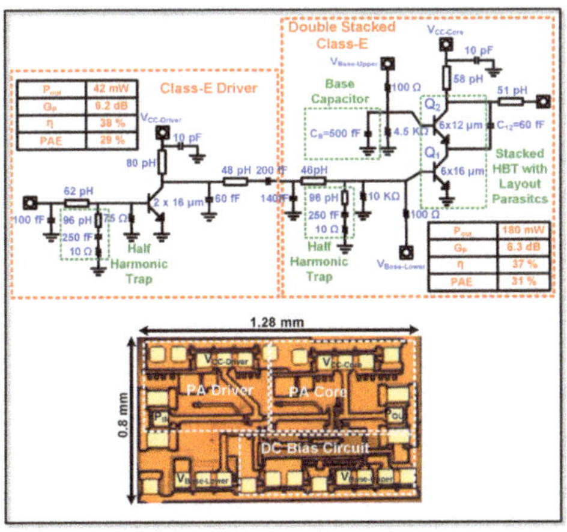

Metric	Measured Performance
Frequency	41 GHz
$P_{out,max}$	23.4 dBm
$G_{p,max}$	14.5 dB
η_{max}	36.1 %
PAE_{max}	34.9 %
P_{-1dB} BW	> 4 GHz
Process	130 nm SiGe BiCMOS

K. Datta, J. Roderick, and H. Hashemi, "Analysis, design, and implementation of mm-wave SiGe stacked class-E power amplifiers," *IEEE RIFC Symposium* Digest, 2013.

The schematic, chip microphotograph, and performance summary table of a two-stage 45-GHz amplifier, consisting of a Class-E amplifier followed by a two-stacked-transistor Class-E amplifier, realized in a 130nm SiGe HBT BiCMOS process, is shown.

Compared to the standard Class-E amplifier design that was shown before, this design generates 2.5 dB higher output power at a similar power efficiency.

40 85-GHz 2-Stacked Class-E SiGeHBT PA

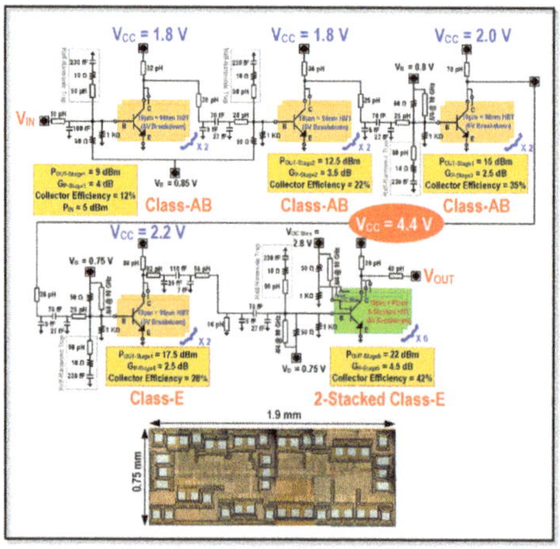

Metric	Measured Performance
Frequency	85 GHz
$P_{out,max}$	22 dBm
$G_{p,max}$	17 dB
η_{max}	19.5 %
PAE_{max}	19.1 %
P_{-1dB} BW	12 GHz
Process	90 nm SiGe BiCMOS

K. Datta and H. Hashemi, "75 – 100 GHz switching power amplifiers using high-breakdown, high-f_{max} multi-port stacked transistor topologies," in *IEEE RFIC Symposium* Digest, 2016.

The schematic, chip microphotograph, and performance summary table of a five-stage 85-GHz amplifier, consisting of three Class-AB amplifiers, a Class-E amplifier, and a two-stacked-transistor Class-E amplifier, realized in a 90nm SiGe HBT BiCMOS process, is shown.

Compared to a five-stage amplifier with a standard Class-E amplifier as the last stage that was shown before, this design generates 3 dB higher output power at a similar power efficiency.

41 Multi-Port Layout of mm-Wave Stacked HBTs

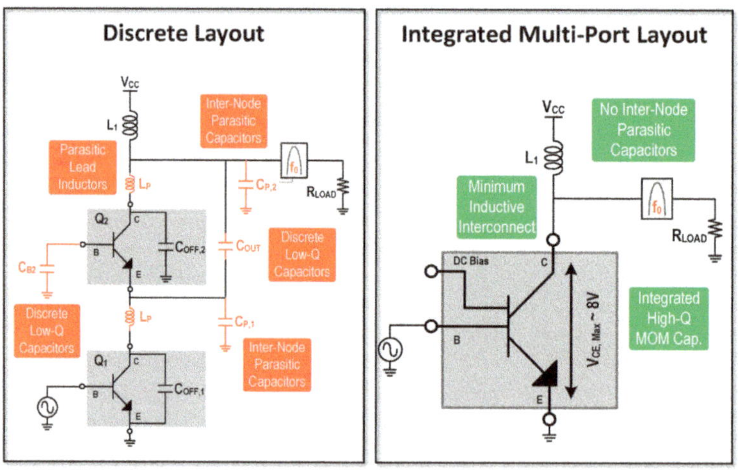

As stated before, the performance of stacked-transistor amplifiers deteriorates with the intrinsic and layout parasitics. Therefore, it is important to layout the stacked transistors in a way that the parasitics are minimized and/or absorbed in the design.

The layout of a two-stacked transistor for the Class-E amplifier of previous example is presented in the subsequent slides.

42 Two-Stacked mm-Wave Layout (I)

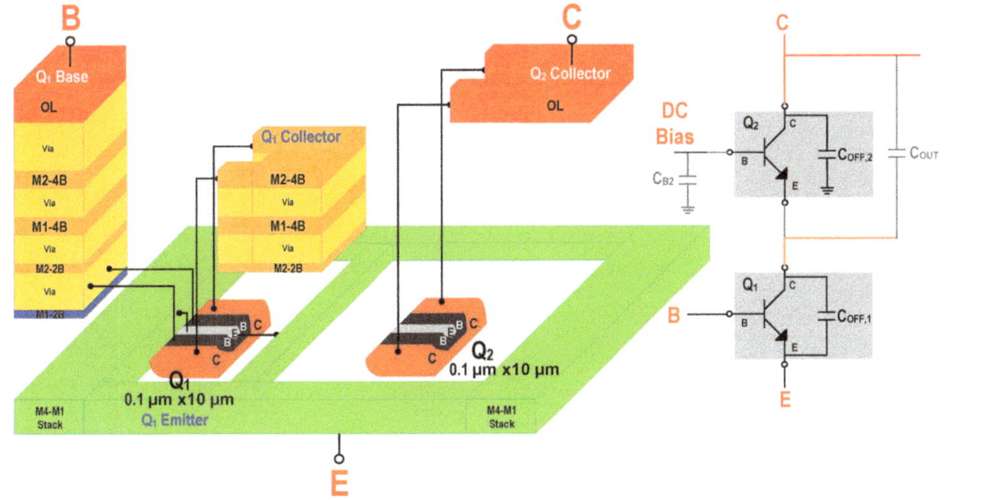

43 Two-Stacked mm-Wave Layout (II)

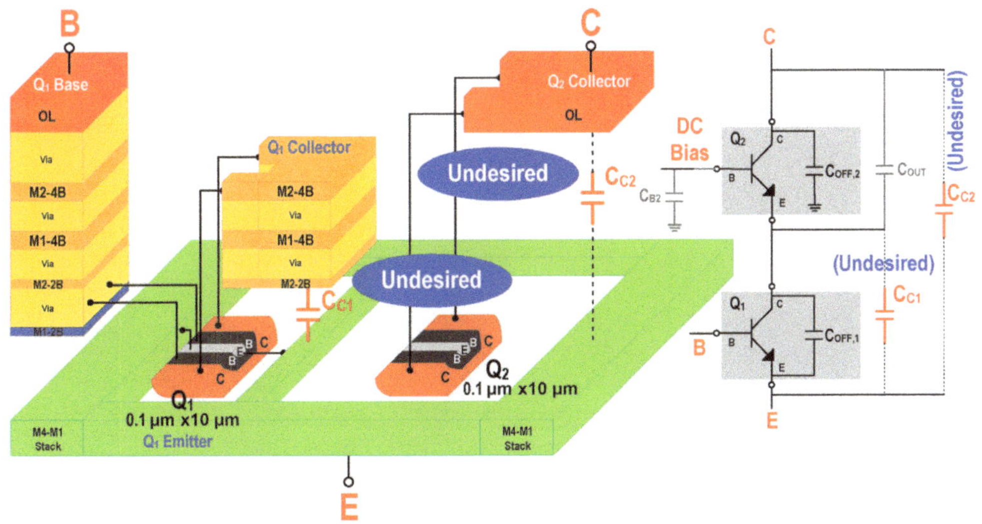

44 Two-Stacked mm-Wave Layout (III)

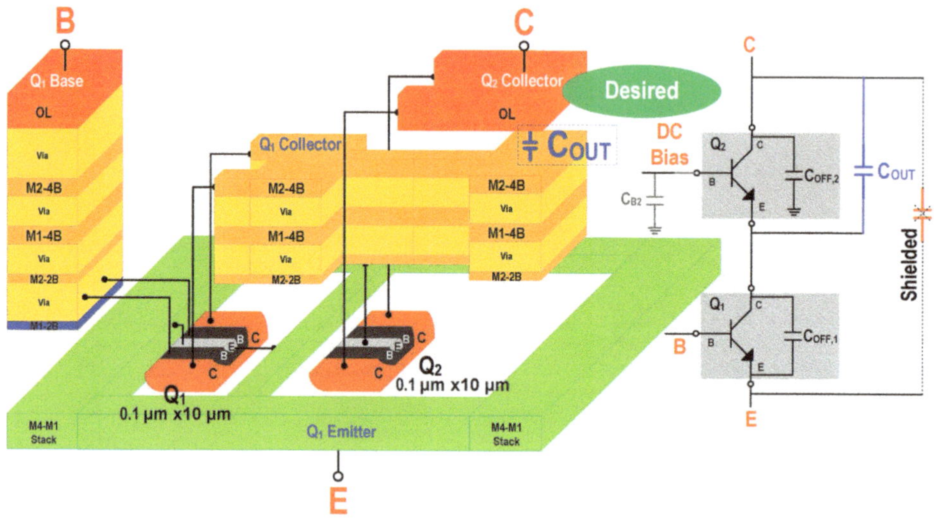

45 Two-Stacked mm-Wave Layout (IV)

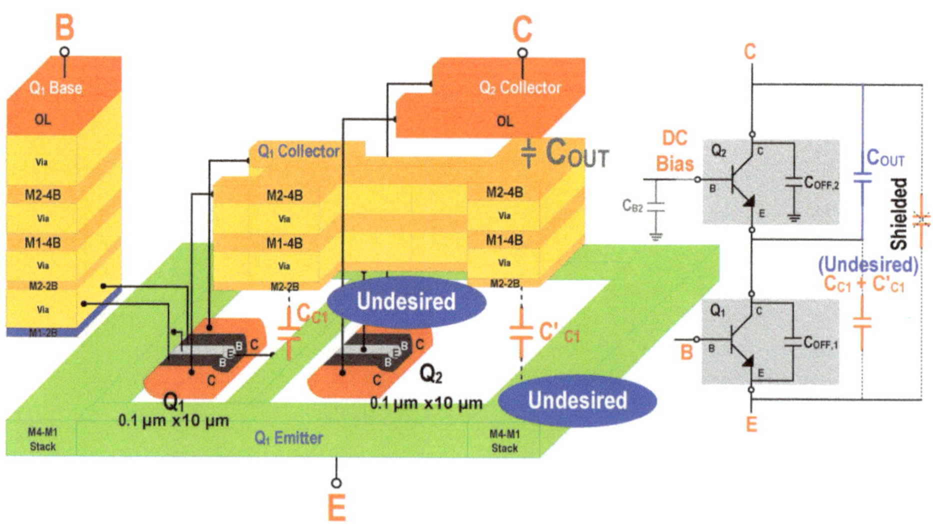

46 Two-Stacked mm-Wave Layout (V)

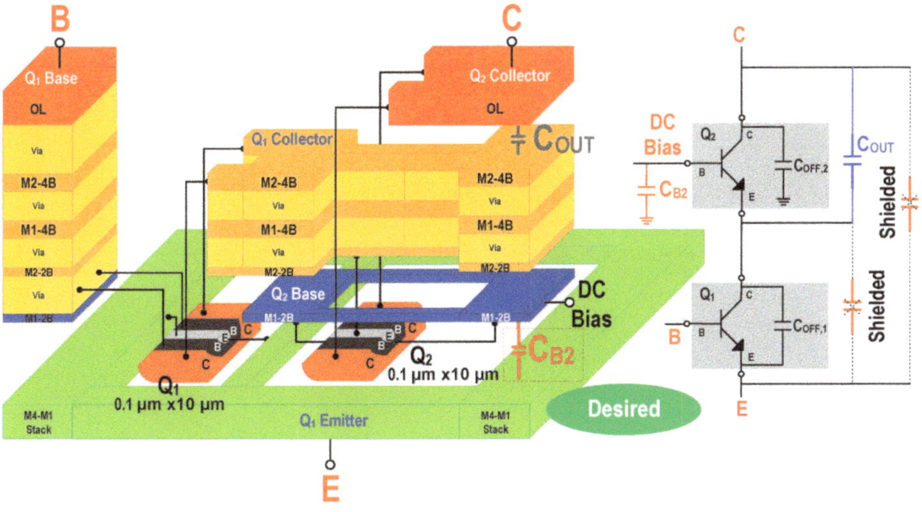

47 Two-Stacked mm-Wave Layout (VI)

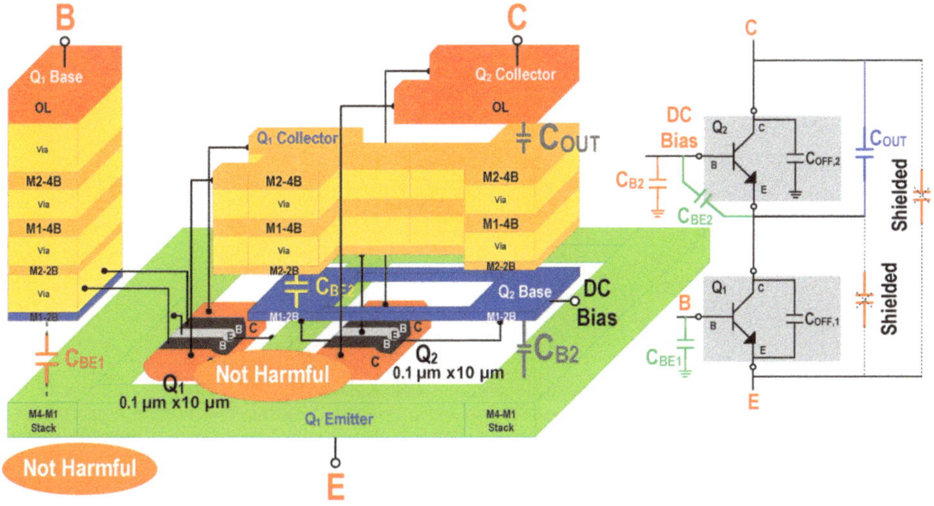

48 Two-Stacked mm-Wave Layout (VII)

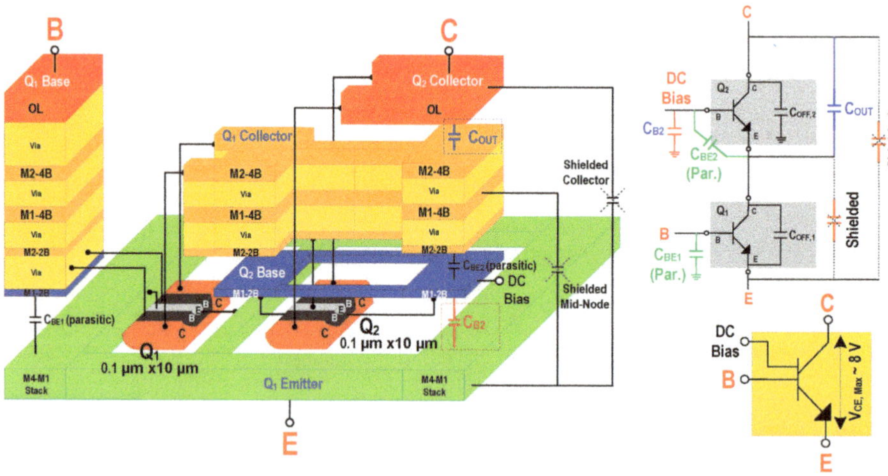

Composite 2-stacked HBT with BV = 8 V and f_{max} = 290 GHz

49 83-GHz 3-Stacked Class-E SiGeHBT PA

The schematic, chip microphotograph, and performance summary table of a six-stage 83-GHz amplifier, consisting of three Class-AB amplifiers, two Class-E amplifiers, and a three-stacked-transistor Class-E amplifier, realized in a 90nm SiGe HBT BiCMOS process, is shown.

Compared to a five-stage amplifiers with a standard Class-E amplifier and a two-transistor-stacked Class-E amplifier as the last stage that were shown before, this design generates around 4 dB and 1.5 dB higher output power, respectively, at a similar power efficiency.

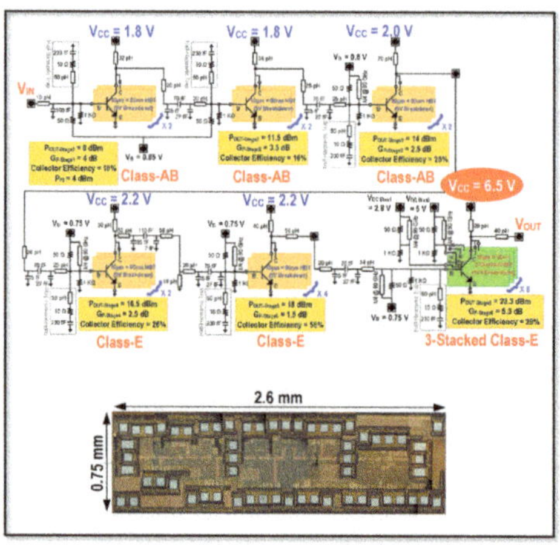

Metric	Measured Performance
Frequency	83 GHz
$P_{out,max}$	23.3 dBm
$G_{p,max}$	18.7 dB
η_{max}	17.4 %
PAE_{max}	17.1 %
P_{-1dB} BW	7 GHz
Process	90 nm SiGe BiCMOS

K. Datta and H. Hashemi, "75 – 100 GHz switching power amplifiers using high-breakdown, high-f_{max} multi-port stacked transistor topologies," in *IEEE RFIC Symposium* Digest, 2016.

50 Three-Stacked mm-Wave Layout (I)

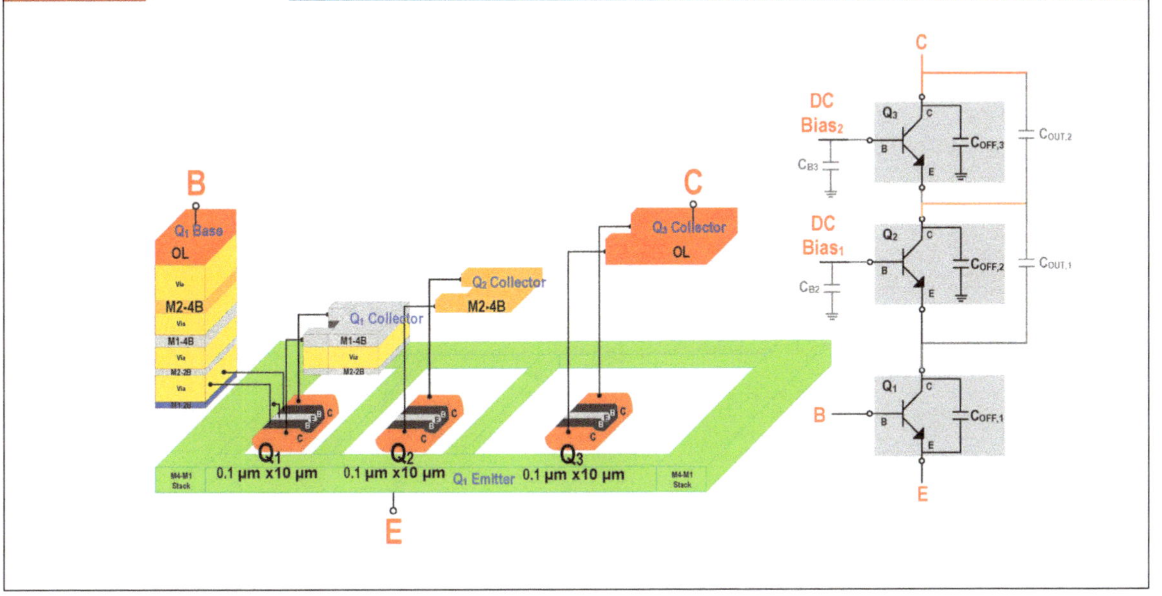

The layout of the three-transistor stacked structure used in the previous example, in which parasitics are absorbed in the design, is shown in the subsequent set of slides.

51 Three-Stacked mm-Wave Layout (II)

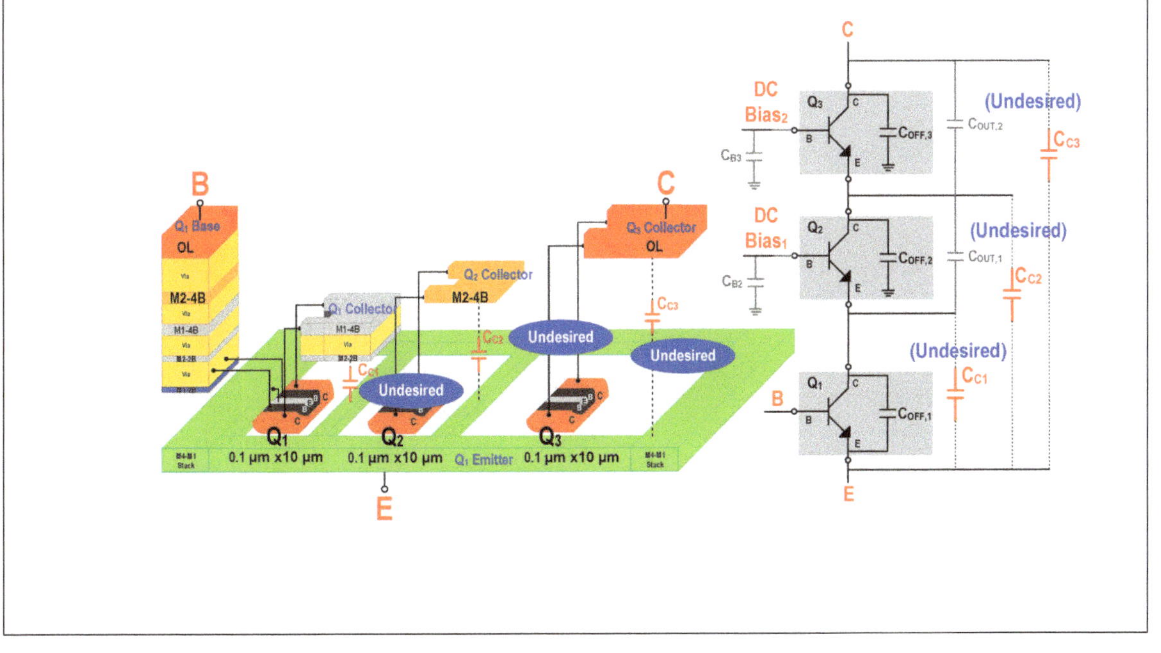

52

Three-Stacked mm-Wave Layout (III)

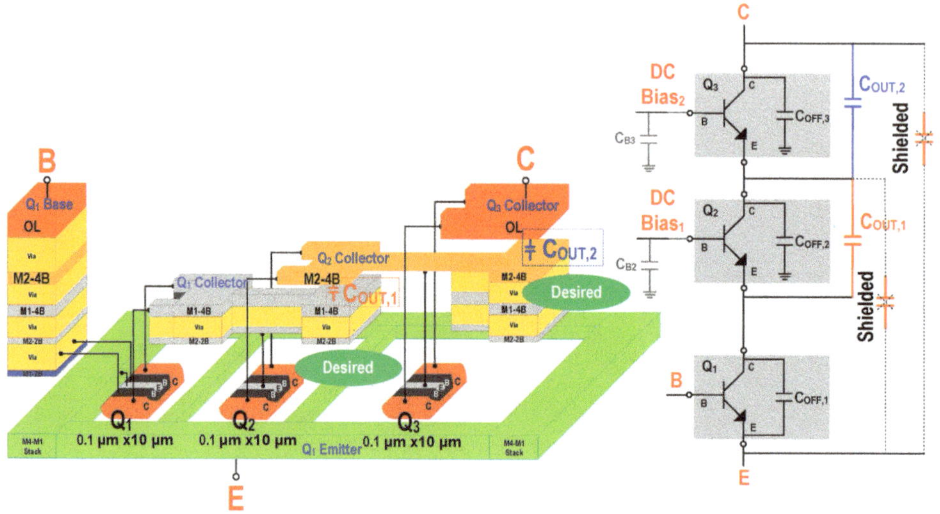

53

Three-Stacked mm-Wave Layout (IV)

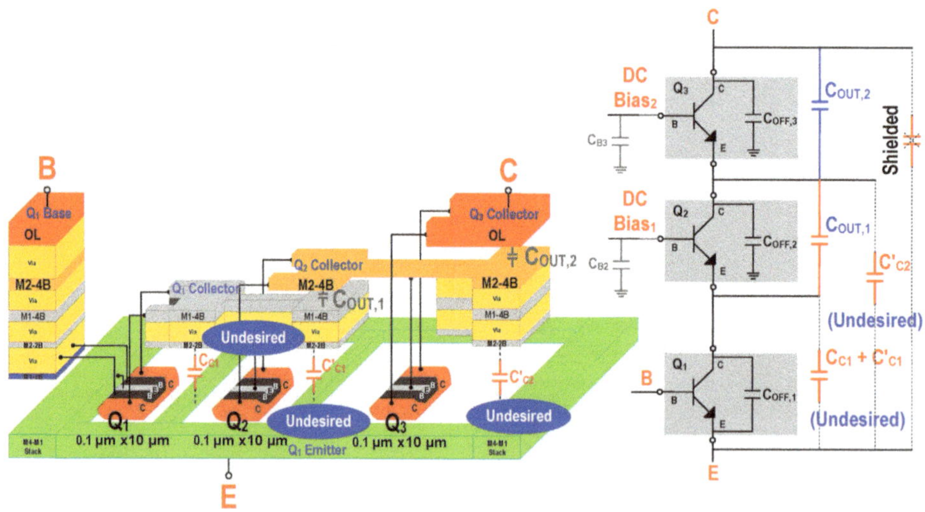

54 Three-Stacked mm-Wave Layout (V)

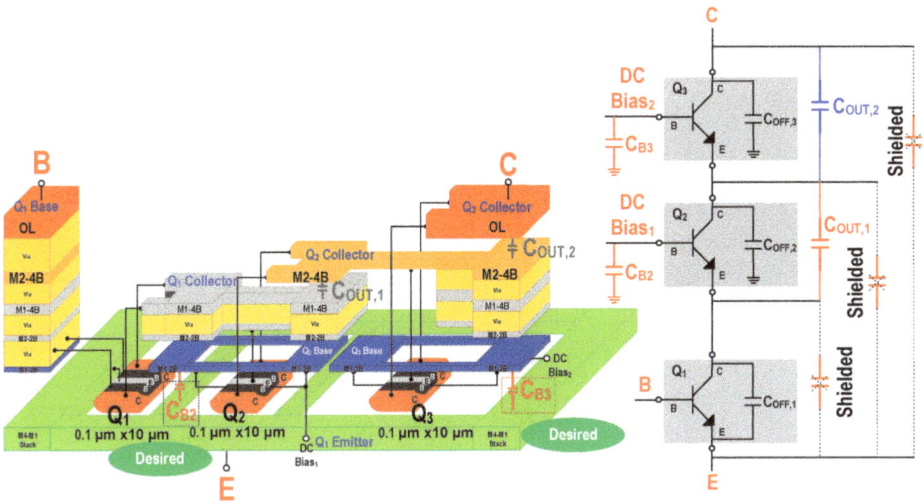

55 Three-Stacked mm-Wave Layout (VI)

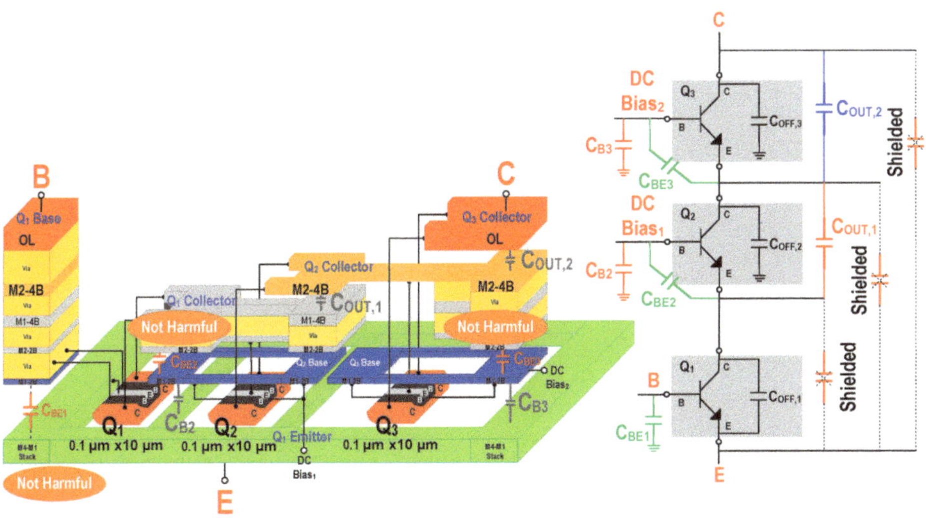

56 Three-Stacked mm-Wave Layout (VII)

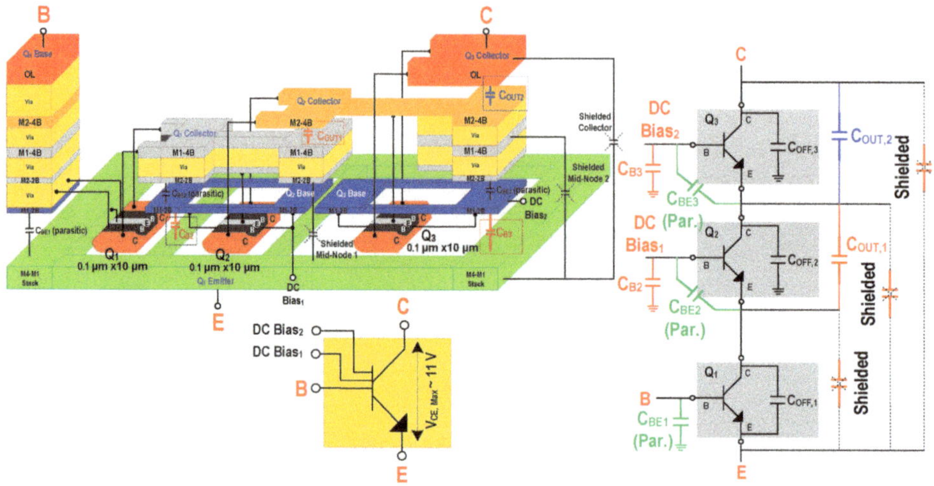

Composite 3-stacked HBT with BV = 11 V and f_{max} = 260 GHz

57 Generalized Stacked-Transistor Switching PAs

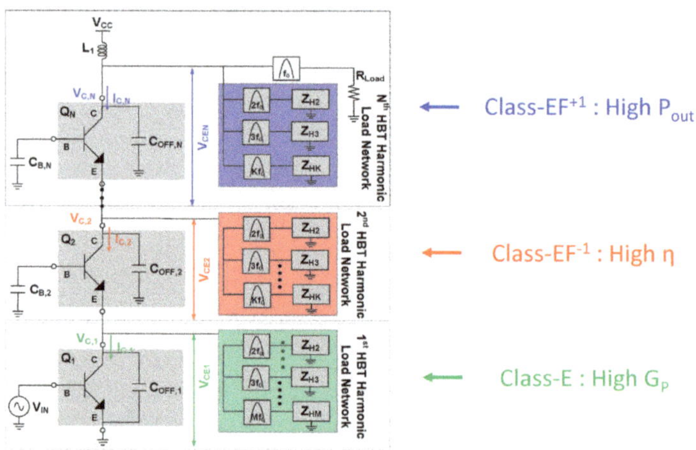

In general, stacked transistors used in an amplifier do not need to have similar current and voltage waveforms. In other words, the collector current and voltage for each transistor can mimic a different amplifier class. The voltage and current waveforms for individual transistors may be set by using proper passive networks connected at each collector (or even between collectors).

For instance, the bottom transistor in the stacked structure can be designed to offer higher power gain, whereas the top transistor can be designed to offer higher output power.

58 Example of a Generalized Two-Stacked PA

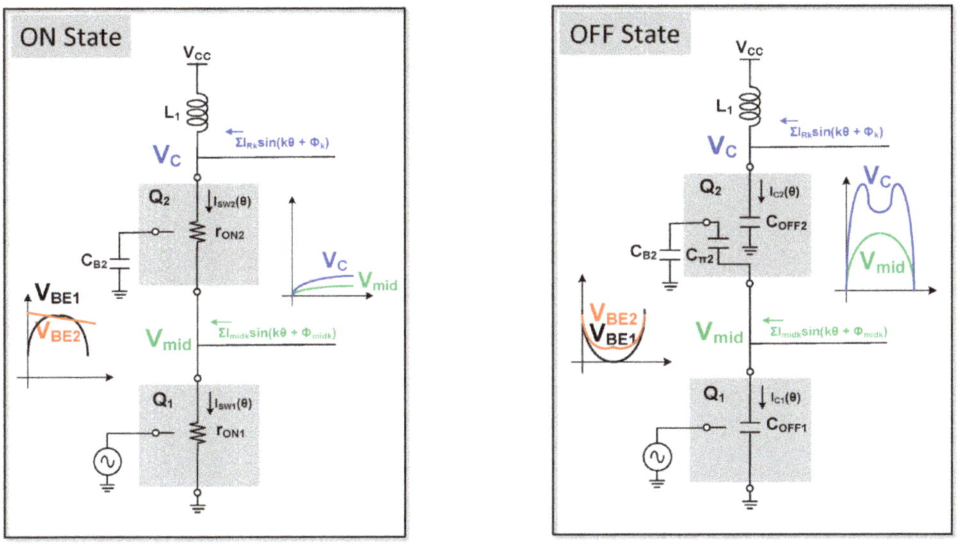

The equivalent circuits corresponding to a two-stacked switching amplifier in the ON and OFF states (assuming synchronous operation) is shown. The passive impedances at each of the collector nodes can be designed so that, for instance, the collector voltage of the bottom transistor (V_{mid}) resembles half sinusoid whereas the collector voltage of the top transistor (V_C) is closer to a square-wave.

59 34-GHz Generalized 2-Stacked SiGeHBT PA

The schematic, chip microphotograph, and performance summary table of a 34-GHz amplifier, consisting of a two-stacked-transistor configuration, with collector impedances so that each transistor collector voltage is set independently, realized in a 130nm SiGe HBT BiCMOS process, is shown.

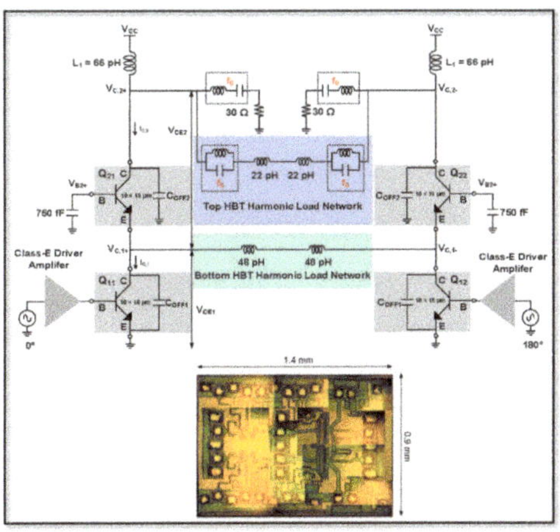

Metric	Measured Performance
Frequency	34 GHz
$P_{out,max}$	25.5 dBm
$G_{p,max}$	13 dB
η_{max}	28 %
PAE_{max}	26 %
Process	130 nm SiGe BiCMOS

K. Datta and H. Hashemi, "Waveform engineering in a mm-wave SiGe HBT stacked class-K power amplifier," in *IEEE RFIC Symposium* Digest, 2017.

Compared to the two-stacked Class-E amplifier design that was shown before (at 41 GHz), this design generates 2 dB higher output power at a lower power efficiency.

60 Survey of mm-Wave Stacked-Transistor PAs

	JSSC 2014	RFIC 2016	RFIC 2016	TMTT 2016	JSSC 2016	TMTT 2014	JSSC 2014	CSICS 2014
Freq. [GHz]	41	85	83	24	29	47.5	91	140
$P_{out,max}$ [dBm]	23.4	22	23.3	25.3	24.8	20.3	19.2	13.2
PAE_{max}	34.9%	19.1%	17.1%	20%	29%	19.4%	14%	2.8%
$G_{P,max}$ [dB]	14.5	17	18.7	13	13	12.8	12.4	9.4
Topology	2-Stack Class E	2-Stack Class E	3-Stack Class E	9-Stack Class AB	4-Stack Class A	4-Stack Class E	3-Stack Class A	3-Stack
Technology	130nm SiGe	90nm SiGe		45nm SOI CMOS				

Stacked-transistor mm-wave power amplifiers in linear and switching modes using SiGe HBT and CMOS SOI transistors have been demonstrated.

Several mm-wave amplifiers based on stacked-transistor configuration, realized in commercial silicon processes, have been reported. The performance summary of some of these reported amplifiers is tabulated in this slide.

61 Outline

- Introduction
- Technology
- Linear Amplifiers
- Switching Amplifiers
- Stacked-Transistor Amplifiers
- **Power Combining**
- Transmitters
- Conclusions

To increase the power delivered to the load, the power of multiple amplifier unit cells may be combined.

62 Transformer-based Combining

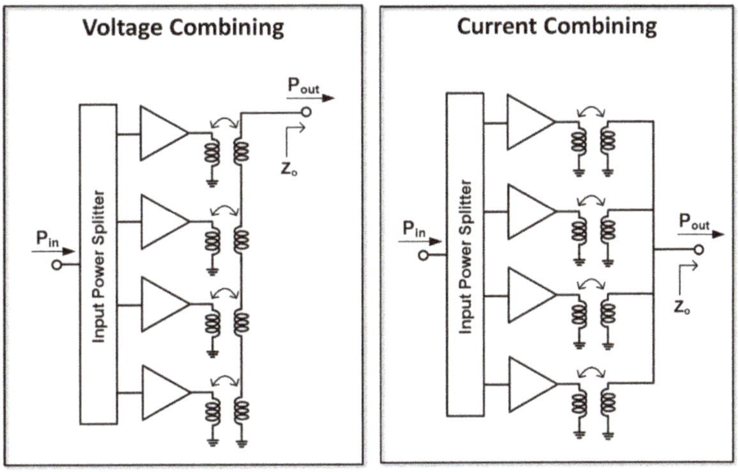

The power delivered to the load can be increased by either increasing the voltage swing across the load or increasing the current swing fed to the load.

The load voltage can be increased by coherent addition of output voltages of multiple amplifier unit cells. This can be done through a series connection of secondaries of transformers at the outputs of individual amplifiers. Assuming 1:1 transformers, in this case, the impedance presented to each amplifier unit cell is "N" times less than the load impedance where "N" is the number of amplifier unit cells whose output voltages are combined.

Alternatively, the load current can be increased by coherent addition of output currents of multiple amplifier unit cells. This can be done through a parallel connection of secondaries of transformers at the outputs of individual amplifiers. Assuming 1:1 transformers, in this case, the impedance presented to each amplifier unit cell is "N" times higher than the load impedance where "N" is the number of amplifier unit cells whose output currents are combined.

63 Corporate Wilkinson Power Combining

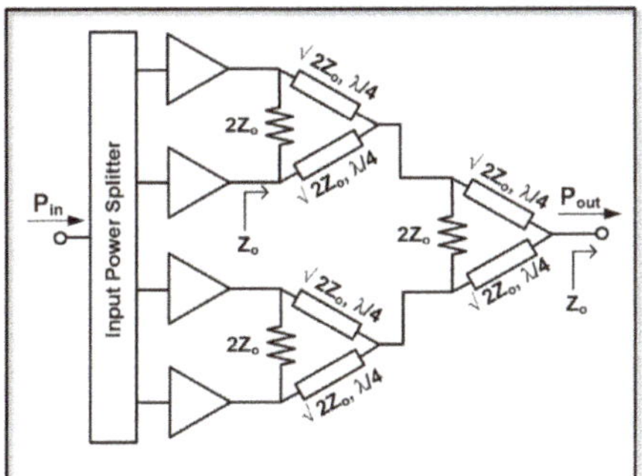

- ✓ Easily scalable
- ✓ Lumped components may be used instead of t-lines
- × Sensitive to mismatches
- × Loss α $\log_2$(# of PAs)

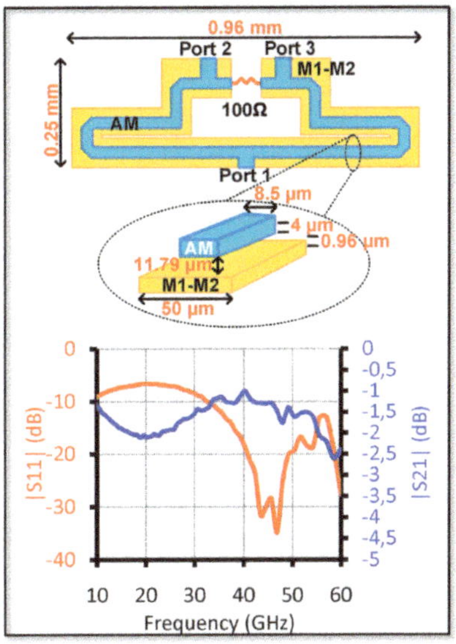

63 Corporate Wilkinson Power Combining

Impedance-matched power combining networks can also be used as power combiners. The three-port Wilkinson structure, is a common 2-way coherent power combiner with an ideal 100% efficiency for identical inputs. An array of two-way Wilkinson combiners may be used in an N-way power combining scheme. The Wilkinson power combiners are somewhat sensitive to the mismatches between the inputs – hence, they may not be desirable to combine the powers of non-identical amplifier unit cells.

An example of the layout and measured performance of an on-chip two-way Wilkinson power combiner is shown.

64 45-GHz Two-Wilkinson-Combined SiGeHBT PA

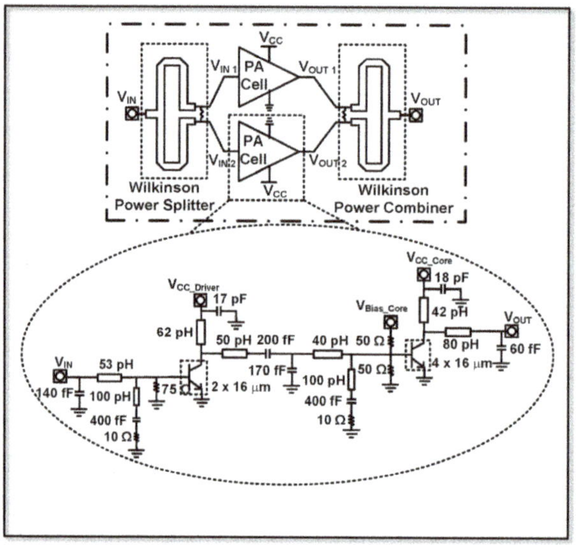

Metric	Measured Performance
Frequency	45 GHz
$P_{out,max}$	22.3 dBm
$G_{p,max}$	8.1 dB
η_{max}	24.5 %
PAE_{max}	21 %
Process	130 nm SiGe BiCMOS

K. Datta, J. Roderick, and H. Hashemi, "A 22.4 dBm two-way Wilkinson power-combined Q-Band SiGe Class-E power amplifier with 23% peak PAE," in *Proceedings of CSICS*, October 2012.

The schematic, chip microphotograph, and performance summary table of a 45-GHz two-way-Wilkinson-power-combined Class-E amplifier, realized in a 130nm SiGe HBT BiCMOS process, is shown. It is interesting to point out that, the output power and efficiency of this design are less compared to the two-stacked-transistor Class-E implementation that was shown before. In other words, purely from the perspective of output power and efficiency, transistor stacking may be more effective compared with Wilkinson power combining.

65

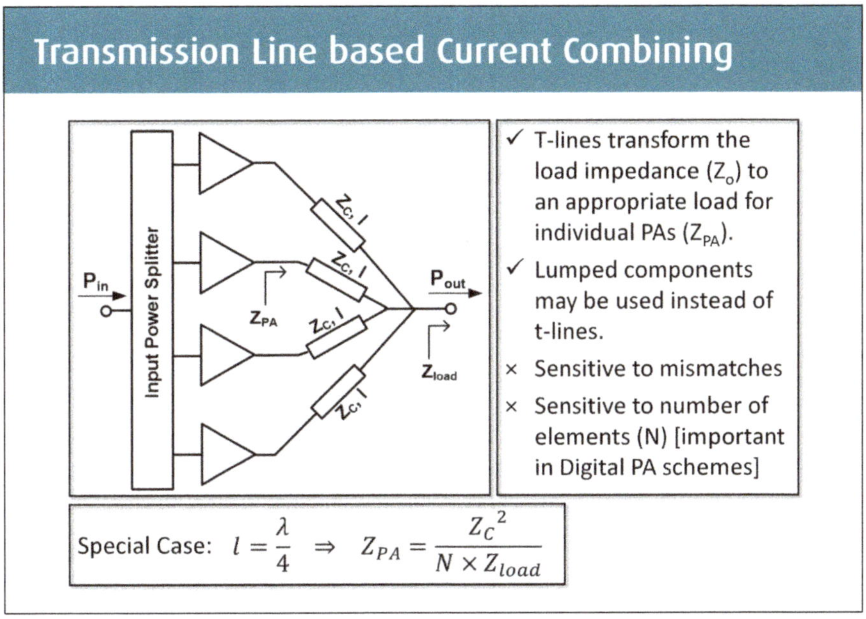

Current combining can use transmission lines instead of transformers. At mm-wave frequencies, transmission line current combining is typically easier to realize when compared with transformer-based combiners. Depending on the frequency, transmission lines or lumped-component approximation of transmission lines may be used. The characteristic impedance and electrical length of transmission lines, and the total number of power-combined amplifiers, determine the impedance that is offered to each amplifier unit cell.

This scheme works well for identical power amplifier unit cells. However, the power combining network is sensitive to the impedances at the outputs of all amplifier unit cells. This can be an issue if amplifier unit cells are turned ON and OFF to create digital amplitude modulation (aka digital power amplifier).

66

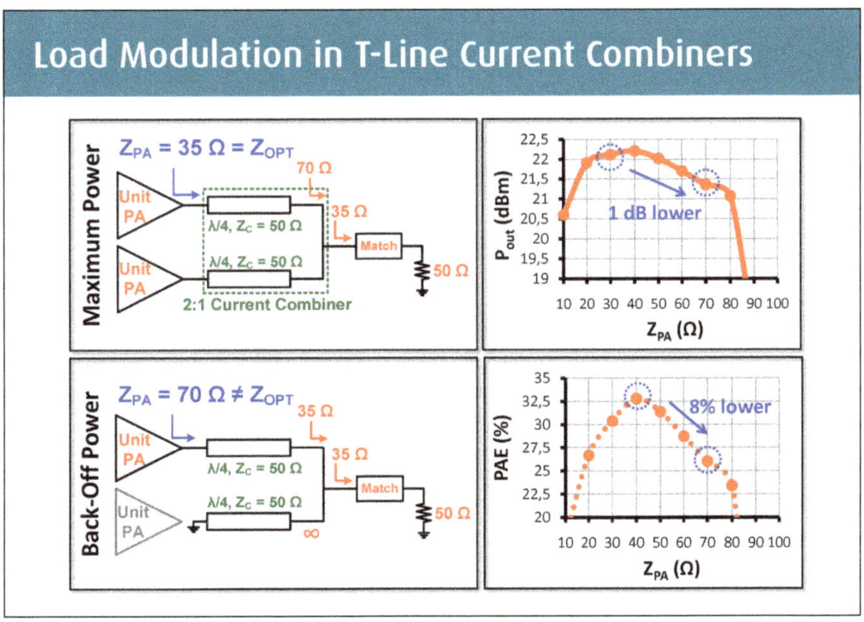

To support power control or amplitude modulation, an array of power-combined switching amplifier unit cells may be used. However, as shown here, the impedance presented to each amplifier unit cell depends on the output impedance of the other amplifier unit cells in the transmission-line-based current-combined network. Therefore, as some of the amplifier unit cells are turned OFF to support lower output power levels (or lower amplitudes for the AM waveform), the impedance presented to all other amplifiers that are turned ON is affected. This deviation of the impedance presented to each amplifier unit cell, aka load modulation, degrades the output power and power efficiency.

67

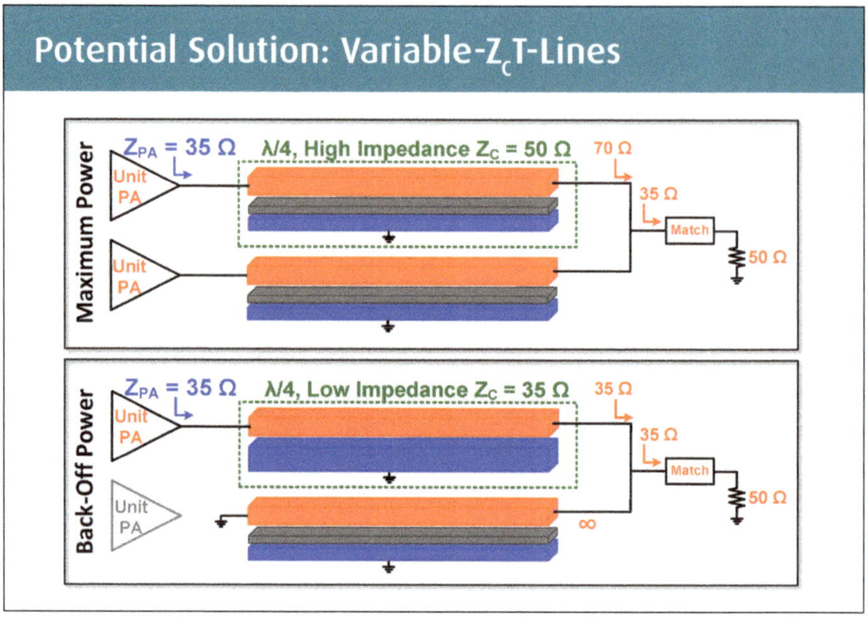

Various methods can be used to ensure that the impedance presented to each amplifier unit cells remains constant as other amplifier unit cells are turned ON and OFF. One approach utilizes transmission lines with variable characteristic impedance.

An example of a two-way transmission-line-base current combiner, using transmission lines with variable characteristic impedance to ensure that the impedance presented to the top amplifier unit cell remains constant (35 Ω in this example) in both cases when the bottom amplifier unit cell is turned ON (top) and OFF (bottom), is shown.

68

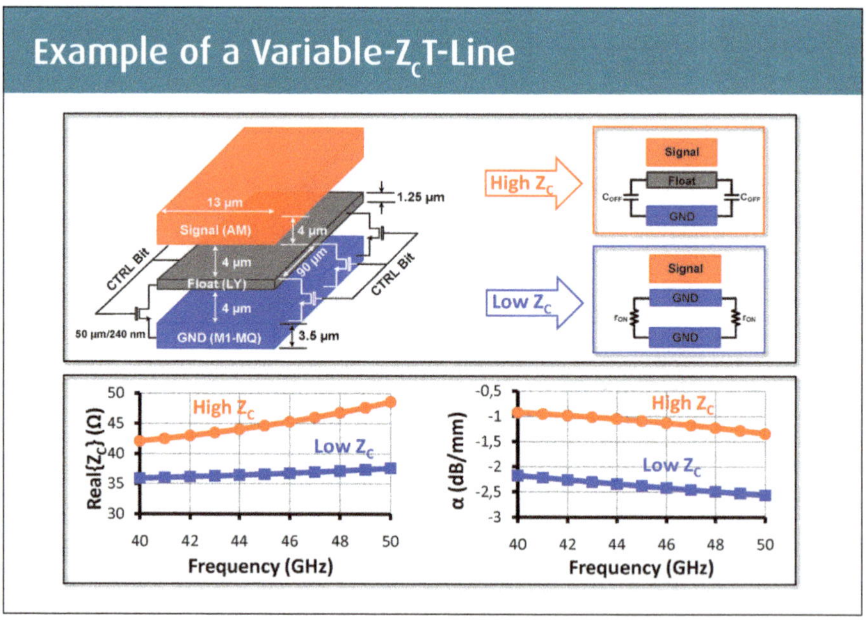

On-chip transmission lines with variable characteristic impedance can be realized in standard silicon processes. One example of such a transmission line is shown here where the distance between the transmission lines signal (realized in a top metal layer called AM) and ground lines (realized in lower metal layers) is changed through MOSFET switches.

The large number of metal layers available in commercial silicon processes enable realization of various transmission lines where the characteristic impedance and/or the propagation constant can be varied by changing the signal and ground properties.

69 45-GHz SiGeHBT Digital Power Amplifier

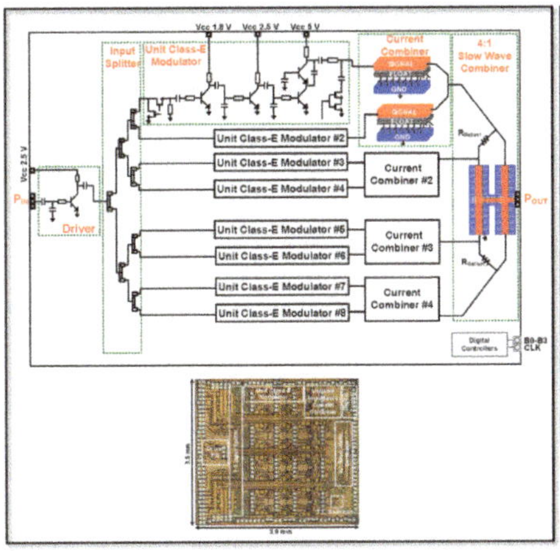

Metric	Measured Performance
Frequency	46 GHz
$P_{out,max}$	28.9 dBm
$G_{p,max}$	13 dB
PAE_{max}	18.4 %
Power Levels	$2^3 = 8$
Efficiency at -6dB power back-off	11%
Process	130 nm SiGe BiCMOS

K. Datta and H. Hashemi, "A 29 dBm, 18.5% peak PAE, mm-wave digital power amplifier with dynamic load modulation," in *ISSCC* Digest of Technical Papers, Feb 2015.

Schematic, layout, and performance summary of a 45-GHz, 29 dBm, SiGe HBT power amplifier with 3-bit amplitude control, based on an 8-way transmission line current combiner with variable-impedance transmission line to mitigate the effect of load modulation, is shown.

70 Spatial Power Combining

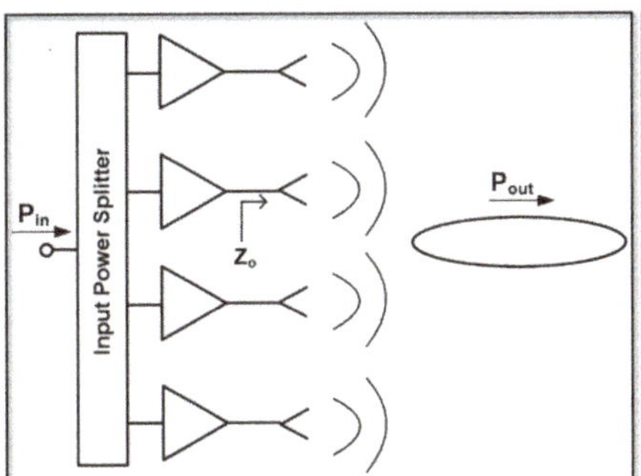

- ✓ Extendible to support beamforming
- ✓ No combining loss
- ✓ Robust to failure
- × Delicate packaging requirement for chip-antenna interfaces (or for on-chip antennas)

It is very likely that antenna arrays are needed in fixed and mobile mm-wave devices to (1) satisfy the link budget requirements, and (2) provide spatial interference mitigation for the future envisioned wireless systems (e.g., 5G standards).

The efficiency of on-chip power combining is limited by the loss of on-chip passive components. Alternatively, the power of multiple amplifier unit cells can be added in space. Although, spatial power combining is lossless, the loss of the radiating structures (antennas) limits the power efficiency. The antennas may be realized on the chip or in a package that houses the silicon chip.

Many of the envisioned mm-wave application, including radar and wireless communications, benefit from beam-forming. In these applications, the power of multiple amplifier unit cells is naturally combined in the desired spatial directions.

71 Survey of mm-Wave Power-Combined Si PAs

	JSSC 2016	ISSCC 2013	ISSCC 2015	RFIC 2014	TMTT 2015	TMTT 2015	RFIC 2014	TMTT 2015
Freq. [GHz]	24	42	46	120	40	45	94	73
$P_{out,max}$ [dBm]	28.7	28.4	28.9	20.8	27.2	40[1]	33[1]	22.6
PAE_{max}	21.9%	18.5%	18.4%	7.6%	10.7%	13.5%[2]	2%[4]	19.3%
$G_{P,max}$ [dB]	16.1	10	13	15	19.4	29[3]	27[3]	25.3
Combining	4-Way On-Chip	16-Way On-Chip	8-Way On-Chip	8-Way On-Chip	8-Way On-Chip	4-Way Spatial	8-Way Spatial	4-Way On-Chip
Technology	0.35mm SiGe	130nm SiGe		90nm SiGe	45nm SOI CMOS			40nm CMOS

[1] Effective isotropic radiated power (EIRP)
[2] PAE of on-chip power amplifiers excluding the loss of radiators
[3] Defined as $EIRP_{max} - P_{in}$
[4] Total efficiency including on-chip power amplifiers as well as radiators

On-chip and spatially power combined mm-wave power amplifiers using SiGe HBT and CMOS SOI transistors have been demonstrated.

Several mm-wave power amplifiers that use power combining have been reported in commercial SiGe HBT and CMOS processes. This table shows the performance summary of a few of such amplifiers.

72 Outline

- Introduction
- Technology
- Linear Amplifiers
- Switching Amplifiers
- Stacked-Transistor Amplifiers
- Power Combining
- **Transmitters**
- Conclusions

Millimeter-wave transmitters should generate the desired modulated waveforms at the required power levels with high power efficiency.

73

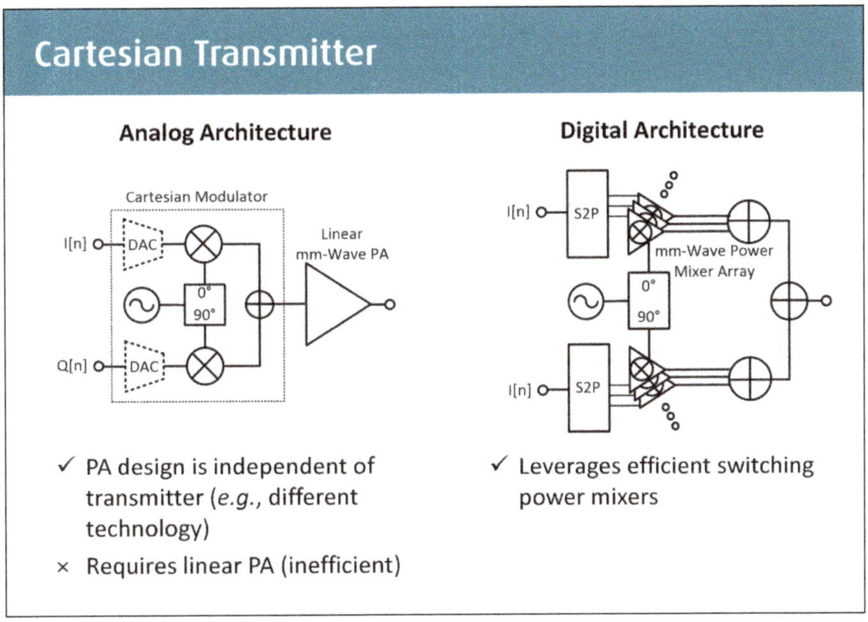

The modulated waveform of the form $I(t)\cos(\omega_{RF}\ t)+Q(t)\sin(\omega_{RF}\ t)$ can be created directly using a Cartesian transmitter scheme.

In an analog Cartesian transmitter, the digital information streams I[n] and Q[n], are converted to analog waveforms and directly modulated with quadrature up-conversion mixers. A linear mm-wave power amplifier is needed to boost the power level of modulated waveform to a desired value without distortion. Similar to RF transmitters, envelope tracking (modulating the PA supply to follow the envelope of modulated waveform) may be used to improve the efficiency of mm-wave PA across all amplitude levels.

Alternatively, in a digital Cartesian transmitter, the digital information data streams I[n] and Q[n] acts as ON/OFF controls of a pair of mm-wave mixer arrays. The mm-wave mixer array may generate the necessary power. The power efficiency of the digital Cartesian transmitter may be higher than its analog counterpart due to the higher efficiency of switching mixer arrays when compared with linear analog mixers.

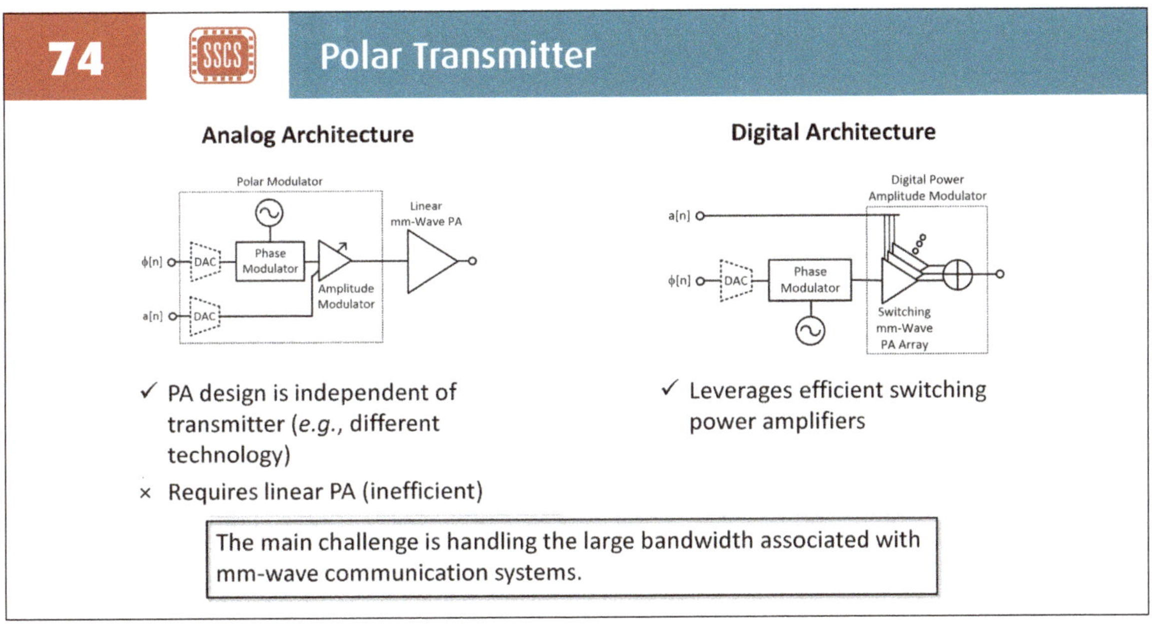

74 Polar Transmitter

The modulated waveform may also be represented in a polar coordinate as $a(t)\cos(\omega\_RF\ t+\emptyset(t))$ enabling realization of polar transmitter architectures. Again, both analog and digital implementations are possible with the digital scheme has the advantage of potentially higher power efficient thanks to the higher efficiency of switching amplifier unit cells that are used for amplitude modulation.

In general, for a given modulated waveform, the bandwidth of $a(t)$ and $\emptyset(t)$ is higher compared with $I(t)$ and $Q(t)$. Therefore, the amplitude and phase modulators of a polar architecture must support a larger bandwidth when compared with the mixers used in the Cartesian scheme.

75 mm-Wave Doherty Amplifiers

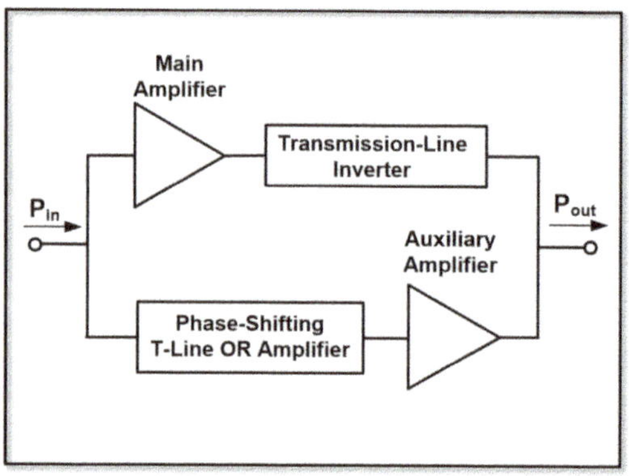

	RFIC 2008	JSSC 2013	TMTT 2015
Freq. [GHz]	60	42	72
$P_{out,max}$ [dBm]	7.8	18	21
Peak PAE	3	20%	19.2%
PAE at -6 dB back-off	1.5%	21%	7%
$G_{P,max}$ [dB]	13.5	8	18.5
Technology	130nm CMOS	45nm SOI	40nm CMOS

Doherty architecture has been applied to mm-waves to improve the power efficiency of power amplifiers across the power levels.

Doherty structure, common at radiofrequencies as a scheme to improve the efficiency of power amplifier across the power levels (e.g., peak power as well as back-off) , is also being applied to mm-wave realizations. The smaller wavelength of mm-wave frequencies enables low-loss compact realizations of transmission-line or quasi-transmission-line inverters.

76 mm-Wave Transmitter Challenges

- Maintaining the linearity of modulators and power amplifiers over a wide bandwidth, needed to maintain high data rate, is a challenge.
- Phase noise of local oscillator is often the major limitation in reducing the error vector magnitude (EVM) of high data-rate transmitters.

One motivation to use mm-wave frequencies in wireless communication schemes is to achieve higher data rate through wider channel bandwidths. Therefore, the mm-wave modulators must support a much higher bandwidth when compared with radiofrequency modulators.

The quality of phase modulated waveforms depends on the phase noise of the local oscillator. The oscillator phase noise degrades with frequency. Therefore, generating complex modulated waveforms at mm-wave frequencies is a challenge.

77 Survey of mm-Wave Silicon Transmitters

	IMS 2015	JSSC 2012	ISSCC 2014	JSSC 2016	JSSC 2015	RFIC 2014
Freq. [GHz]	45	60	60	60	78.5	94
$P_{out,max}$ [dBm]	28	15.6	10.3	10.8	12	N/A
Peak PAE	13.5%	25%	10%	29.8%	15%	N/A
Modulation	64-QAM	16-QAM	16-QAM	16-QAM	16-QAM	64-QAM
Data Rate [Gbps]	0.15	0.5	28.16	6.67	14	1
$P_{out,avg}$ [dBm]	18.5	12.5	N/A	7.2	6	4
η_{avg}	3.1%	15%	N/A	19.8%	N/A	0.1%
EVM [dB]	-28.4	-22	-20	-16.5	-18	-28
Topology	Cartesian	Out Phasing	Cartesian	Digital Polar	Cartesian	Cartesian
Technology	45nm SOI + 130nm SiGe	40nm CMOS	65nm CMOS	40nm CMOS	40nm CMOS	45nm CMOS SOI

The performance summary of selected mm-wave silicon transmitters is highlighted here.

78 Outline

- Introduction
- Technology
- Linear Amplifiers
- Switching Amplifiers
- Stacked-Transistor Amplifiers
- Power Combining
- Transmitters
- **Conclusions**

We conclude this chapter with the following bullet points.

79 Conclusions

- ❑ Technology scaling had led to
  - ❑ increased transistor f_{max},
  - ❑ reduction of breakdown voltages.
- ❑ Efficient mm-wave switching power amplifiers have been demonstrated in silicon. Digital transmitter architectures that leverage these switching amplifiers may be developed.
- ❑ Output power of mm-wave PAs can be increased by
  - ❑ power combining (on-chip and spatial),
  - ❑ transistor stacking.
- ❑ Energy-efficient, high data-rate, integrated, mm-wave, silicon transmitters are still active areas of research.

80 SSCS

Acknowledgements

- Dr. Kunal Datta, formerly at USC and now with Skyworks
- Pingyue Song, USC
- Defense Advanced Research Projects Agency (DARPA)

Kunal Datta, formerly at USC, led the research towards switching mm-wave SiGe HBT amplifiers. PingYue Song at USC supplied the simulations for on-chip mm-wave passives. The research at USC towards silicon mm-wave power amplifiers was supported in part by the Defense Advanced Research Projects Agency.

CHAPTER 02

Millimeter-wave Full-Duplex Wireless: Circuits and Systems

Tolga Dinc

Texas Instruments, USA

Harish Krishnaswamy

Columbia University, USA

In this chapter, we will cover the basic principles of full duplex wireless operation at millimeter-wave frequencies. We will cover potential applications of millimeter-wave full-duplex operation. The main focus of this chapter will be on two different antenna interfaces to enable mmwave full-duplex, namely a wideband reconfigurable T/R antenna pair with polarization-based antenna cancellation and an mm-wave fully-integrated magnetic-free non-reciprocal circulator. Finally, we will briefly discuss some system concepts, including overall self-interference cancellation requirements, its partitioning across different domains and the first demonstration of a millimeter-wave full-duplex wireless link.

1 **Presentation Outline**

- Introduction
- A 60 GHz Full-Duplex Transceiver in 45nm SOI CMOS
- A 25GHz Magnetic-free Circulator in 45nm SOI CMOS
- A 45nm SOI CMOS Circulator
- Conclusion

In this chapter, we will cover the basic principles of full duplex wireless operation at millimeter-wave frequencies. We will cover potential applications of millimeter-wave full-duplex operation. The main focus of this chapter will be on two different antenna interfaces to enable mmwave full-duplex, namely a wideband reconfigurable T/R antenna pair with polarization-based antenna cancellation and an mm-wave fully-integrated magnetic-free non-reciprocal circulator. Finally, we will briefly discuss some system concepts, including overall self-interference cancellation requirements, its partitioning across different domains and the first demonstration of a millimeter-wave full-duplex wireless link.

INTRODUCTION

2 **5G Wireless Capacity**

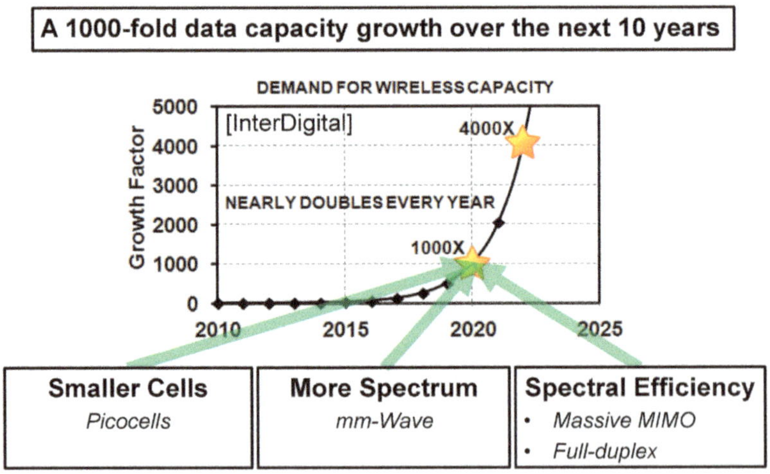

A 1000-fold increase in data traffic is projected over the next 10 years as demand for wireless capacity keeps growing exponentially every year.

Solutions for delivering the 1000-fold increase in capacity fall into three main categories:

1. Deploying smaller cells (especially in urban settings).
2. Allocating more spectrum (This is where moving to mm-wave frequencies becomes promising.)
3. Improving spectral efficiency.

Full-duplex is one of the emergent technologies to improve spectral efficiency and gained a lot of research interest in the recent years.

3 Full-Duplex Wireless

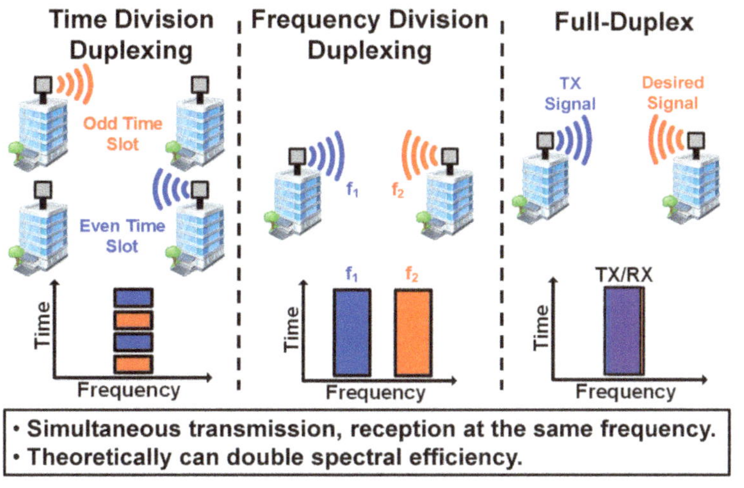

Compared to the traditional duplexing schemes which transmit and receive either in different time intervals (time-division duplexing) or on different frequency channels (frequency-division duplexing), full-duplex can theoretically double the spectral efficiency in physical layer by simultaneous transmission and reception on the same frequency channel. It also offers many other new benefits in the higher layers.

4 mm-Wave Full-Duplex

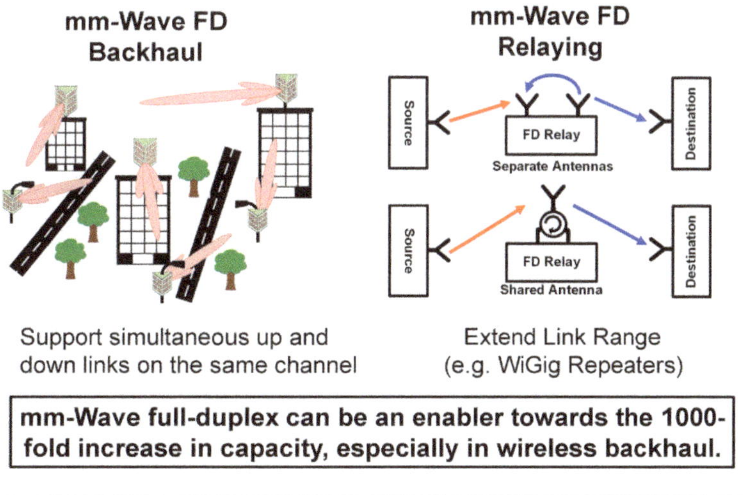

When combined with full-duplex operation, mmWave links can offer wide bandwidths with improved spectral efficiency, improving network capacity. Hence, mm-Wave full-duplex can be an enabler towards the 1000-fold increase in data traffic, especially in mm-Wave backhaul.

Mm-wave full-duplex (FD) can also be useful to extend the link range. For example mm-Wave FD relays can provide significant throughput improvement over existing half-duplex (HD) relays. However, to enable these mm-wave full-duplex applications, low-loss, high isolation, small form-factor, CMOS compatible antenna interfaces are extremely crucial.

5 Self-driving Cars

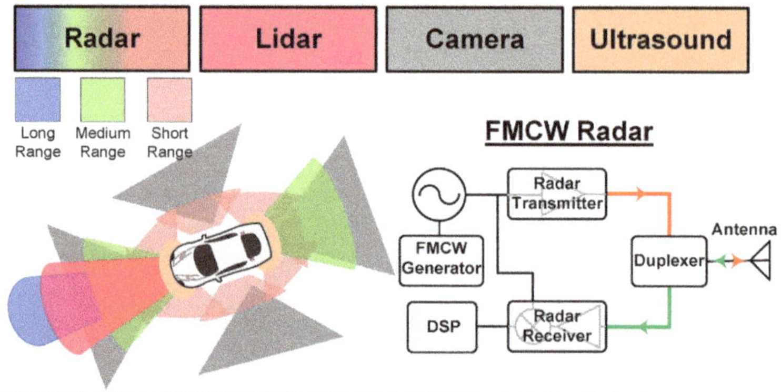

FMCW Radar is the backbone of automotive sensing, covering short/medium/long range in any weather condition, day or night.

Another exciting technology trend is self-driving cars. Among many sensor technologies, FMCW radar forms the backbone of automotive sensing since it can cover short/medium/long range in any weather condition, day or night. As depicted in this simple block diagram, a monostatic FMCW radar requires a duplexer to share a single antenna between transmitter and receiver. As industry moves toward offering low-cost fully-integrated radars, design of low-loss, low noise CMOS-compatible shared antenna interface still remains as an open research problem.

6 5G Base-Stations

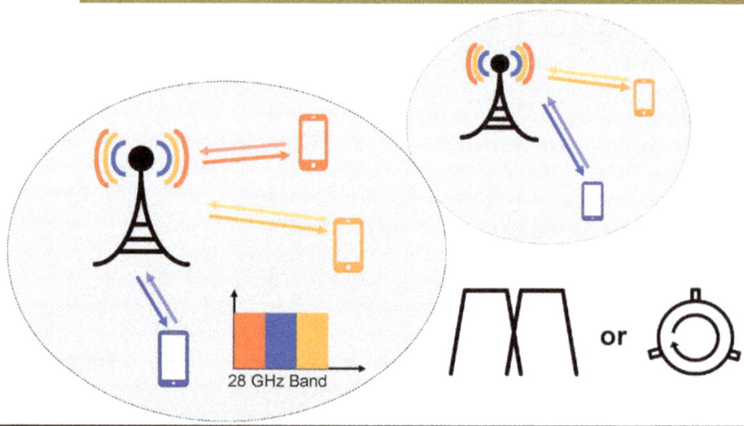

5G base-stations will communicate with multiple users simultaneously on uplink and downlink in adjacent channels.

Furthermore, in the 5G era, millimeter-wave small-cell 5G base-stations are envisioned to communicate with multiple users simultaneously in uplink and downlink. Although the multiple users will occupy adjacent channels in the same band, a low-loss, high-isolation circulator with high power handling capability would be useful to share a single antenna between TX and RX while eliminating the need for high-quality millimeter-wave diplexers.

7 Full-Duplex Challenge (I)

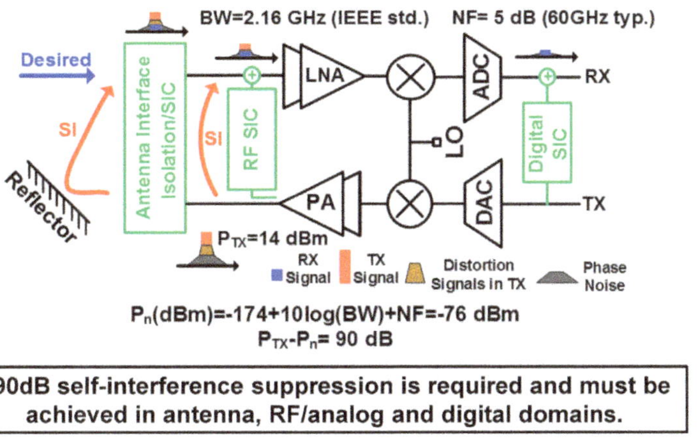

Although the full-duplex concept seems straightforward, strong self-interference from the transmitter to its own receiver, which can be one billion times stronger than the desired signal, forms a fundamental challenge. For example, in a 60GHz full-duplex transceiver with transmit power of +14dBm, receiver noise figure of 5dB and channel bandwidth of 2.16GHz, more than 90 dB self-interference suppression is required. This must be achieved across multiple domains - antenna, RF, analog and digital.

8 Full-Duplex Challenge (II)

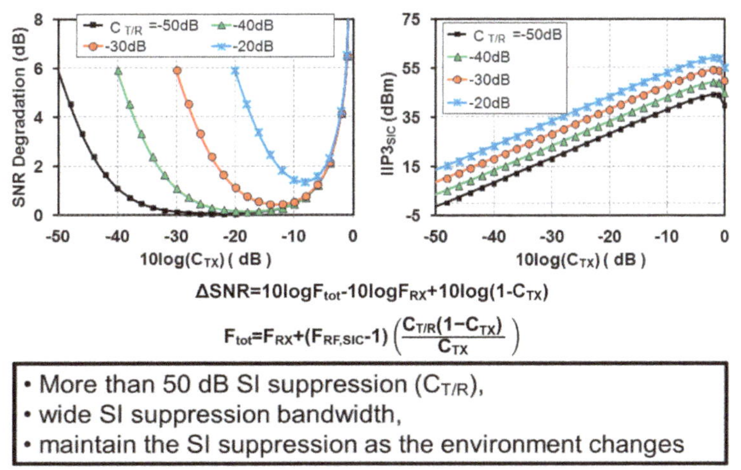

There are several design parameters that trade off with each other. $C_{T/R}$ is the self-interference suppression at the antenna interface. C_{TX} is the coupling coefficient at the transmitter (TX) - the ratio of the transmitter signal that is coupled out for RF self interference cancellation. C_{RX} is the corresponding coupling coefficient at the receiver (RX). F_{RX} is the noise figure of the receiver and F_{tot} is the total noise figure after including the noise contribution of the RF self interference canceller ($F_{RF,SIC}$). While the details of the graphs and calculations are beyond the scope of this chapter, the main message is that a large self-interference suppression within the antenna interface itself is desirable, as it relaxes the dynamic range requirements on the RF, analog, and digital blocks in the receiver chain as well as the RF/ analog and digital SIC circuits.

A 60 GHZ FULL-DUPLEX TRANSCEIVER IN 45NM SOI CMOS

Next, we will discuss a 60GHz full-duplex transceiver enabled by a novel reconfigurable polarization-based cancellation technique in the antenna domain.

9 Polarization-Division Duplex

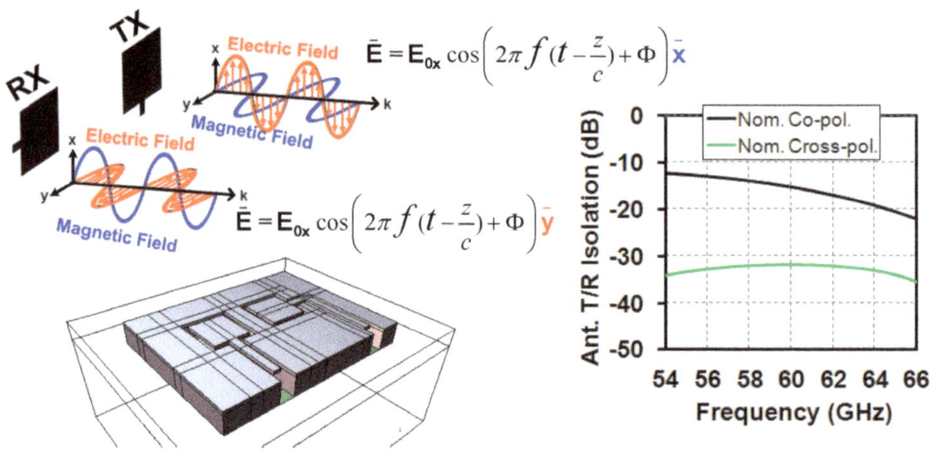

Using different polarizations for TX and RX *improves the T/R isolation by 10 -16dB.*

In RF domain, there are there parameters to characterize a signal : frequency, amplitude and phase. Once we move into the antenna domain, polarization is another degree of freedom. We leverage polarization to increase the initial isolation between the transmit and receive antennas. This provides 10 to 16 dB improvement in the T/R isolation for rectangular slot antennas shown in this figure.

10 Polarization-Based Cancellation

In addition to enhancing the initial isolation, the polarization domain can be used to embed cancellation. This figure shows the SI problem at the antenna interface. The main mechanism is the inherent coupling between the co-located TX and RX antennas. An auxiliary port co-polarized with the transmit port is introduced on the receive antenna. This creates an indirect coupling path from the transmitter output and the receiver input. The auxiliary port is terminated with a reconfigurable reflective termination which sets the amplitude and phase of the indirect signal to cancel the self-interference at the receiver input.

10 Polarization-Based Cancellation

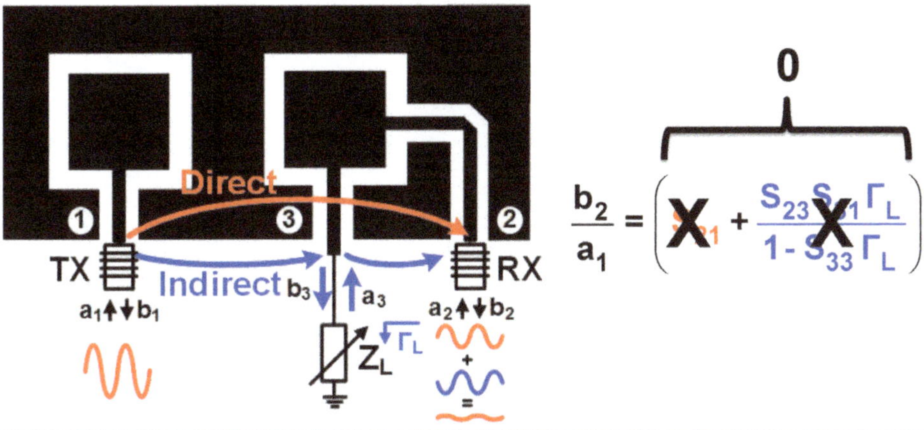

An auxiliary port is introduced on the RX antenna that is co-polarized with TX and terminated with a reflective termination to achieve self-interference cancellation (SIC).

11 60GHz Pol.-Based Cancellation

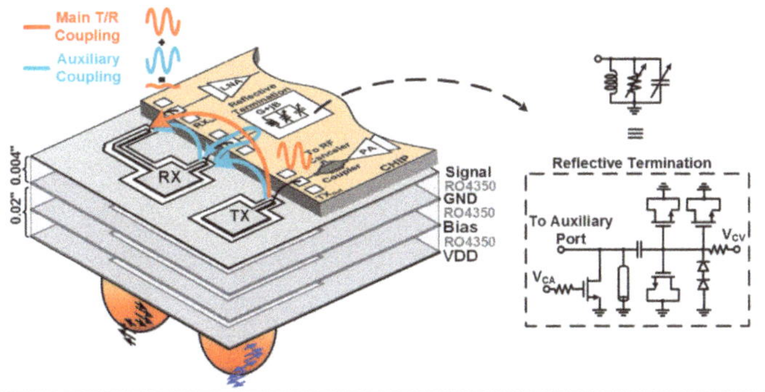

The polarization-based cancellation is implemented at 60GHz and integrated with a 45nm SOI CMOS full-duplex TRX.

This figure illustrates the implementation of the technique at 60GHz and its integration with a CMOS full-duplex transceiver. The T/R antennas are implemented on Rogers 4350B as rectangular slot antennas because of their higher bandwidth. The auxiliary port is terminated with an on-chip reflective termination consisting of variable R, variable C and fixed L in parallel. This enable us to electronically reconfigure the antenna cancellation to combat the SI scattering from the environment.

12

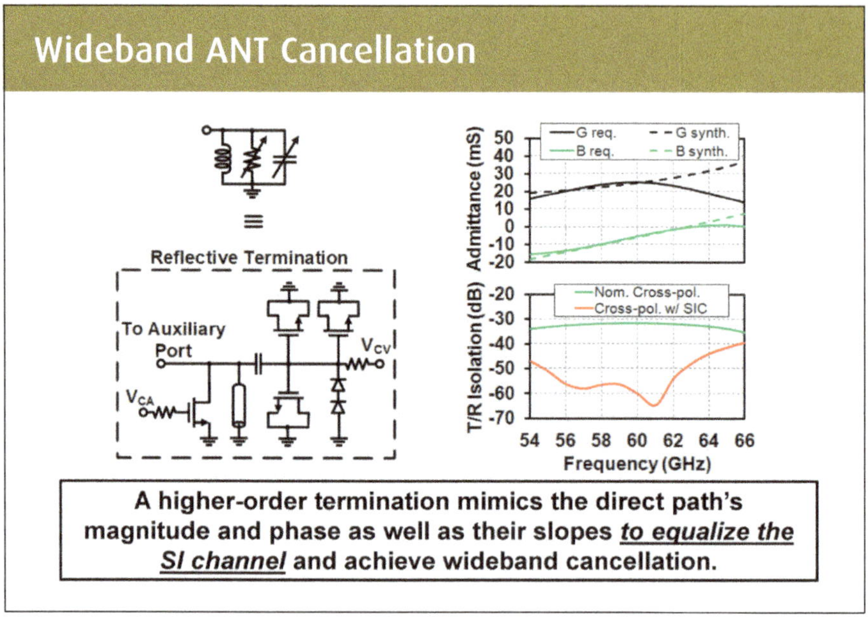

A higher-order termination mimics the direct path's magnitude and phase as well as their slopes to equalize the SI channel and achieve wideband cancellation. This is accomplished by synthesizing the required reflective termination for cancellation across frequency. Solid lines on the figure on the top right show the simulated required conductance and susceptance for perfect cancellation across frequency. The reflective termination explicitly synthesizes both the magnitude and the slope of the required susceptance. The slope of the required conductance is relatively flat and therefore is automatically synthesized by the variable resistance. The synthesized reflective termination provides more than 50 dB isolation over 8 GHz bandwidth in simulation.

13

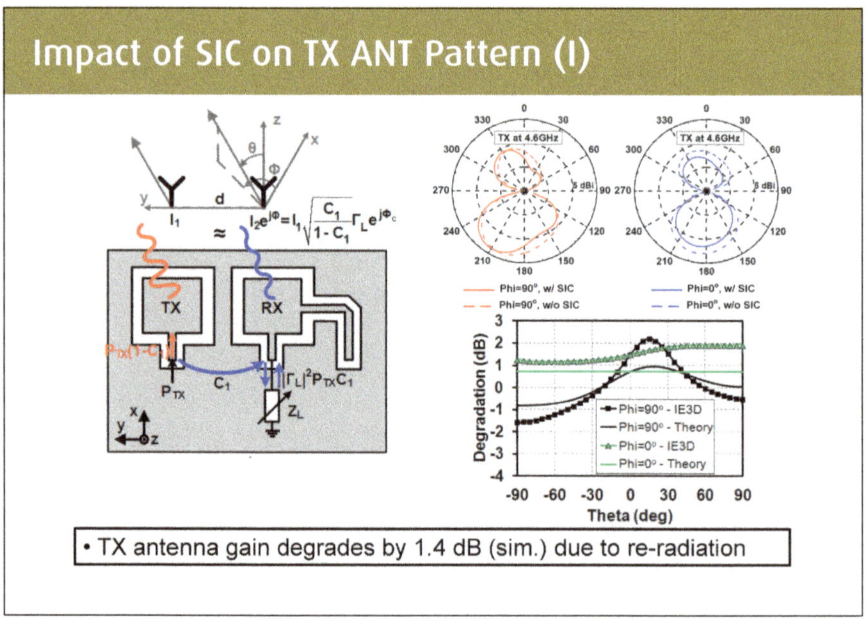

The main concern regarding the self-interference cancellation in the antenna domain is its effect on the antenna patterns. The simulated T/R antenna patterns at 60GHz are shown in this slide. When SIC is enabled, the TX antenna gain degrades by 1.1dB in the broadside direction due to re-radiation from the RX antenna. A small portion of the TX signal couples to the auxiliary port, is reflected by the termination and eventually radiates. This small radiation interferes with the TX radiation in the far field.

14 Impact of SIC on RX ANT Pattern (II)

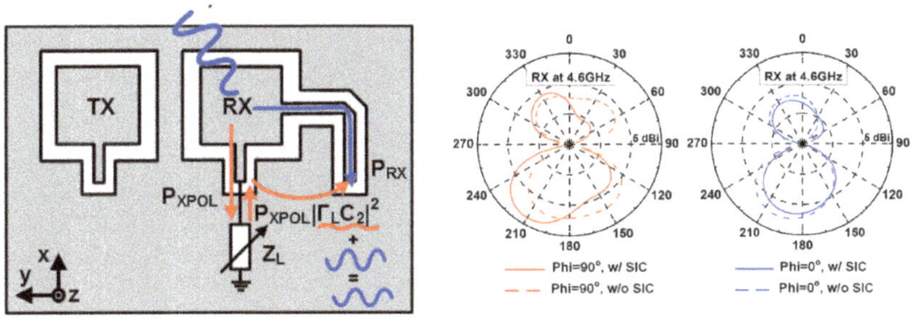

•RX antenna gain degrades by 1 dB in the broadside direction, agrees well with theory.

Similarly, SIC reduces the RX antenna gain by 0.18 dB. These penalties are comparable to the TX efficiency and NF penalties of RF cancellers. Higher initial isolation between the TX and auxiliary ports reduces these penalties.

15 60GHz 45nm SOI CMOS TRX

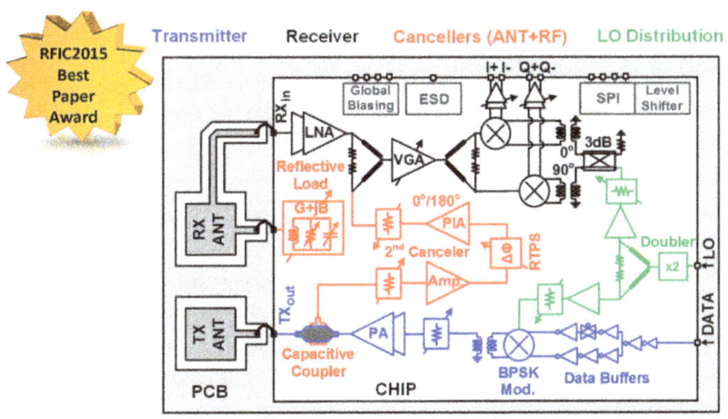

• World's first mm-Wave fully-integrated full-duplex TRX IC.
• RF cancellation further suppresses the self-interference.
• Shared LO to reduce the impact of phase noise on SIC.

A 60 GHz CMOS full-duplex transceiver is designed by using the polarization based antenna cancellation technique. In addition to the antenna cancellation, a second RF cancellation path from the transmitter output to the LNA output is also integrated on chip in order to suppress the self interference further. The transmitter and receiver share the same LO to reduce the impact of phase noise on self interference cancellation. This is the first mm-Wave fully-integrated full-duplex transceiver IC.

16 SIC Distribution across RX (I)

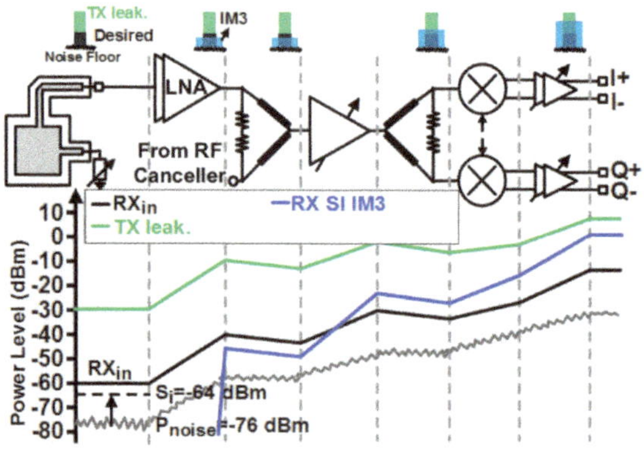

Without SIC, SI and its IM products generated along the RX masks the desired signal.

This graph takes a closer look into the self-interference cancellation distribution along the receiver chain. The gray line shows the noise floor along the receiver. The black line is the desired signal and the green line is the TX leakage, or SI. Blue line tracks the SI intermodulation products generated along the receiver. Without self-interference cancellation, the self interference and its inter-modulation products generated along the receiver mask the desired signal.

17 SIC Distribution across RX (II)

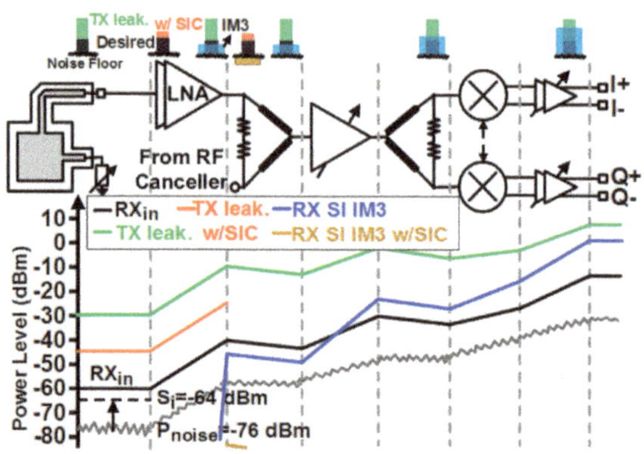

- **ANT cancellation suppress the SI at the LNA input.**
- ***IM products generated by the LNA fall below the noise floor.***

Antenna cancellation suppresses the SI at the LNA input so that inter-modulation products generated by the LNA fall below the noise floor.

18 SIC Distribution across RX (III)

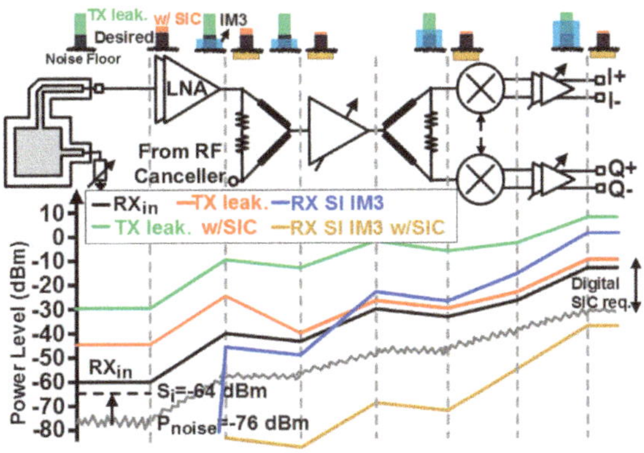

RF cancellation further suppresses the SI so that IM products generated by Mixer and IF Amps fall below noise floor.

RF cancellation should further suppress the SI so that inter-modulation products generated by the mixer and the baseband circuits fall below noise floor. Additional cancellation is required and this can be achieved in digital domain.

19 Chip Photograph

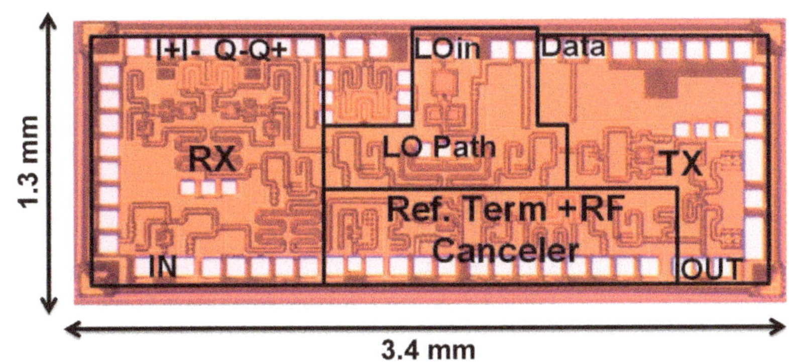

60 GHz Full-Duplex Transceiver in IBM 45nm SOI CMOS.

The transceiver chip is implemented in a GF 45nm SOI CMOS process and occupies 1.3 by 3.4 sq. mm area.

20 RF Canceller Implementation (I)

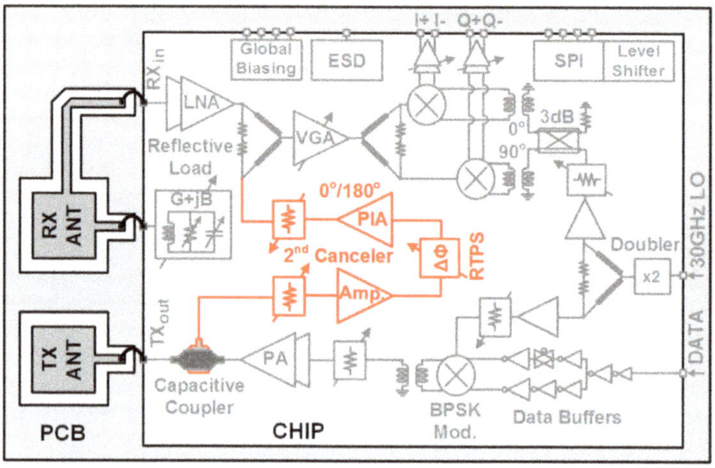

We will now briefly cover the block implementations. The RF cancellation path employs a capacitive coupler at the PA output. The TX copy is fed into an attenuator, a 60GHz amplifier, a reflection-type phase shifter and a phase inverting amplifier. Finally the cancellation signal is injected through a variable attenuator and a Wilkinson combiner to the receiver path so as to not degrade the receiver NF.

21 RF Canceller Implementation (II)

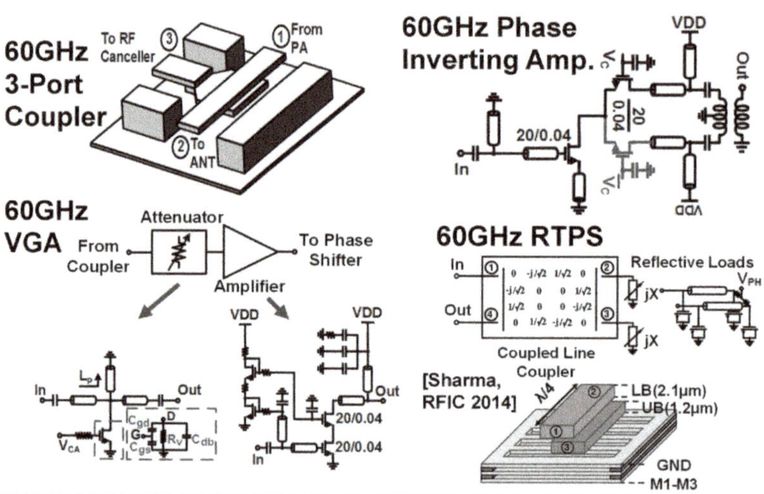

The 3-port 18dB coupler is implemented using the two topmost metal layers. Gain control in the cancellation path is achieved by reflective-type variable attenuators. The reflection-type phase shifter (RTPS) is implemented as a coupled line coupler terminated with variable C-L-C terminations. Phase inverting amplifier is designed essentially as a single-balanced mixer operating in the static mode.

22 RF Cancellation Path Meas. (I)

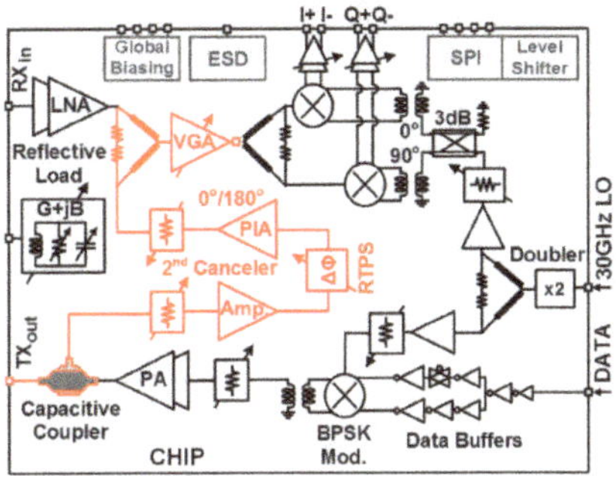

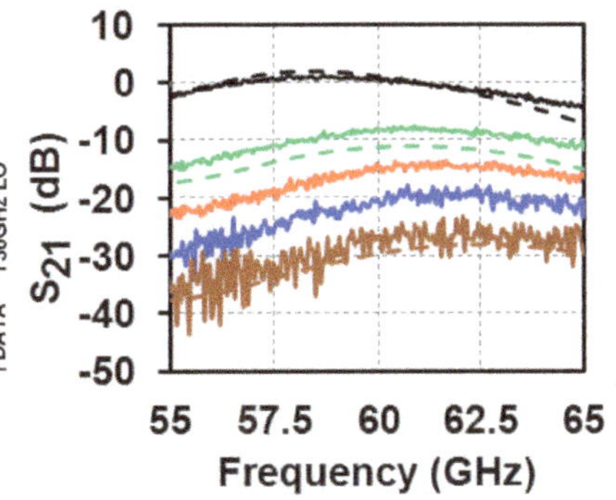

32dB gain control range in the RF Cancellation Path.

The RF canceller is measured from the transmitter output pad to an internal pad after the VGA in the receiver chain. It provides 32 dB gain control range.

23 RF Cancellation Path Meas. (II)

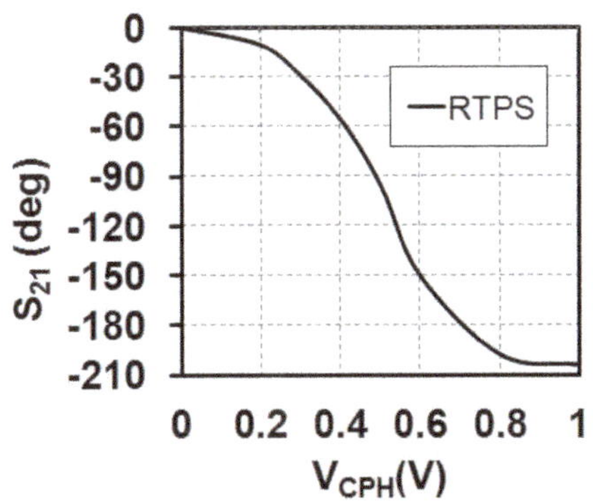

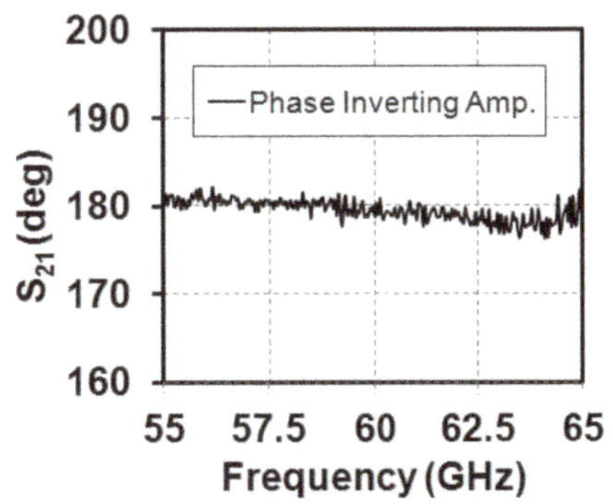

206° analog and 180° discrete phase control (Total > 360°).

The RF cancellation path has 206° continuous and 180° discrete phase control at 60 GHz.

24 Transmitter Implementation (I)

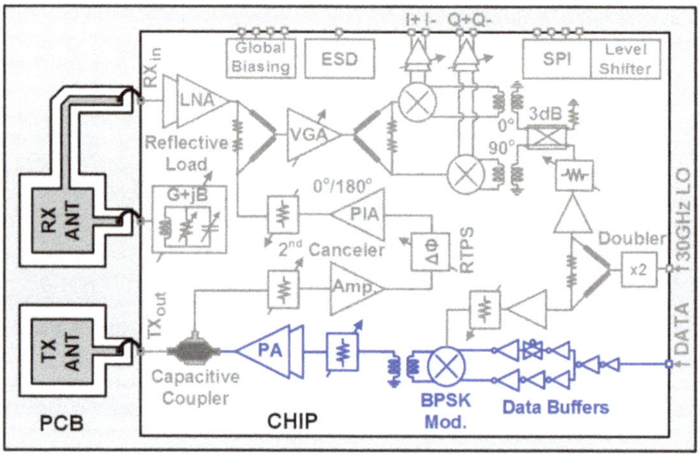

The transmitter includes an inverter-chain data buffer driving a BPSK modulator, a transformer balun, a reflective-type attenuator and a two-stage 60GHz power amplifier.

25 Transmitter Implementation (II)

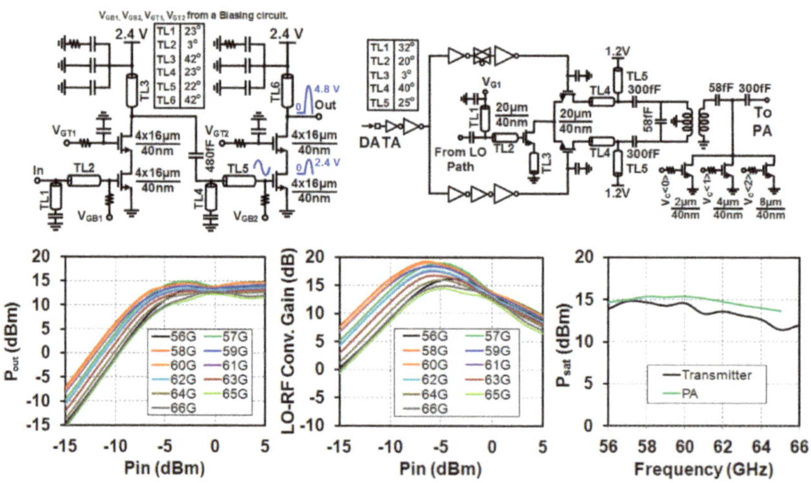

A 60GHz two-stage, two-stacked Class-E-like PA is used to achieve a high output power with high-efficiency. A BPSK modulator (essentially a single-balanced mixer) is implemented on chip for direct modulation of the 60GHz LO signal by a large-swing DATA signal.

26 PA and Transmitter Meas.

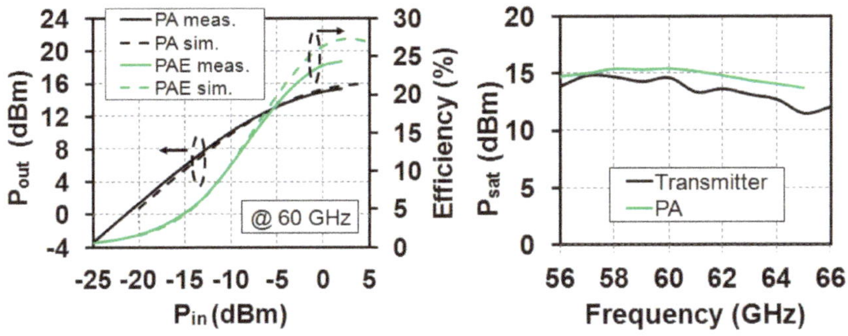

Transmitter achieves a state-of-the-art saturated output power of +15 dBm and peak TX efficiency of 15.3%.

The PA has a saturated output power of 15.4dBm and 24.4% power added efficiency at 60 GHz. The transmitter puts out 15dBm at 57GHz and more than 11.5dBm from 56 to 66 GHz. The peak transmitter efficiency is 15.3% at 57 GHz.

27 Receiver Implementation (I)

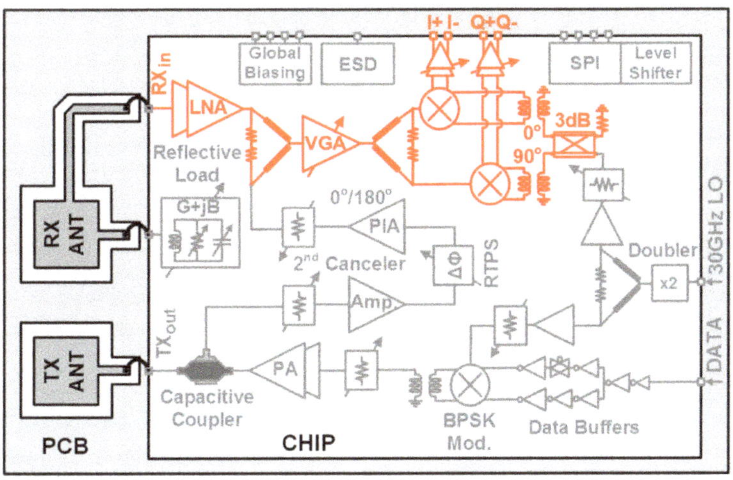

The receiver consists of a 60GHz two-stage LNA, a Wilkinson combiner, a 60GHz variable gain amplifier, a Wilkinson splitter for I/Q split, I/Q down-conversion mixers and differential baseband amplifiers.

28 Receiver Implementation (II)

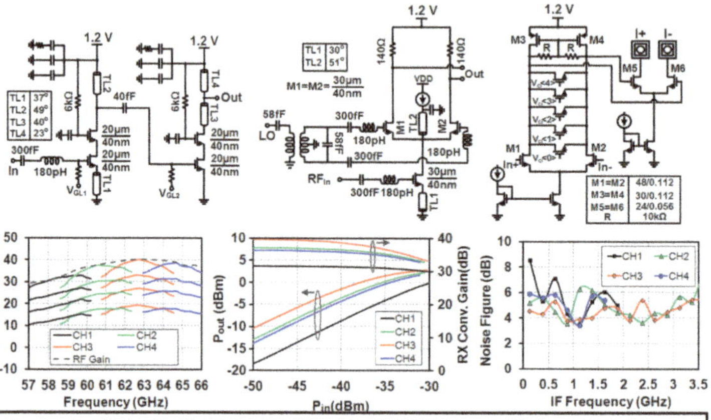

•A peak CG of 40dB in CH3 with 2.25GHz 3dB bandwidth.
•A state-of-the-art NF of 4dB in CH3.

A 60GHz two-stage inductively degenerated cascode LNA is designed for simultaneous noise and impedance matching. I/Q down-conversion mixers are implemented as current stealing single balanced mixers to improve the noise and conversion gain. IF amplifiers include 5-bit NFET resistance banks for gain control. The last stage is designed as an open drain buffer to ease interfacing to 50-ohm measurement equipment.

29 Testing Board

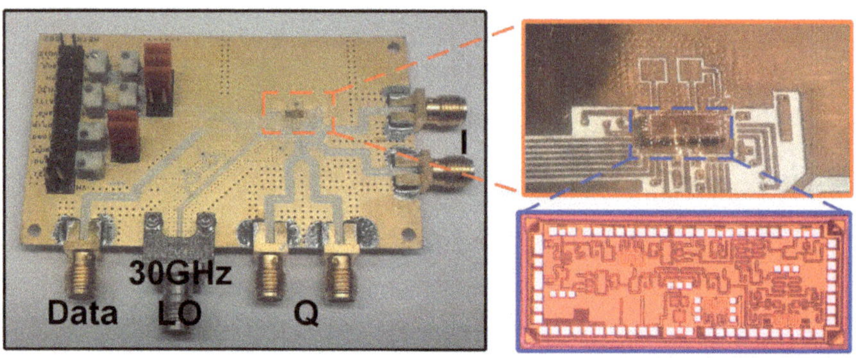

60 GHz Full-Duplex Transceiver PCB used for system level cancellation and link measurements.

The 60 GHz full-duplex transceiver IC is interfaced to the PCB with antennas described earlier.

30 Cancellation Performance (I)

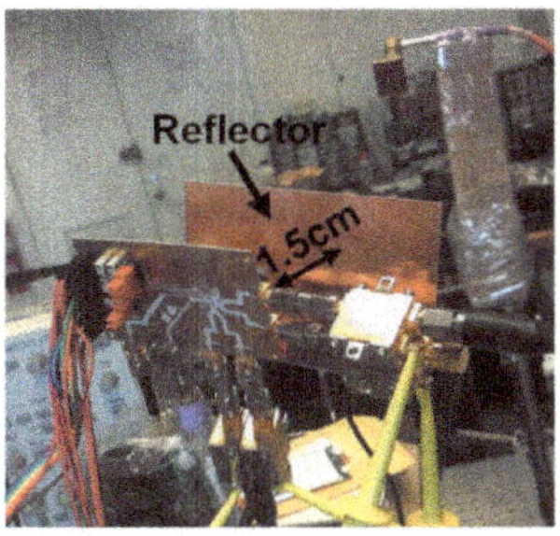

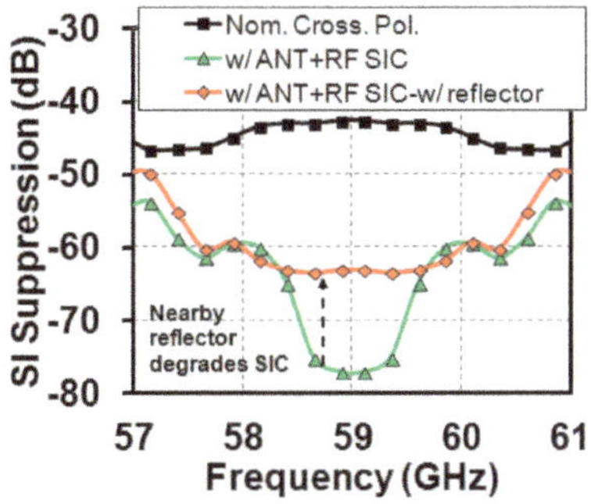

- **Bringing a reflector close to the antenna degrades the SI suppression.**

The self-interference suppression is characterized across frequency. Antenna and RF cancellation together enable more than 70 dB total SI suppression over 1GHz BW. The remaining ~15-20dB cancellation can be achieved in the digital domain. The total SI suppression degrades by about 10 dB when a metallic reflector is placed 1.5 cm away from the antennas.

31 Cancellation Performance (II)

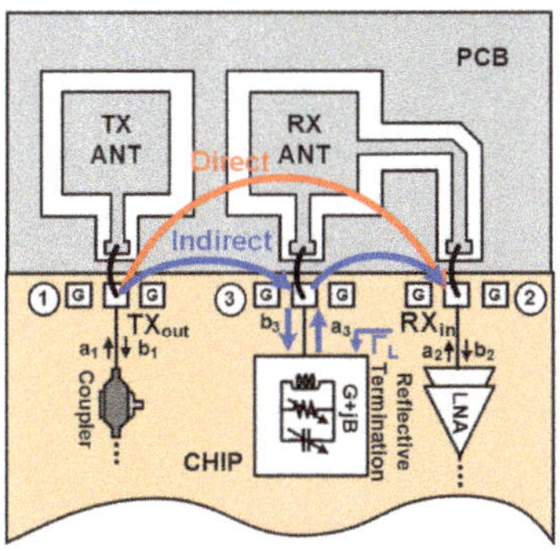

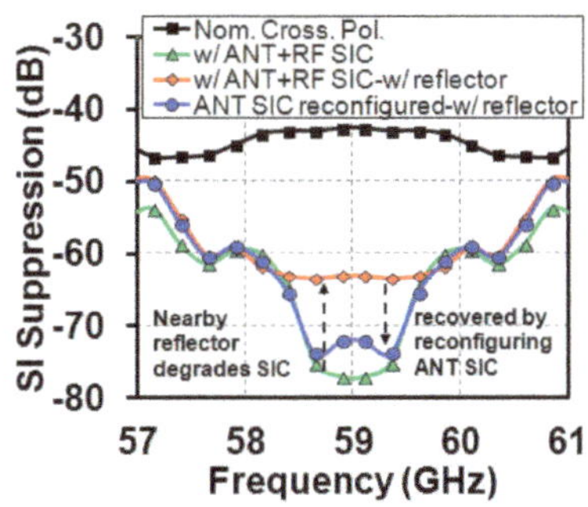

- **Bringing a reflector close to the antenna degrades the SI suppression.**
- **It can be recovered by reconfiguring R and C of the reflective termination.**

The performance can be recovered by reconfiguring only the antenna cancellation, by changing variable resistance and capacitance, while leaving the RF canceller untouched.

32 Demo Setup

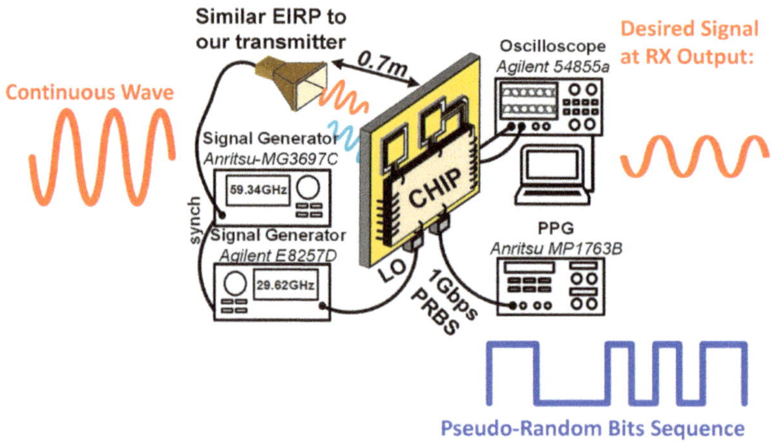

A simple same-channel full-duplex link is demonstrated over 0.7m. A continuous-wave signal at 100MHz offset from the LO frequency is transmitted as the desired signal with a similar EIRP to our transmitter whereas our 60GHz transceiver transmits a 1Gbps BPSK signal.

33 Demonstration

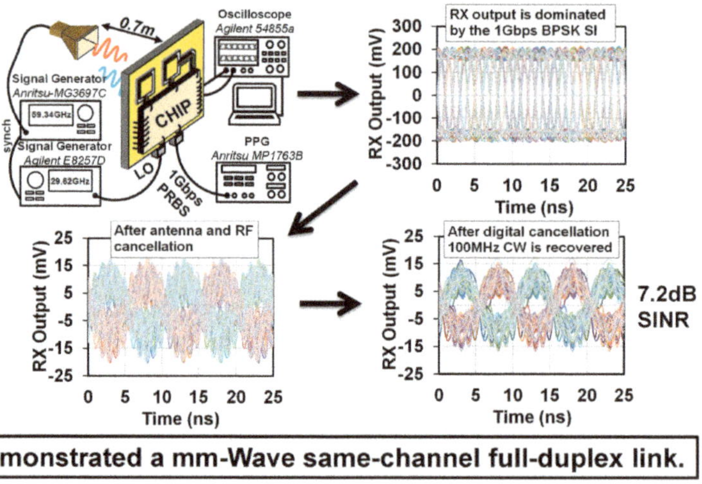

The RX output is dominated by the 1Gbps BPSK SI when the TX is on without SIC. When the antenna and RF SIC are engaged, the desired signal is captured with some residual SI. In this figure, we shift the captured signal by half a cycle to show its quality visually, like an eye diagram. The digitized data is taken from the oscilloscope to perform digital cancellation in Matlab. Digital SIC further suppresses the SI, resulting in an even cleaner received signal in with a signal-to-interference-noise-and-distortion ratio (SINDR) of 7.2dB.

A 25GHz MAGNETIC-FREE CIRCULATOR IN 45nm SOI CMOS

34

Common Challenge: Antenna Interface

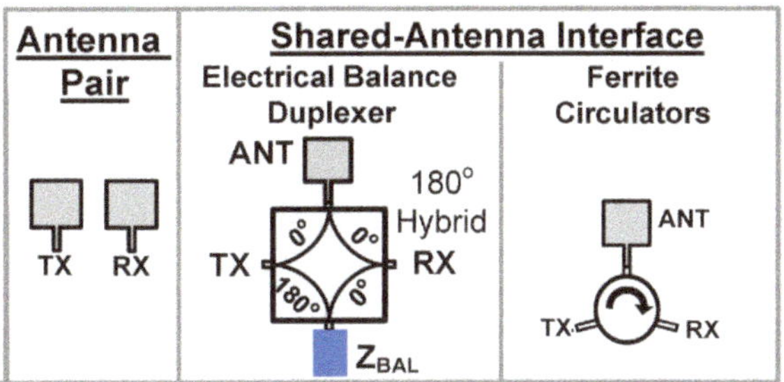

	Antenna Pair	Electrical Balance Duplexer	Ferrite Circulators
Form Factor	NO	YES	NO
Loss	YES	NO	YES
Channel Reciprocity	NO	YES	YES
CMOS Integration	YES	YES	NO

Conventional solutions are either lossy or bulky.

While we have thus far described a two-antenna interface that uses polarization to separate the transmitted and received signal, such solutions are bulky and do not exhibit channel reciprocity. It is therefore desirable to use single-antenna interfaces for full duplex. Current approaches include non-reciprocal magnetic circulators and reciprocal hybrid circuits, such as the electrical balance duplexer. However, magnetic circulators are not CMOS compatible, and hence, are bulky and expensive.

35

Passive Reciprocal Circuits

Reciprocal, passive, matched shared-ANT interfaces have a 3dB fundamental loss (typically around 4dB at RF/mm-wave).

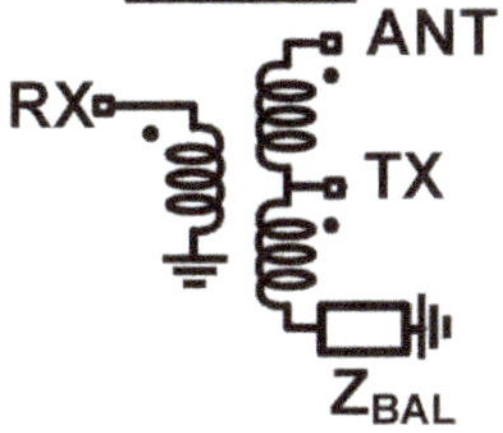

[Ref: B. van Liempd, et al., ISSCC 2015.]

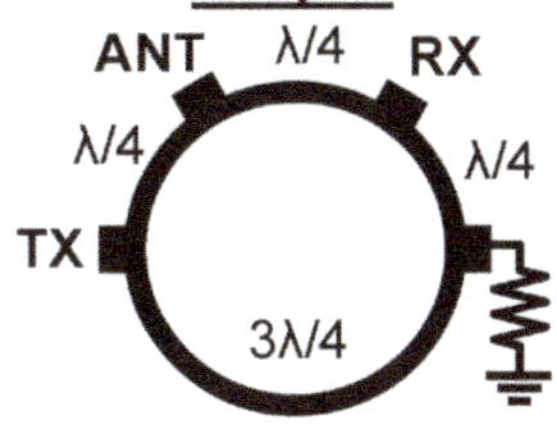

How can we avoid this 3dB fundamental loss?

On the other hand, it is well known fact that a three-port passive network cannot be reciprocal, lossless, and matched at all ports at the same time. As a result, reciprocal hybrid circuits such as the electrical balance duplexer suffer from a 3dB fundamental loss (typically around 4dB at RF and millimeter wave once implementation losses are factored in). The question is: how can we avoid this 3dB fundamental loss?

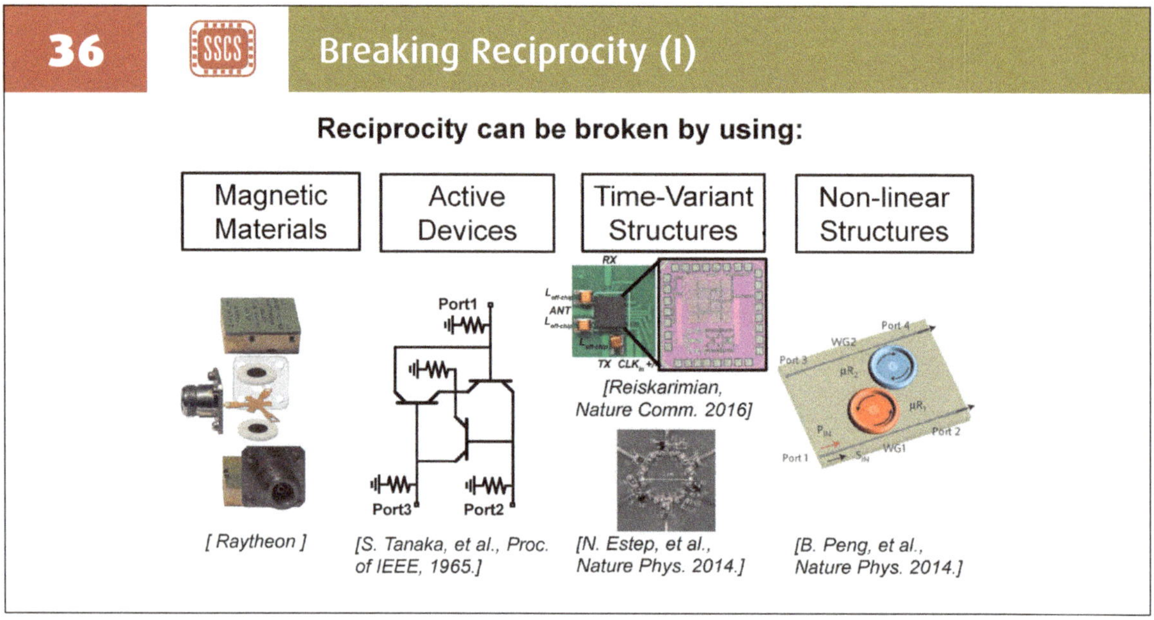

This theoretical loss can be avoided by breaking Lorentz reciprocity. Reciprocity can be broken using magnetic materials with asymmetric permittivity or permeability tensors, active voltage-/current-biased transistors, time-varying structures, or non-linear materials/structures.

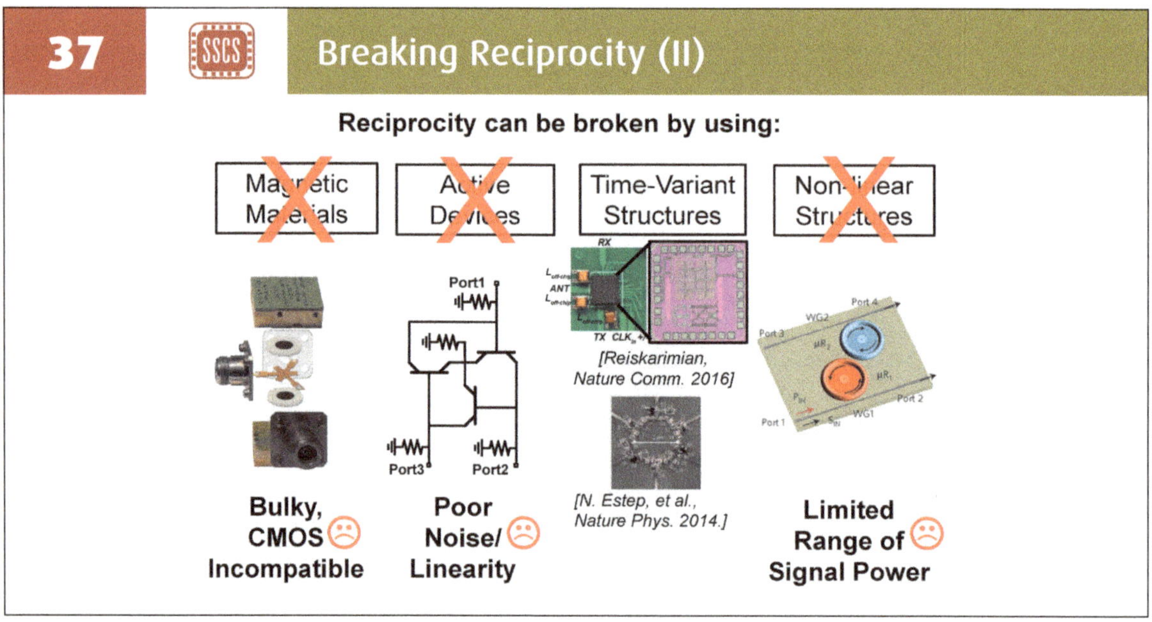

Magnetic circulators are bulky, expensive, and not compatible with CMOS. The use of active transistors severely limits the linearity and noise performance. Nonlinear devices typically exhibit non-reciprocity over a limited range of signal powers. Therefore, recently there has been a strong interest in breaking reciprocity through time-periodic modulation.

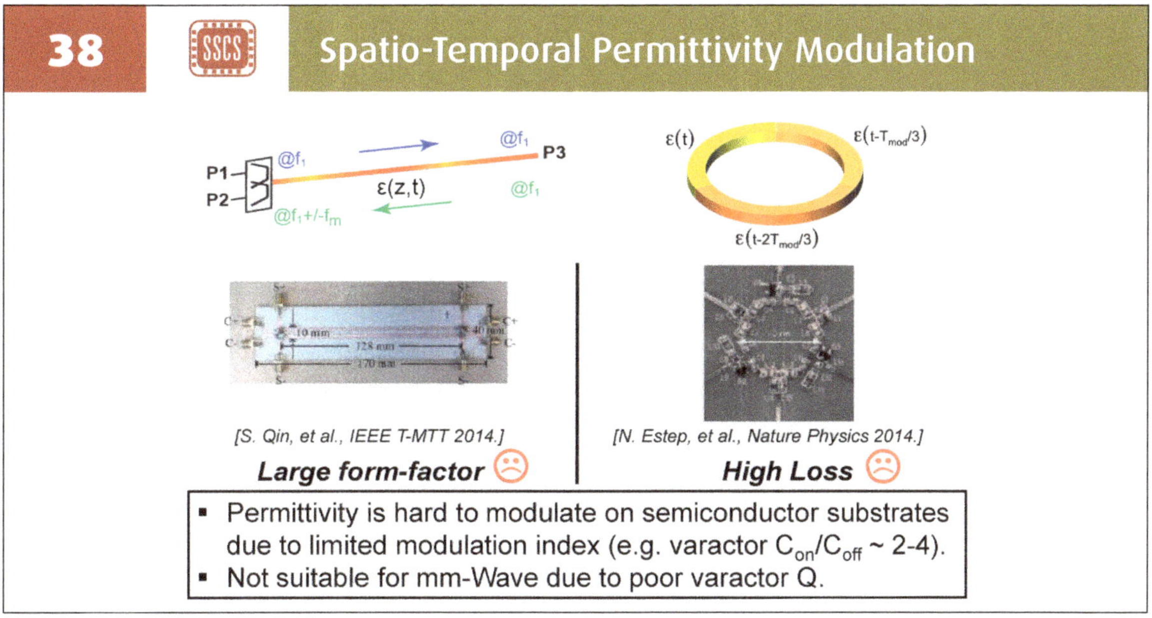

Specifically, there has been recent research on breaking reciprocity through spatio-temporal modulation of material permittivity using varactors. However, permittivity modulation is inherently weak, resulting in either large form factors or high insertion losses and narrow operation bandwidths. Additionally, these techniques are also not suitable for mm-wave due to poor varactor Q. Because of these challenges, there has been no demonstration of an integrated passive mmWave circulator.

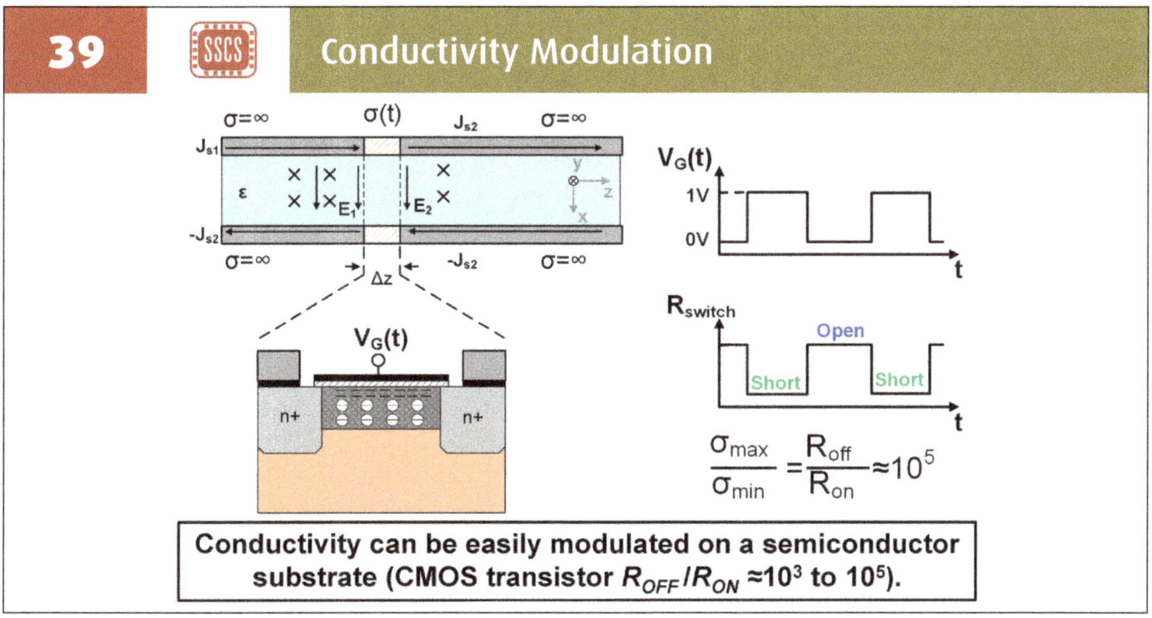

On the other hand, conductivity in semiconductors can easily be controlled using transistor switches, and enables a modulation index several orders of magnitude larger than permittivity. For example, CMOS transistors exhibit ON-OFF conductance ratios as high as 10^3-10^6.

40

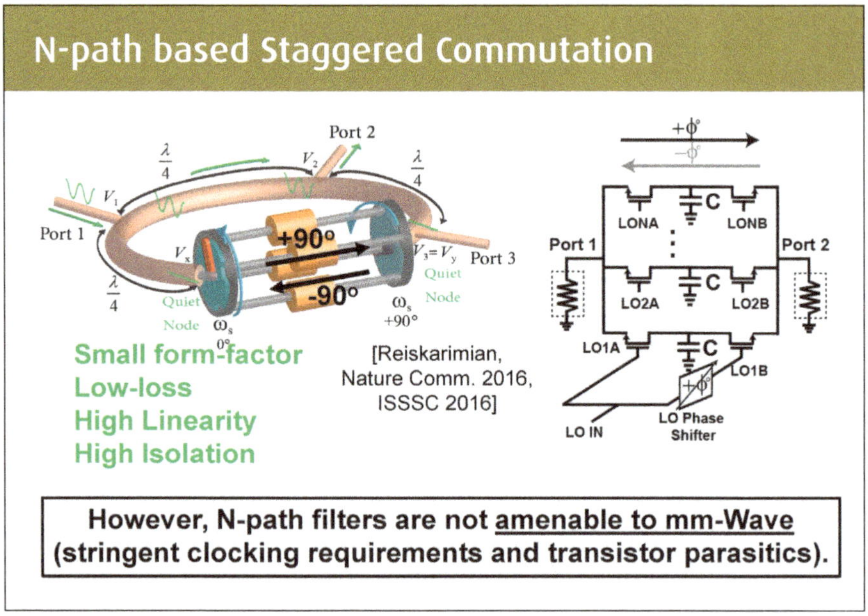

In 2016, using a form of conductivity modulation, we demonstrated a small form factor, very low loss, high isolation and high linearity RF CMOS passive circulator in an N-path filter-based implementation. However, N-path filters are not amenable to mm-wave operation due to stringent clocking requirements and transistor parasitics.

How can we achieve non-reciprocity at mm-Wave?

This brings up the question: How can we achieve non-reciprocity at mm-Wave to enable low loss wideband CMOS circulators?

41

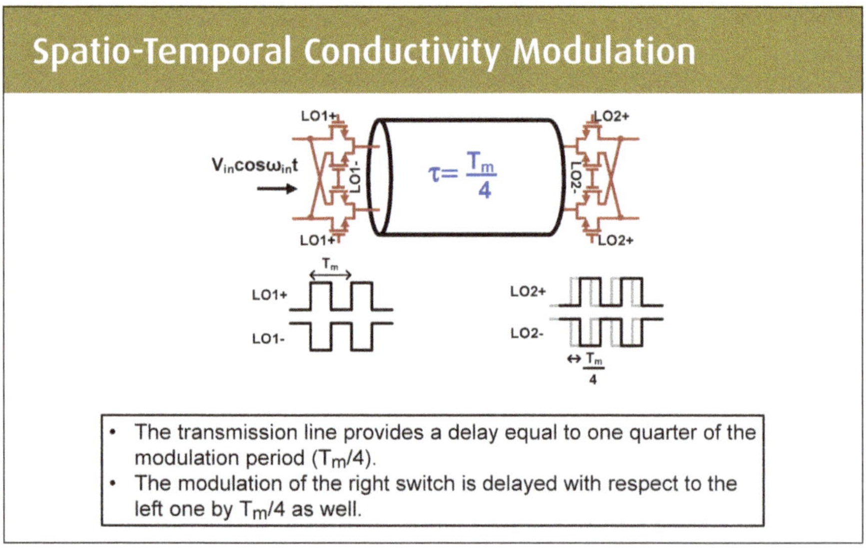

To address this question, we propopose a new concept of spatio-temporal conductivity modulation. The spatio-temporal conductivity modulation concept consists of two sets of switches implemented in a fully-balanced fashion on either end of a differential transmission line delay. The switches are modulated between short and open circuit conditions through periodic square pulses with a 50% duty cycle. The transmission line provides a delay equal to one quarter of the modulation period and the modulation of the right switches are delayed with respect to the left ones by the same amount. Adding this delay between the two sets of switches allows incident signals from different directions to follow different paths.

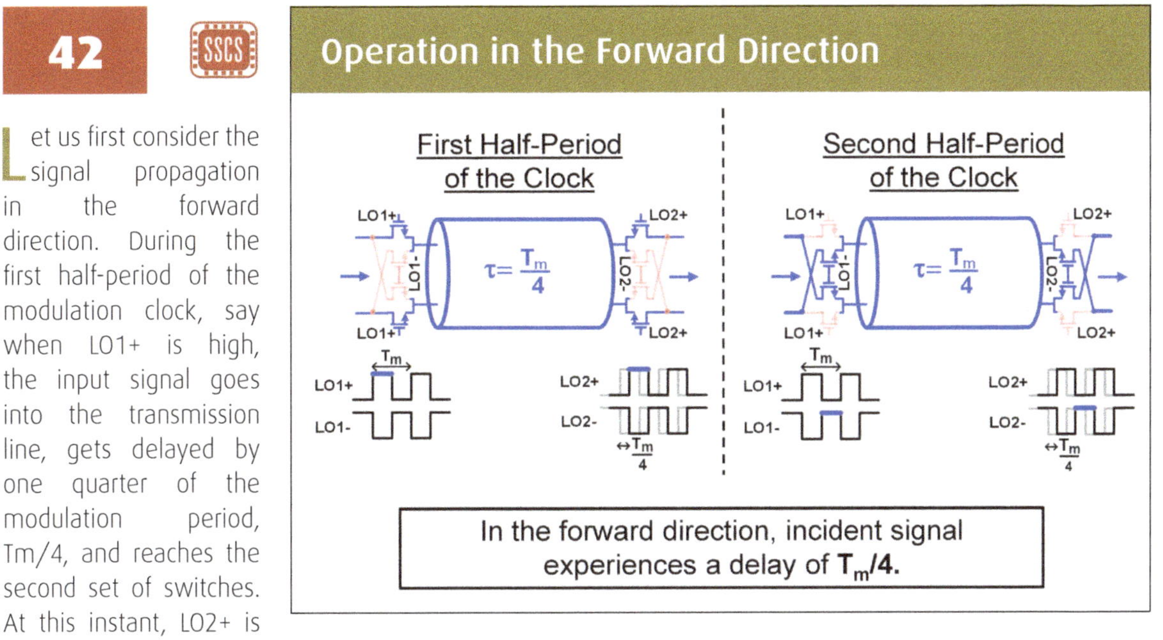

Let us first consider the signal propagation in the forward direction. During the first half-period of the modulation clock, say when LO1+ is high, the input signal goes into the transmission line, gets delayed by one quarter of the modulation period, Tm/4, and reaches the second set of switches. At this instant, LO2+ is high, so that the signal directly passes to the output. A similar explanation also holds for the second half-period of the modulation signal. The signal goes into the transmission line with a sign flip, gets delayed by Tm/4 and the sign flip is recovered by the second set of switches. In short, in the forward direction, the incident signal passes through the structure **without any loss** and **experiences a delay of one quarter of the modulation period**.

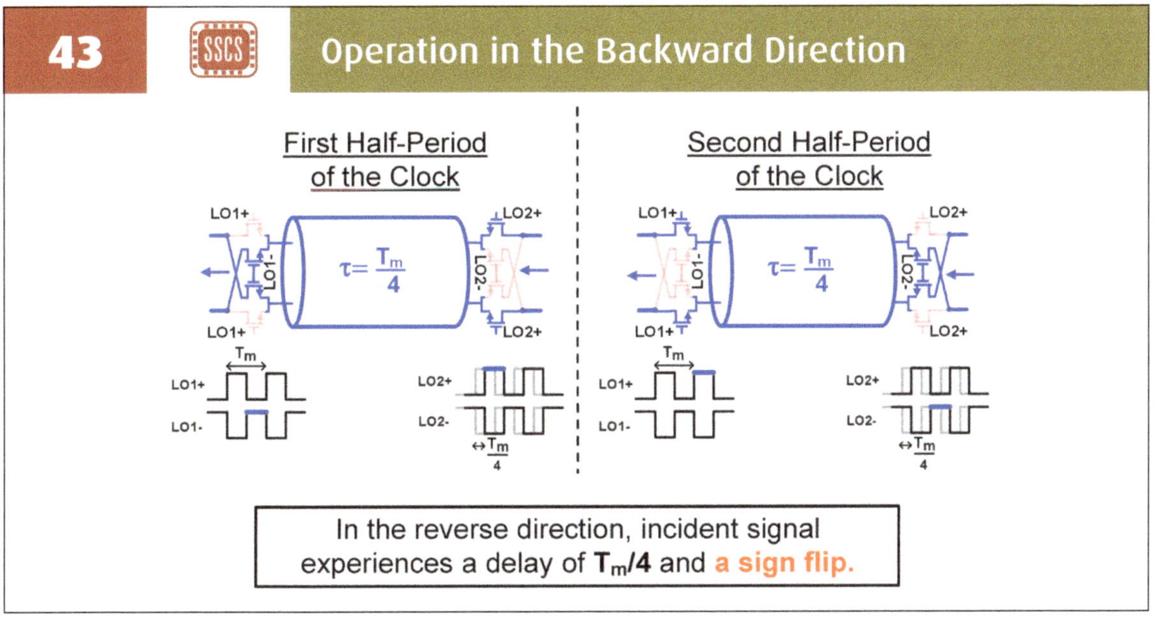

Now, let us look at the signal propagation in the backward direction. During the first half-period of the clock , say when LO2+ is high, the input signal is delayed by the transmisssion line delay of Tm/4 and the second set of switches flips the sign of the signal. A similar operation is also observed for the second half-period. So, in the reverse direction, the incident signal experiences a delay of Tm/4 and a sign flip.

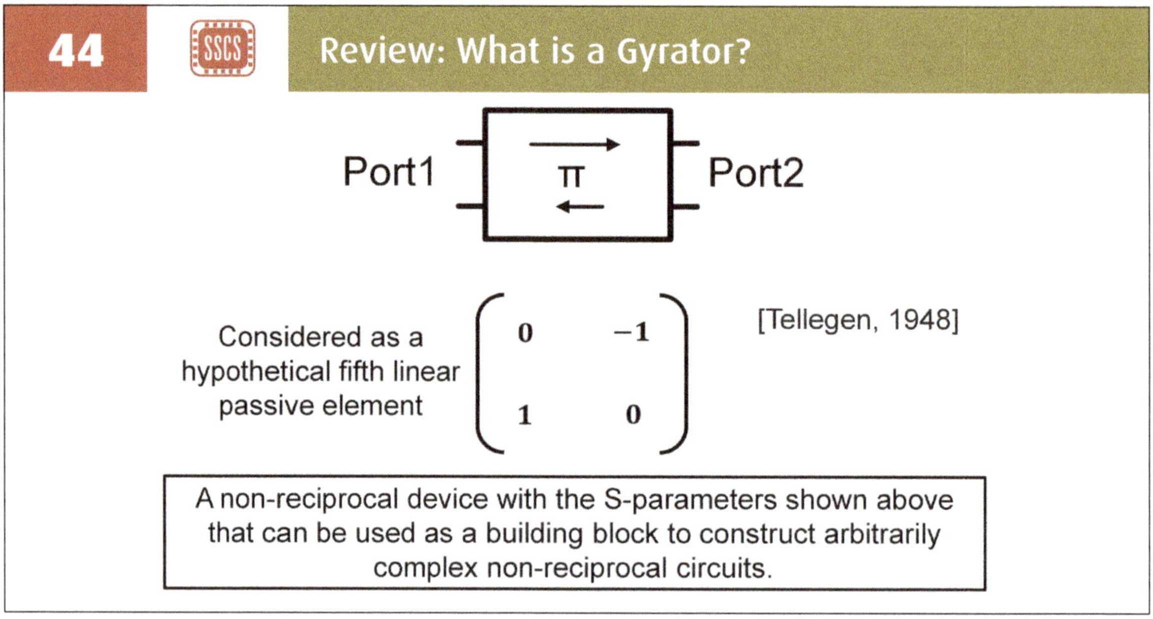

Before proceeding further, we should briefly remind ourselves of what a gyrator is: The gyrator is considered to be the hypothetical fifth passive linear element (after the resistor, capacitor, inductor and transformer) which reverses the polarity of the signal travelling in the backward direction. It enables construction of arbitrarily complex non-reciprocal circuits.

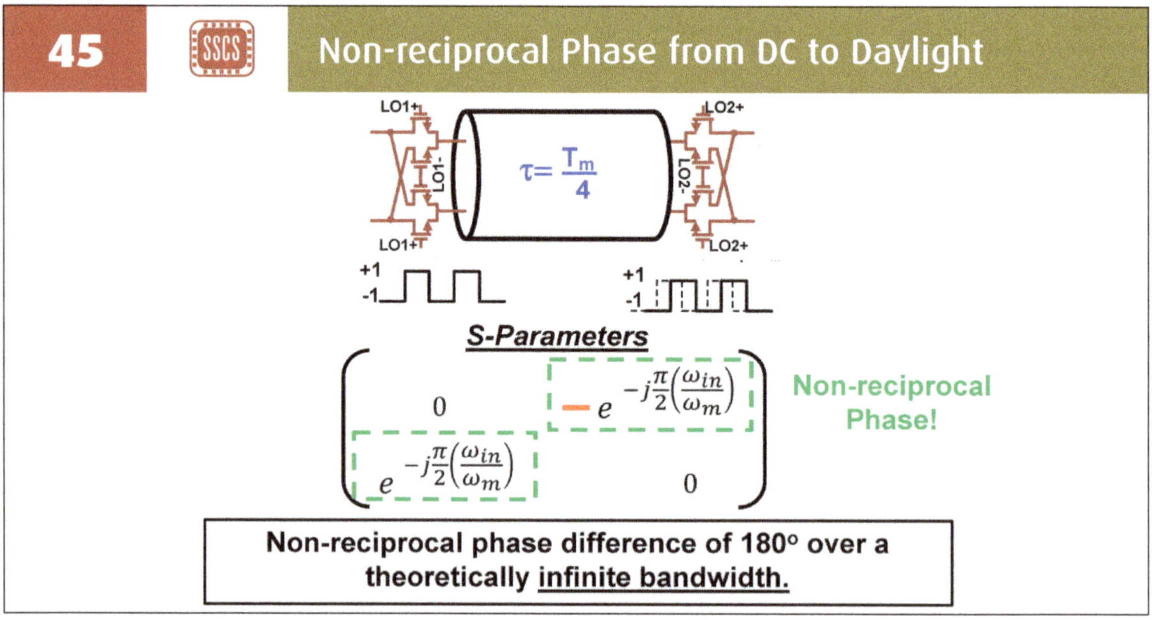

In this slide, we summarize the derived S-parameters of the spatio-temporal conductivity modulation network. As can be seen, spatio-temporal conductivity modulation technique is ideally lossless and breaks phase reciprocity over a theoretically infinite bandwidth. More importantly, it operates as an ideal gyrator over all frequencies, reversing the polarity of the signal travelling in the backward direction.

46

Non-reciprocal Phase Shift of ±90°

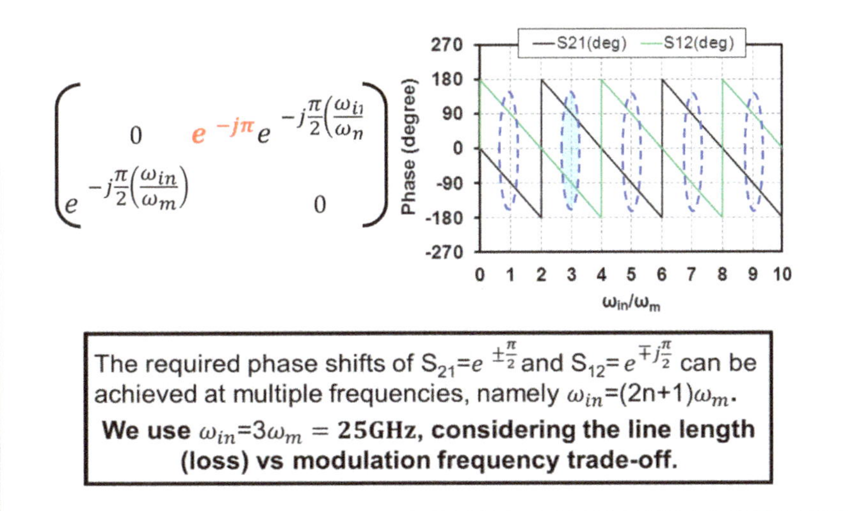

We plot the forward and reverse insertion phase across frequency normalized to the modulation clock frequency. Achieving +90 and -90 degree phase shifts is important to build the circulator in a ring configuration. As can be seen, the spatio-temporal conductance modulation network provides a phase shift of plus or minus 90 degrees at the odd multiples of the modulation frequency. In this work, we chose an operating to modulation frequency ratio of 3. This eases the clock generation and distribution compared to using the same frequency for modulation (for example, using 25GHz square wave clocks vs 8.33GHz clock). It also reduces the delay line length and in return, the loss compared to using other higher odd multiples.

47

Effect of Duty Cycle Impairment

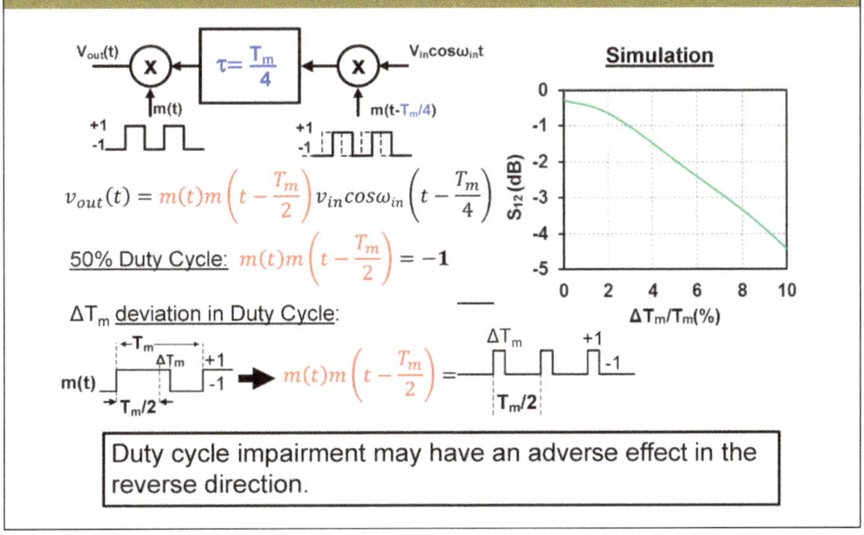

We observed that duty cycle impairment in the modulation clock may have an adverse effect on the operation in the reverse direction. We can model the non-reciprocal structure as multiplication, delay and multiplication in the time domain, and the output voltage can be written shown here, where m(t) is the differential modulation signal toggling between +1 and -1. Here, m(t) is differential modulation signal controlling the left switches. Ideally, m(t) has a 50% duty cycle, and multiplication of m(t) by its half-period delayed version gives us -1 and thus the sign flip. However, if there is a deviation from ideal 50% duty cycle, say by delta Tm, this multiplication will result in a pulse train. This would cause a loss since some portion of the power would be transferred to mixing frequencies. According to our simulations, a deviation from 50% duty cycle to 55% or 45% would degrade the loss by about 2dB in the reverse direction.

48 Use of an Additional Quadrature Path

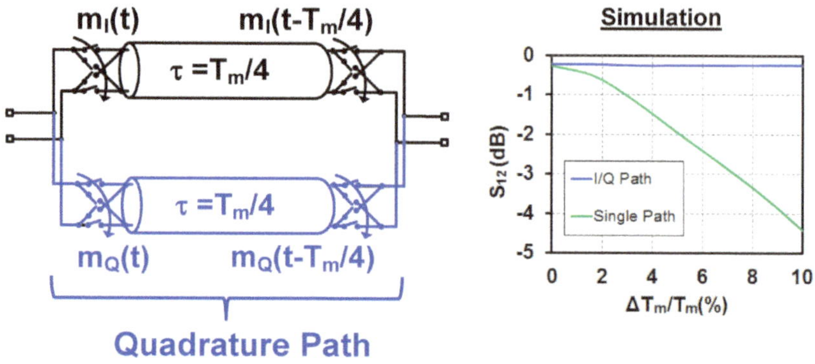

Quadrature path mitigates the effect of duty cycle impairment, ensuring a robust operation.

Addition of a quadrature path mitigates the effect of duty cycle impairment, ensuring robust operation.

A 45nm SOI CMOS CIRCULATOR

We will now discuss the implementation of a 25GHz prototype in 45nm SOI CMOS.

49 Circulator Architecture

[Reiskarimian, Nature Comm. 2016]

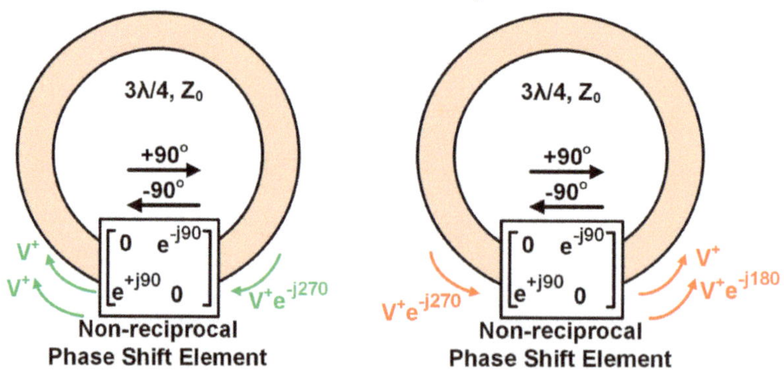

A 3λ/4 line is wrapped around the non-reciprocal phase component to support **non-reciprocal wave propagation**.

49 Circulator Architecture

Similar to the previous RF CMOS circulator work from our group, the mm-wave circulator is realized by wrapping a 3λ/4 transmission-line loop around the non-reciprocal phase element. In the clockwise direction, the −270° phase shift of the transmission line adds to the −90° phase shift through the non-reciprocal part, enabling wave propagation. In the counter-clockwise direction, the −270° phase shift of the transmission line adds to the +90° phase shift of the non-reciprocal part, suppressing wave propagation.

50 Three-Port Passive Circulator

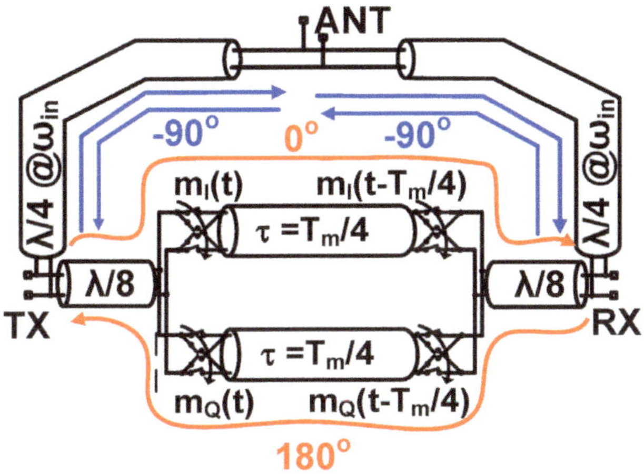

S-parameters at $\omega_{in} = 3\omega_m$

$$S=\begin{bmatrix} 0 & 0 & -1 \\ -j & 0 & 0 \\ 0 & -j & 0 \end{bmatrix}$$

- A circulator is realized by introducing three ports which are λ/4 apart.
- Non-reciprocal phase element is placed symmetrically between the TX and RX ports.

A three port circulator is realized by introducing three ports λ/4 apart from each other. The non-reciprocal phase-element is placed symmetrically between the TX and RX ports. The derived ideal S-parameters show that the signal circulates only in the clock-wise direction.

51 25GHz Three-Port Passive Circulator

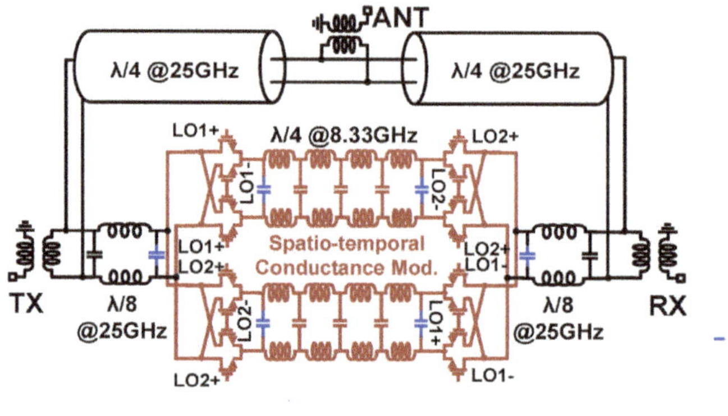

mm-Wave **fully-integrated magnetic-free passive circulator.**

We implemented a magnetic-free passive circulator prototype operating at 25GHz. The circulator is implemented in a fully differential fashion reducing the LO feedthrough and improving power handling. The I and Q delay lines are designed to be a quarter-wavelength at the modulation frequency and are miniaturized using pi-type C-L-C sections. The λ/8 sections on either side are also miniaturized, so that transistor switch capacitive parasitics could be absorbed into the artificial transmission lines.

52 8.33GHz LO Path Implementation

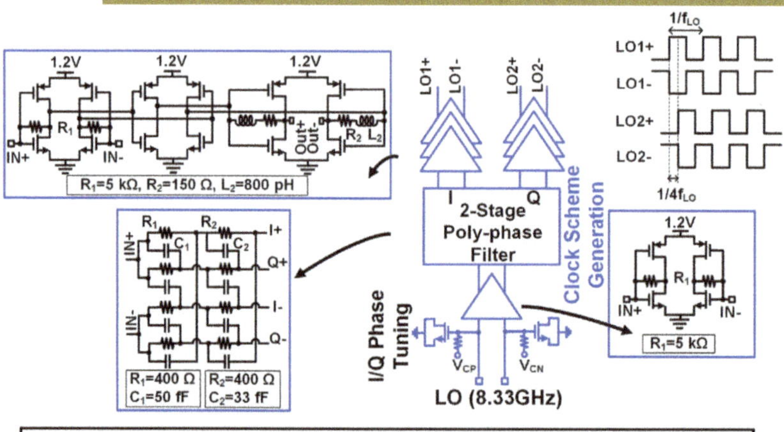

LO path includes an input buffer driving a 2-stage poly-phase filter for I/Q generation and output buffers.

The LO path consists of an input buffer driving a two stage poly-phase filter which generates differential I and Q signals. Self-biased 3 stage inverter chains follow the poly-phase filter to drive the switches. Inductive peaking is used in the last stage to improve the bandwidth. We also included varactors at the differential LO input to compensate for the I/Q imbalance of the poly-phase filter.

53 Chip Photograph

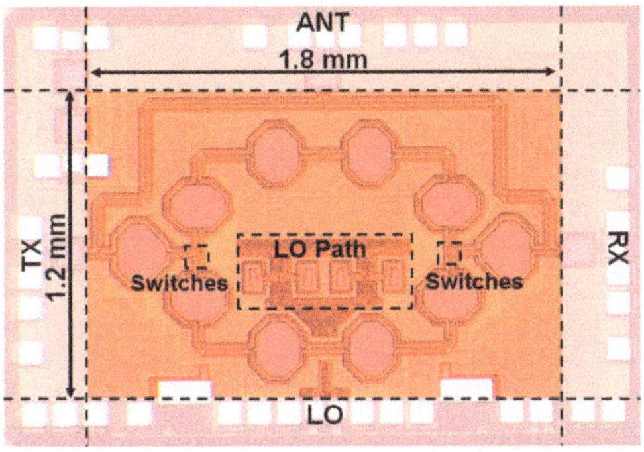

25GHz magnetic-free passive non-reciprocal circulator in GF 45nm SOI CMOS.

The 25GHz magnetic-free passive circulator is implemented in GF 45nm SOI CMOS and occupies 1.2mm x 1.8mm active chip area excluding the on-chip baluns implemented for testing.

54 Small-Signal Measurement Setup

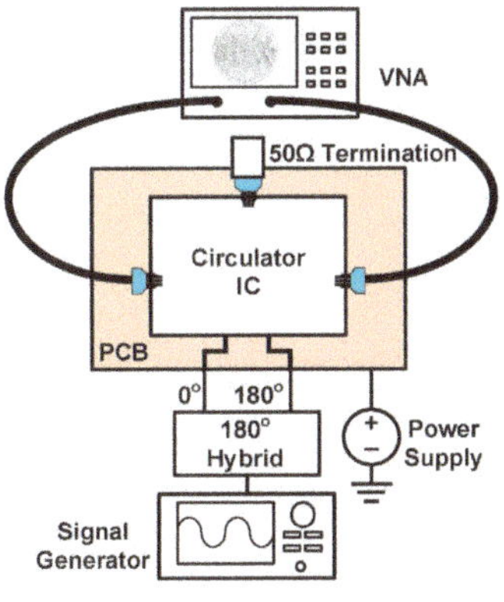

- S-parameters are measured by probing two ports at a time.
- A millimeter-wave probe terminated with a broadband 50Ω termination is landed on the third port.

S-parameters are measured by probing two ports at a time. A millimeter-wave probe terminated with a broadband 50Ω termination is landed on the third port.

55 Small-Signal Measurements

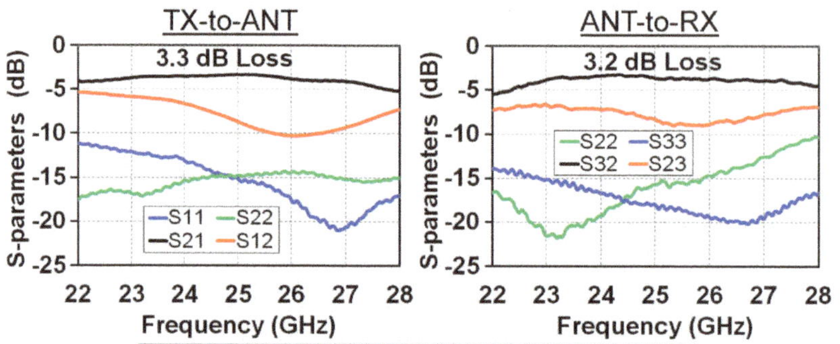

- ANT-to-RX Loss =3.2dB,
- TX-to-ANT Loss=3.3 dB,
- 1dB Insertion Loss BW is 4.6GHz.
- Broadband non-reciprocal behavior.

The small signal TX-to-ANT and ANT-to-RX insertion losses are 3.3dB and 3.2dB, respectively, with a 1dB insertion loss BW of 4.6GHz. A broadband non-reciprocal behavior is observed over more than 6GHz bandwidth.

56 Circulator Loss Distribution

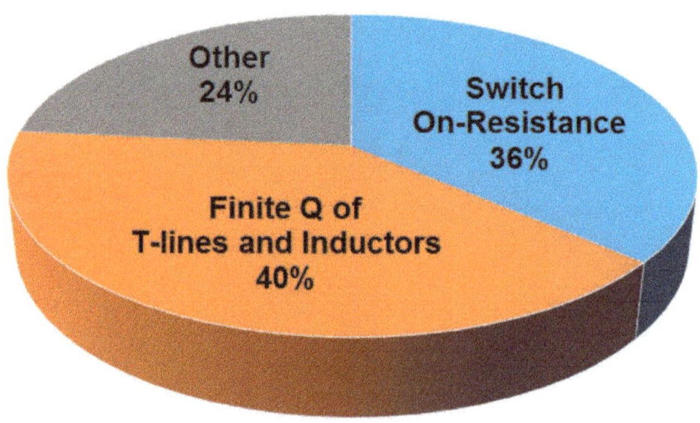

- No 3dB fundamental loss.
- Loss is dominated by switch on-resistance and finite Q of passives.

This graph shows the theoretical distribution of the losses in the circulator, based on our simulations. There is no fundamental loss of 3dB in this structure. Almost 76% of the total loss comes from switch on-resistance and finite Q of passives.

57 Projected Loss in a Better Process

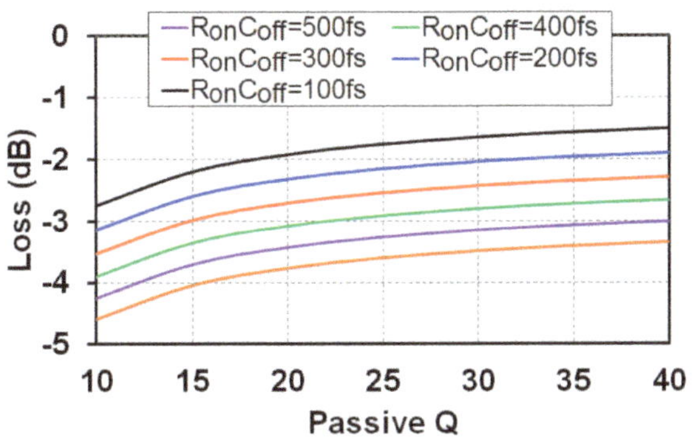

Insertion loss <2dB is possible in a technology with lower $R_{on}C_{off}$ and high resistivity substrate and/or thicker BEOL.

An insertion loss of <2dB is projected in a technology with lower $R_{on}C_{off}$ and high resistivity substrate and/or thicker BEOL.

58 Small-Signal Isolation

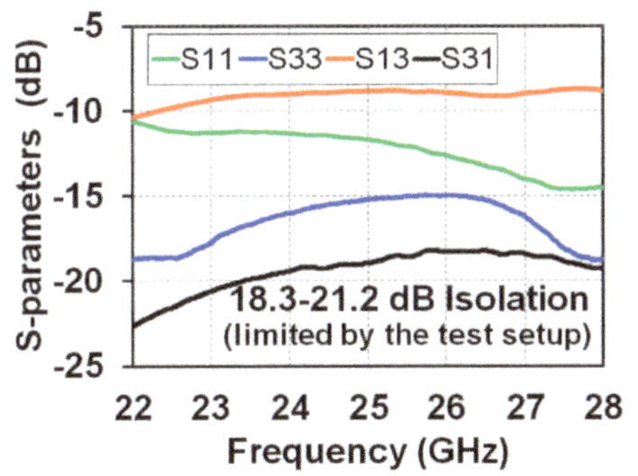

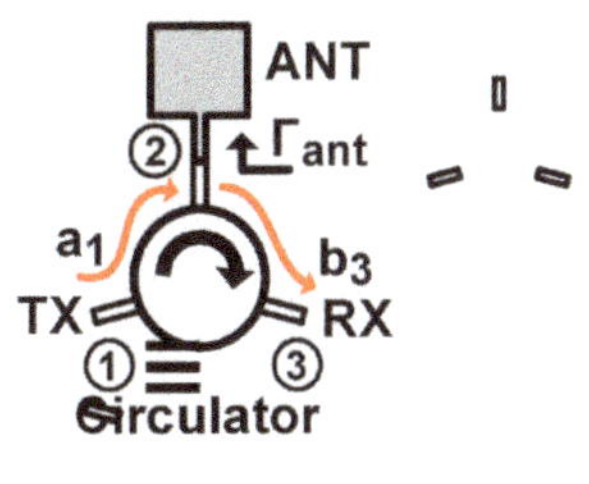

$$\text{Iso} = \frac{b_3}{a_1} = \Gamma_{ant}$$

Isolation is 18.3-21.2 dB (limited by the reflection at the 3rd port, a general limitation for all circulators).

A broadband TX-to-RX isolation of 18.3 to 21.2dB is measured. It is limited by the impedance of the probe and load at the antenna port, which is a challenge for all circulators, as depicted in the right figure. As depicted on the right, TX-to-RX isolation in circulators strongly depends on the matching at the antenna port. This limitation is exacerbated at mm-wave since it is hard to obtain a reflection coefficient better than -20dB from a millimeter-wave probe and termination combination.

59 Isolation with Impedance Tuner

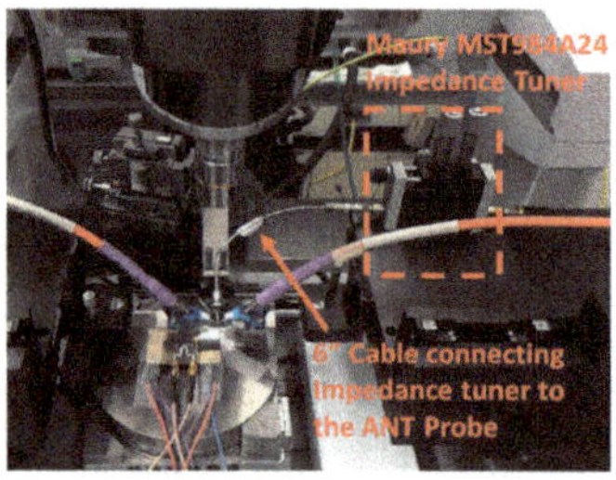

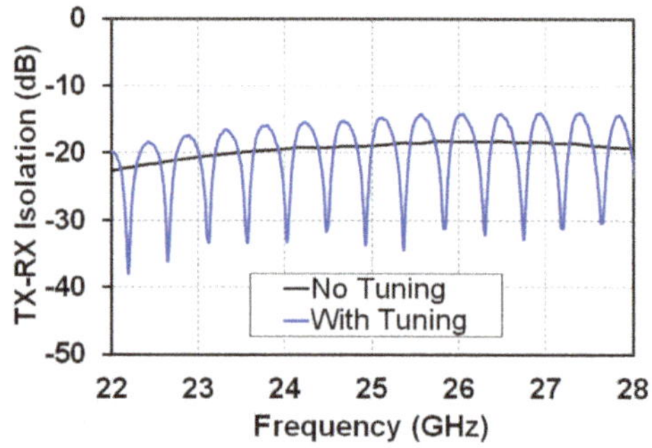

- The third port can be tuned using an antenna tuner.
- However, a 6" cable connecting the tuner to the probe restricts the ability to improve wideband isolation

The antenna port can be tuned using an impedance tuner. However, in our setup, we are limited by a 6" cable connecting the impedance tuner to the probe. With antenna tuning, we can get more than 30dB isolation but this cable causes ripples in the isolation profile, restricting the wideband isolation.

60 Estimated Isolation with On-chip Tuner

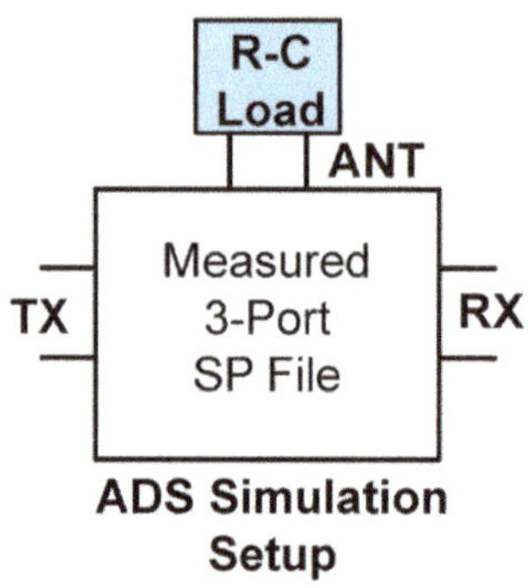

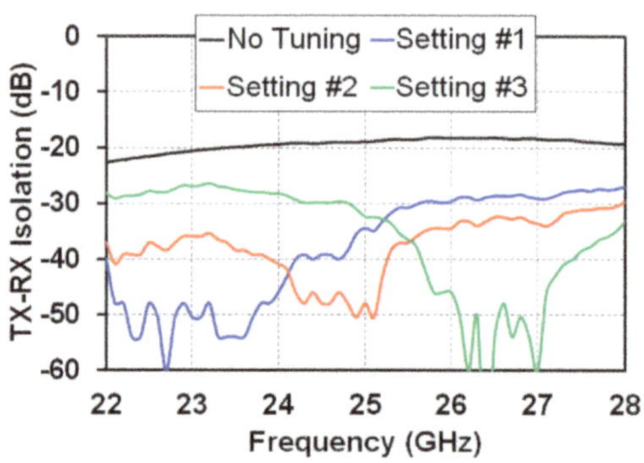

Simulations reveal that >45dB isolation can be achieved over wide bandwidths with fairly simple on-chip tuning.

We have studied wideband antenna tuning in simulation by creating a 3-port touchstone S-parameter file from our measurements. Simulations reveal that >45dB isolation can be achieved over very wide bandwidths with fairly simple R-C tuning, assuming the tuner is integrated on chip.

61 Noise Figure

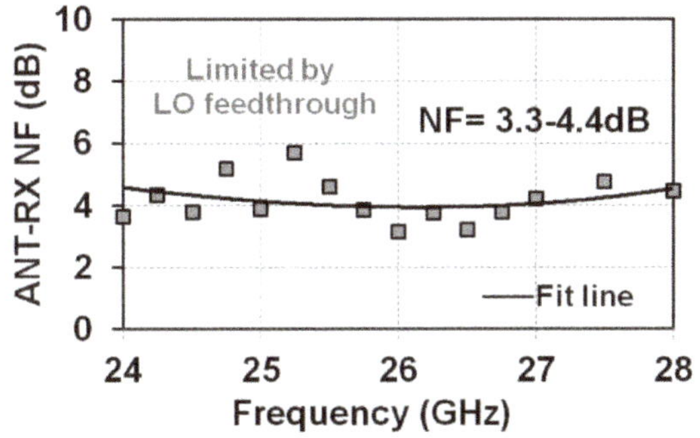

ANT-RX NF is 3.3 to 4.4 dB, consistent with the insertion loss and not degraded by phase noise.

ANT-RX noise figure is 3.3 to 4.4dB, consistent with the insertion loss and not degraded by phase noise.

62 Large-Signal Measurements

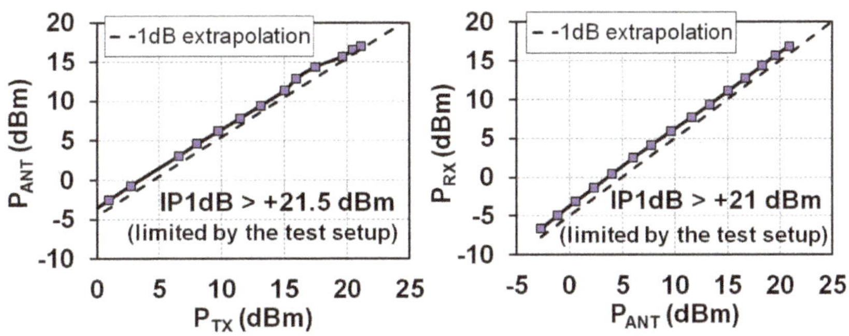

TX-to-ANT and ANT-to-RX IP1dB>21dBm (limited by P_{sat} of the driving amplifier).

The TX-to-ANT and ANT-to-RX input 1dB compression points are higher than +21dBm. They could not be driven into compression due to the limited output power of the driving amplifier in the testing setup.

63 IIP3 Measurements

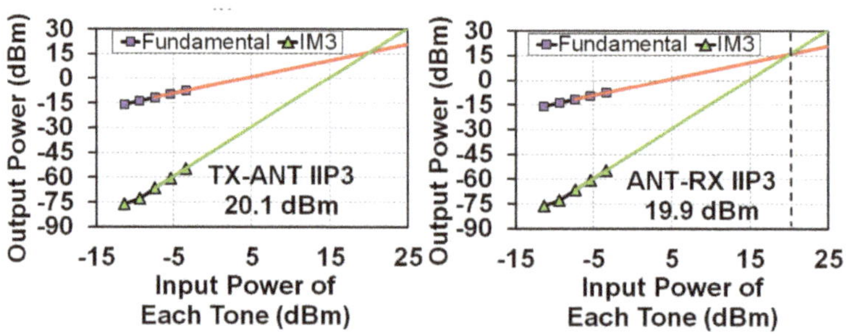

- **TX-to-ANT IIP3=20.1dBm, RX-to-ANT IIP3=19.9dBm**
- The P1dBs are high relative to the IIP3s because the circulator transitions to reciprocal operation under large-signal operation.

The TX-to-ANT and ANT-to-RX IIP3s are around +20dBm. The circulator transitions to reciprocal operation under large-signal operation. Since IIP3 is just an extrapolation from small signal levels, this transition cannot be captured and thus the traditional estimation between P1dB and IIP3 values does not hold.

64 Large-Signal Isolation

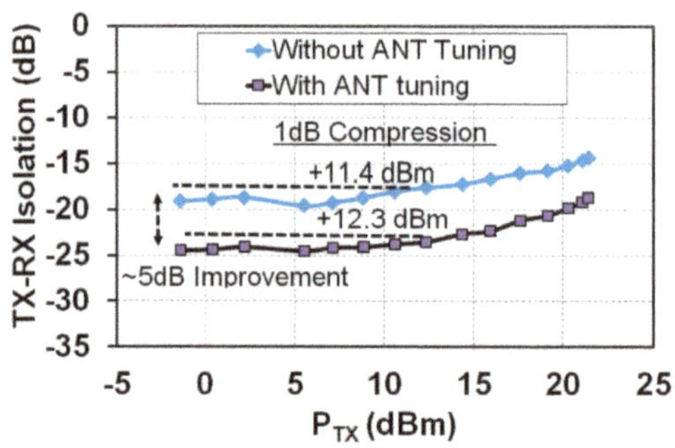

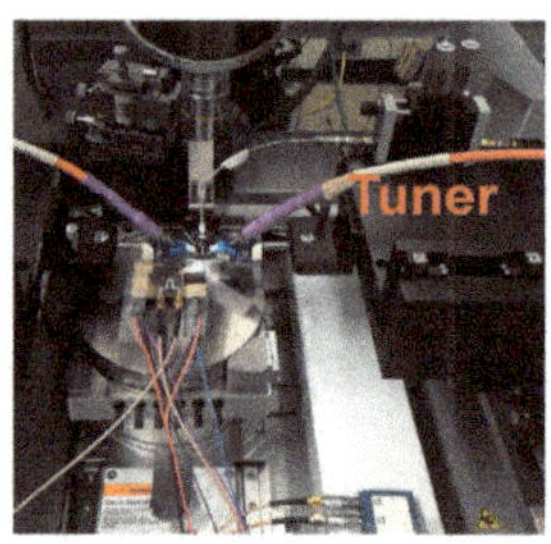

- TX-to-RX isolation degrades by 1dB at 11.4dBm and ~3dB at 21.45dBm.
- This can be compensated by an external ANT impedance tuner, which enables a higher 1dB compression at +12.3dBm.

We also measured the TX-to-RX isolation performance versus the transmit power. TX-to-RX isolation degrades by 1dB and 3dB at +11.4dBm and +21.45dBm, respectively. This can be compensated by an external ANT impedance tuner, enabling a higher 1dB degradation point at +12.3dBm.

65

When compared with prior art, this work is superior to active mm-wave circulators in all metrics: loss, linearity, NF and BW.

Performance Summary & Comparison (I)

	TMTT2010	TMTT2015 [3]	ISSCC2015 [2]	ISSCC2016 [6]	This work
Technique	Active Quasi Circulator	Active Quasi Circulator	Electrical Balance Duplexer	N-Path-Filter-based Magnetic-Free Passive Circulator	Magnetic-Free Passive Circulator Based on Spatio-Temporal Conductance Mod.
Technology	180nm CMOS	180nm CMOS	180nm SOI CMOS	65nm CMOS	**45nm SOI CMOS**
Frequency	24GHz	24GHz	1.9-2.2GHz	0.75GHz	**25GHz**
TX-ANT Transmission	+22.4dB	-5.7dB	-3.7dB	-1.7dB	**-3.3dB**
ANT-RX Transmission	+12.3dB	-5.7dB	-3.9dB	-1.7dB	**-3.2dB**
TX-RX Isolation	>15dB	>20 dB	>40dB	>20dB	**>18.5dB [4]**
Isolation BW [1]	~1%	~1.6%	~15%	4.3%	**18% [5]**
Center frequency/ Modulation frequency	N/A	N/A	N/A	1	**3**
Area	$3.22mm^2$	$0.715mm^2$	$1.75mm^2$	$0.64mm^2/25mm^2$ [2]	**$2.16mm^2$**
ANT-RX NF	17dB	N/R	3.9dB	4.3dB [3]	**3.3-4.4dB**
TX-ANT IP1dB	-19.8dBm	+9.5dBm	N/R	N/R	**>+21.5dBm**
TX-ANT IIP3	-11dBm	N/R	+70dBm	+27.5dBm	**+19.9dBm**
P_{DC}	144.8mW	7.2mW	0	59mW	**78.4mW**

[1] BW over which an isolation better than the value quoted in the row above is maintained.
[2] Includes SMD inductors on PCB.
[3] Includes 2.3dB degradation due to LO phase noise.
[4] Limited by the mmWave test setup.
[5] This is the 1dB insertion loss BW.
N/A: Not Applicable, N/R: Not Reported

TMTT2010: H. S. Wu, C. W. Wang and C. K. C. Tzuang, "CMOS Active Quasi-Circulator With Dual Transmission Gains Incorporating Feedforward Technique at K-Band," in *IEEE TMTT*, vol. 58, no. 8, pp. 2084-2091, Aug. 2010.

66

When compared with low-RF electrical balance duplexers, this work achieves more than 1dB overall advantage in the sum of TX-to-ANT and ANT-to-RX insertion losses while operating at >10x higher frequency.

Performance Summary & Comparison (II)

	TMTT2010	TMTT2015 [3]	ISSCC2015 [2]	ISSCC2016 [6]	This work
Technique	Active Quasi Circulator	Active Quasi Circulator	Electrical Balance Duplexer	N-Path-Filter-based Magnetic-Free Passive Circulator	Magnetic-Free Passive Circulator Based on Spatio-Temporal Conductance Mod.
Technology	180nm CMOS	180nm CMOS	180nm SOI CMOS	65nm CMOS	**45nm SOI CMOS**
Frequency	24GHz	24GHz	1.9-2.2GHz	0.75GHz	**25GHz**
TX-ANT Transmission	+22.4dB	-5.7dB	-3.7dB	-1.7dB	**-3.3dB**
ANT-RX Transmission	+12.3dB	-5.7dB	-3.9dB	-1.7dB	**-3.2dB**
TX-RX Isolation	>15dB	>20 dB	>40dB	>20dB	**>18.5dB [4]**
Isolation BW [1]	~1%	~1.6%	~15%	4.3%	**18% [5]**
Center frequency/ Modulation frequency	N/A	N/A	N/A	1	**3**
Area	$3.22mm^2$	$0.715mm^2$	$1.75mm^2$	$0.64mm^2/25mm^2$ [2]	**$2.16mm^2$**
ANT-RX NF	17dB	N/R	3.9dB	4.3dB [3]	**3.3-4.4dB**
TX-ANT IP1dB	-19.8dBm	+9.5dBm	N/R	N/R	**>+21.5dBm**
TX-ANT IIP3	-11dBm	N/R	+70dBm	+27.5dBm	**+19.9dBm**
P_{DC}	144.8mW	7.2mW	0	59mW	**78.4mW**

[1] BW over which an isolation better than the value quoted in the row above is maintained.
[2] Includes SMD inductors on PCB.
[3] Includes 2.3dB degradation due to LO phase noise.
[4] Limited by the mmWave test setup.
[5] This is the 1dB insertion loss BW.
N/A: Not Applicable, N/R: Not Reported

TMTT2010: H. S. Wu, C. W. Wang and C. K. C. Tzuang, "CMOS Active Quasi-Circulator With Dual Transmission Gains Incorporating Feedforward Technique at K-Band," in *IEEE TMTT*, vol. 58, no. 8, pp. 2084-2091, Aug. 2010.

67

When compared with the 750MHz N-path filter-based passive CMOS circulator from our group, this work most importantly scales to mm-wave and significantly enhances BW.

Performance Summary & Comparison (III)

	TMTT2010	TMTT2015 [3]	ISSCC2015 [2]	ISSCC2016 [6]	This work
Technique	Active Quasi Circulator	Active Quasi Circulator	Electrical Balance Duplexer	N-Path-Filter-based Magnetic-Free Passive Circulator	Magnetic-Free Passive Circulator Based on Spatio-Temporal Conductance Mod.
Technology	180nm CMOS	180nm CMOS	180nm SOI CMOS	65nm CMOS	**45nm SOI CMOS**
Frequency	24GHz	24GHz	1.9-2.2GHz	0.75GHz	**25GHz**
TX-ANT Transmission	+22.4dB	-5.7dB	-3.7dB	-1.7dB	**-3.3dB**
ANT-RX Transmission	+12.3dB	-5.7dB	-3.9dB	-1.7dB	**-3.2dB**
TX-RX Isolation	>15dB	>20 dB	>40dB	>20dB	**>18.5dB [4]**
Isolation BW [1]	~1%	~1.6%	~15%	4.3%	**18% [5]**
Center frequency/ Modulation frequency	N/A	N/A	N/A	1	**3**
Area	3.22mm²	0.715mm²	1.75mm²	0.64mm²/25mm² [2]	**2.16mm²**
ANT-RX NF	17dB	N/R	3.9dB	4.3dB [3]	**3.3-4.4dB**
TX-ANT IP1dB	-19.8dBm	+9.5dBm	N/R	N/R	**>+21.5dBm**
TX-ANT IIP3	-11dBm	N/R	+70dBm	+27.5dBm	**+19.9dBm**
P_{DC}	144.8mW	7.2mW	0	59mW	**78.4mW**

[1] BW over which an isolation better than the value quoted in the row above is maintained.
[2] Includes SMD inductors on PCB.
[3] Includes 2.3dB degradation due to LO phase noise.
[4] Limited by the mmWave test setup.
[5] This is the 1dB insertion loss BW.
N/A: Not Applicable, N/R: Not Reported

TMTT2010: H. S. Wu, C. W. Wang and C. K. C. Tzuang, "CMOS Active Quasi-Circulator With Dual Transmission Gains Incorporating Feedforward Technique at K-Band," in *IEEE TMTT*, vol. 58, no. 8, pp. 2084-2091, Aug. 2010.

CONCLUSION

68

Conclusion

- Mm-Wave fully-integrated same channel full-duplex transceiver is demonstrated at 60GHz in GF 45nm SOI CMOS.
- The first mmWave fully-integrated magnetic-free passive circulator is enabled by the novel concept of spatio-temporal conductivity modulation.
- Topics of research for the future include pushing non-reciprocity and circulators to higher millimeter-wave frequencies (e.g. 77GHz), techniques to further improve linearity and power handling, and incorporation of antenna tuner functionality into the circulator structure.

In conclusion, full duplex operation at millimeter-wave frequencies can be an important enabler for many exciting applications, including automotive radar, 5G mm-Wave base stations and relays and repeaters for range enhancement. A mm-Wave fully-integrated same channel full-duplex transceiver is demonstrated at 60GHz in GF 45nm SOI CMOS. The first mm-Wave fully-integrated magnetic-free passive circulator is enabled by the novel concept of spatio-temporal conductivity modulation. Spatio-temporal conductivity modulation achieves broadband non-reciprocal phase shift (theoretically over infinite BW). The concept is readily scalable across frequency and can be an enabler for 77GHz circulators as well as optical isolators. Topics of research for the future include pushing non-reciprocity and circulators to higher millimeter-wave frequencies (e.g. 77GHz), and techniques to further improve linearity and power handling.

69 Acknowledgements

This work was supported by the DARPA ACT, DARPA SPAR and NSF EFRI programs.

Our thanks to

- Global Foundries for fabrication donation,
- Dr. Troy Olsson of DARPA and Prof. Andrea Alu of UT Austin for feedback and comments,
- Prof. Ken Shepard and Prof. Keren Bergman of Columbia University, and Prof. Arun Natarajan of Oregon State University for equipment support,
- CoSMIC Lab members for discussion.

Thank you for your attention.

70 Selected References

1. T. Dinc, A. Chakrabarti and H. Krishnaswamy, "A 60 GHz CMOS Full-Duplex Transceiver and Link with Polarization-Based Antenna and RF Cancellation," in *IEEE Journal of Solid-State Circuits* (invited paper), vol. 51, no. 5, pp. 1125-1140, May 2016.
2. Tolga Dinc, Anandaroop Chakrabarti and Harish Krishnaswamy, "A 60 GHz Same-Channel Full-Duplex CMOS Transceiver and Link Based on Reconfigurable Polarization-Based Antenna Cancellation," in the *2015 IEEE RFIC Symposium*, May 2015 (Best Student Paper Award – 1st Place).
3. Tolga Dinc and Harish Krishnaswamy, "A T/R Antenna Pair with Polarization-Based Reconfigurable Wideband Self-Interference Cancellation for Simultaneous Transmit and Receive," in the *2015 IEEE International Microwave Symposium*, pp. 1-4, May 2015.
4. T. Dinc, H. Krishnaswamy, " A 28GHz Magnetic-Free Non-reciprocal Passive CMOS Circulator Based on Spatio-Temporal Conductance Modulation," in *ISSCC Dig. Tech. Papers*, Feb. 2017.
5. Tolga Dinc, Aravind Nagulu and Harish Krishnaswamy, "Millimeter-wave Non-Magnetic Non-Reciprocal Circulator Through Spatio-Temporal Conductivity Modulation," (invited) *IEEE Journal of Solid-State Circuits*, vol. 52, no. 4, pp. 3276 - 3292, Dec. 2017.
6. Tolga Dinc, Mykhailo Tymchenko, Aravind Nagulu, Dimitrios Sounas, Andrea Alu and Harish Krishnaswamy, "Synchronized Conductivity Modulation to Realize Broadband Lossless Magnetic-Free Non-Reciprocity," *Nature Communications*, vol .8, no. 795, Oct. 6, 2017.
7. Jin Zhou, Negar Reiskarimian, Jelena Marasevic, Tolga Dinc, Tingjun Chen, Gil Zussman, and Harish Krishnaswamy, "Integrated Full-Duplex Radios," (invited paper) *IEEE Communications Magazine*, vol. 55, no. 4, pp. 142-151, April 2017.

CHAPTER 03

On-chip Transformer Design and Application to RF and mm-Wave Front-ends

John R. Long

University of Waterloo, Canada

The design of monolithic transformers for RF and mm-wave circuit applications are described in this chapter. On-chip transformers are applied to radio front-ends as VCO resonators, low-loss feedback networks, baluns, power combiners/splitters, and even ESD protection networks for wideband inputs. Transformer insertion loss, transformation ratio, and design for a specific bandwidth are outlined. Methods for shielding, which are essential to reduce substrate losses from on-chip transformers at mm-wave frequencies, are also described. Small- and large-signal models for simulation of circuits in a typical design flow are presented. Application-specific aspects of transformer design for VCOs, LNAs, and power amplifiers are detailed throughout using mm-wave case studies as design examples.

1 Outline

- Introduction
- On-chip transformer types and circuits models
- Specifications
- Design for RF and mm-Wave circuit applications (LNA, PA, VCO)
- Summary

2 Transformer Types

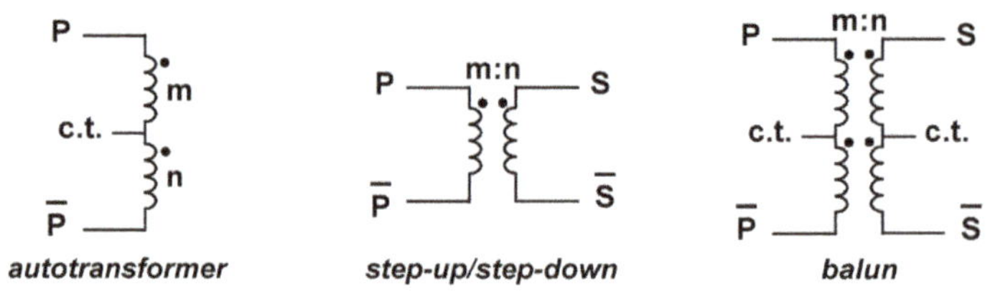

- Autotransformer: primary and secondary are *not* DC-isolated. Conductor losses and leakage inductance may be lower than a multi-winding transformer (optimal when m=n). Primary application is LC-oscillator tanks.
- Step-up/step-down transformer: Bias isolation between primary and secondary. Applied to impedance matching, interstage (AC) coupling, feedback networks, filtering, etc. Can consist of more than two windings.
- Balun: center-tapped (c.t.) transformer used as a phase splitter/combiner. Applications include: on- and off-chip interfacing, and interstage (AC) coupling.

A transformer is designed to couple alternating current from one winding to the other without a significant loss of power, and impedance levels are *transformed* as the ratio of terminal voltage to current flow between windings changes. In addition, direct current is blocked by a transformer, allowing the windings to be biased at different potentials. Three transformer types are typically found on a silicon chip: 1) autotransformers, 2) 1:1 turns ratio, m:n step-up, or step-down transformers, and 3) m:n transformer baluns [2]. An autotransformer synthesizes a relatively high-Q balanced coil in a compact area and is commonly used in L-C oscillator tanks. Power may be fed to the oscillator via the autotransformer's center tap (c.t.). Two-port (1:1 and 1:n turns ratio) and multi-port transformers (e.g., baluns and multi-filar transformers) are used for DC bias isolation and RF interstage coupling, impedance matching, feedback or feed-forward networks, phase splitting, and RF power combining [1].

3 SSCS The Linear Transformer

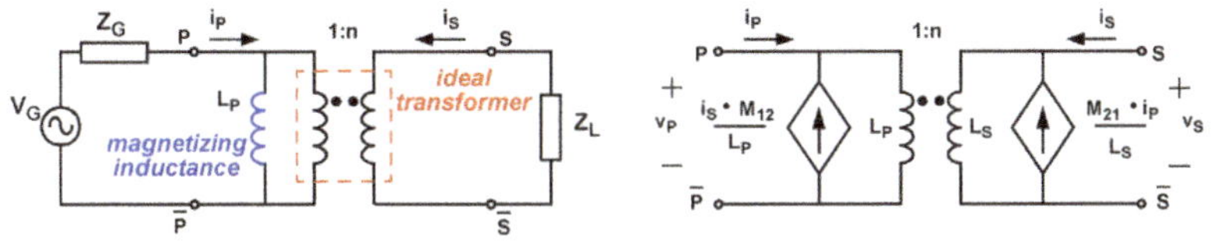

circuit function *circuit model*

- When reactance of the magnetizing inductance (L_P) is much larger than the source impedance (Z_G) and the reflected load impedance (Z_L/n^2):

$$v_S = n \cdot v_P,\ i_S = i_P/n,\ \text{and}\ Z_S = n^2 \cdot Z_P .$$

$$\text{Turns ratio, } n = \frac{v_S}{v_P} = \frac{i_P}{i_S} = \sqrt{\frac{L_S}{L_P}} .$$

- For "perfect" (i.e., 100%) magnetic coupling: $M_{21} = M_{12} = \sqrt{L_P L_S}$.

Operation for a passive transformer relies upon magnetic coupling between two or more conductors, or windings. Magnetic flux produced by current i_P flowing through magnetizing inductance ($k_m{}^2 L_P$) at terminal P induces a current in the secondary winding, which produces a voltage across a load connected between terminals S and $\bar{S}$. The main electrical parameters of interest to a circuit designer are the transformer turns ratio n, and the coefficient of magnetic coupling, k_m. Mutual inductances M_{12} (coil 1 to coil 2) and M_{21} (coil 2 to coil 1) couple the primary and secondary windings. For a passive, reciprocal circuit $M_{12} = M_{21}$. Current and voltage transformations between windings in an ideal transformer are related by the turns ratio, n, where the primary and secondary voltages (v_P , v_S) and currents (i_P , i_S) are defined on Slide 3.

Parameters L_P and L_S are self-inductances of the primary and secondary windings, respectively. The self-inductance is measured across the transformer terminals with all other windings open-circuited. Coupling between the coils may be modeled using current-dependent current sources in any circuit simulation with an appropriate weighting factor, as shown on Slide 3.

4 SSCS Transformer with Magnetic Leakage

Strength of the magnetic coupling between windings is represented by the k-factor, $k_m = M/(\sqrt{L_P L_S})$. The k_m is unity if the magnetic coupling between the windings is perfect (i.e., no leakage of magnetic flux). Uncoupled coils have a k_m of zero. A practical transformer has a k-factor somewhere between these two extremes. Since materials used in the fabrication of an IC chip have magnetic properties similar to air, the confinement of magnetic flux is relatively poor, and $M < \sqrt{L_P L_S}$. Thus, k-factor is always substantially less than unity for a monolithic transformer, however, coupling coefficients above 0.9 are realizable on-chip with overlaid, or 'stacked' windings. At the core of the model shown on this Slide is an ideal linear transformer with magnetizing inductance, L_m and turns ratio $1{:}n/k_m$. Current flowing through magnetizing inductance $k_m L_P$ gives rise to a voltage v_S on the secondary side. Note that the $1{:}n/k_m$ transformation for voltages across the magnetizing inductance and the secondary (i.e., v_S) is "ideal." In other words, leakage does not affect the internal voltage transformation in the model.

4 SSCS Transformer with Magnetic Leakage

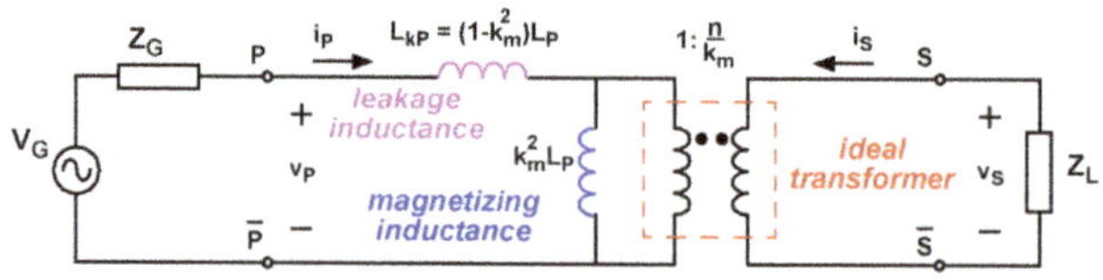

model for magnetic path including leakage

- Imperfect magnetic coupling between primary and secondary windings is called *leakage*, which is characterized by the coefficient of magnetic coupling, k_m

$$k_m = \frac{M}{\sqrt{L_P L_S}}, \text{ and } 0 < k_m < 1.$$

- Leakage prevents signal at the primary input from reaching the secondary output as the frequency increases. Leakage is reduced by shrinking the gap between turns on the windings of an on-chip transformer.

5 SSCS Transformer k-Factor

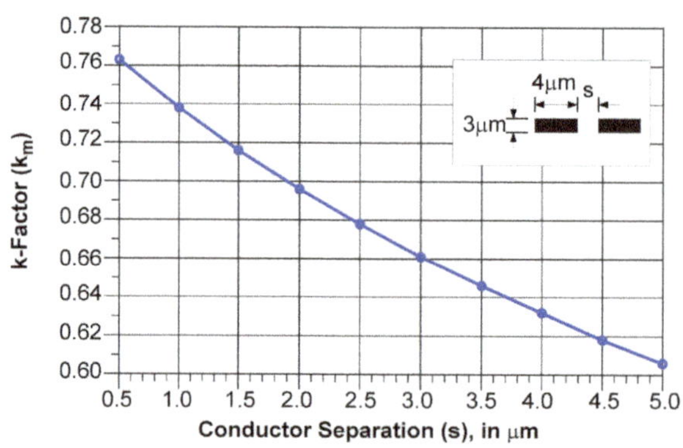

- k-factor is approximately 0.7 for minimum spacing (s) between conductors. Ratio of conductor length to separation diminishes when scaling transformers to mm-wave frequency, which further reduces k-factor.

The relationship between k-factor (k_m) and conductor spacing for a planar transformer (conductor width=4μm and thickness=3μm) as the spacing (s) between windings varies from 0.5μm to 5.0μm is shown on this Slide. Technology limitations restrict coupling to k_m < 0.7 for most silicon technologies when all conductors lie in the same plane. Note that for a conductor thickness of 3μm (as shown on this Slide), the minimum conductor separation imposed by fabrication constraints is approximately 1.5μm. The self and mutual inductances also depend upon the (unwound) length of coupled filaments. For mm-wave applications, where low self-inductance is necessary, smaller values of mutual inductance that accompany shorter length filaments with low self inductance reduce the k-factor that can be realized even further.

6 Transformer L.F. Response

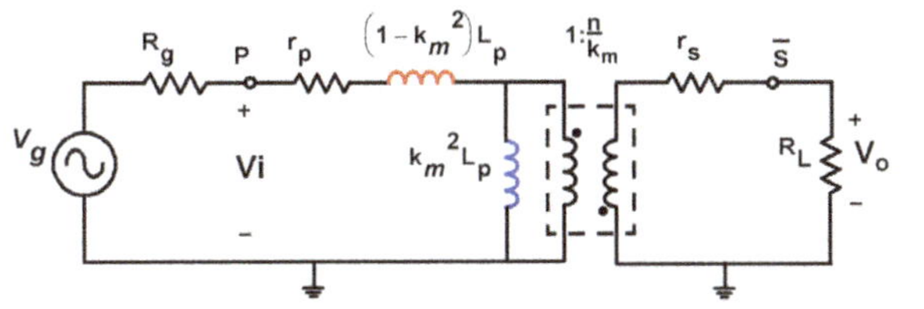

Low-frequency equivalent circuit with leakage k_m

$$\frac{f_U}{f_L} = \frac{(1 + k_m)}{(1 - k_m)} \qquad f_U/f_L = 2, \text{ for km} = 0.33 \qquad f_U/f_L = 9, \text{ for km} = 0.8$$

$$f_L \approx \frac{1}{2\pi}\frac{(R_g + r_p) \parallel R_r}{k_m^2 L_p}$$

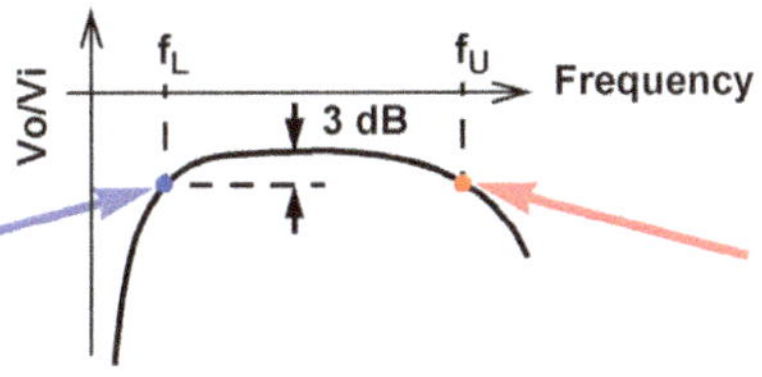

$$f_U \approx \frac{1}{2\pi}\frac{(R_g + r_p) + R_r}{(1 - k_m^2)L_p}$$

Rr = resistance reflected from secondary back to primary

The magnetizing inductance ($k_m^2L_P$) in the primary affects the low end of the frequency response by shunting energy to ground. Also, the leakage inductance ($1\text{-}k_m^2L_P$) in series with generator v_g blocks transmission of the signal from the primary to the secondary as the frequency increases. The terms low, mid and high frequency are relative as implied by the effects of these two key elements in the transformer model.

The series combination of load resistance, R_L and secondary winding loss, r_S may be reflected to the primary side of the transformer as $R_r=(R_L+r_S)/(n/k_m)^2$. Resistance R_r appears in parallel with the magnetizing inductance $k_m^2L_m$ after reflection across the transformer (i.e., from secondary back to primary). The lower (f_l) and upper (f_u) -3dB frequencies of the magnitude response expressed in terms of the generator resistance and secondary winding loss (R_G+r_P), $k_m^2L_m$ and R_r are shown on this Slide. Assuming that reflected and generator resistances are approximately equal and a 1:1 turns ratio, it can be shown that the fractional bandwidth predicted by this low- frequency (L.F.) transformer model is

$$\frac{f_u}{f_l} = \frac{1 + k_m}{1 - k_m}, \text{ for } k_m < 1.$$

Note that this analysis ignores the effects of parasitic capacitances, which are substantial in an on- chip transformer.

7 Transformer M.F. Response

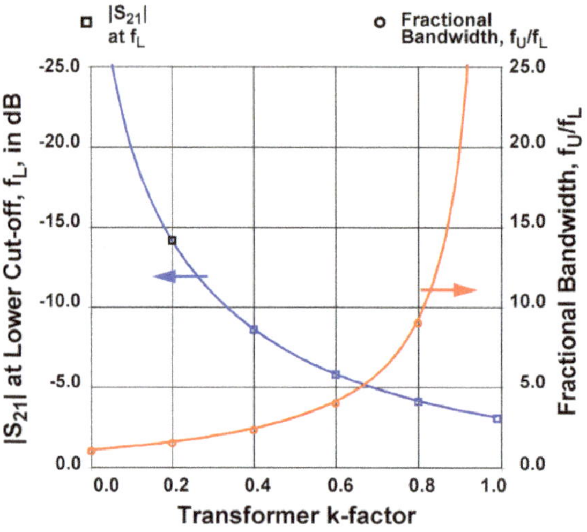

Assuming capacitive parasitics are negligible and $R = R_G + r_P = R_L + r_S$

Max. signal transmission in the passband is:

$$|S_{21}| = \frac{2v_o}{v_G} = \frac{k_m(R_L)}{R}$$

at

$$f_{pk} = \frac{R}{2\pi L_P\sqrt{1-k_m^2}}.$$

As the operating frequency increases, the attenuation through the transformer reaches a minimum in the passband. The series parasitics in the transformer compact model dominate the mid-band (M.F.) response, because the source and load impedances used in RF circuits are typically tens of ohms in most cases, which is an order of magnitude lower than the shunt parasitic reactances.

Assuming a 1:1 transformer with a symmetrically loaded winding where $R = R_G + r_P = R_L + r_S$, it can be shown that the maximum signal transmission in the passband is $|S_{21}| = (k_m(R_L)) / R$, where S_{21} is the forward transmission coefficient (S parameter). The passband peak, f_{pk} occurs at

$$f_{pk} = R/\left(2\pi L_P\sqrt{1-k_m^2}\right).$$

The $|S_{21}|$ varies from -3dB at the lower cut-off frequency, f_L (left axis of Slide 8, in blue) to approximately -6dB when k_m=0.6. Fractional bandwidth for a k_m of 0.8 is approximately 9, indicating that a wide bandwidth can be expected from an on-chip transformer.

8 Transformer H.F. Response

Some qualitative observations can be made from the high frequency equivalent circuit shown in Slide 8. A 1:n turns ratio transformer with k_m=1 is assumed, and the significant high-frequency parasitics of the transformer primary (i.e., C_P and R_P) are shifted to the secondary and scaled by the turns ratio, n. The effect of the magnetizing inductance at high frequency is assumed to be negligible. The value of the interwinding capacitance (C_o) is modified by the turns ratio when its left terminal is moved from primary to secondary loops due the Miller effect. The sign of 'n' can be either positive (non-inverting) or negative (inverting configuration), which leads to different behavior at higher frequencies. For the non-inverting connection, 'n' is positive and the π-type L-C section in the transformer secondary has a

8 Transformer H.F. Response

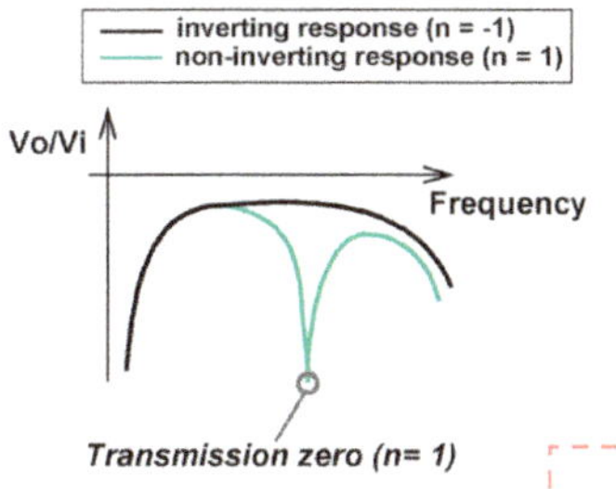

- For n > 0, (Co/n), L_S and r_S introduce a transmission zero.
- For n < 0, (Co/n) is an inductive reactance that decreases in magnitude with increasing frequency. No notch in response.

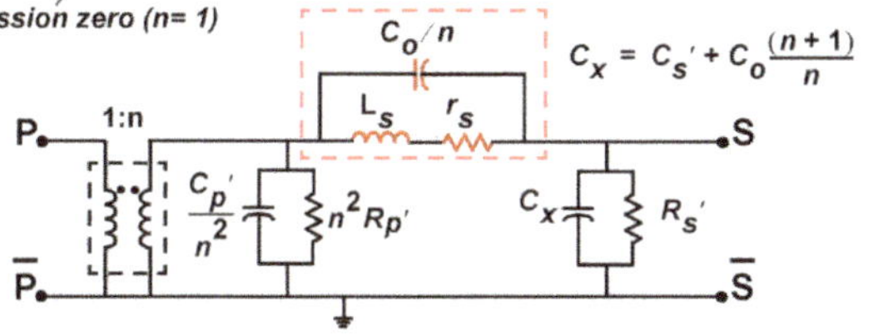

High-frequency equivalent circuit with leakage shifted to secondary

bandpass response with a transmission zero due to the effect of C_o in parallel with L_S. This zero causes a notch in the high frequency response, as seen on Slide 8. However, the inverting connection behaves differently at higher frequencies. The voltage transfer ratio (n) is negative, and the bridging capacitor C_o now has a positive reactance that *decreases* with increasing frequency. Thus, the magnitude response for the inverting connection at high frequencies resembles a bandpass filter with a comparatively higher cut-off than the non-inverting transformer connection.

9 Inverting vs. Non-Inverting

The frequency responses (magnitude and phase) of a 1:1 turns ratio, square-spiral transformer in non-inverting and inverting connections are compared in Slide 9. There is a substantial difference in the magnitude responses of the two connections as seen from the data (measured and simulated), due mainly to parasitic capacitance between the windings that introduces a zero in the response of the non-inverting transformer as described previously (Slide 8). At low frequencies, signal transmission from primary to secondary for a 50-Ohm source and load (i.e., S_{21}) is small, because the input frequency is well below the inductive reactance of the primary and secondary windings. Coupling between the input and output improves as the frequency increases, reaching a maximum in the 1-3GHz range. The phase difference between inverting and non-inverting configurations is 180 degrees at low frequencies, as expected, but deviates significantly at the upper edge of the transformer's operating band (i.e., above 3GHz).

9 SSCS Inverting vs. Non-Inverting

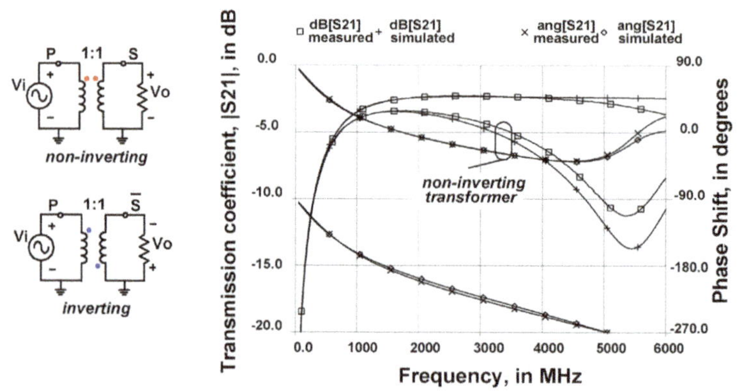

- Frequency response in the inverting configuration (single-ended) is less constrained by effects of interwinding capacitance.

10 SSCS A 2-Wire Planar Transformer

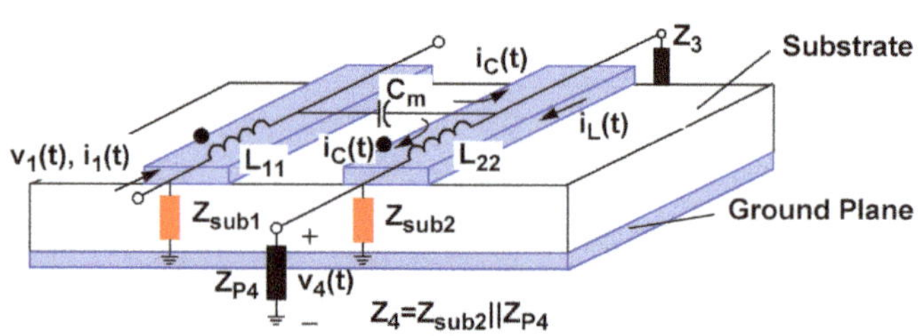

- Parasitic capacitance and substrate dissipation cannot be neglected at RF and mm-wave frequencies. Considering both capacitive and inductive couplings

$$v_4 = \frac{Z_4}{Z_3 + Z_4}\left(Z_3 C_m \frac{dv_1}{dt} + M_{14}\frac{di_1}{dt}\right), \text{ and}$$

$$v_4 = M_{14}\frac{di_1}{dt} \text{ for } Z_3 = 0 \text{ (i.e., magnetic coupling dominant).}$$

Two coupled microstrip lines are shown in Slide 10. A lumped-element circuit model of the mutual electric (capacitive) and magnetic (inductive) couplings is overlaid on the cross section shown. A time-varying voltage (v_1) and current (i_1) applied at terminal 1 induces time-varying voltages and currents via C_m and M_{14}, respectively, at terminal 4. Note that when the impedance at terminal 3 is set to zero, the effect of the (interwinding) capacitance (i.e., C_m) is suppressed, making the magnetic coupling dominant.

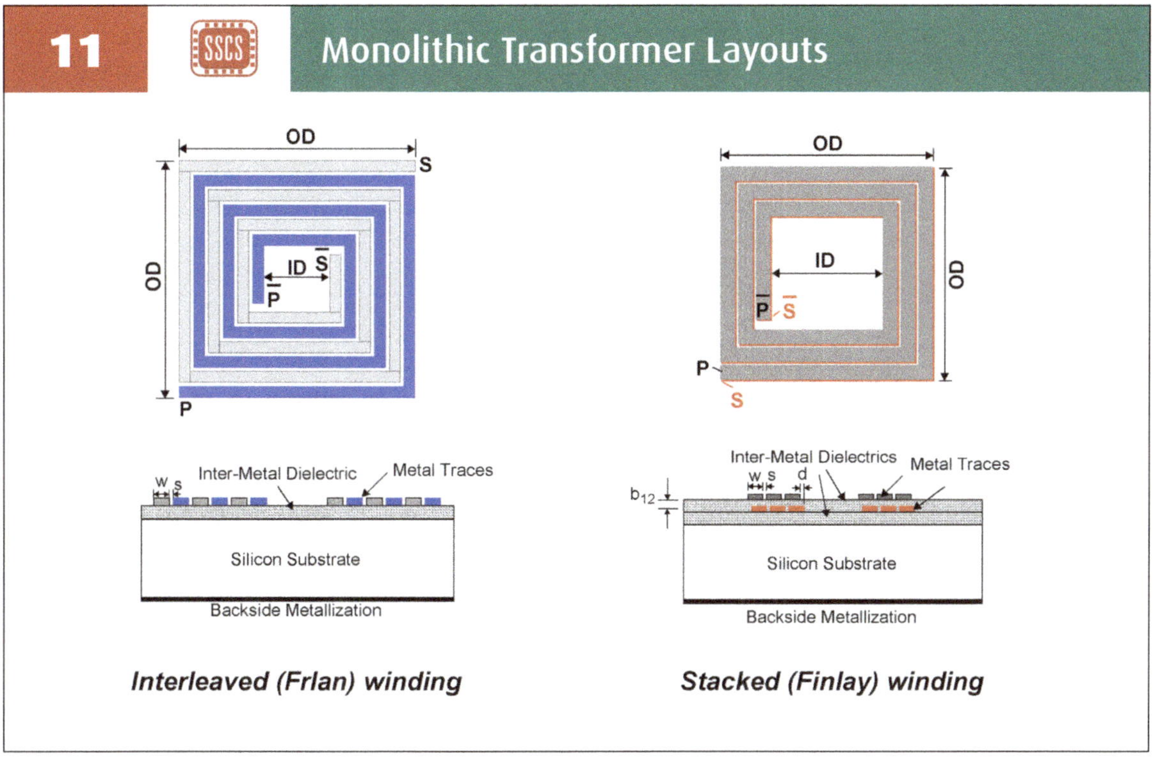

A transformer on a chip is constructed using conductors interwound in the same plane or overlaid as stacked metal, as shown in the plan and cross-sectional views of Slide 11. Conductor losses are reduced when multiple layers of metal are used to construct each winding. The mutual inductance (and capacitance) of the transformer is proportional to the unwound length of each winding.

Interleaving planar metals or conductor stacking promotes mutual inductance at the expense of increased interwinding capacitance. Coupling coefficient, k_m is determined by the mutual and self-inductances, which depend primarily upon the width and spacing of the metal traces and the substrate thickness.

The Frlan-style transformer winding ensures that electrical characteristics of the primary and secondary are identical when they have the same number of turns. Another advantage of this design is that the transformer terminals are on opposite sides of the physical layout, which facilitates connections to other circuit components [3].

Stacked conductor (i.e., Finlay-type) windings utilize broadside magnetic coupling to reduce the overall area required in the physical layout [11]. The dielectric thickness between layers on a silicon chip is less than 1μm in many technologies, giving k_m close to 0.9. Although the windings as shown in the figure are identical, they are implemented on different metal layers causing asymmetry in the electrical response of the transformer. Part of this asymmetry arises from any difference in thickness between metal layers, which results in unequal resistances for the upper (shaded) and lower (solid) windings. Although, the lower winding shields the upper electrically from the substrate, the overlap creates unwanted interwinding capacitance, which limits the transformer frequency response (i.e., upper limit described on Slide 8). However, these impairments are balanced by the savings in chip area when windings are stacked vertically.

12 Example Submicron CMOS Backend

Layer	Thickness, in μm
M9	0.9
M8	0.9
M7	0.3
M6	0.3
M5	0.3
M4	0.1
M3	0.1
M2	0.1
M1	0.07

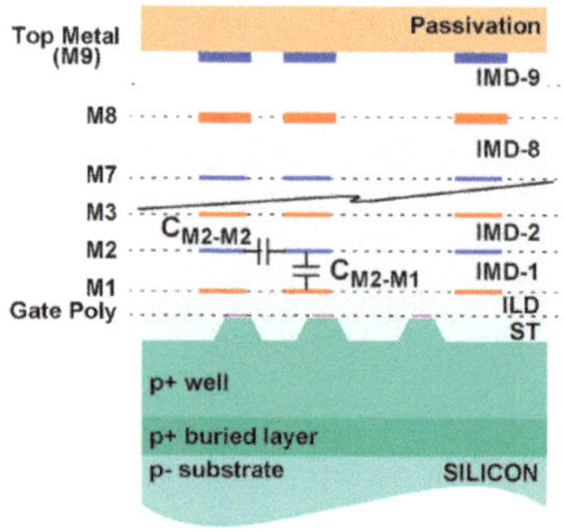

Layer	Thickness, in μm	Effective Permittivity
Passivation	1.8	5.75
IMD-9	1.6	4.3
IMD-8	1.6	4.3
IMD-2 to 7	0.6	3.0
IMD-1	0.25	3.3
ILD	0.45	4.2
ST	0.35	3.9
Silicon	350	11.7

- Example layers for a typical backend metal interconnect stack in deep submicron CMOS technology.

A generic cross-section of a back-end-of-line (BEOL) interconnect scheme is shown on Slide 12. Metal thickness ranges from 70nm (first metal, M1) to 0.9μm for top-metals M8 and M9. The intermetal dielectric (IMD) varies with height above the substrate (ref. table at the right of the Slide), where low-permittivity dielectrics and thin metals close to the substrate are typically used to interconnect dense (digital) circuits. Inductors and transformers are implemented near the top of the interconnect stack, where relatively thick metals reduce ohmic losses, and parasitic capacitance to the substrate is minimized by the greater (total) IMD thickness.

13 Cross-Coupled (Differential) LC-VCO

- The parallel-mode LC tank load is electrically tunable via C_V. Its impedance at resonance (ω_o) is R_T.
- Bandpass filtering by the load removes harmonic energy, making the output voltage almost purely sinusoidal.
- No current is injected by the diff pair into the tank at "zero crossings" (i.e., when the differential output voltage $V_{od}=V_{o+}-V_{o-}=0$). Peak current is injected when V_{od} peaks (conditions for lower phase noise).
- Bias current is regulated by a tail current source and is fed to M_1 and M_2 through the inductor center tap.

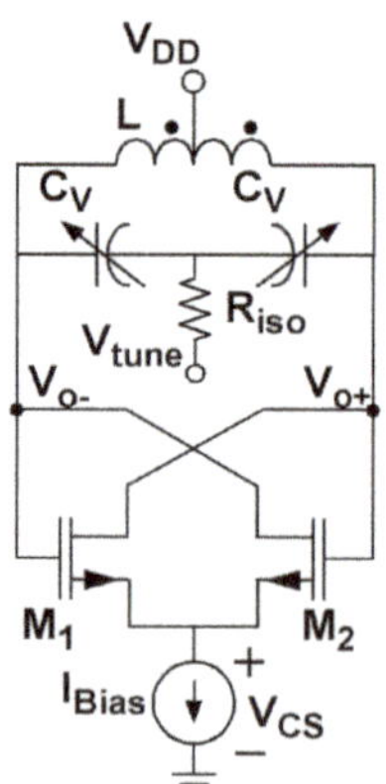

- Drain current of the diff pair is limited by I_{Bias}. The drain voltage can swing from V_{DD} down to almost V_{CS}, but output voltage also depends upon the product of the tank resistance (R_T) and the bias current.

13 Cross-Coupled (Differential) LC-VCO

The voltage-controlled oscillator with an L-C resonant tank (LC VCO) is the first application in this chapter highlighted for transformers on silicon RF ICs [12]. The tank consists of an autotransformer (note magnetic coupling indicated by dotted coils on Slide 13) in parallel with back-to-back varactors, and driven by a cross-coupled MOS differential pair. The autotransformer layout offers a relatively large inductance and quality (Q) factor while consuming minimal chip area. Symmetry in the layout permits biasing of the differential circuit via the virtual ground at the center tap (V_{DD} node). Also, both RF terminals are on the same side of the physical layout, which simplifies connections to the transistors in a differential circuit.

Varactors (C_V) are typically implemented in part as switched capacitors (i.e., fixed value metal- insulator-metal caps with series MOS switches) and in part with accumulation-mode MOS capacitors. The phase noise at any offset frequency ω_d is inversely proportional to the square of the tank Q-factor (i.e., Q_T), making it one of the most important parameters determining the RF performance of an LC VCO [13]. Moreover, the impedance of the tank at resonance (i.e., R_T) affects the amplitude of the VCO (i.e., $A_s \approx (2/\pi)R_T I_{DD}$ at each output) for fundamental frequency, ω_o. Any increase in Q_T increases the VCO output power through an rise in R_T, which may reduce phase noise even further.

14 k-Factor and Common-Mode Rejection

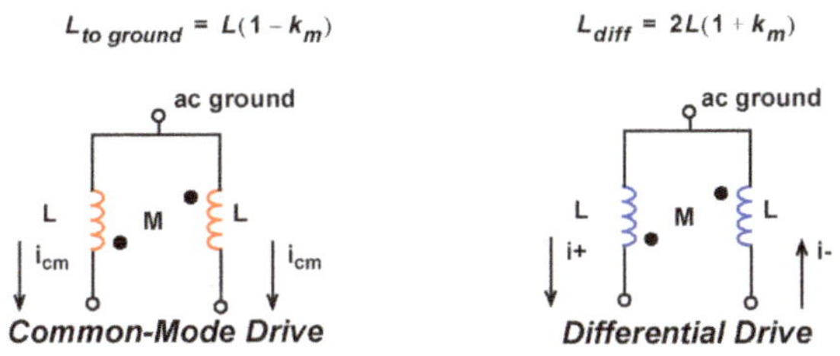

- Total inductance between terminals is 2x larger when driven differentially (i.e., 4L for k_m=1) and smallest for a common-mode signal (i.e., 0 for k_m=1).
- Common-mode rejection improves when tuned to resonate in differential mode, as common-mode components are "off-resonance" and shorted to ac ground.
- Substrate coupling to and from a symmetric inductor is reduced compared to 2 asymmetric inductor loads due to net cancellation at a distance from the coil.

The inductance seen between the inductor's terminals depends upon whether the signal is applied in the differential or the common mode, as illustrated in Slide 14. Positive magnetic coupling between the two halves of the coil gives a total inductance $L_{diff} = 2L(1+k_m)$ when the inductor is driven differentially by AC current i_+ and i_- , as shown at the upper right side of the Slide.

Common-mode currents i_{cm} applied equally to the inductor terminals result in an inductance $L_{to\ ground} = 2L(1-k_m)$ seen between either terminal and the AC ground, as shown on the upper left of the Slide. The center tap is assumed to be ac-grounded in both cases. Because of the difference in inductance between differential and common modes, the resonant frequencies for each mode is not the

14 k-Factor and Common-Mode Rejection

same. For example, common-mode components are shunted to ground when the tank of an LC VCO is tuned to resonate in the differential mode.

The tuned loads for an LC VCO can also be implemented using independent LC resonators (i.e., one connected at each drain). Aside from the penalty in chip area this implies, coupling of RF energy to the substrate increases compared to the autotransformer implementation, where substrate coupling is mainly differential, and is therefore localized.

15 Common-Mode Rejection of an LC Tank

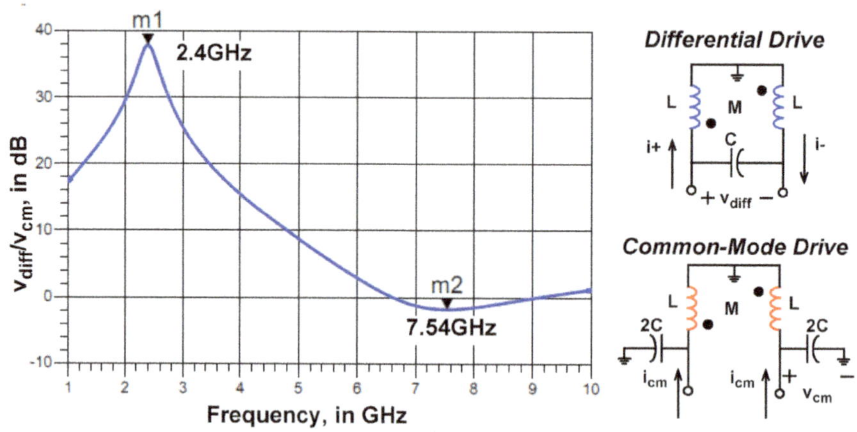

- In this example, a L=2nH inductor (k_m=0.8) resonant in the differential mode has a CMRR of 38dB at 2.4GHz. The common-mode resonant peak is pushed out to 7.54GHz as lower inductance is seen for common-mode drive.

When tuning capacitors (of value 2C) are connected between each RF terminal of the inductor and the AC ground (e.g., as in a VCO), the circuit resonates at $f_d = 1/2\pi\sqrt{CL_{diff}}$ when differential signals are applied. The resonant frequency depends upon the product of C (the differential load capacitance seen between the inductor terminals) and L_{diff}.

Resonance in the common mode occurs at a different frequency: $f_{cm} = 1/2\pi\sqrt{2CL_{cm}}$. The ratio of these resonant frequencies is

$$\frac{f_d}{f_{cm}} = \sqrt{\frac{(1+k_m)}{(1-k_m)}}. \qquad (2)$$

Note that the common-mode resonance (f_{cm}) is higher in frequency than f_d, and for k_m=0.8 (typical for an inductor on a silicon chip in the 1-10 GHz range), f_{cm} is approximately $3f_d$. For the example plotted on Slide 15, f_{cm}=7.54GHz, which is approximately $3xf_d$=2.4GHz from the SPICE simulation shown for the tank. The spread in resonant frequencies implies common-mode rejection, as any signal at f_d is attenuated by about 12dB compared to the differential signal for a single-pole-pair LC resonator. The common-mode rejection added by the tank is very welcome in LC VCO applications, where suppression of the common mode is severely limited by parasitic capacitance at the common (i.e., source) node of a MOS differential pair at RF.

16 Quality Factor

- For the normalized quadratic bandpass function, Z(s)

$$Z(s) = \frac{\omega_P s}{s^2 + \frac{\omega_P s}{Q_P} + \omega_P^2}$$

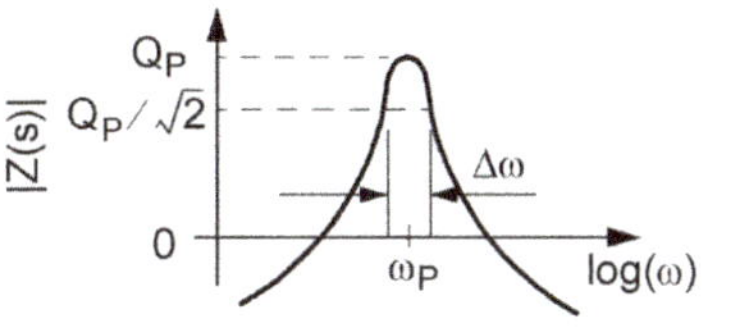

- Q_P relates bandwidth and peak frequency:

$$Q_P = \frac{\omega_P}{\Delta\omega} = \frac{-\omega_P}{2}\frac{\partial\phi}{\partial\omega}\bigg|_{\omega = \omega_P}.$$

- Another interpretation: [energy stored/energy dissipated] per cycle in the steady-state for sinusoidal excitation.

- Energy is dissipated as heat in the conductors and dielectric and is also radiated. For a resonator:

$$\frac{1}{Q_{Total}} = \frac{1}{Q_{Conductor}} + \frac{1}{Q_{Dielectric}} + \frac{1}{Q_{Radiation}} \approx \frac{r}{\omega l} + \frac{g}{\omega c}.$$

The impedance of any inductor at its self-resonant frequency (from simulation or measurement) can be modelled by adjusting the values of L_P, R_P and C_P in its parallel equivalent circuit. The Q- factor is the ratio of total resistance (R_P) to inductive reactance (ωL_P) from the parallel equivalent (i.e., $R_P/\omega L_P$) [4]. Inductor Q defined in this way is consistent with the definition of the Q-factor (Q_P) for a normalized bandpass response Z(s),

$$Z(s) = (\omega_P s)/\left(s^2 + \frac{\omega_P s}{Q_P} + \omega_P^2\right). \quad (3)$$

Q_P is the ratio of center frequency (m_P) to the -3dB bandwidth at resonance from the magnitude response, or the product of the center frequency and slope of the phase responses at resonance,

$$Q_P = \frac{\omega_P}{\Delta\omega} = \frac{-\omega_P}{2}\frac{\partial\phi}{\partial\omega}\bigg|_{\omega = \omega_P}. \quad (4)$$

Analogous to a transmission line, the total Q-factor (Q_{Total}) can also be expressed as $Q_{Total} = Q_{Conductor}||Q_{Substrate}$, where $Q_{Conductor}$ is associated with losses of the conductor metal and any AC current induced by the magnetic field in the substrate. $Q_{Substrate}$ is determined by dissipation of the electric field in the silicon substrate. Typically, $Q_{Conductor} >> Q_{Substrate}$ for ICs fabricated on medium-resistivity silicon substrates.

17 Q-Factor vs. Conductor Spacing

Q-factor varies with the layout parameters of the inductor, and tighter spacing (s) between conductors gives an increase in k-factor, mutual inductance, and peak Q-factor (approx. 20 at 45GHz), as seen from the data plotted on Slide 17 (L=0.4nH, 0.5μm < s < 5μm; conductor width of 4μm and thickness of 3μm). The Q-factor of an inductor at self-resonance is finite, and it is typically well above zero, as seen from the data shown on Slide 17, where self-resonance lies between 70 and 80GHz.

17 Q-Factor vs. Conductor Spacing

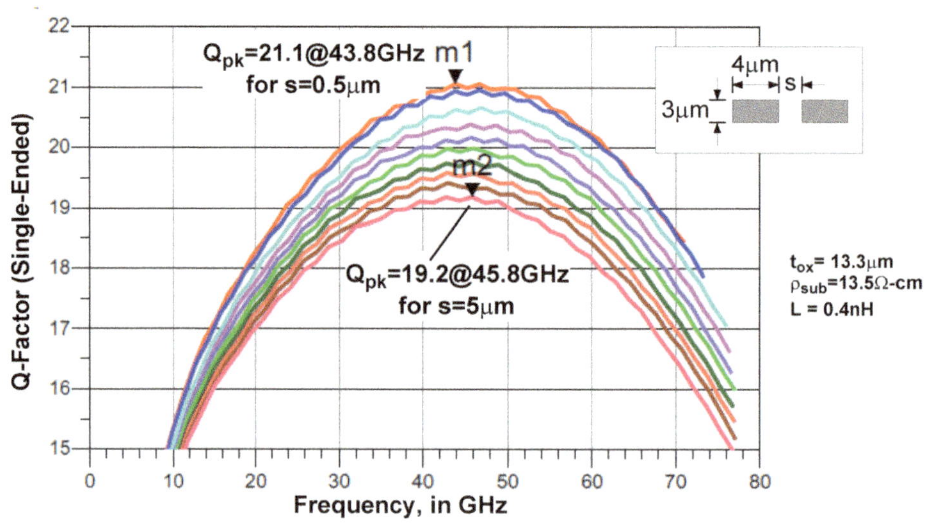

- Single-ended Q for a 2-turn inductor with conductor spacing 0.5µm < s < 5.0µm.

18 Q-Factor for Differential Drive

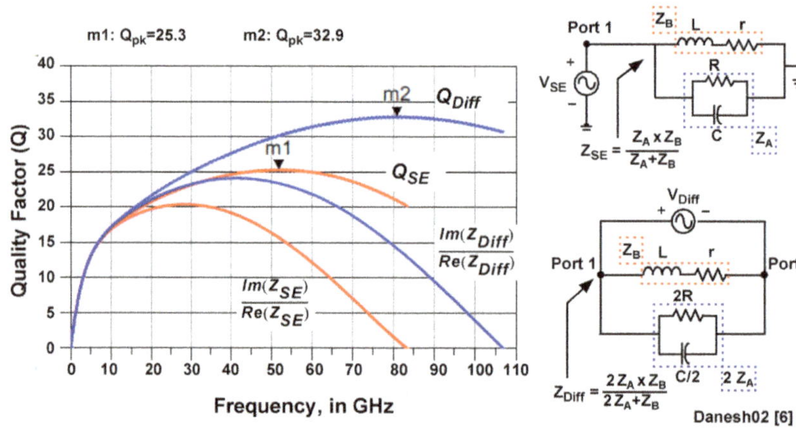

- Differential excitation gives higher Q-factor and a broader band where Q-factor is close to its peak.

The Q-factor of a transformer winding is (ideally) doubled when driven differentially rather than single-ended, because of loading by substrate parasitics on the coil. The substrate parasitics present a higher shunt-equivalent impedance between the RF terminals of the coil when a differential signal is applied (i.e., $Z_{Diff} > Z_{SE}$), as seen on Slide 18. Therefore, the Q-factor for differential drive (Q_{Diff}) improves compared to Q_{SE} for single-ended excitation, without any modification to the IC technology or chip processing. The frequency range below self resonance (which is the usable range as an inductor) also widens due to the reduction in the parasitic capacitance seen between the terminals. The peak Q-factor and self-resonance also occur at higher frequencies due to the reduced effect of

18 Q-Factor for Differential Drive

substrate parasitics when the autotransformer is driven differentially [6].

At lower frequencies (i.e., below 15GHz in this case), the difference in Q between the differential and single-ended excitations is not significant (<1%). This is because the shunt capacitive parasitic components do not affect the low-frequency input impedance, which is dominated by the inductance and resistance of the metal windings. However, the lower substrate parasitics seen when driven differentially result in a higher peak Q-factor and broadening of the Q peak in the vicinity of 80GHz compared with the single-ended connection at 50GHz, as seen from Slide 18.

The inductor Q-factor is often estimated incorrectly by the ratio of imaginary and real parts of the one-port impedance (i.e., "Q" expressed as $\mathrm{Im}[Z_1]/\mathrm{Re}[Z_1]$) . While this is a valid approximation well below the peak of the Q curve (i.e., at low frequencies), it is a poor approximation near the Q peak, and at higher frequencies. The error is largest error when $\mathrm{Im}[Z_1]$ is zero, which occurs at the inductor's self-resonant frequency. At the peak of the Q-factor vs. frequency curve, the error between $\mathrm{Im}[Z_1]/\mathrm{Re}[Z_1]$ and the actual Q-factor is smaller, but the impedance ratio is typically 15-20% pessimistic as an estimate of the inductor Q at the peak. It also cannot accurately predict the frequency where the Q-factor is largest.

19 Optimizing Tank Q-factor

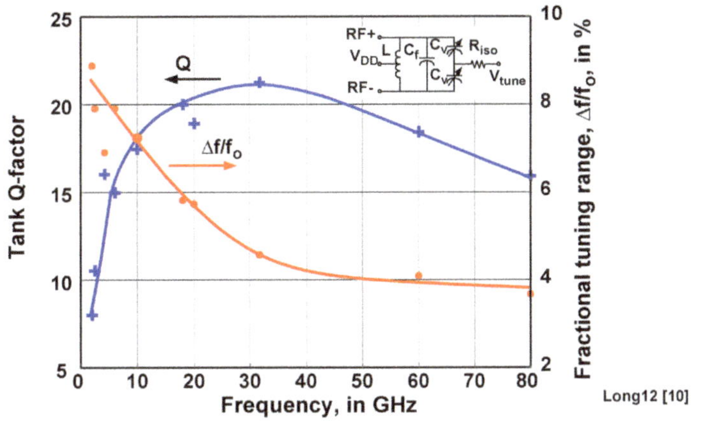

- Q-factor and fractional tuning range for monolithic resonators in 65nm CMOS from simulation (C_f=20fF).

The total Q-factor and fractional tuning bandwidth from simulation for LC tanks optimized at each frequency are shown in Slide 19 (tank schematic shown inset). A 65-nm RF-CMOS technology was used to generate data for the plots on Slide 19 [10]. A fixed portion of the tank capacitance (C_f) accounts for interconnect wiring and transistor parasitics. To construct the tank Q-factor curve shown, an inductor is selected which has the highest peak Q-factor when driven differentially at each frequency. Fixed and varactor capacitances (C_f= 20 fF and n+/n-well thick- oxide C_v, with gate length L = 0.4 um) are added to set the resonance to f_o. The tank Q-factor is then extracted from f_o and the bandwidth of the resonator determined from simulation (i.e., $Q_{tank} = f_o/\Delta f$).

19 Optimizing Tank Q-factor

The tank Q tracks the inductor Q at low frequencies (i.e., below 10 GHz). It is dominated by varactor Q above approximately 40 GHz. Between 20 GHz and 40 GHz, the inductor Q-factor continues to rise with increasing frequency. While the varactor Q is dropping, it is still large enough not to affect the overall tank Q across this range, and therefore the tank Q peaks. However, the tank's fractional tuning range ($\Delta f/f_0$, plotted on right axis of Slide 19) decreases continually with increasing frequency.

A VCO operating in the 15-25 GHz range would have both a high-Q tank and fractional tuning capability better than 5%. If a frequency below 25 GHz were used, a power-efficient frequency multiplier could be used to shift the VCO output to the mm-wave band (e.g., LO > 40 GHz for a 60-GHz receiver).

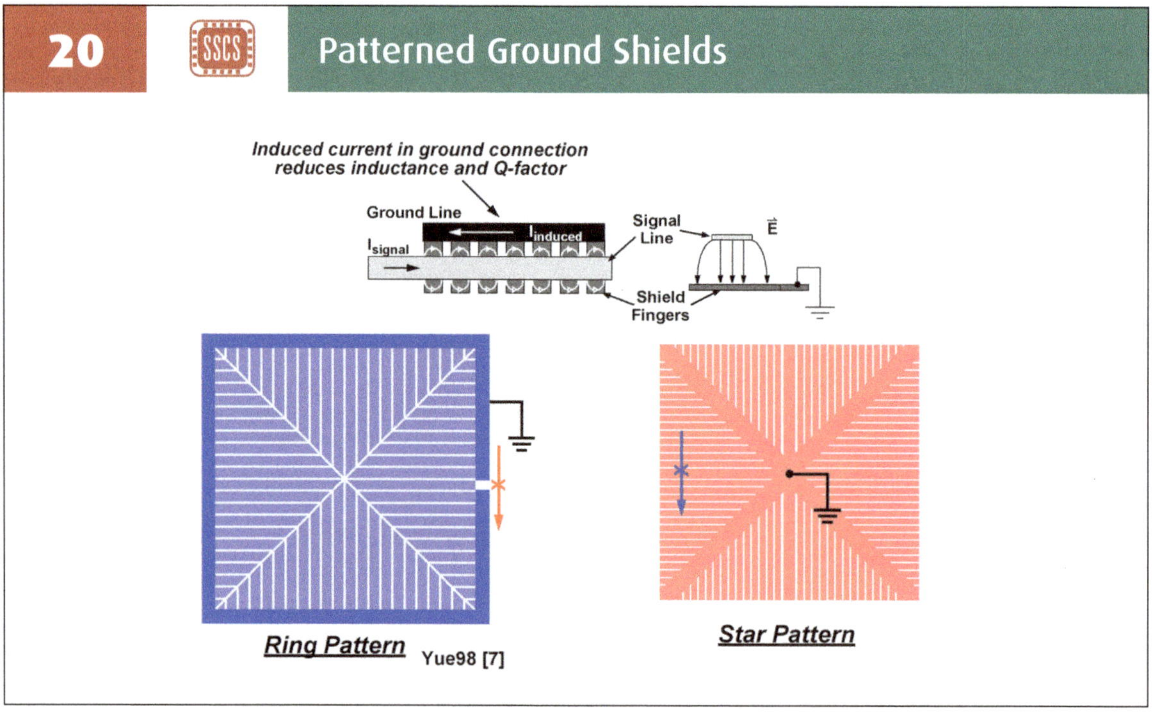

The Q-factor for transformers fabricated in production technologies is limited primarily by energy dissipation in the silicon substrate, which for analog/RF and mixed-signal technologies typically has a resistivity in the 1 Ω-cm to 15 Ω-cm range [5]. Most production silicon-on-insulator (SOI) processes incorporate low-resistivity silicon wafers to simplify handling and lower manufacturing costs, and therefore substrate dissipation can also limit the Q factor in SOI technologies.

Substrate losses are reduced when a patterned shield connected to the on-chip ground is placed between the inductor and the substrate. A simplified illustration of a shielded inductor is shown on Slide 20 [7]. The shield fingers shown in first metal (blue, on the left), are orthogonal to the direction of current flow along the transformer winding in top metal. They block the electric field from entering the underlying substrate. However, the shield adds parasitic capacitance that reduces the self-resonant frequency, and location of the ground point (shown on Slide 20 by the ground schematic symbol) is not unique. Any current induced in the shield by the magnetic flux produced by the inductor winding decreases the inductance and the Q-factor of the shielded inductor. Therefore, gaps between the shield fingers are used

20 Patterned Ground Shields

to inhibit the flow of current induced in the shield by the magnetic field surrounding the inductor (limiting them to small current loops induced in each finger).

The star pattern shield (on the right, in red on Slide 20) is an alternative which requires grounding at one point (i.e., at the center, as shown). Ground on-chip is not well defined due to interconnect parasitics (e.g., wirebond inductance, etc.), so a shielding method (e.g., differential shielding) that does not require an explicit ground is preferred.

21 Differential Shielding

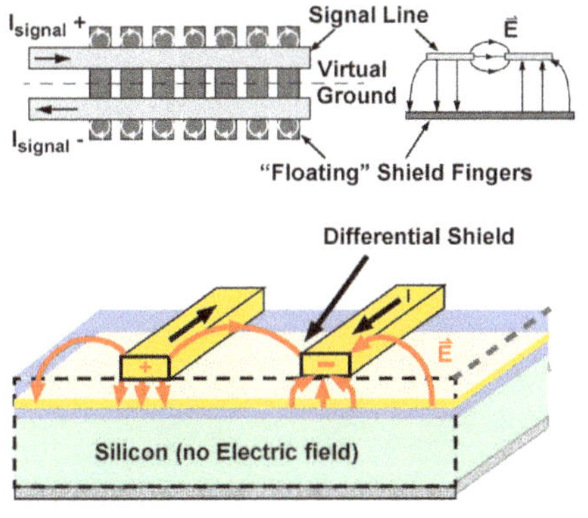

Shielded Balanced Tx Line

Cheung03 [8]

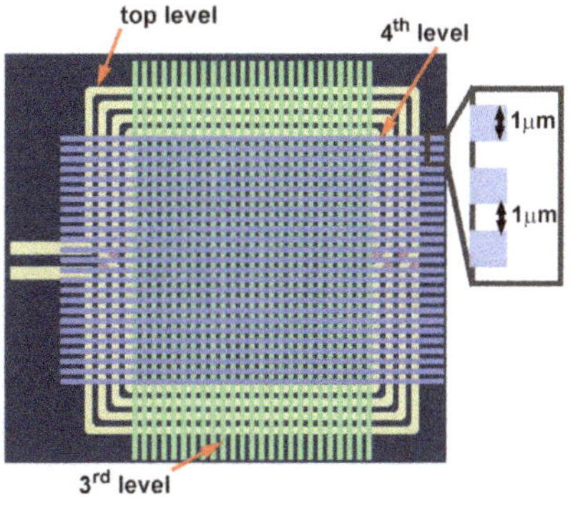

Shielded Symmetric Inductor

Cheung06 [9]

A floating, patterned metal ground shield (shown on the left in Slide 21) does not significantly affect the magnetic field around a balanced transmission line if the dimension of each shield finger in the direction of the signal current flow is kept short. The fingers of this *differential shield* span the transmission line width and shield the signal conductors from the underlying substrate. Because of the patterning, current induced in the shield has little effect on the inductance of the signal conductors. However, shielding the electric field from the substrate increases the capacitance per unit length of the transmission line.

Unlike an inductor shielded by patterned metal connected to ground (as in Slide 20), differential shielding does not require an explicit ground connection, which leads to a simpler implementation. Slide 21 shows (on the right) a differential shield implemented beneath a 4-turn symmetric inductor [8]. The top-metal (yellow) is capacitively coupled to orthogonal metal strips in first (i.e., bottom) and second metals placed beneath it. The fingers oriented vertically in green shield the conductor groups at the top and bottom of the layout. Similarly, fingers oriented horizontally (in blue) shield conductors groups at the top and bottom of the transformer. The shield strips are short in the direction of current flow, but span the entire coil layout to shield each pair of conductor groups electrically. Current flow induced on the metal shield fingers is suppressed by placing them orthogonal to the direction of current flow in the coil winding. With no large-area metal loops to conduct an induced current, the shield metals do not affect the inductance. Due to the balanced (i.e., differential) drive assumed at the inductor input terminals, the net voltage induced on the shield strips is zero (i.e., a virtual ground, ideally) and the autotransformer winding is therefore shielded electrically from the silicon substrate.

22 Differentially-Shielded Inductor

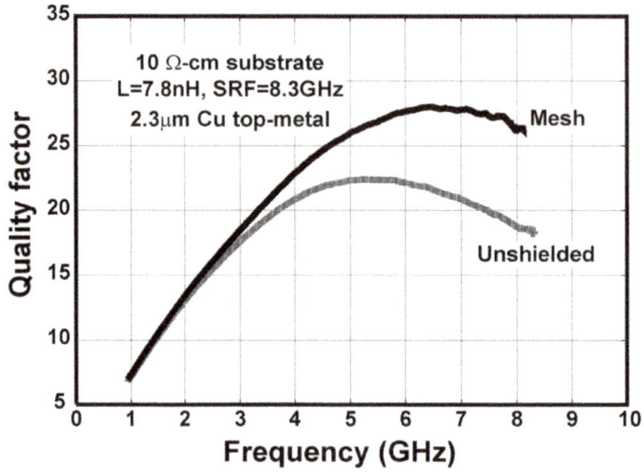

The differentially-shielded inductor (L=7.8nH) demonstrates up to 35% improvement in Q-factor using a production process flow (see the mesh shield vs. unshielded coil Q-factor plots on Slide 22). Each Q-factor sweep begins at 1 GHz and ends at the self-resonant frequency. The self-resonant frequency (SRF=8.3GHz) is reduced by less than 3% when the differential shield is added. The impedance seen between the inductor terminals is capacitive beyond the SRF.

23 Transformers and RF Amplifiers

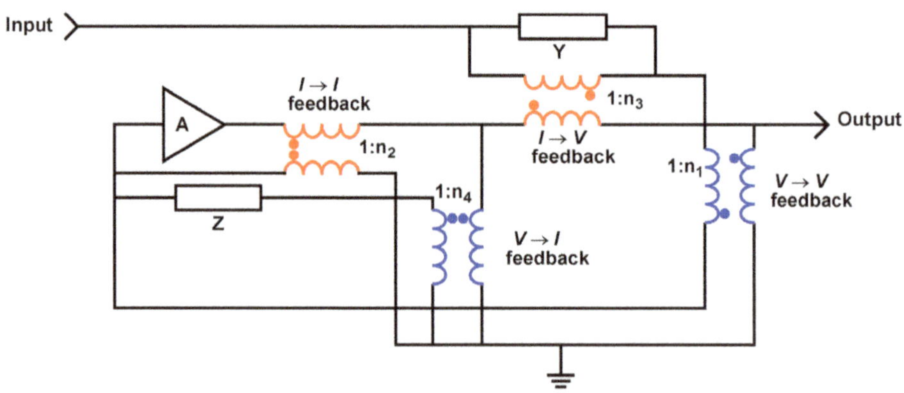

Dual-loop amplifier with 4 feedback loops

- Bandwidth of transformers with floating terminals (i.e., one terminal not at AC ground) is compromised by parasitics. V-I feedback is least affected by substrate parasitics (one terminal on each winding is grounded).

23 Transformers and RF Amplifiers

Four feedback amplifier configurations are shown on Slide 23. A 1:n step-up transformer is an almost ideal feedback element for an RF amplifier, and can be used as an alternative to broadband, resistive feedback networks. The benefits of negative feedback in amplifier design are well-known, however, broadband feedback amplifiers struggle to meet the sub-3dB noise figure required by many wireless applications. Feedback via mutual magnetic coupling in a 1:n transformer, as shown on this Slide, allows for control of amplifier input/output impedances, gain and linearity without introducing excessive noise. Aside from the potential for lower noise, a lower supply voltage is also possible, since ohmic drops can be nearly eliminated when a transformer is used in the DC bias path.

24 Transformer Neutralization

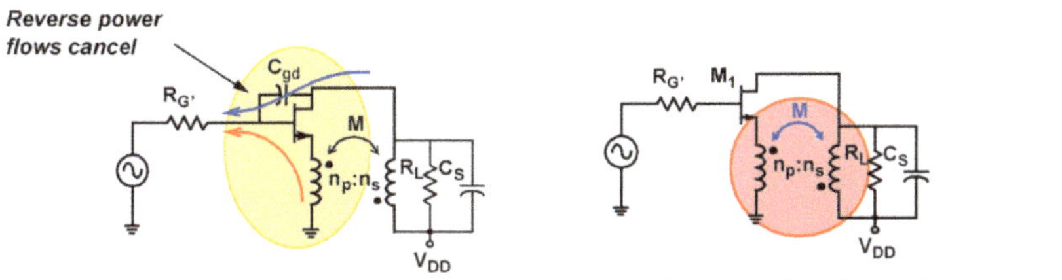

- greater isolation simplifies matching and increases gain

- noiseless feedback stabilizes gain, reduces output impedance

- Neutralizing feedback increases gain and reverse isolation, promotes stability, and raises input and output impedances.
- Neutralization via a step-down transformer cancels signal flow through C_{gd},

$$\frac{C_{gs}}{C_{gd}} \approx \frac{n}{k} \Rightarrow \text{Neutralization condition}.$$

The second application presented in this chapter is feedback around an RF amplifier using a 1:n transformer. Impairments caused by transistor gate-drain overlap capacitance C_{gd} is unavoidable in CMOS technologies. It reduces amplifier gain, and unwanted feedback between input and output degrades that reduces the amplifier's reverse isolation. Its effect on the input impedance in a common-source amplifier is known as the Miller effect (i.e., $C_{in} = C_{gd}(1 + A_V)$, where A_V is the voltage gain).

The Miller effect may be compensated by coupling inductors in series with the drain and source of the MOSFET, as shown on Slide 24. Neutralization cancels signal flow through C_{gd} by adding a second signal path around the amplifier. As shown for the circuit on the upper right of the Slide, signal flow through C_{gd}, which causes the Miller effect, may be cancelled entirely by selecting an appropriate value for the turns ratio 'n' of the transformer, thereby *neutralizing* the amplifier.

When neutralized, the signal flowing through C_{gd} and the signal coupled via the transformer sum to zero at the amplifier input. This increases forward gain and reverse isolation for a given power consumption, but does not reduce the effect of C_{gd} on the input impedance of the amplifier.

25 1:1 Transformer MAG

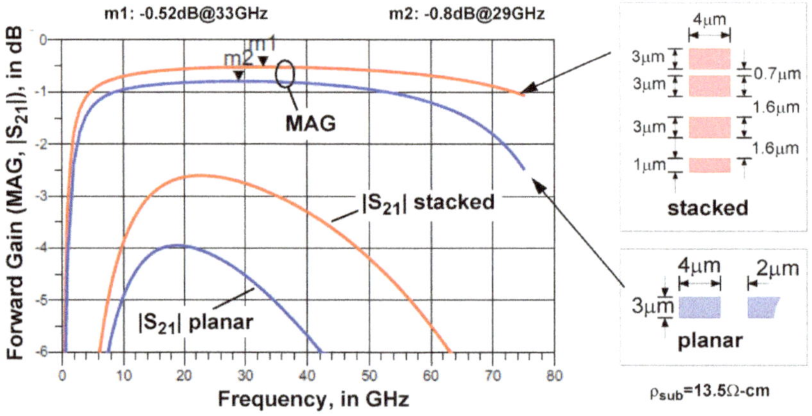

- Maximum available gain (MAG) is realized for conjugate match at source and load ports at each frequency. Untuned (i.e., S_{21}) gain for 50Ω source/load is 2-3 dB lower than MAG. Note that tuning further reduces bandwidth.

The maximum available gain (MAG) for two 1:1 transformer designs are plotted from simulation on Slide 25. A 22-nm SOI-CMOS technology with 11 interconnect metals is used for this study. The first design (called *stacked*) is comprised of 4 metals at the top of the BEOL stack: 3µm topmetal-M11, 3µm metals M10 and M9, and 1µm metal M8. Metals M11 and M9 are used to define one transformer winding, while the other is comprised of metals M10 and M8 (i.e., a Finlay-style transformer). The vertical position in cross-section is determined by intermetal dielectric (IMD) thicknesses in the BEOL stack, which are annotated on the Slide at the upper right. The second, *planar* design, is comprised of four, 3-µm thick topmetal M11 conductors spaced 2µm apart. Primary and secondary metals alternate in a 1:1 configuration (i.e., a Frlan- style transformer). The metals used in both stacked and planar designs are 4µm wide, and reside on a 13.5 Ohm-cm silicon substrate. The self-inductance of each winding is 0.4nH in both cases.

The forward gain in a 50**Ω** system ($|S_{21}|$), and MAG (i.e., matched source and load at every frequency) are plotted and compared on Slide 25. The minimum S_{21} is close to 20GHz for both transformers. However, the stacked transformer has about 1.5-dB lower insertion loss near the peak of the response. Also, the frequency range where the insertion loss is minimum is clearly wider for the stacked design than for the planar 1:1 transformer. This is a consequence of tighter magnetic coupling (i.e., higher k factor) between windings in the stacked configuration.

The difference (in dB) between S_{21} and MAG is the potential improvement in insertion loss possible when the input and output are impedance matched simultaneously. Tuning impedances needed to realize MAG at each frequency are shown on Slide 26. Minimum insertion loss is indicated by marker 'm1' (i.e., stacked: -0.52dB at 33GHz) and marker 'm2' (planar: -0.8dB at 29GHz) annotated at the top of the Slide. While the difference at each frequency is small (i.e., approximately 0.3-dB), the frequency response for the stacked transformer is broader. The frequency range yielding less than 1-dB insertion loss when tuned is approximately 5-70GHz for the stacked transformer, but only 10-60GHz for the planar design. This is again a consequence of tighter magnetic coupling. The minimum insertion loss (i.e., from MAG) in the 28-60GHz range for both transformers is relatively low thanks to the thick metals available in the 22-nm technology, and thick IMD layers separating the upper metals from the silicon substrate.

26 1:1 Tuning Impedances for MAG

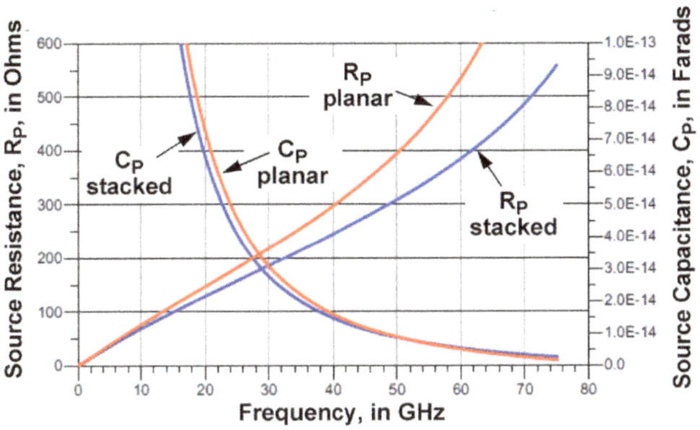

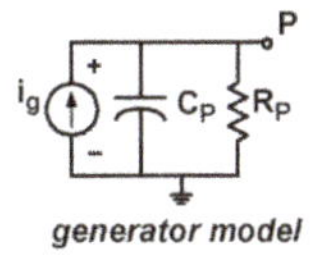

- Source impedances (parallel R-C equivalent, R_P/C_P) required to realize MAG vs. frequency are plotted above (load and source impedances are identical for a 1:1). Circuit loading at terminals must be included in R_P and C_P.

MAG for the 1:1 transformers shown on Slide 25 requires tuning of source and load impedances at each frequency to minimize the insertion loss. For a 1:1 transformer, the tuning impedances at each port are identical. Therefore, only the resistance and capacitance required for MAG tuning on the generator side are plotted on Slide 26. Note that parallel tuning capacitor, C_P is almost identical in value for both stacked and planar transformers, e.g., at about 35fF when tuned to 28GHz. Since the C_P value depends on the self-inductance of the winding (0.4nH for both transformers), this result is expected. However, the shunt resistance (R_P) differs, and the difference between the required tuning resistances grows with increasing frequency, as shown on the Slide. The stacked transformer supports a lower tuning resistance, leading to a lower R_PC_P product and wider bandwidth overall (as seen in Slide 25).

27 Wideband Feedback Amplifier

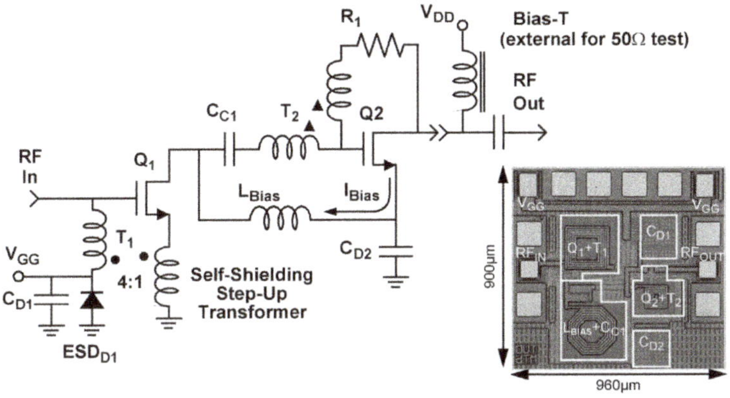

- 2-stages: series/shunt feedback via self-shielding transformer sets input impedance; transimpedance second stage lowers output impedance.

27 Wideband Feedback Amplifier

The wideband low-noise amplifier (LNA) example shown on this Slide is an example application for 1:n transformers on a silicon IC. The circuit is designed to operate across the entire 3.1- 10.6GHz ultra-wideband (UWB), which is open for unlicensed use in the US and many other regions worldwide (although transmit power restrictions apply) [14]. The LNA is implemented in a 0.13-µm CMOS process and consumes just 9mW from a single 1.2-V supply. Bias current reuse (via L_{bias}) enables multi-stage RF performance with single-stage DC power consumption. A noise figure of approximately 2dB across the entire band, as well as broadband gain and port matching without a lossy input filter are achieved through the use of an on-chip, 4:1 feedback transformer.

Current flowing in the source of Q_1 is fed by T_1 back to the RF input where it is summed with generator current at the gate input. The transimpedance second stage also uses a wideband feedback transformer to shape the frequency response. Use of stacked transformer windings for both saves valuable chip area, and this 15-dB gain UWB LNA occupies less than 0.9mm² (ref. chip photo at the lower right) [15].

28 WB-LNA Input Impedance and NF

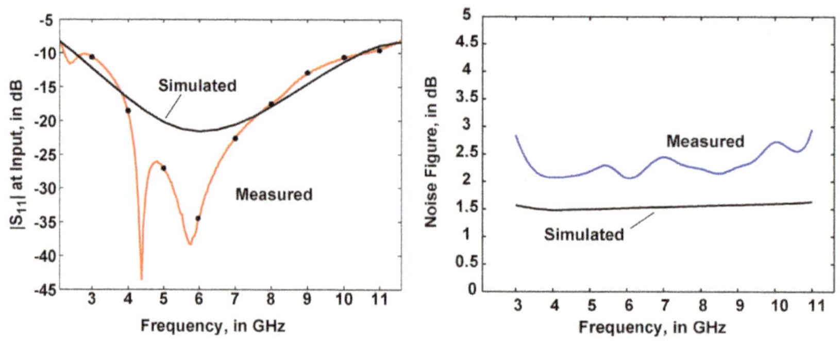

- Broadband gain of approx 15dB while drawing 7.5mA from 1.2V (0.13um CMOS). Noise figure < 2.5dB and relatively flat across > 7GHz. Excellent input match across band (S_{11} < -15dB).

The input reflection coefficient ($|S_{11}|$) measured across the 3-11GHz band is plotted on the left side of this Slide. The input impedance is controlled by feedback transformer T_1 and loop gain of the first stage in the LNA. The lower frequency limit is determined by T_1's magnetizing inductance in combination with the system impedance (i.e., 50Ω.). On-wafer measurements show $|S_{11}|$ < 10dB across 2.2-10.6GHz with nulls of -44dB and -38dB at 4.4GHz and 5.8GHz, respectively. Simulations for $|S_{11}|$ agree very well with the on-wafer measurement, although the deep nulls seen in the measured data are not captured in simulation.

The noise figure (NF) measured on-wafer across a broad bandwidth is compared to simulation on the right in Slide 28. This measured result demonstrates the effectiveness of reactive feedback in realizing a low NF consistently across more than 7-GHz bandwidth. The minimum, measured NF is 2.1dB at 6GHz, while its variation across 3-11GHz is less than 0.9dB. The minimum NF measured at 5GHz for a packaged version of the same amplifier is at 2.04dB, with a variation of 0.94dB over 3-11GHz (not shown on the Slide).

29 Step-Up Transformer Cutaway

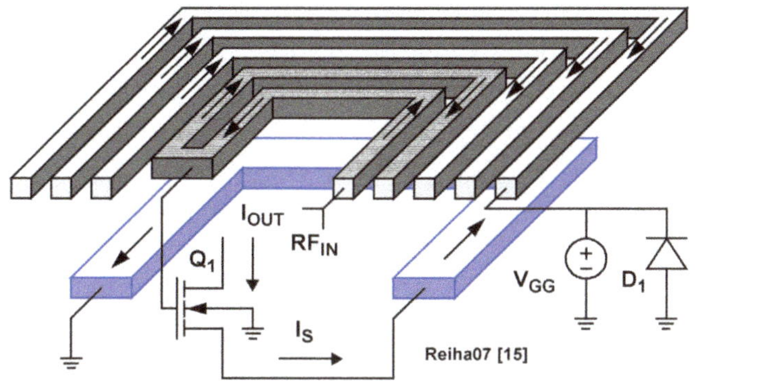

- Turns ratio is determined by the number of turns on each winding and k-factor.
- Stacked layout using separate wiring planes for primary and secondary simplifies implementation and saves chip area.

Slide 29 illustrates the input stage design of the LNA. The Finlay (overlay) transformer windings realize the highest magnetic (and electric) coupling. RF input signal enters at the innermost winding and couples to the adjacent turn. This creates leakage inductance in series with C_{gs} of the MOSFET to peak the frequency response. The gate tie-down diode (D_1) for Q_1 prevents gate oxide failure during the backend manufacturing process and also provides limited ESD protection. High-frequency roll-off in the amplifier response is determined primarily by shunt parasitic capacitance to the substrate at the RF input in combination with the C_{gd} of Q_1. For a higher cut-off frequency, the current amplifier stage should drive a very low-impedance load, which in this case is the input of the second stage. Equally important is the lower cut-off (-3dB) frequency, which is determined by the self and mutual inductances of the secondary coil when driven from a 50 Ω source.

30 1:2 Step-Up Transformer

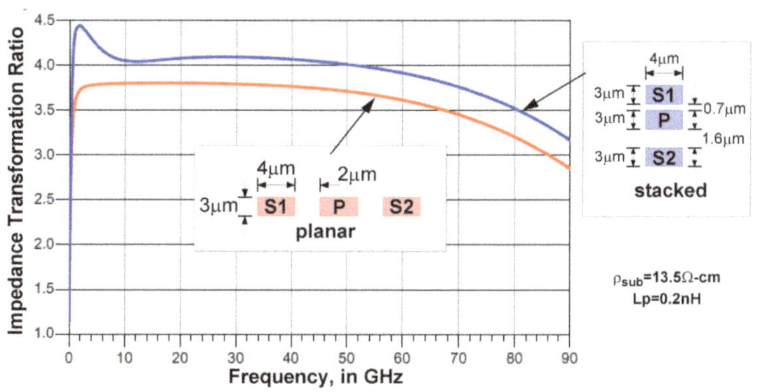

- Impedance transformation ratio is 4x for a 1:2 transformer (ideally). Weaker coupling (lower k_m) for the planar transformer results in an error (n^2=3.75 vs. 4). Tighter coupling and lower substrate parasitics are advantages of the stacked layout.

30 1:2 Step-Up Transformer

The effectiveness of conductor stacking for a 1:2 step-up transformer example is shown on Slide 30. The ideal impedance transformation ratio of a 1:2 transformer is four (i.e., 2^2), however, magnetic leakage affects the magnitude of the reflected impedance as shown on the Slide. The planar 1:2 transformer (w=4μm, t=3μm and s=2μm) supports an impedance transformation of 3.75 from 4GHz to above 50GHz. Stacked windings (layer thicknesses shown at the upper right on Slide 30) improve the magnetic coupling beyond what can be achieved with a planar layout. A near-ideal transformation ratio of 4 is seen from low frequency to almost 60GHz. Tuning at the ports minimizes insertion loss at each frequency across the band, as in the MAG simulations. The primary self-inductance of 0.2nH is suitable for mm-wave applications in bands between 25 and 40GHz, and at 60GHz.

31 Transformer Balun

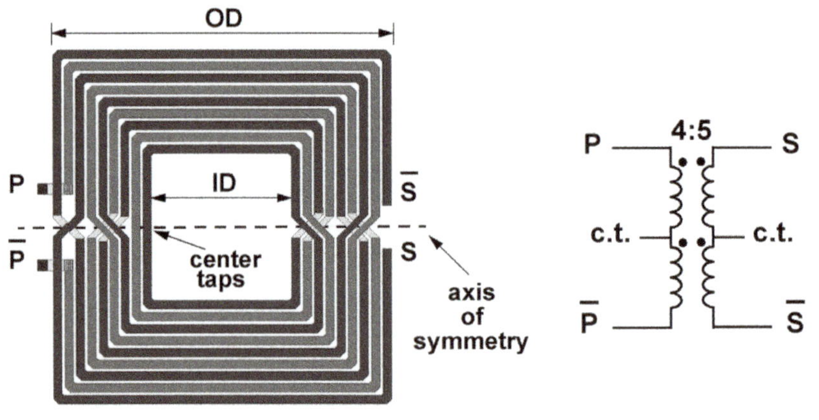

Square symmetric 4:5 (Rabjohn) transformer

The 1:1 and 1:n transformers described previously consist of two independent windings, and are *bifilar* transformers. Multi-filament transformers can also be constructed on-chip. These devices are used to implement power dividers/combiners and baluns. A balun couples a balanced (e.g., differential) circuit to an unbalanced one. There are many structures used to implement baluns at RF and microwave frequencies, although a differential amplifier is the most commonly used method for unbalanced to balanced signal conversion on a chip.

The square symmetric layout shown on Slide 31 was first proposed by Rabjohn [2]. It consists of two groups (i.e., top and bottom) of interwound conductors (dark and light shading) that are divided along a line of symmetry running horizontally. The groups are interconnected in a way which brings all four terminals to the outer edge of the physical layout. This is an advantage when connecting the transformer terminals to other circuitry. Also, the mid-point between the terminals on each winding (i.e., center tap, or c.t.) can be located precisely in the symmetric layout as indicated. The turns ratio for the example shown is 4:5 between primary and secondary.

32 Balun Response

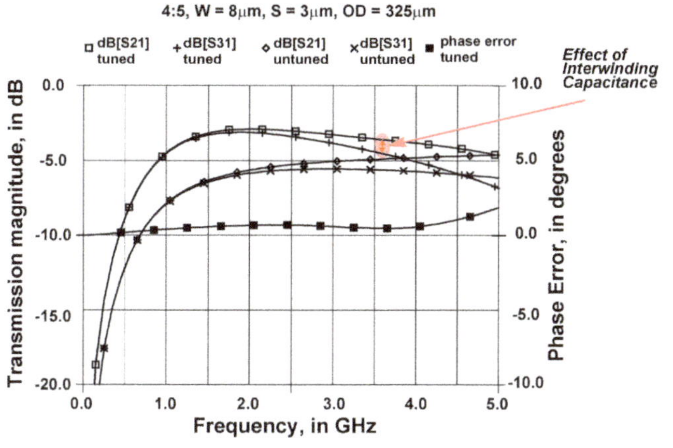

- Output magnitude and phase errors disappear when driven differentially.

The measured and simulated responses for the balun from Slide 31 are compared on Slide 32. The experimental transformer is designed with OD = 325μm, 8-μm metal width, and 3-μm spacing between metals in the planar windings. Slight differences in magnitude response at the inverting and non-inverting secondary ports are clearly seen from the measurements. This is due to the effect of interwinding capacitance, which was described on Slides 8 and 9. The effect is not reduced by adding tuning capacitance in shunt with the transformer ports. Capacitors connected at the input and output ports (425fF across the primary, and 1.7pF across each secondary winding) tune the balun to match the (50Ω) source to the secondary load. The tuned responses are also plotted on the Slide. The measured transmission loss is reduced from over 5dB to very close to the ideal 3dB through tuning. The phase error between secondary ports of the tuned balun is also shown (right axis), where the phase error is the deviation from a 180° phase difference between ports. This error is on the order of 1° in the passband (2 to 3GHz).

33 mm-Wave Balun

MAG and $|S_{21}|$ from simulation for a 1:2, stacked balun aimed at mm-wave frequency applications (self-inductance of 0.2nH at the primary) are shown on this Slide. MAG simulation predicts that an insertion loss of 1dB is realizable across the mm-wave bands proposed for 5G communication. Actual tuning of the balun in the 26-28GHz range to minimize the $|S_{21}|$ insertion loss is also shown. Tuning yields a minimum insertion loss of approximately 1.2dB at 26GHz, and the simulated -3dB bandwidth is 19GHz. An insertion loss of 1dB is acceptable for the power amplifier application, but lower loss would be desired for a balun used in the receive path (e.g., in front of a low-noise amplifier). As noted on the Slide, a fully planar balun layout would have about 0.5dB more attenuation in-band and less bandwidth.

33 mm-Wave Balun

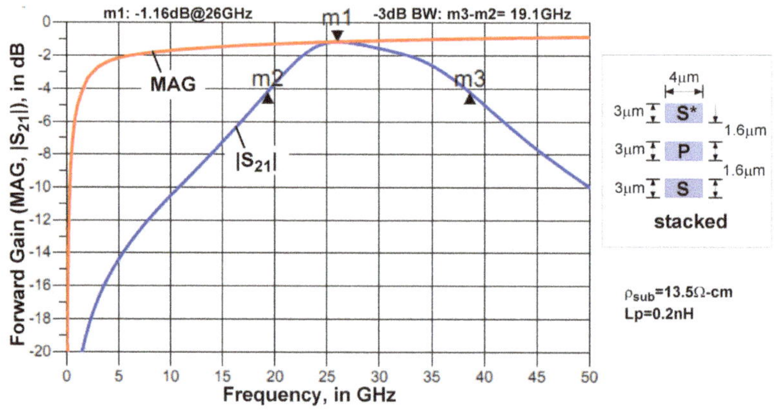

- MAG and $|S_{21}|$ for a 3-conductor balun after conjugate match (i.e., $v_{o\text{-}diff}/v_{in}$). A fully-planar layout has ~0.5dB more attenuation in-band, and 2.4GHz less bandwidth.

34 Optimal Source Impedance

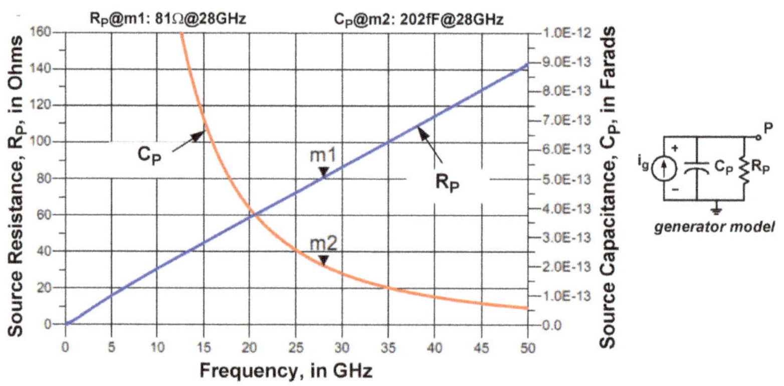

- Optimal source impedance decreases with decreasing frequency and magnetizing inductance. Tuning cap must be large enough to absorb circuit parasitics; can tune on primary or secondary side (or both).

The tuning component values (C_P and R_P) required to realize MAG for the balun at each frequency shown on Slide 33 are plotted on this Slide. Note that at 28GHz (the band selected for $|S_{21}|$ tuning on Slide 33), R_P=81Ω and C_P=202fF. A tuned source with resistance (R_P) closer to 50Ω requires a larger self-inductance in the primary (L=0.2nH is used here). Tuning reduces the upper -3dB bandwidth. A tuning capacitance (C_P) of 202fF is easily realized on-chip and circuit parasitics could be absorbed into that value.

The tuning values shown here are connected at the primary (generator) side only. Tuning could also be applied at the secondary, or at both primary and secondary terminals.

35 Push-Pull Amplifier

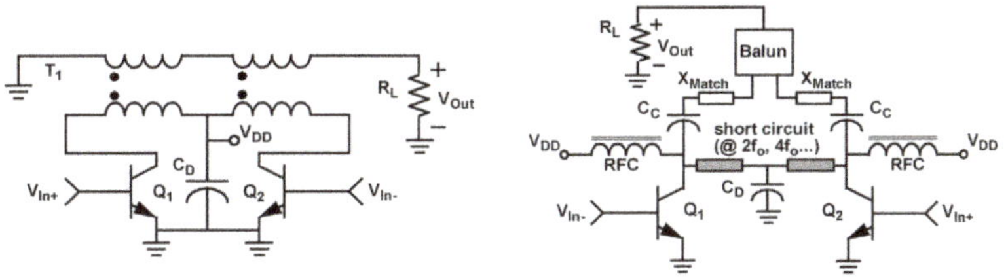

Wideband **2 Single-Ended Stages**

- Push-pull topology shorts even harmonics via symmetry using a transformer balun. Drain sees resistive load at fundamental and all odd-order harmonics.
- Coupling 2 single-ended amplifiers requires harmonic terminations (e.g., short-circuit at $2f_O$, $4f_O$, etc.) at each output and separate bias paths.
- Parasitics such as supply and ground path inductances have less effect on performance for push-pull and balanced amplifiers.

Power combining in push-pull power amplifiers on-chip is advantageous when attempting to realize greater power outputs at mm-wave frequencies on a chip. Slide 35 shows two possible realizations: 1) a wideband output (left side) using series-connected windings at the secondary, and 2) a narrowband, tuned approach to maximizing the power delivered to load R_L (at the right side of the Slide). Implementation of on-chip transformers for use at mm-wave frequencies in a power amplifier application is considered in the remaining Slides.

36 Multi-Path mm-Wave Power Amplifier

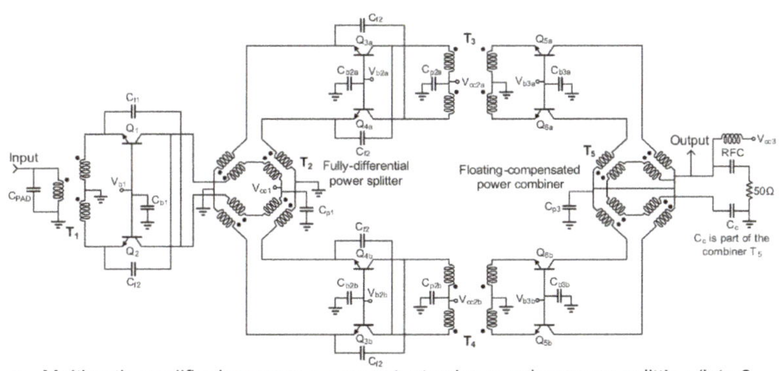

- Multi-path amplifier increases power output using passive power splitting (into 2 paths) and power combining of multiple PA stages.
- Interstage gain realized using current-mode step-up transformers and impedance mismatch between active stages.

The dual-path, multi-stage transformer coupled mm-wave power amplifier shown on Slide 36 is the last application example in this chapter [16]. It consists of identical, 3-stage differential amplifiers running in parallel. An on-chip output power-combining balun sums the (differential) power produced by the two amplifier groups to a single-ended, 50-Ω output. Similarly, an on-chip

36 Multi-Path mm-Wave Power Amplifier

power dividing input balun produces two pairs of differential signals to drive the two amplifier groups from a single-ended, 50-Ω input. Monolithic transformers couple the common-base amplifier stages to each other.

Inter-stage transformers T_2, T_3 and T_4 are designed to present the optimal load to the transistors in each stage for maximum power generation (i.e., Class-A load line matching). Fully-differential power splitter T_2 couples the first stage to each amplifier paths (i.e., a differential 2:4 splitter).

Transformers T_3 and T_4 with a turns ratio of 6:1 (electrical) couple the 2nd and 3rd stages in each path. Power splitting after the Ist stage maintains power gain from the last two stages for efficiency, as transformers T_3 and T_4 in each path are easier to optimize for losses. The compact 4:1 combiner T_5 delivers power to the single-ended load from each power amplifier (PA) output.

37 Self-Shielding Transformers

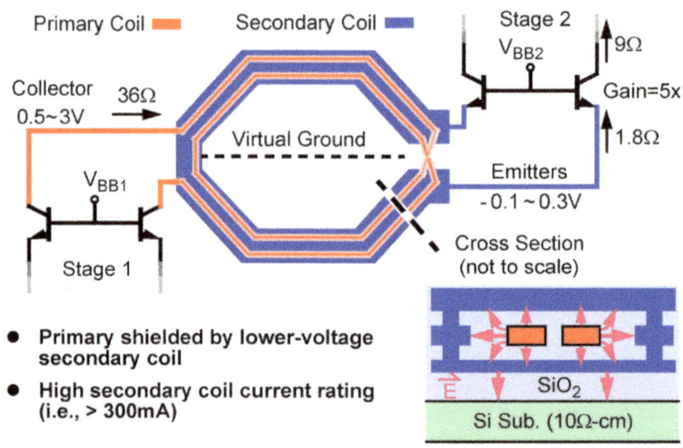

Self-shielding of RF signal energy from the silicon substrate can be realized using multi-filament transformers as shown on Slide 37. Self-shielding minimizes losses to the underlying substrate from a monolithic transformer [9]. For self-shielding, the wider metal in one transformer winding (e.g., carrying a high DC current in the power amplifier) forms a shield around the other winding so that electric field is confined. Self-shielding permits both excellent AC performance and the ability to satisfy DC current rating constraints imposed for on-chip interconnect reliability (e.g., due to electromigration).

For the PA shown on Slide 36, self-shielding limits the losses from each transformer to the substrate. To implement self-shielding, the winding with the lowest voltage swing (i.e., secondary, at emitters of $Q_{5a,\ 6a}$) is placed beneath the winding with the highest swing (i.e., primary, at collectors of $Q_{3a,\ 4a}$) in the physical layout. The secondary is therefore shielded by the primary, thereby minimizing energy loss to the substrate.

In general, self-shielding a transformer consists of the following steps: I) identifying the portion of the circuit connected to low impedance or low voltage node, and 2) using the winding connected to the lowest-voltage to shield the higher-voltage winding, or windings. Self-shielding minimizes losses caused by leakage of the electric field without the use of an explicitly grounded shield layer. As noted on Slide 20, ground shields are difficult to implement properly on a chip as a true ground reference does not really exist on the IC.

38 PA Power Combiner and Balun

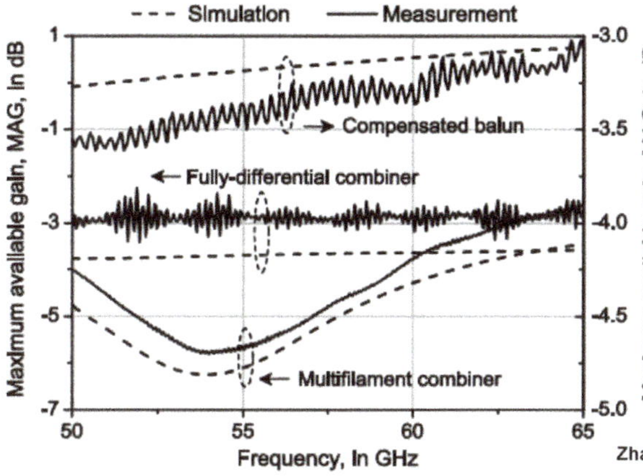

Zhao RFIC 2012 [17]

- MAG in the back-to-back test configuration is twice that of a single combiner plus the losses from coupled transmission lines interconnecting the 4 primary terminals of each transformer.

A laterally-compensated balun and a fully-differential combiner prototype designed for operation in the 60-GHz band were fabricated and characterized using on-wafer measurements. Also, a compensated 60-GHz power combiner prototype with the ac output coupling capacitor was also characterized to benchmark the performance of these component at mm-wave frequencies, and to verify predictions from simulations. The simulated and measured data are shown on Slide 38.

Maximum available gain (MAG) is used to determine the minimum power loss of the transformer combiners. MAG in the back-to-back test configuration is twice that of a single combiner plus the losses from the coupled transmission lines interconnecting the 4 primary terminals of each transformer. Slide 38 shows a comparison of the measured and simulated MAG (back-to-back) for the 60-GHz combiner prototypes used in the mm-wave PA from Slide 36. Good agreement between simulation and measurement is seen in the 50-65GHz frequency range. The single compensated balun and fully-differential combiners with 200-µm long transmission lines at the primary terminals have a measured MAG of 1.62dB and 1.8dB at 60GHz, respectively. Insertion loss for the fully-differential combiner is higher, as the secondary winding is implemented using both top and second (thick) metal layers, leaving the primary in relatively thin (and lossy) third metal (i.e., third from the top in the metal stack).

39 Compensated Power Combiner

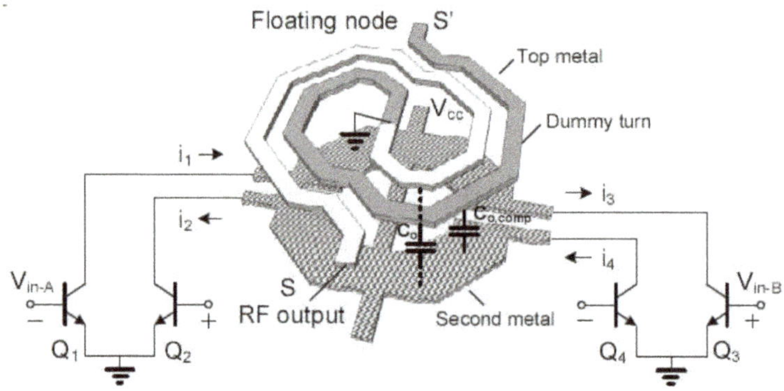

- Interwinding capacitance compensation mitigates amplitude and phase imbalances and maximizes efficiency when combining power from 4 transistors. Insertion loss at 60GHz is <1dB and chip area is <0.015mm^2. Uniformity of impedances reflected from the load to the transistor outputs is better than 3%.

Parasitic capacitance between the primary and secondary windings of a transformer cause impedances reflected from the output back to each input port to differ. This effect is more pronounced when the physical layout is scaled to operate at mm-wave frequencies. An RF balun designed to compensate for amplitude and phase imbalances in the power-combining output is illustrated on Slide 39.

An additional compensating filament called a 'dummy winding' is added to the secondary for compensation. The dummy winding (i.e., shaded in the Slide) replicates the voltage swing seen at the actual RF output (S), and along the length of the winding with decreasing amplitude from S to ground. Electric coupling across the interwinding capacitance (e.g., highlighted by capacitor C_0 on the unshaded winding shown on Slide 39) is equalized by coupling introduced between the dummy and primary windings (e.g., coupling via $C_{0,\ comp}$). Proper compensation ensures that the impedances seen by each transistor at the primary terminals are almost identical. This improves uniformity of the power output of the amplifier across frequency.

40 Dual-Path Power-Combining Amplifier

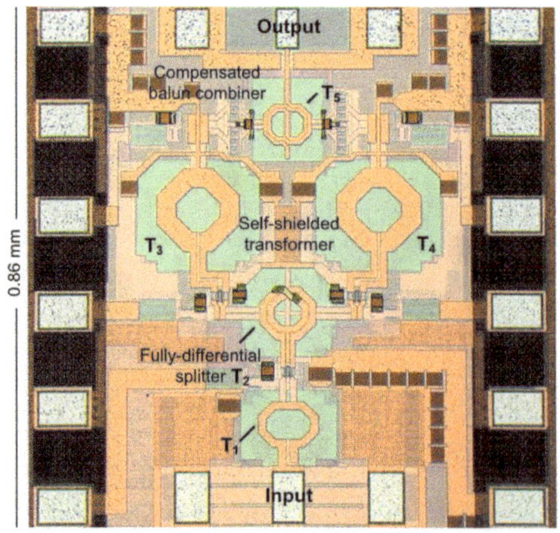

3-stage 60GHz PA with integrated I/O baluns in 130-nm BiCMOS
RF P_{out}=200mW @ 1.8V

Maximum output power > 20dBm and peak-PAE above 20% at 61.5GHz

Small-signal gain > 20dB
-3dB bandwidth > 10GHz

Reverse isolation > 51dB from 50-65GHz

0.25mm^2 active area
353mW quiescent @ 1.8V

40 Dual-Path Power-Combining Amplifier

A frequency-scalable, three-stage, transformer-coupled millimeter-wave power amplifier implemented in 130-nm SiGe-BiCMOS is shown in Slide 40 [17]. Monolithic, self-shielded transformers designed for low insertion loss and compact dimensions on the prototype chip include: a 2:4 input power splitter, a 4:1 output balun combiner, and inter-stage coupling transformers. The balun combiner and fully-differential splitter are compensated for imbalances caused by parasitic interwinding capacitance, and simulations predict better than 3% uniformity between reflected port-to-port impedances at 60GHz. The 0.72-mm^2 60-GHz-band PA realizes a measured small-signal gain higher than 20dB (peak gain) with over 10GHz bandwidth. Reverse isolation is better than 51dB from 50-65GHz, and the PA is unconditionally stable. It consumes 353mW (quiescent) from a 1.8-V supply and the active area is just 0.25mm^2. Maximum output power and peak-PAE are 20.1dBm and 18% at 62GHz, respectively.

41 Summary

- Monolithic transformers are synthesized from transmission lines; they are not "lumped elements" but do behave that way over a limited frequency range. Component parasitics must be captured to optimize RF circuit performance.
- Stacked metal windings offer wider bandwidth and higher k_m than planar windings in some technologies. Metal thickness, intermetal dielectric thickness and metal height above substrate affects the performance of stacked vs. planar transformers.
- Interconnect performance at RF is not well-understood, modeled or captured in conventional CAD tools.
- Passive devices in Si-technology are not readily scalable, and further improvements in process technology and IC-CAD models of passive devices are needed for analog RF/MMIC applications.

Concluding remarks are noted on this Slide.

42 References

1. J.R. Long, "Monolithic Transformers for Silicon RF IC Design" *IEEE Journal of Solid-State Circuits*, vol. 35, no. 9, Sept. 2000, pp. 1368 1382.
2. G.G. Rabjohn, *Monolithic Microwave Transformers*. M.Eng. thesis, Carleton University, 1991.
3. A. M. Niknejad and R. G. Meyer, "Analysis, Design, and Optimization of Spiral Inductors and Transformers for Si RF IC's", *IEEE Journal of Solid-State Circuits*, Vol. 33, No. 10, October 1998, pp. 1470 1481.
4. K.K. O, "Estimation Methods for Quality Factors of Inductors Fabricated in Silicon Integrated Circuit Process Technologies", *IEEE Journal of Solid-State Circuits*, vol. 33, no. 8, August 1998, pp. 1249 1252.
5. H. Hasegawa, M. Furukawa and H. Yanai, "Properties of Microstriplines on Si SiO2 System", *IEEE Transactions on Microwave Theory and Techniques*, Vol. 19, No. 11, November 1971, pp. 869 881.
6. M. Danesh and J.R. Long, "Differentially Driven Symmetric Microstrip Inductors", *IEEE Transactions on Microwave Theory and Techniques*, vol. 50, no. 1, January 2002, pp. 332 341.
7. C. P. Yue and S. S. Wong, "On Chip Spiral Inductors with Patterned Ground Shields for Si Based RF IC's", *IEEE Journal of Solid-State Circuits*, Vol. 33, No. 5, May 1998, pp. 743 752.
8. D. Cheung, J.R. Long, "Differentially shielded monolithic inductors," *Proc. of the IEEE Custom Integrated Circuits Conference (CICC)*, 2003, p. 95 98.
9. T.S.D. Cheung and J.R. Long; "Shielded passive devices for silicon based monolithic microwave and millimeter wave integrated circuits," *IEEE Journal of Solid-State Circuits*, vol. 41, no. 5, May 2006, pp. 1183 1200.
10. J.R. Long, Y. Zhao, W. Wu, M. Spirito, L. Vera and E. Gordon, "Passive Circuit Technologies for mm Wave Wireless Systems on Silicon," *IEEE Transactions on Circuits and Systems - I*, vol. 59, no. 8, Aug. 2012, pp. 1680-1693.
11. H.J. Finlay, U.K. patent application no. 8800115, 1985.
12. P. Andreani, B. Razavi and J.R. Long, "Voltage Controlled Oscillators", *IEEE RFIC Virtual Journal Issue* 1, IEEE, December 2012. URL: http://ieeexplore.ieee.org/virtual-journals/rfic/issue/1/
13. P. Andreani and A. Fard, "More on the 1/f2 phase noise performance of CMOS differential pair LC tank oscillators", *IEEE Journal of Solid-State Circuits*, vol. 41, no. 12, Dec. 2006, pp. 2703 2712.
14. P.-I.Mak, D.Leenaerts, H.C. Luong and J.R. Long, "Low-Noise Amplifiers," in *IEEE RFIC Virtual Journal*, April 2014, URL: http://ieeexplore.ieee.org/virtual-journals/rfic/issue/4/
15. M. T. Reiha andJ.R. Long, "A 1.2 V Reactive Feedback 3.1-10.6 GHz Low Noise Amplifier in 0.13um CMOS," *IEEE Journal of Solid State Circuits*, vol. 42, no. 5, May 2007 Page(s):1023 1033.
16. T.S.D. Cheung and J.R. Long, A 21 26 GHz SiGe bipolar power amplifier MMIC, *IEEE Journal of Solid-State Circuits*, vol. 40, no. 12, pp. 2583 2597, Dec. 2005.
17. Y. Zhao, J.R. Long, "A Wideband, Dual Path, Millimeter wave Power Amplifier with 20 dBm Output Power and PAE above 15% in 130 nm SiGe BiCMOS," *IEEE Journal of Solid-State Circuits*, vol. 47, no. 9, Sept. 2012.

A brief list of references is provided for further reading on the many topics covered in this chapter.

PART II

MILLIMETER-WAVE AND RF RECEIVER CIRCUITS

CHAPTER 04

Millimeter-wave LO Distribution Insights

Amr Fahim

Inphi Corporation, USA

SSCS

In this chapter, highlights into the design and optimization of LO distribution networks for millimeter wave (mmWave) frequencies is illustrated. First, applications and requirements of mmWave LO distribution networks is given. mmWave LO distribution networks is present in a wide variety of applications ranging from phased-array systems to optical communication networks. The requirements on LO distribution networks is multi-faceted as it ranges from clock signal shape, jitter, multi-phase generation, and transmission line requirements. A discussion on each of these requirements is detailed in this chapter, highlighting essential design insights.

Frequency planning is an essential aspect of mmWaveLO distribution network design. Four main frequency plans are illustrated, each with its own trade-offs. This involves the distribution of the main LO generation unit, the multi-phase clock generation and distribution. Hybrid topologies, where a mix of global and local LO generation and distribution are also considered.

The concept of using frequency doublers and triplers are illustrated for mmWave LO distribution. A detailed discussion on the design and implementation of such structures is given. Several examples are highlighted showing how different design techniques can be combined together successfully. Injection locked oscillators as frequency multipliers are also discussed.

Finally, a discussion on the actual buffer design and physical LO distribution is given. The trade-off between different buffer topologies is given. Bandwidth extension techniques such as inductive peaking and neutralization are illustrated. Wilkinson divider to achieve good output port-to-port isolation is described. Finally, the important phenomenon of jitter amplification is described.

A. ROLE OF mmWAVE DISTRIBUTION NETWORK

1 Phased-Array System

- Phase shifting modules to steer signal wavefronts
  - Antenna spacing D=λ/2 for maximum current steering
  - Trend is towards increasing number of array elements → sharper beams, weaker grating lobes, larger antenna gains
- How to route LO signal to all array elements?
  1. Global: LO generation, multi-phase generation
  2. Local: LO generation, multi-phase generation
  3. Hybrid: Global LO generation, local multi-phase generation

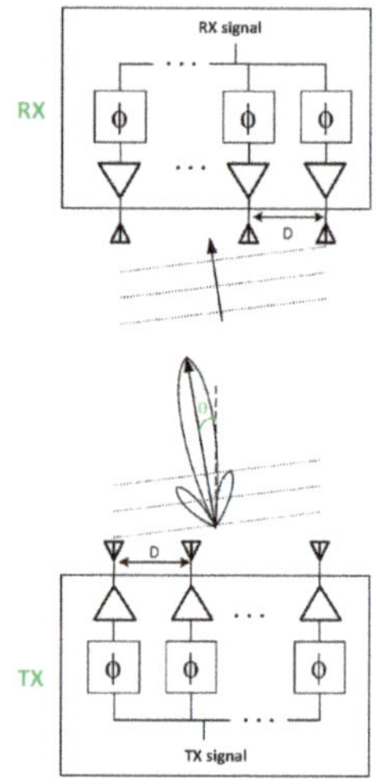

Before we start let's talk about what LO distribution networks are and two main applications where they are used the most. There are two primary applications of interest for mmWave LO distribution networks. One application is a phased-array system where a array of transmit (TX) elements transmit all at the same frequency but at different phases to send a wireless beam with a defined directionality. This directionality is controlled by precisely controlling the phase relationship between all transmit elements. Similarly, a receiver is composed of an array of receive (RX) elements to receive from a fixed direction. The directionality is controlled by precisely controlling the phase relationship between all receive elements. Each RX and TX element requires an LO signal. The challenge then becomes how do you send the LO signal to each element with precisely the same phase and while meeting jitter requirements? Do you generate a single LO signal and distribute it globally, or do you have a more distributed type of network where each element generates its own LO signal, or a hybrid topology where you have a global generation of LO signal and local multi-phase generation?

2 DSP-based systems

Another application is centered more around base station or optical broadband communication applications where the entire received spectrum is digitized. These types of systems employ large arrays of ADCs and DACs. There are three broad categories of techniques for LO generation and distribution: 1) Global generation of all phases of the LO signal, 2) Global generation of one phase, then local phase generation, 3) local LO generation and phase generation. Relative LO skew between the different array elements is important. Duty cycle is also important to control where non-overlapping clocks may be required, or in systems that use both low and high states of the clock.

2 DSP-based systems

- Front-end Topology of DSP-based systems
  - Minimum analog processing
  - Wideband systems:
    - base-station wireless transceivers
    - optical coherent communication systems
- How to route LO to all sub-systems?
  1. Global: LO generation, multi-phase generation
  2. Local: LO generation, multi-phase generation
  3. Hybrid: Global LO generation, local multi-phase generation

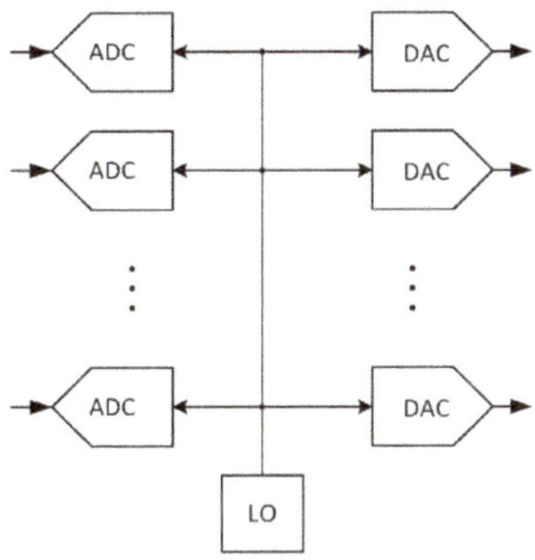

3 Summary of mmWave LO distribution requirements

1. Control shape of the clock (duty cycle)
2. Tight timing between different clock leaf nodes (skew)
3. Spectrally pure clock (low jitter)
4. Generate multiple phases of clock (LO)
5. Support large variation in routing length (tens of microns to several millimeters)

Now, let's talk about the key points to take into account when choosing an architecture of mmWave LO distribution network: 1) Controlling the shape, or duty cycle, of the clock is important; 2) Timing skew between the different leaf nodes (either individual ADCs, DACs, or transceiver elements) must be well controlled, 3) low-jitter or spectrally pure LO source and distribution network, 4) multi-phase clock generation, 5) support large differences in LO route length between the different leaf nodes.

B. ISSUES IN mmWAVE LO DISTRIBUTION

4 Jitter definitions (I)

- Random Jitter (RJ):
  - Jitter assumed to follow a Gaussian distribution
  - RJ is usually expressed as an RMS value
- Deterministic Jitter (DJ):
  - Deterministic Jitter is bounded and expressed as peak-to-peak jitter
- Total Jitter:
  - $T = DJ + 2 \cdot n \cdot RJ_{rms}$
  - Parameter "n" depends on the required BER
  - $\int_Q^\infty Gauss(x)dx = \frac{1}{\sqrt{2\pi}} \int_Q^\infty e^{-\frac{x^2}{2}} dx = \frac{1}{2} erfc\left(\frac{Q}{\sqrt{2}}\right)$

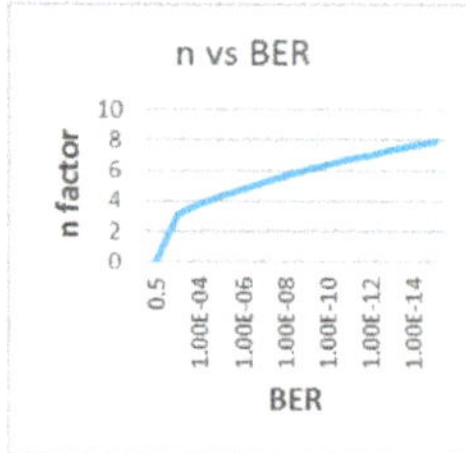

Jitter can be classified into random jitter (RJ) and deterministic jitter (DJ). Random jitter is assumed to have a Gaussian distribution and is expressed as an rms value. Deterministic jitter is jitter that occurs in a deterministic fashion and is hence bounded and is expressed and peak-to-peak jitter. The total jitter, expressed as a peak-to-peak quantity, is the sum of the RJ and DJ. In summing the two the RJ is converted to a peak-to-peak value by multiply by 2n, where n is determined by the BER required by the system.

5 Jitter definitions (II)

- Time Interval Error (TIE):
  - Accumulated timing error between an ideal edge and the actual measured edge

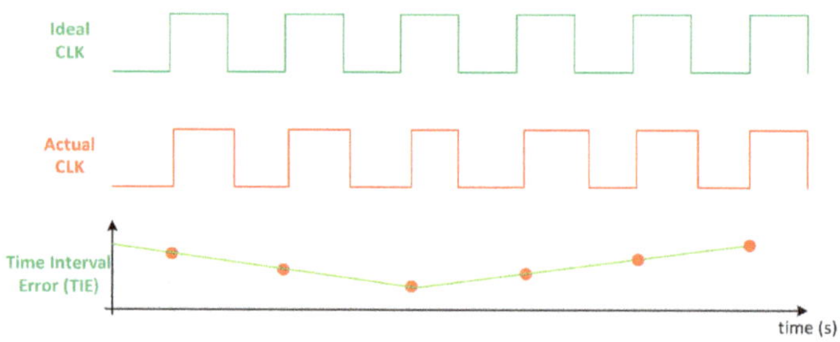

The Time interval error (TIE) metric is more popular in wireline applications. It is defined as the timing error between nth ideal clock edge and the nth real clock edge, for all integer n [1,N], where N is the number of clock edges measured.

6 Jitter definitions (III)

- Phase noise (phase jitter):
  - Frequency domain equivalent of TIE
  - Described in units of dBc/Hz @ f_{offset}
  - PN filtered by timing recover loop (HPF) – sets f_1
  - BW & Reciprocal mixing set f_2

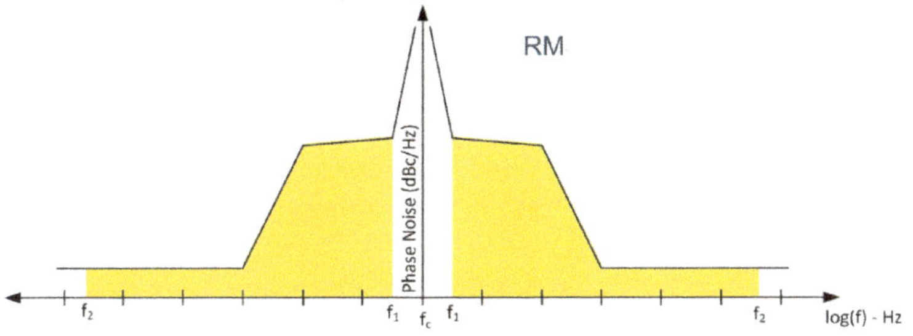

Phase noise (also known as phase jitter) is the frequency domain representation of TIE. The units of jitter is dBc/Hz @ f_{offset}. The low-frequency, or "close-in", phase noise is dominated by the LO generation circuitry, typically a PLL. The high-frequency noise floor is dominated by the LO distribution network, typically buffers and delay elements. The total jitter is obtained by integrating the phase noise profile. The integration interval of $[f_1,f_2]$ depends on the system. The lower bound, f_1, is set by the bandwidth of the timing recover loop in the DSP, which acts as a high-pass filter on the phase noise profile. The upper bound, f_2, depends on the bandwidth of the system.

7 Contribution of jitter

- LO distribution network may dominate phase noise for high-bandwidth systems
  - LO distribution network dominates high offset frequency phase noise.
  - ex. If PLL BW=1MHz, in-band phase noise 40dB higher than out-of-band phase → for BW>10GHz, LO distribution network dominant.

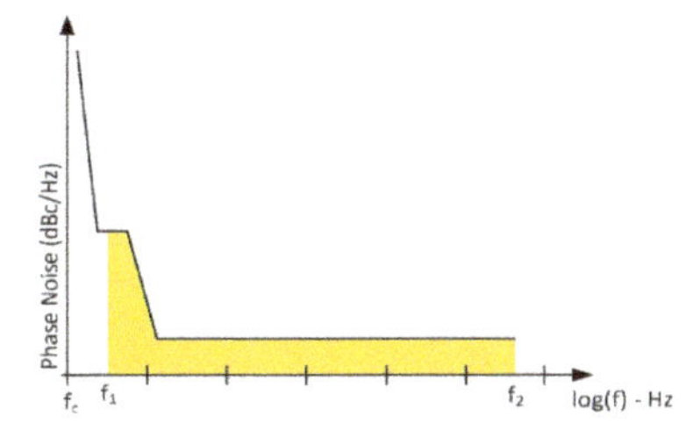

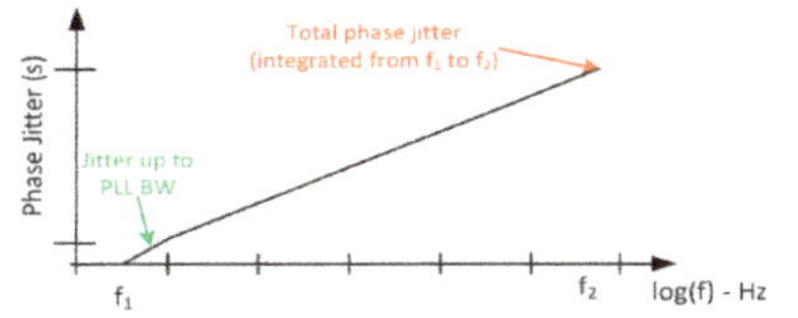

One trend that we see in communication systems, is the increase in bandwidth. This results in increasing f_2, the upper integration bound on phase noise. Another trend that we see in high-performance circuits is that the PLL jitter is gradually being reduced by using higher input reference frequencies, sub-sampling PLL architecture, and other low-jitter PLL architectures. The combination of these two trends results in the jitter of LO distribution network dominating over LO generators in a given system.

8 Why duty cycle? –RF Passive Mixers

- Gain ∝ LO duty cycle

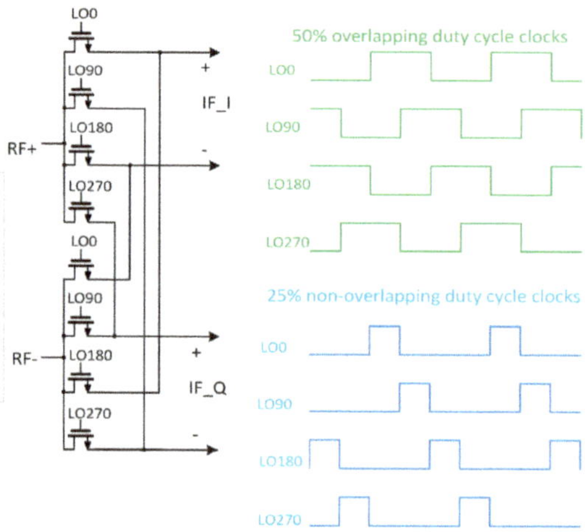

There are many reasons why duty cycle control is important. One application is in passive mixers, which are commonly used in wireless transceivers. If a quadrature mixer is used with 25% duty cycle clocks, this results in high conversion gain.

9 Why duty cycle? –Subharmonic RF Mixers

- Intentionally distort clock to 25% to maximize 2nd order harmonic term
- Fourier series of 25% duty cycle LO:
  - $LO = \frac{\sqrt{2}}{\pi}\left[\cos(\omega_c t) + \frac{1}{\sqrt{2}}\cos(2\omega_c t) + \frac{1}{3}\cos(3\omega_c t) - \frac{1}{5}\cos(5\omega_c t)\right]$
  - Can convert to a double-balanced subharmonic mixer by adding 2 more phases
  - Can expand into I/Q subharmonic mixer by adding more phase slots

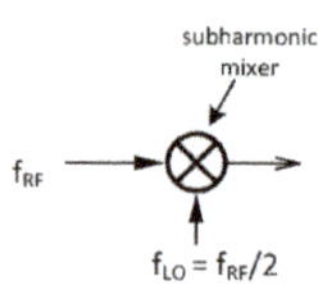

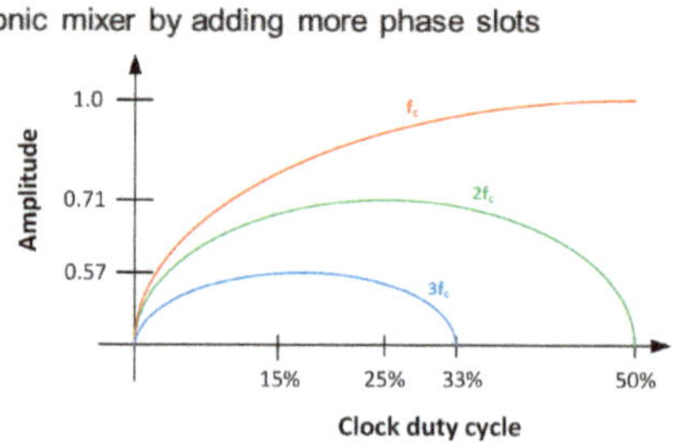

Another example of the use of a 25% duty cycle clock is in a sub-harmonic RF mixer. In a sub-harmonic mixer, the HD2 is intentionally maximized to be used as the main LO mixing tone on the input signal.

10

Why duty cycle?

- Other concerns:
  - Duty cycle control facilitates frequency doubler design
  - Very high-speed digital logic using both phases of the clock (balance of setup & hold time)

Other reasons for controlling the duty cycle include:

1. Facilitation of frequency doubler design,
2. Very high-speed logic where both edges of the clock are used, such as in a high-speed 2:1 Serdes multiplexer.

11

Why is skew important?

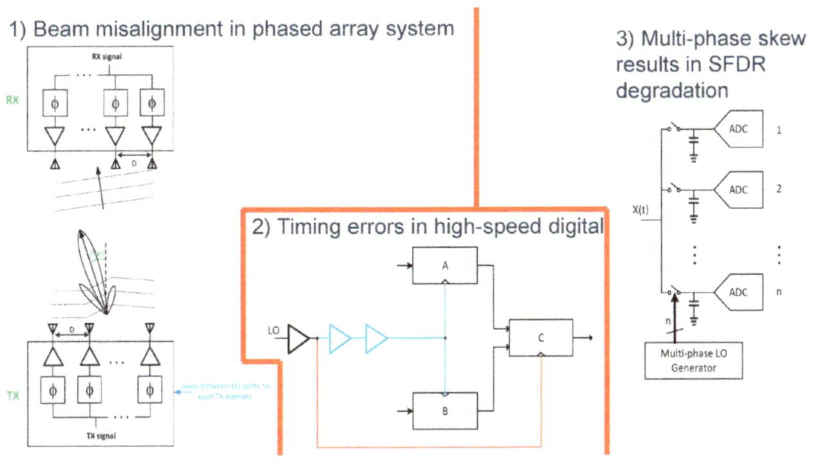

Skew control is also very important for mmWave LO distribution networks. In phased-array systems, undesired skew in the LO distribution network may result in incorrect beam direction and broadening of signal beamwidth. In high-speed digital logic, clock skew can result in reduced hold and setup time margins. In high-speed time-interleaved ADCs, skew between the different ADC elements can result in degraded spurious free dynamic range (SFDR) performance.

12 Noise coupling

- Factor affecting signal integrity:
  1. Power supply ripple induced jitter
  2. Signal noise coupling (magnetic, capacitive)
  3. Amplitude variation due to reflections

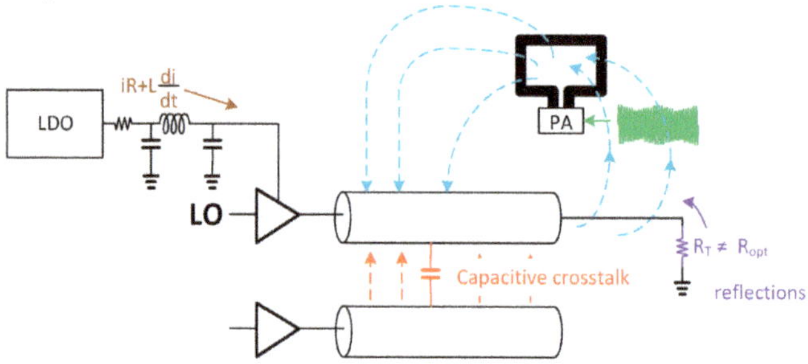

External noise can couple into the LO distribution network and degrade its performance. One such source of the degradation is power supply ripples caused by data toggling. This data toggling can cause IR or L di/dt voltage drops, which are data dependent and hence add jitter to the LO distribution network. Another source of external noise is capacitive coupling of nearby signals, which can be another clock line or data line. The coupling can also be magnetic, especially from a nearby power amplifier (PA), where an inductor with high current can radiate strong magnetic fields, injecting the clock line with data dependent jitter. Lastly, the LO line can be corrupted by the termination network, if the termination impedance is data dependent.

13 LO distribution design space chart

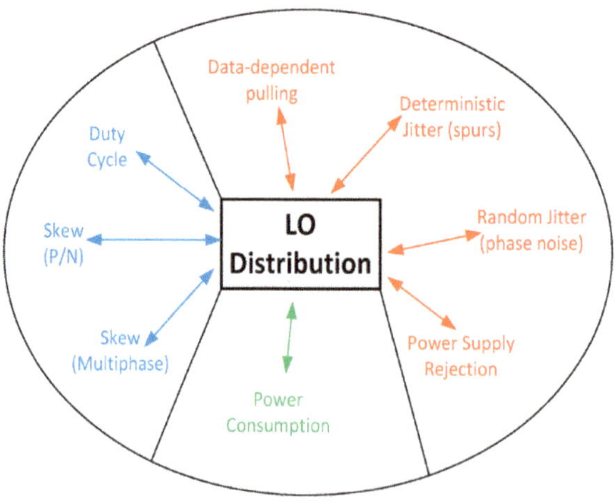

Design space of mmWave LO distribution network encompasses several design parameters. Parameters shown in red are related to jitter performance of the clock signal. Parameters shown in blue are related to quality or shape of the signal, such as duty cycle and clock skew. Finally, the parameter shown in green is power consumption, which can be a dominant source of total power consumption in a mmWave communication system.

C. mmWAVE QUADRATURE LO GENERATION

14 Why Quadrature Generation?

- Communication system: I/Q
  - Two orthogonal spacing enables doubling of number of bits per symbol
  - e.g. 16QAM vs 4PAM: 4-bits/symbol versus 2-bits/symbol

16 QAM constellation

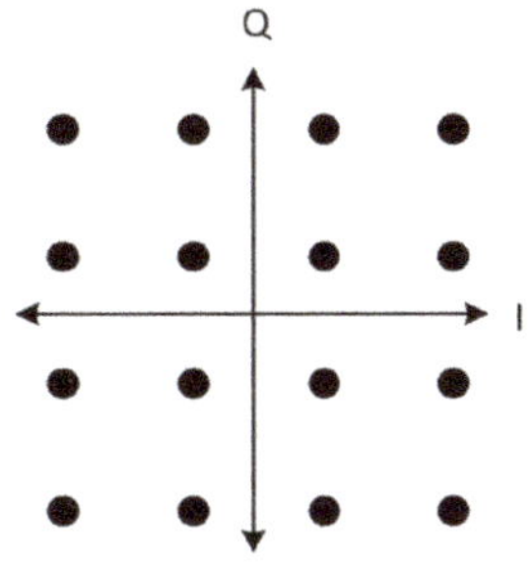

Quadrature (I/Q) LO generation is a special case of multi-phase clock generation. It is especially important in communication systems since it enables quadrature amplitude modulation, which effectively doubles the data rate (bits/symbol) for the same bandwidth when compared to simple amplitude modulation systems.

15 Divide-by-2 method

- Output of both latches of D flip-flop
  - Sensitive to skew between the differential input clock phases
  - Robust, but power hungry to generate I/Q phases

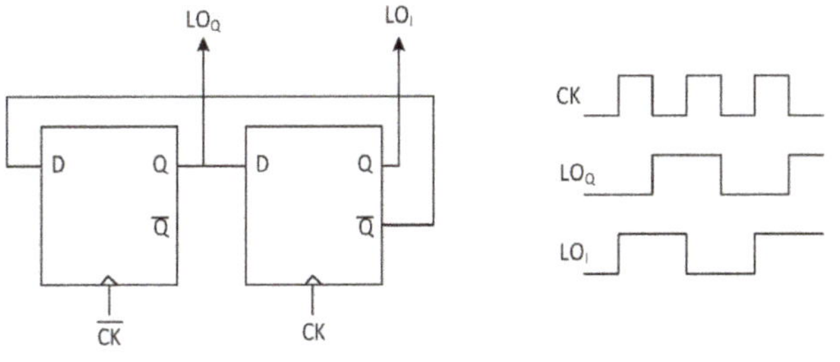

One popular method of I/Q LO generation is by using a D flip-flop connected as a divide-by-2 circuit. The output of both latches provide I/Q clocks. Although a robust method of producing I/Q LO signals, it is fairly power hungry and has limited frequency of operation.

16 I/Q LC Oscillator (I)

- I/Q VCO
  - Essentially two LC tanks connected in a 2-stage ring oscillator fashion
  - Each connection must “inject” sufficient current to modify the phase of the next stage

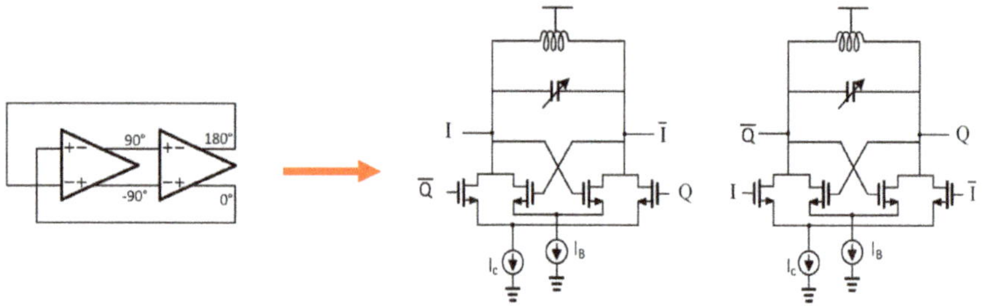

Another method of generating I/Q LO signals is to use a quadrature VCO. A quadrature VCO is essentially a pair of cross-coupled injection locked oscillators.

17 I/Q LC Oscillator (II)

- Critical parameters in quadrature VCOs
  - Condition for each oscillator stage to lock to injected current from previous stage:
    - $\frac{I_{inj}}{I} \geq \frac{4}{\pi}\frac{|\Delta\omega|}{\sqrt{\Delta\omega^2+\left(\frac{\omega_c}{2Q}\right)^2}}$
      - where $\Delta\omega$ is the difference in radial free running frequencies of the two LC VCO cores (due to systematic & dynamic mismatches)
    - Lock Range: $\omega_L = \frac{\omega_c}{Q}\frac{1}{\sqrt{\left(\frac{4I}{\pi I_{inj}}\right)^2-1}}$
    - Excess Phase Noise: $\frac{PN_{QVCO}}{PN_{VCO}} = Q\frac{1}{\sqrt{1+\left(\frac{m\cos\emptyset}{1+m\sin\emptyset}\right)^2}}, m = \frac{I_{inj}}{I}, \emptyset = injected\ phase$
      - Optimal PN @ injected current with phase shift of $\frac{\pi}{2}$

Conditions required for injection oscillator is given by the first equation. As the equation shows, there is a direct trade-off between Q and injection current. The injected current results in excess phase noise, limiting the minimum phase noise of a quadrature VCO. More about operation of this circuit later on when discussing injection locked oscillators.

18 RC Polyphase Filter (I)

- Polyphase filter – RC
  - Quadrature phases occurs when $\omega_c = \frac{1}{RC}$
  - Wideband operation can be obtained by cascading several stages of the RC polyphase stages
  - *Main drawback*: insertion loss, noise due to resistors
    - Can improve noise by lowering resistor values (increase cap) → power and operating frequency limitations
  - Limited use at mmWave frequencies

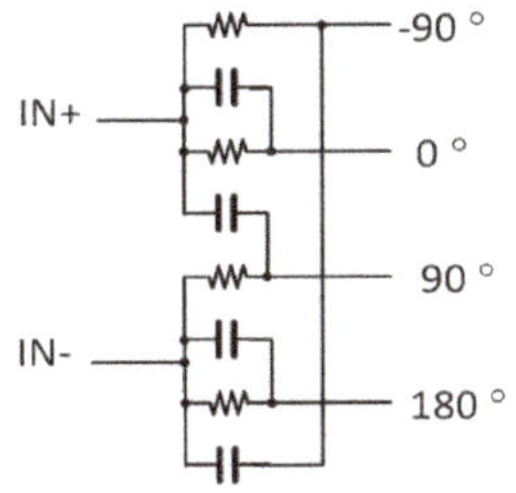

A different method of quadrature LO generation is through the use of RC polyphase filters. Fairly wideband operation is achievable through RC polyphase filters. The main drawback of this approach is insertion loss, or increase in noise. For this reason, this technique is of limited use at mmWave frequencies.

19 RC Polyphase Filter (II)

- LC based polyphase filter
  - Quadrature phases occur when $\omega_c = \frac{1}{\sqrt{LC}}$
  - Very narrow band of operation
    - Can improve BW by cascading several stages, lowering Q

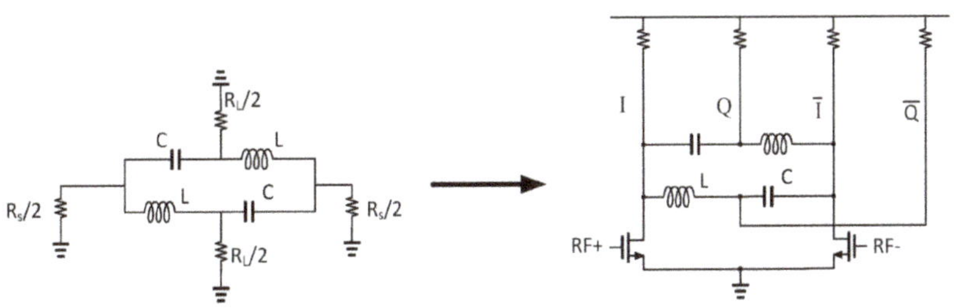

Another approach to polyphase filters is through the use of LC polyphase filter. This results in narrowband operation, but with lower insertion loss than an RC polyphase filter.

20 90° Hybrid Polyphaser Filter

- Utilize routing channel to produce quadrature LO
  - Advantages: low noise (low-insertion loss), dual-use of routing channel
  - Disadvantage: require VERY long routes for moderate mmWave frequencies

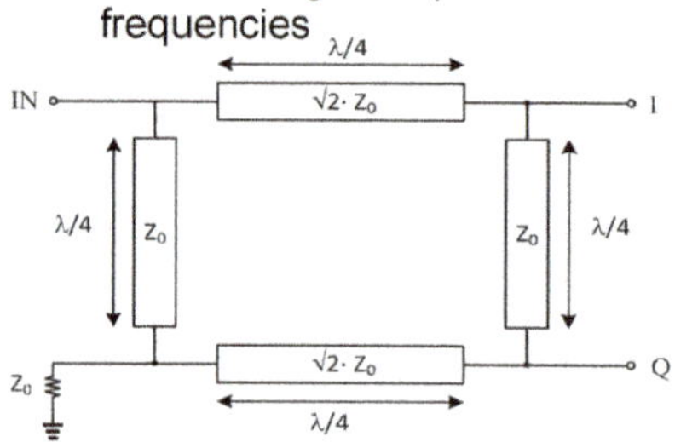

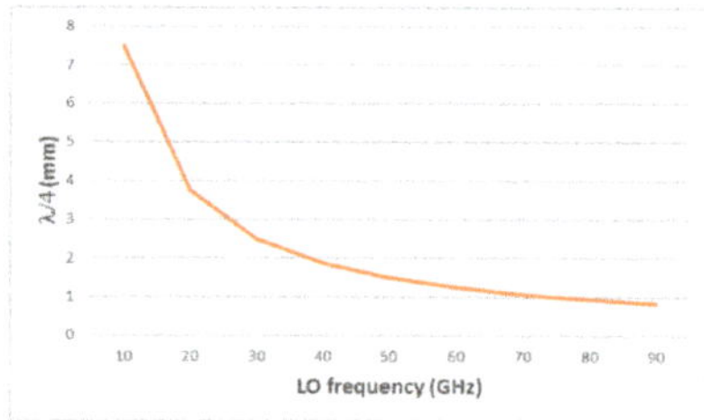

90 degree hybrids are a popular method for I/Q signal generation in microwave circuits and are widely used in board or package level implementations. 90 degree hybrids are difficult to integrate on-chip due to size (and loss) of the required on-silicon transmission lines.

21 Transformer-based Polyphaser Filter

- Transformer-based polyphase filter
  - Basic idea:
    - Monolithic form of 90° hybrid
  - Definitions:
    - If Input port=PORT1, I=PORT2, Q=PORT3, then can define I and Q transfer functions as S_{21} and S_{31}, respectively.
    - $K_c = C_{in}/(C_p+C_{in})$, $K_L=L_M/L$
  - Necessary design equations:
    - $Z_0 = \sqrt{\frac{L}{C}}$
    - $\omega_{\lambda/4} = \frac{1}{\sqrt{LC(1-k^2)}}$
      - This shows that for 90° @ 3dB BW, require that k=0.707.

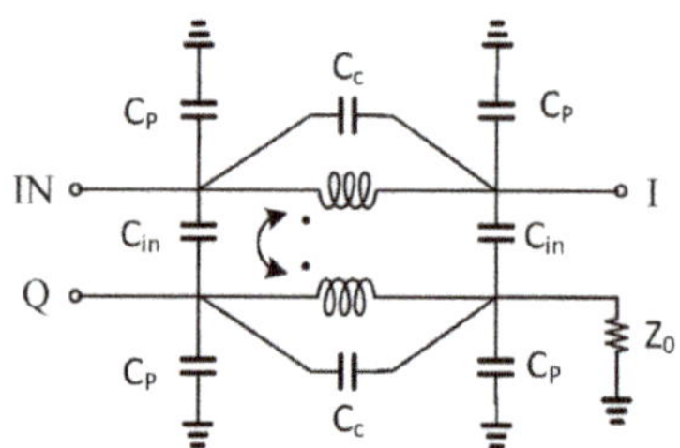

Lumped element equivalent of a 90 degrees hybrid results in a transformer-based polyphase filter. Main disadvantage is that coupling coefficient, k, is required to be exactly 1[2] to avoid gain imbalances.

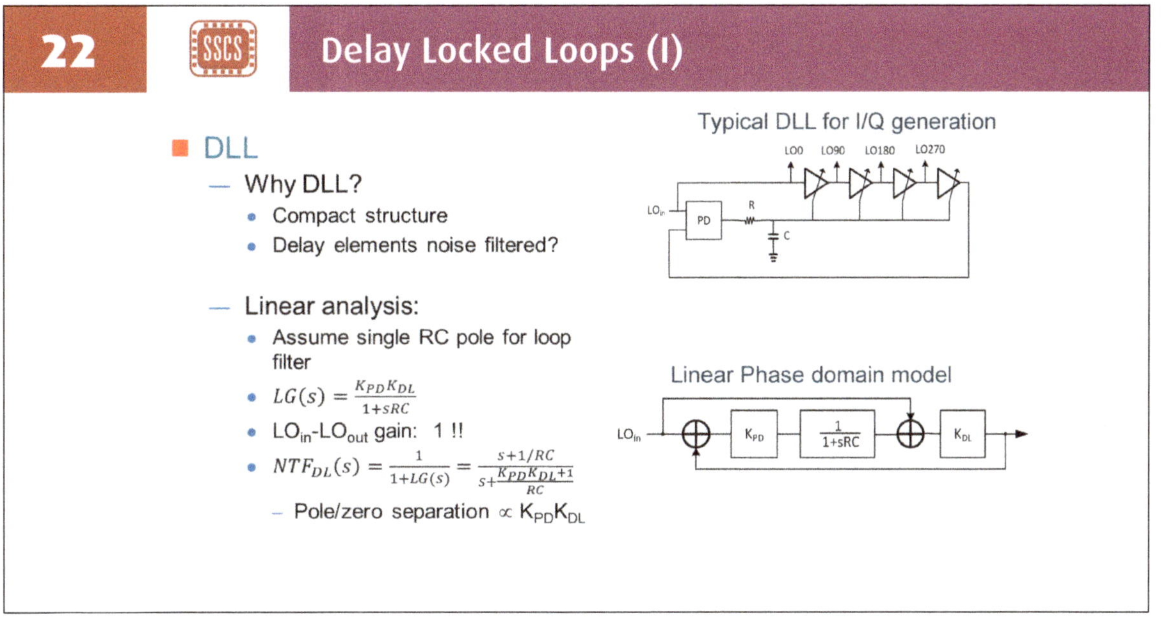

Delay locked loops (DLLs) can be used to generate quadrature clocks. The primary advantage is compact area and some noise shaping of the delay elements (low-frequency noise suppressed to within the loop bandwidth).

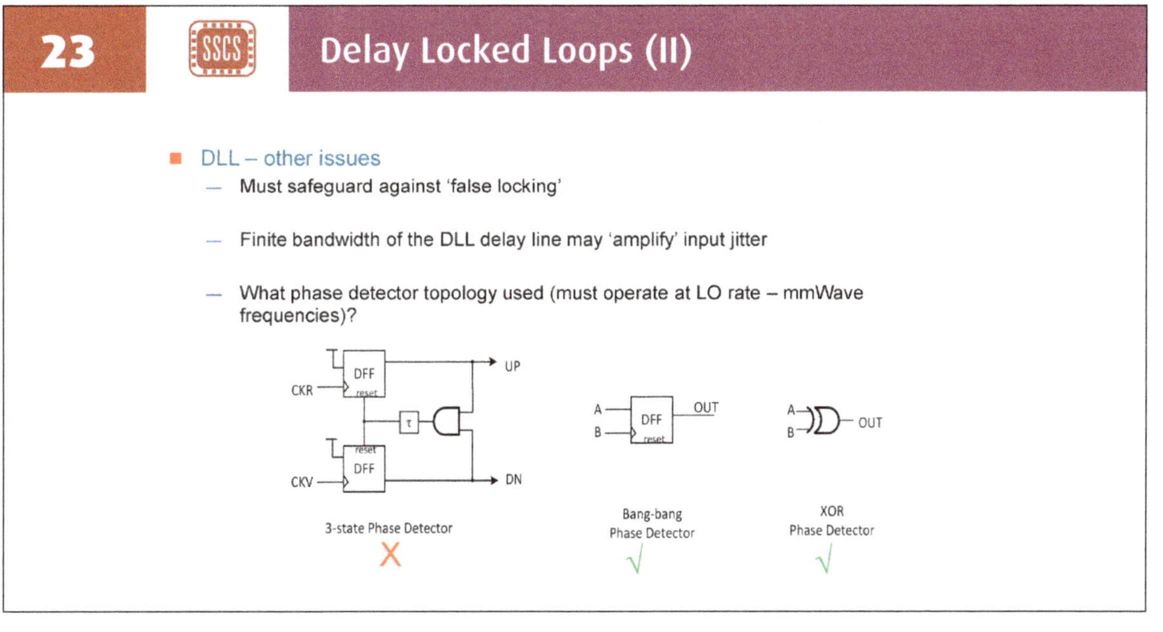

Implementation issues of DLLs include **1)** safeguard against false locking, **2)** avoid excessive number of delay elements to avoid jitter amplification, **3)** requirement of a high-speed low-latency phase detector.

24 Summary

- Summary of Quadrature LO generation approaches:

Quadrature Generation Topology	Accuracy	BW	Power	Area
Divide-by-2	↑	↑↑	↓	↑↑
QVCO	↑↑	↑	↑	↑
RC polyphase	↑	↓	↓	↓
LC polyphase	↓↓	↓	↓	↓
90° hybrid	↓	↓	↓	↓↓↓
Transformer polyphase	↓	↓	↓	↓
DLL	↑	↑	↓	↑↑

Here is a summary of quadrature LO techniques. The nomenclature adopted here is that a down arrow means worse metric. For example, a down arrow for the 'area' metric means 'worse' area, or larger area, and not smaller area.

D. mmWAVE MULTI-PHASE LO GENERATION

25 Skew compensation

- Accuracy limited by mismatches in the wires, devices (buffers) and coupling noise.

Typical LO Chain

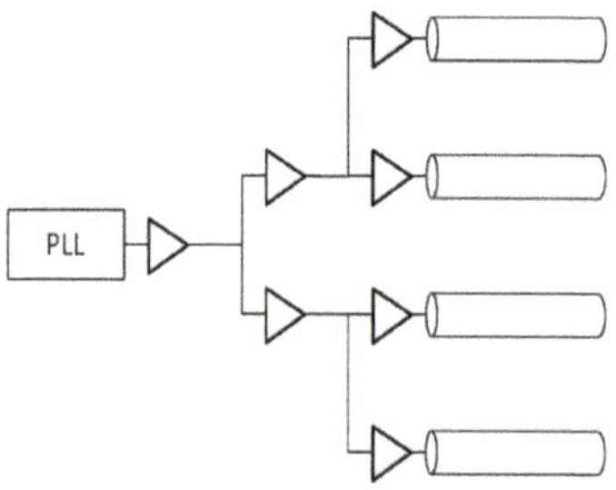

LO route cross section

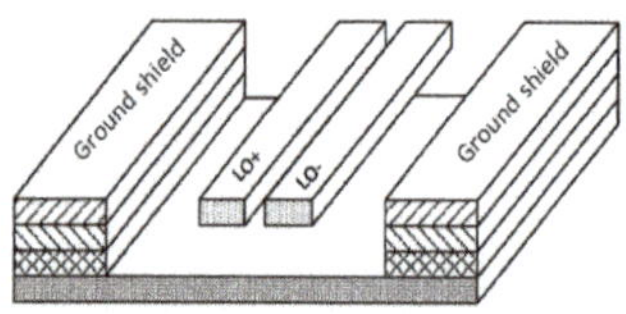

The physical implementation of the actual routing channel is important. Shown here is a commonly used method of routing an LO signal. In order to obtain good matching between different LO paths, the LO signals are shielded in a G-S-S-G configuration, with side and bottom shields.

26 Skew compensation in LO generation

- Conventional technique for skew compensation
  - Require buffer + routing to phase detector → uncorrectable skew error may result
  - Sub-picosecond skew becomes difficult

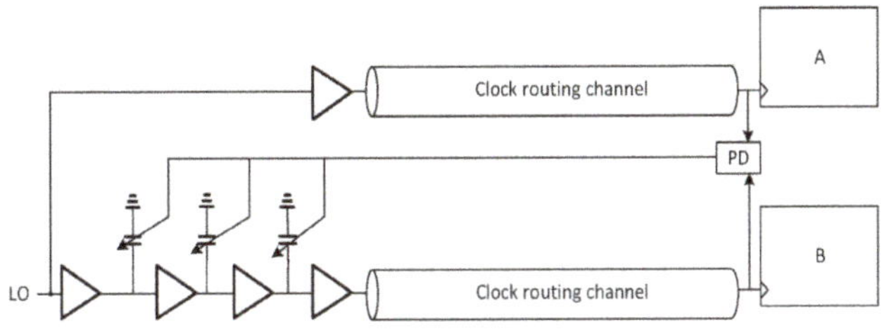

Conventional analog skew compensation schemes are not capable of achieving sub-picosecond resolution where distance between clock leaf nodes can be large. Thus, any mismatch in the long routing path to the actual phase detector (PD) limits the accuracy of the skew compensation scheme.

27 DSP-based skew compensation (I)

- DSP techniques
  - Look at final metric of signal (or statistical measure of signal) to determine skew
  - Require high accuracy calibration ADC for TX mode
  - Suitable for sub-picosecond skew (since directly affect signal metric).

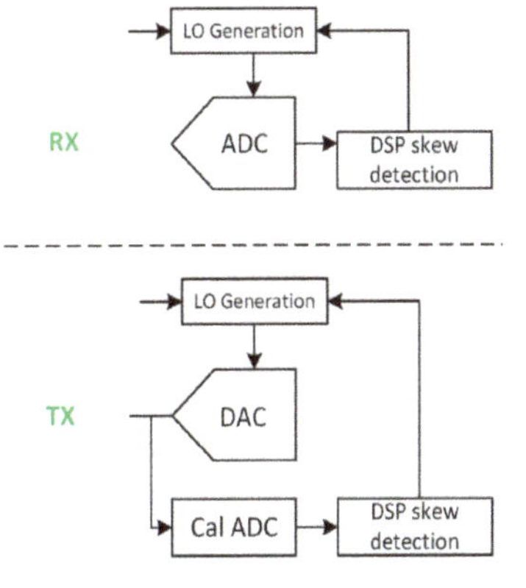

An alternative method of skew compensation is to perform an SNR calculation in the digital domain (through a DSP engine) and skew the LO signal accordingly. This results in a more power hungry, but more accurate method of calibrating the skew.

28 DSP-based skew compensation (II)

- Time-interleaved ADC example:
  - Expect value depends on skew
  - Notch to enhance average value (remove harmonic tones)

[2014 Lei]

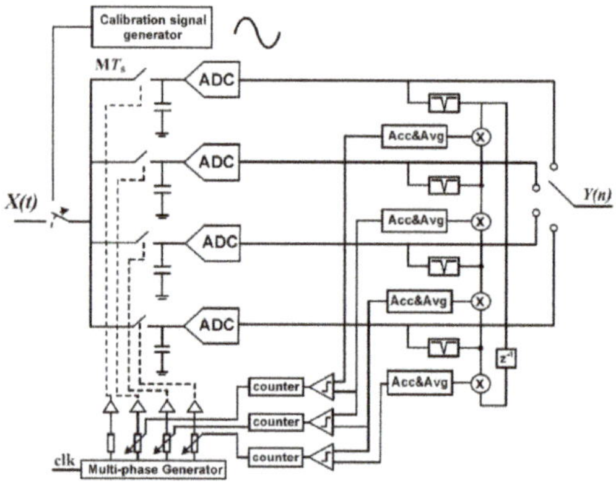

An example of DSP-based skew compensation is illustrated here in the context of a time-interleaved ADC clock skew correction scheme. The cross correlation between the digital signals from adjacent channels are observed. The output of this cross correlation function is then used to adjust the relative clock skew.

E. mmWAVE LO DISTRIBUTION – FREQUENCY PLANNING

29 Frequency Planning (I)

- Frequency Plan A:
  - Central PLL & LO phase generation
  - All required phases generated centrally
    - LO Phase generation power & area minimized
  - All phases distributed globally
    - LO distribution power is highest
    - How to maintain skew & duty cycle over large distance?
  - Estimated power consumption:
    - $P = P_{PLL} + P_{PG} + kNP_{buff}$

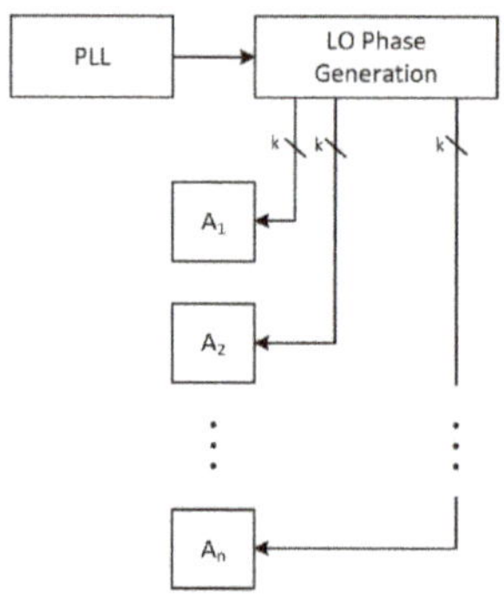

K LO phases
N elements

How do you partition your LO distribution network? In Frequency plan A, a central PLL & LO generation are implemented. All multiphase clocks generated are then distributed to the array elements.

30 Frequency Planning (II)

- **Frequency Plan B:**
  - Central PLL, distributed LO phase generation
  - All required phases generated locally
    - LO Phase generation circuit duplicated for each array element A_i
    - Alleviates routing several phases of clock globally
  - Estimated power consumption:
    - $P = P_{PLL} + NP_{PG} + NP_{buff}$

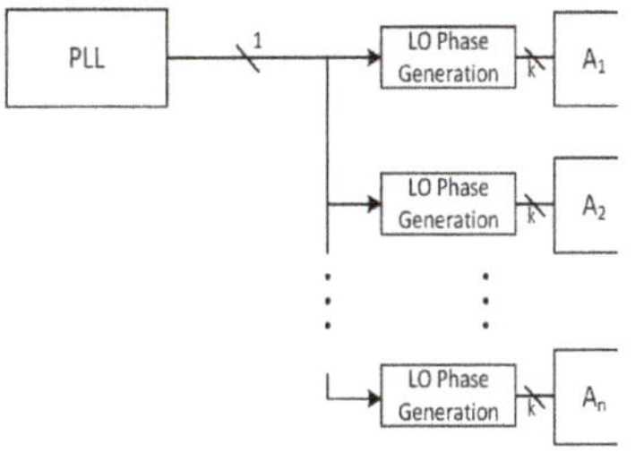

k LO phases
N elements

In Frequency plan B, a central PLL generates the clock and is distributed to all array elements. A local multi-phase clock distribution then produces all required phases of the clock. This is the most common type of LO distribution.

31 Frequency Planning (III)

- **Frequency Plan C:**
  - Distributed PLL & LO phase generation
    - Highest LO generation power consumption
    - Lowest LO distribution power consumption
  - Low reference frequency (f_{xo}) is distributed.
  - Estimated power consumption:
    - $P = NP_{PLL} + NP_{PG} + N\left(\frac{f_{ref}}{f_c}\right)P_{buff}$

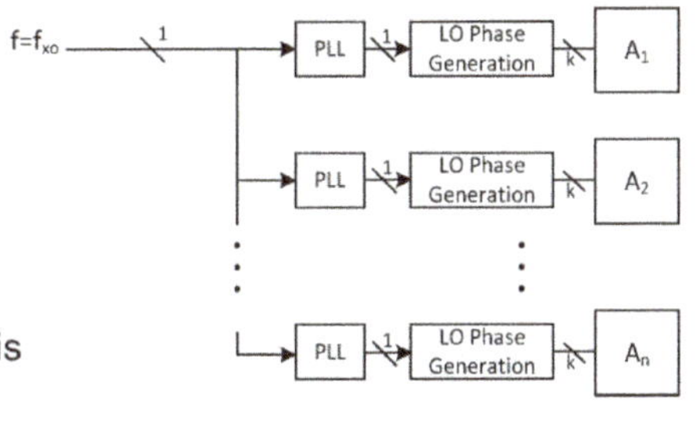

k LO phases
N elements

In Frequency plan C, the low-frequency reference clock is distributed to all array elements. Each array element would generate its own clock (PLL) and locally generate all required phases.

32 Frequency Planning (IV)

- Frequency Plan D:
  - Central PLL, distributed LO phase generation
  - Use of frequency multipliers to reduce power consumption
  - All required phases generated locally
    - LO Phase generation circuit duplicated for each array element A_i
    - Alleviates routing several phases of clock globally
  - Estimated power consumption:
    - $P = P_{PLL}\left(\frac{f}{f_c}\right) + NP_{PG} + NP_{buff}\left(\frac{f}{f_c}\right)$

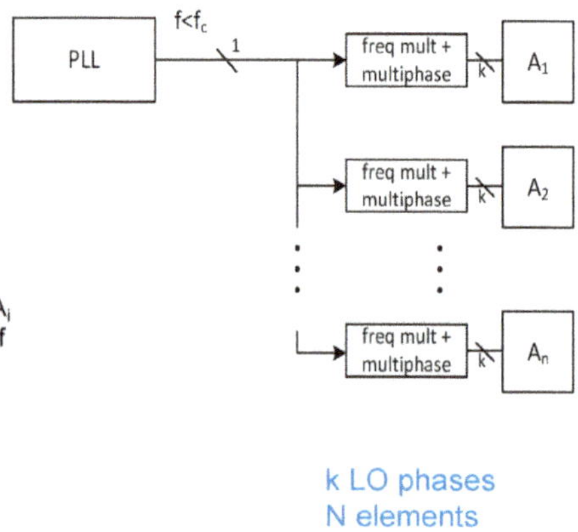

In Frequency plan D, a Central PLL is used that produces an intermediate frequency and is sent to all array elements. A local frequency multiplier is used to generate the required frequency. Local multi-phase clock distribution networks are used per array element.

33 Frequency Planning (V)

- Summary of power consumption of different frequency plan configurations:

Configuration	PLL	PG	Buffers
A	1	1	kN
B	1	N	N
C	N	N	$N\frac{f_{ref}}{f_c}$
D	$\frac{f}{f_c}$	N	$N\frac{f}{f_c}$

Here is a summary of the different frequency plans in terms of power consumption. There are other metrics that must be taken into account, such as phase noise, especially in the context of phased arrays. Common LO blocks to all array elements would end up in correlated noise source which can be canceled out with a carrier recovery loop in the digital domain. LO blocks that are not in common to each array element would result in uncorrelated noise and can potentially be averaged out when summing the individual received components in the phased-array system.

34 Choice of VCO frequency (I)

- Option A:
  - VCO @ f_c – high phase noise due to cap bank Q

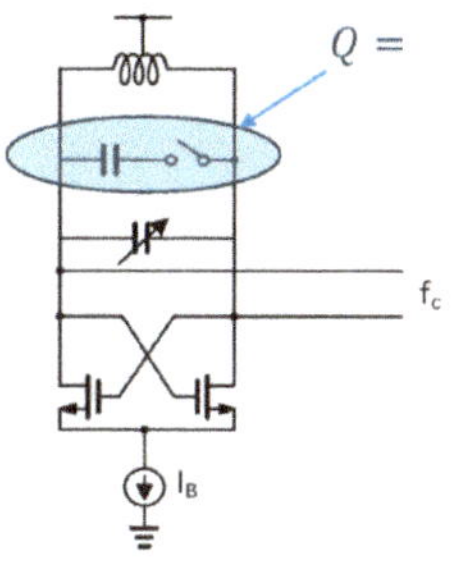

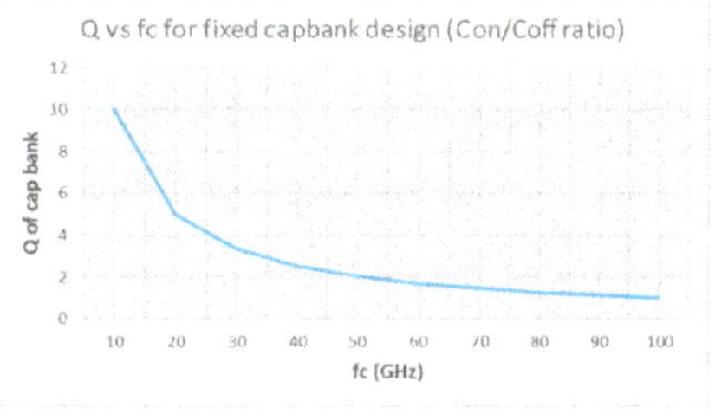

Before moving forward, a note about the VCO frequency in an LO distribution network must be mentioned. Choice of VCO frequency determines the optimal frequency plan choice for the LO distribution network. As the VCO frequency is increased, the quality factor, Q, of the capacitor bank decreases. This fundamentally limits the VCO phase noise.

35 Choice of VCO frequency (II)

- Option B:
  - Tuning range relaxed by 2x (or 3x)
  - Lower phase noise now possible by re-optimizing capbank

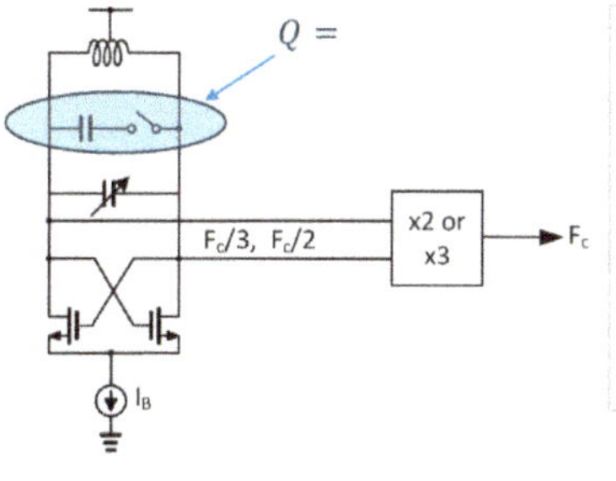

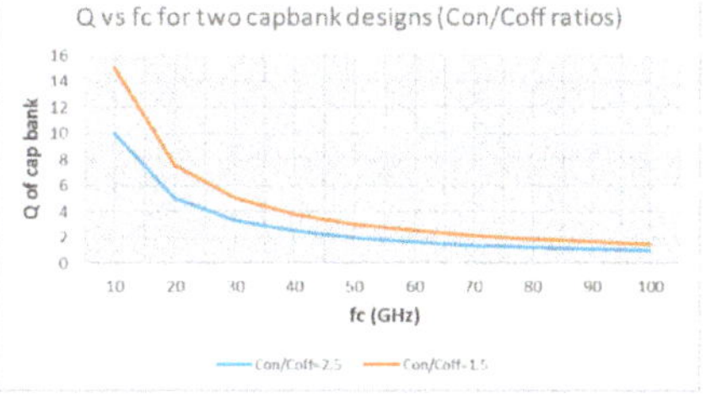

By using a frequency doubler or tripler, the VCO frequency can be lowered. This is a higher power solution, but the overall phase noise is better than simply increasing the VCO frequency.

36 Frequency multipliers

- Metrics & Figure of Merit (FOM)

Metric	Units	Definition
HR1	dBc	Suppression of the input fundamental tone, relative to the second harmonic
Center Frequency (f_c)	GHz	Output frequency of operation
Range (f_r)	GHz	Output frequency range of the frequency multiplier
Power Consumption (Pwr)	mW	Total power consumption

In order to evaluated different frequency doublers and triplers, a figure-of-merit (FOM) is required. One such FOM is reported here that takes into account center frequency, tuning range, fundamental frequency suppression, power consumption and transistor feature length.

37 Frequency Doubler Topologies (I)

- Mixer-based frequency doubling:
  - *Basic idea*: RF & LO ports shorted, output frequency consists of DC term and double frequency term

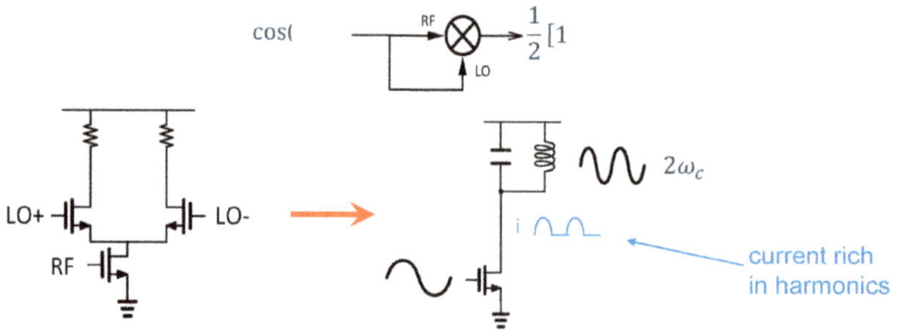

Basic idea of a mixer-based frequency doubler is to short the RF and LO terminals of a mixer. The output results in a DC term and a doubled frequency term. A frequency doubler can be implemented by a single-balanced mixer. Since the LO and RF terminals are tied to the same input, this results in a simple common source amplifier configuration, where the input is hard-switched to produce an output rich in harmonic terms.

38 Frequency Doubler Topologies (II)

- Advantage:
  - Simple
- Disadvantage:
  - Poor HR1 performance (similar issue to single-balanced mixer)
- Several stages can be cascaded for better HR1 performance
- Not very popular doubler topology

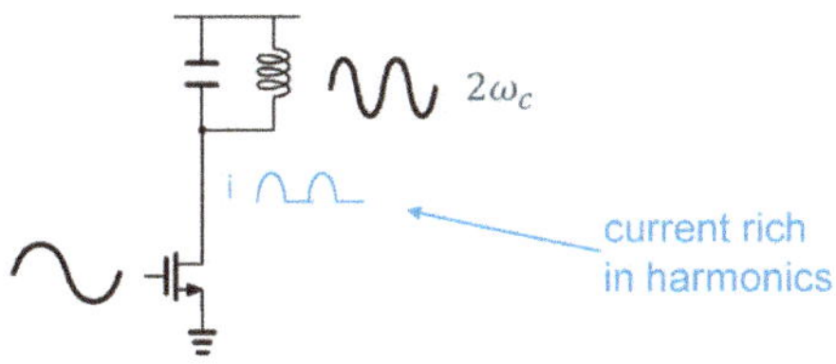

Unfortunately, the output also contains a strong HR1 term, which needs to be filtered out. For this reason, this doubler topology is not very popular.

39 Frequency Doubler Topologies (III)

- Gilbert cell mixer based doubler (aka *push-push* doubler):
  - Basic idea: In order to improve HR1, use a double balanced mixer

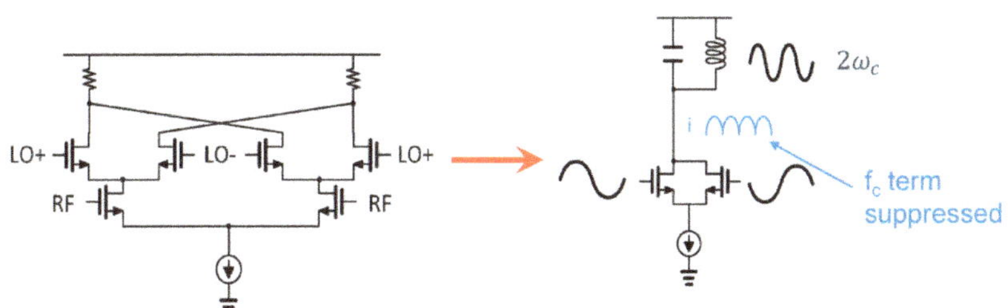

An alternative approach is to use a Gilbert cell mixer, or a double balanced mixer. The resulting output current has HR1 canceled. This topology is commonly used in mmWave LO distribution networks.

40 Frequency Doubler Topologies (IV)

- Advantage:
  - Simple
- Disadvantage:
  - Differential output not easily achieved
- Several stages can be cascaded for even better HR1 performance
- Very wideband operation (100+GHz output frequency has been reported)
- Very common doubler topology

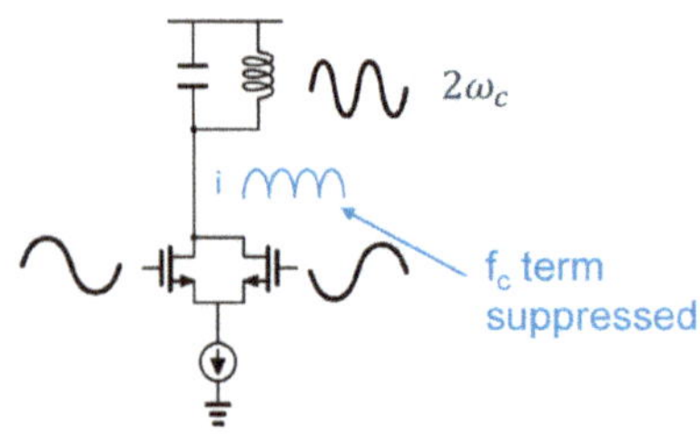

Main advantage of this circuit is its simplicity, although the output is single-ended. This doubler results in 40-50dB of HR1 suppression. Fairly wideband operation is possible (limited by LC tuning network).

41 Frequency Doubler Topologies (V)

- Differential output push-push topology:
  - Differential requires I/Q signals

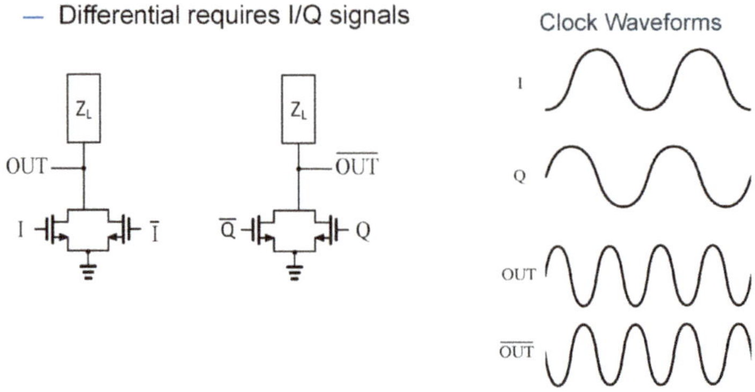

A doubler with differential output is possible if quadrature input clocks are available.

42 Frequency Doubler Topologies (VI)

[C. Yang 2011]

- QVCO + doubler hybrid:
  - Common mode node of the VCO is exploited since it has a large $2f_c$ component.
  - $2f_c$ from QVCO is buffered to a cross-coupled regenerative LC latch centered at $2f_c$
  - The transformer is used as a buffer

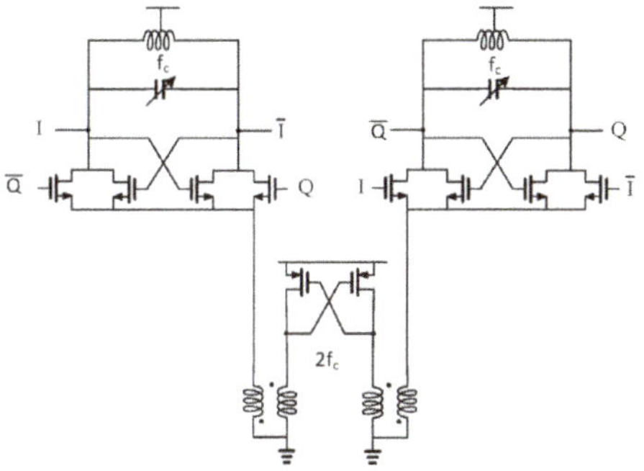

130nm CMOS
8.8-9.2GHz
HR1 = 58dBc
12mW

Shown here is an example of doubler utilizing a quadrature VCO. The tail current terminals of both VCO cores are used to provide the doubled frequency component. The differential signal is then inductively coupled to a regenerative amplifier to increase its amplitude.

43 Frequency Doubler Topologies (VII)

- Pseudo differential:
  - Avoids requirement of I/Q inputs
  - Single-ended output of doubler feeds a CS stage to generate a differential signal: LC resonator for good HR1
  - Both doubler & CS stage "injection locked" to resonate at $2f_c$ set by LC tank, with range of:
    - $\frac{\Delta\omega_{max}}{\omega_c} = \frac{1}{Q}\frac{4}{3\pi}\frac{V_i}{V_0}\frac{1}{\sqrt{1-\left(\frac{8}{3\pi}\frac{V_i}{V_o(\omega)}\right)^2}}$
    - $V_{out}(\omega) = \frac{4}{\pi} I_B \frac{R_P}{\sqrt{1+(2Q\frac{\omega-\omega_c}{\omega_c})^2}}$

[Monaco 2010]

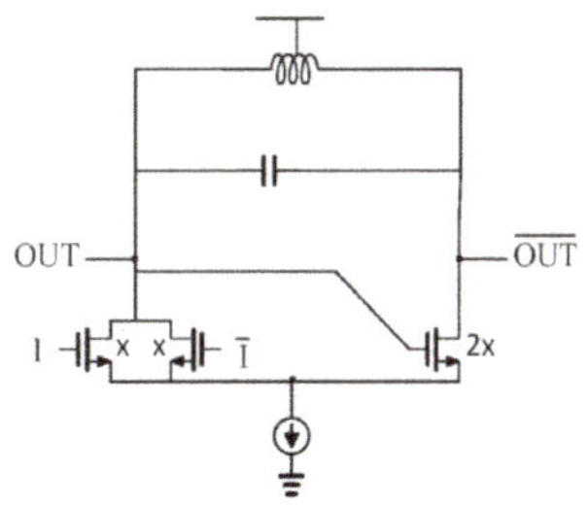

65nm CMOS
11-15GHz range
HR1 = 53dBc
6mW @ 0.75V

Another example of a doubler with differential output is illustrated here. This doubler utilizes a cross-coupled FET pair with LC load. The output is not truly differential due to the internal delay required to produce a complementary output. A skew compensation circuit may be required if truly differential outputs are desired.

44 Frequency Doubler Topologies (VIII)

- Differential push-push by output balun

[Siriburanon 2015]

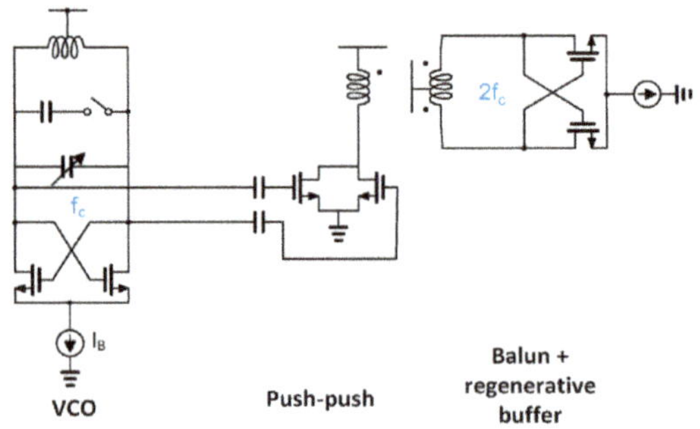

65nm CMOS
27.5-29.6GHz
HR1 > 50dBc
33mW (inc. PLL)

A 'push-push' doubler is shown where a transformer based balun is used to produce a differential output.

45 Frequency Doubler Topologies (IX)

- Survey of recent frequency doublers for mmWave:

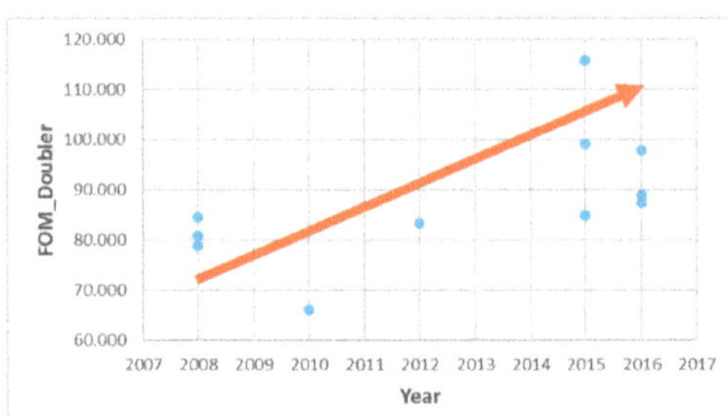

FOI

A survey of recent frequency doublers operating at millimeter wave frequencies shows that the FOM of doublers is steadily increasing approach to a value of 110.

46 Frequency Tripler Topologies (I)

- Tripler topologies – nonlinear type
  - Must bias FET classB/C operation (maximize g_3 term)
  - Optimum class C biasing may require calibration
  - Can cascade several stages for better HR1, HR2

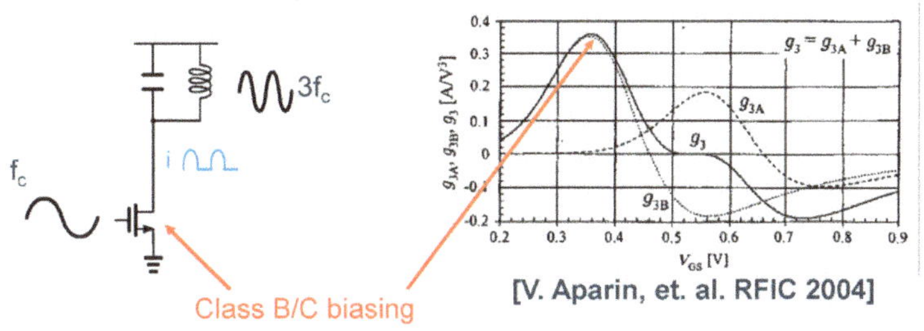

[V. Aparin, et. al. RFIC 2004]

Now, let's turn our attention to frequency triplers. One way to implement frequency triplers is to make use of the nonlinear operation of a FET biased in class C configuration. One issue common to this approach is that a calibration circuit may be required to properly bias the FET device. Passive filters would be required to filter HR1 and HR2 tones.

47 Frequency Tripler Topologies (II)

- Example: Nonlinear type
  - Shunt LC peaking @ 3f$_c$
  - Notch @ f$_c$ for better HR1
  - CG w/ cap fdbk for larger gain
  - Regenerative amplifier @ 3f$_c$

    130nm CMOS
    57-63GHz
    HR1 = 28dBc
    HR2 = 43dBc
    10mW

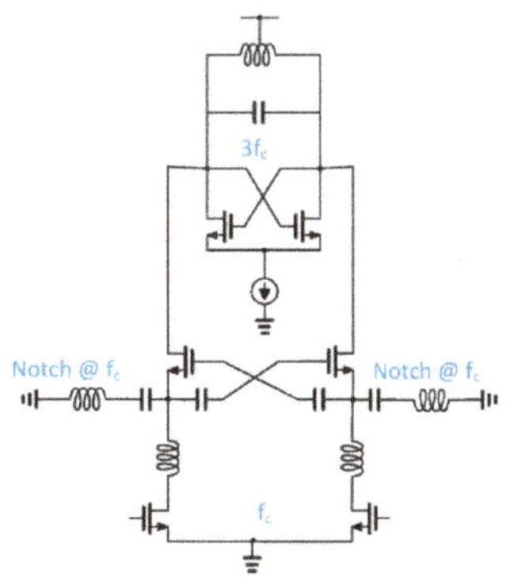

[C. Kuo 2010]

This slide shows an example of tripler using class C biased FET, followed by an LC notch at f_0 (to filter HR1). This is then followed by a regenerative amplifier (resonator) to band-pass filter around $3f_0$. HR2 rejection is achieved in part by fully balanced circuit configuration and also in part by filtering of the regenerative amplifier.

48 Frequency Tripler Topologies (III)

- Tripler topologies – mixer based
  - *Basic idea*: Doubler (x2) followed by mixing, followed by filtering @ $3f_c$
  - Main Issue: low HR2 (since mixer is usually a single-balanced mixer – resulting doubled term is single-ended).

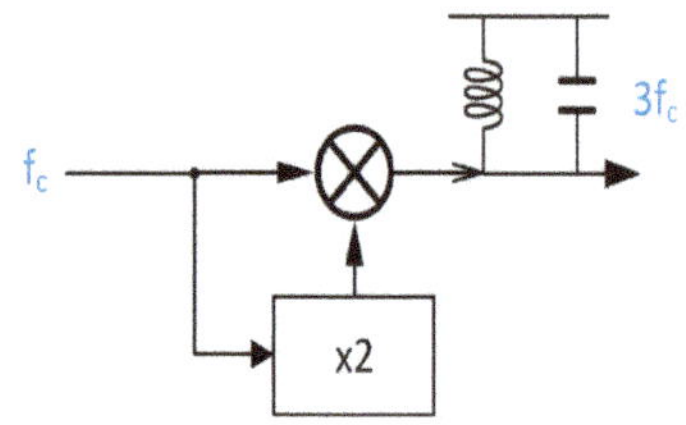

Another method of implementing triplers is to use a mixer-based topology. The idea is to use a doubler and mixer the doubled signal with the original signal to produce a tripled frequency term at frequency $3f_0$. The output LC filter is centered at $3f_0$. Main issue is poor HR2 rejection since the mixer is usually implemented as a single-balanced mixer to simplify the design (output of the doubler is usually single-ended).

49 Frequency Tripler Topologies (IV)

- Example 1: VCO
  - VCO & mixer cascoded to save power consumption
  - LC trap between stages to filter $2f_c$
  - Resonator in mixer around $3f_c$ to improve HR1 (HR2)

[Y. Lee 2016]

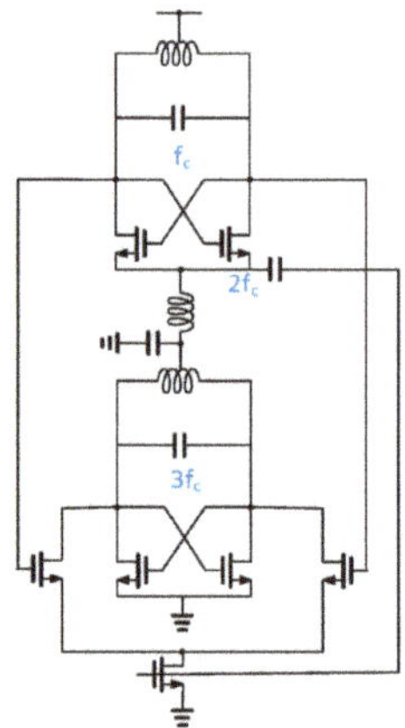

180nm CMOS
20 - 28GHz
HR1 = 30dBc
HR2 = 10dBc
9mW (inc. VCO)

One example of mixer-based tripler topology is shown here. A VCO core is used as the source of the input frequency, f_0. A doubled frequency is obtained by tapping off the center node of the VCO, which toggles at $2f_0$. The two frequencies then go to a single-balance mixer with a regenerative output and load centered at $3f_0$. This circuit suffers low HR2 due to the reuse of the current (stacking of the VCO and the mixer core).

50 Frequency Tripler Topologies (V)

- Example 2: VCO/mixer merging
  - Similar to example 1
  - VCO/tripler cascaded for better headroom

180nm CMOS
19 - 29GHz
HR1 = 32dBc
HR2 = 18dBc
11.1mW (inc. VCO)

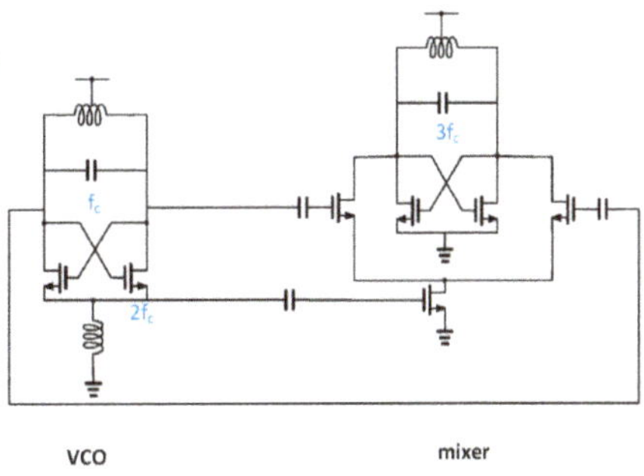

[P. Tsai 2011]

Here is another example of a mixer-based tripler. It is similar to the previous topology, except that the two components are cascaded and not stacked as in the previous slide. This results in better HR2, but still lower than the nonlinearity-based tripler approach. As stated earlier, this is inherently due to the use of a single-balanced mixer core.

51 Frequency Tripler Topologies (VI)

- Example 3: mixing only
  - Compact design
  - Generate 2x term, then mix with 3x term

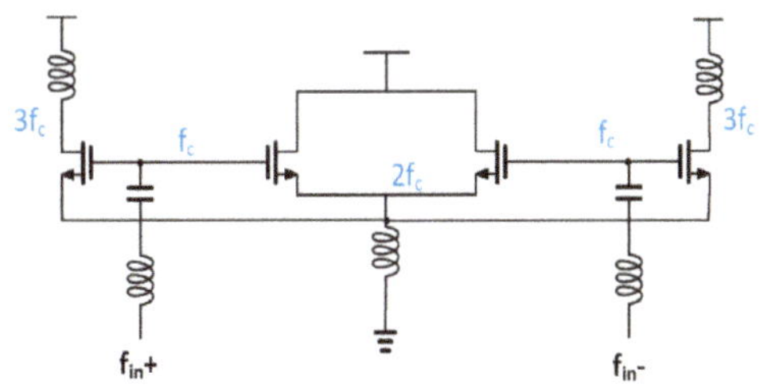

180nm CMOS
15.9 – 21.3GHz
HR1 = 20dBc
HR2 = 25dBc
18.8mW

[Y. Lo 2012]

This is a third example of a mixer-based tripler. This is a fairly compact design. A doubler core in the center is used to produce a $2f_0$ term, which then is fed to the source terminal of the left and right NFET devices and mixed with the f_0 input (which is at the gate of the NFET devices). Again, the HR2 term is not too impressive, simply due to the fact that the $2f_0$ term generated was single-ended.

52 Frequency Tripler Topologies (VII)

- Example 4: Tripler with I/Q outputs

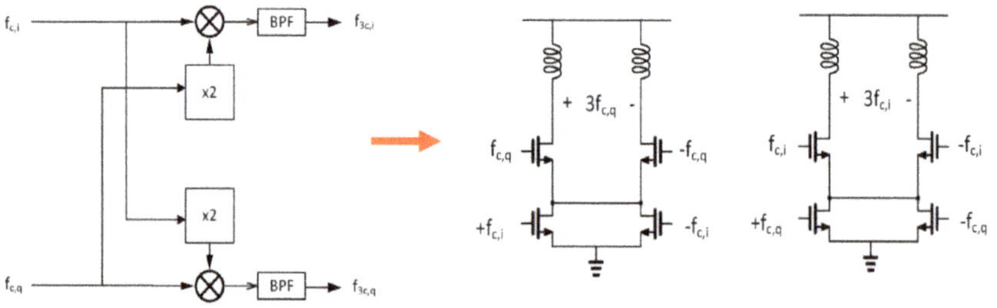

180nm CMOS
56.4-63GHz
HR1 = 22.5dBc
HR2 = 16.1dBc
7.5mW

[C. Kuo 2009]

One example shown here shows how to convert I/Q inputs to I/Q tripled terms. This works by first doubling the I and Q terms and then mixing them with the Q and I inputs, respectively. Then the mixer output is band-pass filtered at around $3f_0$. The schematic on the right-hand side shows a stacked implementation of the doubler and mixer, where the doubler is at the bottom and the output feeds the differential pair directly.

53 Frequency Tripler Topologies (VIII)

- Survey of recent frequency triplers for mmWave:

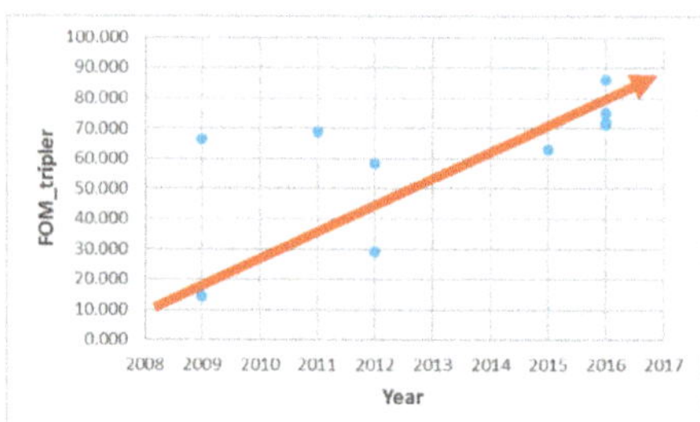

FOI

Here we show a survey of recent mmWave triplers. Again, the figure shows a steady trend towards higher FOM over time, approaching a value of 110.

54 Comparison of Frequency Doublers/ Triplers

- Survey of doublers/triplers over past 8 years
  - Compared in terms of power consumption normalized to center frequency (mW/GHz)

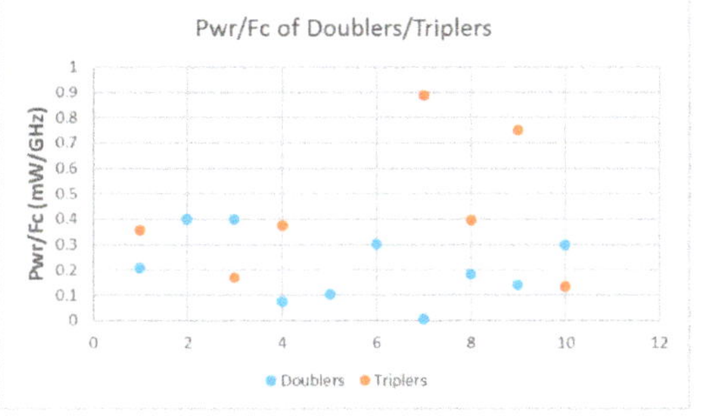

If you compare the doublers and triplers in terms of power consumption per frequency (mW/GHz), it shows that the power efficiency of doublers and triplers are comparable. This points to the conclusion that using a tripler gives you a better trade-off between power and phase noise performance when compared to a doubler. The main drawback of a tripler is higher HR1 and HR2 terms.

55 Injection Locked Oscillators (I)

- What is Injection Locked Oscillators (ILOs)?
  - Phenomenon of tuning an oscillator frequency and phase to follow a reference signal
- Applications of ILOs?
  - Quadrature VCOs (as seen before)
  - Regenerative amplifiers (as seen in doublers, triplers)
  - Low power frequency multiplication (FM-ILO)
  - Low power frequency division (DIV-ILO)

Now we'll shift our attention to injection locked oscillators. We've already seen an example of an injection locked oscillator when discussing quadrature VCOs (QVCOs). In essence, a QVCO is a pair of injection locked oscillators that are injected in quadrature phases. We've also seen them in regenerative amplifiers in the context of doublers and triplers. Other forms of injection locked oscillators are frequency divider and multipliers. In the context of mmWave LO distribution, the frequency multiplier form of injection locked oscillators is of interest. An intermediate frequency can be chosen for global LO distribution (as in Option D discussed earlier), then a local injection locked oscillator would multiply this frequency to the desired frequency.

56 Injection Locked Oscillators (II)

- Basic idea of FM-ILO:
  - ILO Locking range > PLL BW, then have phase noise advantage
  - Injection voltage is a pulsed waveform
  - Reference spur generally higher than PLL reference spur
- Lock Range:
  - $\omega_L \approx \frac{\omega_{out}}{2Q}\frac{I_{inj}}{I_{osc}}\frac{1}{N}$

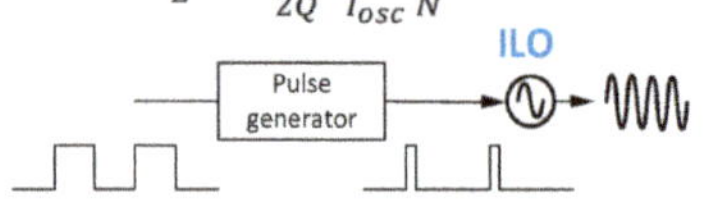

Basic LC-based ILO

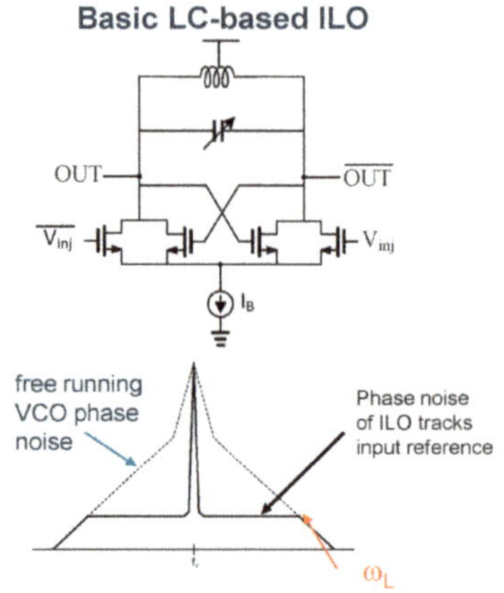

The basic operation of the injection locked oscillator (ILO) is now explained. Consider a basic ILO structure where the input and output frequencies are the same. The input clock frequency is injected into an LC oscillator tank. In order for the oscillator core to reach steady state, the oscillator phase would change to counteract the injected pulses. The effect of this injection is that the output phase noise tracks the input phase noise up to a frequency offset given as w_L. This is advantageous in terms of LO distribution, where a lower LO frequency can be distributed and the frequency can be multiplied locally while avoiding a full phase-locked loop (PLL) implementation. This frequency multiplication occurs by centered the LC tank to the desired output frequency, which would then amplify the harmonic of the input frequency. As its operation suggests, the stronger the input injection signal, the larger the lock range of the ILO. Also, the higher the multiplication ratio, the less the lock range of the ILO (since the amplitude of the harmonic LO signals decrease with frequency. In practice, a frequency locked loop (FLL) would be needed for high multiplication ratios due to limited locking range.

57 Injection Locked Oscillators (III)

- Two types of frequency multiplying ILOs
  - Stand-alone ILOs
    - Limited frequency multiplication
  - PLL / FLL assisted ILOs
    - PLL / ILO loops 'fight' each other resulting in phase error
      - Max phase error tolerable: $\theta_{max} = \frac{\pi}{2} + \sin^{-1}\left(\frac{I_{inj}}{I_{osc}}\right)$

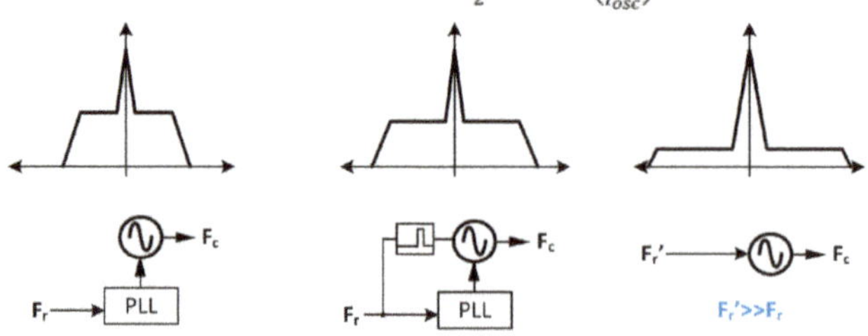

57 Injection Locked Oscillators (III)

In general, there are two classes of ILOs: stand-alone ILOs and PLL/FLL assisted ILOs. Stand-alone ILOs are common for low multiplication ratios (<4). In PLL assisted ILOs, the phase correcting action of the PLL tends to pull the injection locking to a different phase, causing contention on the phase locking operation. In order to minimize this contention, the input injection strength must satisfy the maximum tolerable phase error criteria given by the equation shown above.

F. mmWAVE LO DISTRIBUTION – BUFFER DESIGN

58 Buffer Design (I)

- CMOS versus CML
  - Power, output swing, CMR of noise (vdd, input), noise duty cycle, differential skew

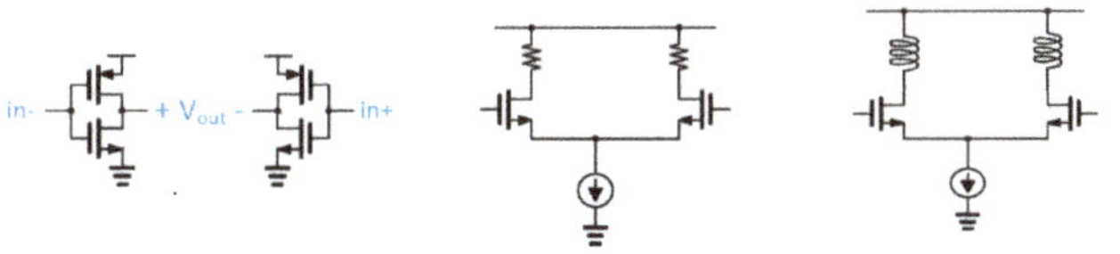

Parameter	CMOS inverter	CML-R	CML-L
Power	$2CV_{DD}^2 f_c$	$I_B V_{DD}$	$I_B V_{DD}$
$V_{out,swing}$	V_{DD}	$I_B R$	$I_B R_P$

The choice of the active buffer topology driving the mmWave LO traces is important. The three choices of buffer design are CMOS, CML-R and CML-L.

59 Buffer Design (II)

- Comparison of CMOS, CML-R, CML-L
  - Assuming a fixed frequency, ω_c and trace capacitance, C, the R and L parameters for CML-R, CML-L, respectively are defined.
    - $\omega = \frac{1}{RC}, \omega = \sqrt{\frac{1}{LC}}$
  - Also, assume that tank Q is dominated by the inductor:
    - $Q = \frac{\omega L}{R_s}$
  - Power / $V_{out,sw}$ for each configuration, is now given as below:

Parameter	CMOS inverter	CML-R	CML-L
$P/V_{out,swing}$	$\frac{V_{DD}\omega C}{\pi}$	$V_{DD}\omega C$	$\frac{V_{DD}\omega C}{Q}$
$P/V_{out,swing}$(norm)	$\frac{1}{\pi}$	1	$\frac{1}{Q}$

59 Buffer Design (II)

Normalizing the power consumption to the voltage swing yields interesting results. In this analysis, some assumptions where taken. For the CML-R, it is assumed that the largest R value given a load C was chosen to maximize the voltage swing. For the CML-L, it is assumed that the optimal L value is used for a load C to resonate at the correct frequency. Also, it is assumed that the LC tank Q is dominated by the inductor. Taking these assumptions, it can be shown that CMOS is ~3x more power efficient than CML-R. The CML-L is shown to be Q times more power efficient than the CML-R.

60 Buffer Design (III)

Buffer design

- CML-R with inductive peaking
  - Useful for increasing BW of CML buffer (lower-Q)

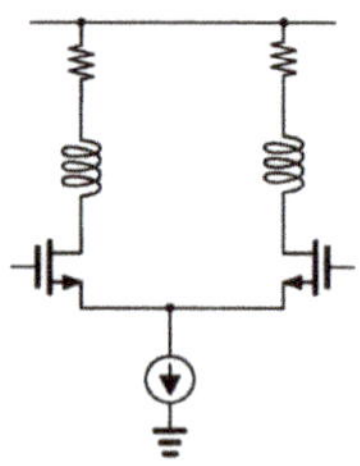

- Neutralized differential pairs
  - Higher operating speed is possible by canceling C_{gd} of input diffpair → important to combat Miller effect
  - Main issue is matching C_{gd} to C_N

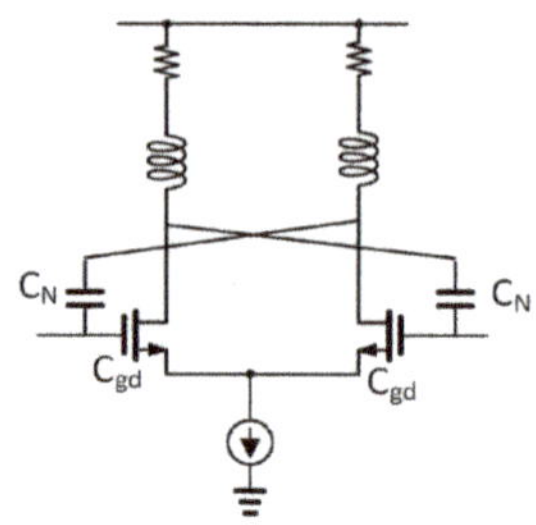

Just a few more points about the buffer design bandwidth improvement. One method of extending the bandwidth of CML-R is to use inductive peaking, where an inductor is placed in series with the load resistor. Another method to improve bandwidth is to use a neutralized differential pair. Neutralized differential pairs operate by canceling the C_{gd} of the input differential pair. This component of the input capacitance is the most dominant, since it is multiplied by the Miller effect. The cross-coupled capacitors, C_N, effectively appears as a negative capacitance which are used to cancel the C_{gd} input capacitance. This is commonly used in low-noise amplifiers (LNAs) and power amplifiers (PAs). Whether this benefit can be realized in LO buffer design depends on the fan-out of the buffer. If the driving buffer is the same strength at the neutralized buffer, current is drawn from the output stage back to the C_N capacitors to sharpen the input rise/fall times. This comes, however, at the expense of the driving capability of the next stage. If the buffers fan-out is high, then the degradation in performance of the output buffer is small and the total cascaded performance can be improved.

G. mmWAVE LO DISTRIBUTION

61 Impedance Matching

- Wilkinson dividers
  - Useful for splitting signal between different paths, while maintaining impedance matching

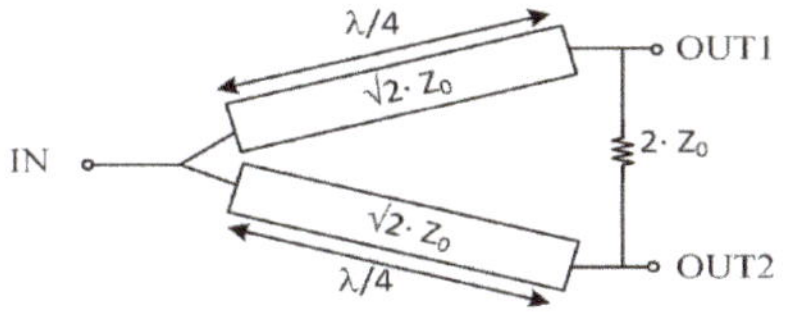

The design of the LO distribution lines is also important. A common method of splitting the LO lines is through the use of Wilkinson dividers. The idea behind Wilkinson dividers is to guarantee a perfectly matched system when splitting the mmWave line into two paths while maintaining good isolation between the two output ports. The isolation is achieved by terminating the two outputs with a differential impedance of $2Z_0$. So, if the length of each trace is l/4, the total distance from one output port to another is l/2. This means that the signal that an output port sees from another output port would be out of phase through the transmission line, and in phase through $2Z_0$, causing a net cancellation of the signal. At mmWave frequencies, Wilkinson dividers provide no more than 20-30dB of output-to-output port rejection. This rejection is especially important in phased-array systems where an RF signal can leak into the LO port, then couple back into an adjacent RF channel.

62 Jitter Amplification (I)

- Def'n: *Jitter amplification is the phenomenon when input jitter of a noiseless buffer chain is amplified as observed at the output of the buffer chain.*
  - Occurs when clk channel bandwidth is less than clk frequency. Most common in DLLs.
  - Not to be confused with jitter accumulation (occurs in VCOs).

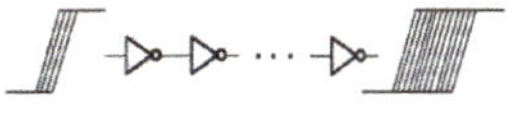

- Jitter amplification can be quantified by looking at the Jitter transfer function (JTF) curves – describes *ratio* of output to input jitter.

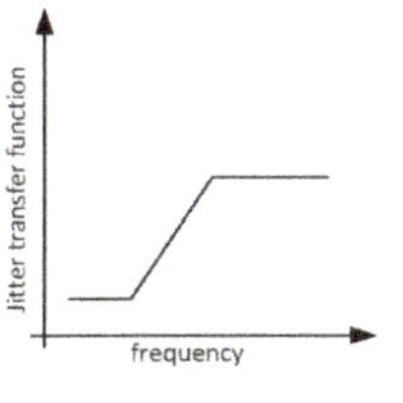

62 Jitter Amplification (I)

One important and often neglected phenomenon that appears in mmWave LO distribution networks is jitter amplification. Jitter amplification is a phenomenon that occurs when the LO signal is higher than the small signal-bandwidth a buffer chain. Its effect is to amplify the input jitter, even if the buffers themselves are noiseless. This is not to be confused with jitter accumulation which occurs in PLLs, where the jitter in a buffer is accumulated over time to produce more jitter than a standalone buffer. The jitter transfer function of a chain of buffers always exhibits a high-pass filter characteristics, signifying that jitter amplification becomes more problematic at higher frequencies.

63 Jitter Amplification (II)

- Jitter amplification effect types:
  - RJA: Random jitter amplification
  - PJA: Periodic jitter amplification
  - DCA: Duty cycle error amplification
- Discrete-time jitter impulse response: time-domain equivalent of JTF – useful in quantifying amplification factors for both duty cycle & random jitter:
  - RJA: $\sqrt{\sum_1^k h_i^2}$
  - PJA: $\sum_1^k h_i$
  - DCA: $\left|\sum_1^{\frac{k}{2}}(h_{2i-1} - h_{2i})\right|$

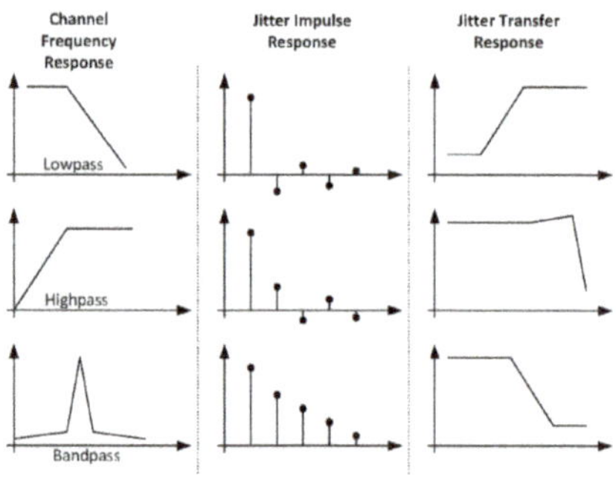

[B. Casper 2009]

The effect of jitter amplification on an LO distribution network is a function of the distribution of the jitter introduced. The jitter may be random, deterministic (or periodic), or it may manifest itself as duty cycle distortion. The jitter transfer function is shaped by the filter response of the LO distribution network. For a low-pass filter (LPF) response, the jitter transfer function is high-pass, meaning jitter is amplified at higher frequencies. If the LO distribution network has a bandpass filter (BPF) response – such as a tune LC line – then the jitter transfer function is a low-pass filter. This means that the jitter amplification effect can be curtailed by narrowband tuning the LO distribution network.

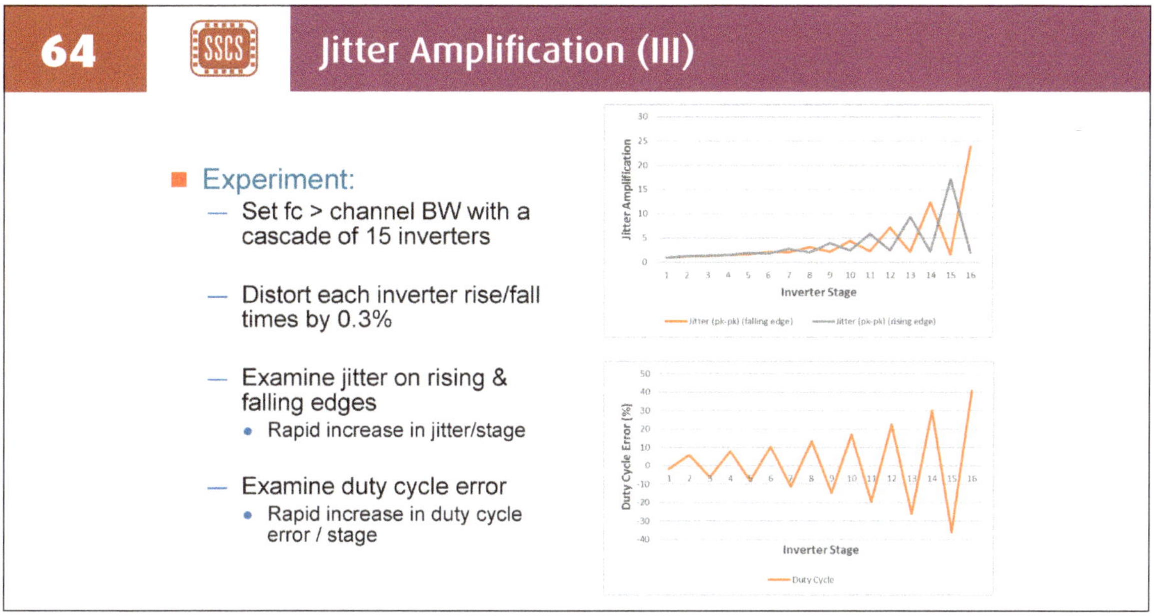

In this slide, the effect of jitter amplification on cascaded jitter and duty cycle distortion is illustrated. A chain of 15 noiseless inverters with finite bandwidth is illustrated. As shown in the top right figure, a 20x amplification of input jitter is possible. This means that if the input jitter is 50fs, the output jitter is 1ps for a chain of noiseless amplifiers! In reality, the total jitter may be significantly higher since the noise of the early stage amplifiers is also amplified by the chain. Duty cycle may also be heavily distorted. As shown, a duty cycle distortion at the input of 0.3% is amplified to 30% after 15 stages of inverters.

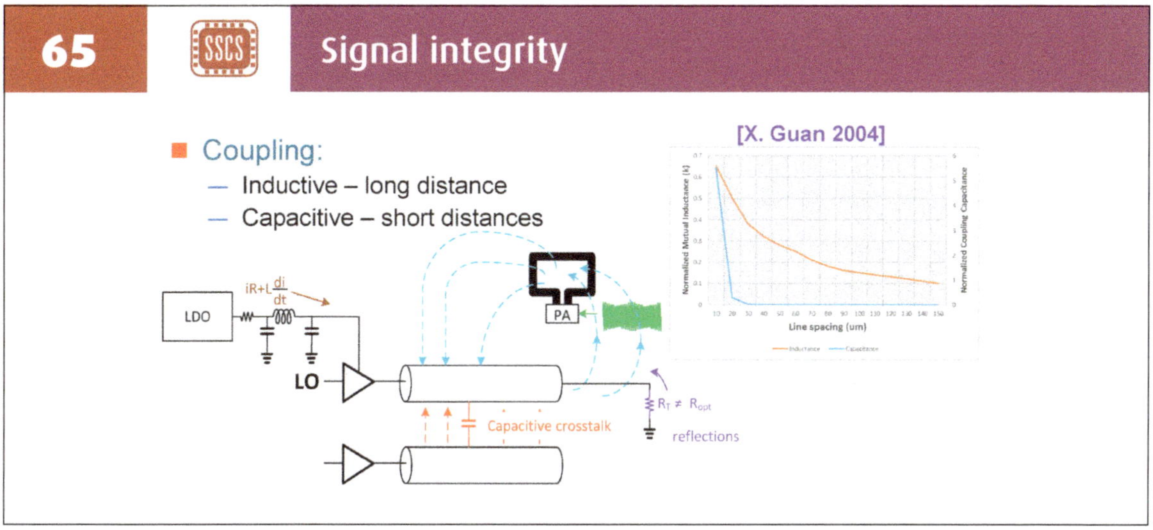

Putting it altogether, this graph shows the various sources of jitter degradation possible in an LO distribution network. Power supply noise and ripples can cause IR and Ldi/dt drops. If these ripples are data dependent, it can degrade the jitter performance of the LO distribution network. Capacitive and inductive coupling can also degrade the noise performance of the LO distribution network. As the figure shows, capacitive coupling decreases quite rapidly over distance, whereas inductive coupling becomes the more dominant form of coupling over larger distances.

66 References

1. C. Yang, "Delay-Locked Loops –An Overview," Phase-Locking in High-Performance Systems, IEEE Press, 2003, pp. 13 –22.

2. A. Mirzaei, M. Heidari, R. Bagheri, S. Chehrazi, A. Abidi, "The quadrature LC oscillator: A complete portrait based on injection locking," *IEEE J. of Solid-State Circuits*, vol. 42, no. 9, pp. 1916-1932, 2007.

3. F. O'Mahony, C. Yue, M. Horowitz, and S. Wong, "A 10GHz global clock distribution using coupled standing wave oscillators," *IEEE J. of Solid-State Circuits*, vol. 38, no. 11, pp. 1813-1820, Dec. 2003.

4. F. Aryanfar, T. Wu, M. Koochakzadeh, C. Werner, K. Chang, "A sub-resonant 40GHz clock distribution network with near zero skew," *IEEE MTT-S International Microwave Symposium (IMS) Digest*, 2010, pp. 1190-1193.

5. A. Elkohly, M. Talegaonkar, T. Anand, P. Hanumolu, "Design and Analysis of Low-Power High-Frequency Robust Sub-Harmonic Injection-Locked Clock Multipliers," *IEEE J. of Solid-State Circuits*, vol. 50, no. 12, pp. 3160-3174, 2015.

6. F. Rao and S. Hindi, "Mechanism of jitter amplification in clock channels," *DesignCon* 2014, San Jose, CA, Jan 2014.

7. S. Chaudhuri, W. Anderson, J. McCall, and S. Darabi, "Jitter amplification characterization of passive clock channels at 6.4 and 9.6 Gb/s," *Proc. of IEEE 15thTopical Meeting on Electrical Performance of Electronic Packaging*, Scottsdale, AZ, Oct 2006, pp. 21-24.

8. F. Rao and S. Hindi, "Frequency domain analysis of jitter amplification in clock channels," *Proc. of IEEE 21stTopical Meeting on Electrical Performance of Electronic Packaging*, Tempe, AZ, Oct 2012, pp. 51-54.

9. A. Mirzaei, M. Heidari, A. Abidi, "Analysis of Oscillators Locked by Large Injection Signals: Generalized Adler's Equation and Geometrical Interpretation," *IEEE Custom Integrated Circuits Conference (CICC)* 2006, pp. 737-740.

10. C. Jany, A. Siligari, J. Gonzalez-Jimenez, P. Vincent, and P. Ferrari, "A programmable frequency multiply-by-29 architecture for millimeter wave applications," *IEEE J. of Solid-State Circuits (JSSC)*, vol. 50, no. 7, pp. 1669-1679, 2015.

11. H. Wang, L. Zhang, D. Yang, D. Zeng, L. Zhang, Y. Wang, Z. Yu, "A 60GHz wideband injection-locked frequency divider with adaptive-phase-enhancing technique," 2011, pp. 17-32.

12. I. Chamasand S. Raman, "Analysis and design of a CMOS phase-tunable injection-coupled LC quadrature VCO (PTIC-QVCO)," *IEEE J. of Solid-State Circuits (JSSC)*, vol. 44, no. 3, pp. 784-796, 2009.

13. J. Chien, L. Lu, "Analysis and design of wideband injection-locked ring oscillators with multiple-input injection," *IEEE J. of Solid-State Circuit (JSSC)*, vol. 42, no. 9, pp. 1906-1915, 2007.

14. J. Lee and H.Wang, "Study of subharmonicallyinjection-locked PLLs," *IEEE J. of Solid-State Circuits (JSSC)*, vol. 44, no. 5, pp. 1539-1553, 2009.

15. B. Afshar, "Millimeter-wave circuits for 60GHz and beyond," *EECS UCB Technical Report* No, UCB/EECS-2010-113, Aug 2010.

16. Q. Ma, "Integrated millimeter-wave broadband phased array receiver frontend in silicon technology," University of Technology, Eindhoven, Netherlands, 2016.

17. A. Agrawal, A. Natarajan, "A scalable 28GHz coupled-PLL in 65nm CMOS with single-wire synchronization for large-scale 5G mm-wave arrays," *IEEE Int'l Soldi-State Circuits Conferences (ISSCC)*, 2016, pp. 38-39.

18. Z. Xu, Q. Gu, Y. Wu, H. Jian, F. Wang, M. Chang, "An integrated frequency synthesizer for 81-86GHz satellite communications in 65nm CMOS," *IEEE Radio Frequency Integrated Circuits (RFIC) Conference*, 2010, pp. 57-60.

19. J. Lee, M. Liu, H. Wang, "A 75-GHz phased-locked loop in 90-nm CMOS technology," *IEEE J. of Solid-State Circuits (JSSC)*, vol. 48, no. 6, pp. 1414-1426, 2008.

20. C. Marcu, "LO generation and distribution for 60GHz phased array transceivers," EECS *UCB Technical Report* No, UCB/EECS-2011-132, Dec. 2011.

CHAPTER 05

Low Noise, High Performance RF Frequency Synthesizers

Foster Dai

Auburn University, USA

SSCS

Modern RF frequency synthesizer designs are driven by ever increasing system requirements such as low power (e.g., for mobile devices, sensors, wearables, and internet of things, etc.), low phase noise (e.g., for wideband modulations such as 64~256 QAM, etc.) and multi-phase clock generations (e.g., for beam forming, phase array, N-path filtering, passive-mixing and interleaved data converters). Traditional LC based PLLs occupy large area and present challenges for technology scaling. On the other hand, area efficient ring oscillators (ROs) often suffer from poor jitter and phase noise performances. Recent techniques including injection locking (IL) and sub-sampling (SS) have achieved impressive in-band noise performance. As a result, the integrated phase noise of inductor-less PLLs can be greatly improved with a widened loop bandwidth. Improving spectral purity is normally obtained at the price of higher power consumption. This talk discusses the techniques to break this trade-off by both circuit and architectural innovations. This talk addresses the design challenges such as the stability issues associated with SSPLLs and the reference spur issues introduced in ILPLLs.

1 SSCS Outlines

- **PLL Fundamentals and Design Challenges**
- Low Noise Multiphase Oscillators
- Multi-ring Coupled Oscillators
- Injection locking PLL with Multi-ring Oscillators
- Sub-sampling PLL with Robust Relocking
- Inductor-less Sub-sampling PLL with Ring Oscillators
- Conclusions

2 SSCS Charge Pump PLL

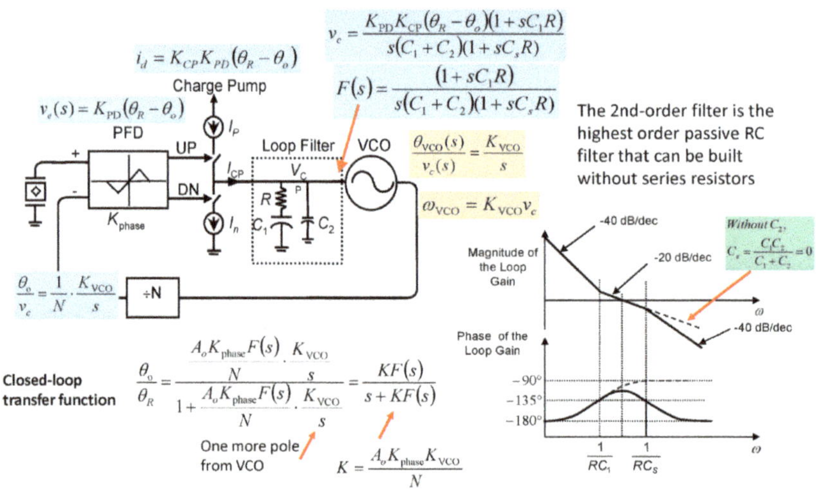

Phase-locked loop (PLL) is a feedback system which forces the divided-down VCO output phase to follow the reference signal phase. The loop is composed of a phase frequency detector (PFD), a low pass loop filter, a voltage controlled oscillator (VCO), and a frequency divider. The up- and down-pulses produced by a tri-state PFD are converted into an analog control signal at the input of the VCO by a charge pump, which is made of two controllable current sources connected to a common output that either charges or discharges capacitors attached to loop filter. The smoothed voltage signal is then used to tune the VCO output frequency. The feedback loop forces the VCO output frequency to be N times larger than the reference frequency for integer-N frequency synthesis.

3 Tri-State PFD and Charge Pump

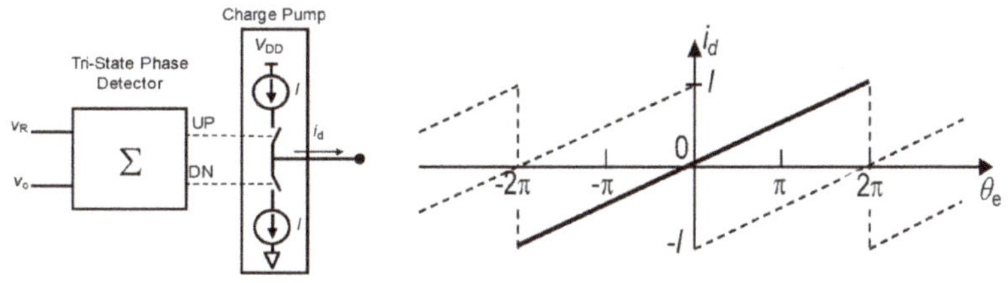

$$i_d = I\frac{\tau}{T} = \left(\frac{I}{2\pi}\right)(\theta_R - \theta_o)$$

$$K_{PFD/CP} = \frac{I}{2\pi}$$

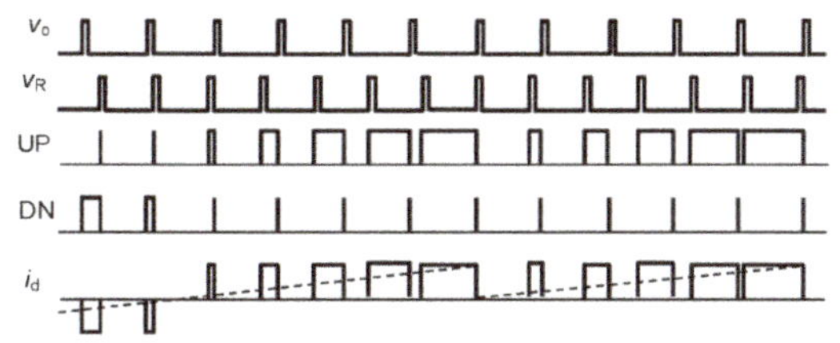

In tri-state PFD operation, the phase errors are represented by the differences of the pulse widths for the up- and down-signals. When the pulses are used to turn on or off the current sources in the charge pump, the phase error corresponds to charge or discharge time, which leads to either increase or decrease of the VCO tuning voltage. As a result, the VCO output frequency will increase or decrease based on the detected phase error.

4 Tri-State PFD with Dead Zone

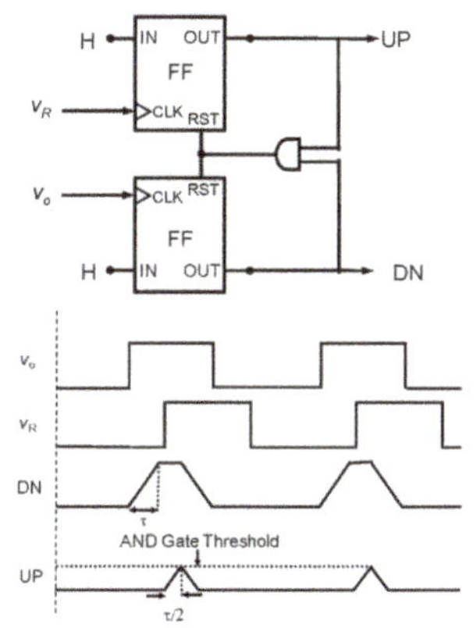

Reset is faster than FF delay → phase error.

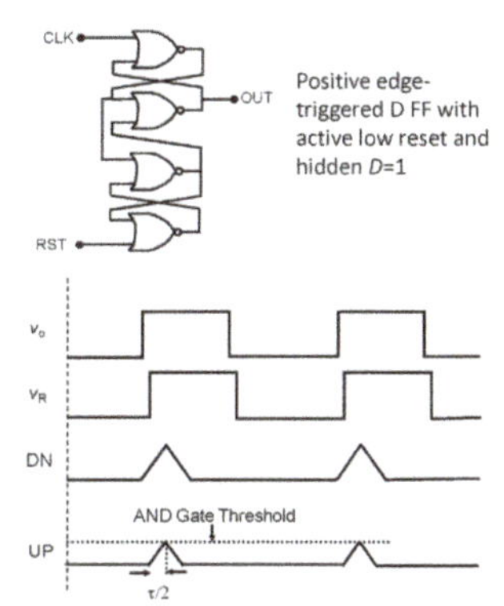

If loop is close to lock (phase difference < τ/2), the charge pump will never be turned on → loop loses gain near lock.

For a small phase difference between the phase detector input signals, narrow pulses are required at the output. However, due to finite rise and fall time, such narrow pulses cannot activate the charge pump so the average output current will not follow the input phase. This region is called "dead zone". Dead zone occurs primarily due to a difference in rise time between the latch outputs and the reset path delay in the tri-state PFD. One way to combat the dead zone is to add delay into the feedback path to ensure that the time for reset is comparable to the delay in the forward path. However, adding delay in the feedback path will cause both current sources in the charge pump to be on simultaneously, leading to undesired charge pump noise, increased reference spur and power consumption.

5 Phase Noise due to PFD Dead Zone

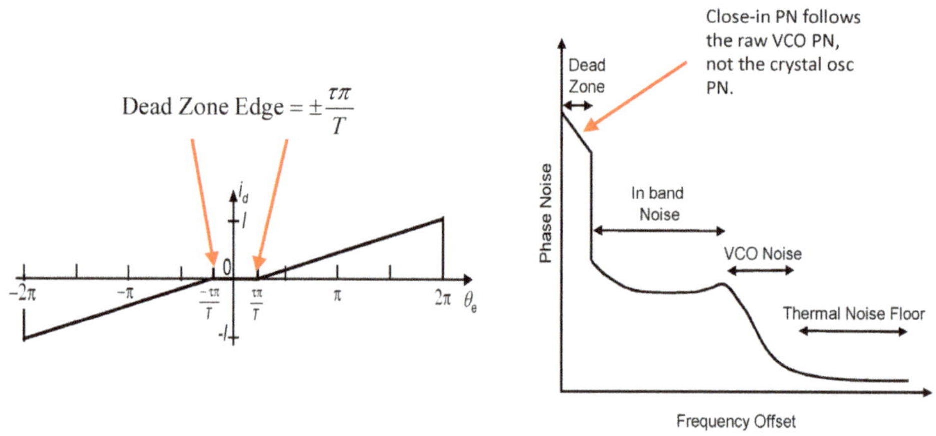

The presence of the dead zone means that, unless the phase difference between input and output reaches a certain finite value, the loop is essentially open. This means that the resultant phase noise (PN) that does not pass this threshold will not be suppressed by the loop. Thus, for noise components very close to the carrier frequency, the loop is essentially open and the PN will be the raw phase noise of the VCO.

6 PLL Phase Noise Sources

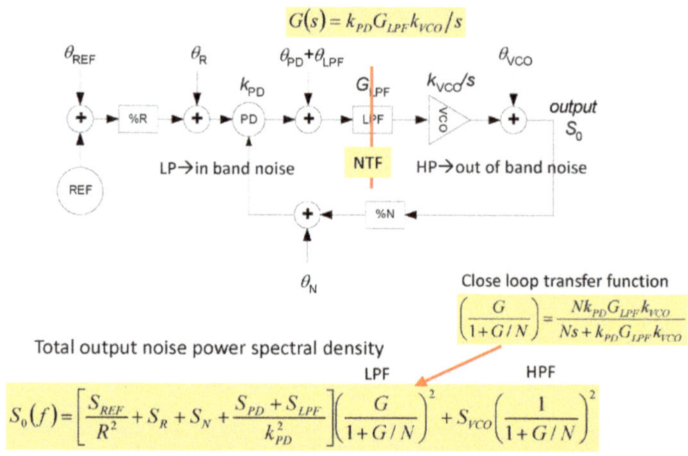

The PLL noise comes from every elements in the loop. The closed loop enforces a high-pass transfer function on the VCO noise and it thus contributes to PLL's out-band noise. On the other hand, the loop exhibits a low-pass transfer function on the noises coming from other building blocks and they thus contribute to PLL's in-band noise. At the lock-state, charge pump noise often dominates the in-band phase noise. The PLL magnifies the in-band noise from the reference, phase detector, charge pump, LPF and the dividers by an amount of 20logN dB. Therefore, reducing N such as the case of a fractional-N PLL or removing N such as the case of a sub-sampling PLL leads to lower in-band phase noise.

7 Frequency Synthesizer Design Challenges

- Tuning range →cover the entire band
- Amplitude → LO (clock) speed and drive strength, mixer noise
- Step size →be able to synthesize every channel
- Settling time → be able to change its frequency in a given time
- Acquisition time → After power on, its frequency must move to a programmed frequency in a given amount of time.
- I and Q matching → amplitude and phase
- **Spectrum purity (phase noise, jitter)** →related to modulation requirement.
- **Spurs** → cause unwanted tones, especially after mixer.
- **Multiphase generation** → for applications such as phase array, beam forming, passive mixing, N-path filtering and interleaved data converters
- **Power** → trade-off with PN and output frequency
- **Area**: VCOs may be the last block in a TRx with off-chip PA that still requires an inductor → inductor-less PLLs

Modern frequency synthesis faces the above mentioned challenges. In particular, phase noise reduction is desired for high data-rate communication links and multi-phase generation is critical for applications such as phase array, beam forming, passive mixing, N-path filtering and interleaved data converters. Moreover, when power amplifiers are often placed off chip, VCOs may be the last block in a TRx that still requires inductors. Can we built an inductor-less PLL with PN performance close to what an LC PLL can do? This talk will present a few techniques that can benefit the implementation of an area efficient PLL without inductors.

8 Outlines

- PLL Fundamentals and Design Challenges
- **Low Noise Multiphase Oscillators**
- Multi-ring Coupled Oscillators
- Injection locking PLL with Multi-ring Oscillators
- Sub-sampling PLL with Robust Relocking
- Inductor-less Sub-sampling PLL with Ring Oscillators
- Conclusions

9

Motivations

1. Multi-phase clock generation is critical for emerging technologies such as phase array, beam forming, passive mixing, N-path filtering and interleaved data converters.
2. Multi-phase signals can be generated using poly-phase filters, which are narrow-band and lossy.
3. Multi-phase signals can be generated by dividing a frequency that is N-times higher than the operation frequency, which consumes large power and is challenge for high frequency implementation.
4. Need techniques that can provide multi-phase clocks with low power, low phase noise and accurate phase across a wide bandwidth.

We start our discussion with the designs of coupled oscillators. Multi-phase clock generation is critical for emerging technologies such as phase array, beam forming, passive mixing, N-path filtering and interleaved data converters. Multi-phase signals can be generated using poly-phase filters, which are narrow-band, or by dividing a frequency that is N-times higher than the operation frequency. Neither technique provides multi-phase clocks with low power, low phase noise and accurate phase across a wide bandwidth. On the other hand, a multi-phase oscillator can be an excellent candidate for multiple phase clock generations because of its low phase noise and accurate phase across a wide frequency range.

10

Oscillator Noise Sources and Sample Process

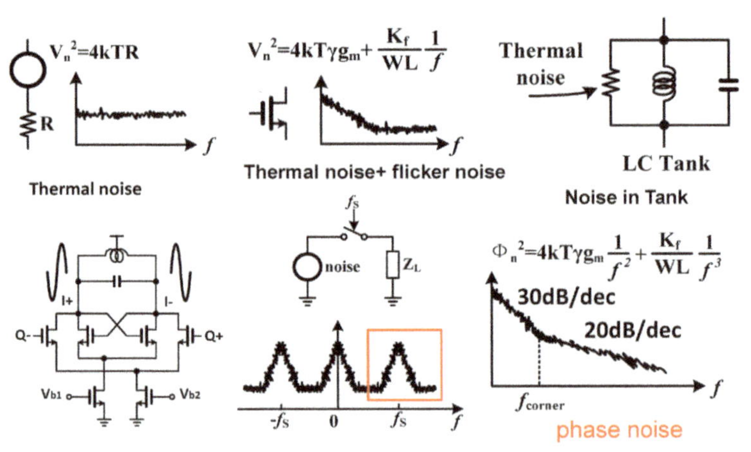

Quadrature voltage controlled oscillator (QVCO) continues to be an attractive research topic. Quadrature signal coupling can be done using active or passive devices. Phase noise in a VCO using MOSFET as switches is dominated by the tank loaded Q-factor as well as the transistor's thermal noise and 1/f noise. VCO switching behavior is a sample-process, which up-converts the phase noise to its oscillation frequency. Conversion of frequency noise to phase noise leads to -30dB/dec roll-off in 1/f noise dominated region followed by -20dB/dec roll-off after the 1/f corner frequency until it reaches the thermal noise floor.

11

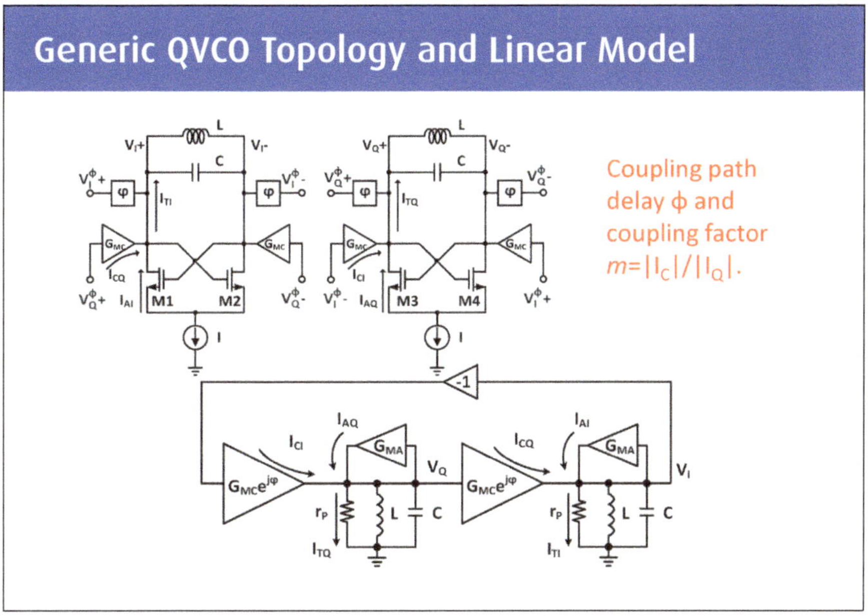

Without losing generality, we start our analysis with phase noise performance of a two-phase coupled oscillator such as a QVCO. A simplified linear model of the two-phase coupled oscillator includes the main transconductance G_{MA} to compensate the loss in the tank, G_{MC} for coupling, and the tanks. A leading phase delay of ϕ is introduced on the coupling path G_{MC} for the reason explained next. The coupling factor m is defined as the ratio between the amount of current for coupling to the oscillation current, namely, $m=|I_C|/|I_Q|$.

12

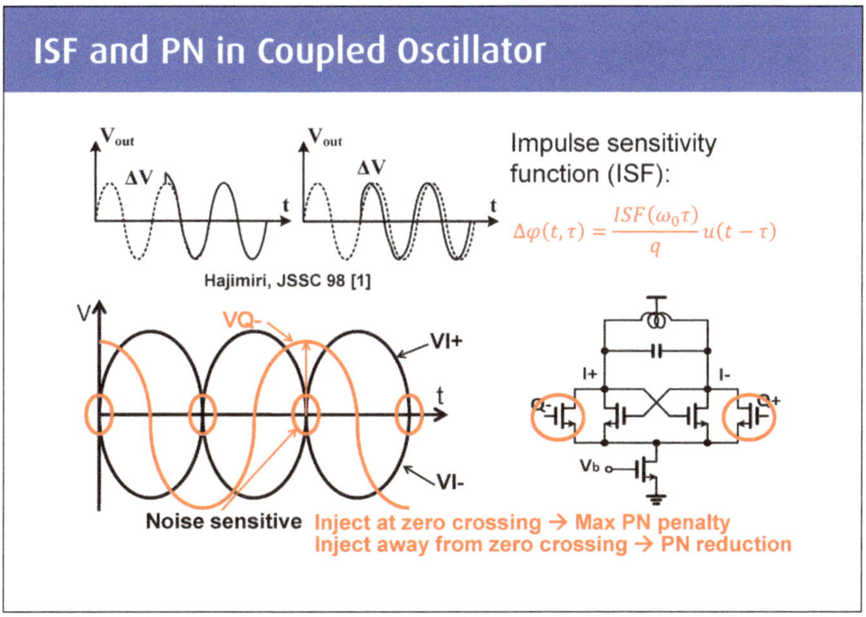

Let's review the impulse sensitivity function (ISF) of a QVCO. Due to quadrature phase relationship, the peak of a quadrature phase signal is coupled to the in-phase core at its most sensitive instance, namely, the zero-cross of the in-phase waveform. Therefore, it causes worst phase noise penalty. However, if we introduce a phase delay on the coupling path, the quadrature phase signal will then be coupled to the in-phase core at a instance away from the zero-cross, avoiding phase noise degradation.

13 SSCS Quadrature Coupling with Phase Shift (I)

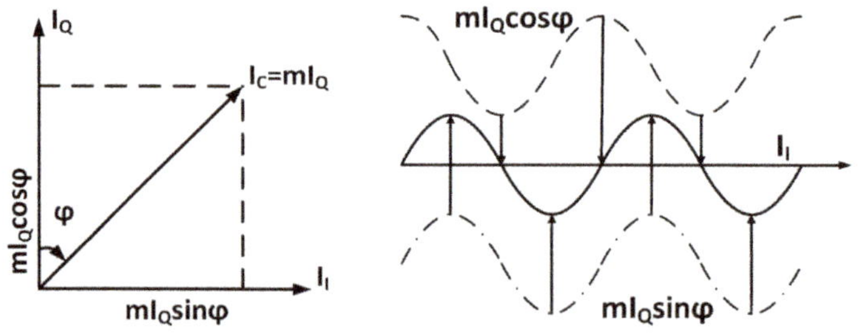

Total in-phase signal swing: $V_{QVCO}=V_0(1+m\sin\varphi)$.

Noise power coupled from Q-core: $\overline{V_{n-c}^2} = V_0^2(m\cos\varphi)^2$

Noise factor: $F_{QVCO} = 1 + \left(\frac{m\cos\varphi}{1+m\sin\varphi}\right)^2$

Assuming the coupling path introduces a phase delay of ϕ, the current coupled from the quadrature core is thus expressed as mI_Q, which is phase-rotated with respect to I_Q. This current vector can be projected into in-phase and quadrature phase axes. Note that the in-phase component injects its maximum current at the peak of I_I and thus has negligible noise contribution, while the quadrature phase component injects its maximum current at the zero-crossing of I_I and thus leads to the maximum phase noise penalty based on ISF analysis. If the VCO output swing is not limited, it will increase to $V_{QVCO}=V_0(1+m\sin\varphi)$ due to coupling. While the noise coupled from another core can be written as $\overline{V_{n-c}^2} = V_0^2(m\cos\varphi)^2$. The resultant noise factor of the QVCO is thus given by $F_{QVCO} = 1 + \left(\frac{m\cos\varphi}{1+m\sin\varphi}\right)^2$, where the first term is the noise factor of the single core and the 2nd term models the noise coupled from the quadrature core.

14 SSCS Quadrature Coupling with Phase Shift (II)

Phase noise power spectral density is given by

$$L(\Delta\omega) = \frac{KTr_p}{V_0^2(1+m\sin\varphi)^2}\left(1+\left(\frac{m\cos\varphi}{1+m\sin\varphi}\right)^2\right)\left(\frac{\omega_0}{Q\Delta\omega}\right)^2.$$

1. This simple analysis yields the same results obtained in [Mirzaei, JSSC 2007].
2. Inserting a 90 degree phase-shift in coupling paths improves the phase noise by a factor of $(1+m)^2$.
3. *m*=1, 90 degree phase shift → PN reduction of 9dB, when compared with a similar QVCO without phase shift.
4. *m*=1, 90 degree phase shift → PN reduction of 6dB, when compared to its single-phase VCO counterpart.
5. Large phase shift does not correspond to the best quadrature phase accuracy considering the mismatches.

With this noise factor of the QVCO, we can derive the phase noise power spectral density as shown above,where K is Boltzman constant and T is absolute temperature, r_p is tank parallel resistance, ω_0 is oscillation frequency, and *Q*is tank quality factor.

15 Quadrature Coupling with Phase Shift (III)

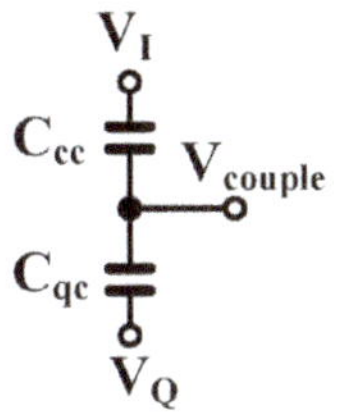

$$m = \frac{C_{qc}}{C_{cc} + C_{qc}}$$

- Coupling with active devices -- extra noise;
  - Parallel coupling: large current consumption
  - Serial coupling: reduced headroom that leads to phase noise degradation
- Coupling with passive components:
  - Inductor: large area
  - Resistor: introduce noise to coupling path
  - Capacitor: small area, no noise degradation and easy to implement phase shifting.
- Noise performance need to be optimized.

$$V_{couple}(t) = mV_Q(t) + (1 - m)V_I(t)$$

Instead of using noisy transistors for coupling, capacitive coupling is utilized to improve the phase noise performance of the QVCO. A phase delay can be easily implemented using capacitive coupling technique. Since we have two perpendicular vectors, in-phase signal V_I and quadrature-phase signal V_Q, we can form another vector V_{couple} with any desired phase that is determined by the capacitance ratio or the coupling strength factor *m*.

16 Capacitive Coupling Technique

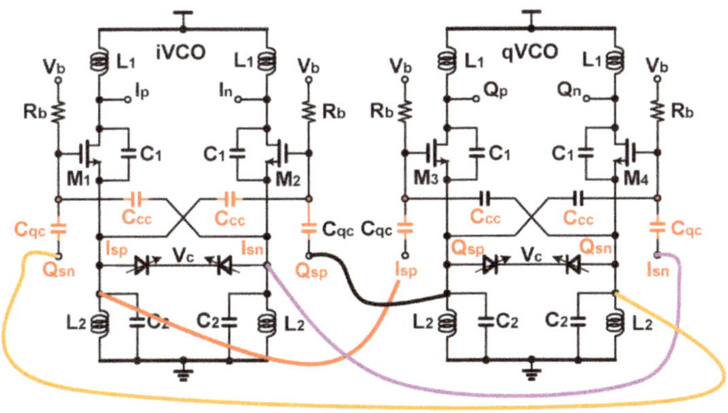

- √ Capacitive coupling for phase noise reduction
- √ Enhanced swing for low voltage supply
- √ Varactor referenced to ground for better noise

Zhao and Dai, TCAS 12, CICC 2011

This paper presents a 0.6-V QVCO design with enhanced swing for low power supply applications. The QVCO comprises a capacitive coupling that is employed for phase noise reduction. Optimized capacitive coupling combined with source inductive enhance-swing technique enables low power and low phase noise simultaneously. Due to the inherent phase shift in the proposed quadrature-coupling path, the problem associated with ±90° phase ambiguity between the quadrature outputs can also be avoided.

17 Optimization of Capacitive Coupling

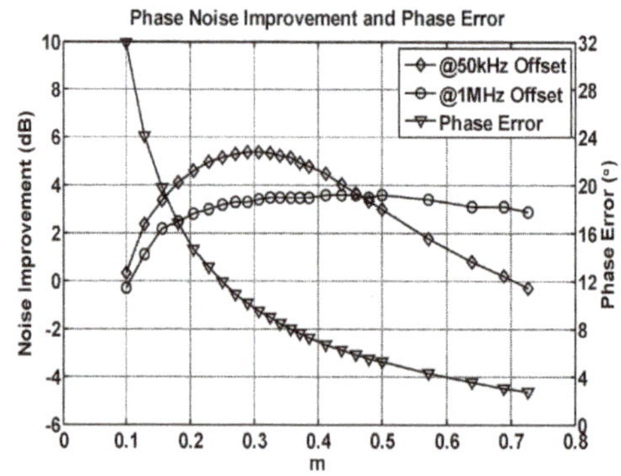

- Coupling-strength factor *m* adjusts the peak location

$$m = \frac{C_{qc}}{C_{qc} + C_{cc}}$$

- Trade-off between FoM and phase error
- QVCO noise: 3.5dB and 5.2dB better than SVCO @ 1MHz and 50kHz offset when coupling factor m=0.4

The choice of coupling-strength factor *m* is determined by the trade-off between phase noise improvement and phase error of the QVCO outputs. The larger the coupling strength is, the smaller the phase error becomes, yet less phase noise improvement is achieved. As *m* approaches to 1, where the proposed QVCO is similar to conventional one without phase delay on the coupling path, the phase noise would be degraded compared to a stand-alone VCO. While when *m* is equal to 0, the two VCOs become independent to each other without phase correlation. Therefore, there is an optimum point of *m* to achieve the best phase noise improvement with acceptable phase error. After careful trade-off between the phase noise and phase error, coupling strength of 0.4 is chosen to implement the QVCO. The corresponding FoM and phase error are 198.1dB and 0.04° in simulation, respectively.

18 Die Photo of the QVCO and SVCO

- Implemented with 0.13um CMOS
- Fout: 5.6 GHz
- core area: 0.6x0.8 = 0.48mm^2
- Supply Voltage : 0.5~0.6V
- Current: total 7mA for both iVCO and qVCO
- Power: 4.2 mW
- FoM: 191.5 dB

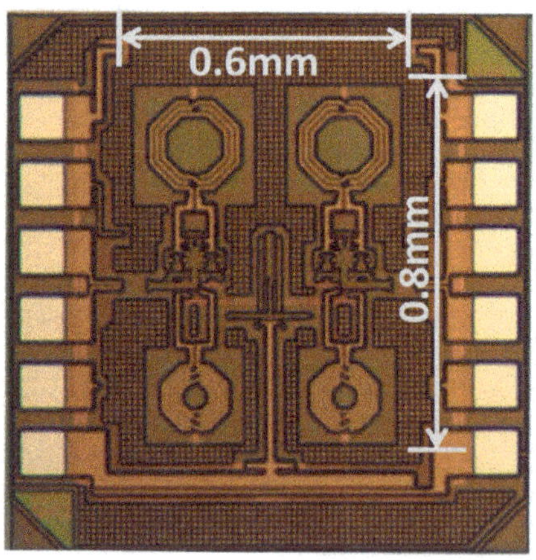

Zhao and Dai, TCAS 12

The prototype QVCO was fabricated in 0.13μm CMOS technology. The QVCO achieves FoM of 191.5dB. The measurement result proves the effectiveness of noise improvement using the proposed noise-coupling technique. The QVCO consumes 3.6mW power with a 0.6 V supply and occupies a die area of 1.2 x 1.2 mm^2 including pads.

19 Measured Phase Noise of QVCO and SVCO

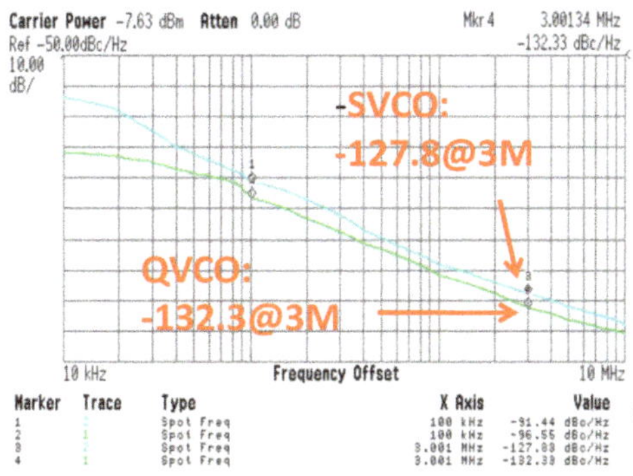

$$\text{FoM} = 10\log\left[\left(\frac{f_0}{\Delta f}\right)^2 \frac{1mW}{P_{tot}}\right] - L(\Delta f)$$

- Single-phase VCO (SVCO) FoM: 190dBc/Hz
- Quadrature-phase VCO (QVCO) FoM: 191.5dBc/Hz
- Noise improvement: 4.5dB

The proposed QVCO can achieve 4.5dB lower phase noise comparing to its single-phase counterpart at 3-MHz offset. The QVCO achieves a measured phase noise of -132.3dBc/Hz @ 3MHz offset with a center frequency of 5.6GHz and consumes 4.2mW from a 0.6V supply. This performance corresponds to a Figure-of-Merit (FoM) of 191.5dB.

20 Noise and Frequency with Tuning Voltage

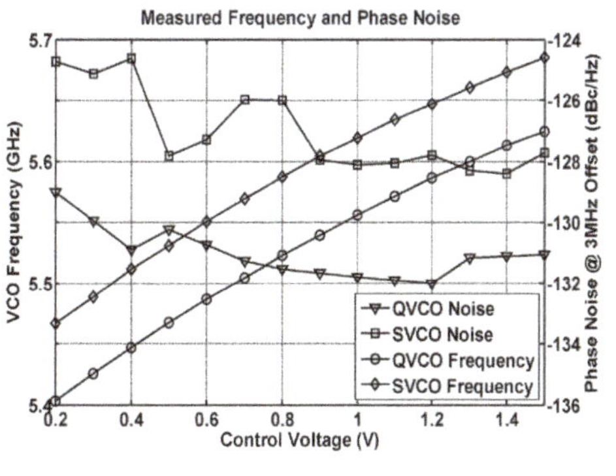

- Frequency tuning range: 200MHz;
- QVCO phase noise: -128.5~-132.3dBc/Hz@3MHz offset
- Improvement over single-phase VCO: 3~5dB

Across the frequency tuning range of 5.46GHz to 5.68GHz, the phase noise of QVCO varies from -129.2dBc/Hz to -132dBc/Hz, which is about 2-6dB lower than the phase noise of its single-phase counterpart SVCO.

21 Outlines

- PLL Fundamentals and Design Challenges
- Low Noise Multiphase Oscillators
- **Multi-ring Coupled Oscillators**
- Injection locking PLL with Multi-ring Oscillators
- Sub-sampling PLL with Robust Relocking
- Inductor-less Sub-sampling PLL with Ring Oscillators
- Conclusions

22 Motivation

- **Ring oscillator**
  - Small area √
  - Wide tuning range √
  - Multi-phase outputs √
  - Poor phase noise ×
  - Low output frequency ×
- **LC oscillator**
  - Large area ×
  - Narrow tuning range ×
  - Difficult to generate multi-phase outputs ×
  - Low phase noise √
  - High output frequency √

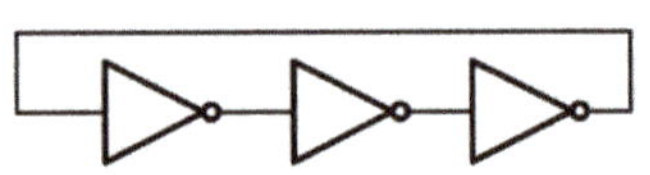

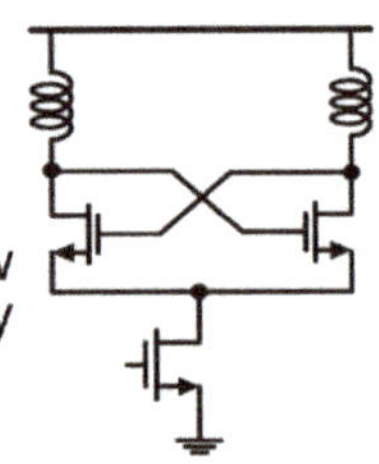

An **inductor-less oscillator** with low phase noise and high output frequency is highly desirable.

Multi-phase signal can be generated by ring oscillators with either resistive load or LC based delay cells. However, ring oscillators without LC tanks normally end up with unacceptable phase noise for most of the wireless applications. Thus, it is highly desirable for low cost applications to push the phase noise of ring oscillators close to what LC-based oscillators can reach. In this section, we present a multi-ring oscillator (MRO) that couples multiple rings with proper phase shift to achieve phase noise improvement greater than $10[\log]_{10}(N_{RO})$ dB, leading to better Figure of Merit (FoM) than its single ring (SRO) counterpart.

23

Challenges in High Frequency and Multiphase Ring-VCO Designs

- **RVCO oscillation frequency:**

$$f_{RO} = 1/(2 \times No.of\ Stage \times Delay)$$

Trade-off between frequency and no. of output phases in single ring:

Achieving high oscillation frequency and large number of output phases require delay stages with increased speed, which leads to increased power consumption and/or degraded phase noise.

- **Basic technique for PN reduction -- increase signal power:**

$$L(f_m) = 10\log\left[\frac{2FkTB}{P_{sig}}\left(1+\left(\frac{\omega_0}{2Q\Delta\omega}\right)^2\right)\left(1+\frac{\Delta\omega_{1/f^3}}{|\Delta\omega|}\right)\right]$$

Disadvantages: Increasing signal power often comes with increased noise, which will not benefit the PN reduction. Faster delay with higher power consumption also degrades the FoM.

Ring oscillators comprise of multiple delay stages connected in a ring configuration. In order to operate at high frequency with large numbers of available output phases, more delay stages are required. However, this will lead to reduced oscillation frequency or increased power consumption in order to reduce the delay between stages. Under a given power consumption, the higher the oscillation frequency is and the more output phases are required, the worse the phase noise becomes. This dilemma can be solved by splitting the delay stages into multiple rings and allowing each ring to oscillate at higher frequency, yet still maintaining correlated phase relationship through coupling among the rings.

24

Coupled Ring Oscillators

- **Coupled Ring oscillator operating frequency:**

$$f_{RO} = M/(2 \times No.of\ Stage \times Delay)$$

where *M* is No. of coupled rings.

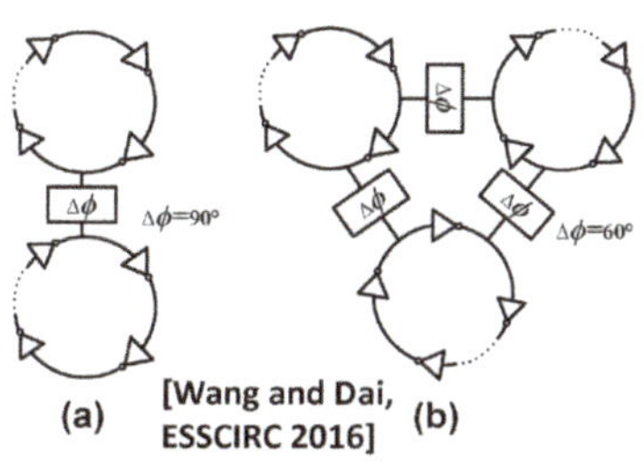

Proposed (a) double ring coupled oscillator (DRO) and (b) triple ring coupled oscillator (TRO)

Coupled ring oscillator:

1. Break the dilemma between high output frequency and number of stages
2. Provide additional output phases
3. Coupling with phase shift provides improved phase noise and FoM in ring oscillator design.

7

Leveraging the capacitive coupling technique discussed in previous section, a double ring coupled ring oscillator (DRO) is formed by connecting two ring oscillators through capacitive coupled paths with 90° phase shift. Similarly, a triple ring coupled ring oscillator (TRO) can be constructed by connecting three ring oscillators through capacitive coupled paths with 60° phase shift. This ensures that the noise coupled from the adjacent rings is not injected at the sensitive instances for best phase noise performance.

25 Common Source Coupling in RVCO

Benefits of CS coupling for PN reduction: Soltanian, ESSCIRC, 2006

- Coupling capacitors provide phase shifting that minimizes the 1/*f* noise up-conversion generated by the current sources.
- Capacitive coupling provides needed phase shifting that minimizes the noise injection from the adjacent cores.

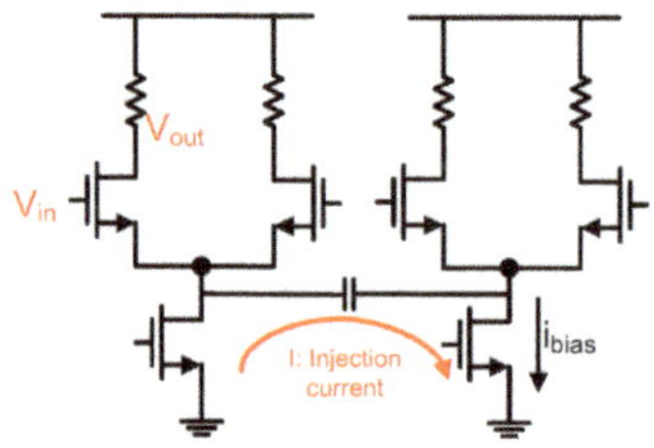

One stage of the proposed multi-ring oscillator (MRO).

- No active devices that add extra noise and power.
- Ignorable impact on the oscillator's frequency response compared with coupling through the output nodes.
- Coupling ensures the phase coherency among multiple rings, providing additional correlated phases without lowering output frequency.

The common-source coupling scheme has two advantages: (i) the capacitive coupling provides needed phase shift that minimizes the noise injection from the adjacent rings; (ii) the coupling capacitors provide phase shift needed for minimizing 1/*f* noise up-conversion generated by the current sources. As a result, the overall phase noise obtained from the TRO is greatly improved compared with its single-ring oscillator (SRO) counterpart. Note that the capacitive coupling does not use any additional active devices, which add extra noise and power, nor does the coupling devices load any output nodes of the delay stages, which is sensitive to the bandwidth.

26 Architecture of Proposed TRO

Proposed triple ring coupled ring oscillator (TRO)

[Wang and Dai, ESSCIRC 2016]

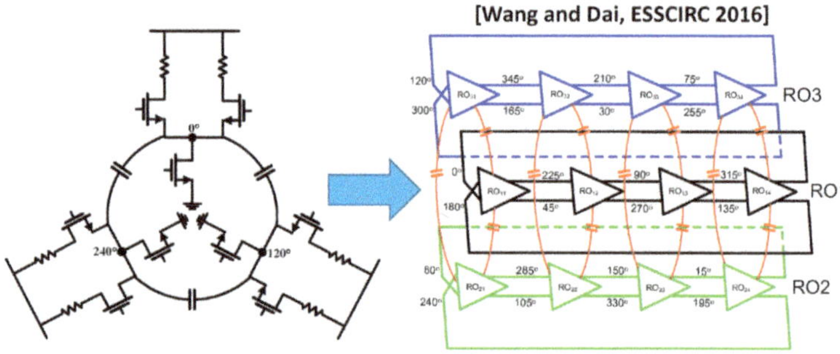

Output frequency of proposed coupled ring oscillator:

The output frequency of the proposed TRO is determined by the total amount of delay experienced in a SRO, while the coupling scheme ensures their relative phase relationship. The increase of output phases does not come with penalty of reduced oscillation frequency, as the conventional SRO would suffer.

The proposed multi-ring oscillator (MRO) is further illustrated with block and circuit diagrams. Three identical rings are implemented with 4 delay cells per ring. The ring oscillator is implemented with differential delay cells with resistive load without inductors. For comparison, the structure can be reconfigured as a SRO, a DRO and a TRO when one ring, two rings and three rings are powered on, respectively. The proposed TRO with 3 rings and 4 delay cells per ring provides 24 non-overlapped multi-phase outputs. In addition, the increase of output phases does not come with penalty of reduced oscillation frequency. Thus, high oscillation frequency and large numbers of output phases can be achieved simultaneously.

27

Proposed TRO Output Phases

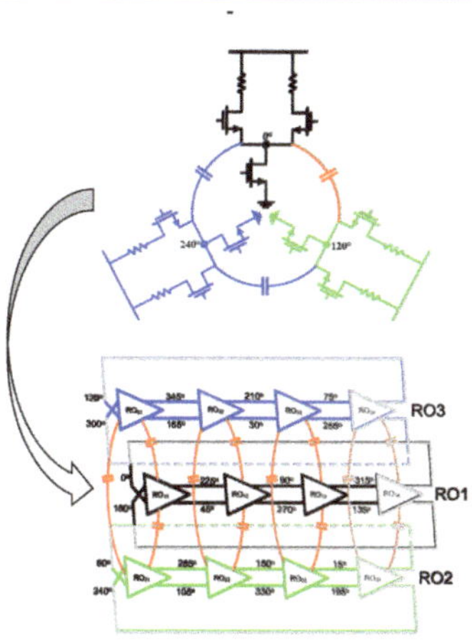

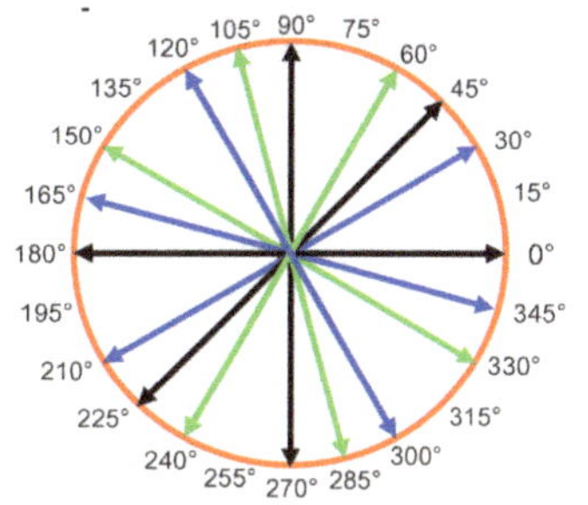

The proposed TRO generates 24 output phases. For M rings with N delay cells, the available output phases are $\varphi_{m,n} = \pi \cdot \left(\frac{m}{M} + \frac{2n}{N}\right), m = 0,1,\ldots,M-1, n = 0,1,\ldots,N-1$.

Some M and N combinations may lead to overlapped phases.

The output frequency of the proposed MRO is determined by the total amount of delay experienced in a SRO, while the coupling scheme between the rings ensures their relative phase relationship. Assuming the MRO contains M rings with N delay cells per ring, the available output phases are $\varphi_{m,n} = \pi \cdot \left(\frac{m}{M} + \frac{2n}{N}\right), m = 0,1,\ldots,M-1, n = 0,1,\ldots,N-1$. Note that output phases may be overlapped for some combinations of M and N, e.g., the DRO case with 4 delay cells per ring. However, the proposed TRO with 3 rings and 4 delay cells per ring provides 24 non-overlapped multi-phase outputs. In addition, the increase of output phases does not come with penalty of reduced oscillation frequency, as the conventional SRO would suffer. Thus, high oscillation frequency and large numbers of output phases can be achieved simultaneously. It should be pointed out that the proposed MRO is not equivalent to duplicating rings and coupling them in-phase, which will not produce any additional output phases, nor will it the subtle noise reduction brought by the proposed multi-ring coupled structure.

28

Prototype of The MRO RFIC

- **Process: 130 nm CMOS. Area: 350×350 μm², inductor-less.**
- **Power 1.3 mW to 4.6 mW per ring under 1.2 V power supply.**
- **Tuning Frequency: 1-1.5 GHz with 1.2V supply, up to 1.8GHz with 1.5V supply.**

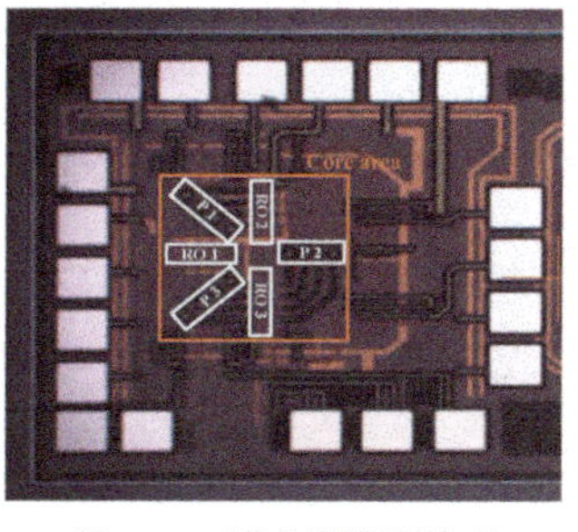

Wang and Dai, ESSCIRC 16

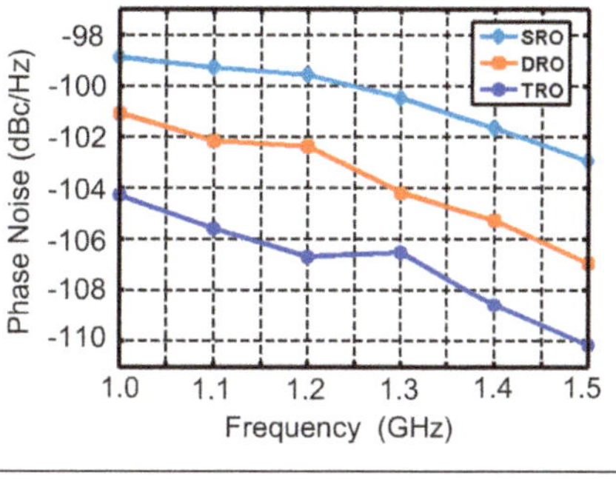

The proposed MRO RFIC was implemented in a 0.13 μm CMOS technology. The ring oscillator circuits can be reconfigured as a SRO, a DRO or a TRO for comparison. The output frequency of the MRO can be tuned from 1 to 1.5 GHz with the power consumption per ring ranging from 1.3 mW to 4.6 mW under a 1.2 V supply, respectively. The measured phase noise @1MHz offset for the MROs across the frequency tuning range of 1~1.5GHz is given above, in which phase noise improvement using the proposed MRO technique has been clearly demonstrated.

29 Phase Noise at 1.5GHz Output

- **SRO -102.98@1MHz**
- **DRO -106.97@1MHz 4dB ↓**
- **TRO -110.37@1MHz 7dB ↓**

FoM 160 dBc/Hz
161 dBc/Hz
162 dBc/Hz

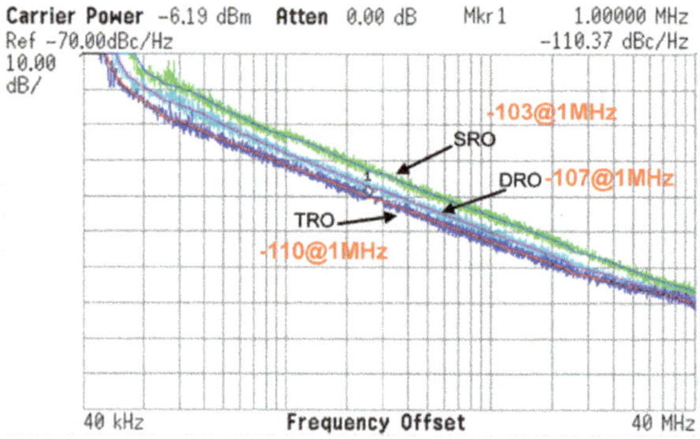

The TRO triples the number of output phases compared to its SRO counterpart. The measured phase noises at 1.5 GHz output @ 1 MHz offset were -102.98, -106.97 and -110.17 dBc/Hz for the SRO, DRO and TRO, respectively. It demonstrates about 7 dB phase noise reduction when comparing the TRO with the SRO and 4 dB phase noise reduction when comparing the DRO with the SRO.

30 Phase Noise at 1.8GHz Output

- **Supply voltage 1.5V**
- **Output frequency 1.8GHz**
- **Power consumption 48mW**

- **SRO -108.5 dBc/Hz@1MHz**
- **DRO -113.2 dBc/Hz@1MHz**
- **TRO -118.5 dBc/Hz@1MHz**

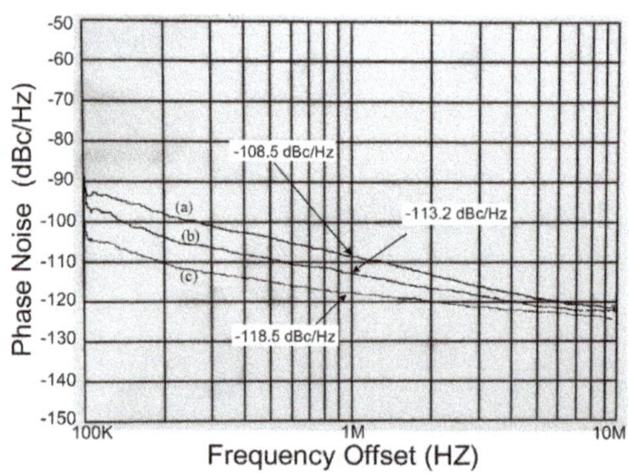

Moreover, the tradeoff between power consumption and phase noise in the MROs has been investigated with higher power. Under an increased 1.5 V power supply, it shows the measured phase noise performance of the TRO reaches -118 dBc/Hz at 1.8 GHz output, approaching the phase noise performance of LC-based oscillators. We expect that reduced power consumption and improved figure-of-merit (FoM) can be achieved in advanced technology nodes with smaller feature sizes.

31 Performance Summary and Comparison

Architecture	This Work		JSSC12	VLSI15	JSSC13
	DRO	TRO	Multi-loop	N-Path	Cyclic
Technology	130 nm		90 nm	90 nm	90 nm
f_o range (GHz)	1.0~1.5		0.63~8.1	0.2~1.8	1~12.8
PN @ 1MHz (dBc/Hz)	-107.0@1.5GHz	-110.2@1.5GHz	-106, 0.63GHz	-110, 1GHz	-105, 7.7GHz
Power (mW)	2.6~9.2	3.9~13.8	7~26	4.7~?	13~200
Output phases	8	24	4	6	3
FoM (dBc/Hz)	161	162	135	163	160
FoM per phase (dBc/Hz)	170	176	141	171	165

$$\text{FoM} = 10\log\left[\left(\frac{f_o}{\Delta f}\right)^2 \frac{1\,\text{mW}}{P}\right] - L(\Delta f), \text{FoM}_\text{P} = \text{FoM} + 10\log(\text{No. of Phases})$$

This work demonstrates an inductor-less ring oscillator with good phase noise performance when compared to state-of-the-art ring oscillator designs summarized in the table. The FoM of the TRO reaches 162 dBc/Hz. Considering multi-phase outputs, the FoM per phase of the MRO reaches 176.8 dBc/Hz. Therefore, the proposed inductor-less MRO provides an effective means for high frequency, low phase noise and low cost multi-phase clock generations.

32 Outlines

- PLL Fundamentals and Design Challenges
- Low Noise Multiphase Oscillators
- Multi-ring Coupled Oscillators
- **Injection locking PLL with Multi-ring Oscillators**
- Sub-sampling PLL with Robust Relocking
- Inductor-less Sub-sampling PLL with Ring Oscillators
- Conclusions

33

Motivation

- **Improving ring PLL's jitter performance by injection locking or sub-sampling.**

Injection locking PLL:
Pro: Wide filter bandwidth for VCO noise rejection and low in-band phase noise. √

Con: High reference spur ×

Subsampling ring PLL:
Pro: Low in-band phase noise √
Con: Stability issue ×

We will address the design challenges in both IL-PLL and SS-PLL.

Let's further investigate the in-band noise reduction techniques for inductor-less PLLs. With lower in-band noise, we can widen the loop bandwidth, which in turn reduces the contribution of oscillator noise. Recently, techniques including injection locking (IL) and sub-sampling phase detector (SSPD) has achieved impressive in-band noise floor with ring PLLs. Injection-locked ring oscillator can greatly reduce the in-band phase noise by resetting jitter accumulation over one reference clock period, however, the high level of reference spur and stability problems hinder its application in the wireless systems. Sub-sampling PLL can achieve excellent in-band noise floor. However, there may be stability issues when loops are switching between the frequency lock loop (FLL) and the sub-sampling loop (SSL).

34

Injection-Locking PLL

- **Injection-locking Oscillator and PLL**

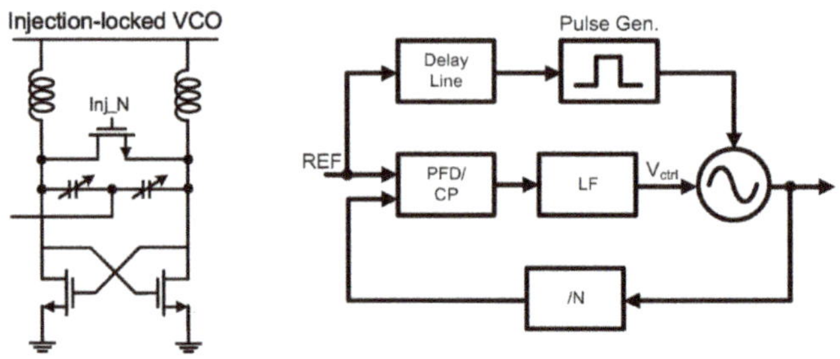

(a) Injection-locked VCO **(b) Injection locking PLL (ILPLL)**

Pulses generated from reference source is directly injected into oscillator to lower the in-band phase noise.

Injection pulses can be applied to the oscillator's outputs using a transistor as the switch. Normally, pulses at reference frequency are used to realign the oscillator edge to the reference edge every reference cycle.

35 Sources of Reference Spur in IL-PLL (I)

- **Frequency error causes spur.**

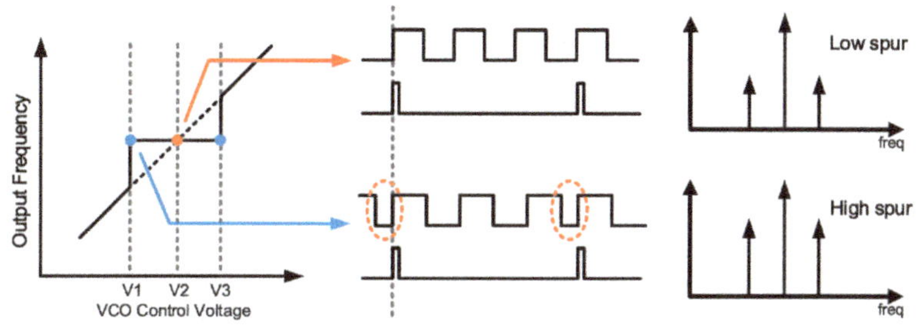

If V_{tune} doesn't correspond to an output frequency that is integer multiple of the reference frequency, frequency errors can occur. Accumulated phase error can be corrected by injection pulses, yet the reference spur will be high.

However, periodically injecting reference pulses into the oscillator will cause high reference spurs. The reference spur in injection locking mainly comes from two sources: (1) timing mismatches between injection signal and oscillation waveform and (2) distortion of the output waveform induced by the injection pulses.

36 Sources of Reference Spur in IL-PLL (II)

- **If frequency is locked to integer multiple of the reference frequency, phase mismatch and finite injection pulse width can still cause reference spurs.**

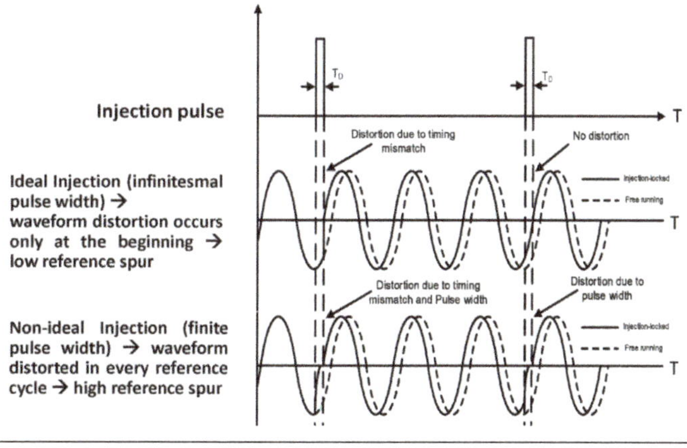

Due to non-ideal injection, distortion of the output waveform exists even if injection happens at waveform's zero crossing. Although injection instance can be calibrated towards the zero-crossing point, injection pulse width T_D sets the lower-bound of spur level in the conventional IL-PLL designs. The distorted waveform in one stage propagates through other stages, leading to reference spurs at every multi-phase output. Some designs use "soft injection" with narrower injection pulses to reduce the reference spur. However, it comes with the penalty of phase noise degradation and requires increased reference frequency or decreased loop division ratio. The dilemma between phase noise and reference spur reductions can be hardly resolved.

37 Use of Auxiliary VCO in IL-PLL

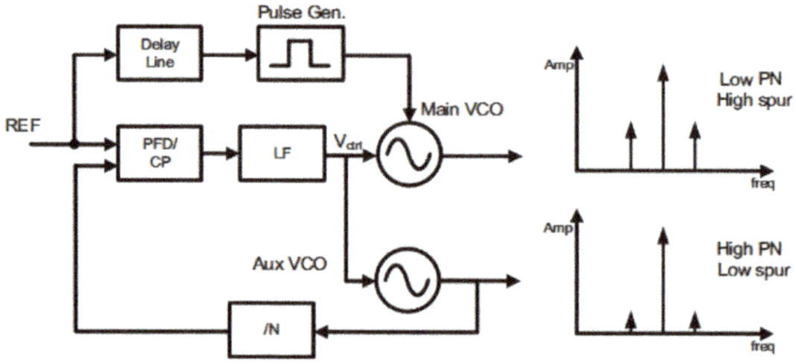

Choi, ISSCC2016

Auxiliary VCO is not affected by the injection pulses, trading PN for spur performance.

This design suggests pulling the loop feedback for frequency control from an auxiliary ring oscillator that is not injection locked and thus has low reference spur. Another main ring oscillator is used for injection-locking with improved phase noise, yet with high reference spur.

38 What if we coupled two Oscillators in an IL-PLL

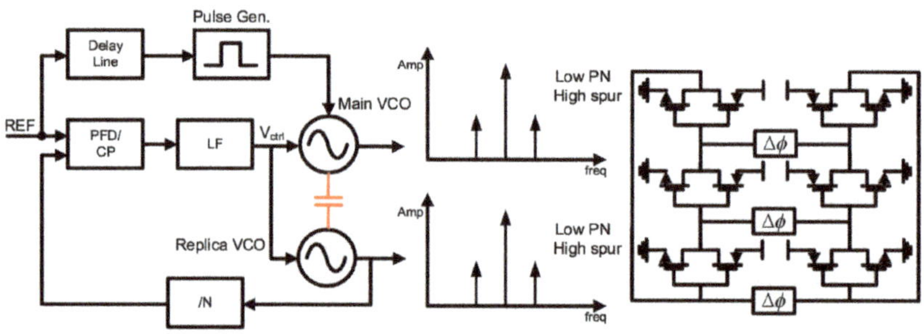

Coupling: The aux VCO tracks the phase changes in the main VCO, leading to low PN. However, the reference spurs remain high if coupling takes place at output nodes of the ring oscillator. The distorted waveform directly coupled into output of the auxiliary VCO.

What if we coupled the two ring oscillators to form an IL-PLL? In the proposed architecture, the auxiliary oscillator tracks the phase changes in the main ring due to coupling, leading to low phase noise. However, the reference spurs in both rings remain high if the coupling takes place at the output nodes of the oscillators since the distorted waveform directly coupled into the outputs of the auxiliary oscillator.

39 Source Coupling in IL-PLL

- **Proposed Multi-ring Oscillators with Source Coupling:**

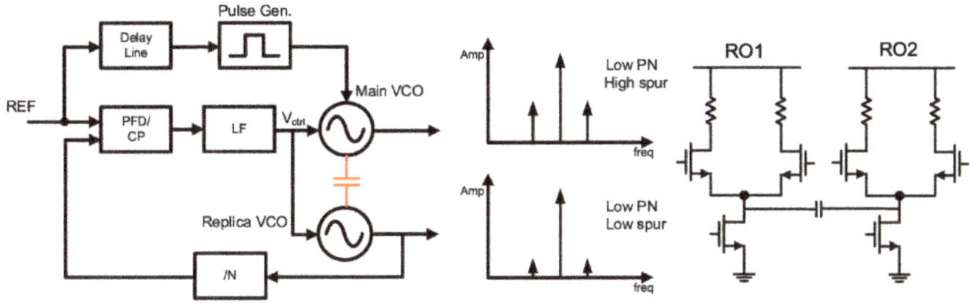

Source coupling at common–mode node leads to less spurs in the auxiliary ring, while achieving low PN in both oscillators.

What about we apply the coupling at the common-mode nodes using the capacitive coupling technique discussed in previous section? Indeed, source coupling at common-mode node leads to less spurs in the auxiliary ring, while achieving low PN in both oscillators. The dilemma between "hard injection" and reference spur can be resolved without degrading phase noise by employing the proposed MRO based IL-PLL.

40 Simulated Waveform of Source Coupled MRO

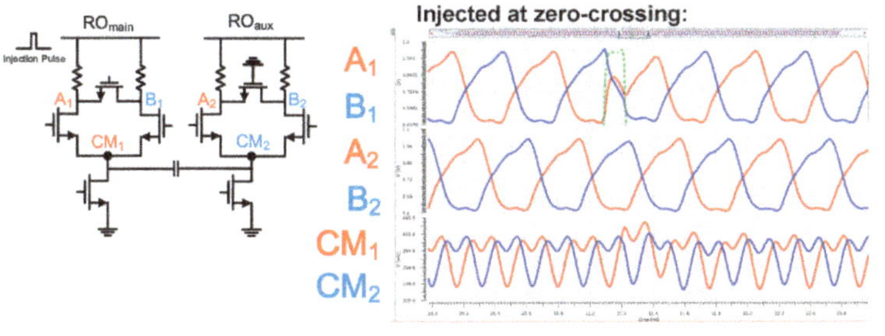

Ideally, CM_1 experiences only phase modulation, no amplitude modulation, in presence of the differential injection. Slightly distorted waveform at CM_1 is coupled to CM_2, which causes slight current fluctuation in the auxiliary ring. The coupled rings maintain their phase relationship, which forces the phase in auxiliary rings to follow that of the main ring, leading to the same PN performance in both rings. However, this process leads to ignorable waveform distortion in auxiliary ring, resulting in much lower reference spurs.

The injection pulse resets the signal in ring oscillator RO_{main} and distorts the output waveform. After injection, the output signals will retain their stable states in a short period usually less than one cycle. The recovering time is determined by the coupling strength. During the recovering period, perturbed current is coupled into auxiliary ring's common-mode point, which forces the phase change in auxiliary ring's outputs. Their phase coherency is guaranteed by the symmetry of the circuits. However, the output waveform in RO_{aux} is much smoother compared with the waveform in RO_{main}, since there is no direct reference signal injected into the RO_{aux}'s output nodes.

41 Proposed Injection-locked PLL with MRO

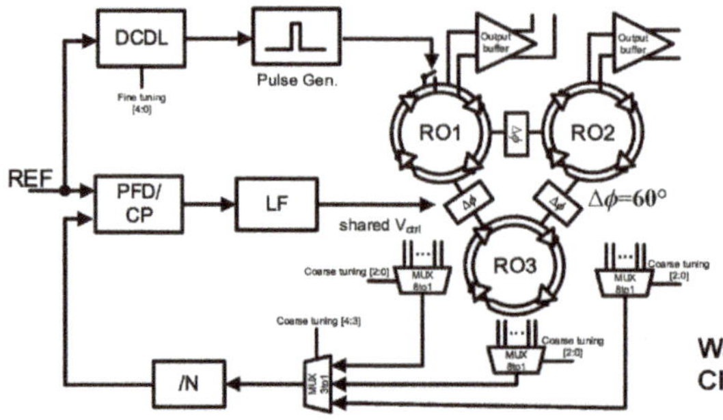

Wang and Dai, CICC 2017

1. Reference pulses is injected in the main ring, forcing all coupled rings to have the same phase noise as the main ring.
2. The reference spurs in the aux rings are much smaller due to less amplitude perturbation.
3. Multiphase outputs are used as a coarse tune and a digitally controlled delay line (DCDL) is employed for fine tune to adjust the injection timing and to achieve factional-N synthesis.

The proposed IL-PLL architecture is composed of TRO, frequency divider, phase frequency detector(PFD), charge pump, pulse generator, phase selector and digital controlled delay line(DCDL). The injection signal from pulse generator is injected at the first stage of the ring oscillator RO1 in a TRO configuration. The outputs in both RO1 and RO2 are buffered for comparison. The prototype of 800MHz-1.3GHz IL-PLL was fabricated using a 130 nm CMOS technology.

42 Injection Timing Adjustment in IL-PLL

Coarse tuning: two stage mux provides wide range of one VCO period with resolution of 2*pi/24

Fine tuning: 5 bits inverter based DCDL in reference path provides 20 ps tuning range

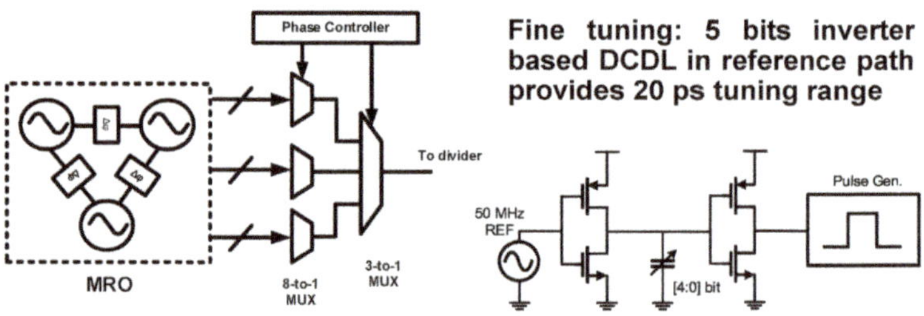

Delay in frequency loop and injection path should be balanced. More delays added in reference path lead to degraded in-band phase noise.

One of the 24 output phases from TRO is selected into feedback loop as coarse calibration of the delay difference between injection and frequency loop by using two stage multiplexers in the phase selector. Multi-phase outputs in TRO reduces the required timing adjustment range in digital controlled delay line (DCDL) that only needs to cover 1/24 period of the output cycle, leading to improved phase noise in the injection signal, since the more delay the injection path has, the worse phase noise we will get. A 5 bit DCDL with a tuning range of 40 ps is used as fine tune in the calibration of loop delay difference, which can cover PVT variations and phase mismatches in the oscillator from the range of 800 MHz to 1.3 GHz.

43 Measured IL-PLL Phase Noise

- **Measured in-band phase noise in main and aux rings demonstrates the same phase noise performance in all rings.**

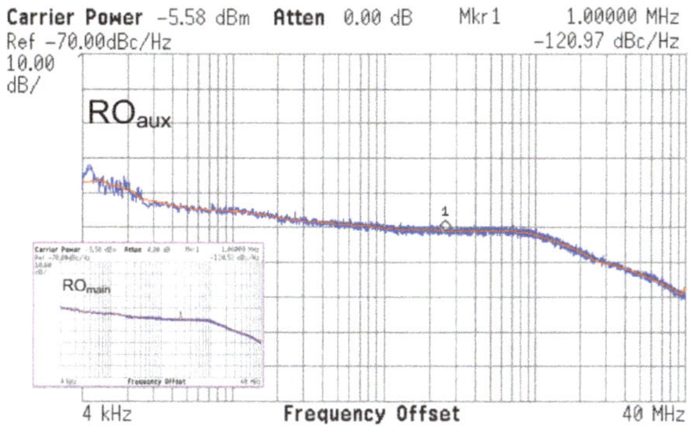

After injection-locking, the phase noise in RO_{aux} reaches -120.97 dBc/Hz @1MHz offset, while the phase noise in RO_{main} is measured as 120.52 dBc/Hz, demonst rating almost identical phase noise performance in both main and auxiliary rings. In another word, the phase noise in RO_{aux} follows the phase noise in RO_{main} even if there is no injection in RO_{aux}.

44 Measured Reference Spur Performance

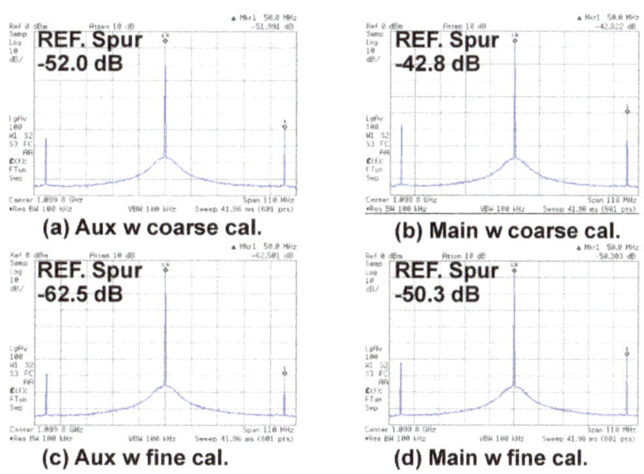

(a) Aux w coarse cal. (b) Main w coarse cal. (c) Aux w fine cal. (d) Main w fine cal.

Without coarse phase calibration, the reference spur level could be as high as -30 dBc and -38 dBc in main and auxiliary rings, respectively. The measured output spectra of RO_{aux} and RO_{main} with only coarse phase calibration are shown in (a) and (b). By selecting the proper output phase for feedback in coarse calibration, their reference spur levels reduce to -52 dBc and -43 dBc, respectively. When both coarse and fine phase calibrations are turned on, the measured reference spur levels drop to -62.5 dBc for RO_{aux} and -50.3 dBc for RO_{main}, respectively. (c) and (d) demonstrates about 10 dB spur rejection by fine phase calibration and about 20 dB spur suppression comparing to non-calibrated IL-PLL case.

45 Measured IL-PLL Reference Spur and PN

- **Tuning range: 800MHz-1.3 GHz.**
- **The reference spur in aux ring is below -60 dBc covering the tuning range, demonstrating >10 dB spur reduction.**
- **In-band phase noise in aux ring -121@1M offset, 1.2GHz output; jitter 513 fs from 10kHz to 50MHz**

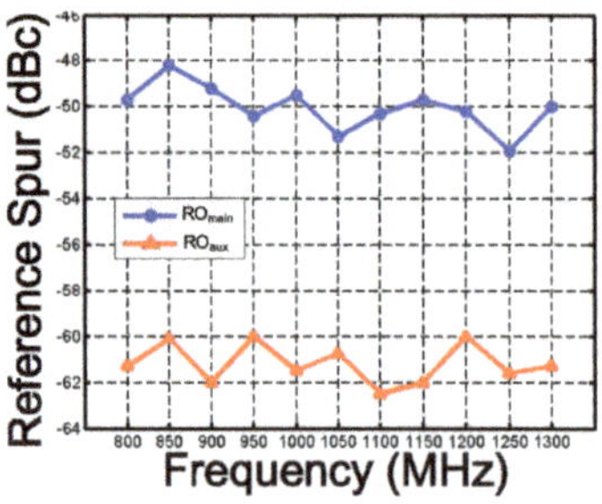

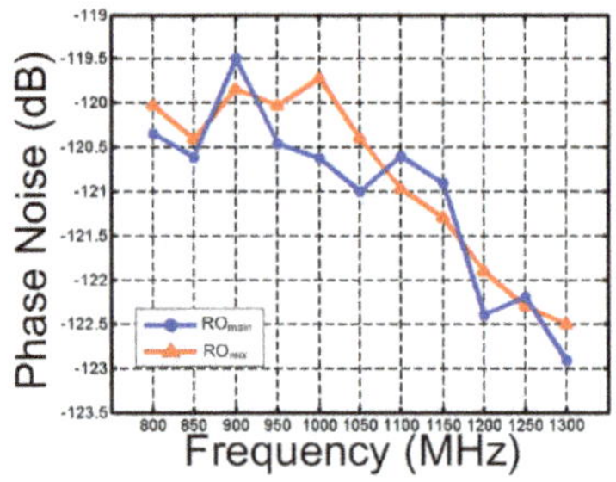

For the entire tuning range from 800 MHz to 1.3 GHz, the reference spur is measured from -48 ≃ -52 dBc in RO_{main} and -60 ≃ -62.5 dBc in RO_{aux}, respectively, demonstrating an average of 12 dB reference spur suppression comparing to the main ring outputs. The measured phase noise @ 1MHz offset ranges from -119.5 dBc/Hz to -123dBc/Hz over the entire IL-PLL tuning range, showing similar phase noise performances for both main and auxiliary rings.

46 Performance Comparison Table

	This work		[6] CICC15	[5] ISSCC16	[7] ISSCC15
Method	Analog		Analog	Analog	Analog
VCO Type	Ring		Ring	Ring	Ring
Frequency (GHz)	0.8~1.3		2	0.96~1.44	0.8-1.7
Reference (MHz)	50		125	120	380
Division ratio	16-26		16	8-12	4
Measured Output	RO_{aux}	RO_{main}	RO_{main}	RO_{main}	RO_{main}
PN (dBc/Hz) @1MHz	-121 @1.1	-121 @1.1	-113 @1.2	-134.4 @1.2	-116 @1.522
Out.Jitter (σ.) *	513fs	508fs	971fs	185fs	3.6ps
Reference Spur(dBc)	-60~ -62.5	-48~ -52	-43	-47~ -53	-63
Power (mW)	13.5		9.5	3.74	3
Area(mm2)	0.27		0.1	0.06	0.1
Process (nm)	130		65	65	65
FOM	-234.5		-234.5	-244.9	-224.5

Here is the performance comparison table, The TRO has the best phase noise at the targeted frequency of 1.5GHz, reaching -110 dBc/Hz. Also it has the most number of output phases as 24 in total. When considering multi-phase outputs, the extra power will be consumed. So we introduce a new FoM formula as FOM per phase to count this extra power. It equals conventional FoM+10log(No. of Phases). FOM per phase of the proposed MRO reaches 176 dBc/Hz outperforming all other designs.

47 Outlines

- PLL Fundamentals and Design Challenges
- Low Noise Multiphase Oscillators
- Multi-ring Coupled Oscillators
- Injection locking PLL with Multi-ring Oscillators
- **Sub-sampling PLL with Robust Relocking**
- Inductor-less Sub-sampling PLL with Ring Oscillators
- Conclusions

48 SSPLL vs IL-PLL

■ IL-PLL can achieve larger filtering bandwidth for VCO noise and can be beneficial for integrated PN or jitter, e.g., when ring oscillator is used.

■SSPLL is more like a traditional PLL with a phase detector and a feedback loop

- More robust bandwidth/loop-dynamic control

- Lower reference spur

Injection-locked ring oscillator based PLL can greatly reduce the in-band phase noise by resetting the jitter accumulated over a reference cycle. Since it is feed forward system, a wider loop bandwidth up to $f_{ref}/2$ is achievable. Sub-sampling PLL (SSPLL) is another technique that can achieve very low in-band noise floor. SSPLL lowers its in-band noise by using a sub-sampling phase detector (SSPD), which has high gain, yet a narrow detectable range of one VCO period. Loop dynamics and stability need to be carefully investigated for the use of SSPLL.

49

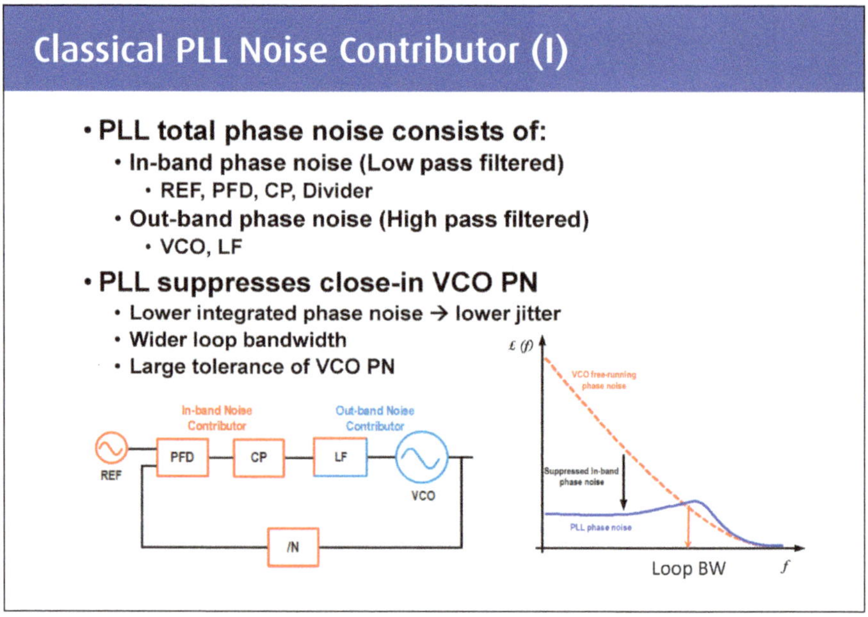

With reduced in-band noise, the SSPLL bandwidth can be widened. The penalty from inferior phase noise of inductor-less ring oscillators can be largely eliminated given a wide PLL bandwidth, while maintaining its benefit of small form factor and scalability with technology. The optimal loop bandwidth with lowest integrated jitter is located at the crossing point where the in-band phase noise floor intersects the oscillator free running phase noise curve. In order to reduce the phase noise contribution from oscillators, it's important to lower the PLL's in-band phase noise such that a larger loop bandwidth can be employed.

50

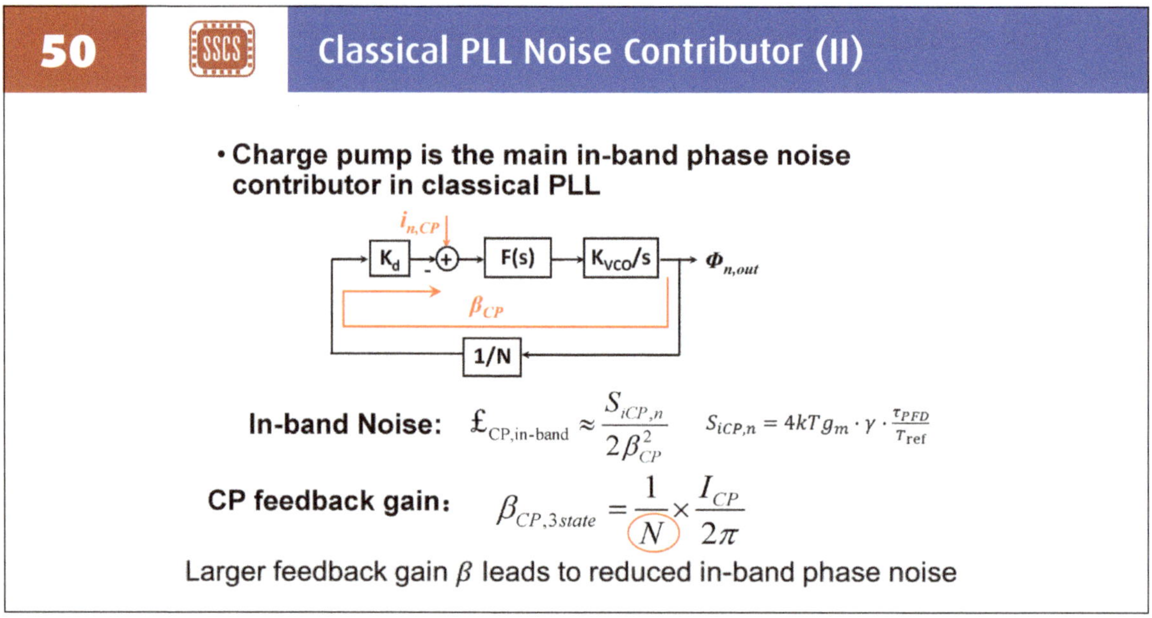

A possible approach to reduce the in-band noise floor is to replace the tri-state phase frequency detector (PFD) with a high gain SSPD. As a result, charge-pump noise, a dominant in-band noise contributor in traditional charge-pump PLL, can be reduced by a factor of N, where N is the loop division ratio.

51 Sub-Sampling Phase Detector (I)

- **Sub-sampling phase detector directly samples VCO waveform**

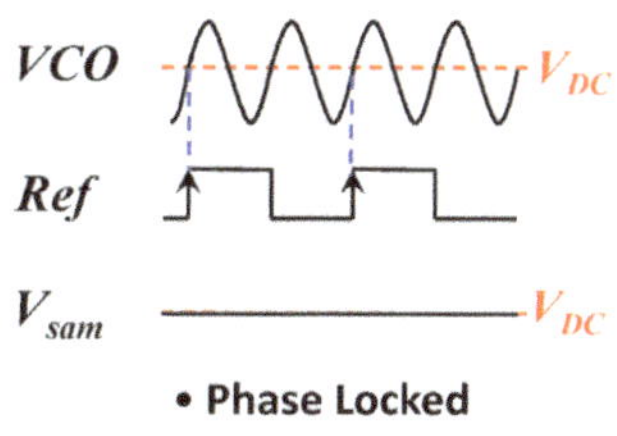

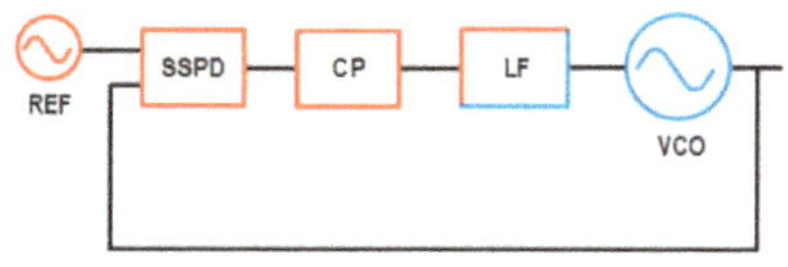

Gao, ISSCC 2009

- **Converting time-domain error to voltage-domain error**
- **No divider required, removing N from the equation!!**

The prior-art sub-sampling PLL (SSPLL) uses the reference edge to sample the VCO waveform by means of a sub-sampling phase detector (SSPD). SSPD converts time-domain error to voltage-domain error and the SS loop (SSL) doesn't need the divider at lock, equivalently removing N from the loop transfer function!

52 Sub-Sampling Phase Detector (II)

- **SSPD can achieve higher gain with sharp VCO slope:**

SS Gain: $\beta_{SS} \approx \frac{SR}{2\pi f_{VCO}} \cdot g_m \approx A_{VCO} \cdot g_m$

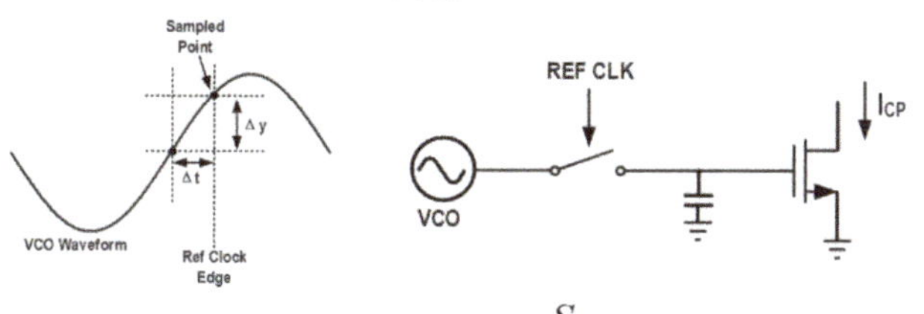

In-band PN due to CP: $\pounds_{SS,\text{in-band}} \approx \frac{S_{iCP,n}}{2\beta_{SS}^2}$ $\Delta\varphi(t,\tau) = \frac{ISF(\omega_0\tau)}{q} u(t-\tau)$

- **Larger β_{SS} leads to lower in-band PN**

Since the VCO slope is N-times larger than the reference slope, the SSPD has N-times larger gain than the traditional phase detector that compares the phase error at reference frequency. As a result, any in-band noise after the SSPD such as the charge pump noise is reduced by a factor of N.

53 Sub-Sampling Phase Detector

- **Narrow phase locking range (one VCO period)**
- **No frequency locking, might lock to harmonics**

SSPD Gain: $K_{SSPD} = \frac{I_{SSPD}}{\Delta\varphi} = A_{VCO} \cdot g_m \cdot \frac{\sin(\Delta\varphi)}{\Delta\varphi}$

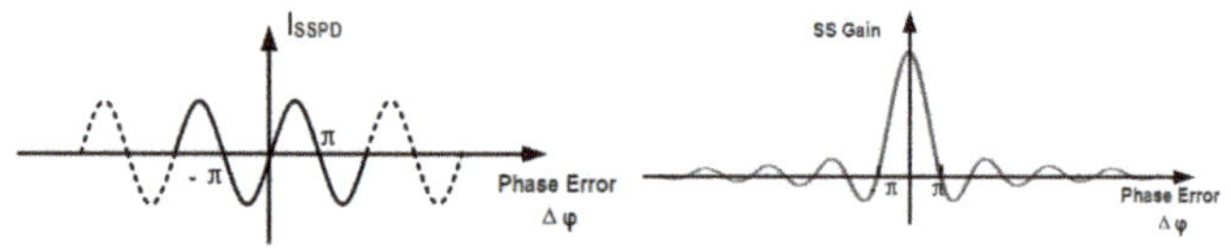

- **Gain follows sinc function within the detectable range of [-π,+ π]**
- **Beyond detectable range of one VCO cycle, may lock to harmonics**
- **Large gain variation within detectable range of [-π,+ π]**

A SSPD samples the VCO waveform without frequency downscaling and maintains its high gain only within a small region around zero crossing of the VCO waveform. If a perturbation causes a relatively large phase error after the loop is locked, the in-band noise may be degraded due to reduced SSPD gain. Furthermore, the sampled voltage of a SSPD operating on a sinusoidal VCO signal only works well within $\pm\frac{\pi}{2}$ phase shift around the zero-crossing. Beyond that, the SSPLL may lock to another VCO zero crossing without regaining lock on its own.

54 Prior Art SSPLL Design (I)

- **Frequency loop uses a PFD with large dead-zone**
  - Low phase noise after lock-in
  - Slow relocking with interference. Need to wait for phase error to be accumulated in order to reactivate the frequency loop.

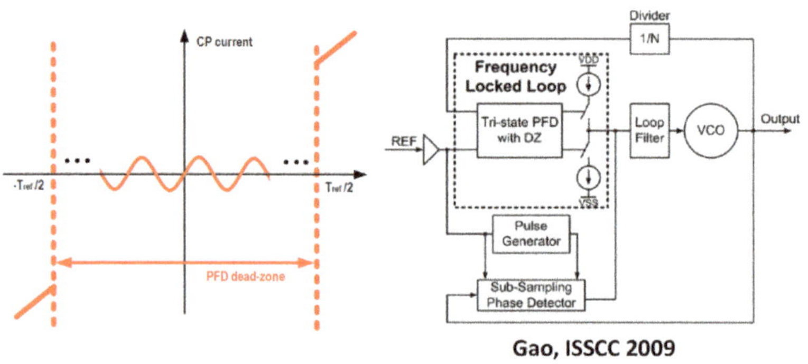

Gao, ISSCC 2009

As a SSPD only detects phase errors, other means are needed for frequency error detection. Furthermore, switching between the two detector outputs is needed to define which detector controls the VCO. To this end, the prior-art SSPLL uses a tri-state PFD with intentionally enlarged dead-zone to switch between the frequency lock loop (FLL) and the sub-sampling loop (SSL).

55 Prior Art SSPLL Design (II)

- **Auxiliary frequency loop with large dead-zone PFD**

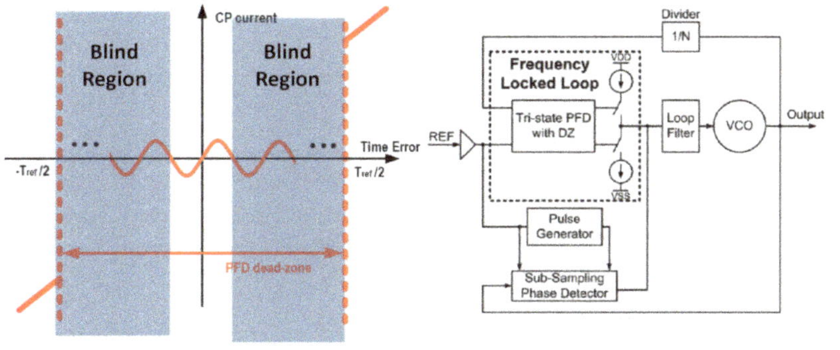

Due to the narrow capture range of the SSL, the SSL may lose lock in the presence of perturbations. Moreover, prolonged relocking time is potentially required as the phase errors need to be accumulated for a quite long time before the dead-zone is passed that triggers the FLL to be switched on.

56 Prior Art SSPLL Design (III)

- **Simultaneously turn on frequency loop and SS loop**
  - **PFD always active for fast relocking**
  - **Large loop gain variation with phase error**

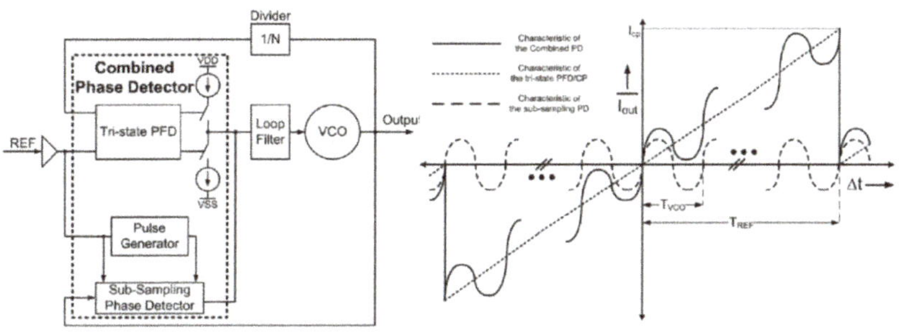

Hsu and Kinget, TCAS. 2015.

This problem was partially solved in the given paper by removing the dead-zone from the FLL. However, the revised FLL is constantly injecting its charge pump current as well as its noise into the loop filter. Depending on the amount of current injected from the FLL, the in-band phase noise of the PLL may be degraded.

57 SSPD and PFD (I)

- SSPD achieves low PN locking
- PFD achieves wide range locking

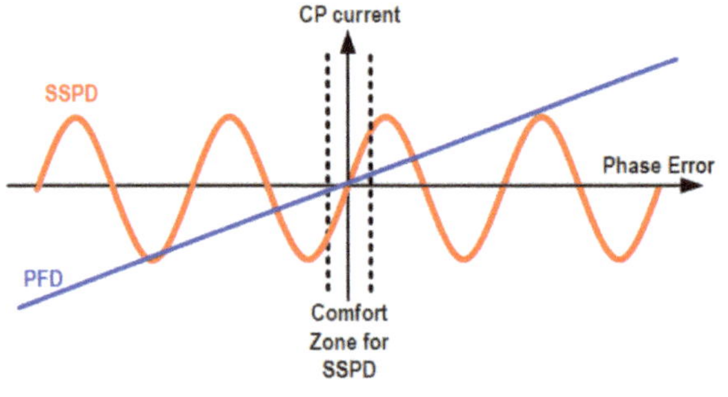

Let's have a close look at the SSPD and PFD. The SSPD achieves high gain within a narrow range of one VCO cycle, while the PFD has low gain across a wide locking range.

58 SSPD and PFD (II)

- Use SSPD for small phase error
  - Low noise locking
- Use PFD for large phase error
  - Large capture range, robust relocking

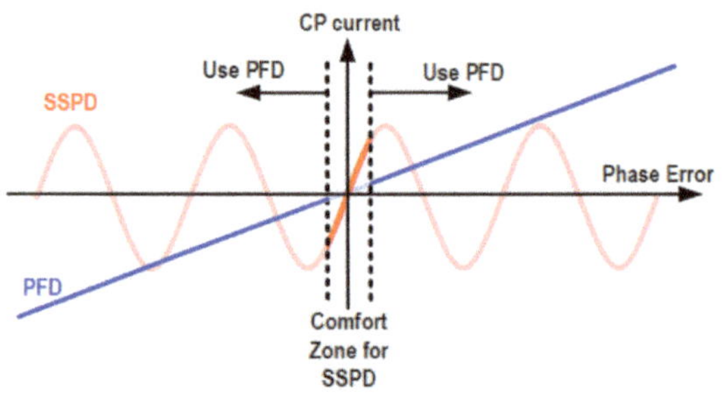

We switch back and forth between the SSL and the FLL by using the SSPD for small phase error detection and PFD for large phase error detection. If they have different gains, the overall transfer function of the PD is not continuous, nor is it monotonic or linear. The loop switching ends up with quite large variation of the open loop gain and thus quite large variations of phase and gain margins, causing stability concern.

59

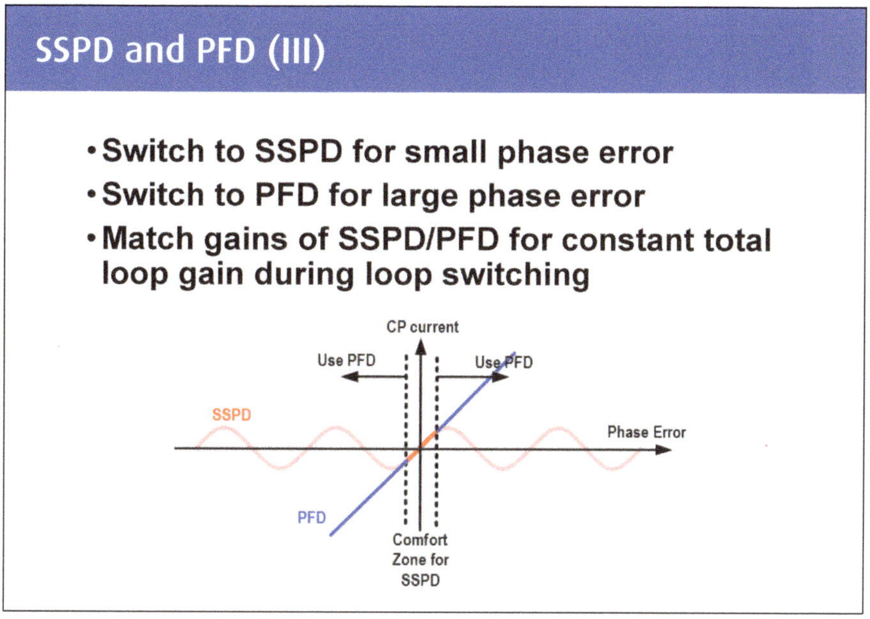

What if we increase the PFD gain such that it matches the SSPD gain when the loop switches from FLL to SSL? Constant gain can be achieved during the loop transition. As a result, the loop is more stable and more robust. However, increase of PFD gain is normally done by increasing the charge pump current, which means more charge pump noise is injected into the loop when the FLL is on. Don't worry. This can be taken care of by turning off the charge pump when the loop is switched to SSL.

60

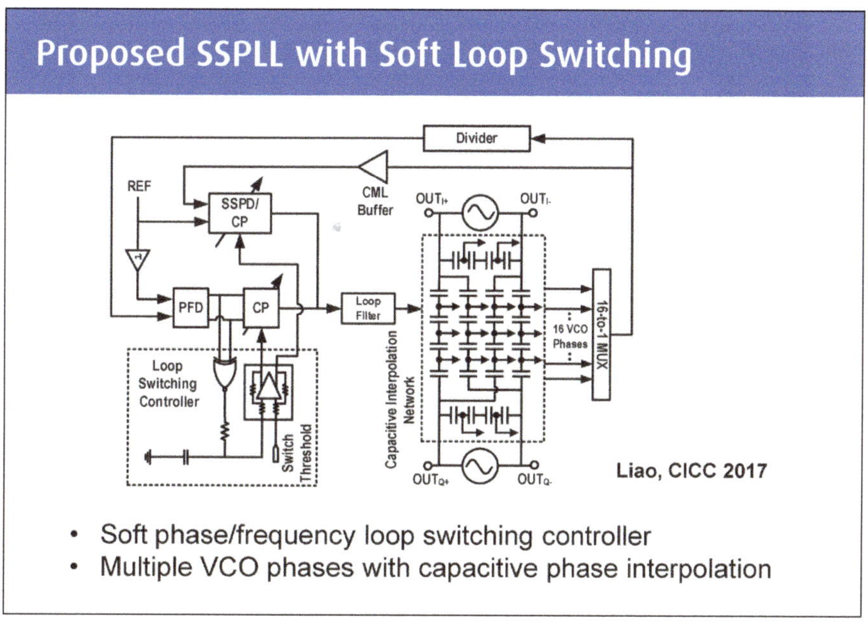

We thus propose an *automatic soft switching* scheme that eliminates FLL noise in lock, but still ensures agile and robust locking. When the phase error is approaching zero, the proposed scheme *gradually increases the SSL gain and decreases the FLL gain,* while maintaining *a constant total loop gain* during loop transition. As a result, the loop dynamics such as loop bandwidth and gain/phase margin will not vary much throughout the transition. When the loop is locked, the gain of the FLL is effectively turned off while the SSL is fully turned on, eliminating the FLL noise contribution to in-band phase noise of the loop.

61

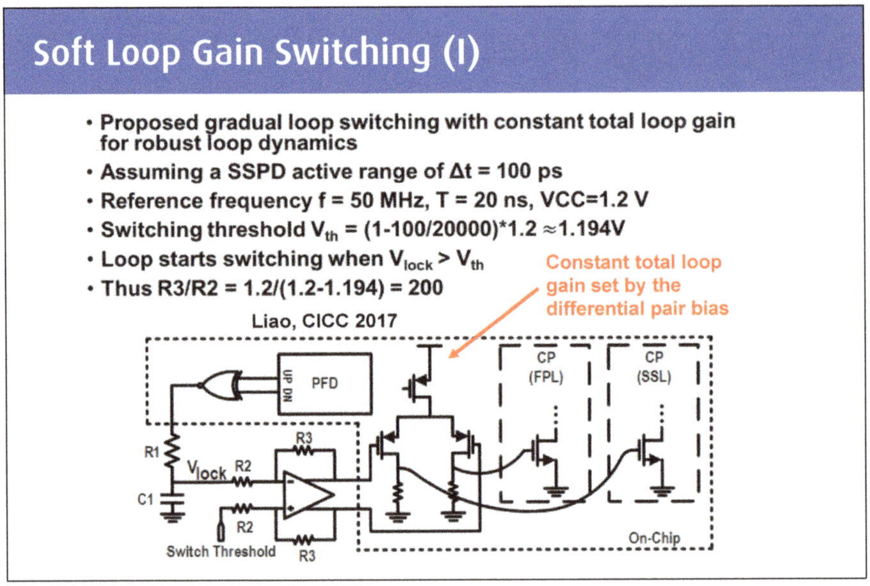

The proposed loop gain switching controller includes 3 parts. Firstly, an XNOR gate is tied to the PFD output in the FLL. Along with a low-pass RC filter, an averaged loop phase error can be measured. The output signal V_{lock} is inversely proportional to the phase error, meaning smaller leads to a higher V_{lock}. In the 2nd part, an amplifier, or a soft comparator, consisting of an operational amplifier is utilized to compare V_{lock} with a programmable switching threshold. This threshold shall be set sufficiently high in order to ensure that the switching from the FLL to the SSL occurs only after the phase error is within the locking range of the SSPD, i.e., one VCO period. Additionally, a high threshold also helps with fast switching from the SSPD to the PFD once the loop loses lock due to perturbation. The 3rd part consists of a PMOS differential pair which directly drives current sources in CPs of the FLL and the SSL. The differential pair ensures a constant sum of a scaled PFD current and the SSPD current for a constant overall loop gain.

62

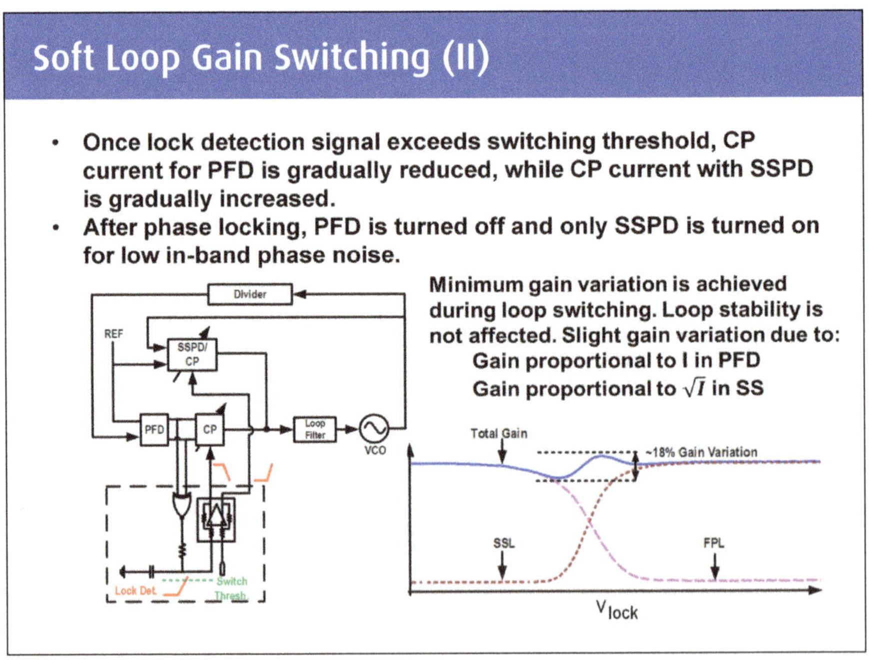

The simulated loop gain normalized by its maximum value versus phase error (V_{lock}) over different process corners and temperatures is shown. A tolerable worst peak-to-peak gain variation of 18% is observed. To compensate for different process corners, the bandwidths (i.e., open-loop gain) of the two loops are calibrated to be equal before normal operation. As a result, the loop only needs to tolerate the variations from temperature and voltage.

63 Loop Stability Comparison

- **Illustration of loop dynamic variation compared with other approaches:**

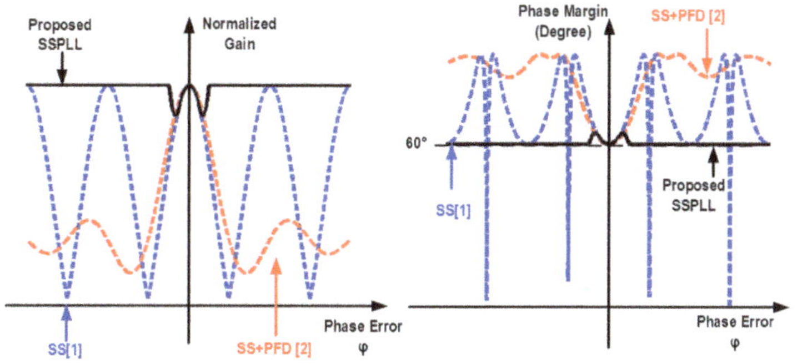

Proposed: D. Liao, JSSC, 2018
[1] X. Gao, JSSC, 2009. [2] C.-W. Hsu, TCAS I, 2015.

Figure above presents the total loop gain normalized by its maximum value versus static phase errors for some prior art designs and our proposed soft switching scheme. The SSPD assisted with dead-zone PFD [1] shows a periodic behavior because it can lock to any zero crossing of the VCO waveform. Since the phase error is still within the dead-zone, the FLL is not activated in this case. Each null corresponds to a large drop in phase margin, which causes potential stability issues. Since G_{SSL} follows a sinc function with respect to the phase error and G_{FLL} shows a constant gain, directly summing these two terms as suggested in [2] leads to a gain profile of a sinc function with a DC offset. Although this combined SSPFD approach [2] can avoid harmonic locking, the loop experiences large gain variation as the phase error changes from ±π to 0. Consequently, loop bandwidth and phase margin also vary dramatically, causing stability concern. The problem can be further exacerbated in this architecture since the gain of the FLL needs to be very small in order to reduce the extra noise from the FLL. With the minimum FLL gain, the loop barely maintains a positive total gain for arbitrary phase error, making the total loop gain close to 0 when the phase error approaches ±π.

64

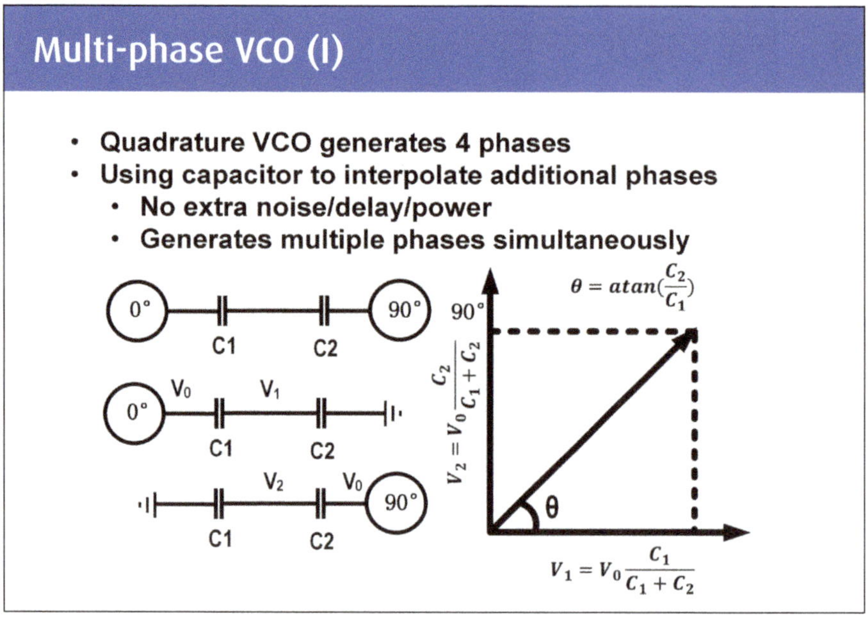

In integer-N mode, the VCO zero crossing will always be aligned with the reference edge after phase locking since the VCO frequency is an integer multiple of the reference frequency. Extending the SSPLL to fractional-N mode requires that the VCO or the feedback edge moves periodically around the reference edge, creating instantaneous phase errors even after phase locking. In our proposed fractional-N SSPLL, edge alignment is achieved through utilizing multiple interpolated VCO phases uniformly spanning from 0° to 360°. The generation of multiple clock phases is achieved through capacitive interpolation with a quadrature LC oscillator. Consider a simple case of two capacitors connected in series between the in-phase (*I*) and the quadrature (*Q*) component of a quadrature VCO (QVCO) output. By tuning the ratio of two capacitors, arbitrary phase between 0° and 90° can be interpolated.

65

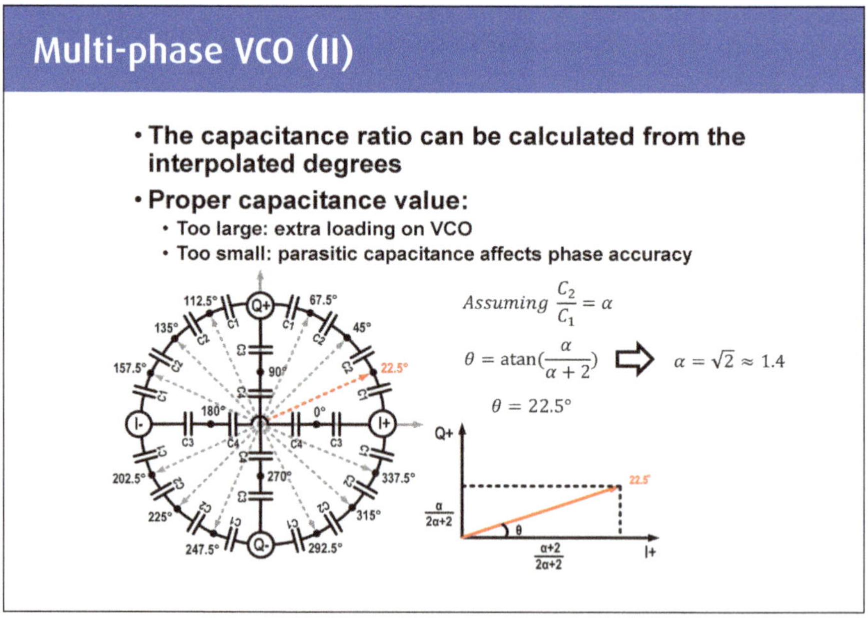

In this SSPLL, the QVCO output is further extended into interpolating 16 phases. Four capacitors are connected in series between 0° and 90° from the QVCO to generate 3 additional sub-phases of 22.5°, 45° and 67.5° respectively. Let us define a capacitor ratio $\alpha = \frac{C_2}{C_1}$. From *I*+ to *Q*+, the phase at the first node can be found to be $\tan\theta = \alpha/(\alpha+2)$. Using θ of 22.5°, α can be calculated to be $\sqrt{2}$ which is approximated with 1.4 in the actual implementation.

66

Measurement Results (I)

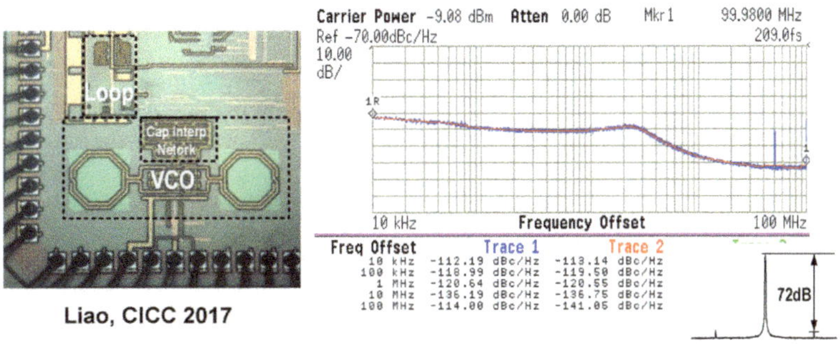

- An integrated jitter of 209 fs was achieved for integer mode at 2.4 GHz with a reference spur of -72 dBc
- -120 dBc/Hz in-band noise floor with soft loop switching

The proposed SSPLL is implemented in a 130 nm CMOS technology with the total active area of 0.43 mm². The system consumes 21 mW under a 1.3 V power supply. Most of the power is consumed by the QVCO which delivers a phase noise of -121 dBc/Hz and -140 dBc/Hz at 1 MHz and 10 MHz, respectively. In integer-N mode, a low in-band noise floor of -120 dBc/Hz has been measured as expected due to using SSPD. The loop bandwidth is set to around 1.5 MHz where the in-band noise floor intersects the VCO free-running phase noise for minimal PLL noise. An integrated jitter of 158 fs (10 kHz-10 MHz) has been measured at 2.4 GHz. With careful circuit and layout design, a very low reference spur of -72 dB has been measured.

67

Measurement Results (II)

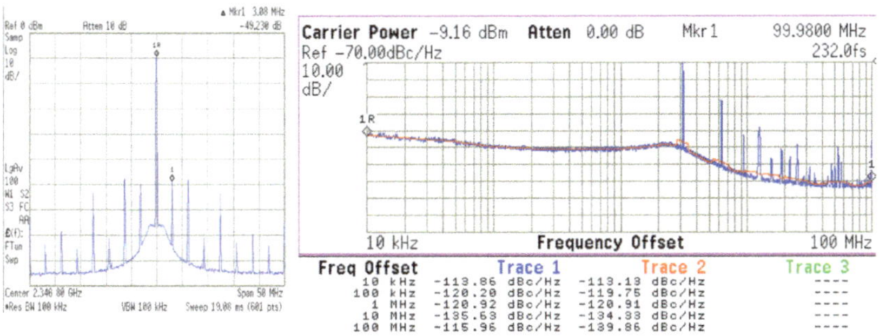

- An integrated jitter of 232 fs achieved for fractional mode at 2.397 GHz
- -120 dBc/Hz in-band noise floor, same as in integer mode
- Closest fractional spur of -49 dBc was measured at 2.397 GHz

In the fractional-N mode, the measured phase noise at 2.4GHz is shown, maintaining an in-band noise floor around -120 dBc/Hz with an integrated jitter of 169 fs (10kHz-10MHz) and fractional spur of -49 dBc.

68 Measured SSPLL Loop Dynamics

- A step voltage of 150 mV was injected onto VCO's supply voltage after lock-in as a disturbance.

By setting a low switching threshold, loop showed responses similar to [Gao, 2009] using large dead-zone where FPL needs to wait for phase error accumulation before being triggered.

By setting a high switching threshold, the loop could instantly switch to frequency loop to regain locking. Relocking time is decided by loop bandwidth and switching speed.

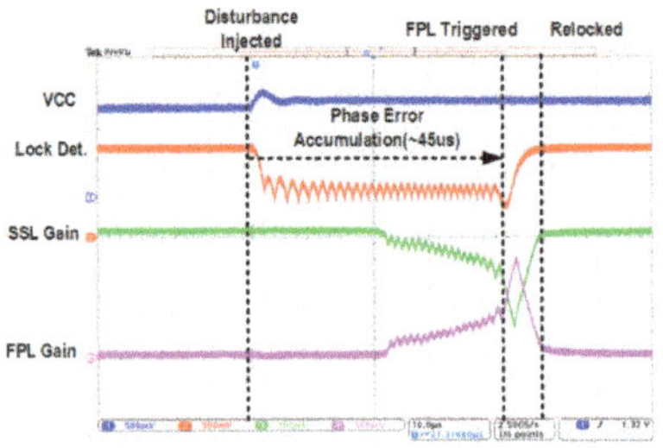

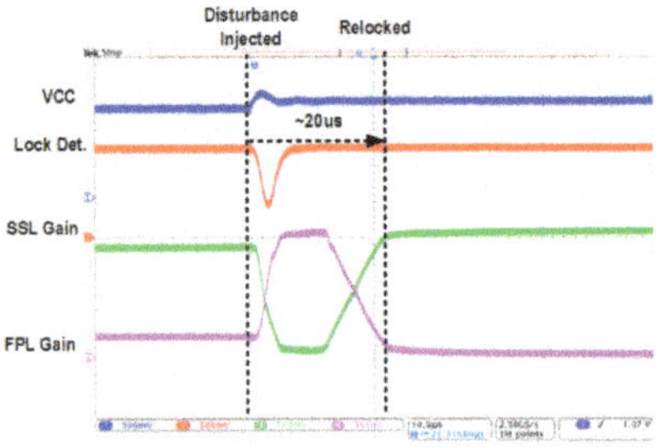

A step voltage of 150 mV was injected onto VCO's supply voltage after lock-in as a disturbance. Prior-art SSPLL using large dead-zone needs prolonged locking time for phase error accumulation before the loop switching can be triggered (left). Proposed soft loop switching scheme quickly regains lock (right). After the loop is relocked, it is switched back to the SSL. Since the FLL is completely disconnected from the loop filter after phase lock, the interference and noise from the FLL can be avoided.

69 Outlines

- PLL Fundamentals and Design Challenges
- Low Noise Multiphase Oscillators
- Multi-ring Coupled Oscillators
- Injection locking PLL with Multi-ring Oscillators
- Sub-sampling PLL with Robust Relocking
- **Inductor-less Sub-sampling PLL with Ring Oscillators**
- Conclusions

70 Proposed inductor-less SSPLL with MRO

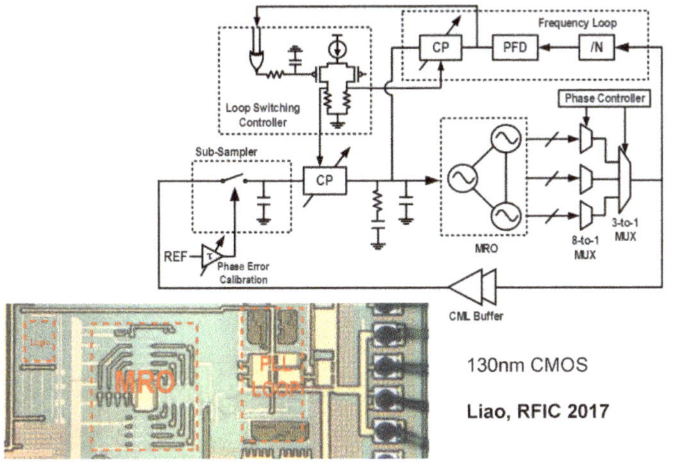

130nm CMOS

Liao, RFIC 2017

Finally, the proposed SSPLL with soft-loop switching can be employed to reduce the total integrated phase noise in an inductor-less PLL. This design illustrates a compact factional-N inductor-less sub-sampling phase-locked loop using multiple coupled ring oscillators. As a result of the improved in-band noise of the SSPLL, the loop bandwidth can be widened to suppresses the phase noise of the ring oscillator.

71 Fractional-N Mode using MRO

- Different VCO phases are selected in each reference cycle for fractional mode
- Fractional frequency equals (N-n/24)*f_{ref}

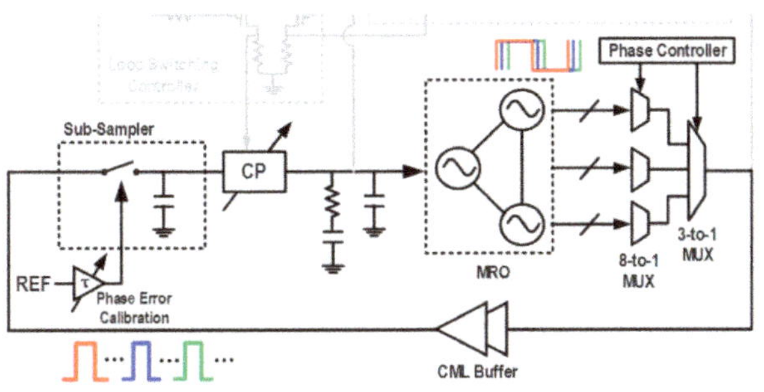

Unlike prior-art single ring topology, multi-ring coupled oscillator (MRO) allows generating more output phases without compromising oscillation frequency. The MRO with proper phase shift also achieves reduced phase noise comparing to their single-ring counterpart. Fractional-N mode is implemented utilizing the multiple output phases from the MRO oscillator. The inductor-less 24 phase SSPLL was implemented in a 0.13um CMOS technology, consuming 22 mW from a 1.3 V power supply.

72 Measurement PN and Jitter of SSPLL with MRO

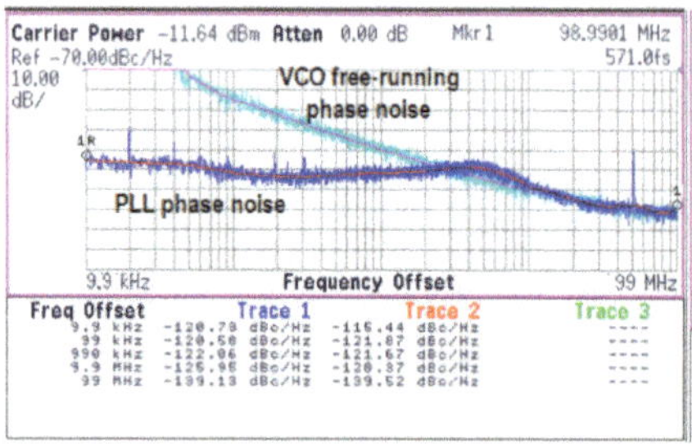

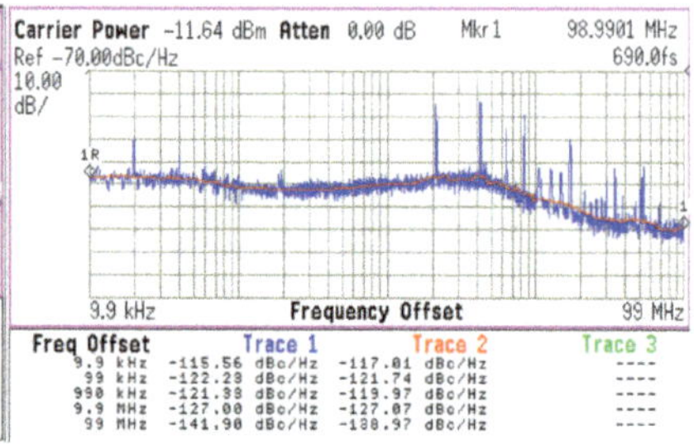

The ring VCO phase noise is -110 dBc/Hz@1MHz offset and loop bandwidth is 3 MHz. Achieved an integrated jitter of 571 fs in integer-N mode at 1.2 GHz.

With 50 MHz reference and 24 phases, the fractional resolution is 2.08 MHz and the integrated jitter from 10K to 100MHz is 690 fs.

The measured largest in-band fractional spur at 2.08 MHz is -42 dBc. The measured free-running phase noise for the triple ring coupled oscillator reaches -110 dBc/Hz at 1 MHz offset and -131 dBc/Hz at 10 MHz offset, respectively. The in-band phase noise of the SSPLL is measured as -121 dBc/Hz and its integrated jitters were measured as 571 fs and 690 fs at 1.2GHz output in integer mode and fractional-N mode, respectively, achieving a FoM of -230 dB. Note that the design was implemented in a large feature size CMOS process without using any high frequency reference injection as other IL-PLL would require. The proposed inductor-less SSPLL with the innovative MRO thus provides a low-cost and high performance frequency synthesis means for applications such as IoT and wireless communications.

73 Outlines

- PLL Fundamentals and Design Challenges
- Low Noise Multiphase Oscillators
- Multi-ring Coupled Oscillators
- Injection locking PLL with Multi-ring Oscillators
- Sub-sampling PLL with Robust Relocking
- Inductor-less Sub-sampling PLL with Ring Oscillators
- **Conclusions**

74 Conclusions

- Area efficient, low jitter, low power multi-phase frequency generation is highly desirable for applications such as beam forming, phase array, N-path filtering, passive-mixing and interleaved data converters.
- Capacitive source coupling technique with proper phase shift minimizes the noise injection from the adjacent cores and the current source noise up-conversion.
- The proposed multi-ring coupled oscillators provides low noise multi-phase generation with improved phase noise comparing to its single ring oscillator. Its auxiliary rings are also immune from reference perturbation when used in IL-PLLs, leading to greatly reduced reference spurs.
- The proposed soft loop gain switching scheme improves the stability of a SSPLL with minimum loop dynamic variation and robust relocking.
- The combination of sub-sampling technique and multi-ring coupled oscillators provides an excellent solution for inductor-less and low noise fractional-N frequency synthesis.

75 Acknowledgement

Thanks to my students:

Dongyi Liao, Ruixin Wang and Feng Zhao

for their contributions to this talk.

Thanks to professors:

Bram Nauta and Eric Klumperink

for valuable discussions on SSPLL designs.

Thanks to sponsors:

Globalfoundries and MOSIS for chip fabrication support and Intel Research Labs for gift fund support.

76 References

1. A. Hajimiri and T. H. Lee, "A general theory of phase noise in electrical oscillators," *IEEE Journal of Solid-State Circuits (JSSC)*, vol. 33, no. 2, pp. 179-194, Feb 1998.

2. F. Zhao and F. F. Dai, "A 0.6-V Quadrature VCO With Enhanced Swing and Optimized Capacitive Coupling for Phase Noise Reduction," *IEEE Transactions on Circuits and Systems I*, vol. 59, no.8, pp. 1694-1705,Aug.2012.

3. A. Hajimiri, S. Limotyrakis and T. H. Lee, "Jitter and phase noise in ring oscillators," *IEEE Journal of Solid-State Circuits*, vol. 34, no. 6, pp. 790-804, Jun 1999.

4. A. Mazzanti and P. Andreani, "Class-C Harmonic CMOS VCOs, With a General Result on Phase Noise," *IEEE Journal of Solid-State Circuits*, vol. 43, no. 12, pp. 2716-2729, Dec. 2008.

5. B. Soltanian and P. Kinget, "A Low Phase Noise Quadrature LC VCO Using Capacitive Common-Source Coupling," *the 32nd European Solid-State Circuits Conference*, 2006.

6. Erik Pankratz, and Edgar Sanchez, "Multiloop High Power Supply Rejection Quadrature Ring Oscillator," *IEEE Journal of Solid-State Circuits*, vol 47, No. 9, Sep 2012.

7. Chunyang Zhai, Jeffrey Fredenburg, John Bell, and Michael P. Flynn, "An N-path Filter Enhanced Low Phase Noise Ring VCO," *IEEE Symposium on VLSI Circuits*, 2014

8. Mohammed M. Abdul-Latif, and Edgar Sanchez-Sinencio, "Low Phase Noise Wide Tuning Range N-Push Cyclic-Coupled Ring Oscillators," *IEEE Journal of Solid-State Circuits*, vol. 47, No. 6, June 2012.

9. A. Mirzaei, M. E. Heidari, R. Bagheri, S. Chehrazi, and A. A. Abidi, "The quadrature LC oscillator: a complete portrait based on injection locking," *IEEE JSSC*, vol. 42, no. 9, pp. 1916–1932, Sep. 2007.

10. X. Gao, E. Klumperink and B. Nauta, "Sub-sampling PLL techniques," *IEEE Custom Integrated Circuits Conference (CICC)*, San Jose, CA, 2015.

11. A. Elkholy, M. Talegaonkar, T. Anand and P. Kumar Hanumolu, "Design and Analysis of Low-Power High-Frequency Robust Sub-Harmonic Injection-Locked Clock Multipliers," *IEEE JSSC*, vol. 50, no. 12, pp. 3160-3174, Dec. 2015.

12. S. Choi, S. Yoo, Y. Lim and J. Choi, "A PVT-Robust and Low-Jitter Ring-VCO-Based Injection-Locked Clock Multiplier With a Continuous Frequency-Tracking Loop Using a Replica-Delay Cell and a Dual-Edge Phase Detector," *IEEE JSSC*, vol. 51, no. 8, pp. 1878-1889, Aug. 2016.

13. D. Lee, T. Lee, Y. H. Kim, Y. J. Kim and L. S. Kim, "An injection locked PLL for power supply variation robustness using negative phase shift phenomenon of injection locked frequency divider," *IEEE CICC*, 2015.

14. W. Deng et al., "14.1 A 0.048mm2 3mW synthesizable fractional-N PLL with a soft injection-locking technique," *IEEE International Solid-State Circuits Conference (ISSCC)*, 2015.

15. R. Wang and F. F. Dai, "A 1~1.5 GHz capacitive coupled inductor-less multi-ring oscillator with improved phase noise," *42nd European Solid-State Circuits Conference*, pp. 377-380, Lausanne, 2016.

16. Dongyi Liao, Ruixin Wang and Fa Foster Dai, "A Low-Noise Inductor-less Fractional-N Sub-Sampling PLL with Multi-Ring Oscillator," *IEEE Radio Frequency Integrated Circuits Symposium*, June, 2017

17. Dongyi Liao, Fa Foster Dai, Bram Nauta, and Eric Klumperink, "Multi-Phase Sub-Sampling Fractional-N PLL with Soft Loop Switching for Fast Robust Locking," *IEEE CICC*, Austin, TX, April, 2017

18. Ruixin Wang and Fa Foster Dai, "A 0.8~1.3 GHz Multi-phase Injection-locked PLL Using Capacitive Coupled Multi-ring Oscillator with Reference Spur Suppression," *IEEE CICC, Austin*, TX, April, 2017

19. Dongyi Liao, Fa Foster Dai, Bram Nauta, and Eric Klumperink, "A 2.4 GHz 16-Phase Sub-Sampling Fractional-N PLL with Robust Soft Loop Switching," *IEEE JSSC*, vol. 53, no. 3, pp. 715-727, March, 2018

20. John W.M. Rogers, Calvin Plett, and Fa Foster Dai, *Integrated Circuit Design for High-Speed Frequency Synthesis*, ARTECH HOUSE PUBLISHERS, INC., ISBN: 1-58053-982-3, Norwood, MA, February, 2006.

CHAPTER 06

N-Path Filters and Mixer-First Receivers – A Review

Eric Klumperink
Bram Nauta

University of Twente
IC Design group Enschede
The Netherlands

SSCS

Given the plethora of low GHz radio standards and frequency bands, multi-band software defined radio receivers are coming up. Such digital intensive radios are implemented in CMOS and require a digitally reconfigurable analog RF front-end with high linearity and low noise. Moreover, flexibly programmable high-Q RF-filtering is wanted. The N-path filter concept, also known as commutated filter or frequency translated filter, can offer high-Q and high-linearity RF-filtering with acceptable noise. Moreover, a digital clock-frequency defines the center frequency of the RF-filter. This filtering can also be combined with frequency conversion in mixer-first switch-R-C receivers. This chapter reviews recent developments in the field of N-path filters and mixers, focusing on ways to improve the selectivity, linearity and blocker tolerance.

1 Outline

- **Introduction**
- Basic N-path Filters & Mixers – the good
- N-path filters: the bad
- LPTV Wonderland
- Gm-Assisted approaches
- Improve Selectivity Passively
- Conclusions

The rationale behind N-path filters and mixer-first receivers will first be introduced. Then properties of N-path filters will be reviewed (both good and bad ones and other issues you may wonder about). Finally, examples of published circuit implementations will be discussed both exploiting transconductors (Gm) and fully passive solutions which can achieve better linearity.

2 Trend to Software Radio

- Armstrong 1915

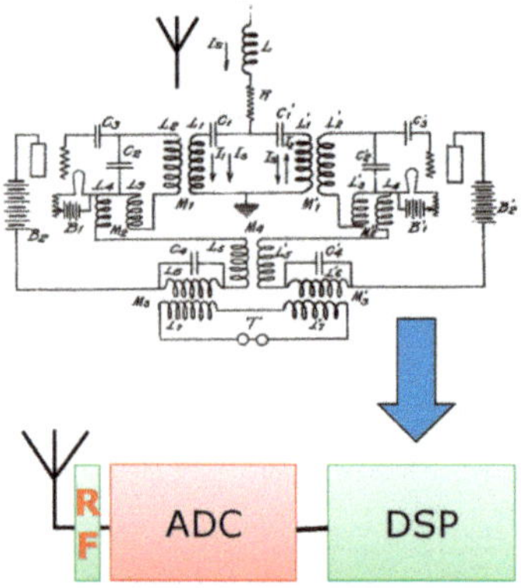

- Trend Analog ⇒ Digital
- ADC feasibility & power:
  - RF-filtering
  - Mix down
- Integration in CMOS
- Inductors:
  - Large area, poor Q
  - Limited tuning-range
- Challenges:
  - Selectivity ⇔ high Q
  - Linearity ⇔ passive R, L, C
  - Digitally Programmable
  - "SAW-less" CMOS receivers

Traditional radio receivers heavily rely on inductors (L) and capacitors (C) to realize high quality factor (Q) LC-tanks with high selectivity to select the wanted radio channel. Nowadays LC-filters and Surface Acoustic Wave(SAW) filters are still commonly used in radio front-ends, but this approach is increasingly problematic due to the numerous radio standards and bands to be supported in wireless devices. Hence the trend is to move functionality to the digital signal processing (DSP), i.e. remove dedicated fixed filters and replace them by software defined selectivity. To make the Analog to Digital Conversion (ADC) feasible and power efficient, a radio front-end is still needed for down-conversion and filtering to relax speed and dynamic range requirements. Our goal is to integrate this front-end entirely in CMOS, together with the DSP, without external Surface Acoustic Wave (SAW) filters, but this comes with several challenging listed.

3 Optimize Dynamic Range

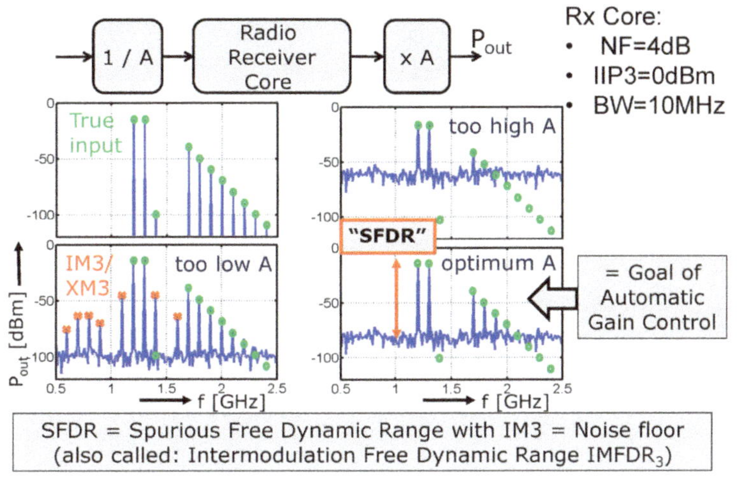

Radio receivers have a dynamic range that is limited at the low end by noise and at the high end by distortion, most notably intermodulation distortion (IM3) and cross-modulation (XM3). By doing an experiment and putting attenuation 1/A in front of a receiver core, and amplification A behind it, the nominal output remains the same, but noise and distortion contributions change. SFDR of IMFDR3 is defined as shown on the slide as the best compromise.

4 Challenge: High SFDR

Goal: high SFDR for Software Defined Radio

Spurious Free Dynamic Range (SFDR):

- Limited both by noise and nonlinearity
- Assume noise figure NF and bandwidth B
- Assume linearity is limited by IIP3

$$\mathbf{SFDR} = \tfrac{2}{3}(\mathbf{IIP3} - \mathbf{NF} - \mathbf{10logB} + \mathbf{174dBm})$$

High SFDR ⇨ low NF, high IIP3

Note: NF already close to 0dB, more room for increasing IIP3

There is more room at the top!

Focus of this talk: high-linearity …

As the total noise power in a radio channel with bandwidth B depends on Noise Figure NF and B, while IM3 can be related to the Input referred Intermodulation Intercept point, SFDR can be expressed as shown. For a given channel bandwidth B, increasing SFDR requires either lowering NF or increasing IIP3. Given that the NF is always >0dB, while typical radio receivers achieve 2-4dB, there is not much room for NF improvement, i.e. IIP3 improvement is the way to go.

5 Outline

- Introduction
- **Basic N-path Filters & Mixers – the good**
- N-path filters: the bad
- LPTV Wonderland
- Gm-Assisted approaches
- Improve Selectivity Passively
- Conclusions

Let's now see how we can achieve a high SFDR using mixer-first receivers and how this relates to N-path filters.

6 Mixer-First Architecture

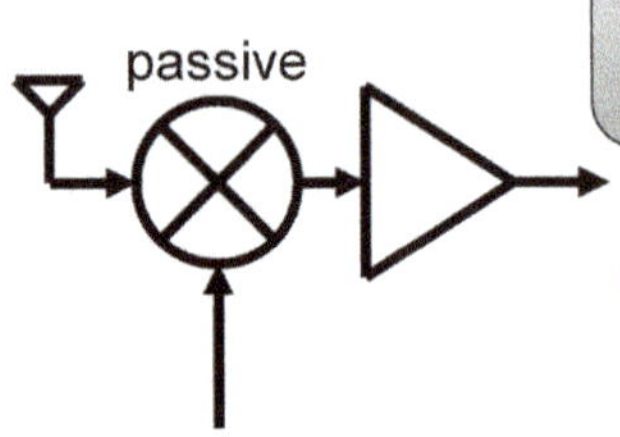

"Mixer first":
- High linearity passive mixer
- Baseband: linear by Feedback

No voltage gain before mixer:
- Noise folding?
- Conversion loss?

Solve with:
- Harmonic cancellation
- Optimized mixer duty cycle

Passive mixers can achieve very high IIP3, much higher than feasible with low noise amplifiers working at RF. Baseband amplifiers on the other hand can be very linear due to negative feedback. Hence putting a mixer first, without voltage gain before it, seems attractive for SFDR, but there are several worries (listed). We will show however, that there are solutions for them even in a very simple circuit.

7

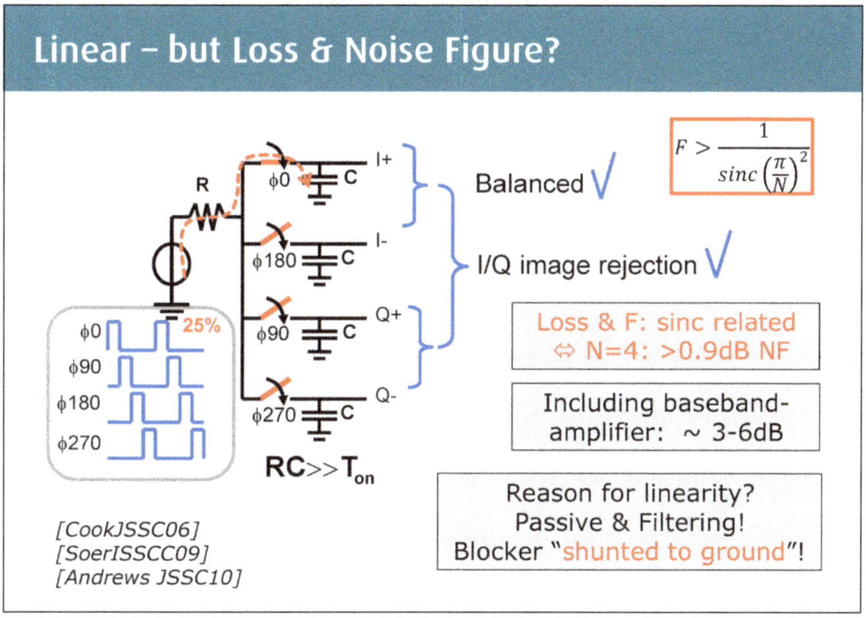

Just four capacitors and four switches driven by non-overlapping 25% duty-cycle polyphase clocks can realize a 4-path (N=4) mixer-first receiver. Balanced I and Q baseband voltage-signals are available allowing for image rejection, as required for zero-IF or low-IF receivers. Detailed analysis shows that noise factor F can be low (see formula and example numbers). High linearity is also possible as capacitors are extremely linear and can shunt strong RF-blockers to ground, while MOSFET switches with low resistance can result in very high IIP3 numbers.

8

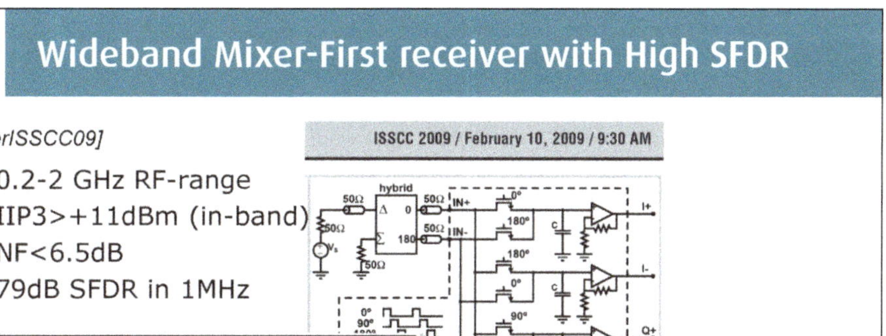

	This Work	[5]	[1]	Tektronix RSA2200A [6]	
RF Frequency	0.2 - 2.0	0.8 - 6	0.6 - 10	0 - 3	GHz
IF -3dB Bandwidth	25	20	264	~10	MHz
Gain	19	20	14		dB
DSB NF	6.5	5	7	24	dB
IIP3	+11	-4	0	+30	dBm
IIP2	+65	+60	+20		dBm
SFDR	79	70	71	80	dB
Power Dissipation	67	60	90		mW

This slide shows an early mixer-first receiver, from the University of Twente including base band amplifiers, with NF<6.5dB demonstrating that high SFDR can be achieved over a decade of RF-frequencies. Note that in-band IIP3 and IIP2 numbers are reported here, and that out-of-band IIP3 is even higher due to the capacitive filtering (typically out-of-band-IIP3=20-30dBm depending on the switch-size).

9 Mixer-first Receiver with Z-matching

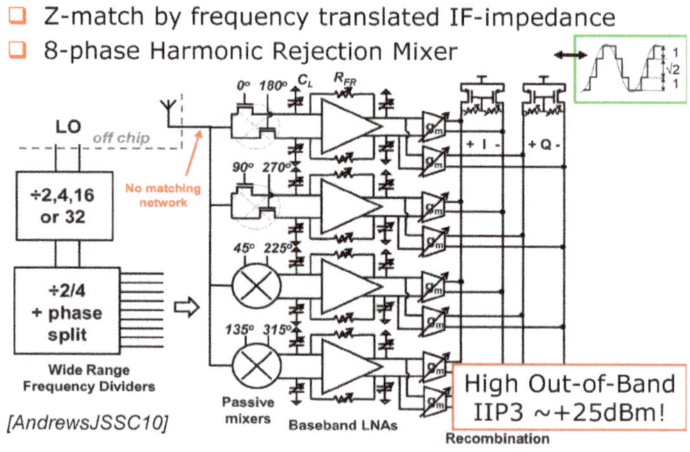

One year later, an 8-path mixer including RF-impedance matching was published at ISSCC by researchers from Cornell University. Baseband transimpedance amplifiers with resistive feedback (R_{FR}) realize a finite input resistance (note NOT virtual ground!). The passive mixers mix this baseband impedance up to RF and allow for realizing impedance matching to the antenna. Moreover, 8-path polyphase mixers with non-overlapping clocks of 1/8 duty cycle, allow for realizing poly-phase harmonic rejection mixing, exploiting 1:sqrt(2):1 weighing of three output signal contributions (Gm-weighing).

10 Observation: input has Band-Pass nature @ fs

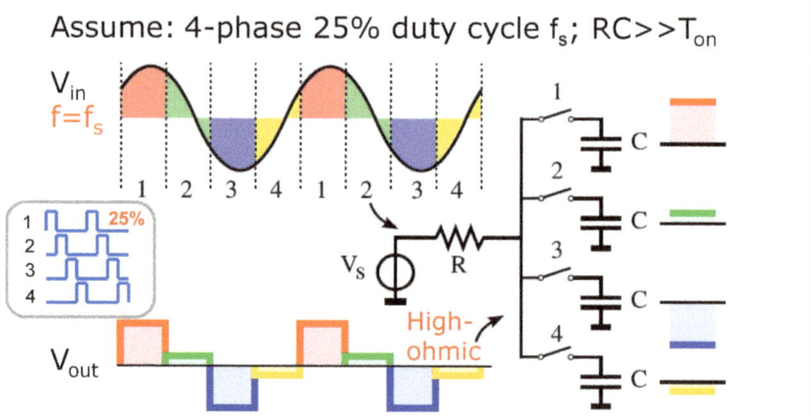

Thinking about the input impedance of the 4-path mixer with non-overlapping clocks, the input voltage is important. During each clock phase the resistive source "sees" one capacitor. For RC>>T_{on} (the on-time of the switch), the capacitor voltage will approximate the long-term average value of the input signal during that on-time. If the RF-frequency is equal to the switching frequency f_S, the contribution of each period of the sine is equal, and a stair-case approximation of the input sine wave results at the right side of resistor R. In other words: the input impedance is high-ohmic. Apart from the fundamental, harmonics are also present and using more clock-phases results in a better approximation of the input sinewave, i.e. less attenuation.

11

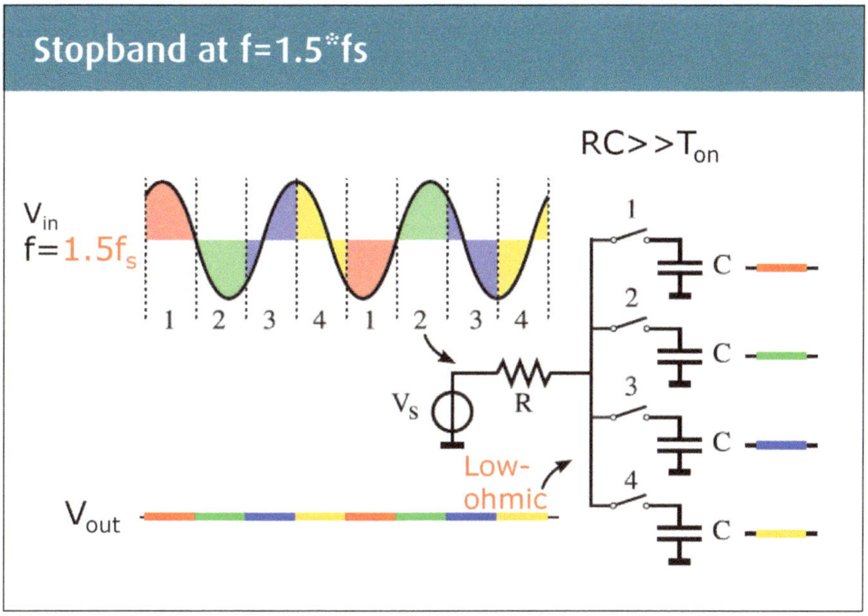

In case we choose an RF-frequency of 1.5x the switching frequency, contributions of alternating periods cancel each other resulting in zero input voltage. As no input voltage is developed now, while the input current is significant, the input impedance of the network is low-ohmic now. Combining the results from the previous and this slide, we conclude that frequency selectivity occurs around the switching frequency which is often referred to as N-path filtering (explained later). Note that an N-path filter results if we choose the RF-input node behind the resistor as output, while choosing the baseband capacitor voltages renders a polyphase N-path down-conversion mixer (with implicit N-path filtering).

12

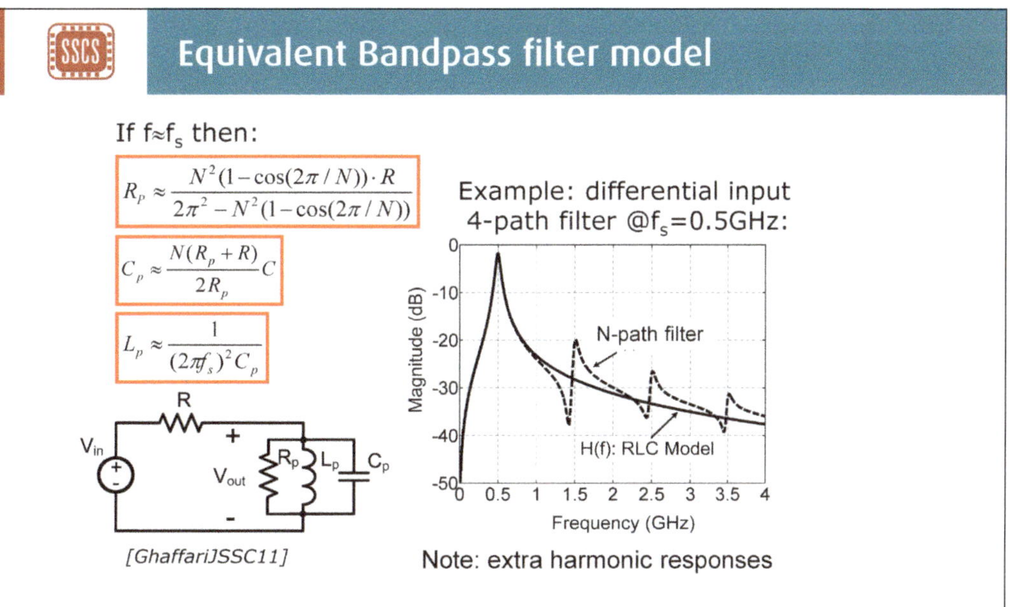

Detailed mathematical analysis shows that indeed band-pass behavior results, not only around the fundamental of the switching frequency but also around harmonics of it. When exploiting a differential input, even order responses can be cancelled (see figure). Around the fundamental, it appears possible to approximate the amplitude transfer with a parallel R-L-C tank model with the values given (N=number of paths, R the source resistance, C one baseband capacitor).

13

Low pass – Bandpass transformation

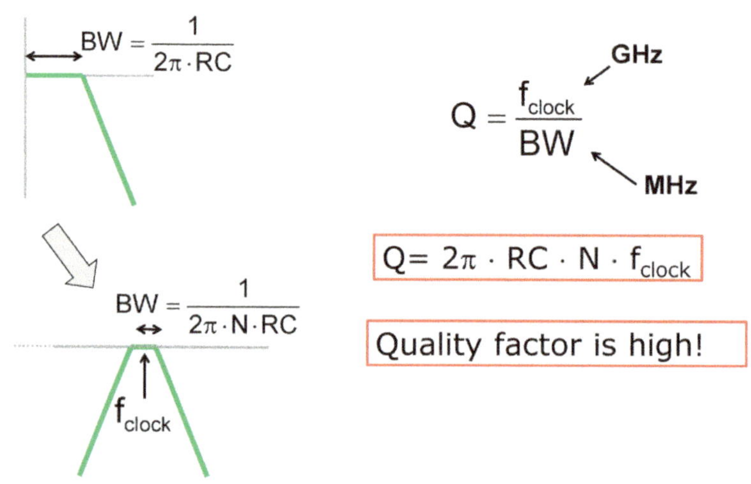

Taking a frequency domain view, we can conclude that periodic switching leads to frequency shifting (e.g. down- and up mixing). RF signals are mixed down, low-pass filtered and then mixed up again around the switching frequency. The resulting filter shape is essentially defined by the low-pass R-C network, but the transfer is shifted by the switching frequency f_s=f_{clock}. Apart from a simple frequency shift, the bandwidth is also reduced by N times, as each capacitor in the switched network sees the resistor only during 1/N of the period. The resulting Q of this filter can be very high as GHz switching frequencies are feasible now.

14

Wide Tuning Range Linear 4-path band-pass filter

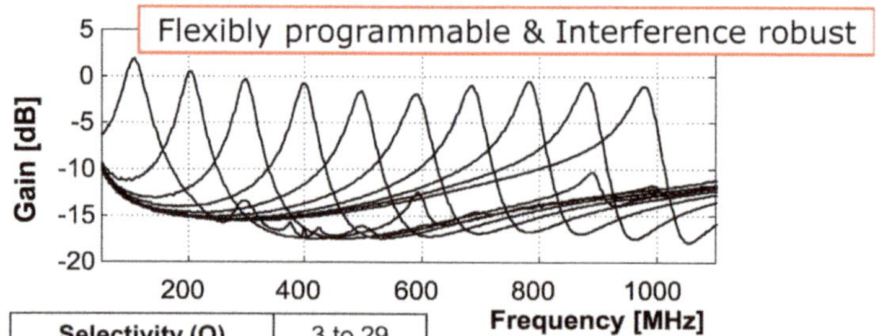

Selectivity (Q)	3 to 29
Compression (P_{1dB})	+2dBm
Linearity (IIP3)	+19dBm
Noise Figure	3-5dB*
Power	2-16 mW

* Theory: ~1dB; meas. setup dominates

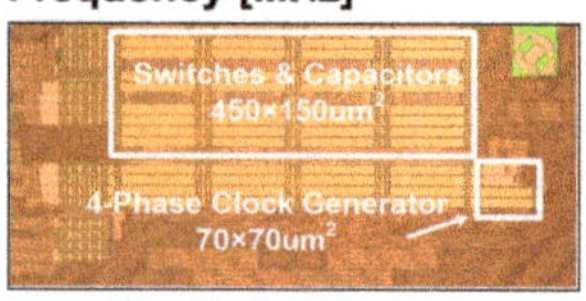

[GhaffariJSSC11]
(first: RFIC'10)

Taking the RF-input node as output, the 4-path switched capacitor mixer-chip from slide 9 was used as 4-Path filter, and the listed measurement results were obtained. Although Q was not pushed, values higher than possible with on-chip inductors are realized. Linearity and compression numbers are attractive, while moderate noise figure is achieved. Only dynamic power is consumed to drive the gates of the switch-transistor with 1/4 duty-cycle clocks.

15 Two-stage 4-Path Blocker filtering I/Q Receiver

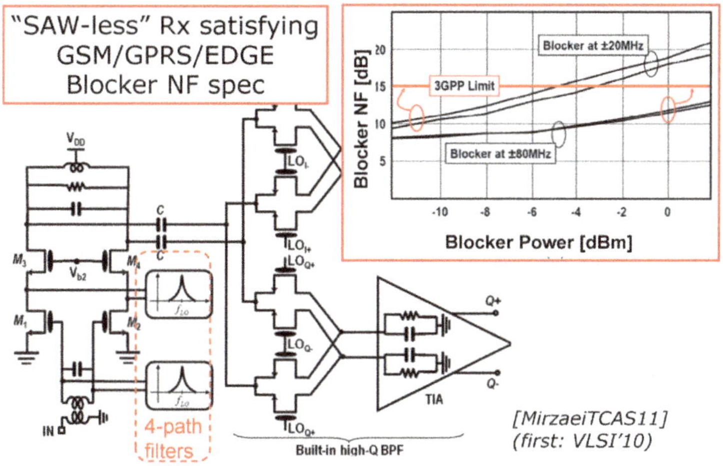

[MirzaeiTCAS11]
(first: VLSI'10)

Almost at the same time, Broadcom researchers published a chip designed to demonstrate that a multi-standard CMOS radio receiver without SAW-filter is feasible. To handle the strong out-of-band blocker two 4-path filters are added, one in parallel to the input transistors M1-M2 and the second in parallel to M3-M4. In the presence of strong out-of-band blockers (indicated in frequency offset and power level), noise figure can be kept below the 15dB 3GPP-specification.

16 Not a new idea ...

- Downconvert + LPF + Upconvert = BPF

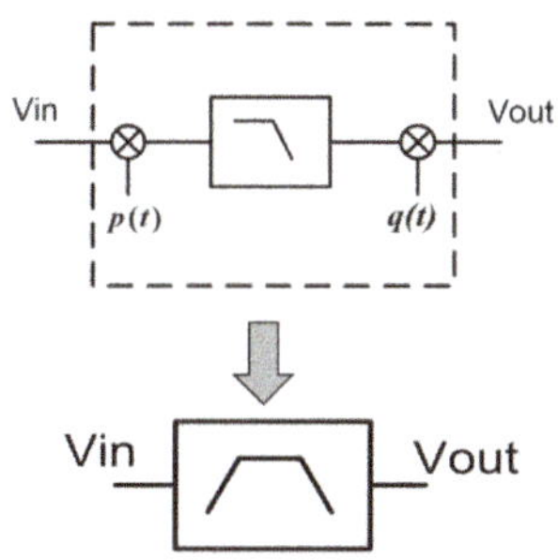

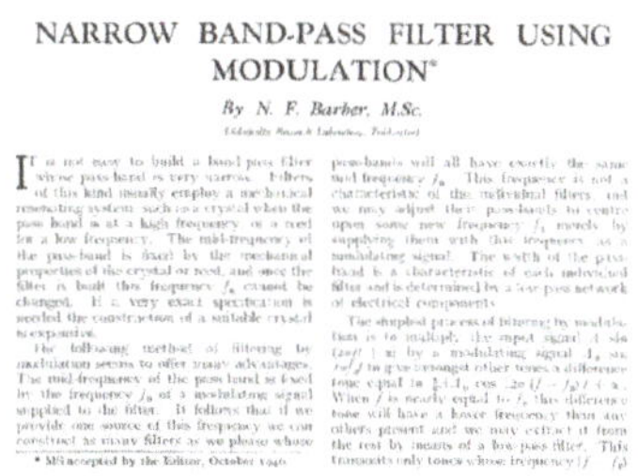
NARROW BAND-PASS FILTER USING MODULATION*

By N. F. Barber, M.Sc.

[Barber, Wireless Engineer, May 1947]

It turned out that these N-path filter ideas were not new, but rather are a re-invention or earlier work that can be traced back to at least 1947. Barber proposed to convert a low-pass filter into a band-pass filter using two modulators or mixers.

17 "Commutated Networks" [Smith1953][LePage1953]

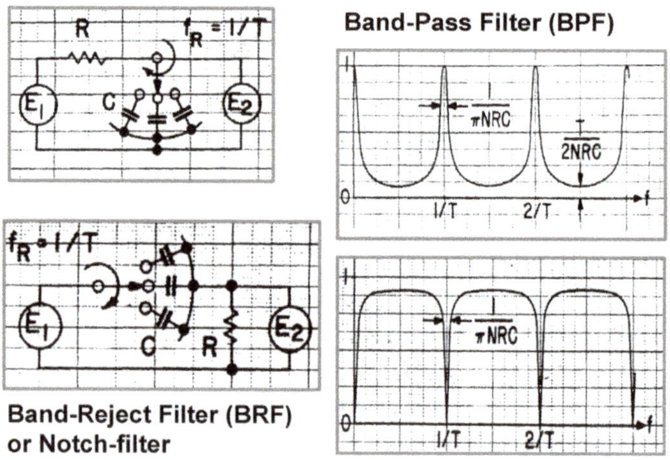

Already in the 50s of the previous century the basic filter shape of switched-R-C filters was analyzed. Mechanically rotating switches or "commutators" were used to implement such filters. Both band-pass and band-reject "commutated networks" were analyzed.

18 ISSCC 1960: Go solid-state!!

6.4: Solid-State Sampled-Data Bandpass Filters

L. E. Franks and F. J. Witt

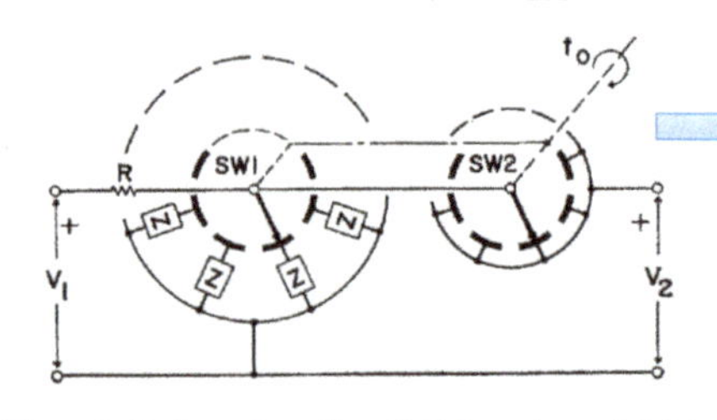

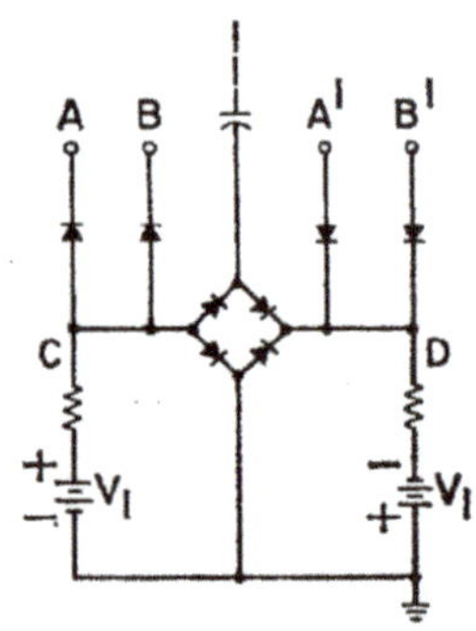

Figure 1—Basic configuration of N-element sampled-data network for the realization of narrow-band filter characteristics. The switch rotors revolve in synchronism at f_o revolutions per second. With solid-state switches, SW_2 can be eliminated.

[FranksISSCC60]

70 • 1960 International Solid-State Circuits Conference

At ISSCC 1960, Franks proposed a "solid-state" implementation of a commutated network, exploiting diodes as passive mixer switches. From the name of the paper it is clear that the authors realized the sampling nature of the network, but here the time-continuous output of the filter was still used (later work in the 60s and 70s gradually moved to using the time-discrete sampled output, and over time using a time-continuous output was kind of forgotten.

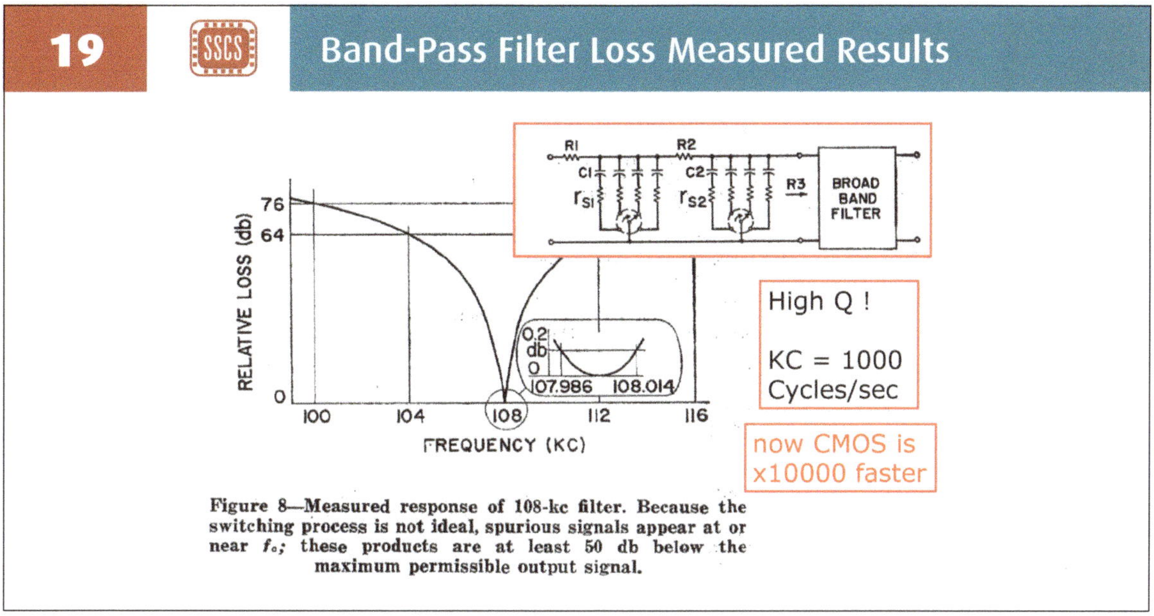

The paper demonstrated a notch filter with two cascaded switched-R-C 4-path filter sections. High Q was demonstrated and a notch-frequency (=switching frequency) in the order of 100kHz. Note that CMOS technology nowadays allows 10000x higher switching frequencies, allowing use at RF frequencies.

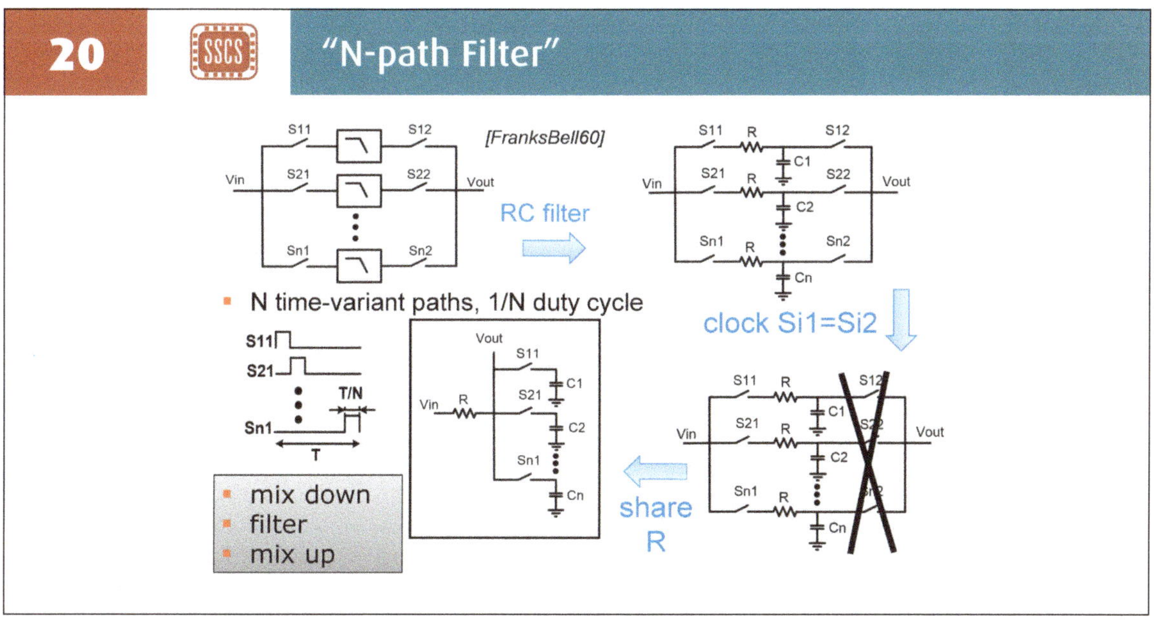

Franks generalized the concept and used the name "N-path filter". Each of the N paths has a switch before and after a low-pass filter, where the switching clock-phase can be different. If chosen equal, while a simple R-C low-pass filter is used, the N-path filter network can be simplified as shown.

21 Outline

- Introduction
- Basic N-path Filters & Mixers – the good
- **N-path filters: the bad**
- LPTV Wonderland
- Gm-Assisted approaches
- Improve Selectivity Passively
- Conclusions

N-path filters have some performance limitations and challenges that will be discussed next.

22 N Path Filters: the bad

- Phase Noise:
  - essentially an N-path filter is a set of mixers...
  - so: reciprocal mixing with phase noise is an issue

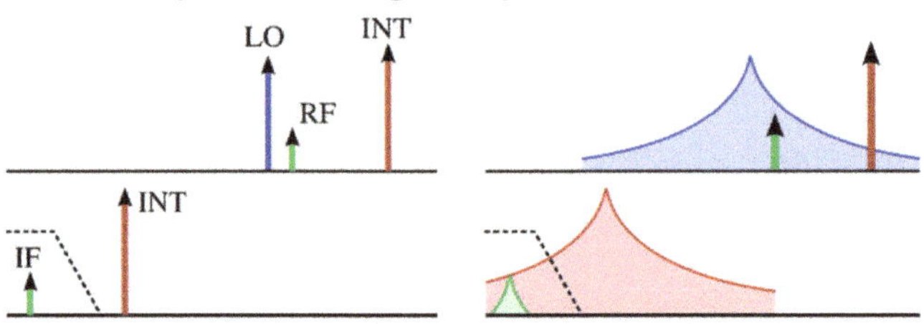

- Harmonic responses
- Folding

As in any mixer, phase noise around the Local Oscillator signal (LO) can mix with interferers (INT) to the same intermediate frequency (IF) than the wanted RF signal and disturb it. Moreover, the time-variance introduced by the periodic switching results in harmonic responses and folding, as will be discussed next.

23 N=8 fundamental response

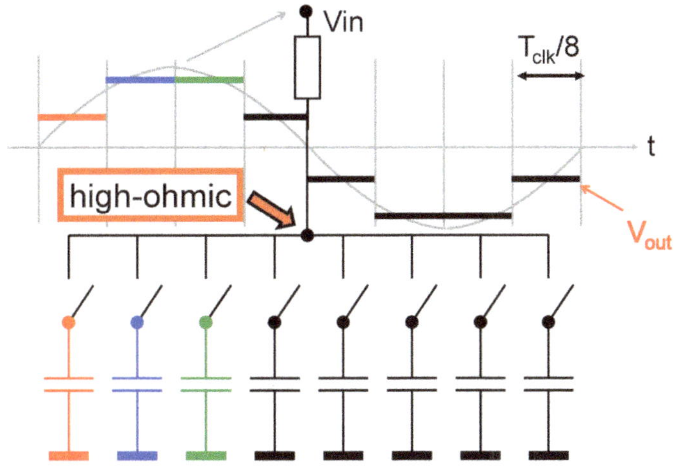

Let us return to the time-domain view of slide 11 for intuitive insight on harmonic responses and folding. Consider now an 8-path filter with a sinewave input a frequency equal to the clock-frequency (pass-band). During each of the 8 non-overlapping clock phases, one capacitor is connected to input signal V_{in} for 1/8th of the clock-period, via a shared resistor, developing a voltage equal to the average of V_{in} over "its" timeslot. Switching multiplexes these voltages to V_{out}, the stair-case approximation of V_{in} (signal passes).

24 N=8, f_{in} = 2 x f_{clock}

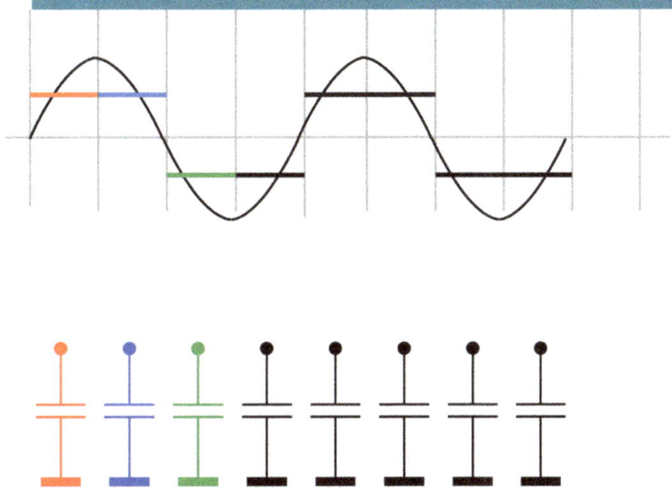

Now consider the response at two times higher the input frequency. Again the input voltage is passed, albeit with some extra attenuation.

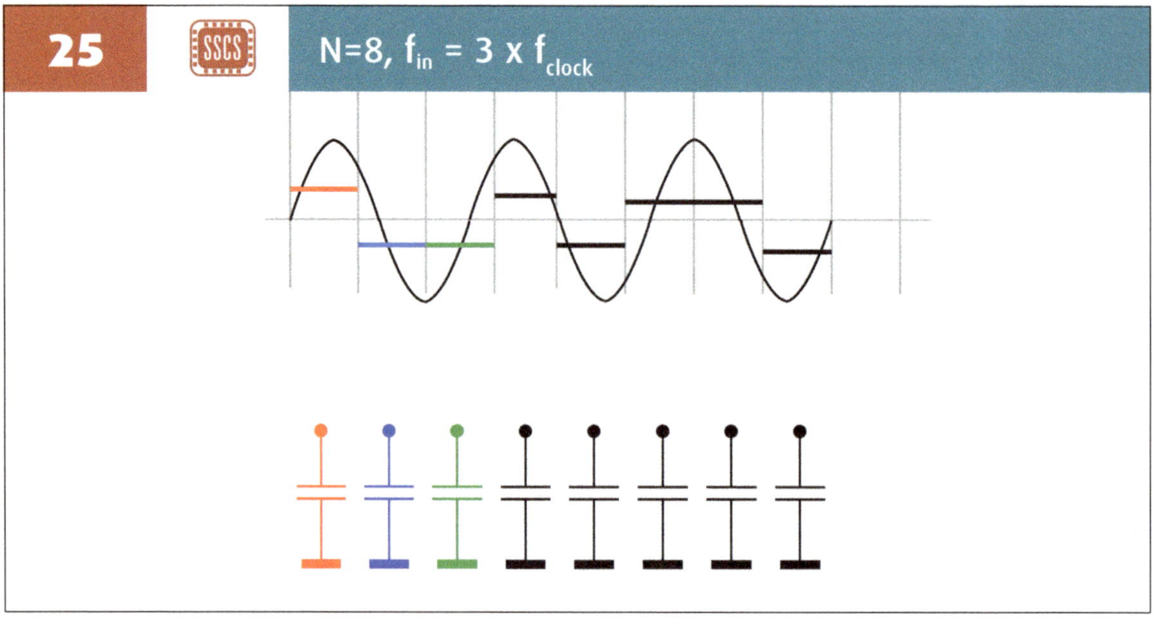

For an input frequency equal to 3x the clock frequency again a harmonic pass-band occurs.

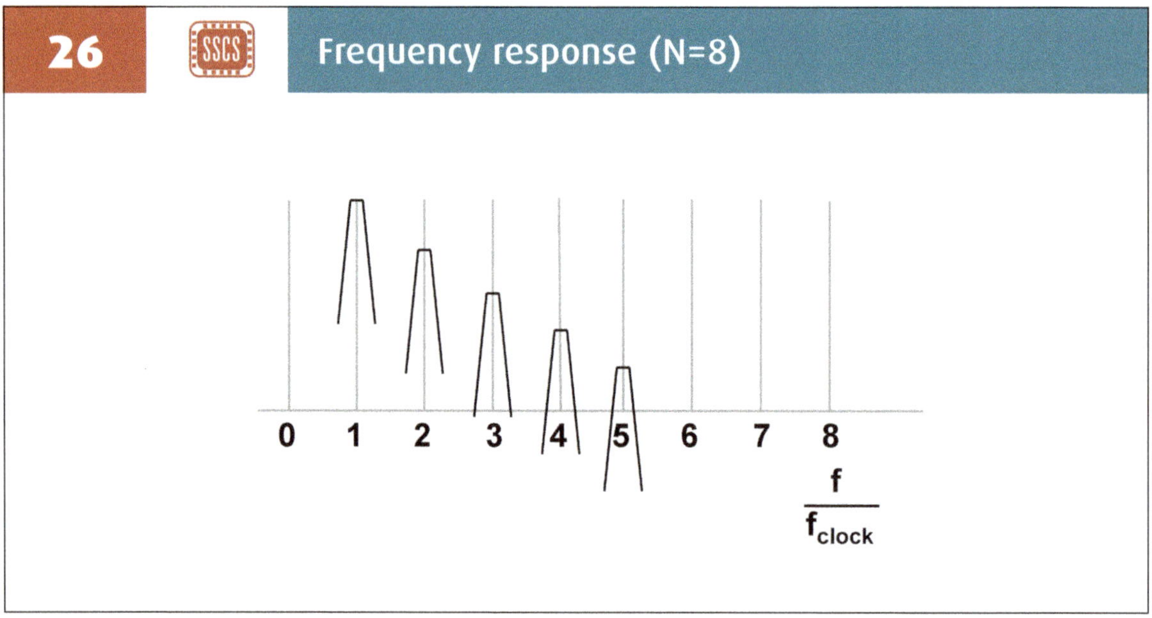

Combining these responses, we see that the N-path filter not only responds around the fundamental of the clock but also at its harmonics, albeit with attenuation (like a "comb-filter" with attenuation).

27 Differential version

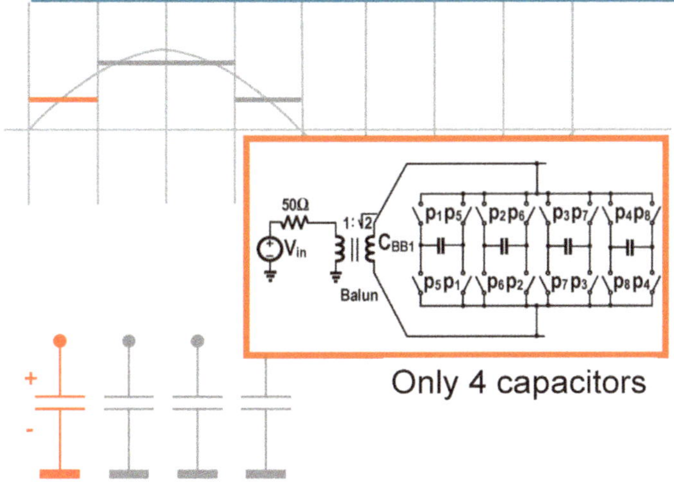

When using an RF-balun, a differential RF-signal is available. Now an 8-path filter can be implemented using only 4 capacitors.

28 Differential Version (after flipping)

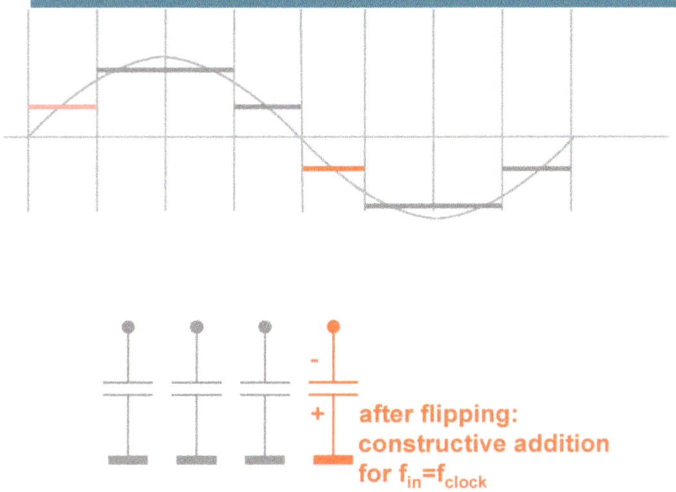

The capacitor that was used during the first clock-phase is now flipped and then re-used in the 5th time slot. Flipping after half a clock-period is also applied to the other 3 capacitors.

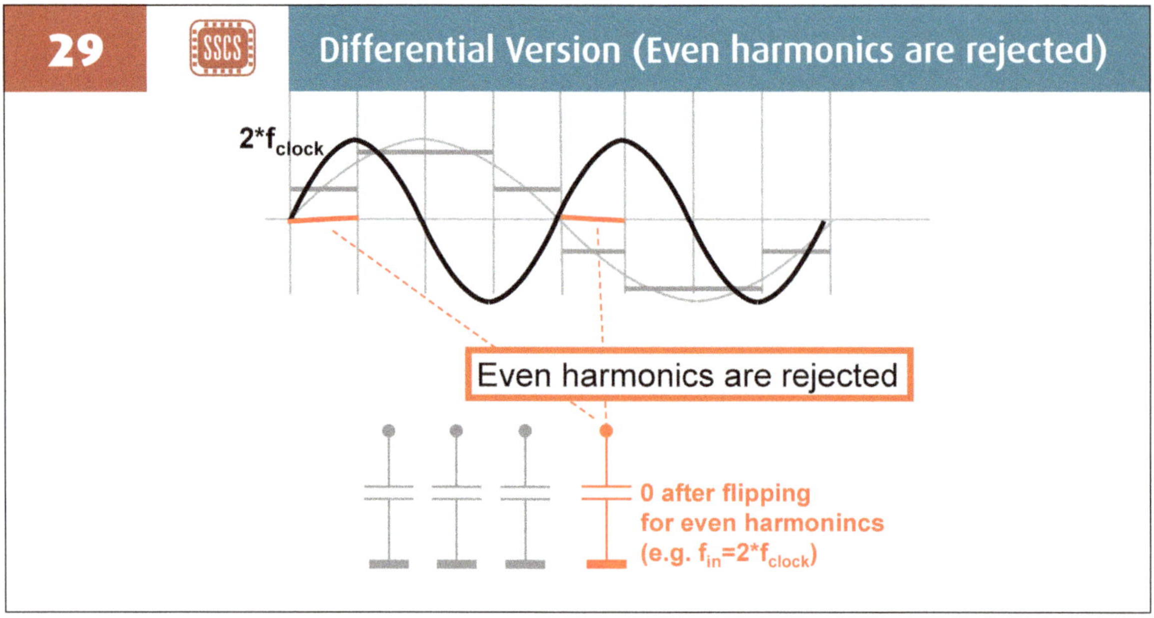

For the fundamental frequency (grey waveforms), there is still a pass-band response after flipping, as V_{in} changes polarity after half a clock-period. However, for the even harmonics the polarity of V_{in} remains the same, while the capacitor is flipped, so that the average voltage is zero (rejection).

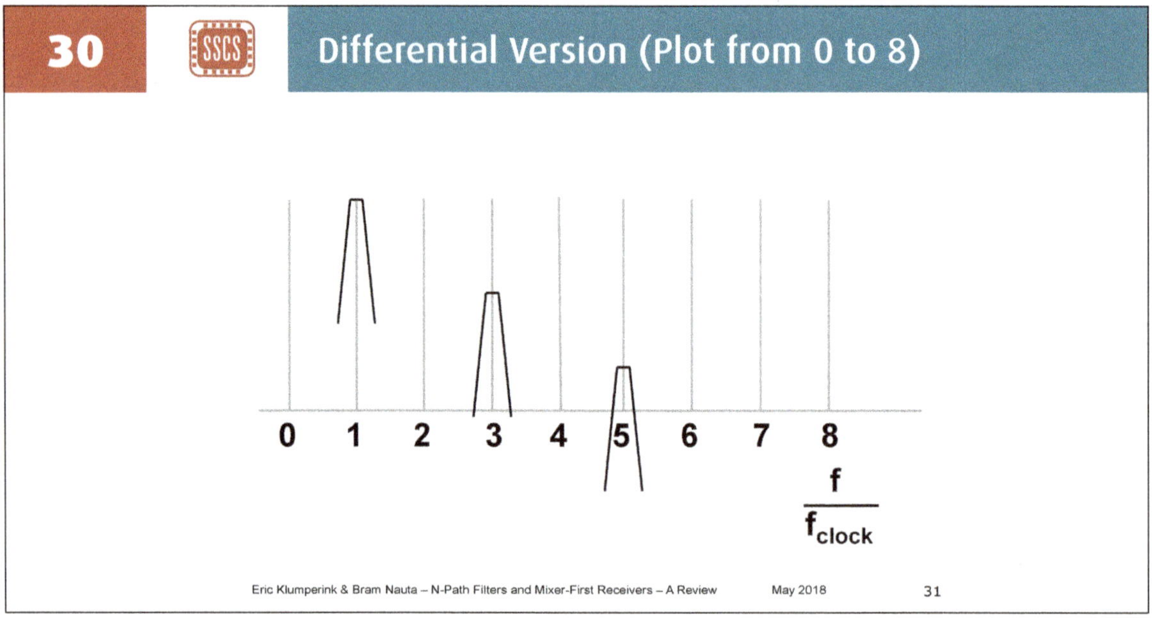

The resulting wideband filter response now has no even harmonics anymore.

31 Folding (N=8)

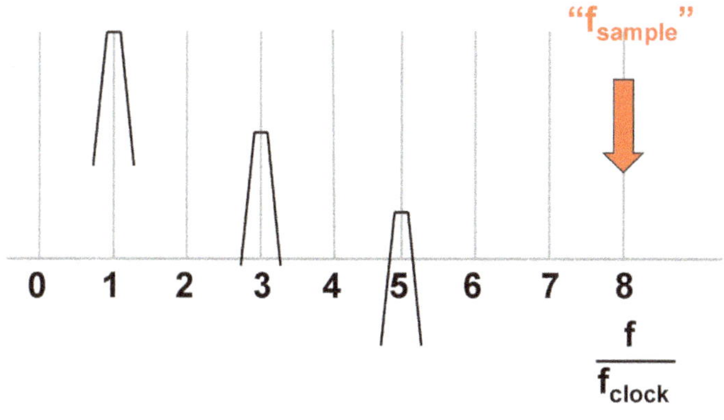

Another issue is signal folding, which is somewhat similar to aliasing for sampling, but with extra attenuation due to the R-C low-pass filtering (note that T_{on}>>RC, while fast settling samplers use T_{on}<<RC). If 8 clock-phases are used, the effective sampling frequency is N times the clock-frequency (here 8x).

32 Folding (N=8)-attenuation and folding from 7 to 1

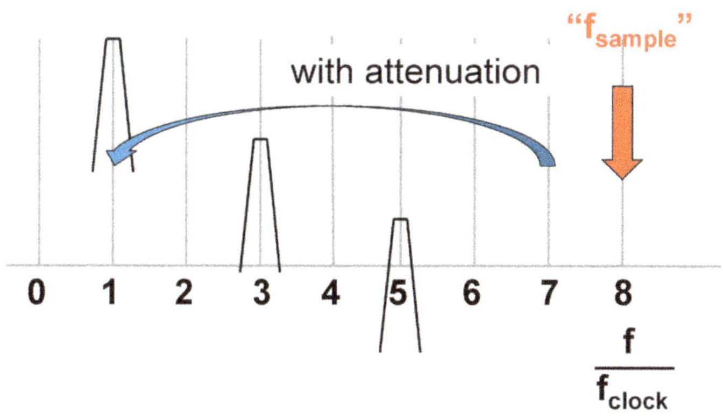

RF-frequency content around 7x f_{clock} is now folded on top of content around 1xf_{clock}, with attenuation.

33 Folding (N=8)-attenuation and folding from 9 to 1

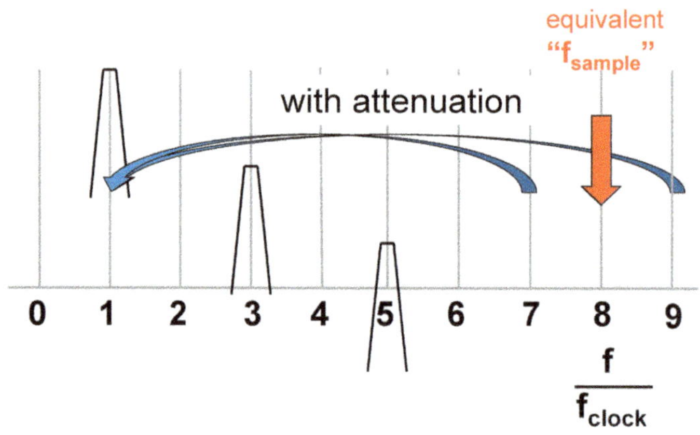

Similarly, RF-frequency content around $9xf_{clock}$ also folds on top of content around $1xf_{clock}$. Hence, some form of time invariant low-pass or band-pass pre-filtering before high-Q flexibly programmable N-path filtering is needed to mitigate folding products sufficiently.

34 Outline

- Introduction
- Basic N-path Filters & Mixers – the good
- N-path filters: the bad
- **LPTV Wonderland**
- Gm-Assisted approaches
- Improve Selectivity Passively
- Conclusions

We have now discussed some limitations of N-path filters related to their Linear Periodic Time Variant (LPTV) behavior. However, there is more to wonder about when studying LPTV circuits.

35

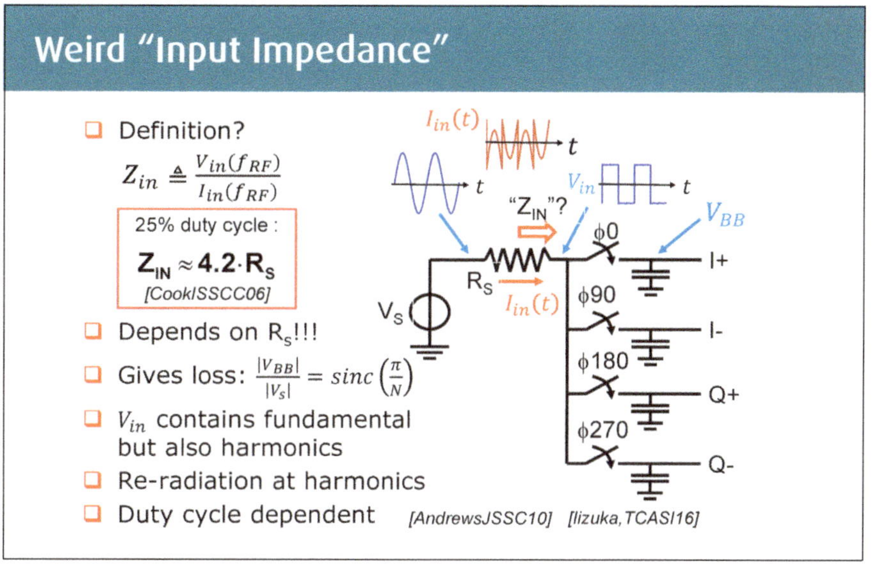

We talked loosely about "input impedance", but one might wonder how to define it exactly? Commonly a voltage source excitation with a sine is assumed, while evaluating the resulting current at the same frequency and dividing them. Curiously, the result depends on R_S! Driving with a current source also gives a puzzling result: integration results in potentially unbounded capacitor voltages! When driving with a voltage source, voltages have bounded values, but the stair-case shape of V_{in} results in an I_{in} waveform with many harmonics. Hence, if the source is an antenna with incident sinewave at $f_{RF}=f_{clock}$, the N-path filter/mixer circuit re-radiates content at harmonics. The amplitude of current I_{in} depends on the value of the capacitor voltage divided by R_S, which explains why the input impedance depends on R_S.

36

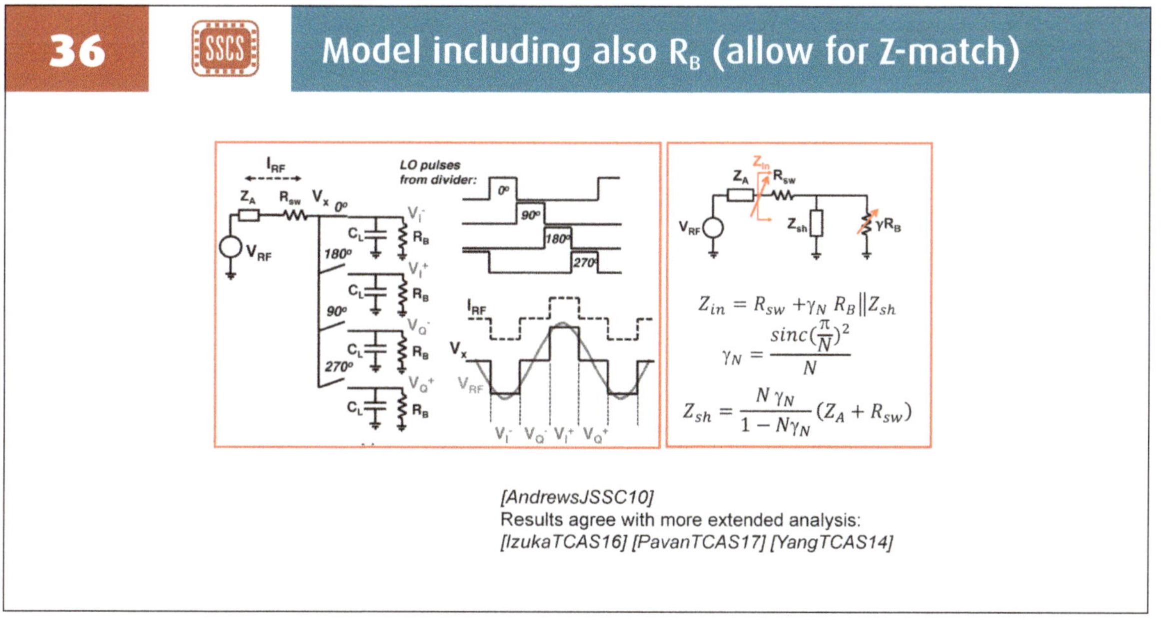

If a mixer with input-impedance matching is needed, baseband resistor R_B can be added and this slide shows the relevant equations to model the effective input impedance seen from the RF-antenna port (Linear Time Invariant model with a contribution due to R_B and due to the switched capacitor (Z_{sh} part)).

37 Q: Is this Mixing or Sampling?

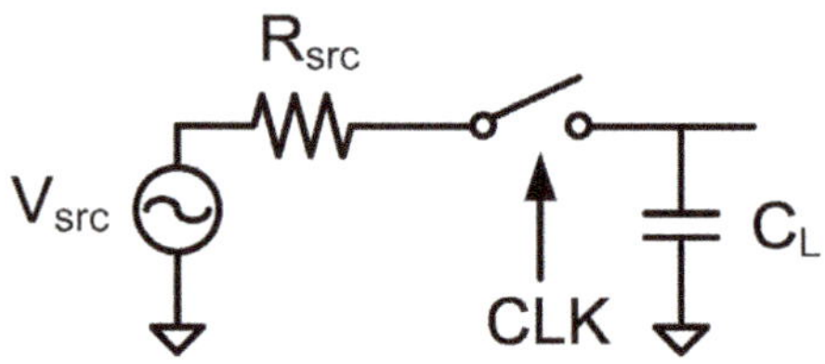

Can be both !

One might also wonder whether the basic switched-R-C circuit shown here is a sampler or mixer? It turns out that it actually can be both! We will illustrate this, taking two viewpoints: a signal processing view (how are the signal used) and a detailed circuit analysis view (signal waveform and noise behavior).

38 Mixing or Sampling: Use of output? Rate?

[RuJSSC10]

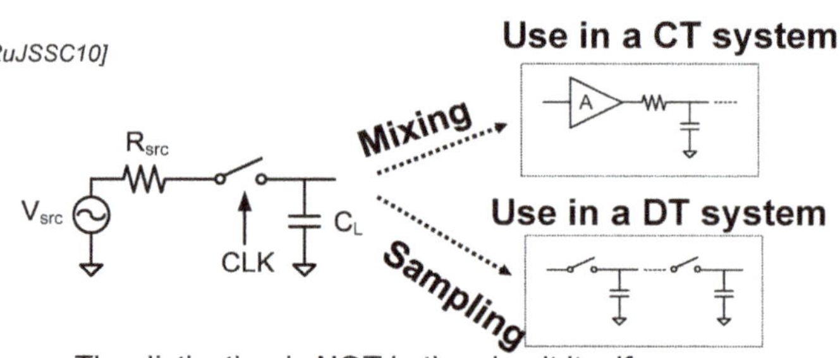

The distinction is NOT in the circuit itself

Key question: how is the output signal used??

Does "observation rate" change or not?

- CT Mixing: RF(t) ➔ IF(t) — No
- CT-to-DT Sampling: RF(t) ➔ IF(k) — Yes!

Consider first the signal processing view and ask yourself how the output signal is used. Based on the "observation rate" of the output signal, now two cases can be distinguished [RuJSSC10]: the case where the observation rate changes (sampling) and the case it does NOT (mixing). Sampling reduces the observation rate of a time-continuous input signal (infinite observation rate) to a limited rate, while mixing doesn't alter the rate.

39 Another view: Small or Big Γ=Ton/RC ??

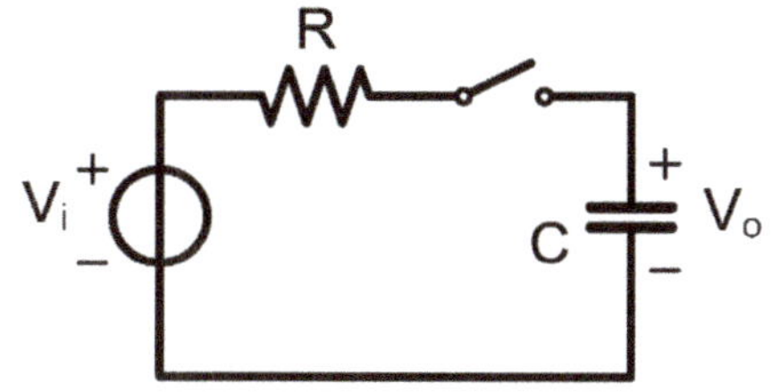

[SoerTCAS10]

Another distinction between sampling and mixing can be made based on the value of the on-time of the switch, compared to the RC time-constant (the ratio is labelled Gamma).

40 Crucial point: Small or Big Γ=Ton/RC

- Both show a hold, only big Γ tracks and samples input well
- For small Γ: simplification for Γ→ 0 and $f_o = f_i + n \cdot f_s$:

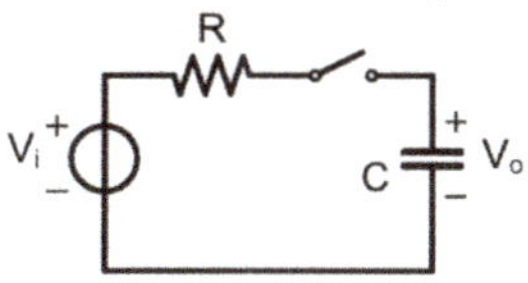

$$H_{n,SE}(f_o) \approx \frac{\mathrm{sinc}(Dn)}{1 + j\frac{f_o}{D f_{rc}}} e^{-j\pi Dn}$$

Harmonic mixing including a first-order Low-pass filter defined by RC and D

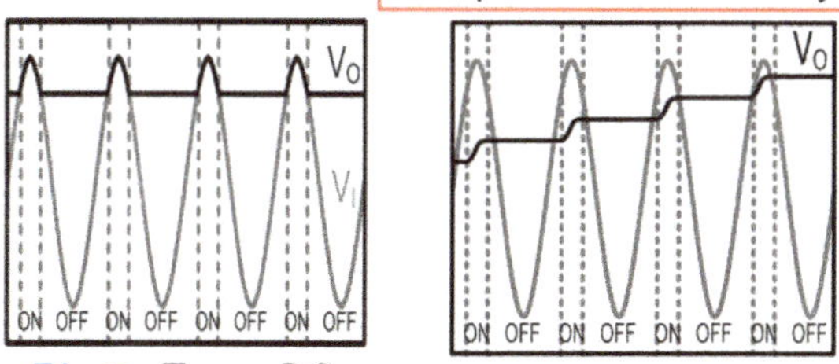

will ultimately become Periodic Steady State (PSS) for sine input

Big Γ: $T_{on} >> RC$ Small Γ: $T_{on} << RC$

For large Gamma, the capacitor voltages settle well within the on-time of the switch, providing a wideband track-and-hold (or also sometimes called sample-and-hold) with fast settling and small sampling error. For small Gamma, the output voltage is heavily R-C low-pass filtered and only slowly responds to changes in the envelope of the RF-waveform. This behavior makes the circuit useful as a down-conversion mixer or N-path filter with slow-settling high-Q behavior. Also, noise behavior appears to be attractive (next slide).

41 "Mixing" and "Sampling" Region

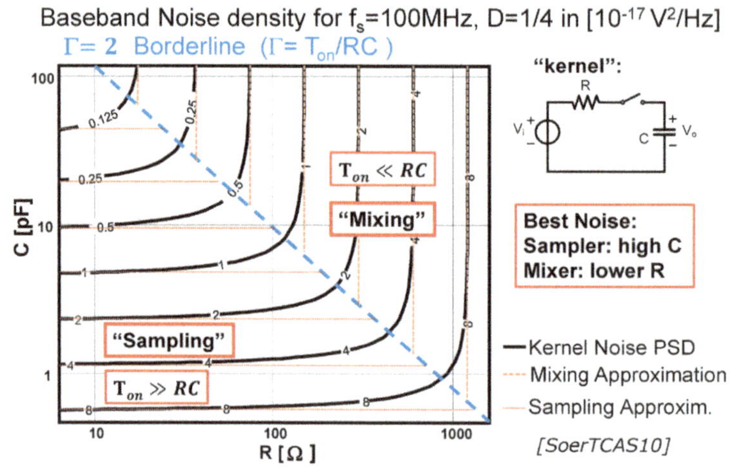

Using detailed time-variant circuit analysis the baseband noise density was calculated for a single switched-R-C "kernel". The plot shows equal noise density curves in which symmetry around Gamma=2 occurs, providing a basis for defining two regions: the "sampling" and "mixing" region.

42 Noise Figure (NF) versus RC and Duty Cycle

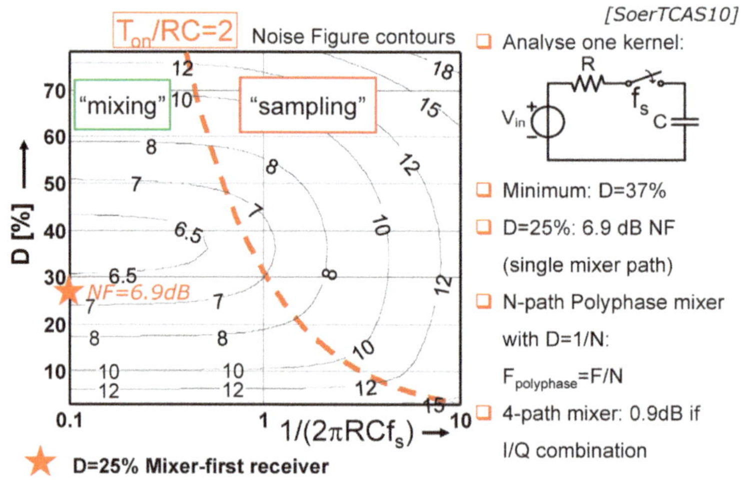

Expressing the noise performance as noise figure, equal noise figure (NF) contour plots for variable RC and duty-cycle were made. The Gamma=2 borderline is also indicated. Clearly, the mixing region gives the lowest noise figure (makes intuitive sense as a wideband sampler aliases a lot of noise to baseband, while for big RC-time signal and noise bandwidth are both limited and are roughly the same). The ¼ duty cycle mixer shown earlier is not far from the optimum NF-point. It achieves a theoretical NF=6.9dB for a single kernel, and 6dB less after 4-path signal combining when applied in a zero-IF mixer with image rejection.

43 Parasitic Capacitance at the input?

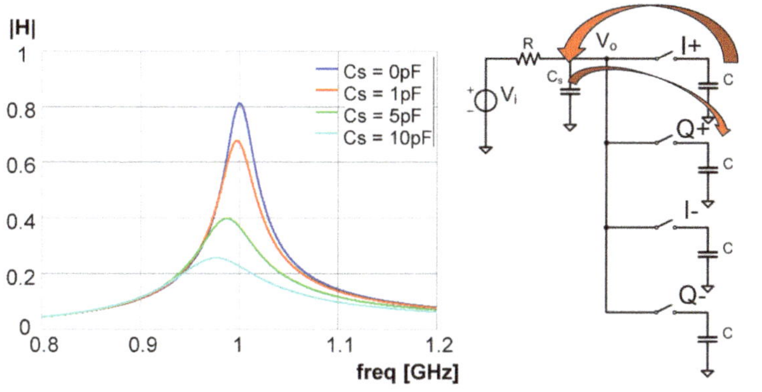

- C_s shares charge between N-path Caps ⇔ loss
- C_s also shifts the passband centre (I↔Q)
- Modeling, also with series-L: *[YangTCAS15] [PavanTCAS18-1]*

When a parasitic capacitance C_S is present at the shared RF-node, RF-signal will partly be shunted to ground. Moreover, this capacitor C_S introduces interaction between the four grounded capacitors. As indicated with the arrows, charge sharing of the I+ capacitor with C_S will take place, while in the next clock phase, the charge of C_S will be shared with the Q+ capacitor. This renders a kind of quadrature cross-talk leading to an asymmetric N-path filter RF-response (like in polyphase filters) and shifts the point of optimum input impedance matching to lower RF-frequencies than the switching frequency.

44 Outline

- Introduction
- Basic N-path Filters & Mixers – the good
- N-path filters: the bad
- LPTV Wonderland
- **Gm-Assisted approaches**
- Improve Selectivity Passively
- Conclusions

After introducing the N-path filter and mixer concept and its properties, lets now look at different implementation examples that have been proposed over the last decade. The focus will be on improving frequency selectivity on the one hand and improving linearity on the other. We will start with concepts exploiting transconductors in an assisting role, and then focus on passive-only approaches.

45 Gm-Assistance to lower power: Increase Z-level!

- MHz Bandwidth @50ohm ⇔ big C, Rsw<<50Ω ⇔high LO-power
- Solution: higher Z-level!

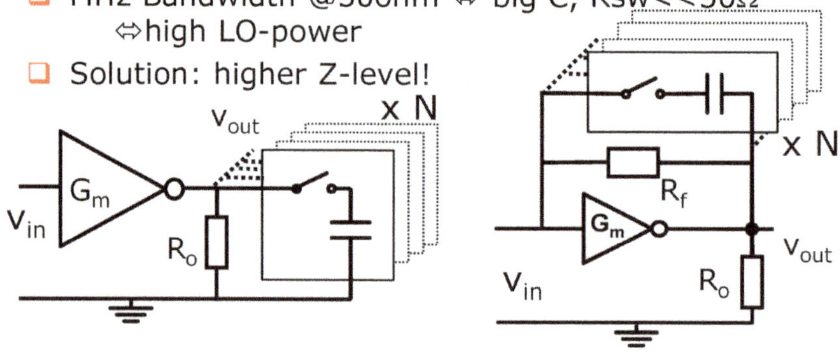

a) Increase R-level > 50Ω
e.g. R_o = 500Ω ⇔ C/10

[BorremansJSSC11]
[Liempd,JSSC14][SoerISSCC14]

b) Notch in feedback ⇔ BPF
Miller effect "Gain Boosted"

[Lin/MakJSSC14]; [LinISSCC15]
[LinTCAS14][ParkJSSC14]

In order to realize high-Q filtering for a given RF center-frequency, a narrow bandwidth is needed. For a given driving-resistance this translates in requiring more capacitance and lower switch resistance, i.e. wide switch-transistors. Driving such wide transistors requires strong clock-drivers which increase dynamic power consumption. Increase the driving resistance is a solution direction to save both chip area (capacitance) and power dissipation. Two main approaches have been proposed, both exploiting a transconductor G_m. V_{in} can be amplified to V_{out} with voltage gain G_mR_o, where R_o is the driving resistance for the N-path filter capacitors and R_o is typically chosen 5-10x higher than 50ohm. It is also possible to put an N-path notch filter (see slide 18) in the feedback path of the amplifier, to obtain band-pass filtering. This results not only in filtering of V_{out}, but also introduces some filtering at the input. The benefits can be explained in different ways, e.g. as exploiting the Miller effect or as a gain-boosted N-path filter.

46 Filter Shape and Stop-Band Rejection

- Desired: low loss flat in-band shape, high Roll-off

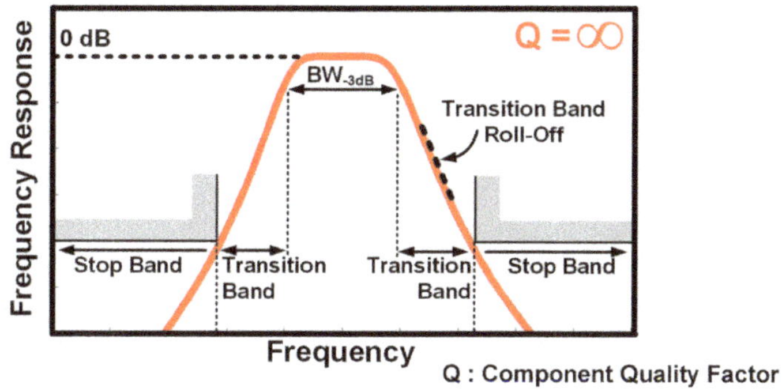

Courtesy Pingyue Song & Hossein Hashemi [SongISSCC18]

The basic N-path filter shows a roll-off related to 1storder RC-filtering. We would like to improve selectivity, not by just reducing the bandwidth but also by improving the filter shape. The plot shows the shape of a higher order LC-filter with infinite-Q resonators, and key specifications like low pass-band insertion loss, a flat-toppass-band, large transition-band roll-off and high stop-band attenuation.

47 High-order N-path BPFs

- Use N-path filter as parallel LC tank
- Synthesize a high-order BPF with gyrator coupling
- All-pole singly-terminated 6th-order BPF

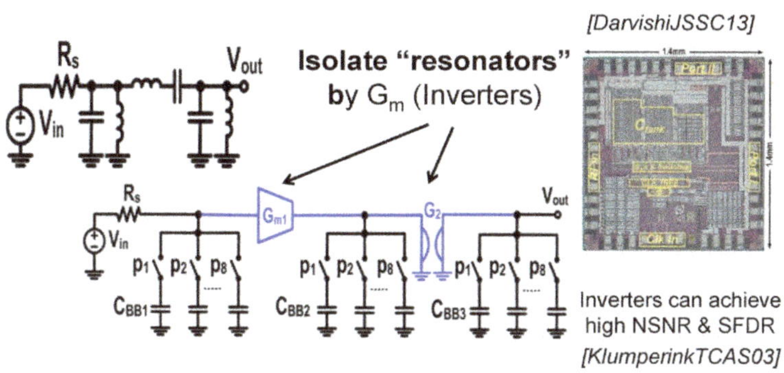

A 6th order LC prototype filter can be approximated using 3 switched-capacitor N-path band-pass filter sections. To avoid unwanted mutual loading effects, high-ohmic voltage sensing and current injection is used, i.e. transconductors are used. CMOS Inverters are used because of their high SFDR, which is related to Normalized Signal to Noise Ratio (NSNR, see [KlumperinkTCAS03]).

48 Linearity and NF

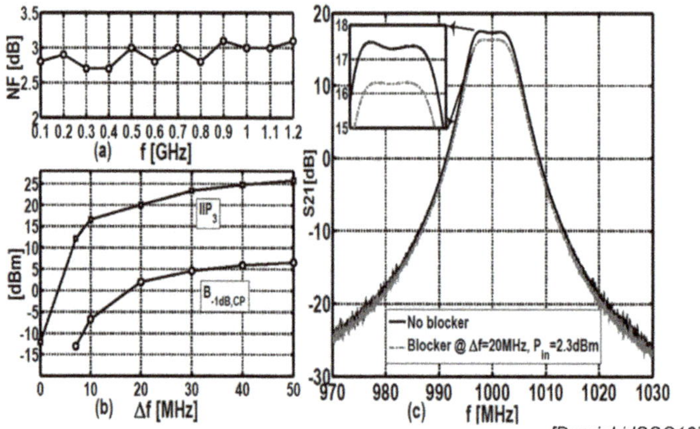

[DarvishiJSSC13]

This slide shows that the achieved filter shape is close to a flat-top shape, while steep roll-off and large out-of-band attenuation is achieved. The filter can be tuned from 0.1 to 1.2 GHz by the clock frequency, while its noise figure remains close to 3dB. Out-of-band IIP3>25dBm and a blocker compression point >6dBm is achieved, where the first fully passive N-path filter stage plays an important role.

49 Outline

- Introduction
- Basic N-path Filters & Mixers – the good
- N-path filters: the bad
- LPTV Wonderland
- Gm-Assisted approaches
- **Improve Selectivity Passively**
- Conclusions

Transconductors help to improve filter shape but are significantly less linear than passive circuits. Hence several ideas have been proposed to improve selectivity exploiting passive circuits.

50 Extra filter roll-off by Passive IIR-filtering

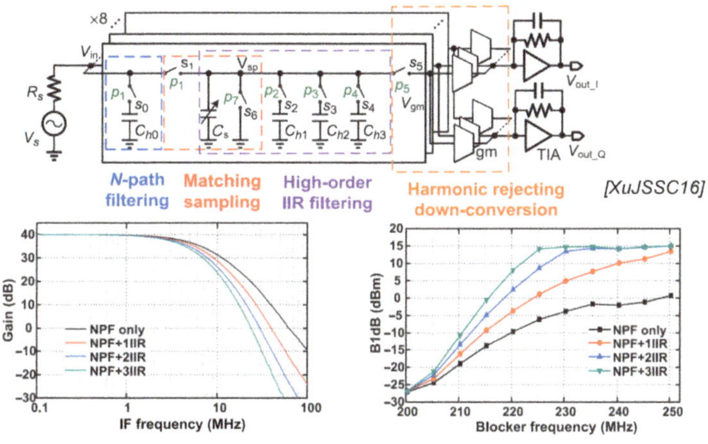

Extra filter roll-off can be implemented by cascading an N-path filter with switched-capacitor Infinite Impulse Response (IIR) filter stages. Here up to 3 extra orders are added, resulting in improved roll-off and out-of-band blocker compression points. A disadvantage is extra loss, i.e. noise figure degradation.

51 Passive RX BPF & Mixer architecture

- 1st V-V BPF attenuates OOB voltage signal
- 2nd V-I BPF bypasses OOB current signal

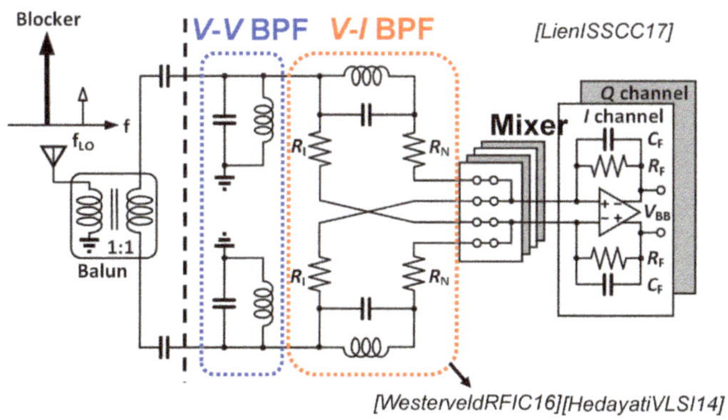

To improve the filter shape targeting high linearity, consider this concept with LC-Band-Pass filter modeling N-path filters. The first voltage-mode stage implements band-pass voltage filtering (V-V BPF). The resulting voltage is converted to current by resistors R_I and mixed to baseband, while a cross-coupled notch-filter current is subtracted (hence overall extra band-pass filtering is implemented).

52 Current Path for In-band signals

- LC tanks act as "open"

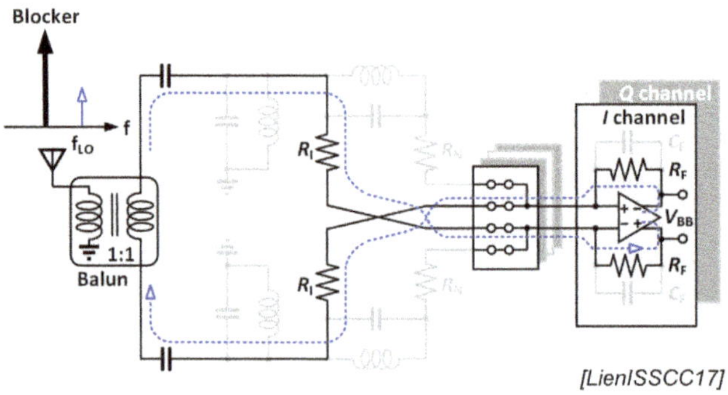

[LienISSCC17]

This slide shows the in-band signal flow. The LC-tanks are high-ohmic and hardly affect the signal flow. The wanted signal is V-I converted, mixed down, and converted to voltage again by R_F in the base-band transimpedance amplifiers. The 4-Path filtering and mixing is compatible with an I/Q output.

53 for Out-of-band (OOB) signals

- LC tanks act as "short"
- Blocker current circulates in R-network

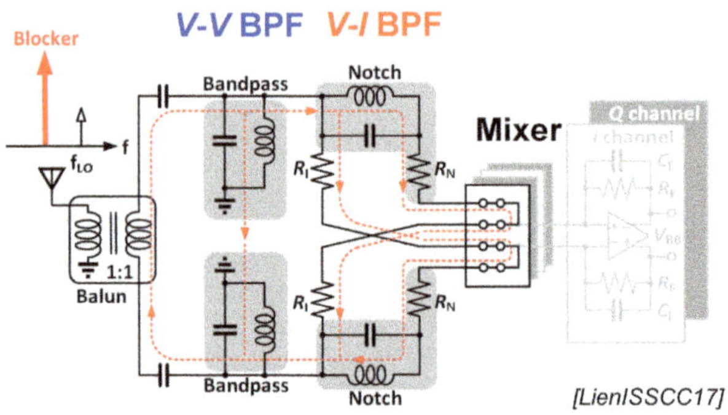

[LienISSCC17]

Out-of-band, the LC-tank conduct significant current and blocker current is bypassed. The V-I filter has a series-resistance R_N, which is chosen equal to R_I, so that far out-of-band its current cancels the current in R_N (check the current-flow including the cross-coupling). Hence the out-of-band blocker current is largely by-passed and does hardly affect the I/Q baseband outputs.

54 Measured gain and S11

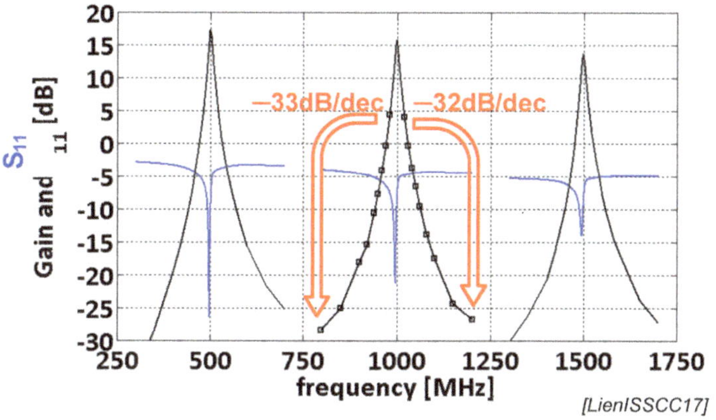

This slide shows the measured gain from RF to the I-output and the narrowband S_{11} matching. Filter roll-off is indeed higher than -20dB/decade as achieved in a traditional N-path filter, although less than -40dB/decade, as the two N-path filter sections load each other.

55 N-path filter (differential)

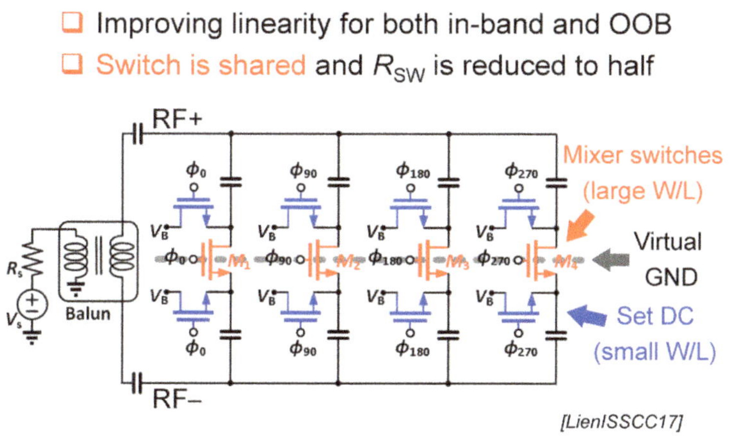

This slide shows how the first V-V filter was actually implemented. To improve linearity one shared switch (red) is used for two capacitors connected to differential RF-input voltages RF+ and RF-. Much smaller (blue) switches are used to set the bias level periodically to V_B. For pure differential input signals, the potential at the middle point of the red switching MOSFET remains almost constant (virtual ground).

56 Linearity simulation

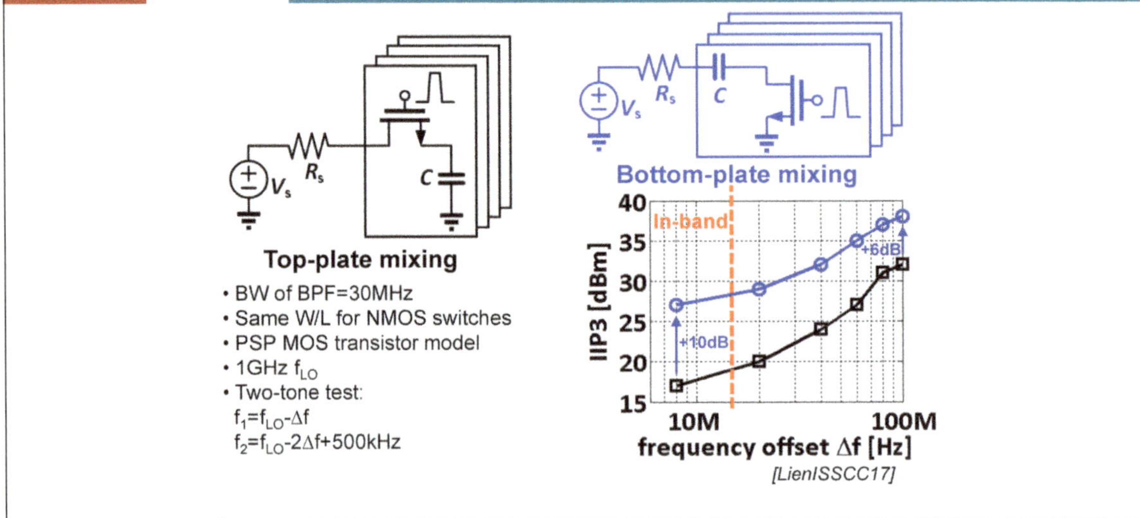

Exploiting the virtual ground, a half-circuit model of the mixer can be drawn, with a grounded switch and a capacitor directly connected to the signal source. This "bottom-plate mixing" is compared here to the traditional "top-plate" mixing, assuming identical switch and capacitor sizing. IIP3 simulations show 10dB improvement for in-band linearity and 6dB for out-of-band. Intuitively this can be understood as the top-plate mixing produces much more gate-channel voltage modulation.

57 Linearity improvement simulation

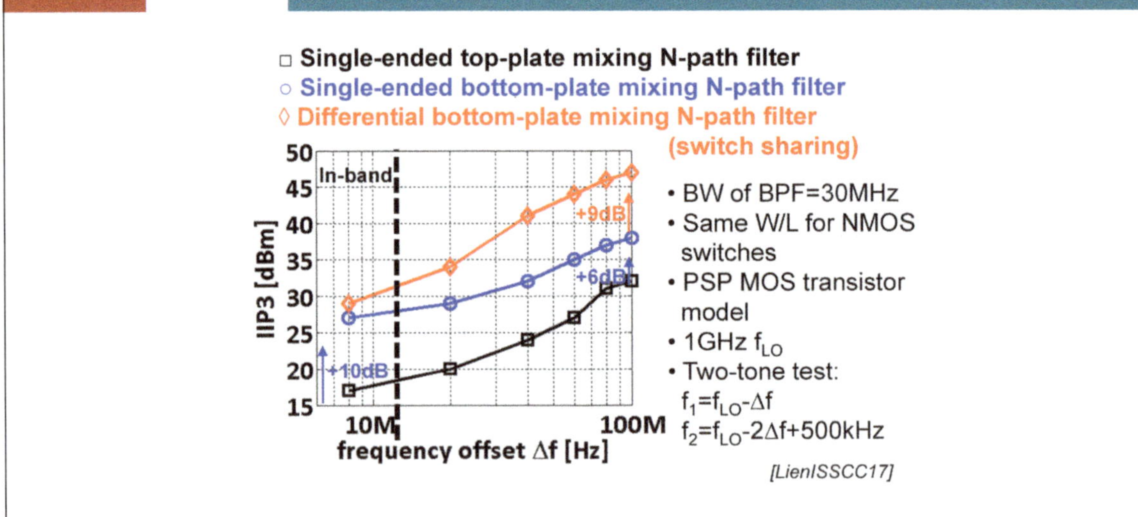

This slide shows the single-ended simulations from the previous slide as well as the IIP3 simulation of full differential bottom-plate mixer with switch sharing, which halves the effective switch-resistance. This leads to less out-of-band voltage modulation and 9dB better IIP3 (can be verified analytically).

58 Measured IIP3 and IIP2

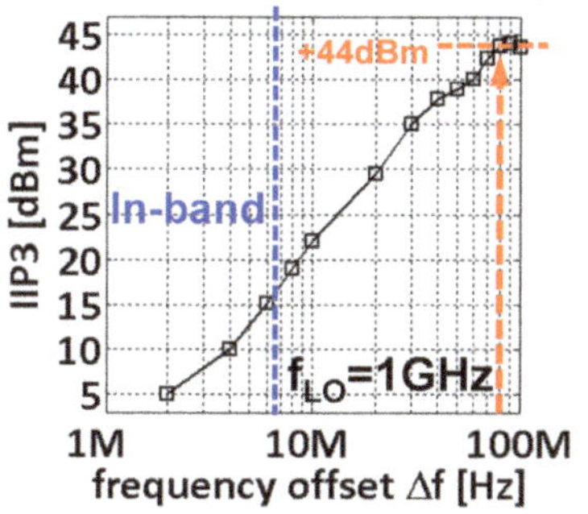

$f_1=f_{LO}-\Delta f$

$f_2=f_{LO}-2\Delta f+500kHz$

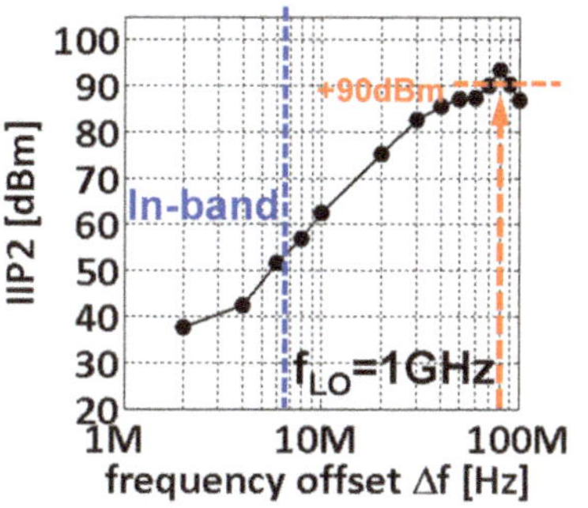

$f_1=f_{LO}-\Delta f$

$f_2=f_{LO}-\Delta f+500kHz$

[LienISSCC17]

This slide shows measured 2nd and 3rd order intercept point measurements versus offset-frequency. An IIP3 of 44dBm and IIP2 of 90dBm is achieved out-of-band.

59 Measured blocker Noise Figure

[LienISSCC17]

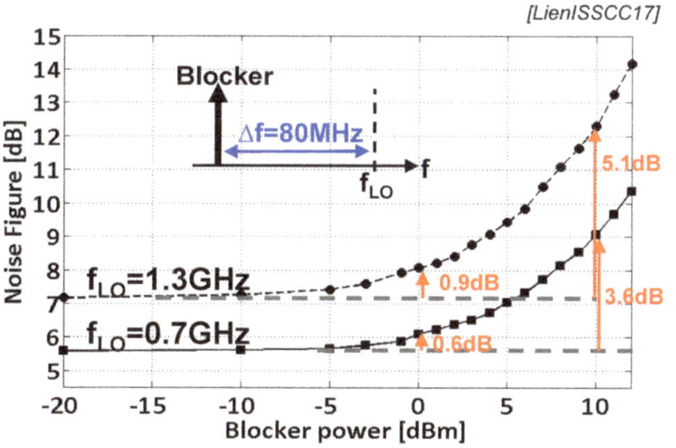

Note: NF degrades due to Parasitic-C @ RF ("Switched-capacitor Resistor" loss)

Noise figure in the presence of a blocker with variable blocker power was also measured. At 0dBm blocker power and 80MHz blocker offset-frequency, noise figure only degrades by less than 1dB.

60 Synthesize complex poles via feedback

Add positive feedback C_2

+

negative feedback C_1

complex poles !

while also:
Low Noise
High Linearity

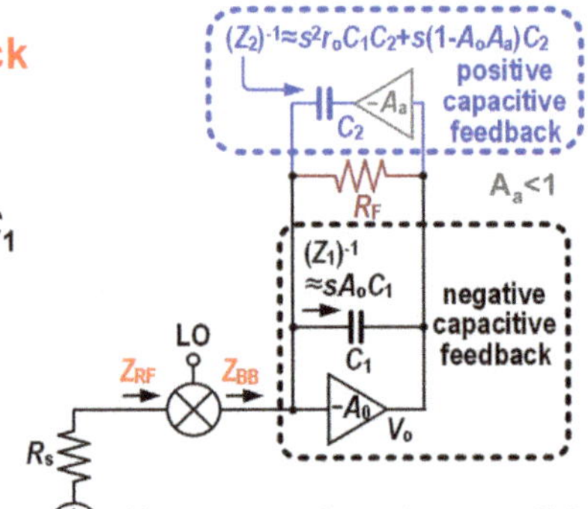

Note: output impedance: r_o // C_o

[LienJSSC18]

In order to further improve selectivity, a positive capacitive feedback technique has been proposed for the baseband amplifier in a mixer-first receiver. Together with the regular negative feedback network in a transimpedance baseband amplifier, this allows for realizing complex poles to improve the filter shape, while also achieving low noise and high linearity (to be shown next).

61 I/Q Receiver with the Attenuator

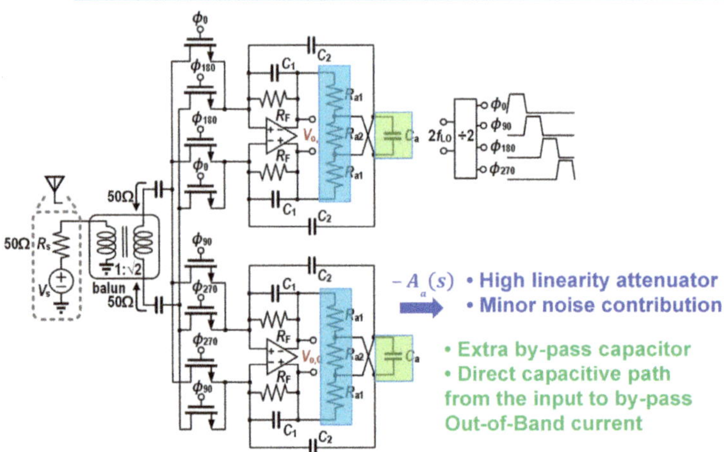

[LienJSSC18]

In order to achieve high linearity and low noise, resistive attenuator circuits are used behind the baseband amplifier to drive the positive feedback capacitors. Bypass capacitors provide an out-of-band blocker shunting path, as illustrated on the next slide.

62 Out-of-Band Current Path

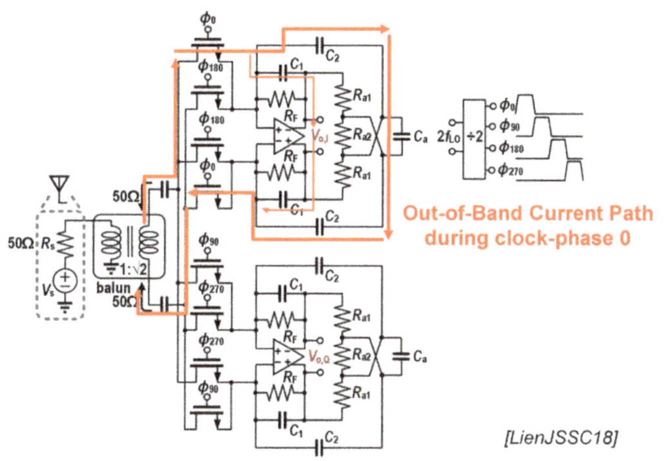

With red arrows the out-of-band current path of the receiver is illustrated. Apart from a path via the amplifier, a fully passive path also exists, which helps to improve linearity.

63 Measured/simulated gain and S_{11}

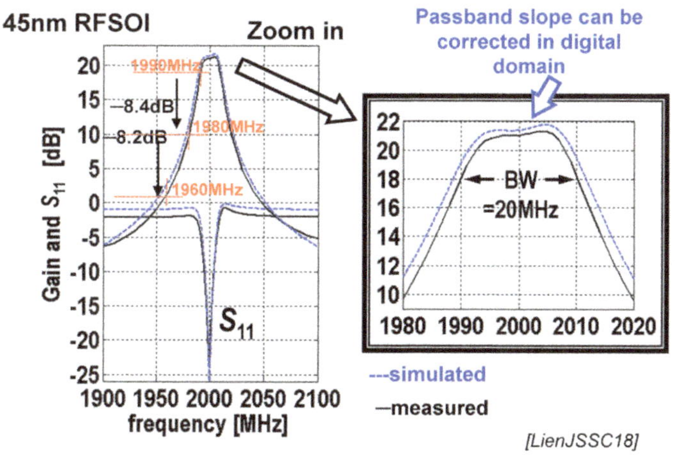

This slide shows the measured pass-band gain and S_{11}. The filter close-in roll-off of the receiver indeed benefits from the positive feedback and the filter shape corresponds to a complex-pole response. Due to complex resistive feedback exploiting the quadrature base-band signals, the S_{11} is nicely centered around the switching frequency of 2GHz. On the other hand, some tilt in the pass-band response occurs, but this can be compensated in the digital domain.

64 Measured IIP3 and IIP2

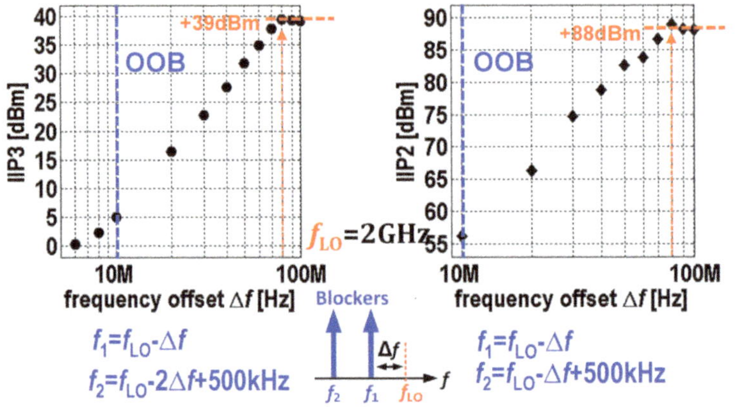

[LienJSSC18]

Measured IIP3 and IIP2 are shown here and are only a few dB worse that achieved with the bottom-plate mixing technique.

65 Measured gain and DSB NF

- NF=2.4dB(f_{LO}=1GHz), NF=7.2dB(f_{LO}=8GHz)
- NF<5.4dB(f_{LO}<6GHz)

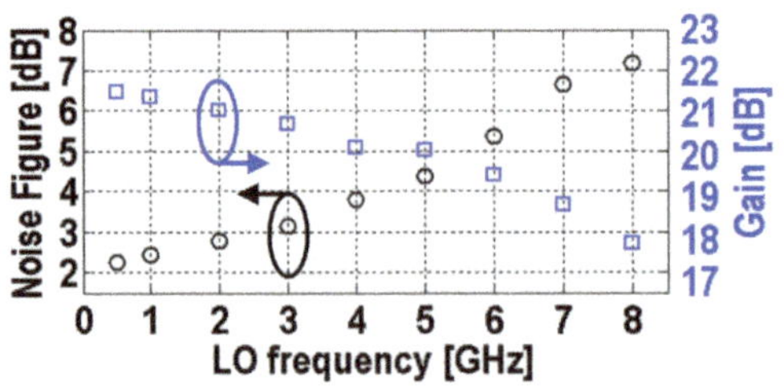

[LienJSSC18]

The receiver covers a very wide range of RF-frequency bands from a 500MHz to well beyond 6GHz. Noise figure can be as low as 2dB and remains below 5.5dB up to 6GHz, while the gain drops from 21.5dB to 19.5dB over this 6GHz range.

66 Even more roll-off needed? Add TZs

- Increase filter slope adding transmission zeros (TZs):

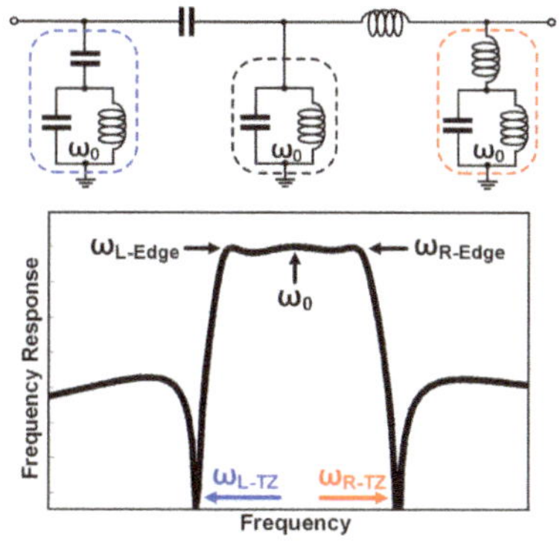

Courtesy Pingyue Song & Hossein Hashemi [SongISSCC18]

Even steeper roll-off can potentially be achieved adding transmission zeros (TZs) as shown in this example.

67 Implementation with extra inductors and $-R_{QB}$

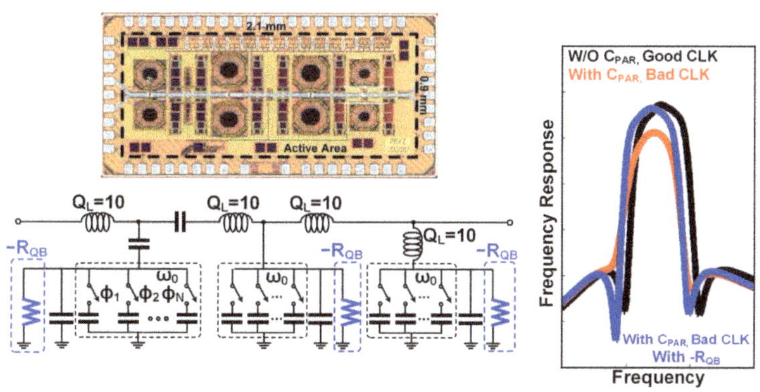

Courtesy Pingyue Song & Hossein Hashemi [SongISSCC18]

The filter can be implemented using N-path filters for the high-Q resonators. In order to limit interaction between N-path filter sections by capacitive charge sharing, inductors were added. Their finite Q degrades pass-band shape but still steep roll-off can be achieved due to the transmission zeros. Negative resistors are added to compensate for losses due to finite-Q inductors and charge sharing.

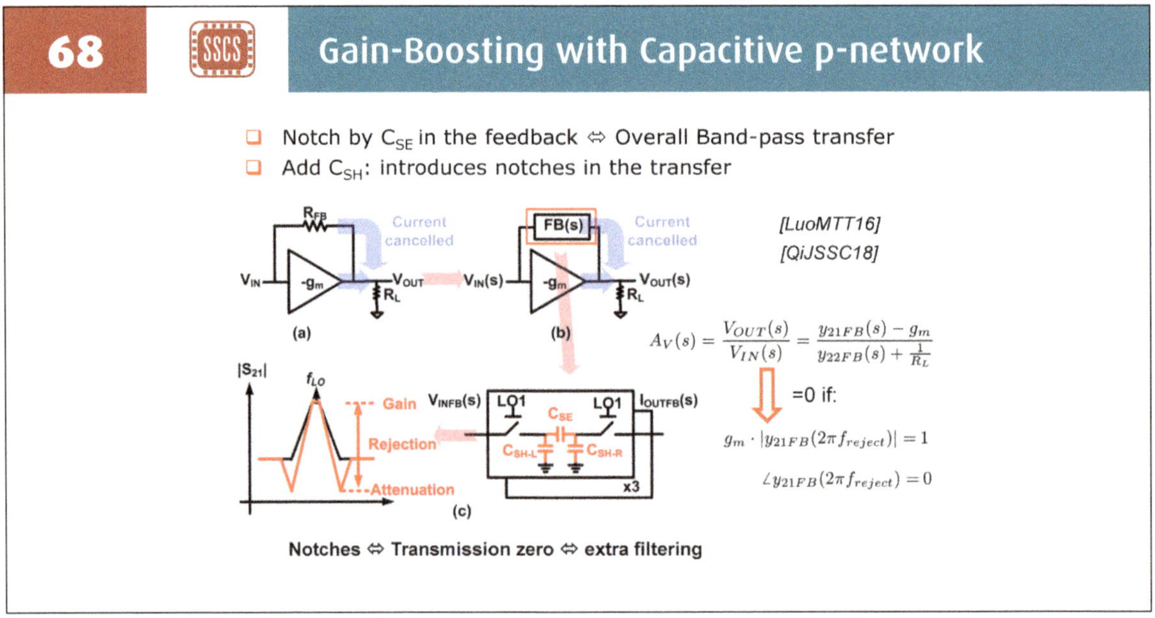

Instead of explicitly implementing transmission zeros, it has also been proposed to use a capacitive PI-network instead of simple switched capacitors in the feedback path of a gain-boosted N-path filter. Modelling the switched capacitors to ground as LC-tanks, phase and amplitude conditions for zero voltage gain (i.e. a transmission zero) can be found, and it is possible to improve filter shape (see the next slide).

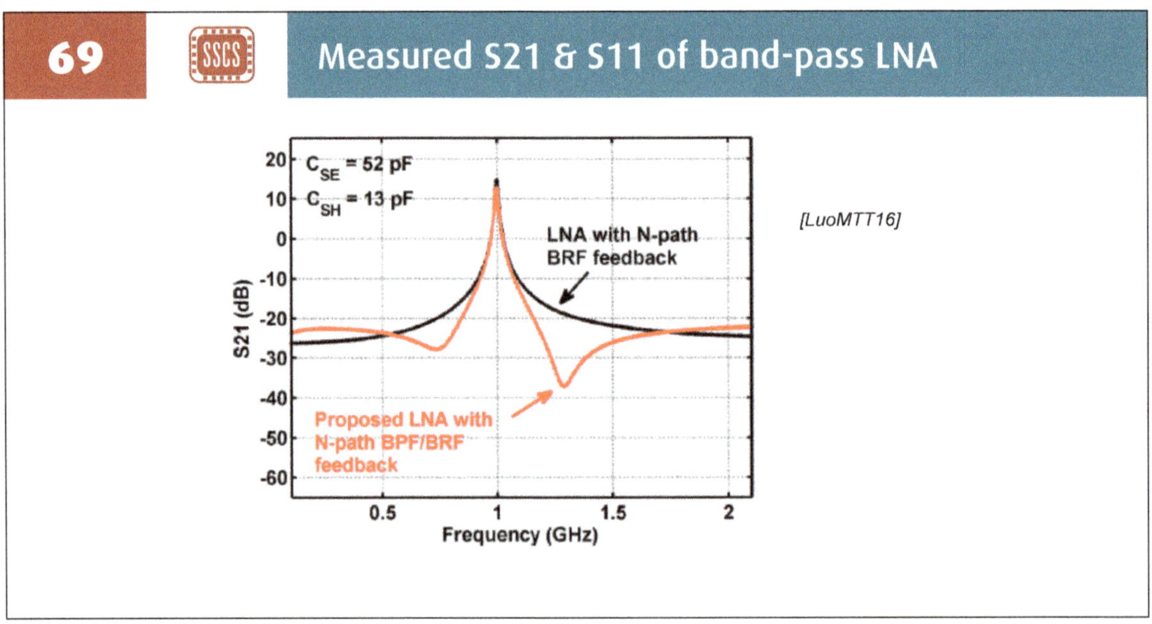

This slide shows the measured filter shape and compares it to the case without transmission zeros. Close to the pass-band the shape is similar, but the transmission zeros provide 8-17dB extra rejection.

70 Conclusions

- N-path filter and mixers allow for high-Q, high-linearity receivers with a digitally programmable center-frequency
- N-path Filter suppression and shape can be improved by:
  - Higher order filter via Transconductor/Gyrator coupling
  - Gain-Boosting
  - Adding extra switched-capacitor IIR-filtering
  - Exploiting positive capacitive feedback
  - Adding transmission zeros
- Achievable linearity and compression can be pushed by:
  - Passive-first approaches with a large Rs/Rsw ratio
  - High-swing switch drivers (thick oxide SOI)
  - Bottom-plate mixing

This slide summarizes the main techniques discussed and the next slides benchmark key performance parameters of the discussed designs.

71 IIP3 versus normalized offset-frequency

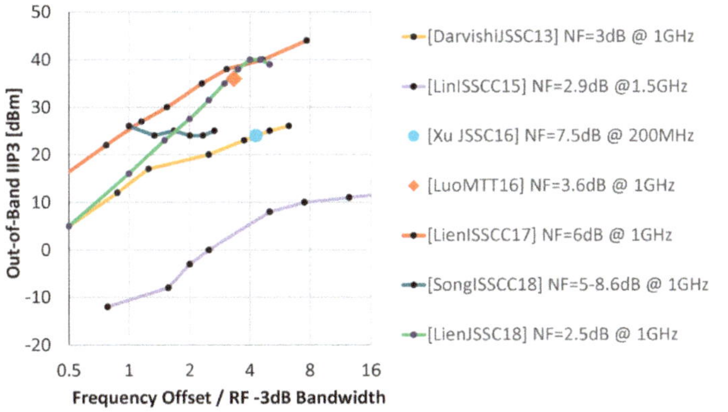

Note: RF-3dB BW = Equivalent RF-referred (double side-band) -3dB bandwidth

Out-of-band IIP3 versus offset frequency of published designs is plotted here (note that normalization is done versus RF -3dB bandwidth, i.e. 0.5 corresponds to the -3dB baseband roll-off point). Noise Figure and nominal receiver center frequency are indicated in the figure caption. The best IIP3 is achieved by the bottom-plate mixing [LienISSCC17], but the positive capacitive feedback [LienJSSC18] and gain-boosted LNA with capacitive PI-network [LuoMTT16] are not much worse. Note that the slope of improvement of out-of-band IIP3 varies per design, indicating a benefit for [LienJSSC18].

72 B1dB vs. normalized offset-frequency

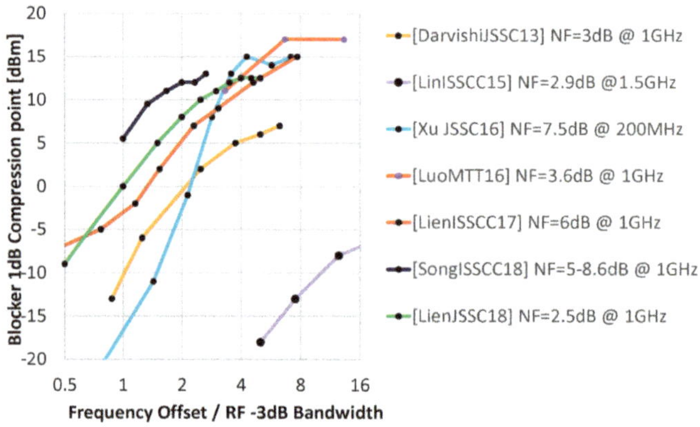

Note: RF-3dB BW = Equivalent RF-referred (double side-band) -3dB bandwidth

For the same designs, the blocker compression point versus offset frequency is shown here. Clearly the fully passive design with transmission zeros helps close-in B1dB [SongISSCC18]. [XuJSSC16] shows the steepest curve (N-path filter + 3rd order IIR), and achieves 15dBm out-of-band B1dB, but for close-in interferers up to 3x the channel bandwidth, other designs still achieve better results.

73 Benchmark table N-Path Filters/Mixers

	Darvishi JSSC13	Lin/Mak ISSCC15	Xu JSSC16	Luo MTT16	Lien ISSCC17	Lien JSSC18	Song ISSCC18
N-Path Filter Technique	6th order 8-Path	Gain-Boost 4-Path	8-Path BP 3x IIR BB	Gain Boost PI-network	2-stage 4-Path	Positive C-fdb 4-Path	13th order 4-Path+TZ
Output:	RF & BB	RF & BB	BB	RF	BB	BB	RF
RF freq. [GHz]	.1 to 1.2	.1 to 1.5	.1 to .7	**.4-6**	.1 to 2	**0.2-8**	0.8-1.1
BW_{-3dB} @RF [MHz]	8	4	7	15	13	20	30
BW_{-20dB} @RF[MHz]	20	36	24	115	86	85	60
BW_{-20dB}/BW_{-3dB}	2.5	9	3.4	7.7	6.6	4.3	**2**
NF [dB] @ RF-freq [Hz]	3 1G	2.9 @1.5G	7.5 @200M	3.6 @1G	6.3 @1G	**2.5 @1.4**	5.0 – 8.6 .8-1.1 G
NF_{blk}@0dBm [dB]	N.A.	13.5@80M	9@30M	11@200M	8@80M	**4.7@80M**	N.A.
OOB B1dB [dBm]	+7	-6	**+15**	**+17***	+15	+12	**12@2xBW**
OOB IIP3 [dBm]	+26	+13	+24	+36*	**+44**	+39	24
Area [mm²]	.27 @65nm	**.028** @65nm	2 @40nm	.28 32nm SOI	0.49 28nm	.8 45nm SOI	1.2 65nm
Power @ f [mW]	22-69 @.1-1.2G	**11 @1.5G**	59-106 @.1-.7G	81-210 @.4-6G	38-96mW 0.1-2G	50mW + 30mW/G	80-97 @.8-1.1G

* 2V LO-Drivers (thick-oxide SOI); all others: thin-oxide devices with regular supply

Detailed performance numbers of the discussed design can be found here, with strong points in bold.

74 SSCS

Acknowledges

- PhD students:
  - Michiel Soer, Amir Ghaffari, Milad Darvishi, Remko Stuiksma, Hugo Westerveld, Yuan-Ching Lien
- Colleague Staff members in Twente:
  - Ronan v.d. Zee, Frank van Vliet, Gerard Wienk, Henk de Vries, Anne Johan Annema, Pieter Tjerk de Boer, Andre Kokkeler
- Industrial Support by:
  - NXP, STMicroelectronics, Mediatek, STARS
  - Dutch Science Foundation STW, NWO/TTW

The authors would like to acknowledge the following for their contributions to this work.

75 SSCS

References

[AndrewsJSSC10] C. Andrews and A. C. Molnar, "A Passive Mixer-First Receiver With Digitally Controlled and Widely Tunable RF Interface," *IEEE JSSC*, pp. 2696-2708, 2010.

[Barber1947] N. F. Barber, "Narrow Band-Pass Filter Using Modulation," *Wireless Engineer*, pp. 132-134, 1947.

[BorremansJSSC11] J. Borremans et al., "A 40 nm CMOS 0.4-6 GHz Receiver Resilient to Out-of-Band Blockers," *Solid-State Circuits, IEEE Journal of*, vol. 46, no. 7, pp. 1659-1671, 2011.

[CookJSSC06] B. Cook, A. Berny, A. Molnar, S. Lanzisera, K. Pister, "Low-power 2.4-GHz transceiver with passive RX front-end and 400-mV supply", *IEEE JSSC*, pp. 2757–2766, Dec. 2006.

[DarvishiJSSC13] M. Darvishi, R. van der Zee, B. Nauta, "Design of Active N-Path Filters," *IEEE JSSC*, pp. 2962-2976, Dec. 2013.

[FranksISSCC60] L. Franks and F. Witt, "Solid-state sampled-data bandpass filters," in *Solid-State Circuits Conference. Digest of Technical Papers. 1960 IEEE International*, 1960, vol. III, pp. 70-71.

[FranksBell60]] L. Franks and I. Sandberg, "An alternative approach to the realizations of network functions: The N-path filters," *Bell Syst. Tech. J.*, pp. 1321–1350, Sep. 1960.

[GhaffariJSSC11] A. Ghaffari, E. Klumperink, M. Soer, and B. Nauta, "Tunable high-Q N-path band-pass filters: Modeling and verification," *IEEE JSSC*, pp. 998–1010, May 2011.

[GhaffariJSSC13] A. Ghaffari, E. A. M. Klumperink, B. Nauta, "Tunable N-Path Notch Filters for Blocker Suppression: Modeling and Verification," *IEEE JSSC*, vol. 48, no. 6, pp. 1370, 1382, June 2013

[HedayatiVLSI14] Hedayati, V. Aparin, and K. Entesari, "A +22dBm IIP3 and 3.5dB NF wideband receiver with RF and baseband blocker filtering techniques," in *2014 Symposium on VLSI Circuits Digest of Technical Papers*, 2014, pp. 1-2.

[IzukaTCAS16] T. Iizuka and A. A. Abidi, "FET-R-C Circuits: A Unified Treatment - Part I: Signal Transfer Characteristics of a Single-Path," *IEEE Transactions on Circuits and Systems I: Regular Papers*, vol. 63, pp. 1325-1336, 2016; Part II: Extension to Multi-Paths, Noise Figure, and Driving-Point Impedance,",--, vol. 63, pp. 1337-1348, 2016.

[KlumperinkTCAS03] E. A. M. Klumperink and B. Nauta, "Systematic comparison of HF CMOS transconductors," in *IEEE Transactions on Circuits and Systems II: Analog and Digital Signal Processing*, vol. 50, no. 10, pp. 728-741, Oct. 2003.

[LiempdJSSC14] B. van Liempd et al., "A 0.9 V 0.4–6 GHz Harmonic Recombination SDR Receiver in 28 nm CMOS With HR3/HR5 and IIP2 Calibration," in *IEEE Journal of Solid-State Circuits*, vol. 49, no. 8, pp. 1815-1826, Aug. 2014.

[LePage1953] W. R. LePage, C. R. Cahn, and J. S. Brown, "Analysis of a comb filter using synchronously commutated capacitors," *American Institute of Electrical Engineers, Part I: Communication and Electronics*, Tr. of the, vol. 72, pp. 63-68, 1953.

[LinTCAS14] Zhicheng Lin; Pui-In Mak; Martins, R. P., "Analysis and Modeling of a Gain-Boosted N-Path Switched-Capacitor Bandpass Filter,", *IEEE TCAS-I*, vol. 61-9, pp. 2560, 2568, Sept. 2014

[LinISSCC15] Zhicheng Lin; P.-I. Mak; R. P. Martins, "2.4 A 0.028mm2 11mW single-mixing blocker-tolerant receiver with double-RF N-path filtering, S11 centering, +13dBm OB-IIP3 and 1.5-to-2.9dB NF" *ISSCC*, 22-26 Feb. 2015.

[LienISSCC17] Y. Lien, E. Klumperink, B. Tenbroek, J. Strange, and B. Nauta, "A high-linearity CMOS receiver achieving+44dBm IIP3 and +13dBm B1dB for SAW-less LTE radio," *ISSCC*, pp. 412-413 , 2017.

[LienJSSC18] Y. C. Lien, E. A. M. Klumperink, B. Tenbroek, J. Strange, and B. Nauta, "Enhanced-Selectivity High-Linearity Low-Noise Mixer-First Receiver With Complex Pole Pair Due to Capacitive Positive Feedback," *IEEE Journal of Solid-State Circuits*, vol. PP, no. 99, pp. 1-13, 2018.

[LuoMTT16] C. k. Luo, P. S. Gudem, and J. F. Buckwalter, "A 0.4 - 6-GHz 17-dBm B1dB 36-dBm IIP3 Channel-Selecting Low-Noise Amplifier for SAW-Less 3G/4G FDD Diversity Receivers," *IEEE MTT*, vol. 64, no. 4, pp. 1110-1121, 2016.

[MirzaeiTCAS11] A. Mirzaei, H. Darabi, "Analysis of Imperfections on Performance of 4-Phase Passive-Mixer-Based High-Q Bandpass Filters in SAW-Less Receivers," *IEEE TCAS-I*, pp.879, 892, May 2011.

[ParkJSSC14] J. W. Park and B. Razavi, "Channel Selection at RF Using Miller Bandpass Filters," *Solid-State Circuits, IEEE Journal of*, vol. 49, no. 12, pp. 3063-3078, 2014.

[PavanTCAS17-1] S. Pavan, E. A. M. Klumperink, "Simplified Unified Analysis of Switched-RC Passive Mixers, Samplers, and N -Path Filters Using the Adjoint Network, *IEEE TCAS-I*, Vol. 64, Issue 10, pp. 2714-2725, 2017.

[PavanTCAS18-1] S. Pavan, E. A. M. Klumperink, "Analysis of the Effect of Source Capacitance and Inductance on N-Path Mixers and Filters, *IEEE TCAS-I*, vol. 65, no. 5, pp. 1469-1480, May 2018..

[PavanTCAS18-2] S. Pavan, E. A. M. Klumperink, "Generalized Analysis or Higher Order N-Path Mixer and Filters using the Adjoint Network", accepted for *IEEE TCAS-I*, 2018 (prepublication doi: 10. 1109/TCSI. 2018. 2816342) .

[QiJSSC18] G. Qi, B. van Liempd, P. I. Mak, R. P. Martins and J. Craninckx, "A SAW-Less Tunable RF Front End for FDD and IBFD Combining an Electrical-Balance Duplexer and a Switched-LC N-Path LNA," *IEEE JSSC*, doi: 10. 1109/JSSC. 2018. 2791477

[RuJSSC10] Z. Ru, E. A. M. Klumperink, and B. Nauta, "Discrete-Time Mixing Receiver Architecture for RF-Sampling Software-Defined Radio," *Solid-State Circuits, IEEE Journal of*, vol. 45, no. 9, pp. 1732-1745, 2010.

[Smith1953] B. D. Smith, "Analysis of Commutated Networks," *Aeronautical and Navigational Electronics, Transactions of the IRE Professional Group on IRE Trans.*, vol. PGAE-10, pp. 21-26, December 1953.

[SoerISSCC09] M. C. M. Soer, E. A. M. Klumperink, Z. Ru, F. E. van Vliet, B. Nauta,"A 0.2-to-2.0GHz 65nm CMOS Receiver Without LNA Achieving >11dBm IIP3 and <6.5dB NF," *ISSCC*, pp. 222-223, Feb. 2009.

[SoerTCAS10] M. C. M. Soer, E. A. M. Klumperink, P. T. de Boer, F. E. van Vliet, B. Nauta, "Unified Frequency-Domain Analysis of Switched-Series-RC Passive Mixers and Samplers," *IEEE TCAS-I*, pp. 2618-2631, 2010.

[SongISSCC18] Pingyue Song, Hossein Hashemi, "A 13th-Order CMOS Reconfigurable RF BPF with Adjustable Transmission Zeros for SAW-Less SDR Receivers", *ISSCC*, pp. 416-418, Feb. 2018.

[XuJSSC16] Y. Xu and P. R. Kinget, "A Switched-Capacitor RF Front End With Embedded Programmable High-Order Filtering," *IEEE Journal of Solid-State Circuits*, vol. 51, no. 5, pp. 1154-1167, 2016.

[WesterveldRFIC16] H. Westerveld, E. A. M. Klumperink, B. Nauta, "A cross-coupled switch-RC mixer-first technique achieving41dBm out-of-band IIP3", *Radio Frequency Integrated Circuits Symposium* (RFIC) pp. 246-249, 2016.

[YangTCAS15] D. Yang, C. Andrews, and A. Molnar, "Optimized Design of N-Phase Passive Mixer-First Receivers in Wideband Operation," *IEEE Transactions on Circuits and Systems I: Regular Papers*, vol. 62, no. 11, pp. 2759-2770, 2015.

CHAPTER 07

Doherty Architecture for mm-wave/RF Digital and Analog Power Amplifiers

Hua Wang

Georgia Institute of Technology, USA

Spectrally efficient modulations with large peak-to-average-power ratios (PAPR) are widely employed in many high-performance wireless communication systems from low-GHz radios to mm-Wave links. This growing trend has posed stringent requirements on next-generation power amplifier (PA) designs and transmitter architectures. Going beyond more conventional PA designs that mostly focus on peak output power and efficiency, next-generation PA solutions should achieve both power efficiency and linearity in the entire power back-off region (PBO) to amplify the large-PAPR signals with high efficiency and fidelity. Doherty architecture is recently gaining an increasing interest due to its PBO efficiency enhancement and its compatibility with broadband and linear amplifications. This tutorial presentation will first introduce the Doherty PA and its operation principle. Multiple popular architectures will be presented, and their pros and cons will be discussed. Several recently published PAs using standard CMOS/SiGe processes will be shown as state-of-the-art design examples to demonstrate the use of Doherty architecture in low-GHz mixed-signal power amplifiers and mm-Wave 5G multi-band analog power amplifiers.

1 Outline

- **Introduction**
- Doherty PA Configurations
- Analog Doherty PAs at RF Frequency
- Digital Doherty PAs at RF Frequency
- Mm-Wave Doherty PAs
- Conclusion

Here is the outline of the chapter. I will first give an introduction on Doherty Power amplifiers and explain why we need them and when we need them. Next, I will delve deeper into the Doherty Power amplifier configurations. We will briefly review the original vacuum tube designs proposed by William Doherty in 1936, and I will explain how those original designs now evolve into modern Doherty PA architectures. I will also show several unique advantages and design challenges of modern Doherty PAs. In the next three sections, I will present multiple Doherty amplifier design examples to demonstrate their use in analog RF PAs, digital RF PAs, and mm-Wave PAs. A summary session will conclude my talk.

2 Introduction (I)

- **Wireless electronics are everywhere in our daily life.**
  - Wireless communication: from 1G to 4G and future 5G

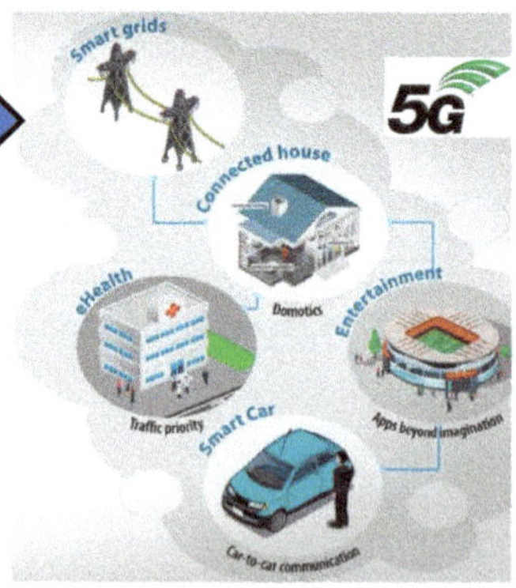

  - Internet of things (IoT)
  - Wireless sensor networks
  - Smart cities
  - **Ubiquitous computing**
  - **Augmented reality (AR)**
  - **Virtual reality (VR)**
  - **Autonomous vehicles**

Wireless communication is now ubiquitous in our society. A study in 2016 showed that the total energy to move digital information wirelessly in one year has now exceeded the total energy to fly all the commercial airlines that transport people from one place to another. Moreover, the upcoming 5G deployment is expected to enable numerous new applications, such as ubiquitous computing, augmented reality, virtual reality, and autonomous driving.

3 Introduction (II)

- **High-performance wireless systems with high data rates**
  - Spectrum-efficient modulations → OFDM and high-order QAM

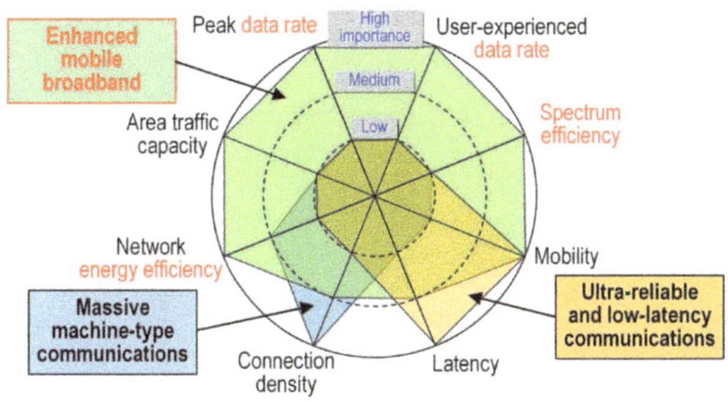

• [1] ITU IMT-2020 vision

To achieve higher data rate within a fixed channel bandwidth, spectrum-efficient modulation schemes are widely used, such as OFDM and high-order QAM.

4 Introduction (III)

- **<u>Spectrum-efficient</u> modulations → challenges for <u>energy-efficient</u> RF/mm-wave signal generation.**
  - OFDM and high-order QAM → large peak-to-average power ratios (PAPR)

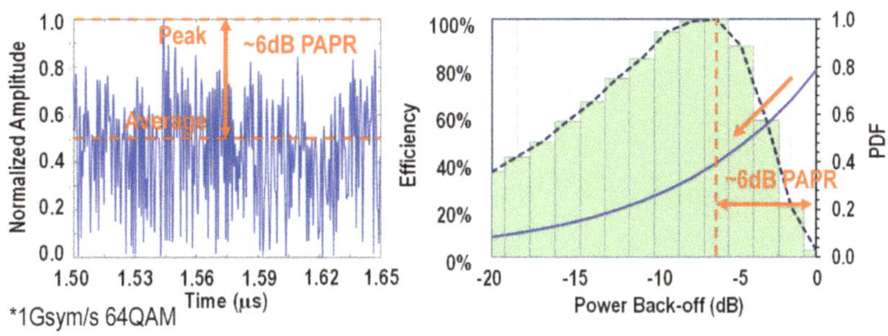

However, such spectrum-efficient modulation schemes normally exhibit large peak-to-average-power-ratio (PAPR). This is making energy-efficient RF/mm-wave signal generation and amplification very challenging. The figure on the lower left shows the normalized time-domain envelope of a 64QAM modulated signal. The time-domain envelope changes drastically and a peak-to-average is about 6dB. Using OFDM multi-carrier scheme will further increase this PAPR. However, most power amplifiers are designed in such a way that their peak efficiency is achieved only at their peak output power. The figure on the right shows an idealistic class-B PA drain efficiency back-off curve and the probability-density-function (pdf) for the envelope of a typical 64QAM modulation signal. It is very clear that the envelope of the 64QAM actually happens at a lower power, called power back-off, for most of the time. Only in rare cases, the output signal reaches its maximum envelope. However, the efficiency of a class-B PA degrades rapidly with power back-off, which will be reduced to only half of the peak efficiency value at the 6dB power back-off.

5

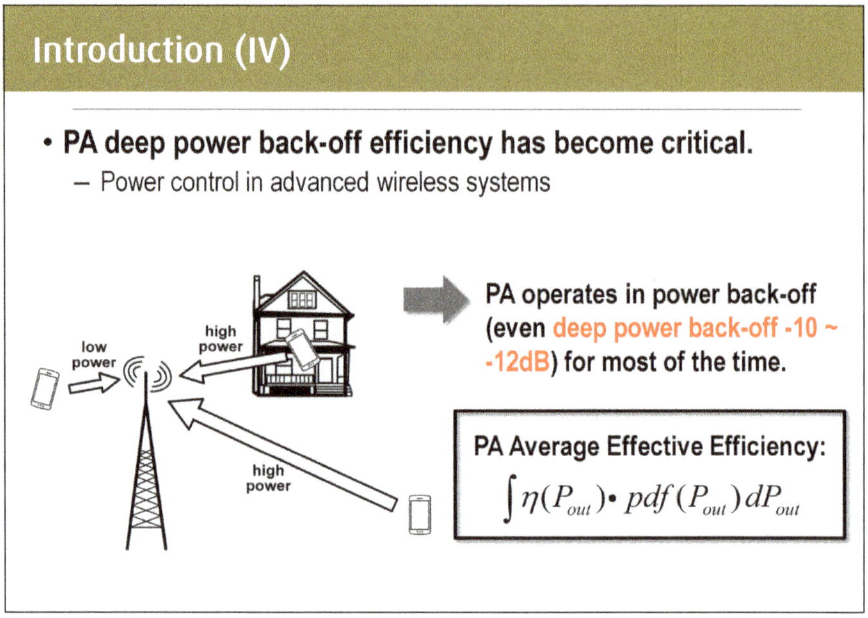

Moreover, besides high PAPR modulations, sophisticated power control schemes in practice will further complicate the need of PA power back-off levels. As a result, practical PAs often operate in power back-off or even deep power back-off like -10~-12dB for most of the time. Considering the high-PAPR complex modulation signals and power control schemes, the PA average efficiency can be expressed as the integration of the PA back-off efficiency curve scaled by the probability density function of the actual envelop of the transmitted signal. This equation again emphasizes that PA back-off efficiency is extremely important.

6

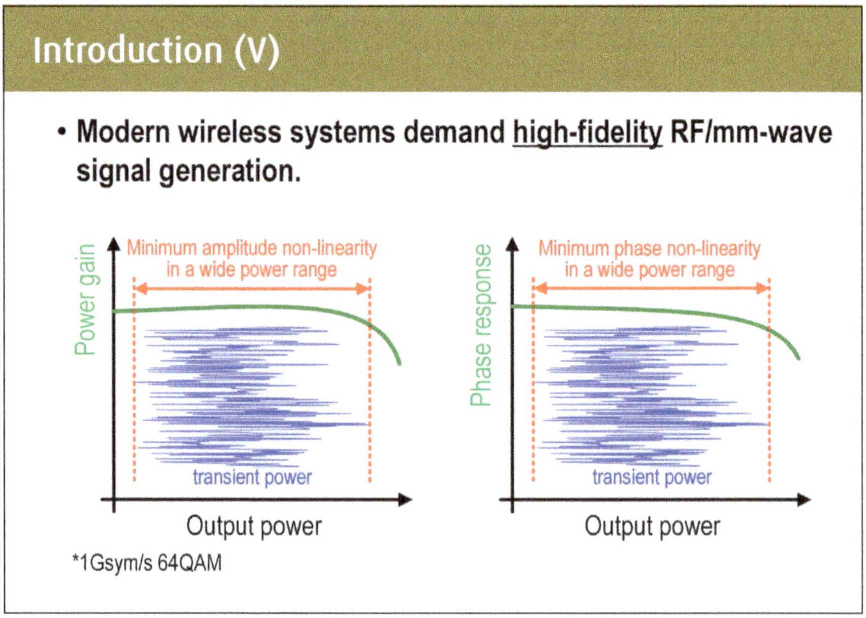

In addition, signal-fidelity is equally important for modern wireless communications. PAs must preserve high linearity that will impact its in-band linearity (such as, Error Vector Magnitude, EVM) for successful signal demodulations at the receiver end and out-of-band linearity (such as, Adjacent Channel Leakage Ratio, ACLR) to satisfy spectrum mask. It is known that all the practical PA architectures and designs invariably exhibit trade-offs between the linearity and efficiency. For wireless communication applications, it is not very meaningful to talk about efficiency without mentioning linearity. From a memoryless large-signal perspective, PAs should exhibit good AM-AM and AM-PM performance through the entire power back-off range to ensure high linearity. Memory effect due to biasing networks or thermal effects should also be considered when amplifying wideband modulation signals.

7 SSCS Introduction (VI)

- **Power amplifier (PA) governs the energy efficiency, linearity, and bandwidth of wireless links.**
  - *Output power*: wireless link budget
  - *Efficiency (peak, PBO, and deep PBO)*: battery life and thermal handling
  - *Linearity*: signal fidelity and quality of service
  - *Bandwidth*: data rate and frequency agility

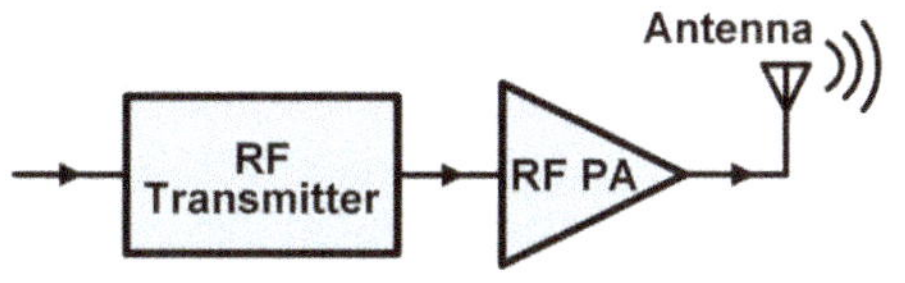

- **High-efficiency, high-fidelity, and broadband RF/mm-wave power generation in silicon is still an unmet challenge.**

• [2] H. Wang, S. Kousai, K. Onizuka, S. Hu, "The Wireless Workhorse: Mixed-Signal Power Amplifiers Leverage Digital and Analog Techniques to Enhance Large-Signal RF Operations," *IEEE Microwave Magazine*, vol. 16, no. 9, pp. 36-63, Oct. 2015.

To summarize, power amplifiers actually govern many performance aspects of wireless links in practice, including its output power, efficiency, linearity, carrier bandwidth, and modulation bandwidth. However, achieving all of these specs in low cost silicon platforms is still an unmet challenge. Thus, there is a strong need for new PA architectures instead of incremental changes.

8 SSCS Introduction (VII)

- **Loadpull and Optimum Load — "Power Amplifier Design 101"**

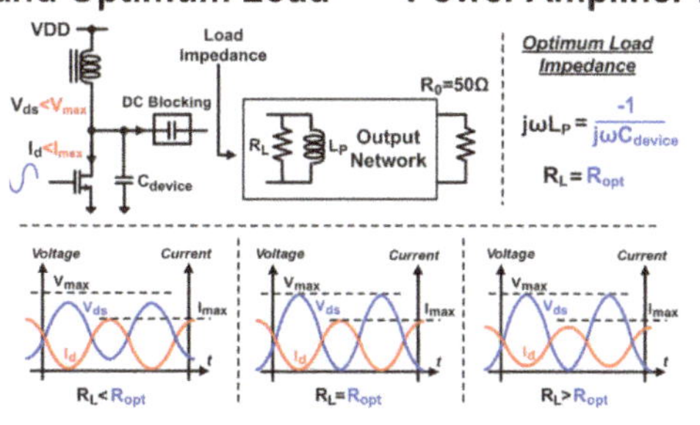

- **$V_{max} = I_d \cdot R_L$ for maximum PA efficiency — "Ohm's Law"**
- **During PA power back-off (PBO), V_{max}(PBO) = I_d(PBO) • R_L(PBO) for maximum PA efficiency — "Ohm's Law" in PBO**

• [3] H. Wang, C. Sideris, and A. Hajimiri, "A CMOS Broadband Power Amplifier With a Transformer-Based High-Order Output Matching Network," *IEEE J. of Solid-State Circuits*, vol. 45, no. 12, pp. 2709 – 2722, Dec. 2010.

Before we delve deeper into advanced PA architectures for back-off efficiency enhancement, let us first review the concept of load pull and optimum load impedance, which is considered as the basic PA designs 101 knowledge. Assume that we have one power device in a common source configuration whose drain is connected to the supply VDD via an inductive choke and a DC blocking capacitor to the output. The input driving signal is fed at the device gate. Assume that we ignore harmonic tuning, device knee voltages, and finite device output impedance. Let us also assume that the device's

8 Introduction (VII)

transconductance, as the ratio between the output current and the input gate voltage, is completely linear, when the device is turned on. For a current-mode PA with a given device sizing, an optimum complex load impedance is needed to achieve the maximum output power and efficiency of the PA. The reactive part of the optimum load impedance should resonate with the device output capacitance at the frequency of operation, while the real part of the optimum load impedance should be set at the value so that the PA's power device achieves its maximum output voltage swing right at its maximum RF output current. In short, we can say $V_{max}=I_d*R_L$, which is essentially like the ohm's law that can be used to describe the PA load pull or load line behaviors.

Obviously, during the PA power back-off, something should be changed to allow a lower amplification power yet still with high efficiency. It is clear that to maintain the peak efficiency, the PA operation should still satisfy the large signal ohm's law, so that the maximum output voltage swing during PBO equals the output current and the load at various PBO levels. Therefore, to maximize the PA efficiency during PBO, we only need to control two of the three fundamental quantities, PA device RF voltage swing, RF output current, or the load impedance.

9 Introduction (VIII)

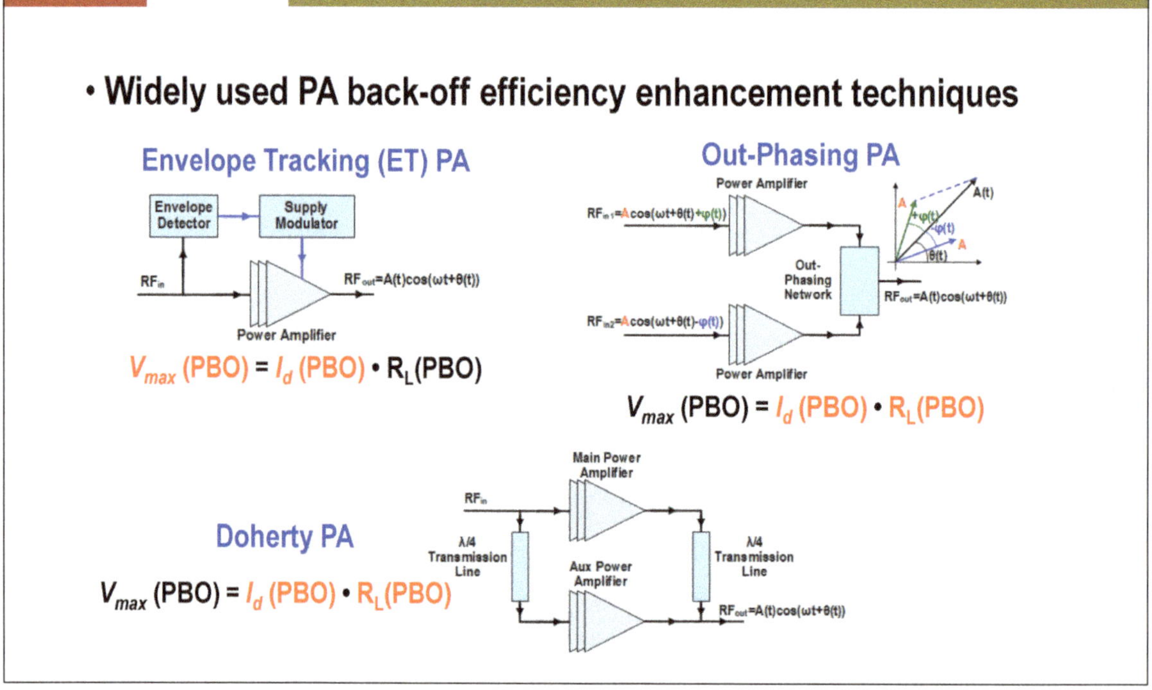

There are three very popular basic PA architectures can enable PA back-off efficiency enhancement. Many existing PA architectures can be considered as the variants of these three architectures or their hybrid combinations. The first one is the envelope tracking (ET) power amplifier. It extracts the real-time amplitude of the modulation signal, based on which the PA supply voltage is modulated, so that the PA always maximizes its RF voltage swing during the back-off. In the perspective of our large signal ohm's law for PAs, the V_{max} and I_d are varying, while the load resistance stays the same. The Envelope Elimination and Restoration (EER) PA is a similar architecture, except that the input signal is a phase-modulated

9 Introduction (VIII)

constant envelope signal, and the PA output envelope is completely modulated by the supply voltage. The second one is the Outphasing power amplifier. It uses two constant envelope signals whose phase difference is real-time modulated, so that the vector summation of the two signals provides the desired envelope-varying modulation signal. In the perspective of our large signal ohm's law for PAs, the I_d and R_L are varying, while the amplifiers' output voltage stays the same. The third one is the Doherty power amplifier. It typically has two PA paths, i.e., the main and auxiliary PAs, which will perform active load modulation to ensure that the main PA output voltage stays the same during the power back-off to maximize its efficiency. Therefore, in our PA ohm's law perspective, the I_d and R_L are varying, while the amplifiers' output voltage stays the same. Next, let us look at the pros and cons of each PA architecture.

10 Introduction (IX)

• **Widely used PA back-off efficiency enhancement techniques**

Envelope Tracking (ET) PA

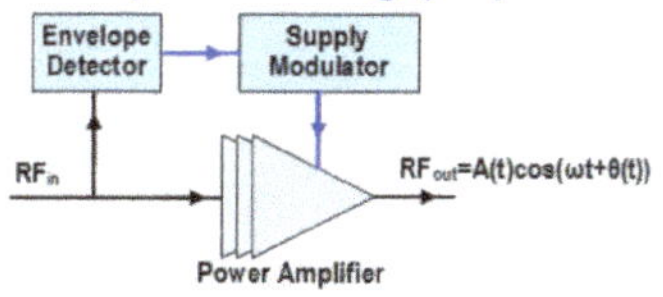

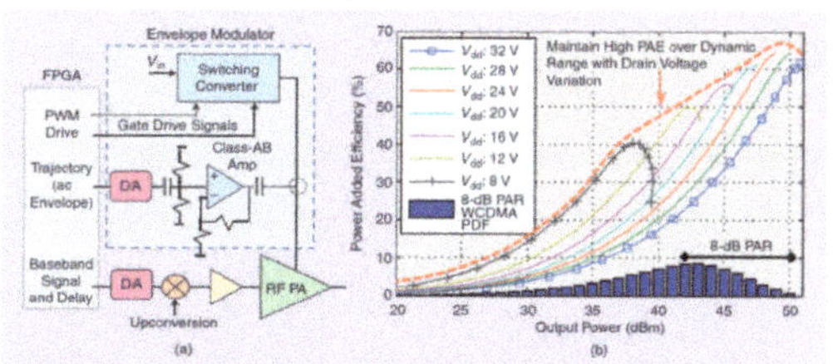

• PA efficiency enhancement over a very large PBO range

• Envelope modulator = an efficient but slow switching-mode converter + a less-efficient but faster feed-back linear amplifier

• Envelope modulator has practical trade-offs among efficiency, speed, accuracy, and power.

• Limiting the modulation rate (~130MHz modulation bandwidth with GaN buck converter + GaN PAs)

• [4] Zoya Popovic, "Amping Up the PA for 5G: Efficient GaN Power Amplifiers with Dynamic Supplies," *IEEE Microwave Magazine*, vol. 18, no. 3, pp. 137 – 149, May 2017.

• [5] Y. Zhang, M. Rodríguez, D. Maksimović, "Very high frequency PWM buck converters using monolithic GaN half-bridge power stages with integrated gate drivers", *IEEE Trans. Power Electron.*, vol. 31, no. 11, pp. 7926-7942, Nov. 2016.

For the ET PA, its advantage is that it can perform PA efficiency enhancement over a very large PBO range, enabling highly efficient power amplification at deep power backoff. Its main challenge is the design of the envelope amplifier. Normally, the supply modulator is composed of a slow but high-efficiency switching mode convertor and a fast but less-efficient linear feedback amplifier to track the real-time signal envelope in an accurate yet efficient manner. In practice, supply modulators often havestringent trade-offs among its own energy efficiency, speed, accuracy, and power dynamic range. Moreover, the supply modulator should handle the envelope signal, which typically exhibits a 3×-5× bandwidth expansion beyond the modulated signal. As a result, reported ET PA often has limited modulation bandwidth. A recently published GaN ET PA with GaN buck convertor supports up-to 130MHz modulation. This may be acceptable for many wireless applications, but is certainly insufficient for GHz-speed modulation data, such as those in 5G systems.

11 Introduction (X)

• Widely used PA back-off efficiency enhancement techniques

Outphasing PA

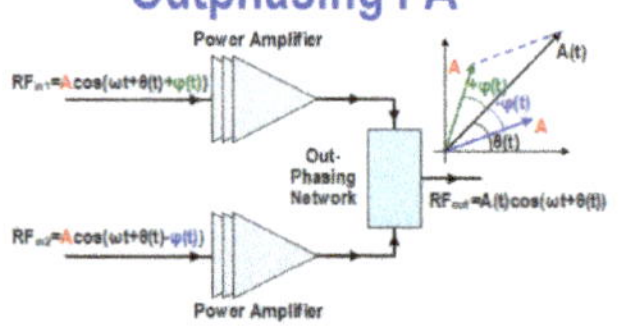

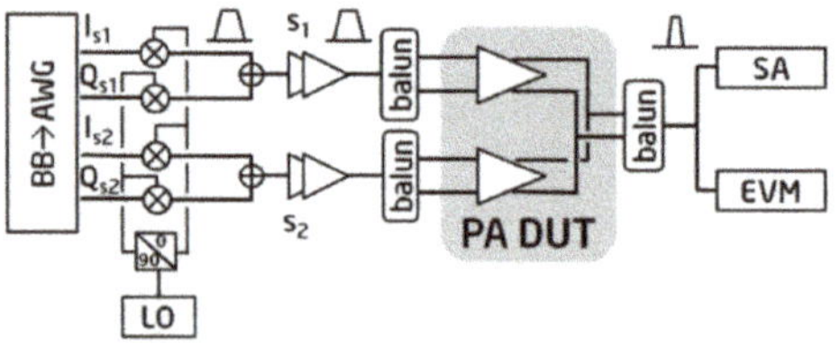

- High PA peak efficiency

  $\varphi(t)=\cos^{-1}[A(t)/2A]$

- Large baseband overhead and complexity to generate the out-phasing signals
- Often requiring extensive digital pre-distortion (DPD)
- Out-phasing network loss, distortion, and carrier BW limiting
- Limiting the modulation rate (<100MHz modulation bandwidth)

• [6] Hongtao Xu, *et al.*, "A Flip-Chip-Packaged 25.3 dBm Class-D Outphasing Power Amplifier in 32 nm CMOS for WLAN Application," *IEEE J. Solid-State Circuits*, vol. 46, no. 7, pp. 1596 - 1605, May 2011.

• [7] T.W. Barton, *et al.*, "Multi-way lossless outphasing system based on an all-transmission-line combiner." *IEEE Trans. Microw. Theory. Tech*, vol. 64, no. 4, pp. 1313-1326, March 2016.

For the outphasing PA, since it is typically built based on multiple nonlinear switching-mode or highly compressed PAs, it can achieve a very high peak efficiency. However, the outphasing angle is determined by the inverse cosine of the target envelope divided by two times of the constant amplitude. This is a very nonlinear computation, and any nonlinear computation will result in signal bandwidth expansion. Since the outphasing signal is at the phase path of the PA, it typically has a 5×-7× bandwidth expansion beyond the modulated signal. Therefore, outphasing signal generation normally requires large baseband overhead and complexity, as well as extensive DPD compensation. Moreover, the outphasing network also presents passive loss, signal distortion due to non-ideal power combining, and also carrier bandwidth limitation. In summary, practical outphasing PA only support limited modulation rate. It is not straightforward to extend outphasing PA to GHz-speed modulation without consuming large power on baseband signal generation.

12 Introduction (XI)

Finally, let us take a look at the Doherty PA. Ideally, it can support a very large modulation bandwidth. It also requires little or even no overhead for baseband processing. It is an RF-in-RF-out amplifier, so it is often employed as a drop-in replacement for class-AB PAs. It is also inherently linear and does not exhibit signal bandwidth expansion or require digital pre-distortion with excessive bandwidth. However, the challenges of Doherty PAs include the large, lossy, and band-width limiting Doherty passive network, and it also require careful controls on the cooperation of the main/auxiliary PAs to ensure its large signal linearity.

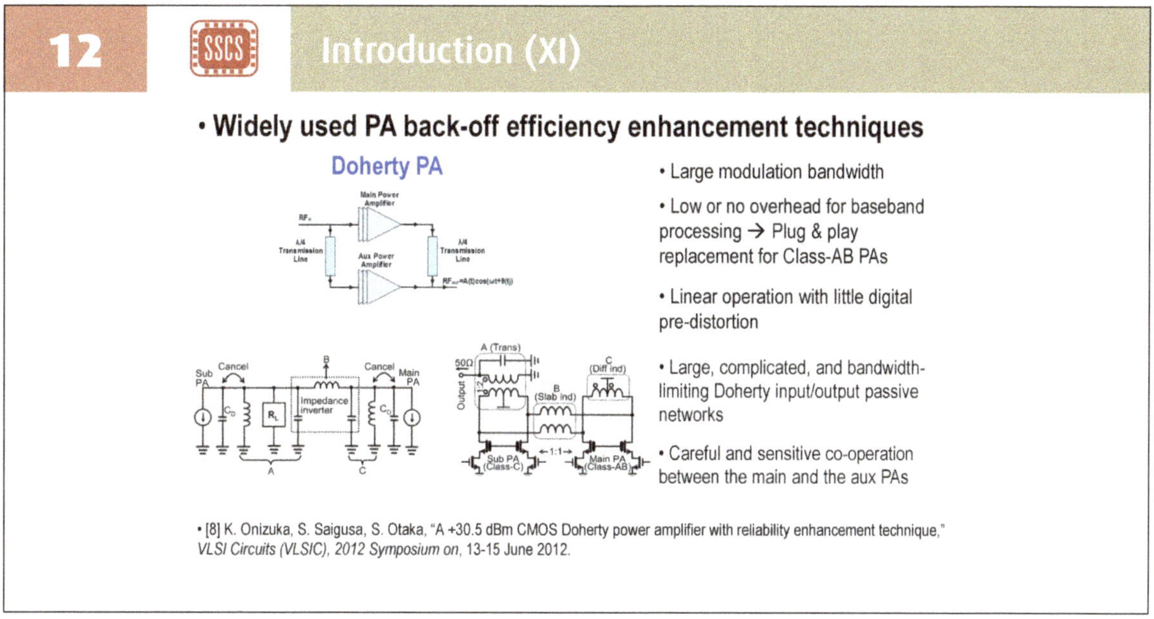

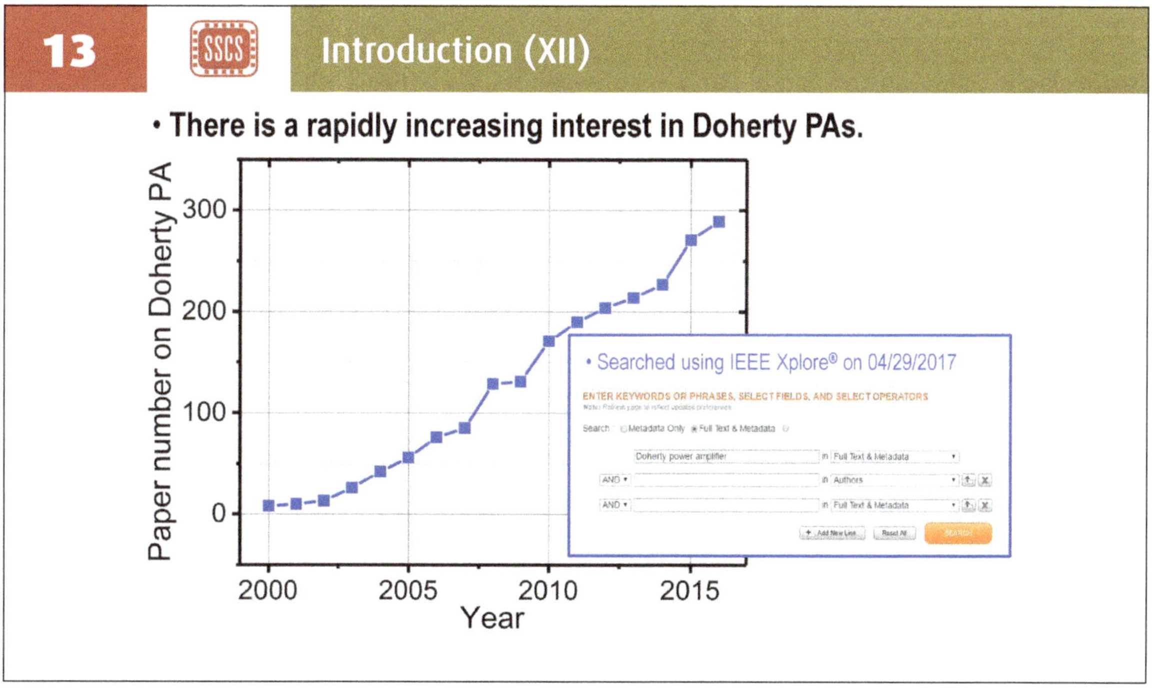

Nevertheless, the advantages of Doherty PA clearly outweigh its technical challenges, and this results in a rapid increasing interest in Doherty PA recently. This plot summarizes conferences/journals published in each year that mentioned Doherty power amplifier from 2000 to 2016. This search is performed using IEEE Xplore yesterday. The search criteria is shown here. We can see that in the last 17 years, the related papers increases from just a few per year to almost 300 per year. Considering Doherty PA is just one advanced and very specialized architecture of power amplifiers, this really shows the increasing interest among the technical field.

14 Outline

- Introduction
- **Doherty PA Configurations**
- Analog Doherty PAs at RF Frequency
- Digital Doherty PAs at RF Frequency
- Mm-Wave Doherty PAs
- Conclusion

Next, I would like to give an introduction on Doherty PA configurations with more details.

15 Doherty PA Configurations (I)

- **First proposed by William H. Doherty in 1936 [9]**

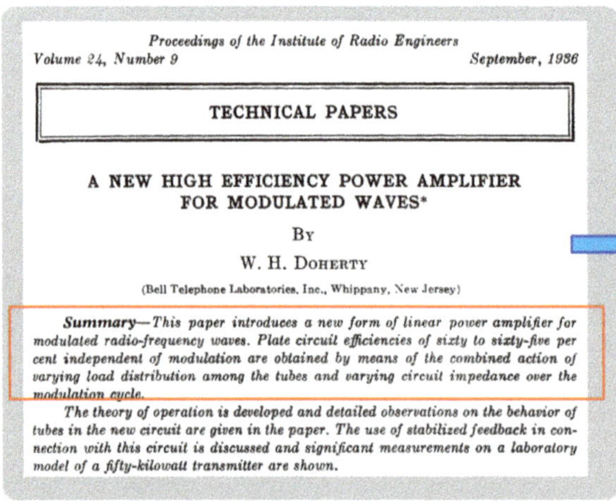
Proceedings of the Institute of Radio Engineers
Volume 24, Number 9 *September, 1936*

TECHNICAL PAPERS

A NEW HIGH EFFICIENCY POWER AMPLIFIER FOR MODULATED WAVES*

By

W. H. Doherty

(Bell Telephone Laboratories, Inc., Whippany, New Jersey)

Summary—*This paper introduces a new form of linear power amplifier for modulated radio-frequency waves. Plate circuit efficiencies of sixty to sixty-five per cent independent of modulation are obtained by means of the combined action of varying load distribution among the tubes and varying circuit impedance over the modulation cycle.*

The theory of operation is developed and detailed observations on the behavior of tubes in the new circuit are given in the paper. The use of stabilized feedback in connection with this circuit is discussed and significant measurements on a laboratory model of a fifty-kilowatt transmitter are shown.

- Multiple tubes (amplifiers)
- Active load modulation
- Constant amplifier efficiency over back-off
- Linear operation

• [9] W. H. Doherty, "A new high efficiency power amplifier for modulated waves," *Proc. Institute of Radio Engineers*, vol. 24, no. 9, pp. 1163-1182, Jun. 2012.

To better understand the Doherty PA configurations, we should re-visit the original paper published by William Doherty in 1936. If we read the first paragraph of its abstract, it essentially captures all the necessary parts of a Doherty PA, including multiple tubes (amplifiers), active load modulation between the tubes, constant amplifier efficiency over the power back-off, and linear operation.

16

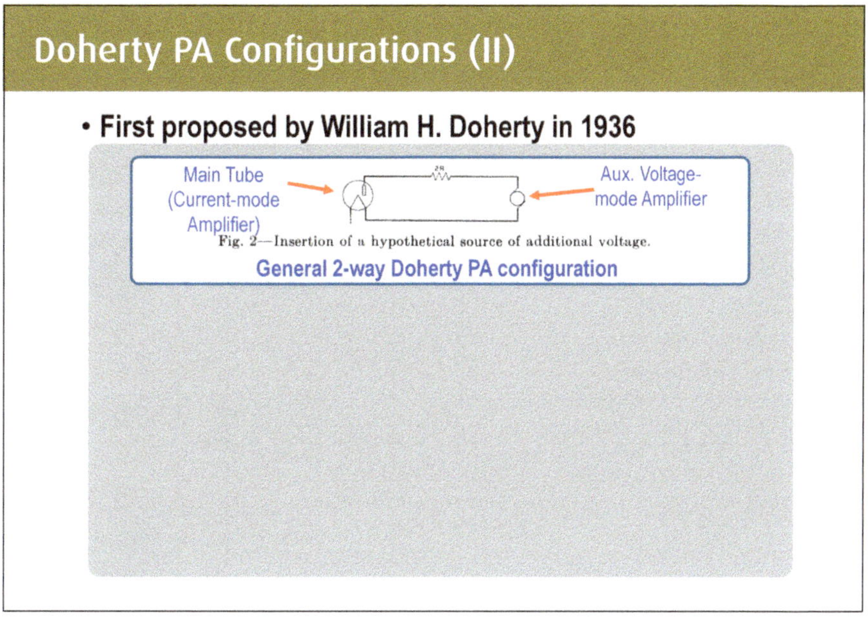

Moreover, in this 1936 paper, William Doherty proposed this general 2-way configuration for Doherty PA, which is composed of a main tube (that can be viewed as a current-mode amplifier) and an auxiliary voltage-mode amplifier connected in series by a resistor 2R. The current-mode amplifier means amplifier with high output impedance, while the voltage-mode amplifier means amplifier with low output impedance, both compared to the value of the load impedance. Therefore, at low output power, the auxiliary voltage amplifier is off, and the main amplifier sees the entire 2R load. At high output power, the auxiliary voltage amplifier contributes equally with the main amplifier, so each of them will see a load of R.

17

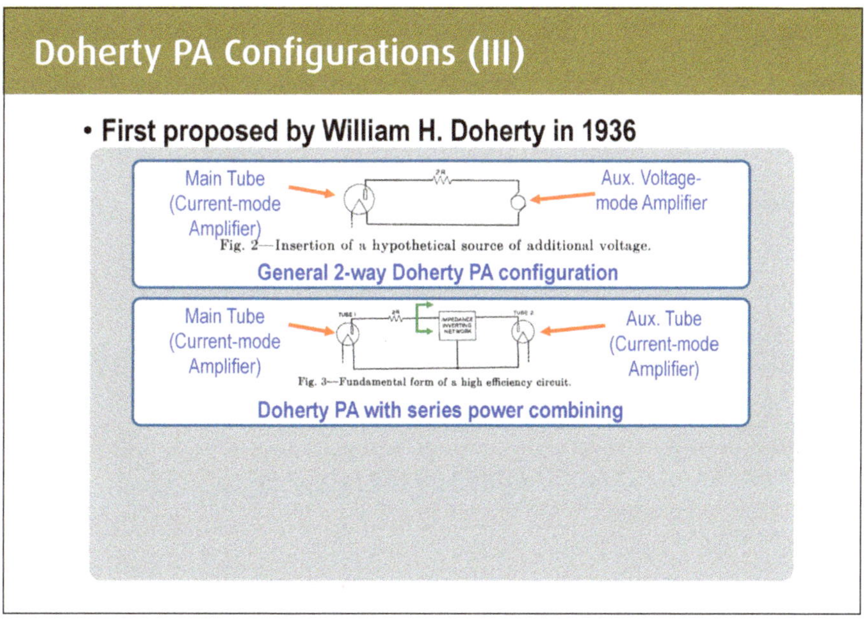

At that time, it was challenging to implement a voltage-mode amplifier. But an impedance inverter network, e.g., a lamda/4 transmission line at the carrier frequency of operation, can be used to convert a current-mode amplifier to an equivalent voltage-mode amplifier. Thus, when the auxiliary PA is off, it shows a high output impedance which then leads to a low impedance in series with the 2R and will not load the 2R. This leads to the Doherty PA with the series power combining configuration.

18

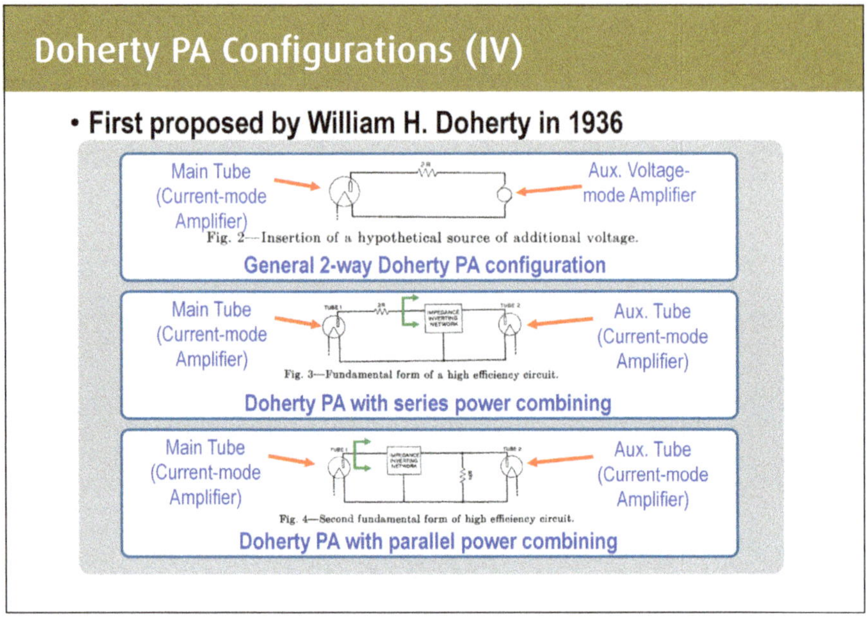

An alternative configuration also exists; we can connect two current-mode amplifiers in parallel with a load R/2 but with an impedance inverter at the output of the main amplifier. Therefore, at low output power when the auxiliary PA is off, the auxiliary PA exhibits high output impedance and will not load the load R/2. The main PA will see its load of 2R due to the impedance inverter. However,at the peak output power level, the auxiliary PA is turned on and contribute equally as the main PA. So the main PA branch will see a load of R at the parallel junction with the auxiliary PA branch. Then, the main PA will see its load of R at its output after the impedance invertor. This results in the Doherty PA with parallel power combining configuration.

19

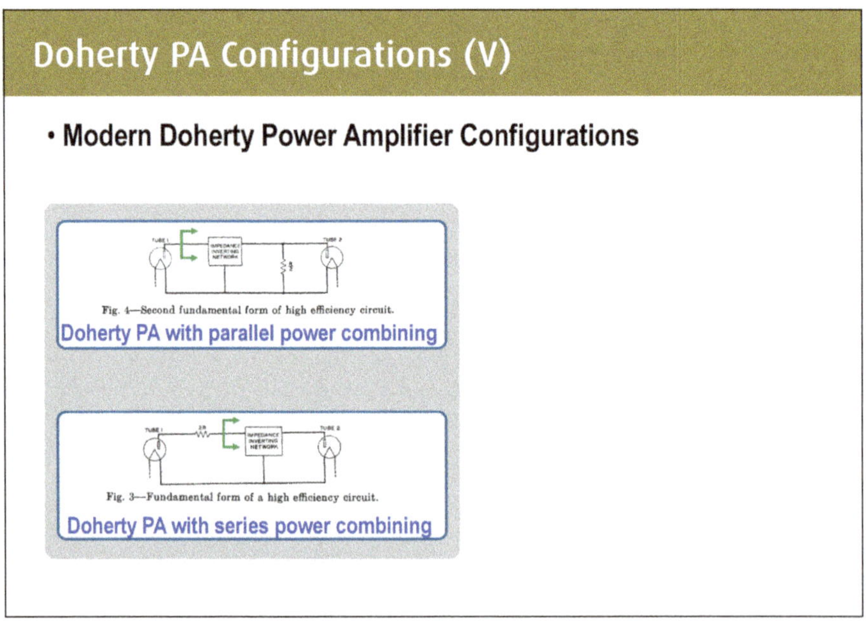

Of course, most of us here probably do not design vacuum tube amplifiers any more. We are more interested in learning how these classic vacuum tube Doherty amplifier configurations have evolved to our modern Doherty PA configurations?

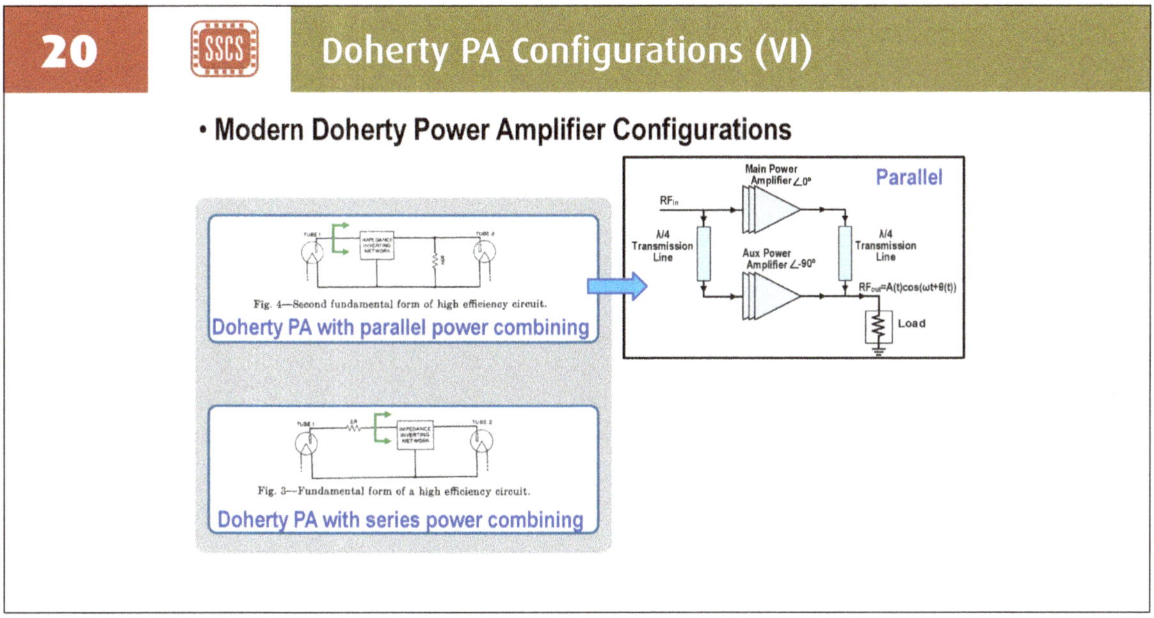

The modern parallel Doherty power amplifier is essentially composed of two current-mode amplifiers connected in parallel with the load. A lamda/4 transmission line is placed at the main PA output as the impedance invertor, and another one at the auxiliary PA input to balance the delay between the two paths.

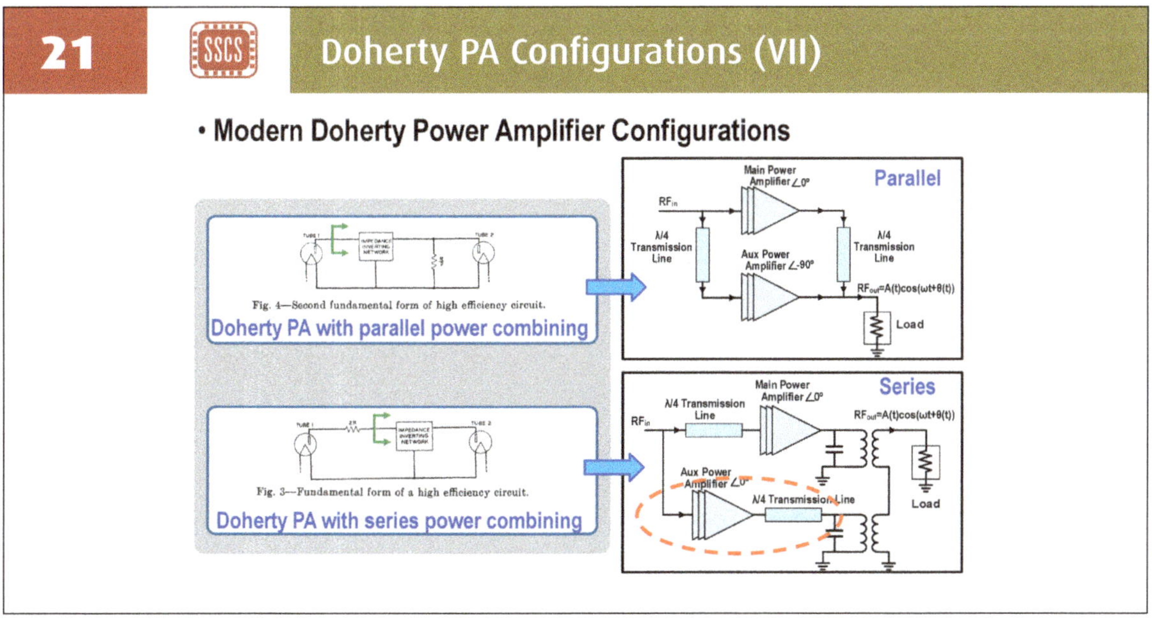

The modern series Doherty power amplifier is essentially composed of two current-mode amplifiers connected in series by a transformer with the load. A lamda/4 transmission line is placed at the auxiliary PA output as the impedance invertor and provide low-impedance, and another one at the main PA input to balance the delay between the two paths. Note this impedance inverter network at the auxiliary PA output is essential and can be a major source of confusion in the design.

22 Doherty PA Configurations (VIII)

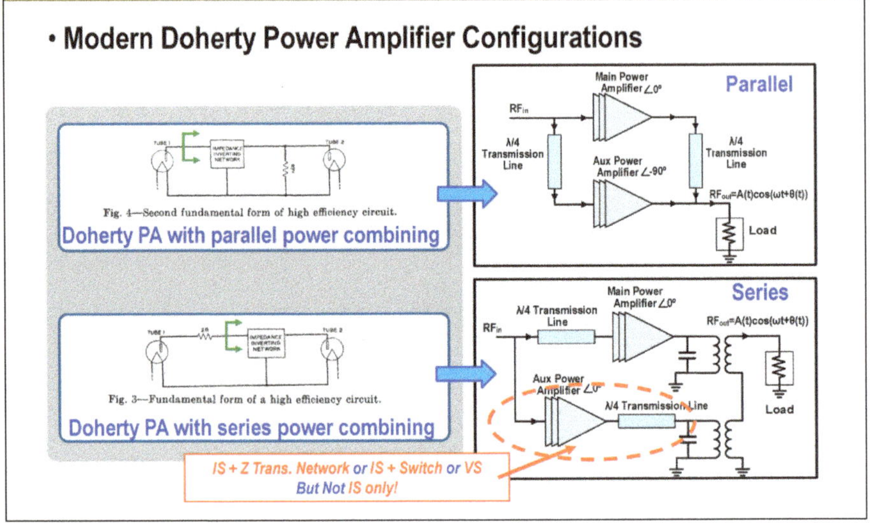

If the auxiliary PA is current-mode with high output impedance, such as a common-source PA, it must have this lamda/4 line, so that the auxiliary PA output will not load the transformer combiner. If you don't want to have a lamda/4 line here, you can use a switch at the auxiliary PA output to force it to be a short circuit when the auxiliary PA is off. Another solution is to use voltage-mode amplifier as the auxiliary PA, such as class-D digital PA or switched-cap digital PA. However, if we just use a current-mode amplifier here without additional circuit to lower its output impedance, it will substantially load the output and degrade the PA efficiency.

23 Parallel Doherty PA Operation (I)

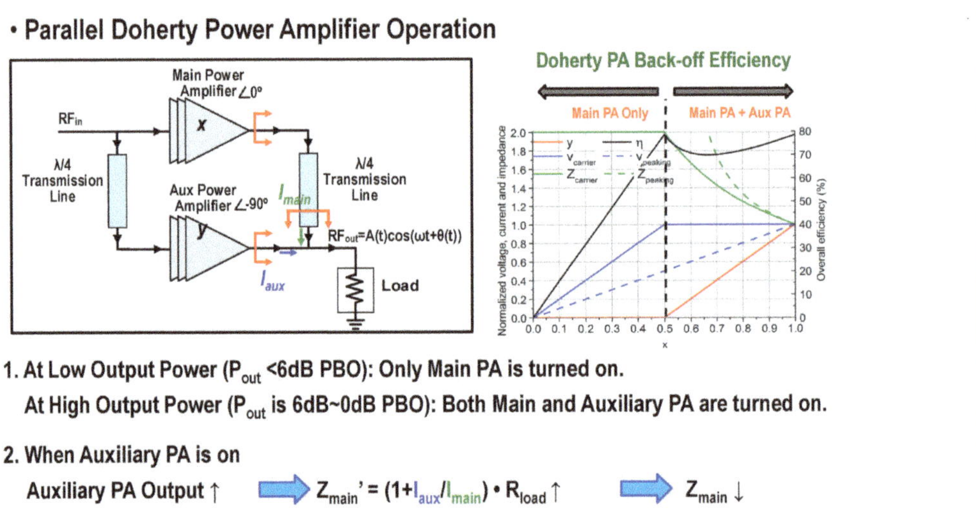

Now, let us look at the detailed operation of the parallel Doherty PA, which is probably the most popular Doherty PA architecture. The series Doherty PA can be derived in a similar way. At low output power, when the P_{out} is more than 6dB lower than the peak P_{out}, only the main PA is on. So it just behaves like a single-branch PA. The main PA should start to saturate at 6dB PBO by itself. Then, at high output

23 Parallel Doherty PA Operation (I)

power P_{out}, the auxiliary PA is turned on. If there is more auxiliary output current, the main PA branch at the main/auxiliary junction will see an increasing active load modulation impedance $Z'_{main} = (1 + I_{aux}/I_{main}) \times R_{load}$. The impedance inverter then ensures that the main PA output will see a decreasing load impedance. With an increasing input drive signal, the main PA should exhibit an increasing output current, which together with the decreasing main PA load impedance will result in a constant voltage swing at the main PA output. This then extends the main PA linear operation range and allows a higher output power P_{out} with high efficiency, i.e., optimum efficiency from the main PA.

24 Parallel Doherty PA Operation (II)

- **Parallel Doherty Power Amplifier Operation**

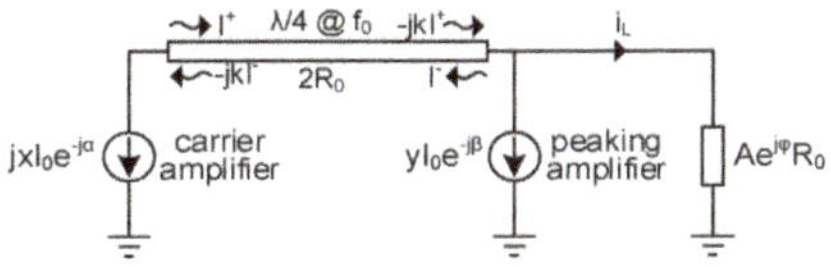

Ideal case: x, y = [0,1], α=β=0°, k=1, A=1, and φ=0°

$$v_{carrier} = -j(2x-y)V_{DD} \qquad v_{peaking} = -xV_{DD}$$

$$Z_{carrier} = \frac{2R_0}{x}(2x-y) \qquad Z_{peaking} = \frac{2x}{y}R_0 \qquad P_{out} = 2x^2 I_0^{\,2} R_0 \qquad \eta = \frac{\pi}{2}\frac{x^2}{x+y}$$

Doherty PA Back-off Efficiency

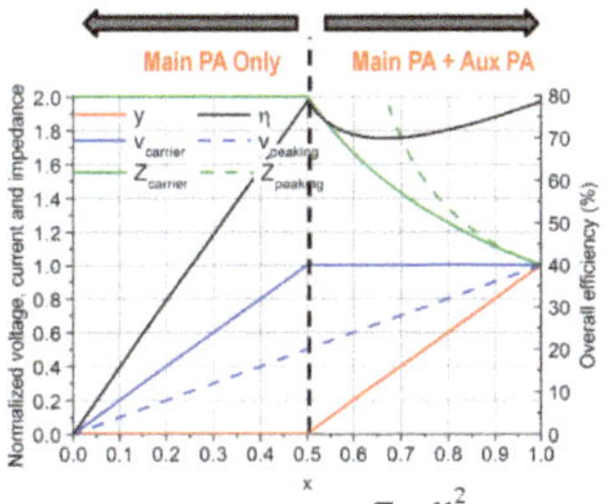

To ensure a constant main PA output voltage → 2x-y=1 ⇒

$$y = \begin{cases} 2x-1, & 1/2 < x \le 1 \quad \textit{High Power Mode} \\ 0, & 0 \le x \le 1/2 \quad \textit{Low Power Mode} \end{cases}$$

- [10] S. Hu, S. Kousai, and H. Wang, "Antenna Impedance Variation Compensation By Exploiting A Digital Doherty Power Amplifier Architecture," *IEEE Trans. Microw. Theory. Tech*, vol. 63, no. 2, pp. 580-597, Feb. 2015.

We can delve deeper into the operation of a parallel Doherty PA. Assuming that the main and auxiliary PAs are both modeled as current sources, the complete PA behavior including the main/auxiliary PA current mismatch, output combiner loss and antenna load variation has been analytically solved in our T-MTT paper in 2015. For the simplest case, if main and auxiliary PAs have the same weighting at their peak output power, called a symmetric 2-way Doherty, assuming there is no additional phase mismatch, the combiner is lossless, and the antenna load has no VSWR mismatch. x stands for the output current weighting from the main PA, and y is the output current weighting from the auxiliary PA. All the voltage swings, active load modulation impedances, the PA output power, and the PA efficiency can be mathematically formulated. The RF voltage swing at the main PA output equals $-j(2x-y)^*V_{DD}$. Since an optimum Doherty PA needs to ensure that the main PA output voltage swing is constant during the active load modulation by the auxiliary PA in order to maximize its efficiency, i.e., satisfying the PA large-signal ohm's law, we then must have 2x-y=1. Therefore, the auxiliary PA output current y equals twice of the main PA output current minus one. All the current values are normalized values. This is the governing equation for the operation of an ideal 2-way symmetric Doherty PA. With this main/auxiliary current relationship, the PA can achieve the desired Doherty PA back-off efficiency curve.

25

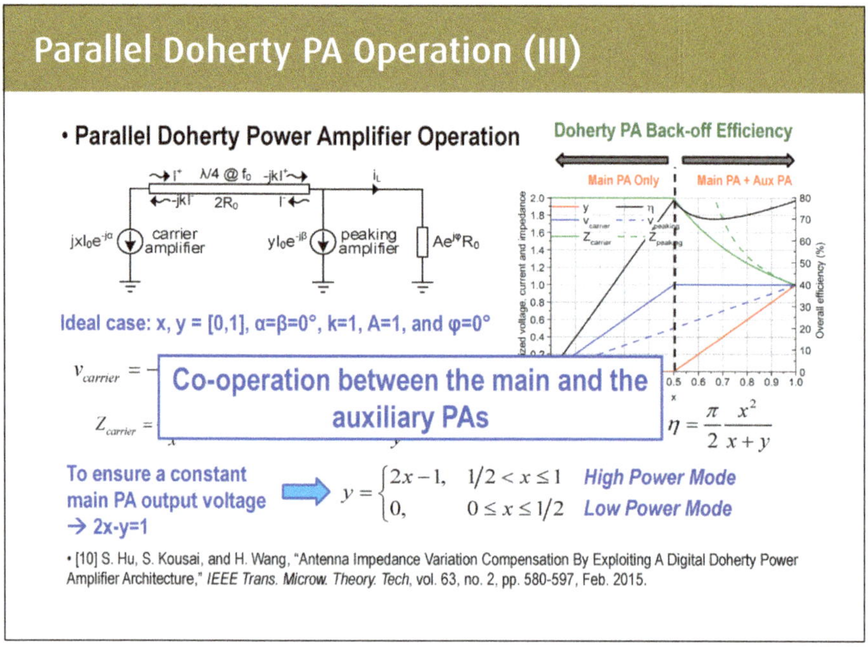

Therefore, this means it is essential to achieve this cooperation between the main and auxiliary PAs.

26

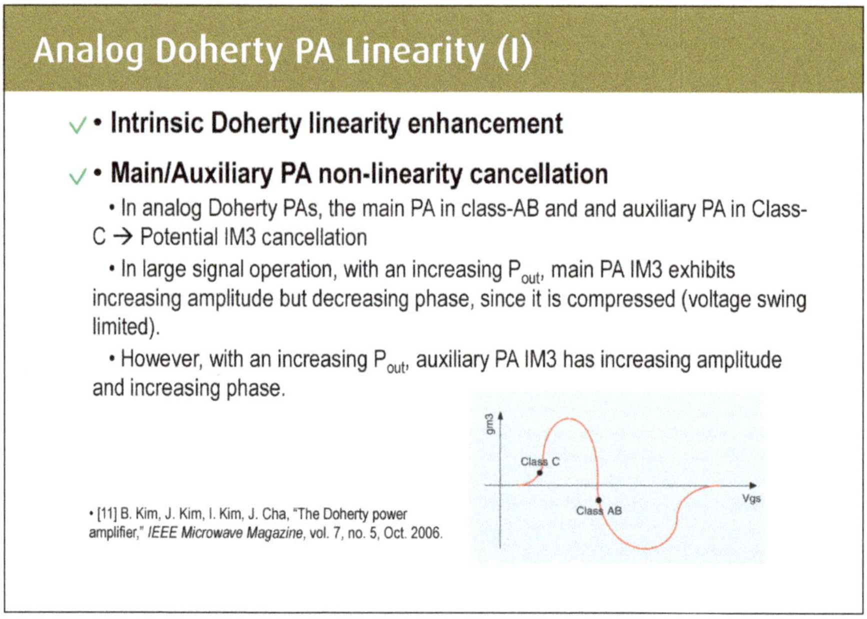

What about the linearity behavior of a Doherty PA? Well, this topic itself can be a tutorial talk, so we cannot cover all the details here. However,it should be emphasized that the linearity of a Doherty PA often exhibits complicated behaviors. There can be some pleasant surprise. First of all, we know that the active load modulation directly extends the linearity range of the main PA, so this is the intrinsic Doherty PA linearity enhancement. Moreover, in practice for analog Doherty PAs, we often bias the main PA in a class-AB mode and he auxiliary PA in class-C mode to allow their sequential turning on operation. Recall the MOS transistor gm3 behavior that is positive at class-C and negative at class-AB, so potentially one can design the Doherty PA to achieve IM3 cancellation between the main/auxiliary PAs. Of course, in reality it is not that simple, since we need to make sure the two IM3 tones are in 180-degree output phase. This can be achieved through careful designs. On other hand, if the tuning-on point of the auxiliary PA is not appropriately designed, or the desired output current relationship between the main and the auxiliary PAs cannot be satisfied, this will lead to substantially worsened large-signal linearity.

27

Analog Doherty PA Linearity (II)

- **Intrinsic Doherty linearity enhancement**
- **Main/Auxiliary PA non-linearity cancellation**
  - In analog Doherty PAs, the main PA in class-AB and and auxiliary PA in Class-C → Potential IM3 cancellation
  - In large signal operation, with an increasing P_{out}, main PA IM3 exhibits increasing amplitude but decreasing phase, since it is compressed (voltage swing limited).
  - Howe[...]sing amplitude and incre[...]

Co-operation between the main and the aux PAs

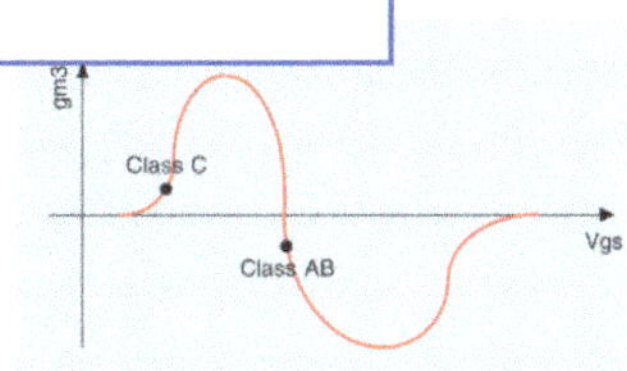

- [11] B. Kim, J. Kim, I. Kim, J. Cha, "The Doherty power amplifier," *IEEE Microwave Magazine*, vol. 7, no. 5, Oct. 2006.

Again, this means the co-operation between the main/auxiliary PAs are extremely important.

28

Digital Doherty PA VSWR Robustness (I)

- **Digital Doherty PA → Fully reconfigurable amplitude/phase relationship between Main/Aux PAs.**

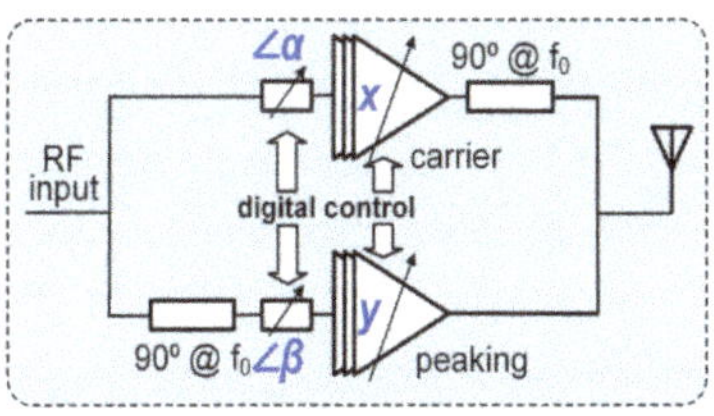

- **Type-I Doherty PA:** Tunable RF currents (x, y) and phase difference (α-β) of the main/auxiliary PAs
- **Type-II Doherty PA:** Only tunable RF currents (x, y); no (α-β) phase tuning
- **Type-III Doherty PA:** No tunable RF currents (x, y); only tunable (α-β) phase tuning
- **Type-IV Doherty PA:** Baseline analog Doherty PA with no RF currents/phase tuning

- [10] S. Hu, S. Kousai, and H. Wang, "Antenna Impedance Variation Compensation By Exploiting A Digital Doherty Power Amplifier Architecture," *IEEE Trans. Microw. Theory. Tech*, vol. 63, no. 2, pp. 580-597, Feb. 2015.

Finally, it should be emphasize that if we implement digital Doherty PA, which will be discussed extensively later on, one can achieve fully reconfigurable amplitude and phase relationship between the main and auxiliary PA paths. Such reconfigurabilty can enhance the PA robustness and restore its efficiency against antenna VSWR impedance variations. This result was published and experimentally verified in our recently published T-MTT paper in 2015. Here we can define four types of reconfigurable Doherty PAs. The type-I Doherty PA has both its amplitude and phase reconfigurable. The type-II Doherty PA only has its amplitude reconfigurable. The type-III Doherty PA has its phase reconfigurable. The type-IV Doherty PA is a baseline analog Doherty PA with no RF current/phase tuning.

29 Digital Doherty PA VSWR Robustness (II)

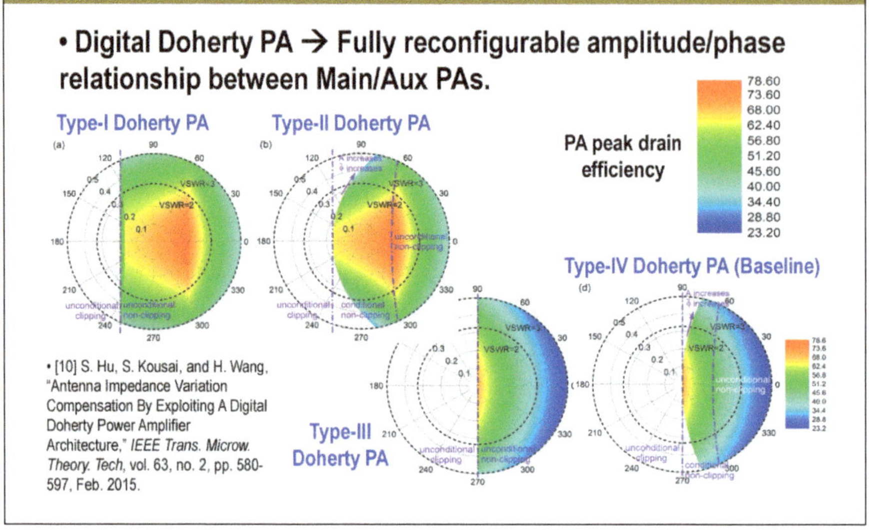

This slide shows the peak PA drain efficiency under VSWR 3:1 antenna load variation and Doherty PA main/auxiliary re-optimization. Since we assume ideal class-B PA, so the peak efficiency is 78% (pi/4). The type-IV Doherty PA exhibits large efficiency degradation, while the type-I digital Doherty PA can greatly restore the peak PA drain efficiency by re-configuring the relative amplitude/phase of the main/auxiliary PAs. Additional detailed results can be found in our T-MTT paper published in 2015.

30 N-Way and N-Stage Doherty PAs

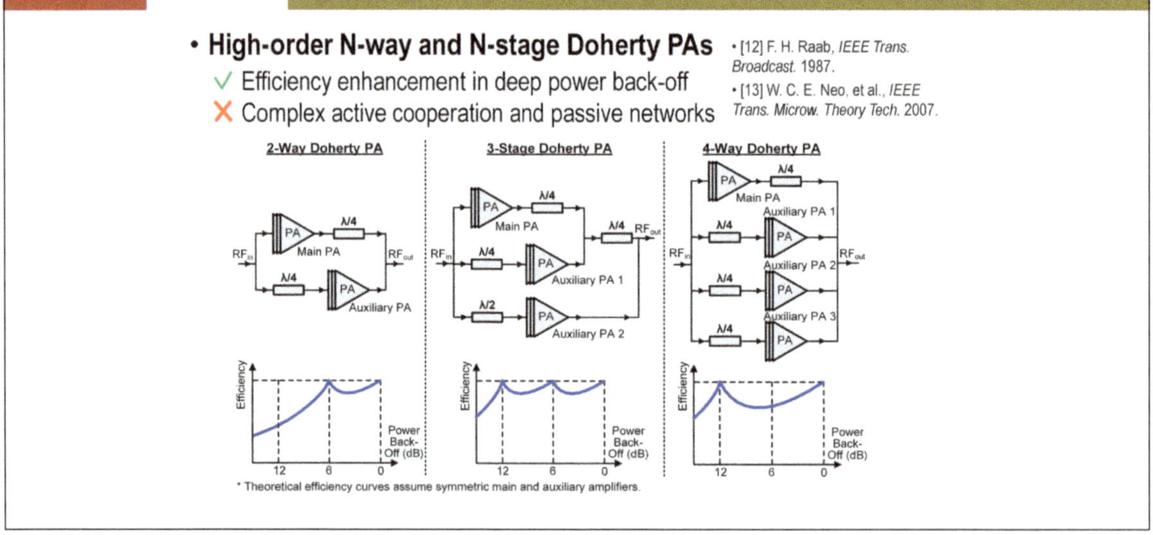

In addition, we know that a symmetric 2-way Doherty PA will only perform PA efficiency enhancement up-to 6dB PBO. Therefore, multiple N-way and N-stage Doherty PA architectures have been proposed to substantially extend this efficiency enhancement range. In generation, an N-way Doherty PA can be viewed as an asymmetric Doherty that provides only one efficiency peaking during the PBO. However, an N-stage Doherty PA will provide N-1 efficiency peaking points and outperform the N-way Doherty PA. These architectures are well suited for amplifying signals with large PAPR at deep power back-off. Of course, the resulting passive networks may also become more complicated, bulky and lossy.

31 Doherty PA Summary

- **Pros:**
  - ✓ Large modulation bandwidth
  - ✓ Low or none baseband overhead (→ Plug & Play replacement of Class-AB PA)
  - ✓ Robustness against VSWR

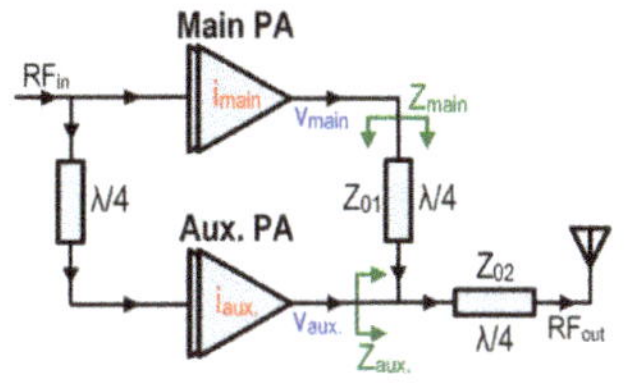

- **Cons:**
  - × Complicated power combiner (→ Loss, BW, impedance transformation ratio)
  - × Careful cooperation between main and auxiliary PAs (→ PA efficiency, linearity)

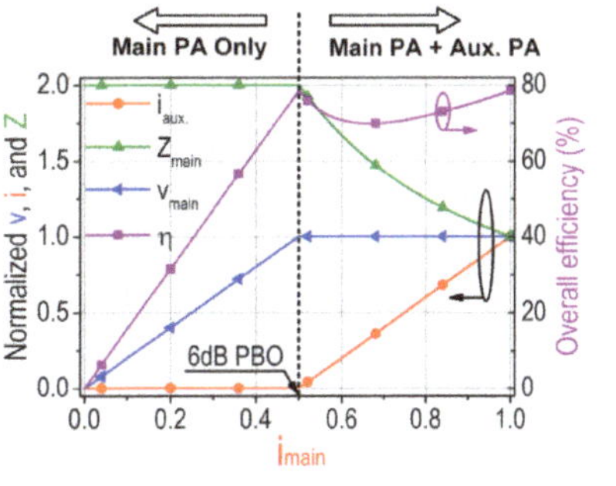

In summary, Doherty PAs have unique advantages of large modulation bandwidth, low or none baseband overhead so that a Doherty PA can be direct replacement of class-AB PA. Doherty PA also shows unique robustness against antenna VSWR. However, Doherty PA requires large and complicated power combiner and careful cooperation between the main/auxiliary PA paths. These challenges lead to extensive research in Doherty PAs, and multiple design examples will be shown in the following sections.

32 Outline

- Introduction
- Doherty PA Configurations
- **Analog Doherty PAs at RF Frequency**
- Digital Doherty PAs at RF Frequency
- Mm-Wave Doherty PAs
- Conclusion

In this section, a recently published analog Doherty PA at RF frequency will be presented as a design example.

33 Parallel Analog Doherty PA (I)

- **Designed CMOS Doherty PA**

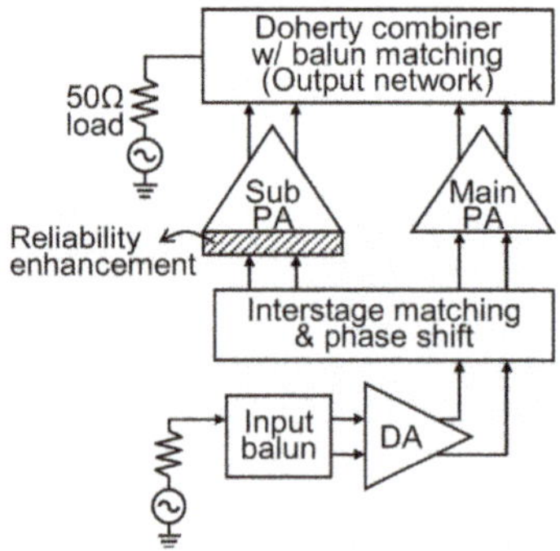

- Slab inductor based parallel Doherty combiner
- Reliability enhancement for Aux. PA

- [8] K. Onizuka, S. Saigusa, S. Otaka, "A +30.5 dBm CMOS Doherty power amplifier with reliability enhancement technique," *VLSI Circuits (VLSIC), 2012 Symposium on*, 13-15 June 2012.

This design was presented at the VLSI conference in 2012. The simplified circuit schematic is shown on the left. It is a parallel Doherty PA, where a slab inductor is used to realize the parallel Doherty output combiner by leveraging its low passive loss but at the expense of extra chip area. Additional reliability enhancement technique is employed for the auxiliary PA.

34 Parallel Analog Doherty PA (II)

- **Parallel Doherty combiner with matching network**

Initial network design | Actual implementation

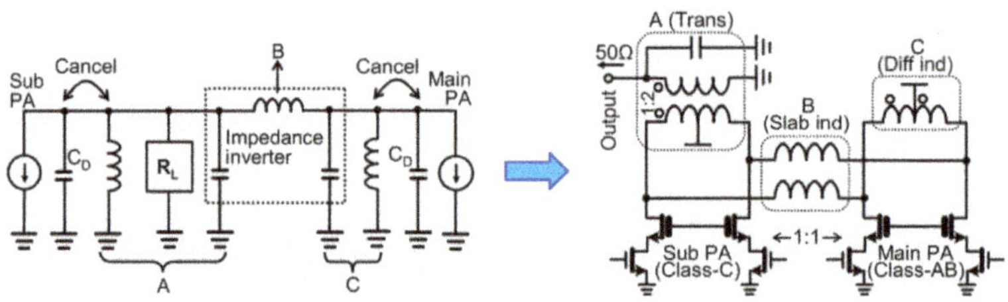

- Synthesized impedance inverter using slab inductors

- [8] K. Onizuka, S. Saigusa, S. Otaka, "A +30.5 dBm CMOS Doherty power amplifier with reliability enhancement technique," *VLSI Circuits (VLSIC), 2012 Symposium on*, 13-15 June 2012.

The slab inductor in conjunction with some output capacitance of the main and auxiliary PAs form a C-L-C pi network to synthesize the required lamda/4 transmission line for impedance inverting. The remaining device output capacitances are resonated out by one additional differential inductor at the main PA side and a 1:2 transformer balun at the Doherty PA output.

35 Parallel Analog Doherty PA (III)

- **Reliability enhancement technique for sub-PA (Aux. PA).**

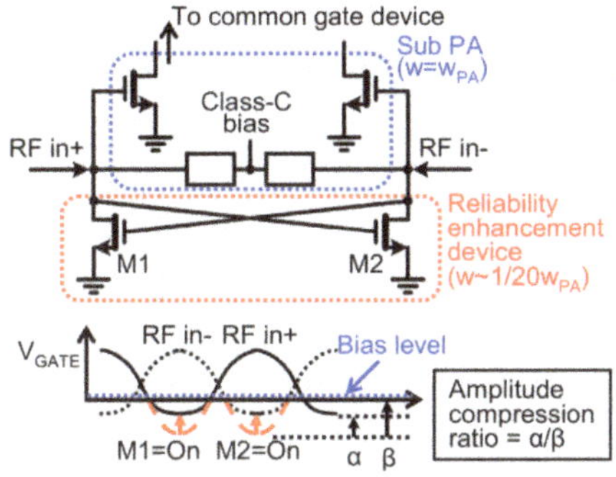

- The negative gate-amplitude of the sub-PA (Aux. PA) is relaxed.
- When the positive input node RF in+ is high, a small NMOS transistor M2 turns on and compresses the negative voltage swing, and vice versa.

- [8] K. Onizuka, S. Saigusa, S. Otaka, "A +30.5 dBm CMOS Doherty power amplifier with reliability enhancement technique," *VLSI Circuits (VLSIC), 2012 Symposium on*, 13-15 June 2012.

The reliability enhancement technique for the class-C biased auxiliary PA is achieved by adding a differential cross-coupled pair in parallel at the input of the auxiliary PA. The device sizes of this differential cross-coupled pair is only 1/20 compared to the power devices in the auxiliary PA. When one side of the input voltage is high, e.g., the positive input RF in+, it turns on the device M_2, which presents a low impedance and pulls the negative input RF in- closer to the ground. This largely attenuates the negative swing at the auxiliary PA, enhancing its reliability.

36 Parallel Analog Doherty PA (IV)

- **Measured Results**

Die photo (Toshiba 65nm Bulk CMOS)

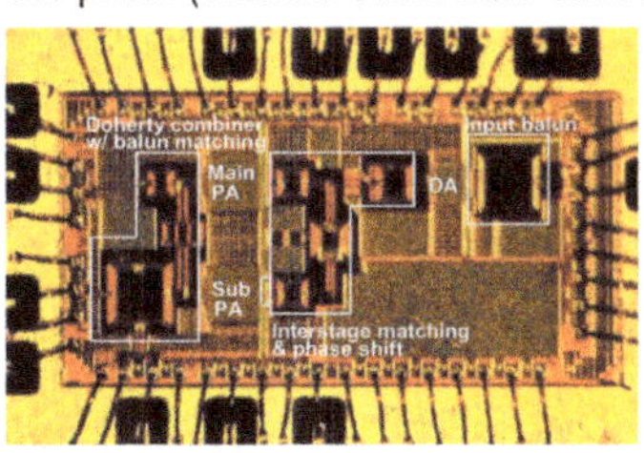

Measured CW performance at 2.48GHz

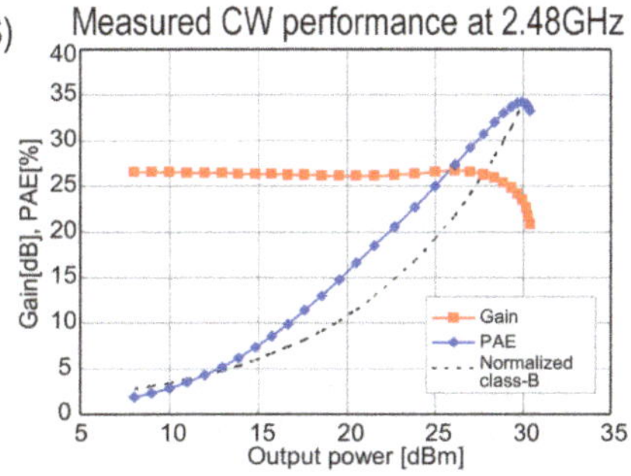

- 27 dB linear gain and 34% peak PAE at 2.48GHz. The back-off efficiency improvement compared with a normalized class-B PA is 1.4x at 6dB back-off.

- [8] K. Onizuka, S. Saigusa, S. Otaka, "A +30.5 dBm CMOS Doherty power amplifier with reliability enhancement technique," *VLSI Circuits (VLSIC), 2012 Symposium on*, 13-15 June 2012.

This analog Doherty PA is implemented in the Toshiba 65nm bulk CMOS process for the operation at 2.48GHz. It achieves a 27dB linear power gain, 30dBm peak output power, 34% peak PAE, and 1.4× PAE enhancement over a normalized class-B PA.

37 Parallel Analog Doherty PA (V)

- **Measured Results**

Measured (a) IMD3/5 and (b) peak and rms EVM for 802.11b/g signals.

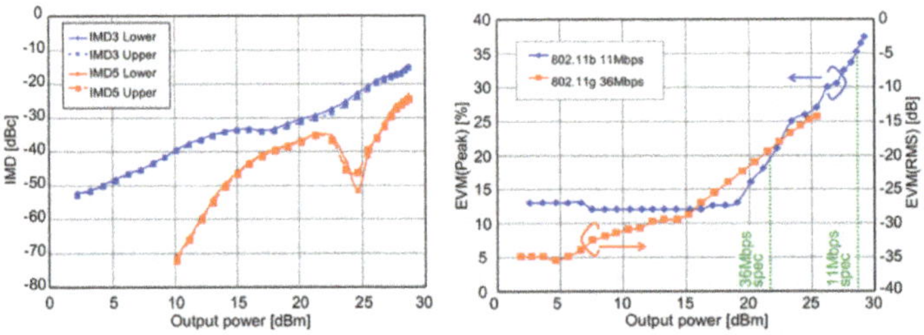

(a) (b)

- (a) shows the measured IMD performances for 4 MHz-spacing two-tone signals. (b) describes the measured peak and RMS EVMs for 802.11b 11Mbps and 802.11g 36Mbps signals.

• [8] K. Onizuka, S. Saigusa, S. Otaka, "A +30.5 dBm CMOS Doherty power amplifier with reliability enhancement technique," *VLSI Circuits (VLSIC), 2012 Symposium on*, 13-15 June 2012.

The PA is carefully design for its linearity. The measured intermodulation distortion (IMD) performance for a two-tone test with 4MHz spacing is shown on the left. Symmetric upper/lower IMD tones show little memory effect. The measured peak and rms EVMs for 802.11b (11Mbps) and 802.11g (36Mbps) signals are shown on the left.

38 Parallel Analog Doherty PA (VI)

- **Measured Results**

Output spectra for (a) 802.11b and (b) 802.11g signals.

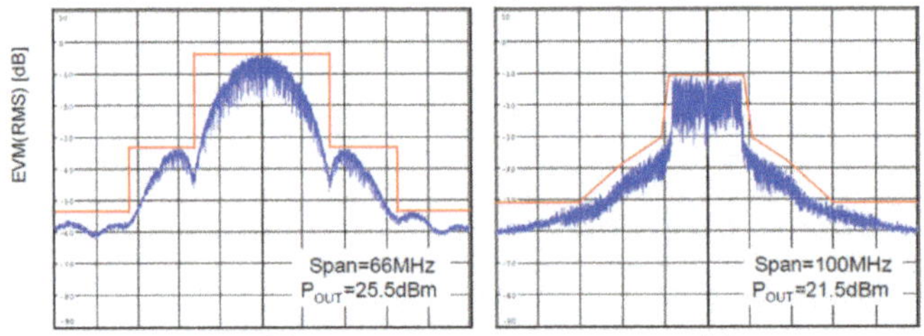

(a) (b)

- (a) and (b) shows the output spectra at the PA output power levels of 25.5 and 21.5 dBm respectively, and both satisfy the spectrum masks.

• [8] K. Onizuka, S. Saigusa, S. Otaka, "A +30.5 dBm CMOS Doherty power amplifier with reliability enhancement technique," *VLSI Circuits (VLSIC), 2012 Symposium on*, 13-15 June 2012.

The output spectra of the PA for 802.11b and 802.11g signals are also presented to illustrate the PA's out-of-band linearity performance. With the spectrum masks being satisfied, the 802.11b signal can be amplified at an average P_{out} level of 25.5dBm, while the 802.11g signal can be amplified at an average P_{out} level of 21.5dBm.

39 Outline

- Introduction
- Doherty PA Configurations
- Analog Doherty PAs at RF Frequency
- **Digital Doherty PAs at RF Frequency**
- Mm-Wave Doherty PAs
- Conclusion

In this section, several recently published digital Doherty PAs at RF frequency will be presented as design examples.

40 Current-Mode Parallel Digital Doherty PA (I)

- **Digital Doherty polar PA architecture**
  - The main and aux. PAs are implemented as digitally controlled RF Power DACs, resulting in precisely and flexibly controlled RF gains of the two paths.
    - ✓ Enhanced PA back-off efficiency
    - ✓ In-field reconfigurability to enhance robustness against antenna mismatch
  - Polar architecture facilitates the adoption of switching-mode amplifier.
    - ✓ Enhanced PA peak efficiency

- **Conventional analog Doherty PA** • **Digital Doherty polar PA**

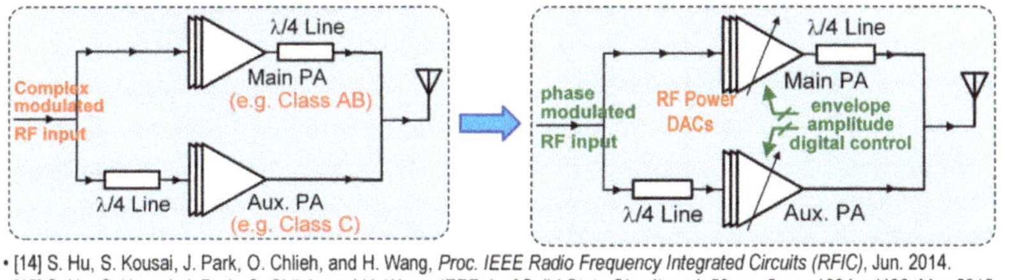

- [14] S. Hu, S. Kousai, J. Park, O. Chlieh, and H. Wang, *Proc. IEEE Radio Frequency Integrated Circuits (RFIC)*, Jun. 2014.
- [15] S. Hu, S. Kousai, J. Park, O. Chlieh, and H. Wang, *IEEE J. of Solid-State Circuits*, vol. 50, no. 5, pp. 1094 – 1106, May 2015.

Different from conventional analog Doherty PA, digital Doherty PA implements its main and auxiliary PAs as digitally controlled RF power DACs, so that the onset point of the auxiliary PA and the relative relationship between the main/auxiliary PAs can be precisely and flexibly controlled in real-time using digital amplitude control codes. In addition, it is possible to further add independent and real-time phase controls between the main and auxiliary PA paths. This allows optimum enhancement of the PA back-off efficiency and in-field reconfigurability to improve the PA robustness against antenna mismatches. Moreover, the digital Doherty PA naturally lends itself to the polar transmitter

40 Current-Mode Parallel Digital Doherty PA (I)

architecture, where switching-mode PAs can be employed to enhance the PA peak efficiency. This digital Doherty PA concept was originally proposed and demonstrated by my group at a conference paper in IEEE RFIC 2014 (RFIC 2014 Best Student Paper Award 1st Place) and its later journal extension in IEEE JSSC 2015.

41 Current-Mode Parallel Digital Doherty PA (II)

- **Digital Doherty polar PA architecture**
  - Main/Auxiliary PAs as 5-bit RF Power DACs (Binary-Weighted Inverse Class-D PAs)
  - Transformer-based Input/Output Passive Networks
  - Polar PA architecture

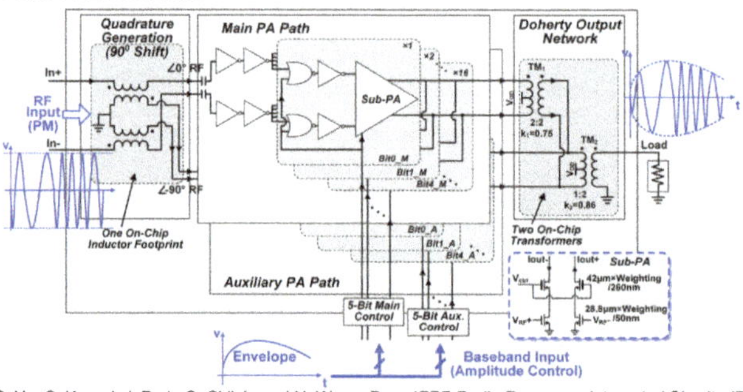

• [14] S. Hu, S. Kousai, J. Park, O. Chlieh, and H. Wang, *Proc. IEEE Radio Frequency Integrated Circuits (RFIC)*, Jun. 2014.
• [15] S. Hu, S. Kousai, J. Park, O. Chlieh, and H. Wang, *IEEE J. of Solid-State Circuits*, vol. 50, no. 5, pp. 1094 – 1106, May 2015.

An example Doherty polar PA is shown on this slide. Both the main and auxiliary PAs are implemented as a 5-bit RF power DAC composed of binary-weighted invers class-D differential PA cells. The PA also employs a transformer-based Doherty output network that simultaneously achieves parallel Doherty load modulation, impedance transformation, power combining, and differential to single-ended conversion. The input of the Doherty PA adopts a transformer-based quadrature signal generation network to divide the RF input into two paths with 90-degree phase-shift to feed the main and auxiliary PAs. It adopts a polar PA architecture, where constant-envelope phase-modulated signal feeds the PA (as the PM path) and the real-time envelope is digitized to control the turning/off of the main and auxiliary PA paths (as the AM path). The information in the AM and PM paths is combined at the PA output to synthesize the desired envelope-varying complex-modulated signal.

42 Current-Mode Parallel Digital Doherty PA (III)

- **Chip microphotograph and assembly**
  - Realized in a standard 65-nm bulk CMOS process.
  - "Staircase" bypass chip capacitor structure minimizes the effective resistance and inductance due to the bonding wires for the PA supply.

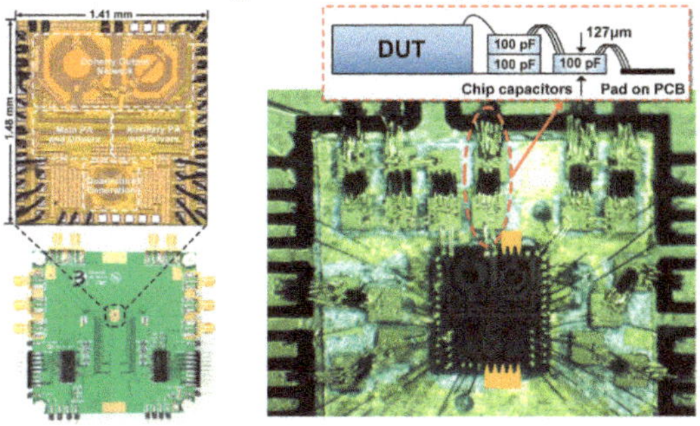

- [14] S. Hu, S. Kousai, J. Park, O. Chlieh, and H. Wang, *Proc. IEEE Radio Frequency Integrated Circuits (RFIC)*, Jun. 2014.
- [15] S. Hu, S. Kousai, J. Park, O. Chlieh, and H. Wang, *IEEE J. of Solid-State Circuits*, vol. 50, no. 5, pp. 1094 – 1106, May 2015.

A proof-of-concept design is implemented in a standard 65nm bulk CMOS process. The total PA chip size is 1.41mm×1.48mm including all the input and output passive networks for full PA integration. Bypass capacitors are placed around the PA chip in a "staircase" style for the supply and biasing lines to minimize the effective resistance and inductance due to wirebonds.

43 Current-Mode Parallel Digital Doherty PA (IV)

- **Continuous wave measurement with the 50ohm load**
  - +27.3 dBm peak output power at 3.82 GHz
  - 32.5% peak PA drain efficiency at 3.60 GHz
  - 3.10-3.98GHz (24.9%) 1dB P_{out} bandwidth
  - Maximum 7.0% absolute efficiency enhancement at 6dB PBO and ×1.48 relative efficiency enhancement over Class-B operation

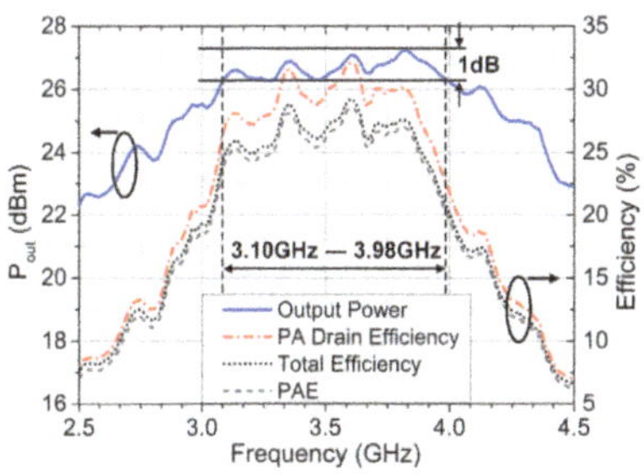

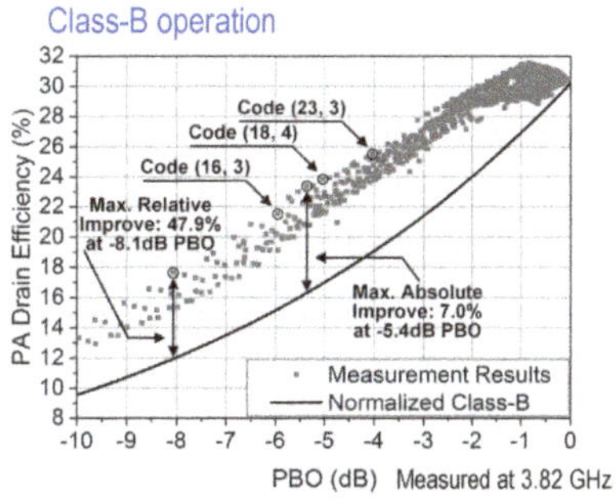

- [14] S. Hu, S. Kousai, J. Park, O. Chlieh, and H. Wang, *Proc. IEEE Radio Frequency Integrated Circuits (RFIC)*, Jun. 2014.
- [15] S. Hu, S. Kousai, J. Park, O. Chlieh, and H. Wang, *IEEE J. of Solid-State Circuits*, vol. 50, no. 5, pp. 1094 – 1106, May 2015.

The PA is first characterized for its continuous-wave performance with a standard 50ohm load. It achieves +27.3dBm peak output power at 3.83GHz carrier frequency and 32.5% peak PA drain efficiency at 3.6GHz. The PA 1dB P_{out} bandwidth is 3.1-3.98GHz at a fractional bandwidth of 24.9%. In terms of its back-off performance, the PA achieves ×1.48 enhancement over class-B operation at its 6dB back-off at 3.82Gz.

44 Current-Mode Parallel Digital Doherty PA (V)

- **Modulation tests (dynamic performance)**
  - QPSK (1MSym/s, PAPR=3.7dB): 3.5% rms EVM, +23.5dBm average Pout, 26.8% PA drain efficiency, **×1.37 relative η improvement compared with Class-B.**
  - 16QAM (500kSym/s, PAPR=5.4dB): 4.4% rms EVM, +22.2dBm average Pout, 20.2% PA drain efficiency, **×1.20 relative η improvement compared with Class-B.**

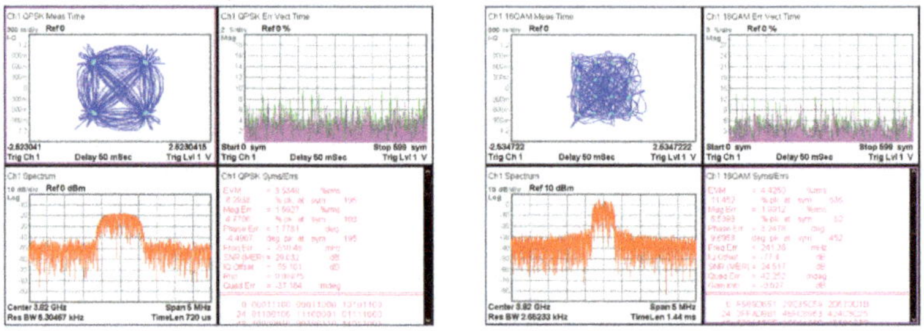

- 1MSym/s QPSK signal with 10MHz sampling rate.
- 500kSym/s 16QAM signal with 10MHz sampling rate.

- [14] S. Hu, S. Kousai, J. Park, O. Chlieh, and H. Wang, *Proc. IEEE Radio Frequency Integrated Circuits (RFIC)*, Jun. 2014.
- [15] S. Hu, S. Kousai, J. Park, O. Chlieh, and H. Wang, *IEEE J. of Solid-State Circuits*, vol. 50, no. 5, pp. 1094 – 1106, May 2015.

The PA is then tested using modulation signals. For a QPSK signal with 1MSym/s and 3.7dB PAPR, the PA achieves 3.5% rms EVM, +23.5dBm averaged Pout, 26.8% PA drain efficiency, ×1.37 relative efficiency improvement compared with Class-B. For a 16QAM signal with 500kSym/s and 5.4dB PAPR, the PA achieves 4.4% rms EVM, +22.2dBm averaged Pout, 20.2% PA drain efficiency, ×1.20 relative efficiency improvement compared with Class-B.

45 HybridClass-G Current-Mode Parallel Digital Doherty PA (I)

- **A classic symmetric Doherty PA only covers up to 6dB PBO**
  - Classic N-way/N-stage Doherty extends PBO but adds substantial complexity, area, and loss on the input/output passive networks.

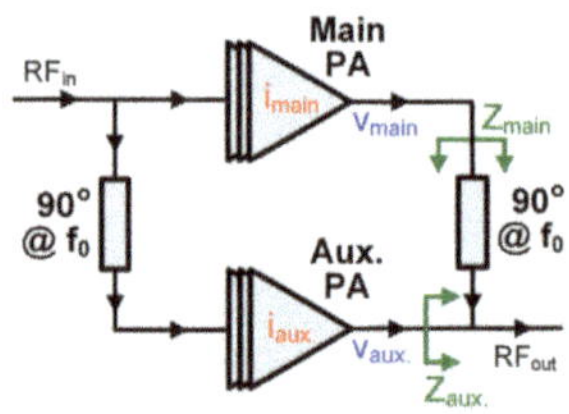

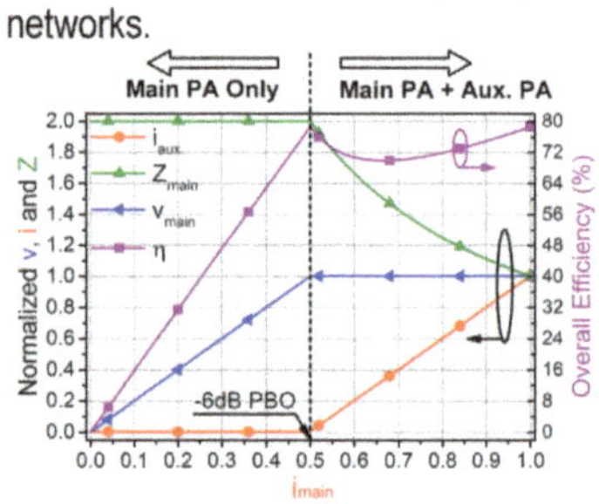

Can we extend the PBO range by adding some simple digital operation?

To further extend the back-off efficiency enhancement to deeper power back-off, classic approaches involve the use of N-way or N-stage Doherty PAs that will add substantial complexity, area, and loss on the input/output passive networks. An interesting question to ask is whether we can extend the back-off efficiency enhancement range by only adding some simple digital operation?

46

HybridClass-G Current-Mode Parallel Digital Doherty PA (II)

- **A classic symmetric Doherty PA only covers up to 6dB PBO**
  - Classic N-way/N-stage Doherty extends PBO but adds substantial complexity, area, and loss on the input/output passive networks.

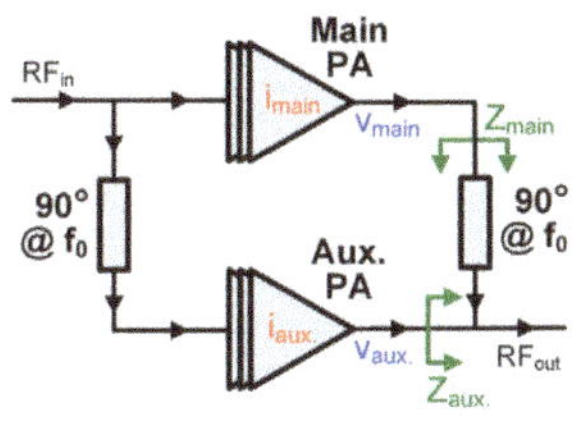

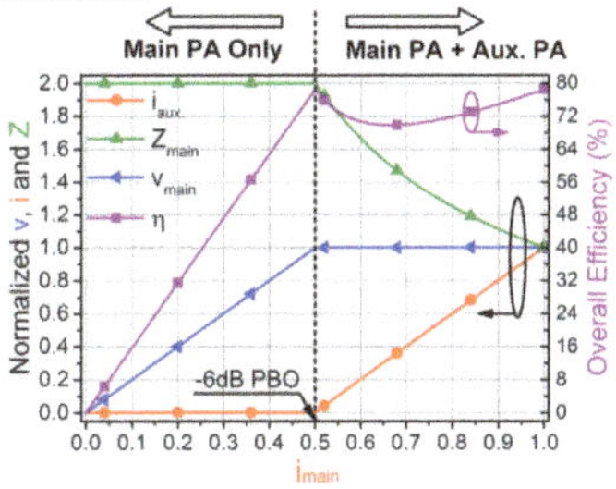

Can we extend the PBO range by adding some simple digital operation? ➡ Hybrid Class-G Doherty PA architecture

- [16] S. Hu, S. Kousai, and H. Wang, *IEEE ISSCC Dig. Tech. Papers*, Feb. 2015.
- [17] S. Hu, S. Kousai, and H. Wang, *IEEE J. of Solid-State Circuits*, vol. 51, no. 3, pp. 598 - 613, Mar. 2016.

This motivates the research on the hybrid class-G Doherty PA architecture as well as general hybrid PA architectures. Our proof of concept work on this topic was published as a conference paper in IEEE ISSCC 2015 and its journal extension in JSSC 2016.

47

HybridClass-G Current-Mode Parallel Digital Doherty PA (III)

- **Class-G operation – Supply Modulation with Discrete Levels**
  - ✓ Reduced complexity. ✓ Potentially large modulation bandwidth.
  - ✗ No efficiency enhancement within each supply mode.
  - ✗ Substantial linearity degradation due to supply switching.

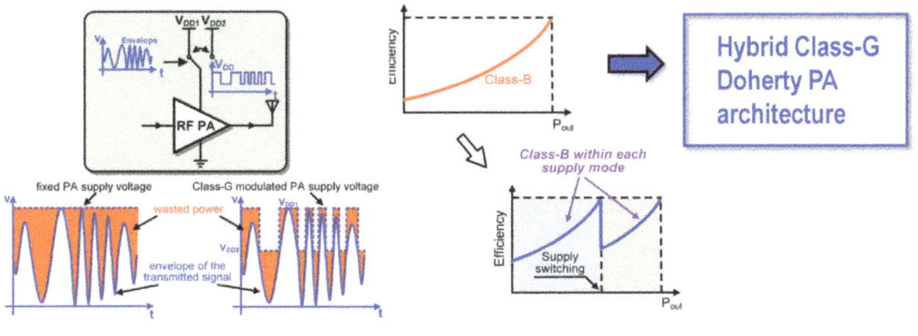

Before describing the details of the hybrid class-G Doherty PA, let us first review the basics of Class-G PA operation. Class-G operation can be viewed as supply modulation with discrete levels, for example VDD and VDD/2. Compared to analog ET operation, Class-G has reduced complexity and potentially larger modulation bandwidth. However, it does not support any efficiency enhancement in-between each supply mode, and it often exhibits substantial linearity degradation due to supply switching. We propose the hybrid Class-G Doherty PA architecture, which achieves Doherty-based efficiency enhancement between supply levels, and uses supply switching to extend the symmetric Doherty operation from 6dB PBO to 12dB PBO.

48

HybridClass-G Current-Mode Parallel Digital Doherty PA (IV)

• **Our Hybrid Class-G Doherty – Full-V_{DD} mode**

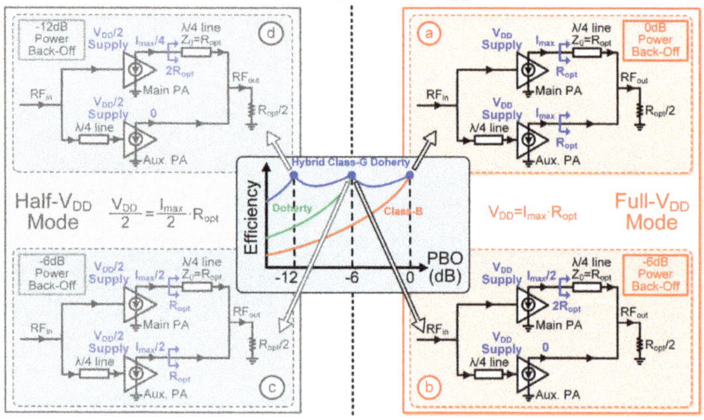

In the Full V_{DD} mode, the main and auxiliary power DACs are scaled accordingly to achieve its optimum Doherty back-off efficiency enhancement to 6dB power back-off. In particular, at the 0dB PBO, both main and auxiliary PAs output their maximum output current I_{max}, while at the 6dB BPO, the main PA outputs $I_{max}/2$ and the auxiliary PA is turned-off. In this process, the optimum PA large-signal load-pull/load-line equation, $V_{DD}=I_{max}\cdot R_{opt}$, is maintained at the main PA output to ensure high efficiency.

49

HybridClass-G Current-Mode Parallel Digital Doherty PA (V)

• **Our Proposed Hybrid Class-G Doherty – Half-V_{DD} mode**

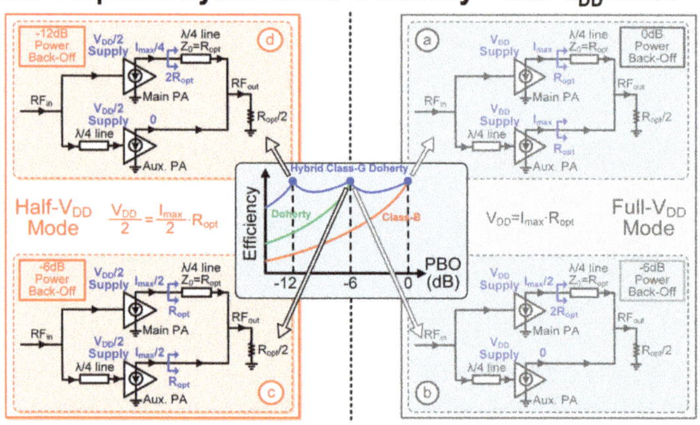

For deep power back-off, the PA is configured to its half-V_{DD} mode, where the supply voltage is switched to $V_{DD}/2$. At 6dB PBO, to keep the continuity of the output power P_{out}, the main and the auxiliary PAs are both configured to output $I_{max}/2$. Then, the two digital PAs are scaled to perform another Doherty-based back-off efficiency enhancement for 6dB to 12dB PBO with a supply voltage of $V_{DD}/2$. At the 12dB PBO, the main PA is configured to output $I_{max}/4$, and the auxiliary PA is turned-off. As a result, the hybrid Class-G Doherty PA can achieve backoff efficiency enhancement to 12dB PBO with two Doherty operations at the full-V_{DD} and the half-V_{DD} mode, essentially realizing a 3-stage Doherty PA by using only one supply switch and a 2-stage Doherty PA output passive network.

50 HybridClass-G Current-Mode Parallel Digital Doherty PA (VI)

- **Hybrid Class-G Doherty – Circuit implementation**

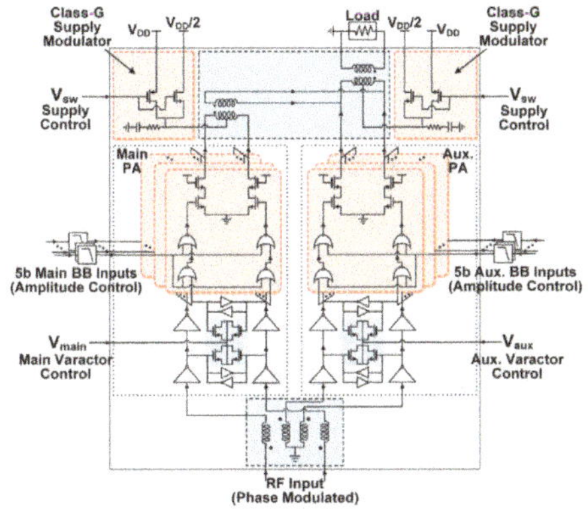

- A 5-bit Class-D^{-1} power DAC driven by 4-stage drivers for each PA path.
  - Precise and flexible control of the currents from two PA paths.
- Class-G supply modulator
  - 12mm/360nm PMOS for V_{DD} with R_{on} = 0.34 Ω.
  - 6mm/360nm NMOS for $V_{DD}/2$ with R_{on} = 0.38 Ω.
  - Minimized shoot through current
  - R-C damping legs for smoothened supply switching

- [16] S. Hu, S. Kousai, and H. Wang, *IEEE ISSCC Dig. Tech. Papers*, Feb. 2015.
- [17] S. Hu, S. Kousai, and H. Wang, *IEEE J. of Solid-State Circuits*, vol. 51, no. 3, pp. 598 - 613, Mar. 2016.

In terms of the circuit implementation of the hybrid Class-G Doherty, 5 bit class-D^{-1} power DACs are implemented for the main and auxiliary PAs. The Class-G supply modulator is implemented using complementary NMOS/PMOS switches that are carefully designed to minimize their on-resistance, shoot-through current, and supply ringing during switching.

51 HybridClass-G Current-Mode Parallel Digital Doherty PA (VII)

- **Chip microphotograph**
  - Fully integrated in a standard 65 nm bulk CMOS process.

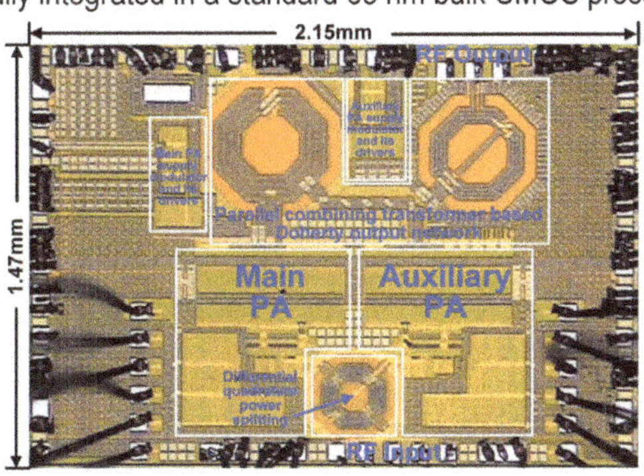

- [16] S. Hu, S. Kousai, and H. Wang, *IEEE ISSCC Dig. Tech. Papers*, Feb. 2015.
- [17] S. Hu, S. Kousai, and H. Wang, *IEEE J. of Solid-State Circuits*, vol. 51, no. 3, pp. 598 - 613, Mar. 2016.

A proof-of-concept design is fully integrated in a standard 65nm bulk CMOS process with the total chip area of 2.15mm×1.47mm, dominated by the I/O pads.

52 HybridClass-G Current-Mode Parallel Digital Doherty PA (VIII)

• **Hybrid Class-G Doherty – CW measurement results (3.7GHz)**

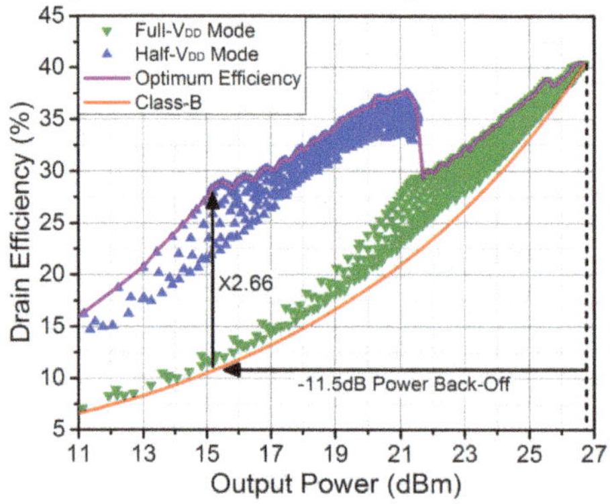

- +26.7 dBm peak output power
- 40.2% peak drain efficiency
- 37.0% drain efficiency at -6 dB power back-off (x1.84 over Class-B)
- 26.2% drain efficiency at -12 dB power back-off (x2.61 over Class-B)
- Maximum x2.66 (266%) over Class-B at -11.5 dB power back-off

→ World record of efficiency enhancement in deep back-off in Si PAs without switches at RF outputs.

• [16] S. Hu, S. Kousai, and H. Wang, *IEEE ISSCC Dig. Tech. Papers*, Feb. 2015.
• [17] S. Hu, S. Kousai, and H. Wang, *IEEE J. of Solid-State Circuits*, vol. 51, no. 3, pp. 598 - 613, Mar. 2016.

The hybrid Class-G Doherty PA is characterized for its continuous-wave performance. At 3.7GHz carrier frequency, it achieves 26.7dBm peak output power, 40.2% peak drain efficiency, 1.84× drain efficiency enhancement over class-B PA at 6dB PBO, 2.62× drain efficiency enhancement over class-B PA at 12dB PBO, and its maximum 2.66× drain efficiency enhancement over class-B PA at 11.5dB PBO, attaining the world record of efficiency enhancement in deep back-off among reported silicon PAs without switches at RF outputs.

53 HybridClass-G Current-Mode Parallel Digital Doherty PA (IX)

• **Hybrid Class-G Doherty – AM-PM distortion**

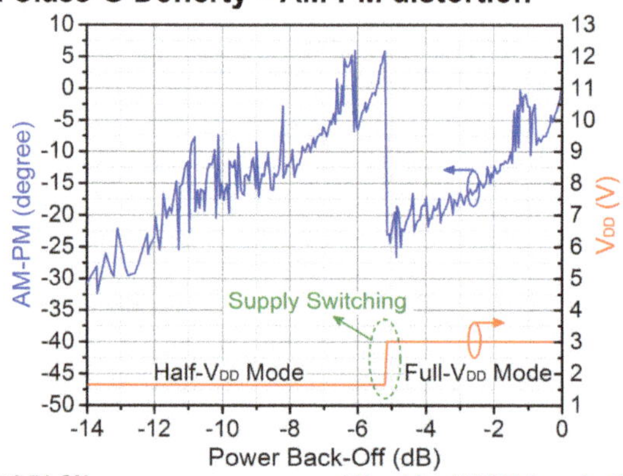

*** Measured at 3.71 GHz.** *** Need for AM-PM linearization.**

However, due to the Class-G supply switching and digital Doherty operation, the PA exhibits large AM-PM distortion, particular at the power level where Class-G supply switching happens.

54 HybridClass-G Current-Mode Parallel Digital Doherty PA (X)

- **AM-PM linearization – Analog-Assisted Technique**

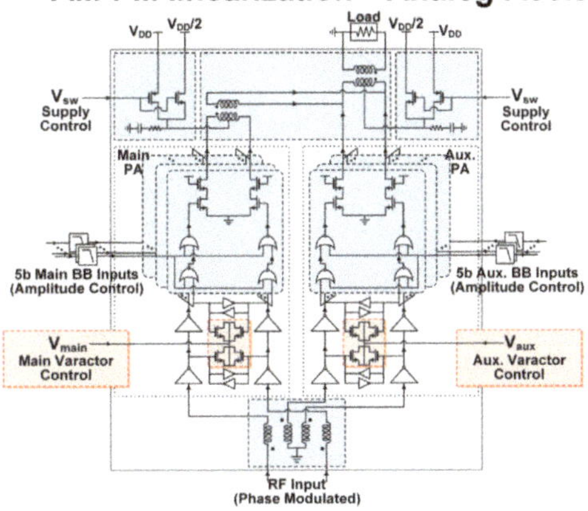

- Two-stage phase compensation varactors are employed in the digital drivers.
- V_{main} and V_{aux} can be decomposed into the common-mode V_{cm} and differential-mode V_{dm}.

$$V_{cm} = \frac{1}{2}(V_{main} + V_{aux})$$
$$V_{dm} = V_{main} - V_{aux}$$

To linearize the AM-PM distortion, the PA implements an analog-assisted technique, which is composed of two-stage compensation varactors in the main and auxiliary digital drivers. The main and auxiliary path control voltages can be decomposed to their common-mode (for compensation of the AM-PM distortion) and differential-mode voltages (for reconfiguring and extending the carrier frequency of the PA).

55 HybridClass-G Current-Mode Parallel Digital Doherty PA (XI)

- **Modulation measurement results by V_{cm} – AM-PM linearization**
  - 3.3 dB EVM and 2.8 dB ACLR improvement.

Achieved mainly by tuning V_{cm}

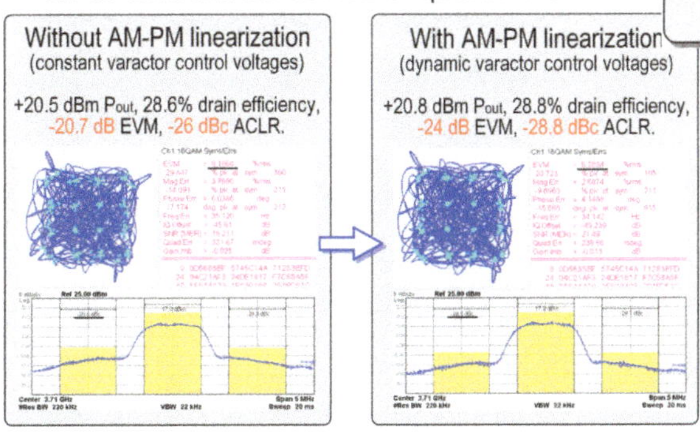

* Measured with 1 MSym/s 16QAM at 3.71 GHz using x10 oversampling.

By real-time adjusting the common-mode control voltages for the compensation varactors, the PA AM-PM can be clearly improved, with 3.3dB EVM and 2.8dB ACLR improvement for a 1MSym/s 16QAM signal. The averaged output power and PA drain efficiency stay almost constant with or without the varactor-based real-time AM-PM compensation.

56 HybridClass-G Current-Mode Parallel Digital Doherty PA (XII)

- **RF bandwidth extension by V_{dm} – CW measurement results**
  - 1dB bandwidth: 1.08 GHz (32%) ⇨ 1.80 GHz (48%)
  - Peak efficiency: @ 4.3 GHz: 25.5% ⇨ 33.3%
    @ 5.0 GHz: 5.3% ⇨ 24.0%

Achieved mainly by tuning V_{dm}

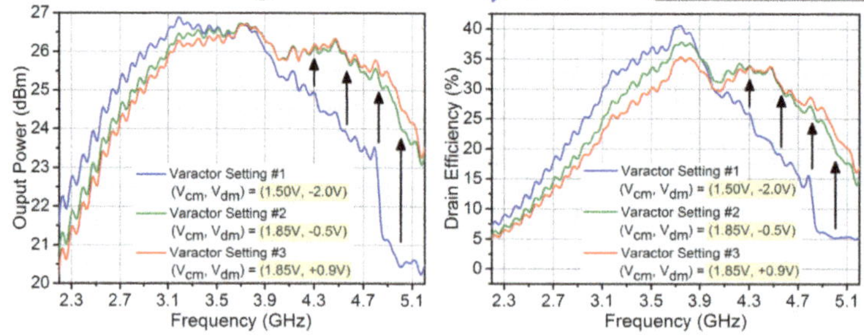

* Measured with all the power cells on.

By adjusting the differential-mode control voltages for the compensation varactors, the PA carrier frequency is substantially extended. The P_{out} 1dB bandwidth is extended from 1.08GHz (32%) to 1.8GHz (48%). The PA peak efficiency is improved from 25.5% to 33.3% at 4.3GHz, while at 5GHz it is improved from 5.3% to 24%. This is because the differential-mode control voltages adjust the relative phase difference between the main/auxiliary PA paths, compensating the frequency-dependent behaviors of the input/output networks of the Doherty PA.

57 HybridClass-G Current-Mode Parallel Digital Doherty PA (XIII)

- **Modulation results – Total PA efficiency at power back-off**
  - Significant efficiency enhancement is achieved at deep power back-off.

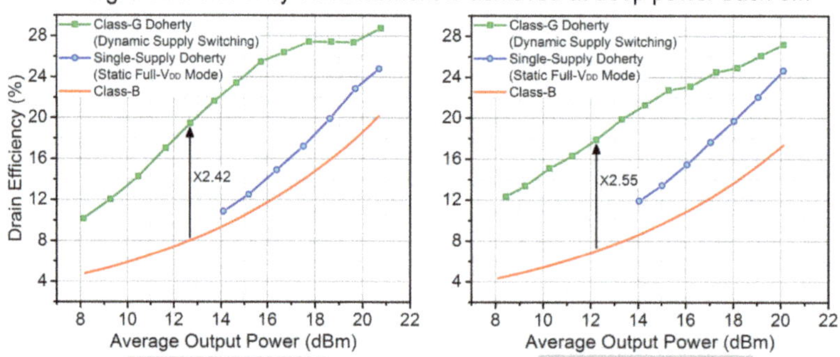

Measured at 3.71 GHz

Measured at 4.3 GHz (16.2% freq. offset)

*1 MSym/s 16QAM with x10 oversampling.

Comprehensive modulation tests are performed versus the PA power back-off. The hybrid class-G Doherty operation clearly achieves drastic efficiency enhancement over a conventional single-supply Doherty PA or a normalized class-B PA.

58

Voltage-Mode Series Doherty PA (I)

- **Switched-capacitor PA as the main/auxiliary PA core**

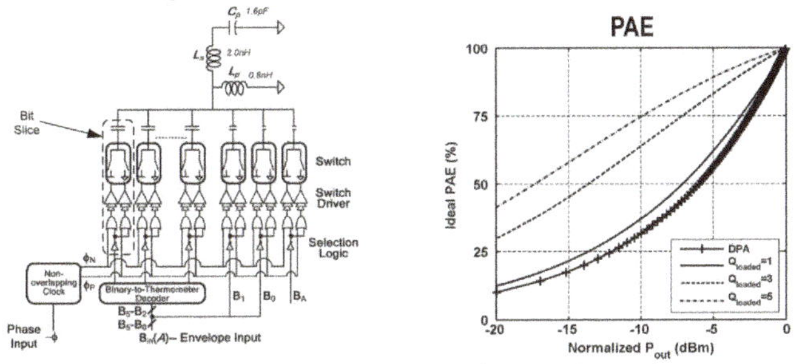

- High linearity due to the capacitance mismatch.
- Low output impedance → Voltage source approximation
- PAE is determined by the Q_{loaded}

- [18] S. Yoo, et al., "A Switched-Capacitor RF Power Amplifier," *IEEE J. of Solid-State Circuits*, vol. 46, no. 12, pp. 2977 - 2987, Sep. 2011.

As an alternative to the current-mode parallel Doherty PA, voltage-mode series Doherty PA is also gaining increasing interest in the technical community. Typical voltage-mode series Doherty PA employ switched-capacitor PAs as the core for the main and auxiliary PAs. It exhibits high linearity due to the low mismatch among switched capacitors and naturally presents low output impedance that can enable voltage-mode series Doherty power combing.

59

Voltage-Mode Series Doherty PA (II)

- **Voltage mode Doherty PA**

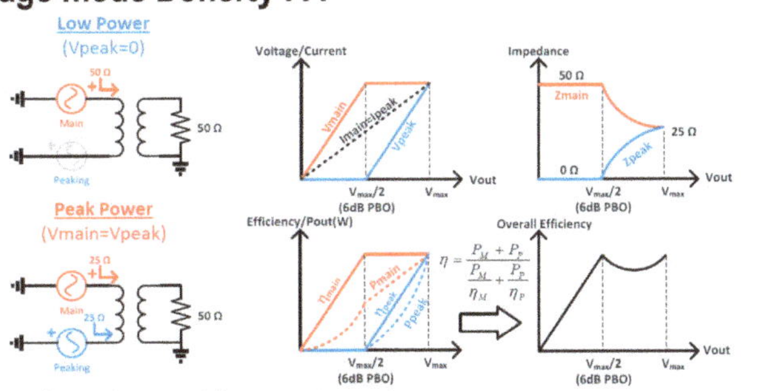

- Low impedance at low power region.
- No additional quarter-wave T line
- Limited output voltage swing

→ Smaller size
→ Wideband modulation

- [20] V. Vorapipat, C. Levy, P. Asbeck, "A wideband voltage mode Doherty power amplifier," *Proc. IEEE Radio Frequency Integrated Circuits (RFIC)*, May. 2016.

The operation principle of the Voltage mode series Doherty PA can be explained as follows. At low power levels (more than 6dB power back-off), the auxiliary PA is off presenting a low output impedance to the summation transformer. The main PA then exhibits a load of 50ohm. For higher power levels, the auxiliary PA is gradually turned on with reversed polarity compared to the main, resulting an increasing differential voltage as the PA output. At the peak output power P_{out}, the auxiliary PA delivers same output voltage amplitude as the main PA but with a reversed phase, resulting in the maximum PA output. Simultaneously, the main and auxiliary PAs achieve series power combining, and then each exhibits a load impedance of 25ohm (50ohm/2) as the desired Doherty active load modulation. The unique advantages of the voltage-mode series Doherty PA include its elimination of the lamda/2 inverter, smaller size, large carrier bandwidth, and wideband modulation.

60 Voltage-Mode Series Doherty PA (III)

- **Voltage mode class-G Doherty PA**

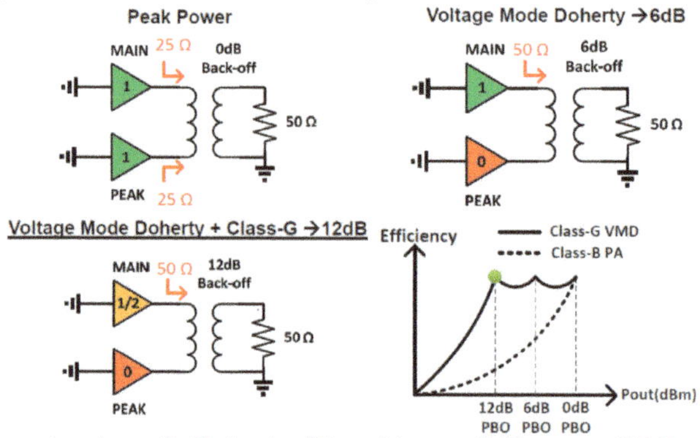

- Voltage mode class-G Doherty PA achieves 6dB more PBO compared to a symmetric Doherty PA.
  - [21] V. Vorapipat, C. Levy, P. Asbeck, IEEE Int. Solid-State Circuits Conf. (ISSCC) Dig. Tech. Papers, pp. 46-47, Feb. 2017.
  - [22] V. Vorapipat, C. Levy, P. Asbeck, *IEEE J. of Solid-State Circuits*, vol. 52, no. 12, pp. 3348 - 3360, Dec. 2017.

The voltage mode Doherty PA can be combined with Class-G supply switching to achieve hybrid voltage-mode class-G Doherty PA, which also extends the Doherty operation to 12dB PBO through a one-bit supply modulator, just like the current mode Doherty PA.

61 Voltage-Mode Series Doherty PA (IV)

- **Voltage-mode class-G Doherty PA → Soft $V_{dd}/2$ & V_{dd} transition**

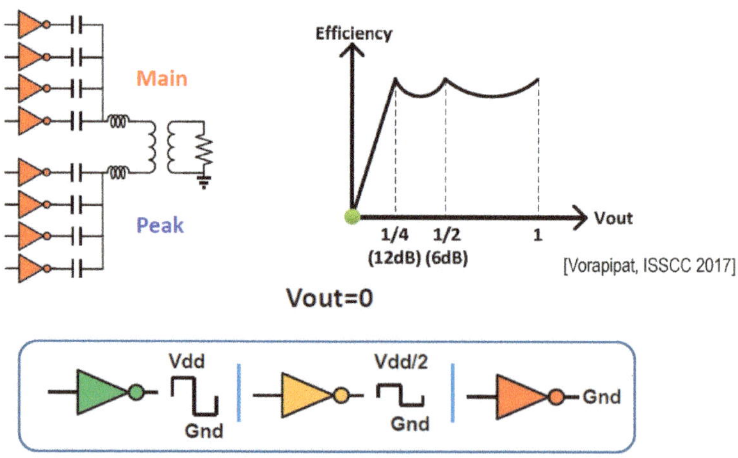

Moreover, the voltage mode Class-G Doherty PA can potentially achieve soft supply switching transition to minimize the PA nonlinearity.

62 Voltage-Mode Series Doherty PA (V)

- **Voltage-mode class-G Doherty PA → Soft $V_{dd}/2$ & V_{dd} transition**

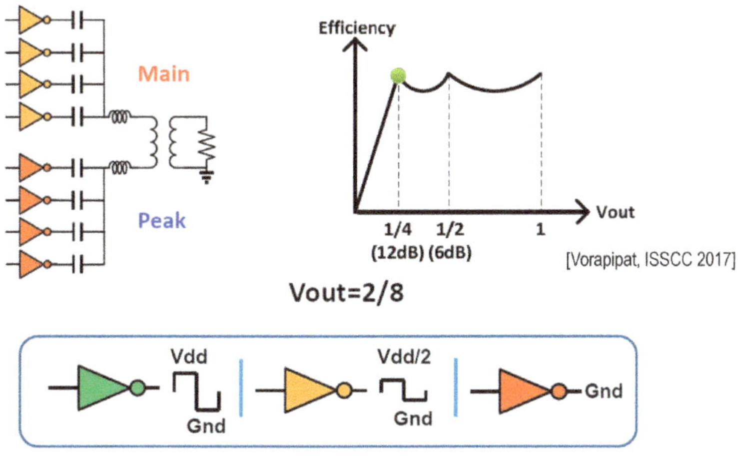

At low power level, i.e., more than 12dB PBO, only the main PA array is turned on and switching between $V_{DD}/2$ and GND, while the auxiliary PA array is kept off.

63 Voltage-Mode Series Doherty PA (VI)

- **Voltage-mode class-G Doherty PA → Soft $V_{dd}/2$ & V_{dd} transition**

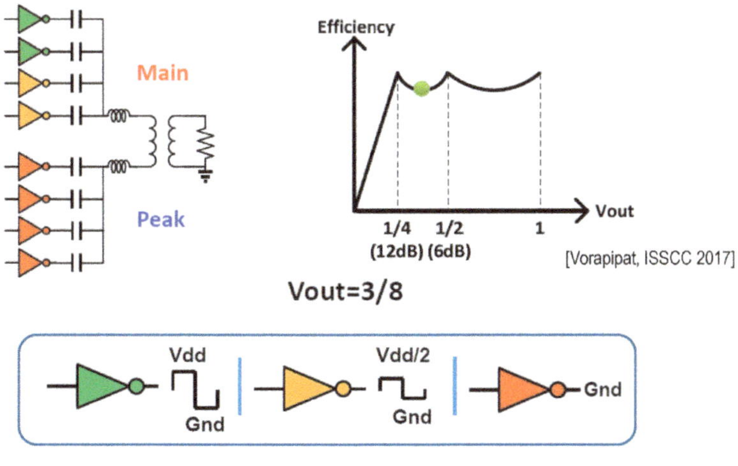

Between 12dB and 6dB PBO, part of the main PA array is gradually changed to V_{DD}, while the auxiliary PA array is still kept off.

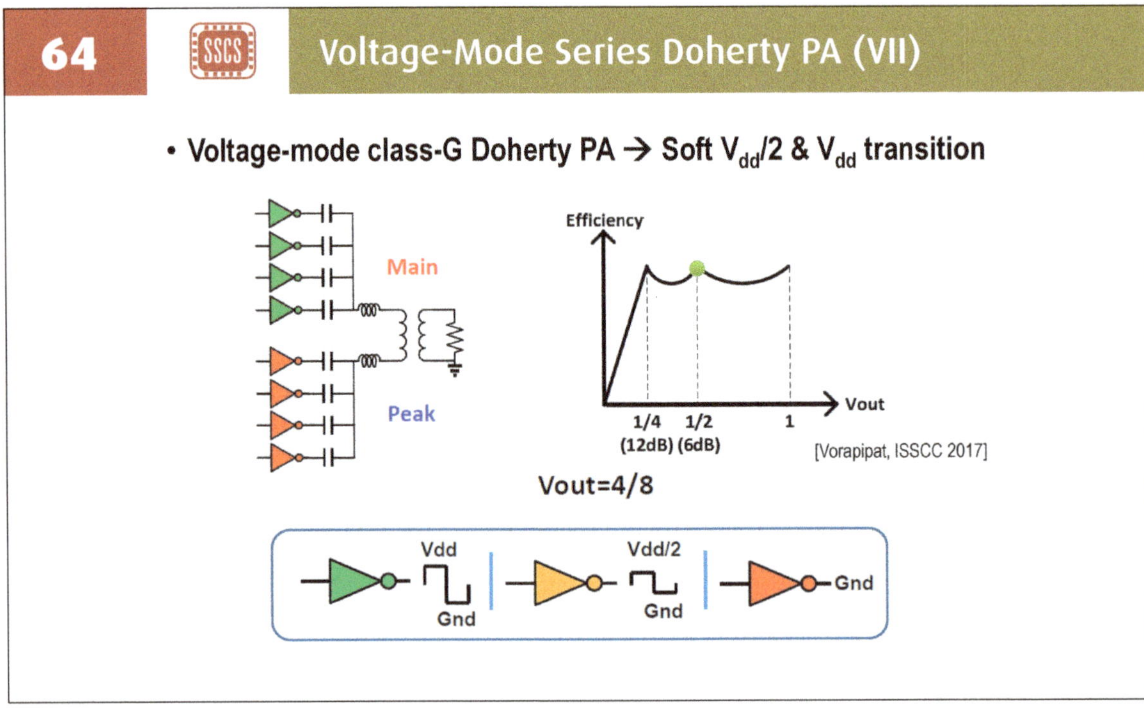

At 6dB PBO, all the main PA array is changed to V_{DD}, while the auxiliary PA array is still off.

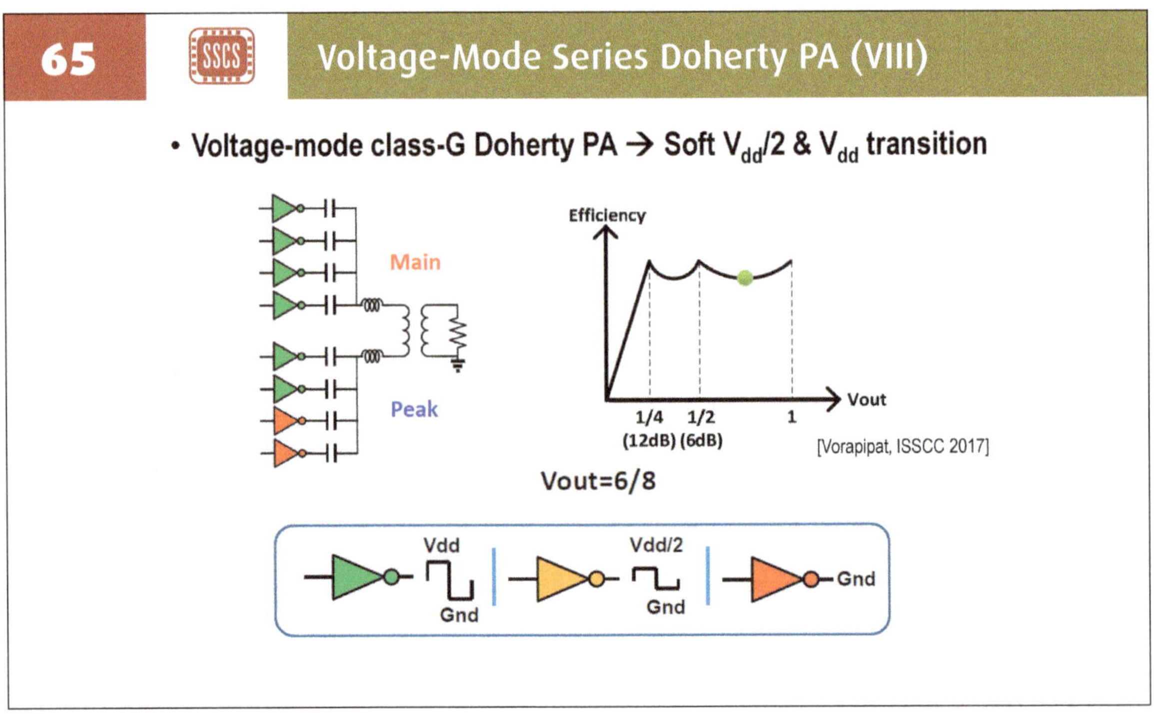

Between 6dB PBO and the peak output power, the auxiliary PA array is gradually turned on, and the on power cells are switching between V_{DD} and GND.

66 Voltage-Mode Series Doherty PA (IX)

- **Voltage-mode class-G Doherty PA → Soft $V_{dd}/2$ & V_{dd} transition**

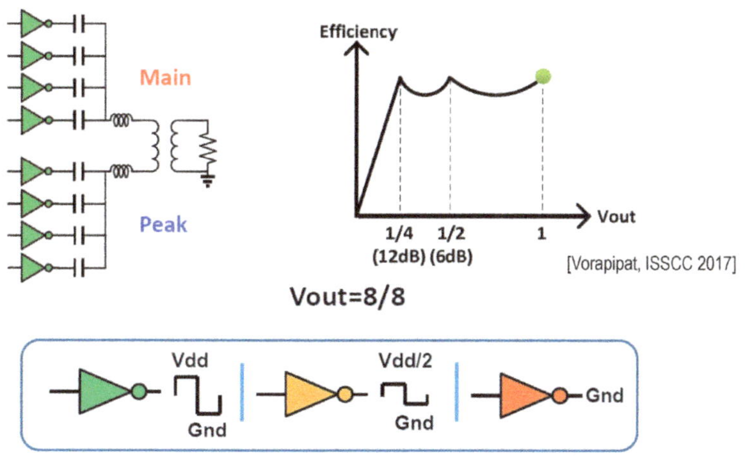

At the peak output power, all the power cells in the main and auxiliary PA arrays are turned on and are switching between V_{DD} and GND, delivering the maximum output power.

67 Voltage-Mode Series Doherty PA (X)

- **Voltage mode class-G Doherty PA**

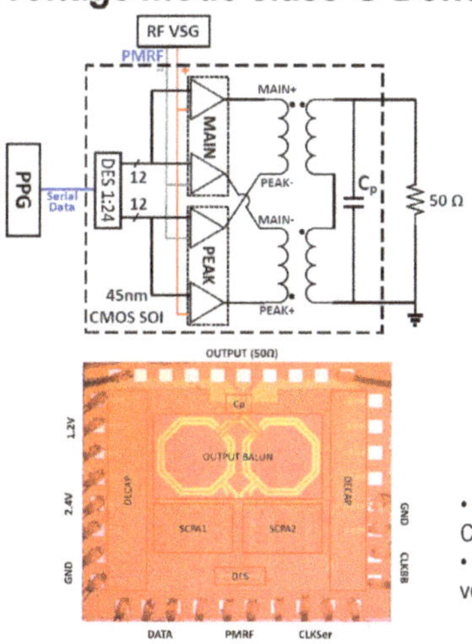

- Low impedance
- Deep power back-off
- Class-G soft supply transition
  → Glitch free
  → High AM-AM & AM-PM linearity

- Limited output power.
- Dual power supply.

• [21] V. Vorapipat, C. Levy, P. Asbeck, IEEE Int. Solid-State Circuits Conf. (ISSCC) Dig. Tech. Papers, pp. 46-47, Feb. 2017.
• [22] V. Vorapipat, C. Levy, P. Asbeck, *IEEE J. of Solid-State Circuits*, vol. 52, no. 12, pp. 3348 - 3360, Dec. 2017.

A proof-of-concept voltage-mode Class-G Doherty PA is implemented in a standard 45nm CMOS SOI process leveraging a two-coil transformer-based series Doherty power combiner.

68 SSCS Voltage-Mode Series Doherty PA (X)

- **Measurement results**

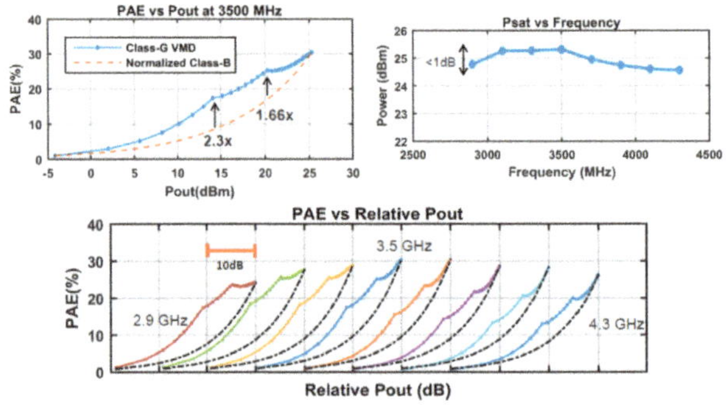

- Back-off improvement is 2.3X at 12PBO and 1.66X at 6PBO over class-B PA.
- 1dB bandwidth is 2.8GHz to 4.3GHz

The continuous-wave measurement of the PA shows a substantial back-off efficiency enhancement of 2.3× at 12dB PBO and 1.66× at 6dB PBO over a normalized idealistic class-B PA. The PA output power P_{out} 1dB bandwidth is from 2.8GHz to 4.3GHz, achieving 42.25% fractional bandwidth.

69 SSCS Voltage-Mode Series Doherty PA (XI)

- **Measurement results**

Static AM-AM and AM-PM

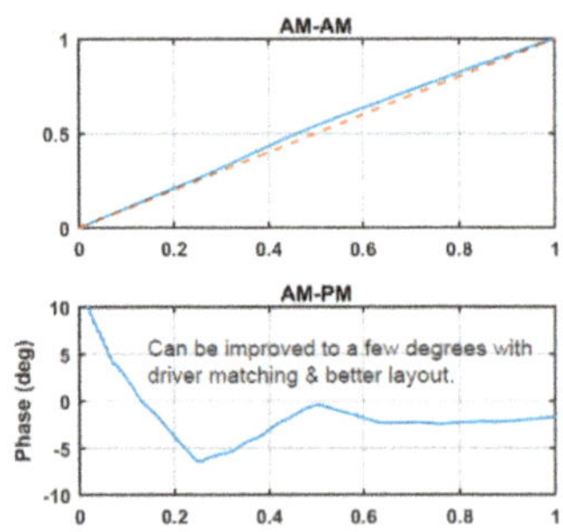

Dynamic AM-AM and AM-PM

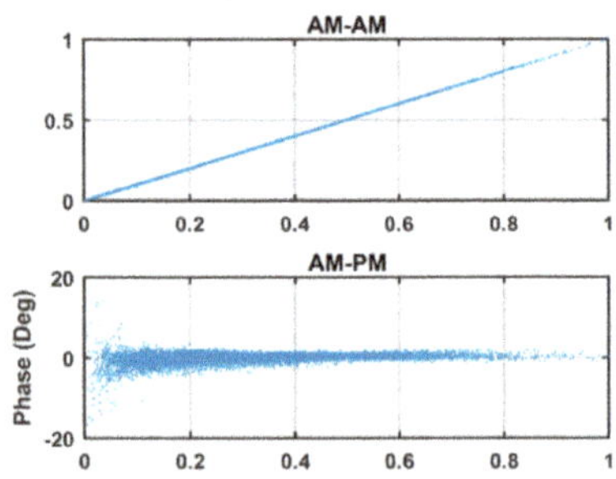

- Class-G supply switching glitch free
- 10MHz 32 Carriers 256-QAM OFDM using memoryless look-up table

The nonlinearity of the PA is also carefully characterized. For the static AM-AM and AM-PM performance, the switched capacitor PA configuration and the soft supply transistion ensures low AM-AM and AM-PM distortion. The dynamic AM-AM and AM-PM performance is tested using 10MHz 32 carrier 256-QAM OFDM signal, showing high linearity with negligible memory effect at such modulation speed.

70 Voltage-Mode Series Doherty PA (XII)

- **Measurement results**

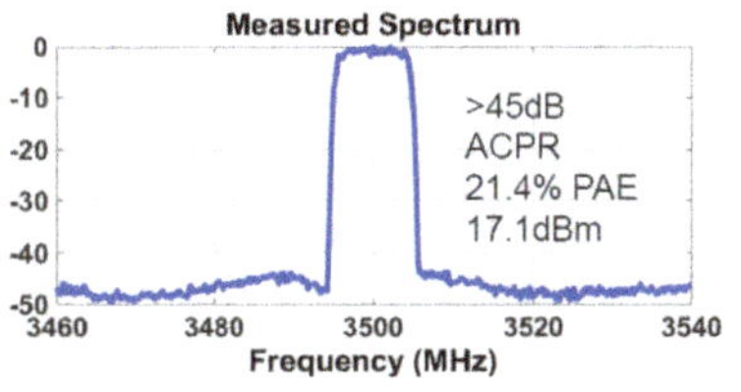

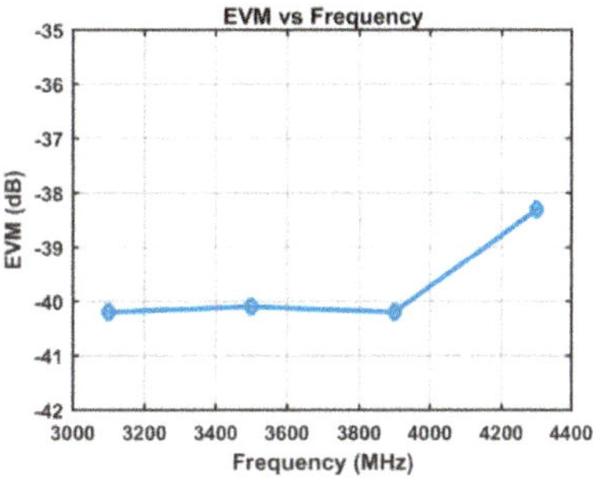

- 10MHz 32-Carrier 256-QAM OFDM using a memoryless LUT.
- Wideband operation.

With a memoryless look up table, the PA supports a 10MHz 32-carrier 256-QAM OFDM signal with a better-than 45dB ACPR, 21.4% PAE, and 17.1dBm averaged output power. A low EVM is kept across the carrier frequency from 3.1GHz to 4.3GHz, showing the capability of wideband operation.

71 Outline

- Introduction
- Doherty PA Configurations
- Analog Doherty PAs at RF Frequency
- Digital Doherty PAs at RF Frequency
- **Mm-Wave Doherty PAs**
- Conclusion

In the next section, state-of-the-art mm-Wave Doherty PAs will be presented show the back-off efficiency enhancement to enable high-speed and high efficiency power amplification with mm-Wave carriers.

72 Multi-Band/Broadband PA for Future 5G Systems

- **Modern wireless systems demand <u>broadband</u> RF/mm-wave power generation.**

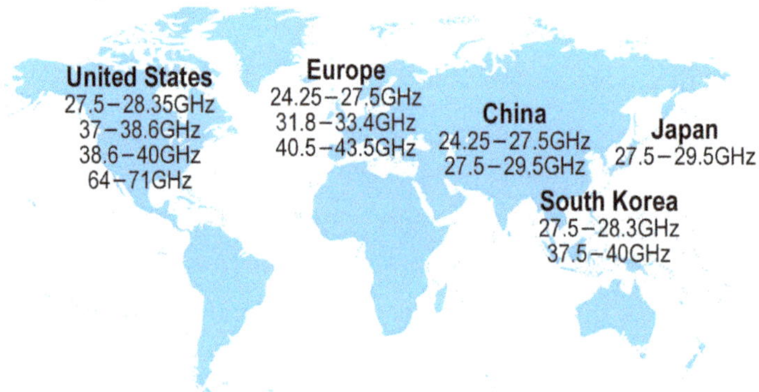

- [23] FCC 16-89; "Spectrum for 4G and 5G", Qualcomm, Oct. 2016.

There is an increasing interest in multi-band and broadband power amplifiers. This is particularly relevant for future mm-Wave 5G systems that have noncontiguous frequency bands, spanning from 24GHz to 71GHz, for different countries and regions. If a mm-wave power amplifier covers multiple of these noncontiguous carrier bands, it then supports cross-network or international roaming and multi-standard 5G communication.

73 Mm-wave Series Doherty Power Amplifiers (I)

- **Transformer-based series combiner with an additional auxiliary PA matching network**

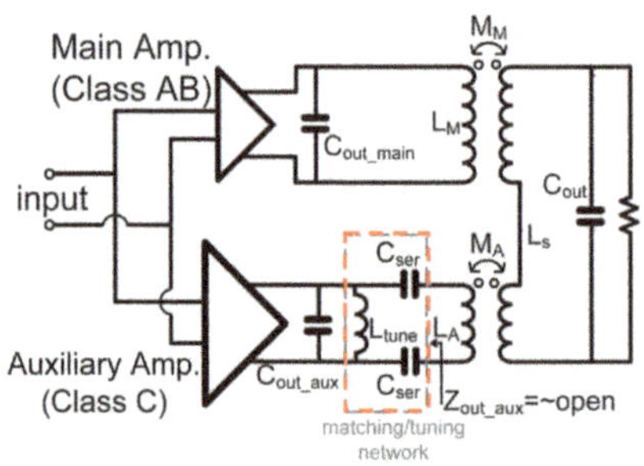

- Additional matching network → improved back-off efficiency
- Asymmetric main/auxiliary PAs → Larger auxiliary amplifier → efficiency enhancement in deeper back-off

- [24] E. Kaymaksut, D. Zhao, P. Reynaert, "Transformer-Based Doherty Power Amplifiers for mm-Wave Applications in 40-nm CMOS," *IEEE Trans. Microw. Theory Tech.* vol. 63, no. 4, pp. 1186 - 1192, Mar 2015.

As an example of mm-Wave Doherty power amplifiers, series Doherty power amplifiers have been proposed and demonstrated that utilize transformer-based series power combiner. To complete the Doherty operation and improve the back-off efficiency, an additional matching and impedance inverting network is needed at the auxiliary PA path. Asymmetric main/auxiliary PAs can be employed to boost the efficiency enhancement in the deep power backoff and compensate for the passive loss of the power combining network.

74 Mm-wave Series Doherty Power Amplifiers (II)

- **PA schematic with adaptive biasing for the auxiliary PA**

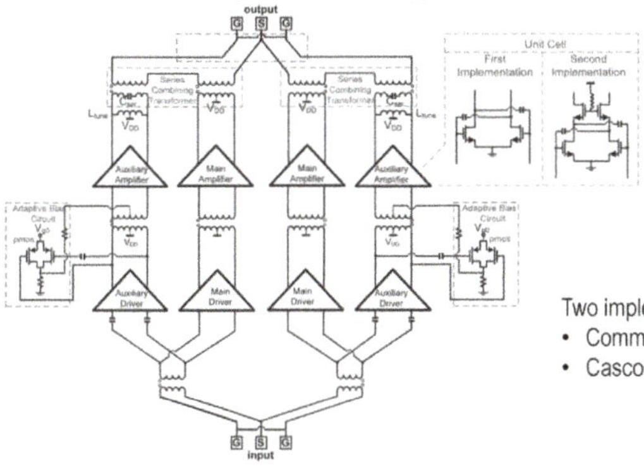

Two implementations
- Common-source
- Cascode

- [24] E. Kaymaksut, D. Zhao, P. Reynaert, "Transformer-Based Doherty Power Amplifiers for mm-Wave Applications in 40-nm CMOS," *IEEE Trans. Microw. Theory Tech.* vol. 63, no. 4, pp. 1186 - 1192, Mar 2015.

The series Doherty PA architecture can be extended to multi-path power combining. Adaptive biasing circuit can be added to the auxiliary PA to facilitate the turning on the auxiliary PA path. Broadband capacitive neutralization is implemented to boost the gain of the PA and enhance its reverse isolation and stability.

75 Mm-wave Series Doherty Power Amplifiers (III)

- **Chip microphotographs**
  - **Common-source implementation**
  - **Cascode implementation**

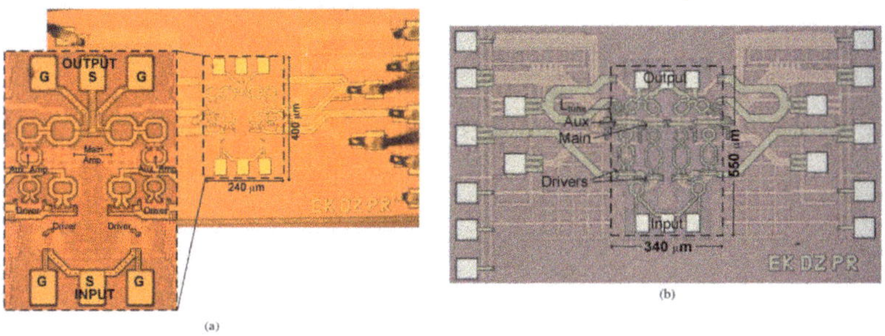

(a) (b)

- [24] E. Kaymaksut, D. Zhao, P. Reynaert, "Transformer-Based Doherty Power Amplifiers for mm-Wave Applications in 40-nm CMOS," *IEEE Trans. Microw. Theory Tech.* vol. 63, no. 4, pp. 1186 - 1192, Mar 2015.

The common-source implementation and the cascade implementation are both demonstrated at an IEEE T-MTT journal paper in 2015. Both designs achieve a compact core chip area.

76 Mm-wave Series Doherty Power Amplifiers (IV)

- **Continuous-wave PA measurement results**
  - Common-source implementation
  - Cascode implementation

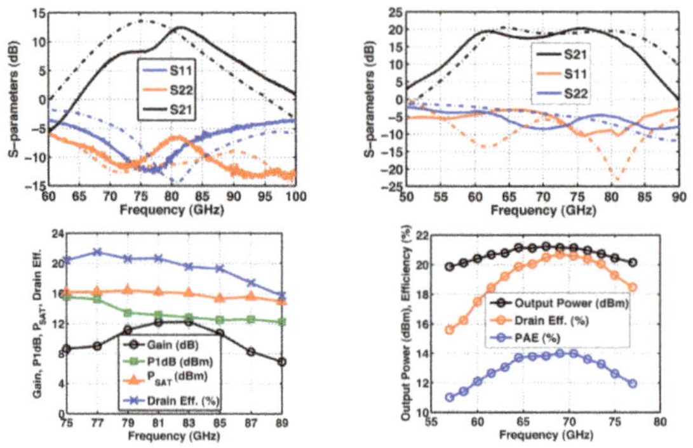

- [24] E. Kaymaksut, D. Zhao, P. Reynaert, "Transformer-Based Doherty Power Amplifiers for mm-Wave Applications in 40-nm CMOS," *IEEE Trans. Microw. Theory Tech.* vol. 63, no. 4, pp. 1186 - 1192, Mar 2015.

The PAs are characterized with continuous-wave testings. The small signal performance as S-parameters and large signal performance as saturated output power, drain efficiency, and PAE, are reported over the carrier frequency, showing a wideband PA performance.

77 Mm-wave Series Doherty Power Amplifiers (V)

- **Continuous-wave PA measurement results**

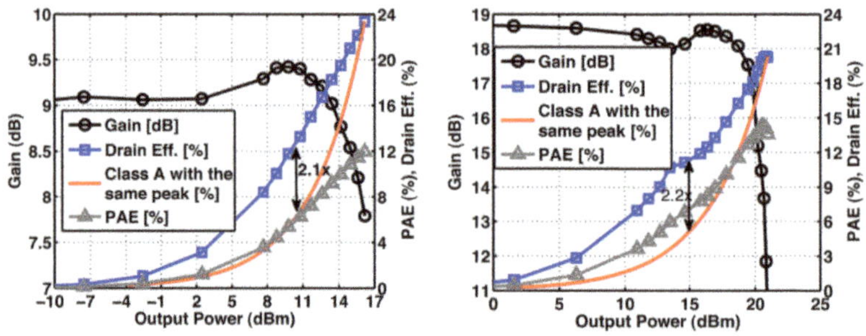

- Common-source implementation
- 1.05× over class-B and 2.1× over class-A at 6dB PA PBO

- Cascode implementation
- 1.1× over class-B and 2.2× over class-A at 6dB PA BPO

- [24] E. Kaymaksut, D. Zhao, P. Reynaert, "Transformer-Based Doherty Power Amplifiers for mm-Wave Applications in 40-nm CMOS," *IEEE Trans. Microw. Theory Tech.* vol. 63, no. 4, pp. 1186 - 1192, Mar 2015.

The PAs' continuous-wave back-off performance is next characterized. The common-source PA shows its drain efficiency with 1.05× enhancement over class-B and 2.1× enhancement over class-A at 6dB PA PBO. The cascode PA shows its drain efficiency with 1.1× enhancement over class-B and 2.2× enhancement over class-A at 6dB PA PBO.

78 SSCS Mm-wave Series Doherty Power Amplifiers (VI)

- **Modulation measurement results**

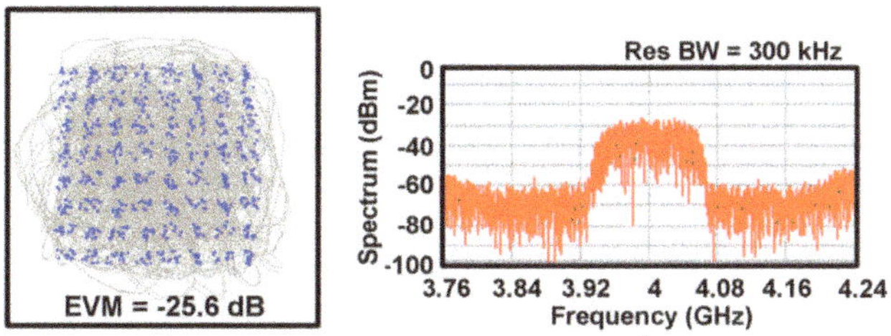

(a) (b)

- +15.9dBm, 100MSym/s 64QAM for the cascode implementation at 72GHz
- The modulation bandwidth is potentially limited by the adaptive biasing circuit.

- [24] E. Kaymaksut, D. Zhao, P. Reynaert, "Transformer-Based Doherty Power Amplifiers for mm-Wave Applications in 40-nm CMOS," *IEEE Trans. Microw. Theory Tech.* vol. 63, no. 4, pp. 1186 - 1192, Mar 2015.

The cascade PA achieves +15.9dBm averaged output power with a 100MSym/s 64QAM at 72GHz. The modulation bandwidth is potentially limited by the adaptive biasing circuit.

79 SSCS Mm-wave Multi-band Parallel Doherty PA (I)

- **Conventional parallel Doherty power combiner in silicon**

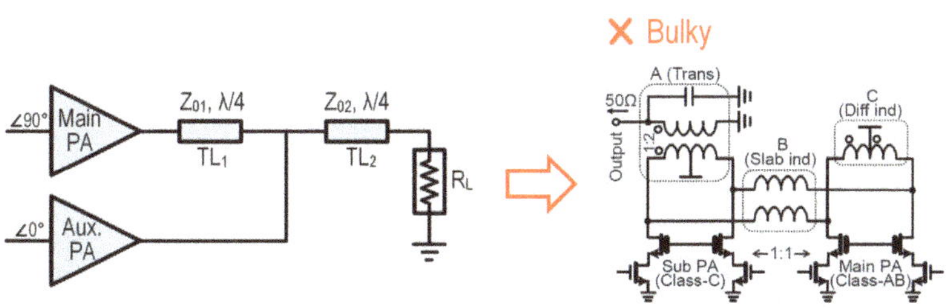

$Z_{01} = R_{opt}$, $Z_{02} = \sqrt{R_L R_{opt}/2}$

➢ π-networks using slab inductors

- [8] K. Onizuka, S. Saigusa, S. Otaka, "A +30.5 dBm CMOS Doherty power amplifier with reliability enhancement technique," *VLSI Circuits (VLSIC), 2012 Symposium on*, 13-15 June 2012.

On the other hand, it is highly desired to explore multi-band mm-Wave Doherty operation for parallel Doherty combiners with current-mode PAs, since most mm-Wave PAs are current-mode PAs in their nature. A conventional paralleled Doherty output network is composed of lamda/4 impedance inverter transmission lines. Although they can be equivalently implemented using C-L-C pi-networks, they still exhibit bulky formfactor in practice.

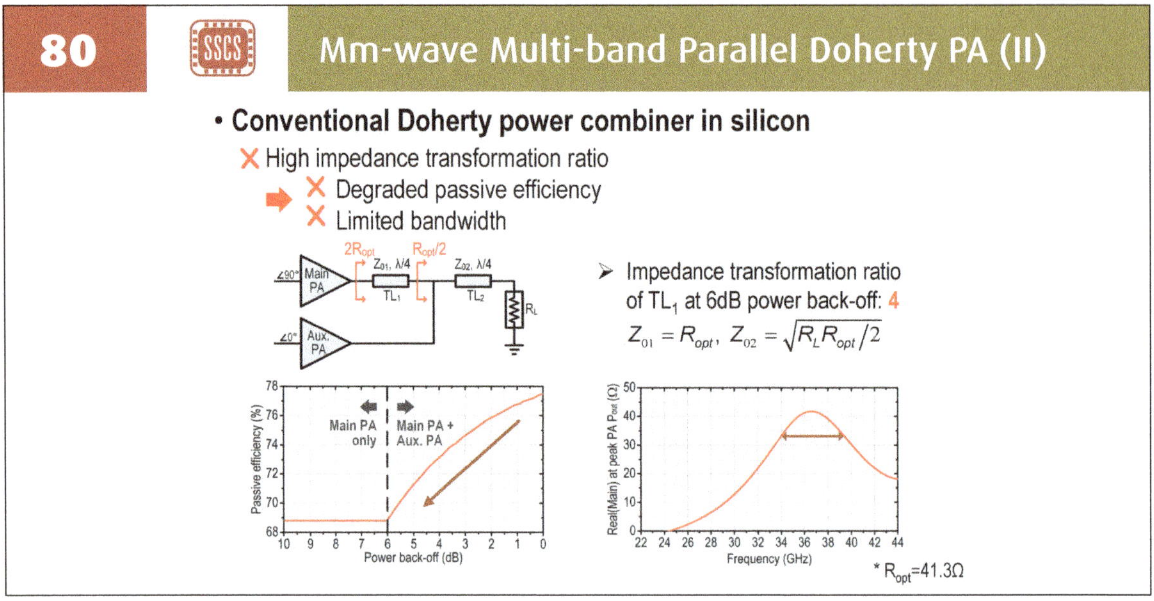

Moreover, conventional implementation of parallel Doherty output network exhibits high impedance transformation ratio of a factor of 4 at the 6dB PBO. Such a high impedance transformation ratio raises the loaded quality factor of the network and results in degraded passive efficiency and limited bandwidth of the network. In particular, the passive loss of a typical parallel Doherty output network degrades with power back-off, which greatly attenuates the achievable PBO efficiency enhancement of the Doherty PA.

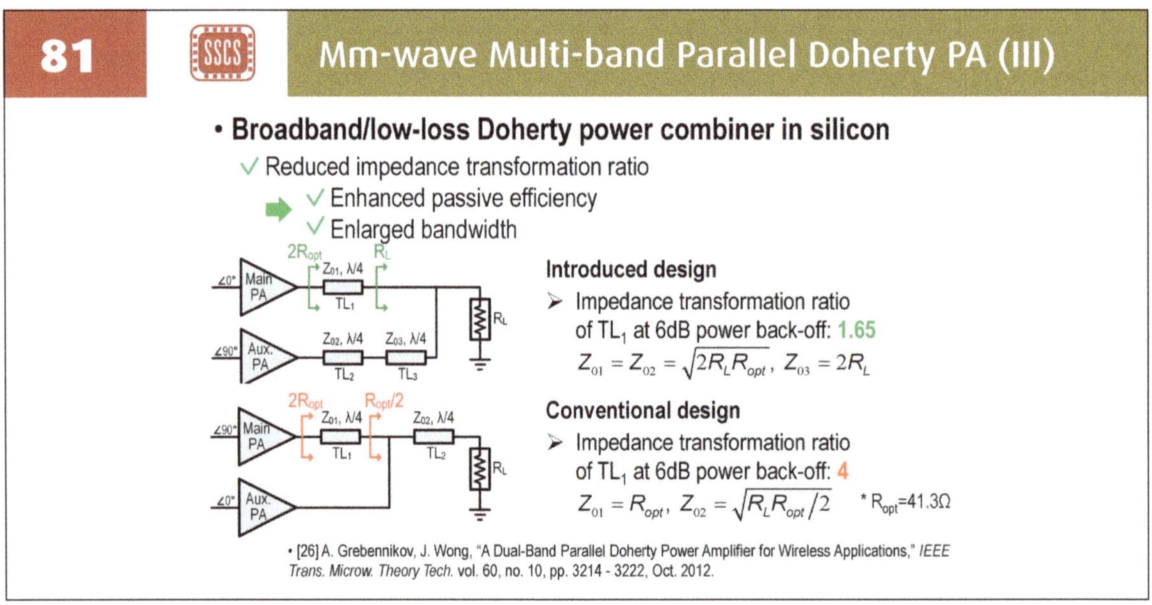

It can be shown that inserting additional impedance transformation network at the auxiliary PA path can greatly reduce the impedance transformation ratio from 4 to 1.65 at 6dB PBO. Such a reduced impedance transformation ratio will directly lead to an enhanced passive efficiency and enlarged carrier bandwidth. However, the additional auxiliary path impedance inverter networks will result in large implementation area and additional loss in practice.

82 Mm-wave Multi-band Parallel Doherty PA (IV)

- **Broadband/low-loss Doherty power combiner in silicon**
  - New transformer-based combiner
  - ✓ Compact (two-transformer footprint)
  - ✓ Absorption of device parasitic capacitors

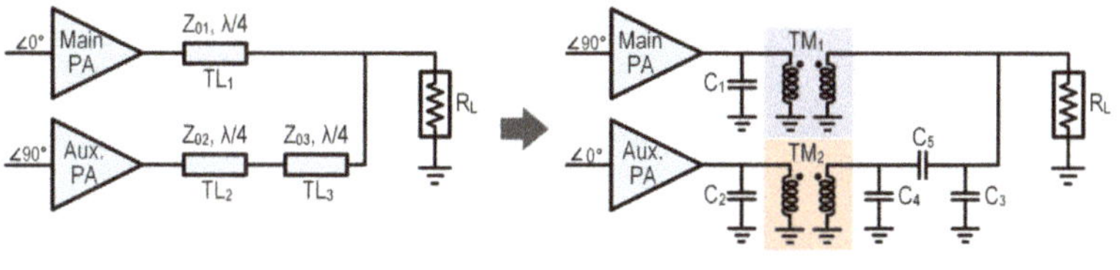

- [25] S. Hu, F. Wang, and H. Wang, "A 28GHz/37GHz/39GHz Multiband Linear Doherty Power Amplifier for 5G Massive MIMO Applications," *IEEE ISSCC Dig. Tech. Papers*, Feb. 2017.

It is shown that the three-T-line Doherty output network can be synthesized using two physical transformers with non-ideal magnetic coupling. Such a transformer-based Doherty output network achieves compact form-factor, differential power combining, and absorption of device parasitic capacitors. This design is reported in an IEEE ISSCC conference paper in 2017.

83 Mm-wave Multi-band Parallel Doherty PA (V)

- **Broadband/low-loss Doherty power combiner in silicon**
  - New transformer-based combiner: two transformers absorbing three λ/4 lines
  - ✓ Closed-form design equations

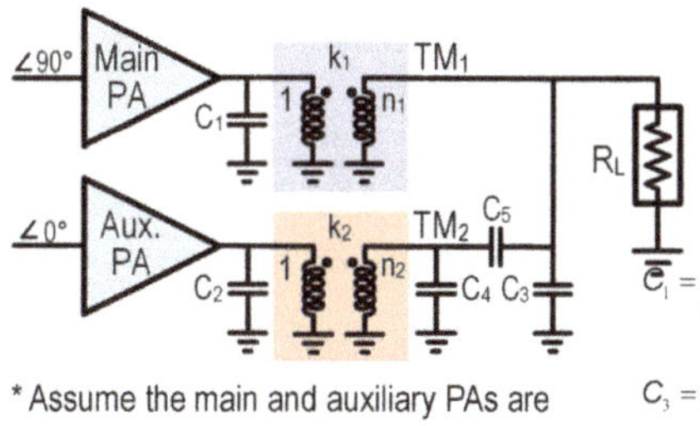

* Assume the main and auxiliary PAs are symmetric.

➢ For given n_1, k_1, n_2, and load-pull impedance R_{opt},

$$k_2 = \frac{\sqrt{n_2^2 R_{opt}/(2R_L)+4} - n_2\sqrt{R_{opt}/(2R_L)}}{2},$$

$$L_{P1} = \frac{k_1\sqrt{2R_L R_{opt}}}{\omega n_1 (1-k_1^2)}, \qquad L_{P2} = \frac{n_1 k_1 \sqrt{2R_L R_{opt}}}{\omega n_2^2 (1-k_1^2)},$$

$$C_1 = \frac{n_1}{\omega k_1 \sqrt{2R_L R_{opt}}}, \qquad C_2 = \frac{n_2^2(1-k_1^2)}{\omega n_1 k_1 (1-k_2^2)\sqrt{2R_L R_{opt}}},$$

$$C_3 = \left(\frac{k_1}{n_1}\right)^2 C_1, \quad C_4 = \left(\frac{k_2}{n_2}\right)^2 C_2, \quad C_5 = \frac{1-k_1^2}{\omega n_1 k_1 \sqrt{2R_L R_{opt}}}.$$

- [25] S. Hu, F. Wang, and H. Wang, "A 28GHz/37GHz/39GHz Multiband Linear Doherty Power Amplifier for 5G Massive MIMO Applications," *IEEE ISSCC Dig. Tech. Papers*, Feb. 2017.

The closed-form design equations for the transformer-based network synthesis of the three lamda/4 transmission lines are derived and presented, assuming the main and auxiliary PAs are symmetric. The design equations also naturally absorb the PA device output capacitance into the synthesized transformer-based Doherty output network.

84 Mm-wave Multi-band Parallel Doherty PA (VI)

- **Broadband/low-loss Doherty power combiner in silicon**
  - ✓ **Compact**
  - ✓ Enhanced passive efficiency in power back-off
  - ✓ Increased carrier bandwidth

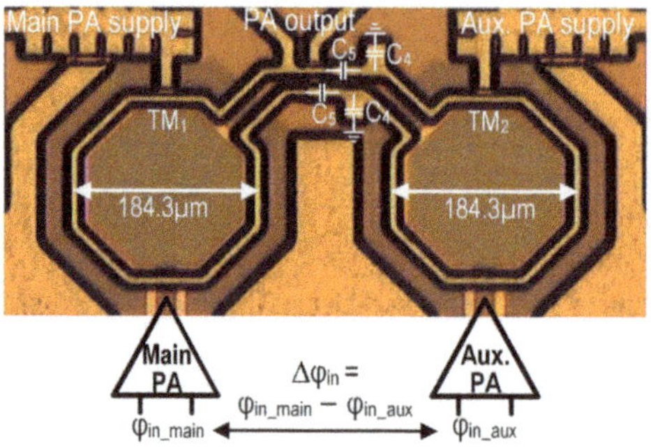

• [25] S. Hu, F. Wang, and H. Wang, "A 28GHz/37GHz/39GHz Multiband Linear Doherty Power Amplifier for 5G Massive MIMO Applications," *IEEE ISSCC Dig. Tech. Papers*, Feb. 2017.

A proof-of-concept parallel Doherty power combiner is implemented in a standard 45nm CMOS SOI process. It is composed of only two on-chip transformers with diameter of 184.3μm, achieves a compact passive network form-factor.

85 Mm-wave Multi-band Parallel Doherty PA (VII)

- **Broadband/low-loss Doherty power combiner in silicon**
  - ✓ Compact
  - ✓ **Enhanced passive efficiency in power back-off**
  - ✓ **Increased carrier bandwidth**

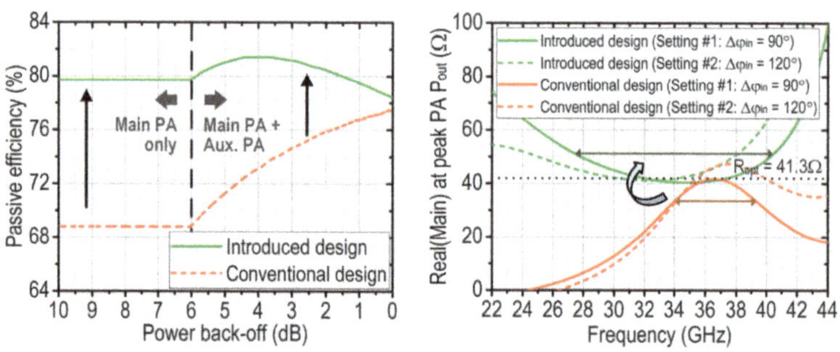

3D EM modeling shows that the synthesized transformer-based Doherty power combiner indeed achieves an improved passive efficiency at the power back-off with almost 80% passive efficiency at 6dB PBO. In contrast, the conventional design only attains 69%. In addition, the synthesized transformer Doherty power combiner also realizes an increased carrier bandwidth that is much wider than the conventional design.

86 Mm-wave Multi-band Parallel Doherty PA (VIII)

- **Conventional design** – Class-AB/-C with adaptive biasing
  - ✗ Limited modulation bandwidth
  - ✗ Weakened Class-AB/-C nonlinearity compensation

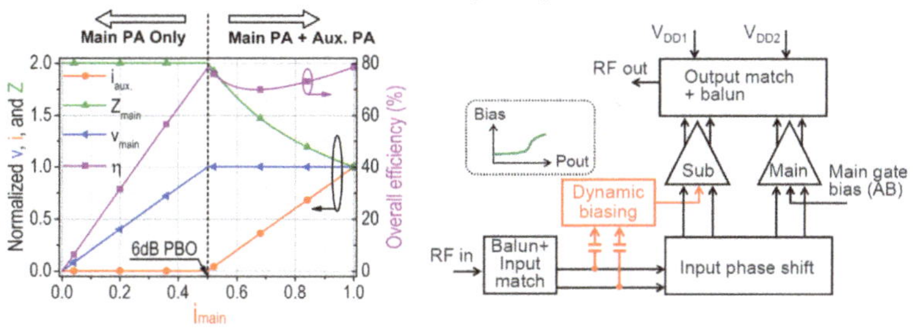

• [24] E. Kaymaksut, D. Zhao, P. Reynaert, "Transformer-Based Doherty Power Amplifiers for mm-Wave Applications in 40-nm CMOS," *IEEE Trans. Microw. Theory Tech.* vol. 63, no. 4, pp. 1186 - 1192, Mar 2015.
• [27] K. Onizuka, K. Ikeuchi, S. Saigusa, S. Otaka, "A 2.4 GHz CMOS Doherty power amplifier with dynamic biasing scheme," *Proc. IEEE Asian Solid State Circuits Conf.*, Nov. 2012.

On the other hand, Doherty PAs often employ adaptive biasing to facilitate the turning-on of their auxiliary path. However, conventional adaptive biasing circuit requires envelope amplifiers to real-time drive the input biasing of the auxiliary PA according to the signal envelope. Since the envelope amplifiers should accommodate the 3×-5×bandwidth expansion of the envelope signal compared to the complex modulated signal, the adaptive biasing path often becomes the bandwidth limiting factor that may weaken the desired Doherty operation and even result in a substantial memory effect.

87 Mm-wave Multi-band Parallel Doherty PA (IX)

- **Introduced design** – Power-dependent uneven-feeding scheme
  - Leverage the power-dependent input impedances of Class-AB/-C Doherty PA final stages
  - Introduce a "driver-PA co-design" method

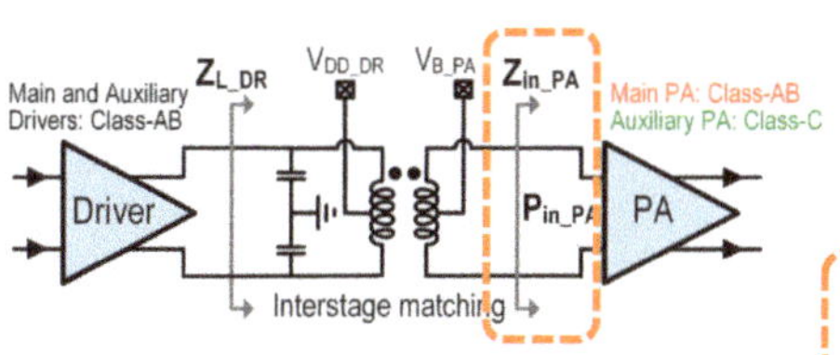

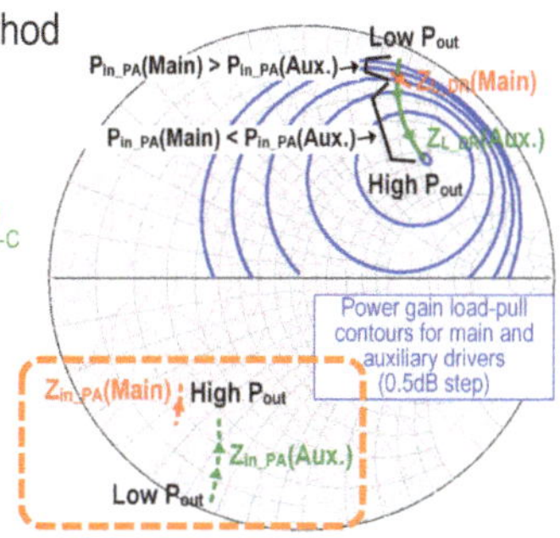

• [25] S. Hu, F. Wang, and H. Wang, "A 28GHz/37GHz/39GHz Multiband Linear Doherty Power Amplifier for 5G Massive MIMO Applications," *IEEE ISSCC Dig. Tech. Papers*, Feb. 2017.

It is shown that utilizing the nonlinear behavior of the power device input impedance, one can achieve a power-dependent uneven-feeding scheme. This is particularly true for SiGe bipolar power devices whose input impedance will vary significantly versus the input power, if the device is biased in class-C mode, typically for the auxiliary PA. In contrast, the input impedance of SiGe device is almost independent of input power if the device is biased in class-AB mode.

88 Mm-wave Multi-band Parallel Doherty PA (X)

- **Introduced design** – Power-dependent uneven-feeding scheme
  - Leverage the power-dependent input impedances of Class-AB/-C Doherty PA final stages
  - Introduce a "driver-PA co-design" method

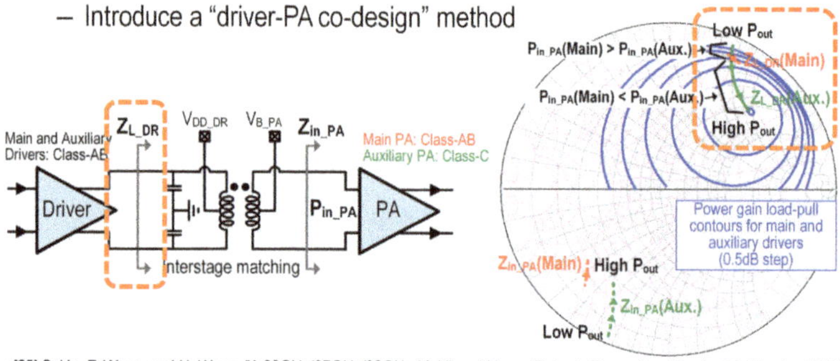

- [25] S. Hu, F. Wang, and H. Wang, "A 28GHz/37GHz/39GHz Multiband Linear Doherty Power Amplifier for 5G Massive MIMO Applications," *IEEE ISSCC Dig. Tech. Papers*, Feb. 2017.

Then, the inter-stage matching network of the auxiliary PA can be designed, so that the load impedance for the auxiliary driver can travel from medium gain circle to high gain circle, which achieves effective gain peaking of the auxiliary PA with respect to the input power.

89 Mm-wave Multi-band Parallel Doherty PA (XI)

- **Introduced design** – Power-dependent uneven-feeding scheme
  - ✓ Enhanced Doherty cooperation → Rapid auxiliary PA ramp-up
  - ✓ No hardware overhead
  - ✓ No sacrifice of modulation bandwidth

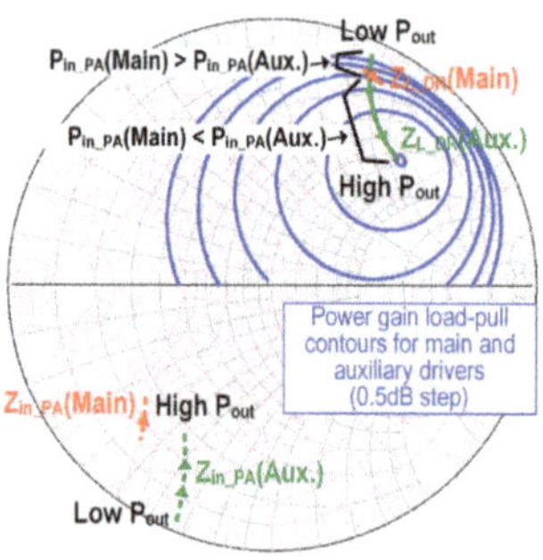

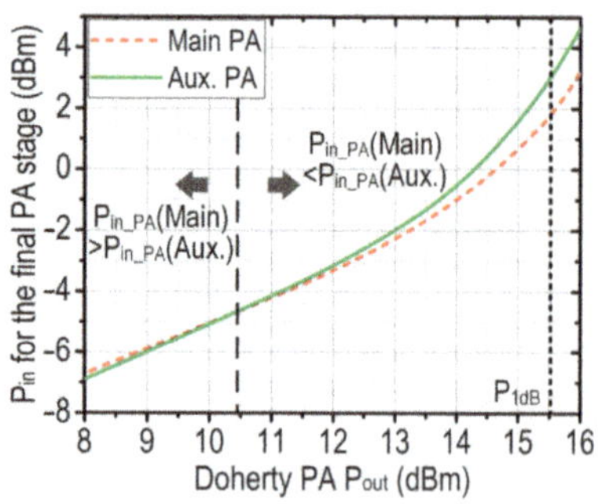

Overall, such a power-dependent uneven-feeding scheme can enhance the desired Doherty operation by rapidly ramping-up the auxiliary PA yet without causing additional hardware overhead or sacrifice of modulation bandwidth.

90 Mm-wave Multi-band Parallel Doherty PA (XII)

- **A 28/37/39GHz multiband Doherty PA for 5G massive MIMO**
  - Prototype in GlobalFoundries 130nm SiGe BiCMOS

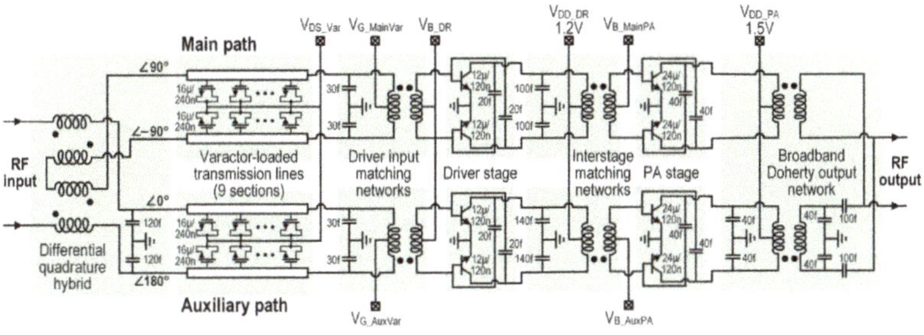

- [25] S. Hu, F. Wang, and H. Wang, "A 28GHz/37GHz/39GHz Multiband Linear Doherty Power Amplifier for 5G Massive MIMO Applications," *IEEE ISSCC Dig. Tech. Papers*, Feb. 2017.

A proof-of-concept PA design is implemented in the Globalfoundries 130nm SiGeBiCMOS process, covering the 28/37/39 GHz carrier bands for multi-band 5G massive MIMO application. The design is presented at an IEEE ISSCC conference paper in 2017. The detailed schematic including the transistor sizing information is shown on this slide.

91 Mm-wave Multi-band Parallel Doherty PA (XIII)

- **A 28/37/39GHz multiband Doherty PA for 5G massive MIMO**
  - Chip microphotograph (130nm SiGe BiCMOS)
  - World-first multiband mm-wave linear Doherty PA in silicon for 5G

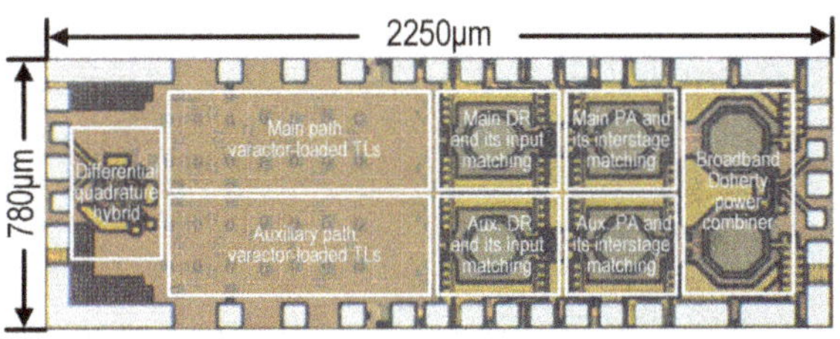

- [25] S. Hu, F. Wang, and H. Wang, "A 28GHz/37GHz/39GHz Multiband Linear Doherty Power Amplifier for 5G Massive MIMO Applications," *IEEE ISSCC Dig. Tech. Papers*, Feb. 2017.

The chip microphotograph is shown with all the major circuit blocks highlighted.

92 Mm-wave Multi-band Parallel Doherty PA (XIV)

- **Continuous-wave measurement results at 37GHz**
  - +17.1dBm P_{sat}, +15.5dBm P_{1dB}, and 27.6% peak collector efficiency
  - Excellent amplitude and phase linearity

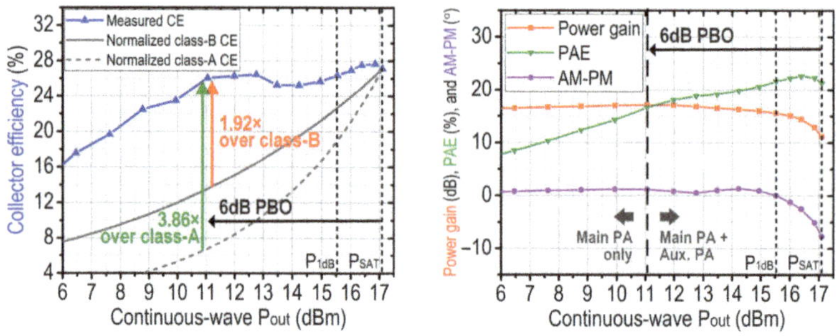

The PA is first characterized for its continuous-wave performance. At 37GHz, it achieves +17.1dBm saturated output power, +15.5dBm OP1dB, and 27.6% peak collector efficiency. The collector efficiency at 6dB PBO is over 26%, resulting in 1.92× efficiency improvement over class-B PA and 3.86× efficiency improvement over class-A PA at 6dB PBO. The PA also achieves very liner AM-AM and AM-PM performance at 37GHz.

93 Mm-wave Multi-band Parallel Doherty PA (XV)

- **Modulation measurement results at 37GHz**
  - 500MSym/s 64QAM (3Gb/s) without any predistortion

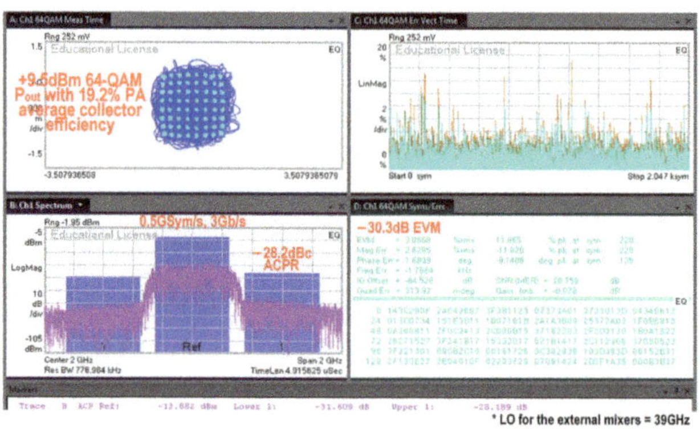

The PA is then characterized using high-speed complex modulation signals. At 37GHz, the PA supports 500MSym/s 64QAM modulation (3Gbit/s) without any digital predistortion. The PA achieves +9.5dBm averaged output power with 19.2% PA averaged collector efficiency.

94 Mm-wave Multi-band Parallel Doherty PA (XVI)

- **Modulation measurement results at 39GHz**
  - 500MSym/s 64QAM (3Gb/s) without any predistortion

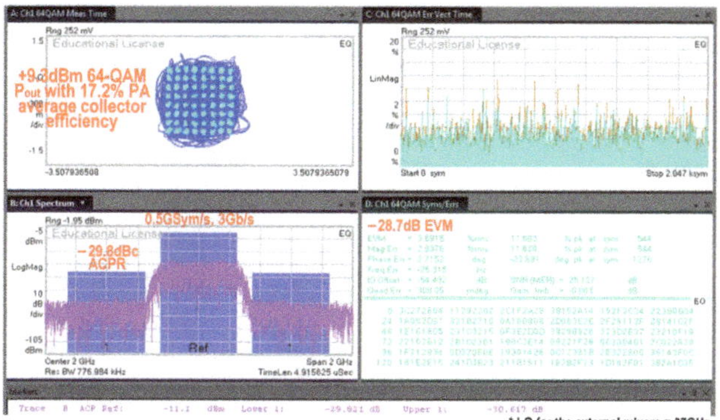

* LO for the external mixers = 37GHz

The same high-speed modulation of 500MSym/s 64QAM modulation (3Gbit/s) is supported at 39GHz, achieving +9.3dBm averaged output power with 17.2% PA averaged collector efficiency.

95 Mm-wave Multi-band Parallel Doherty PA (XVII)

- **Modulation measurement results at 28GHz**
  - 1GSym/s 64QAM (6Gb/s) without any predistortion
  - Fastest data rate among 28GHz Doherty PAs in silicon

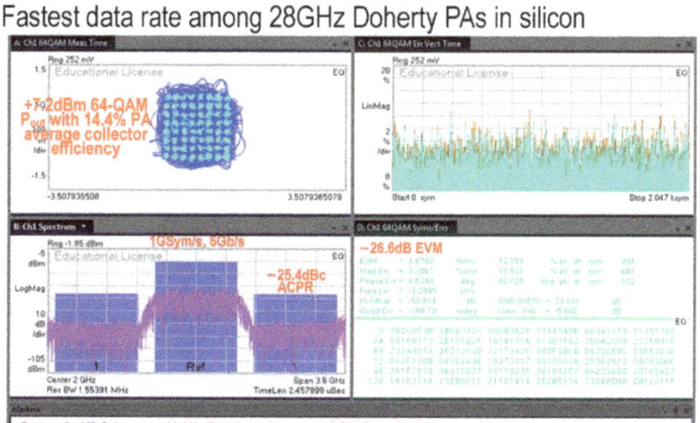

* LO for the external mixers = 26GHz

At 28GHz, the PA supports 1GSym/s 64QAM modulation (6Gbit/s) without any digital predistortion, which was the fastest data rate among reported 28GHz Doherty PA in silicon by then. The PA achieves +7.2dBm averaged output power with 14.4% PA averaged collector efficiency.

96 Conclusion

- Introduction of Doherty PA configurations and operations
- Multiple Doherty PA design examples (**Parallel/Series Doherty configurations, Analog vs. Digital vs. Hybrid Digital Doherty PAs, and RF/Mm-Wave Doherty PAs**)
- Future directions (potential):
  - PBO efficiency enhancement
  - Linearity
  - Mm-wave Doherty PA for 5G applications
  - Modulation bandwidth
  - Robustness

In conclusion, we first introduce the Doherty PA popular configurations and basic operations in this tutorial presentation. Multiple state-of-the-art Doherty PA design examples are presented in detail. Future research directions potentially include further enhancing the PA PBO efficiency, improving the Doherty PA linearity, innovating mm-Wave Doherty PAs particularly for multi-band 5G applications, increasing the modulation bandwidth, and improving the robustness of the Doherty PA in practical application environments. We would like to acknowledge many people and parties for their support of the related research studies.

97 Reference

1. IMT Vision – Framework and overall objectives of the future development of IMT for 2020 and beyond, https://www.itu.int/dms_pubrec/itu-r/rec/m/R-REC-M.2083-0-201509-I!!PDF-E.pdf, retrieved on 08/15/2018.
2. H. Wang, S. Kousai, K. Onizuka, S. Hu, "The Wireless Workhorse: Mixed-Signal Power Amplifiers Leverage Digital and Analog Techniques to Enhance Large-Signal RF Operations," *IEEE Microwave Magazine*, vol. 16, no. 9, pp. 36-63, Oct. 2015.
3. H. Wang, C. Sideris, and A. Hajimiri, "A CMOS Broadband Power Amplifier With a Transformer-Based High-Order Output Matching Network," *IEEE J. of Solid-State Circuits*, vol. 45, no. 12, pp. 2709 – 2722, Dec. 2010.
4. Zoya Popovic, "Amping Up the PA for 5G: Efficient GaN Power Amplifiers with Dynamic Supplies," *IEEE Microwave Magazine*, vol. 18, no. 3, pp. 137 – 149, May 2017.

5. Y. Zhang, M. Rodríguez, D. Maksimović, "Very high frequency PWM buck converters using monolithic GaN half-bridge power stages with integrated gate drivers", *IEEE Trans. Power Electron.*, vol. 31, no. 11, pp. 7926-7942, Nov. 2016.

6. Hongtao Xu, et al., "A Flip-Chip-Packaged 25.3 dBm Class-D Outphasing Power Amplifier in 32 nm CMOS for WLAN Application," *IEEE J. Solid-State Circuits*, vol. 46, no. 7, pp. 1596 - 1605, May 2011.

7. T.W. Barton, et al., "Multi-way lossless outphasing system based on an all-transmission-line combiner." *IEEE Trans. Microw. Theory. Tech*, vol. 64, no. 4, pp. 1313-1326, March 2016.

8. K. Onizuka, S. Saigusa, S. Otaka, "A +30.5 dBm CMOS Doherty power amplifier with reliability enhancement technique," *VLSI Circuits (VLSIC)*, 2012 Symposium on, 13-15 June 2012.

9. W. H. Doherty, "A new high efficiency power amplifier for modulated waves," *Proc. Institute of Radio Engineers*, vol. 24, no. 9, pp. 1163-1182, Jun. 2012.

10. S. Hu, S. Kousai, and H. Wang, "Antenna Impedance Variation Compensation By Exploiting A Digital Doherty Power Amplifier Architecture," *IEEE Trans. Microw. Theory. Tech*, vol. 63, no. 2, pp. 580-597, Feb. 2015.

11. B. Kim, J. Kim, I. Kim, J. Cha, "The Doherty power amplifier," *IEEE Microwave Magazine*, vol. 7, no. 5, Oct. 2006.

12. F. H. Raab, "Efficiency of Doherty RF power-amplifier systems," *IEEE Trans. Broadcast.*, vol. BC-33, no. 3, pp. 77-83, Sep. 1987.

13. W. C. E. Neo, et al., "A Mixed-Signal Approach Towards Linear and Efficient -Way Doherty Amplifiers," *IEEE Trans. Microw. Theory Tech.* vol. 55, no. 5, pp. 866-879, May 2007.

14. S. Hu, S. Kousai, J. Park, O. Chlieh, and H. Wang, "A +27.3dBm Transformer-Based Digital Doherty Polar Power Amplifier Fully Integrated in Bulk CMOS," *Proc. IEEE Radio Frequency Integrated Circuits (RFIC)*, Jun. 2014.

15. S. Hu, S. Kousai, J. Park, O. Chlieh, and H. Wang, "Design of A Transformer-Based Reconfigurable Digital Polar Doherty Power Amplifier Fully Integrated in Bulk CMOS," *IEEE J. of Solid-State Circuits*, vol. 50, no. 5, pp. 1094 – 1106, May 2015.

16. S. Hu, S. Kousai, and H. Wang, "A Broadband CMOS Digital Power Amplifier with Hybrid Class-G Doherty Efficiency Enhancement," *IEEE International Solid-State Circuits Conference (ISSCC) Dig. Tech. Papers*, Feb. 2015.

17. S. Hu, S. Kousai, and H. Wang, "A Broadband Mixed-Signal CMOS Power Amplifier with A Hybrid Class-G Doherty Efficiency Enhancement Technique," *IEEE J. of Solid-State Circuits*, vol. 51, no. 3, pp. 598 - 613, Mar. 2016.

18. S. Yoo, et al., "A Switched-Capacitor RF Power Amplifier," *IEEE J. of Solid-State Circuits*, vol. 46, no. 12, pp. 2977 - 2987, Sep. 2011.

19. S. Hu, S. Kousai, J. Park, O. Chlieh, and H. Wang, "A +27.3dBm Transformer-Based Digital Doherty Polar Power Amplifier Fully Integrated in Bulk CMOS," *Proc. IEEE Radio Frequency Integrated Circuits (RFIC)*, Jun. 2014.

20. V. Vorapipat, C. Levy, P. Asbeck, "A wideband voltage mode Doherty power amplifier," *Proc. IEEE Radio Frequency Integrated Circuits (RFIC)*, May. 2016.

21. V. Vorapipat, C. Levy, P. Asbeck, "A class-G voltage-mode Doherty power amplifier", *IEEE Int. Solid-State Circuits Conf.(ISSCC) Dig. Tech. Papers*, pp. 46-47, Feb. 2017.

22. V. Vorapipat, C. Levy, P. Asbeck, "A Class-G Voltage-Mode Doherty Power Amplifier," *IEEE J. of Solid-State Circuits*, vol. 52, no. 12, pp. 3348 - 3360, Dec. 2017.

23. FCC 16-89; "Spectrum for 4G and 5G", *Qualcomm*, Oct. 2016.,

24. https://www.qualcomm.com/media/documents/files/spectrum-for-4g-and-5g.pdf, retrieved on 08/15/2018.

25. E. Kaymaksut, D. Zhao, P. Reynaert, "Transformer-Based Doherty Power Amplifiers for mm-Wave Applications in 40-nm CMOS," *IEEE Trans. Microw. Theory Tech.* vol. 63, no. 4, pp. 1186 - 1192, Mar 2015.

26. S. Hu, F. Wang, and H. Wang, "A 28GHz/37GHz/39GHz Multiband Linear Doherty Power Amplifier for 5G Massive MIMO Applications," *IEEE International Solid-State Circuits Conference (ISSCC) Dig. Tech. Papers*, Feb. 2017.

27. A. Grebennikov, J. Wong, "A Dual-Band Parallel Doherty Power Amplifier for Wireless Applications," *IEEE Trans. Microw. Theory Tech.* vol. 60, no. 10, pp. 3214 - 3222, Oct. 2012.

28. K. Onizuka, K. Ikeuchi, S. Saigusa, S. Otaka, "A 2.4 GHz CMOS Doherty power amplifier with dynamic biasing scheme," *Proc. IEEE Asian Solid State Circuits Conf.*, Nov. 2012.

PART III

DATA CONVERTERS

CHAPTER 08

A/D Converter Fundamentals and Trends

Hui Pan

Broadcom Inc, USA

Many new Analog-to-Digital Converters (ADCs) techniques have emerged recently, enabling advanced applications from direct RF/IF sampling and software defined radio to coherent optical receivers. To help the readers understand the wide variety of ADCs and the development trends, without being overwhelmed with circuit implementation details, this chapter presents an effort to derive a unified framework for understanding ADCs with focus on the fundamentals.

Introduced as applied in advanced receiver examples, ADCs are viewed as signal processing systems that quantize analog inputs. The system basics are covered such as antialiasing filtering, sampling and DFT/FFT, quantization characteristics, spectral properties, and specifications. Time-interleaving, pipelining, oversampling, and noise shaping are discussed as examples of general signal processing techniques, as opposed to specific ADC architectures, applied to enhance ADC performance in speed and resolution. Insights are thus revealed with ways to reshape and innovate ADCs by adding, dividing, combining, merging, and permuting the signal processing operations.

With the general signal processing components separated, the description of ADCs is narrowed down to the topic of quantization, which is considered for implementation purpose as a search process to find in a predefined full-scale (FS) range the digitally coded interval that encompasses the analog input. Various search algorithms and the core search elements of zero-crossing (ZX) generation and detection are identified and used to derive quantization architectures, leading to signal-folding based flash and pipeline quantizers in addition to the simple ZX based counterparts. The derivation provides a unified view about ADCs with different tradeoffs between efficiency and performance.

Circuit examples are presented to give a flavor of real ADC implementations including the sampler, the core elements of ZX generators and detectors, and some auxiliary signal-folding circuits, along with a brief coverage of circuit imperfections and remedies such as bootstrapping of T/H switches, bottom-plate sampling, auto-zeroing, offset averaging and spatial filtering, comparator metastability and mitigation, reference generation and ripple compensation, thermometer bubble correction, and so on. Important ADC metrics such as INL/DNL, SNDR, ERBW, NSD, NPR, SFDR, and metastability, the Walden & Schreier FoMs, and the measurement methods are discussed in connection with applications.

The chapter concludes with an illustration of architecture evolution using an example of multi-bit/stage pipeline ADC that transforms to an efficient SAR ADC with all the auxiliary pipelining and folding circuits removed, followed by the closing remarks about the trend of departing from the traditional signal-folding based flash and pipeline ADCs to simple ZX-based quantizers coupled with sophisticated signal-processing and digital error corrections in advanced technologies.

1 Outline

- **Applications and Systems**
- Algorithms and Architectures
- Circuit Implementations
- Imperfections and Corrections
- Metrics and Measurements
- Evolutions and Trends
- Concluding Remarks

APPLICATIONS AND SYSTEMS

2 Direct RF/IF Sampling and Software-Defined Radio

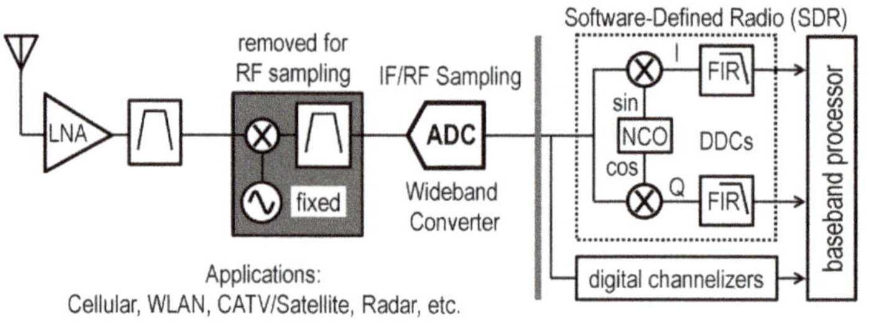

- Multi-GS/s ADCs for direct IF/RF sampling of hundreds of channels
- High ADC dynamic range for DSP to sort out weak signals from in-band blockers
  - SFDR ~ 80dB & SNR ~ 60dB for GSM base station applications *[Pan, JSSC 2000]*
- High power, but gap closing with heterodyne for $f_t >> f_{nyquist}$ *[Gomez, TCAS 2016]*
  - Direct sampling (DS) more efficient than heterodyne with carrier aggregation

2 Direct RF/IF Sampling and Software-Defined Radio

Technology scaling has enabled many advanced applications for data converters. A prominent example in wireless applications is direct RF sampling where the received RF signals are all digitized with a wideband ADC and then sorted out with a digital signal processor (DSP). This eliminates the costly RF counterparts while enabling highly programmable software-defined radio (SDR). For 5G base station applications, the sample rate could go beyond 10GS/s to capture most of the sub-6GHz bands, and the spurious-free dynamic range (SFDR) requirement could reach 80dB or higher to accommodate the large in-band blockers and the fast fading signals from many user equipment units (UEs). Due to processing gain from oversampling for each channel, the ADC SNR requirement could be 20dB lower than the SFDR for the case of capturing 100 channels in the Nyquist band.

3 DSP-Based Wireline Receivers

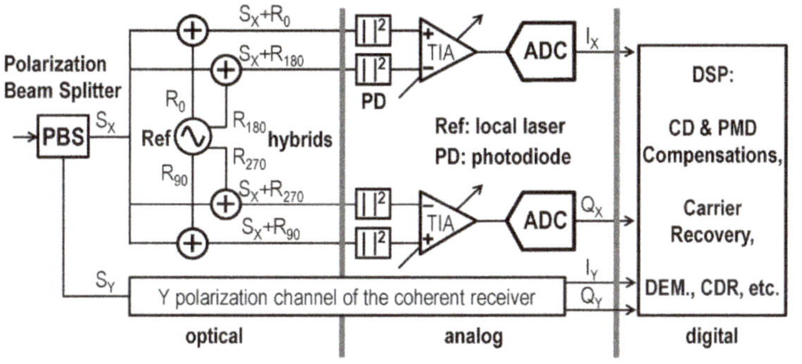

Applications: coherent optical communications (100G and higher)

- Ultra high-speed (>10GS/s) 5~8b ADCs and DSPs enable long-haul optical transmission beyond 10Gb/s per lambda [Crivelli, TCAS-I 2014].

A prominent wireline example is the DSP-based coherent optical receiver where the four (2 polarizations X 2 quadratures) DP-QPSK optical signals are down-converted through the photo diodes and TIAs to baseband electrical signals and then digitized by four ADCs for digital correction of the optical impairments including chromatic dispersion (CD) and polarization mode dispersion (PMD). For 100Gbs/s transmission, the baud rate is 100GS/s/4 = 25GS/s. Given 12.8% channel coding overhead for feed forward error correction (FEC) and 2X oversampling for clock and data recovery, each ADC runs at 64GS/s. In contrast to the wireless example, ENOB (effective number of bits) of 5 ~ 8 suffices for this application with low-order modulation and slow-varying signal strength.

4 A/D Converters: A System Perspective

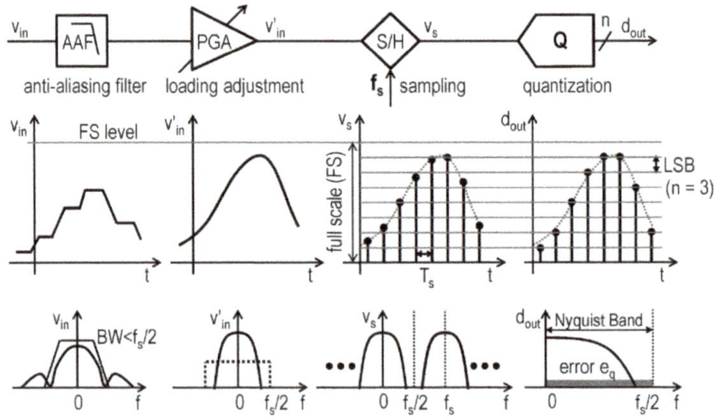

- Sampling and quantization are permutable.

From a system perspective, anti-aliasing filtering (AAF) and variable (or programmable) gain amplification (VGA or PGA) are important parts of ADCs, because they directly impact the ADC output performance. All the out-of-band noise, harmonic spurs, and interferences that are not cleaned up with an AAF fold back to the Nyquist band (from DC to $f_S/2$), consuming the ADC dynamic range and degrading the Effective Number of Bits (ENOB). The VGA tracks out the (slow) signal strength variation to load the ADC optimally for the best SNR, preventing ENOB from degrading. It is worth noting that sampling and quantization are theoretically permutable. This can be visually confirmed by swapping the order of the graphical presentations of sampling and quantization in the slide and checking if the ADC output dots move.

5 Sampling Theorem

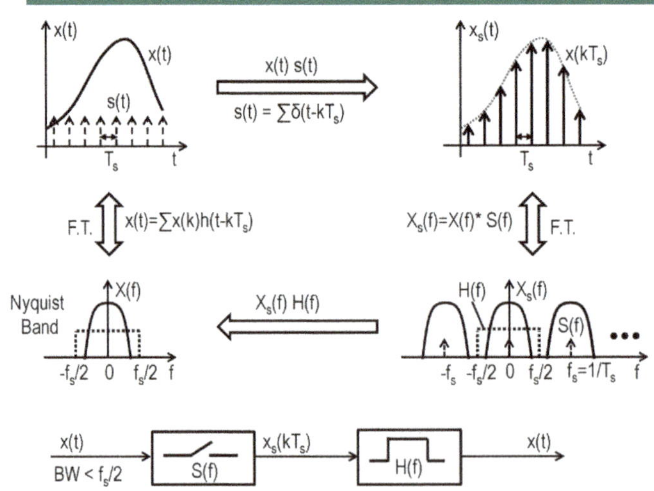

5 Sampling Theorem

Most ADCs are clocked to generate high speed digital samples for discrete signal processing and storage. Loss of information does not happen as long as the Nyquist sampling condition is met, that is, the sampling frequency f_S is higher than 2x the signal bandwidth BW. This can be understood by treating sampling as modulation by an infinite impulse train with spacing $T_S = 1/f_S$ or by mixing with all the harmonics of f_S. To avoid information loss as a result of aliasing or overlapping of the frequency spectrum replicas, the replica spacing f_S must be larger than the double-sided bandwidth, $f_S > 2BW$. Careful readers may notice that the finite energy of the finite (duration) samples $x(kT_S)$ appears to contradict the infinite replicas of the spectrum energy. This is because the time-domain energy is NOT summation of $x^2(kT_S)$; instead, it is summation of all the products of $x^2(kT_S)$ and integration of $\delta^2(t-kT_S)$, which is infinite or undefined.

6 Discrete Fourier Transform (DFT)

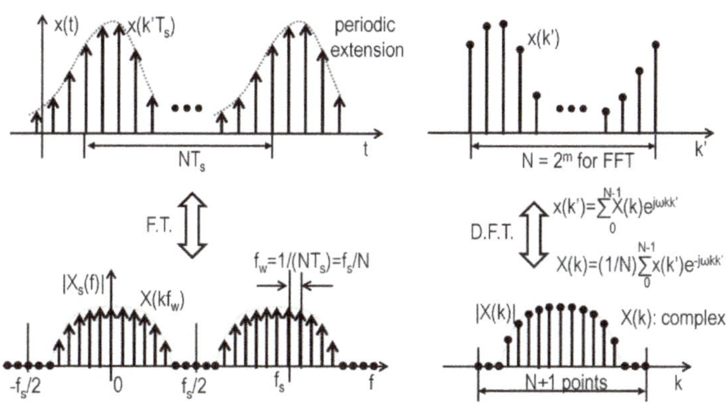

- Important tool for spectrum analysis of finite signal samples
- Fast Fourier Transform (FFT) algorithms for $N = 2^m$

It is also desirable to store and process discrete spectrum samples. The continuous energy spectrum of Fourier transform (FT) is turned to discrete power spectrum by periodical extension of the finite energy signal. Now that the signal is discrete and periodic in both time and frequency domains, we get an N to N mapping of one-period samples between the time and frequency domains, resulting in the so-called discrete Fourier transform (DFT). The number of DFT points N is usually chosen as the power of 2 to implement the fast FT (FFT) algorithm. It is noted that the spectrum samples are mirror symmetric about DC (and multiple sampling frequencies), and therefore only half of them are independent. The spectrum mirror image makes sense because it halves the number of independent spectrum samples to satisfy N to N mapping of the independent variables between the N/2 complex spectrum samples and the N time-domain real samples.

7 Coherent Sampling of Periodical Signals

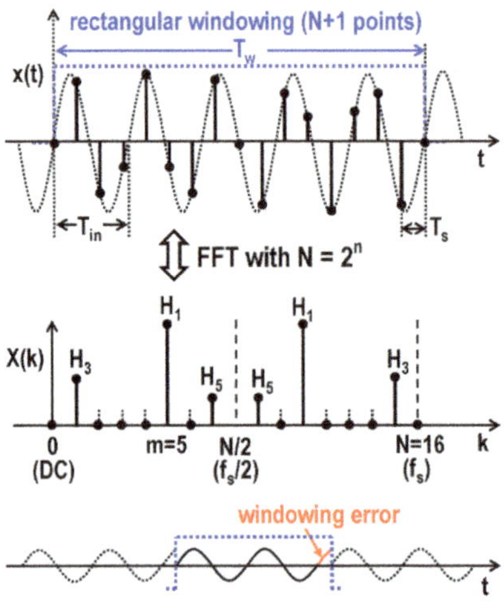

N-Point Coherent Sampling:

$$T_w = N\,T_s = m\,T_{in}$$

$$\Downarrow$$

$$f_{in} = (m/N)\,f_s = m\,f_{bin}$$

$$(f_{bin} = 1/T_w = f_s/N)$$

- Each of the harmonics H_i falls within one bin f_{bin}.
- Less H_i's fall in the same bin with prime ratio $m = f_{in}/f_{bin}$.
- Windowing error causes spectral leakage to all bins.

Since the N points for FFT are one period of the function for Fourier expansion or transformation (FT), the FT period NT_S should match to one or multiple signal period mT_{in} when the signal has period T_{in}. This condition is coherent sampling where the FFT window T_W includes integer number of signal periods, that is, $T_W = NT_S = mT_{in}$. Otherwise, windowing error arises and the fundamental of the signal changes from $f_{in} = 1/T_{in}$ to $f_{bin} = 1/T_W = f_S/N$. As a result, the signal power would spread all over the harmonics of f_{bin}, that is, all the N/2 frequency bins in the Nyquist band. Since the signal harmonic components still dominate the spectrum, the spectrum leakage from the windowing error causes (phase-noise like) skirt shape surrounding each signal harmonic.

8 Quantization and Error Characteristics

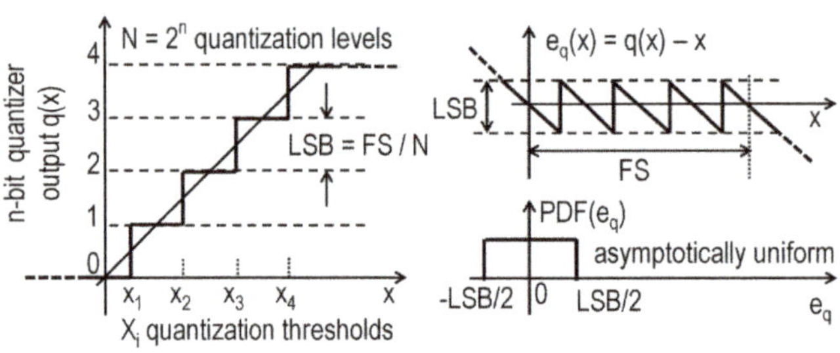

Error Power $P_e(x) = \int (e_q)^2\, PDF(e_q)\, de_q = LSB^2 / 12$

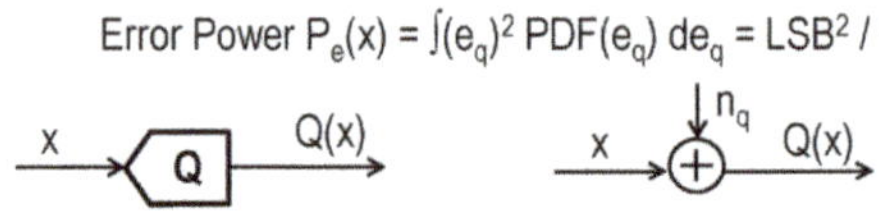

- Quantization error e_q is usually approximated as additive noise n_q
  - Given randomness (e.g. dither) in x on the order of LSB or more
  - Asymptotically uniform PDF independent of x for x(rms) >> LSB

8 Quantization and Error Characteristics

Now we move to the ADC core topic - quantization. For analysis purpose, quantization is characterized with a staircase input-output transfer function or a sawtooth error function. The trip points are determined by the quantization thresholds, which are ideally distributed with uniform least significant bit (LSB) quantization interval LSB = $FS/2^n$ across the quantization full scale (FS) range for an n-bit quantizer. The sawtooth error characteristic could cause very high-order harmonic spurs under stimulation of CW input signals but can be simplified as additive noise for most applications where there is always some randomness in the input signal on the order of one LSB or more. Given LSB << signal (rms), the error probability density functions (PDF) for ideal uniform quantization are uniform between 0 and LSB, and the resulted total error or noise power of $LSB^2/12$ is independent of the signal.

9 Error Waveform & Spectrum of Quantized Sinewave

$e_q(t) = q(A\sin(2\pi f_{in}t)) - A\sin(2\pi f_{in}t)$

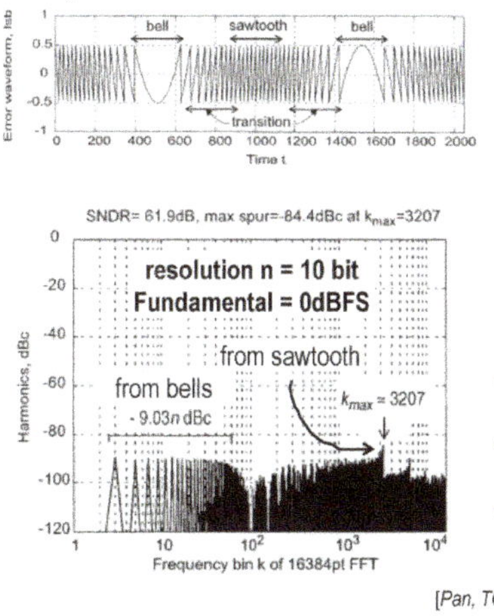

Sawtooth period ~ LSB / $(2\pi f_{in}A) = 1/(2^n\pi f_{in})$

Max harmonic freq. ~ $(2^n\pi)f_{in}$

Max harmonic order $k_{max} \sim 2^n\pi$

Max harmonic level ~ –9n+c (dBFS)

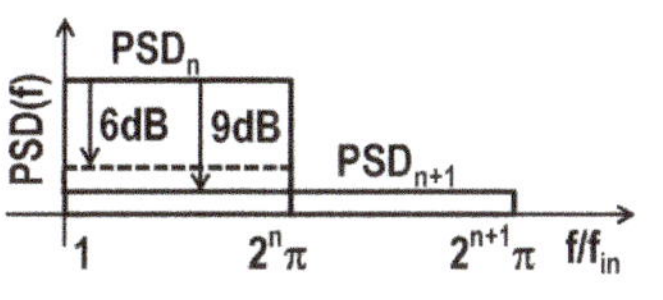

- Quantization causes very high-order distortion (spurs).
- **Spur level actually rises as signal amplitude decreases**.
- All spurs fold within Nyquist band as a result of sampling.

[Pan, TCAS-I 2004]

For wideband ADC applications where SFDR is of paramount importance as shown in the wireless example, quantization spurs should be considered instead of being simplified as additive noise. An intuitive approach to estimation of the quantization spurs under CW input stimulation is based on the error waveform of a quantized sinewave. The sawtooth error around the sinewave zero-crossing dominates the error waveform and therefore contributes the max spur around the sawtooth frequency = $2\pi f_{in}$*A/LSB = $2^n\pi f_{in}$. The spur level should drop 9dB for every bit increase, with 6dB from halved LSB and 3dB from the doubled number of spurs up to the order of $2^n\pi$. As signal amplitude halves, the total spur power remains unchanged, but the spur level should increase by about 3dB as the effective resolution n drops by one bit with half quantization thresholds traversed. This is in stark contrast to the "common sense" that harmonics should diminish with reduced amplitude.

10 Ideal ADC SNR & SFDR, and Real ADC ENOB

Single-tone Signal Power P_s
$= FS^2 / 8$

Noise & Distortion Power P_e
$= LSB^2 / 12$

Single-tone Spur Level P_d
$\approx -9n$ dBFS

SNDR $\equiv P_s / P_e$
$= 6.02n + 1.76$ (dB)

SFDR $\equiv P_s / \max. P_d$
$= 9.03n + c$ (dBFS)
$(c \approx -6)$

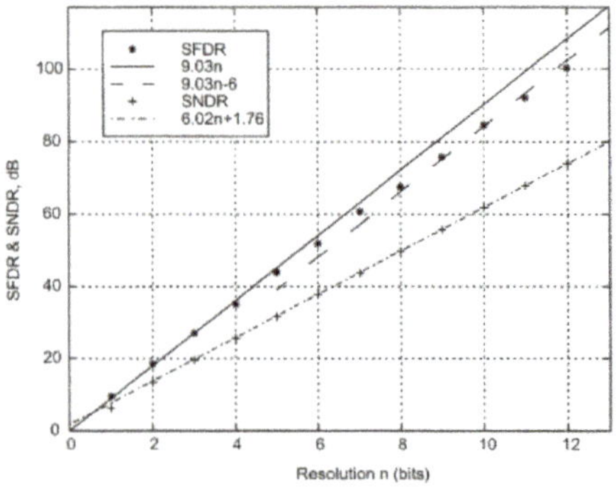

For real quantizers,
Effective Number of Bit (ENOB) = (SNDR – 1.76) / 6.02

ADCs are mostly tested with sinewave signals especially for communications applications. The FS single-tone output SNDR and SFDR are the two most important ADC specs. They are about 6n and 9n dB, respectively, for an ideal n-bit quantizer based on the aforementioned quantization error analysis. Real quantizer SNDR could drop below the 6n dB level by more than 6dB, causing more than 1b reduction of the effective number of bits. As to the real SFDR performance, it is usually limited by the low order distortion of the front-end sampling circuit, that is, the track-hold amplifiers (THAs). It is noted that the sampling effect can be applied after the quantization, which folds all the spurs back to the Nyquist band without changing the SNDR and SFDR.

11 A/D Converter Definitions

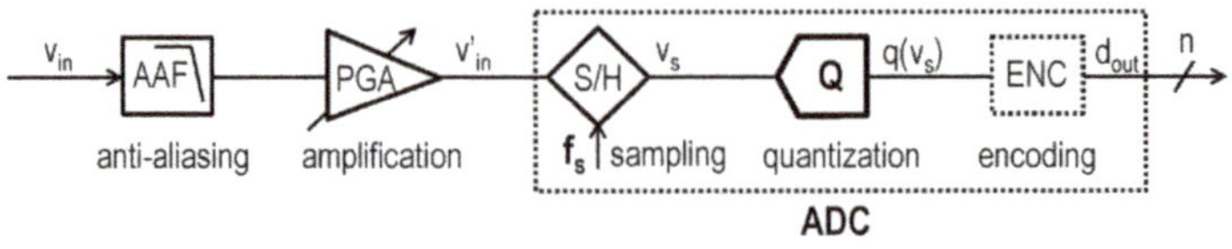

- In a common sense, ADC consists of sampling, quantization, and possibly encoding.
  - AAF & PGA in different design categories can be merged with S/H
  - Sequence usually appears as shown, but some operations are permutable
- In a narrow sense, an ADC consists of only quantization.
  - S/H merged with Q or eliminated for clockless ADCs
  - Encoder merged with Q or included in succeeding DSP
- In a wide sense, an ADC is a compound system that consists of one or more quantizer cores and auxiliary functionalities that enhance conversion speed, resolution, efficiency, system performance, etc.

11 A/D Converter Definitions

It is very beneficial to view A/D converters as a signal processing system that includes at least one quantizer for analog to digital conversion that best meets the required system performance. This view opens the possibility of innovations through addition, division, and recombination of many signal processing and quantization functionalities to enhance conversion rate, resolution, efficiency, etc. This is to be illustrated in the following examples.

12 Sampling Rate Multiplication by Time-Interleaving

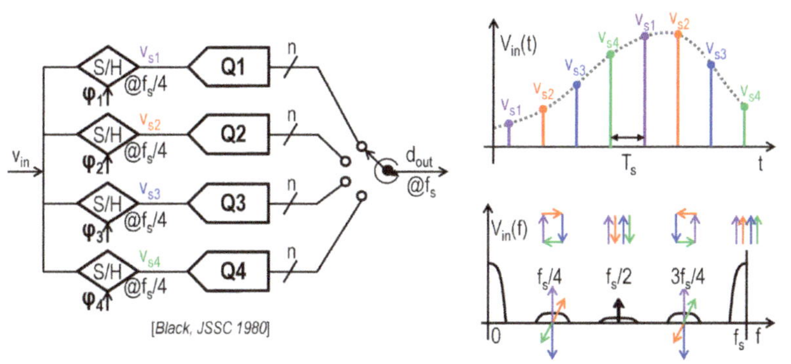

- Conversion rate linearly increases with *L* = number of time-interleaved ADC lanes.
- S/Hs are bottleneck for Nyquist operation as the tracking BW must scale with *L*.
- Spurs arise around i^*f_s/L, $i=1,\ldots,L-1$, for lane mismatch in offset, gain, BW, skew, etc.
- Interleaving spurs from random mismatch drop by 10log*L* due to static scrambling.

Time-interleaving technique can be applied to multiply the ADC sample rate from the lane rate f_S/L to the overall rate f_S, where L is the number of interleaved ADC lanes. In the frequency domain, each lane output spectrum consists of signal spectrum replicas centered at i^*f_S/L, where i is the lane index starting from 0. When all the lane outputs are combined, the spectrum replicas from all the lanes centered at i^*f_S/L for i = 1, 2, ..., L-1 cancel each other because of the multiple $2\pi/L$ phase shifts between the L replicas. However, the L replicas have zero phase shift at i = 0 and L and therefore add constructively. Now that the remaining replicas at i = 0 and L are spaced by $L^*f_S/L = f_S$, the effective sample rate becomes f_S. Interleaving spurs centered around i^*f_S/L for i = 1, 2, ..., L-1, arise when there is any mismatch among the lanes that stops the replicas from cancelling each other completely. The interleaving spur level drops with respect to the signal by 3dB as the number of lanes doubles because the random mismatch errors add in power while the signals add linearly.

13 Conversion Rate Multiplication using Pipelining

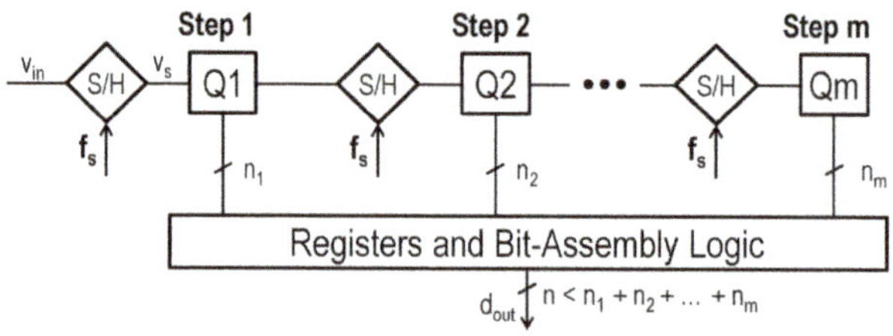

- Sample rate multiplied by up to m = number of pipeline stages with throughput decoupled from latency.
- Sample-and-holds (S/H) for pipeline storage run at full accuracy and full speed in the signal path, causing large overhead in power and area.
  - Accurate settling needs large op-amp DC gain, an issue with process scaling.
  - kT/C noise and distortion accumulate with the multiple S/H stages.
- Best suited for implementation with switch capacitor circuits in CMOS technology [Lewis, JSSC 1987].

Pipelining technique can also be applied to speed up the A/D conversion by dividing the quantization to many small steps of low-resolution quantization Q1, Q2, ..., Qm. The intermediate quantization results or the residues are sampled and held for next step quantization by the inter-stage S/Hs so that each small and fast quantizers can start quantizing the next sample before quantization of the current sample completes. In contrast to the parallel operation of the time-interleaved ADCs that involve some limited overhead in multiphase clock generation, the pipelined operation tends to worsen the power efficiency and the noise due to the repetitive sampling by the full-speed inter-stage S/Hs. Inter-stage gain can be applied to mitigate the tail stage overheads, but this causes severe speed bottleneck in the head stages. In addition, the full accuracy requirement on the S/H is an issue in advanced CMOS because the high DC gain required for precision analog operation is not readily available at low supply voltages.

14 Hierarchical Parallel and Pipelined A/D Conversion

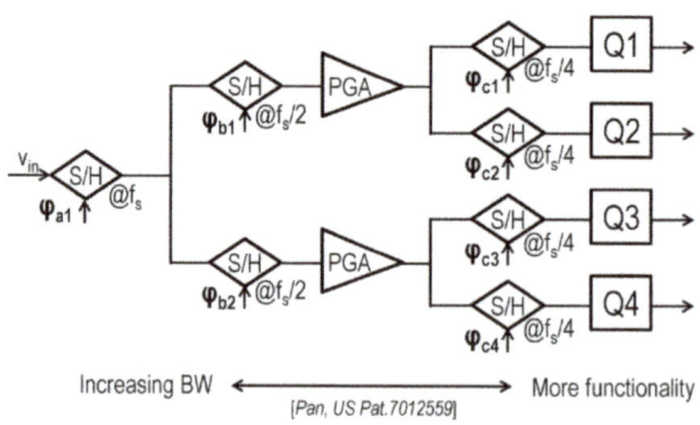

[Pan, US Pat.7012559]

- BW and functionality grow in opposite directions in the hierarchy.
- Number of head-end channels is minimized for max. BW.
- Clock jitter and skew cause little error after first stage S/H(s).
- Phase alignment necessary among multi-rate multi-phase clocks.
- Linearity, kT/C noise, and power consumption get worse.

14 Hierarchical Parallel and Pipelined A/D Conversion

For ultra high speed applications that are bottlenecked by the ADC input BW, hierarchical time-interleaving architecture is a suitable choice, where the head stage complexity and lane number are minimized to realize the highest possible input BW with the complexity and lane number growing towards the tail stages. The drawback is that the multi-stage sampling degrades the noise and linearity performance, which could be an issue for the aforementioned wireless example that requires very large SNDR and SFDR. But for the ultra high speed DSP-based wireline example, this architecture fits very well because the input BW reaches tens GHz but the resolution is only up to 8b.

15 Resolution Boost by Oversampling & Noise Shaping

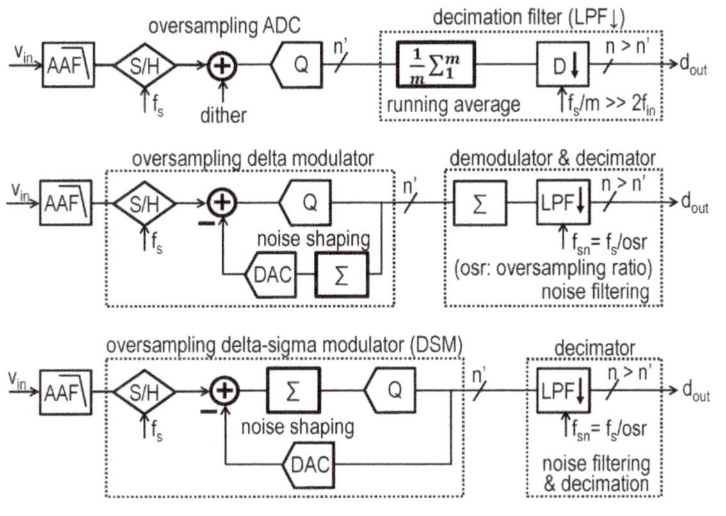

The effective ADC resolution can be enhanced by adding digital processing to the low-resolution ADC output as shown on the top of the slide. The dithered quantization error is suppressed with respect to the static or relatively slow varying signal by sampling the signal m times and taking the average of the m quantized samples as an output sample of the ADC system. By the same token, an integrator or accumulator can be added as shown in the slide to filter out the high frequency quantization error while passing all the low frequency signal. To unwind the integration effect on the signal, a delta operation or modulation is introduced preceding the integration, which further suppresses the low frequency quantization error in the signal band due to the infinite DC loop gain from the accumulator. The two integrators can be merged in front of the quantizer as shown at the bottom of the slide to derive the well-known delta-sigma ADC.

16 Noise and Signal Transfer Functions (NTF & STF)

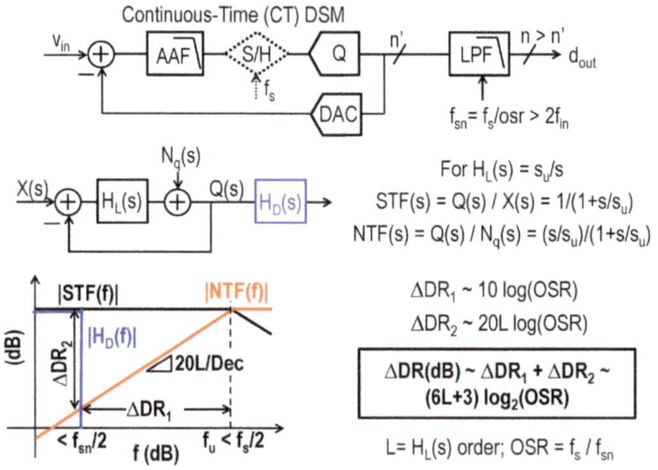

The integrator can be generalized as a low-pass (or even a bandpass) noise shaping filter. The preceding AAF and S/H can be moved within the loop with the AAF merged with the noise shaping loop filter and the S/H absorbed in the clocked regenerative comparators. These transformations lead to continuous time delta-sigma ADCs. The low-pass noise shaping filter in the loop pushes the quantization error power injected at this output to the high frequency end of the Nyquist band while passing the low frequency signal injected at the input. As the signal and the spectrum shaped quantization error pass the low-pass decimation filter, the SNR and resolution are greatly enhanced. The low frequency part of the low resolution quantizer output tracks the analog input due to the high DC loop gain. The accurate low-frequency component is extracted as the high resolution delta-sigma ADC output with a low-pass decimation filter.

ALGORITHMS AND ARCHITECTURES

17 Quantization Q(x) as a Search Process

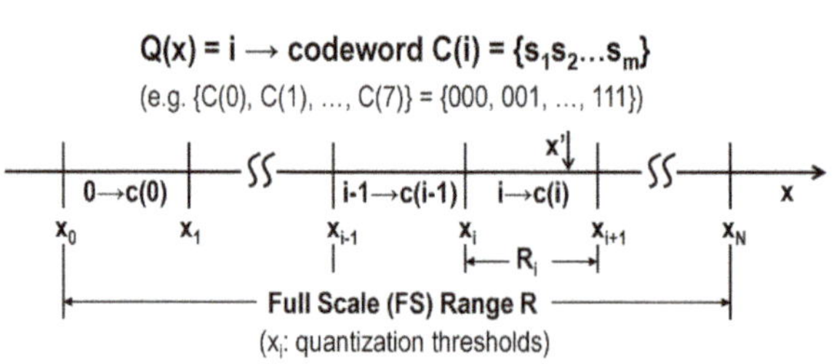

- $Q(x) = i$ for $x \in R_i$, $i = 0, 1, \ldots, N-1$; $N = 2^n$ for n-bit quantizer.
- Sub-range $R_i = x_{i+1} - x_i = R/N$ = LSB for uniform quantization.
- A quantizer searches for the R_i in which a given x falls.
- An algorithm drives the search based on (x, x_i) comparison.
- An optional encoder C maps $Q(x) = i$ to a codeword $C(i)$.
- Q and C can be merged to resolve each code symbol s_k directly.

17 Quantization Q(x) as a Search Process

From an implementation perspective, quantization is a search process that finds the subrange R_i hit by the analog sample x. The search algorithm tends to dominate the quantizer architecture, performance-efficiency tradeoff, and the output code format. In fact, each quantizer output code symbol represents a different way to divide the FS range R into a set of coarse subranges consisting of different combinations of contiguous or non-contiguous fine subranges. Each symbol value resolves a coarse hit subrange and the AND of all the symbol values resolves the fine hit subrange. The search algorithm determines the subrange divisions and the procedure to resolve the hit subrange. For example, each bit of a binary code is a symbol representing a different binary division of the FS, and collection of all the bit values resolves the hit LSB subrange.

18 Search Algorithms for Quantization

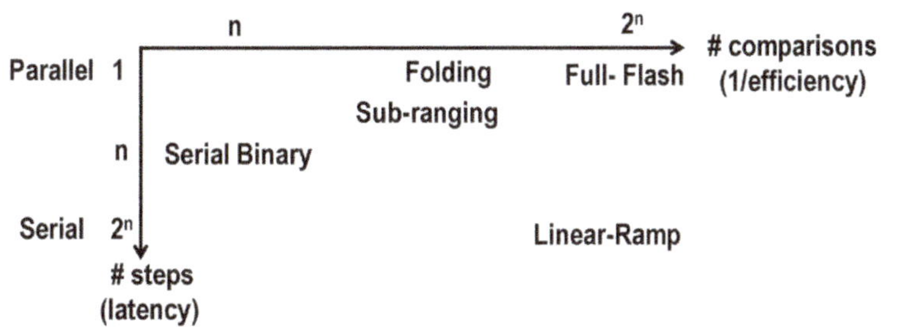

- Parallel brute-force (full-flash) : one step, ~2^n comparisons
- Serial brute-force (linear ramp up/down): ~2^n steps and comparisons
- **Serial binary search (SAR): n steps & comparisons (most efficient)**
- Coarse-fine search (sub-ranging): 1< steps <n, n< comparisons <2^n
- Search efficiency (# comparisons) dominates power efficiency

Every search algorithm involves a different efficiency-latency tradeoff. The efficiency is measured with the total number of decisions that must be made to resolve the finest hit subrange or the hit LSB; the latency is the total number of serial steps of decisions. To resolve the hit LSB in one step for minimum latency, all the LSB subranges must be examined simultaneously, costing the maximum number of decisions. In serial steps, less decisions need to be made because the preceding decisions helps narrowing down the search. The most efficient search is serial binary search which takes n steps to resolve $N = 2^n$ LSB subranges. This accounts for the recent popularity of successive approximation register (SAR) ADCs that implement the most efficient serial binary search algorithm.

19 Quantization Elements and Generic Architecture

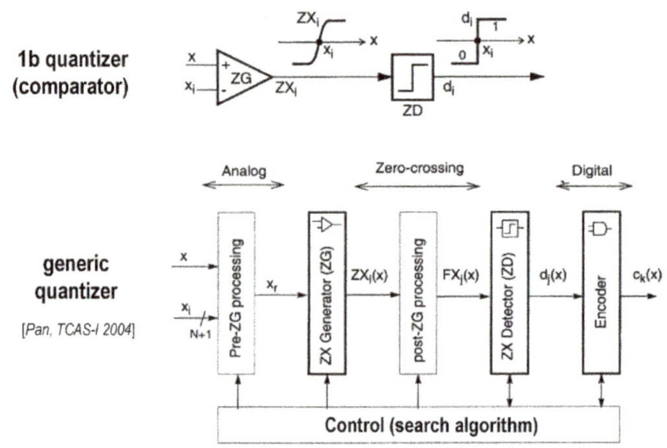

- Quantizer core elements: ZX generation (ZG) and ZX detection (ZD)

The core elements for the search are zero-crossing (ZX) generation and detection. Subrange divisions are implemented with ZX generations (ZG) at quantization thresholds; decision on a hit subrange is based on ZX detection (ZD) at the boundaries of the subrange. A control block like the SAR could implement the search algorithm. Auxiliary blocks like signal folding or multiplexing could facilitate or speed up the search. The ZX signals can be generated with difference amplifiers or even inverters with the signal x converted to time delay. Ideal ZD sharpens the ZX slope to infinity to produce crispy output of 1 or 0. Since the infinite gain is not realizable, real ZX detectors could fall in undecisive metastable (unknown) state causing occasional sparkle error in the ADC output or bit error in the system.

20 Parallel Brute-Force Search: Full-Flash

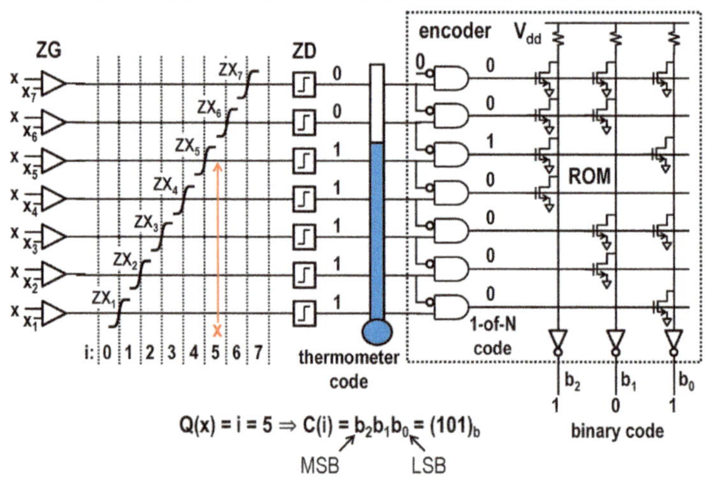

20 Parallel Brute-Force Search: Full-Flash

In full-flash quantizers, the FS range is completely divided with all the (2^n – 1) ZXs signals generated and detected at once. The ZX detector (ZD) output thermometer code can be directly used by the system as the ADC output or converted to a compact binary format using a digital encoder. When the ZDs are implemented with clocked regenerative latches, the signal is sampled at the moment the latches strobe, though a standalone S/H can be added to mitigate the SNR degradation caused by the timing skew among the distributed sampling. The conversion latency is only one clock period, but the complexity (= 2^n ZGs and ZDs) degrades exponentially with resolution n.

21 Serial Binary Search: Successive Approximation

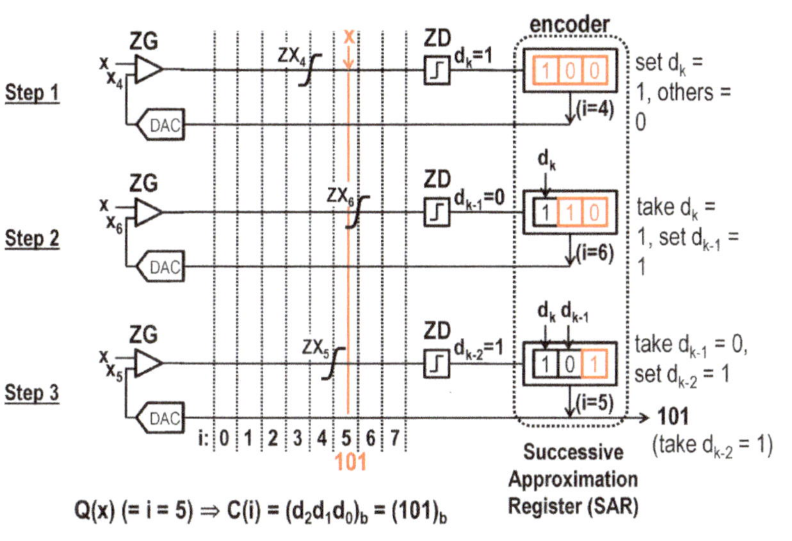

The opposite extreme of the flash quantizers is the SAR quantizers that have the least complexity with only a single ZG and ZD, though they suffer much longer latency and lower throughput. The SAR register is initiated to the mid code corresponding to the mid-point of the FS. Only one ZX signal is generated and detected at each step and the ZD result is pushed into the SAR to direct the ZX generation and detection for the next step to narrow down the search. The D/A converter (DAC) providing the quantization threshold for each step is the most important part of the ZX generator. It is usually implemented in CMOS with a binary weighted array of MOM capacitors that can achieve 10b accuracy in reasonable area.

22 Parallel Binary Search: Analog Encoding

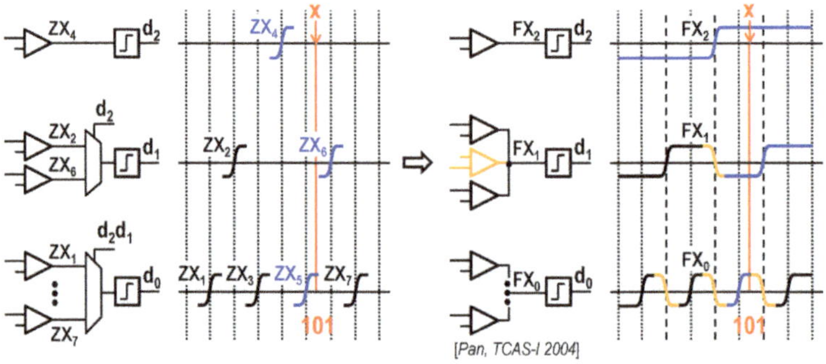

- Signal folding merges ZG and encoding → analog encoding.
- Signal folding is equivalent to automatic multiplexing of ZXs.
- However, there is algorithmic conflict among redefined ZXs.

To shorten the latency of SAR quantizers, one may consider to make all the candidate ZX signals of each search step available to be mux selected for detection. However, the mux control still depends on the ZX detection results of the preceding steps. To remove this serial dependence, one can add all the preceding ZXs to the current step with their polarities flipped, and stitch all the ZXs to form an analog folding signal as shown on the right hand side. By signal folding, each serial search step is converted to an independent parallel channel with the right ZX signal automatically multiplexed for detection. Since ZX detections of the folding signals directly produce the desired output code, which is binary in this example, signal folding is also known as analog encoding.

23 Potential Algorithmic Conflict in Parallel Binary Search

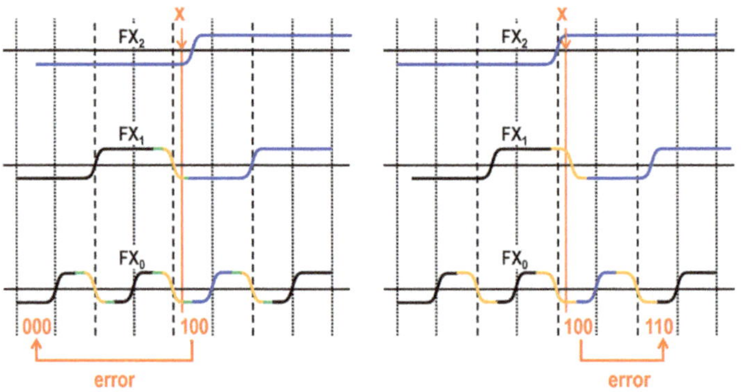

- Midpoint ZX redefined in each parallel channel could be misaligned, causing algorithmic conflict and large quantization error.

23 Potential Algorithmic Conflict in Parallel Binary Search

In the binary analog encoding example, the redundant ZXs added to each channel for signal folding (automatic multiplexing) not only cause overhead, but also incur a major issue of algorithmic conflict as a result of multiple definitions of the same ZX point. Any mismatch among the redundant ZX points in different channels could cause quantization error as large as half FS as shown in the slide. Methods to correct the error or effectively align the redundant ZXs are called "bit-sync".

24 Parallel Binary Search Without Algorithmic Conflicts

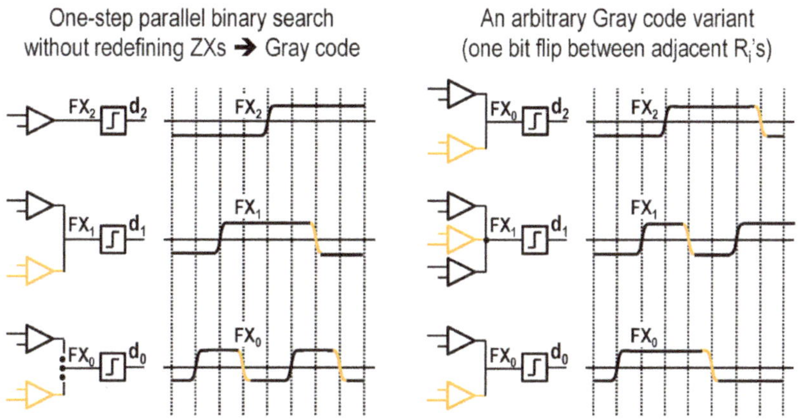

- Gray codes are robust without redefining ZXs.
- One step and total 2^n ZGs, but n instead of 2^n ZDs.

The bit-sync can be avoided by flipping and stitching the candidate ZXs of each step without adding redundant ZXs to any channel. The resulted analog coding generates Gray code. Without any redefinitions of the same ZX point, only one ZX is passed at a time causing only one ZD output bit flip among all the parallel channels as the signal sweeps across the FS. Gray code variants that possess this property of no simultaneous toggling of multiple bits can be realized by redistributing the non-redundant ZXs among the parallel channels to form different folding signals. It is noted that each parallel channel resolves a code symbol representing a different way of dividing the FS range by the corresponding folding signal.

25 Parallel Coarse-Fine Search: Modulus-Residue Encoding

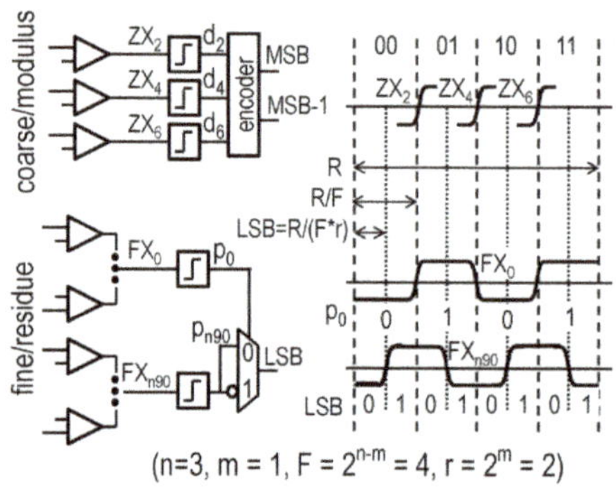

Bit-Sync (align redefined ZXs):

$(MSB\text{-}1)_s = p_0$

$(MSB)_s = \overline{MSB}$, if s = 1,
$= MSB$, if s = 0.

where $s = ((MSB\text{-}1) \oplus p_0)\ \overline{p_{n90}}$

Output Binary Code:

$C = (MSB)_s(MSB\text{-}1)_s(LSBs)$

- (F-1) ZXs (e.g. $ZX_{2,4,6}$) in coarse channels divide R into F folds.
- Fine folding (e.g. FX_0, FX_{90}) divide each fold (R/F) into r LSBs.
- Bit-sync aligns coarse ZXs to fine channels. *[van de Grift, JSSC 1987]*

Each code symbol and corresponding subrange division need not be binary, as is shown in this modulus-residue coding case. The non-redundant ZX signals can be distributed evenly across all the parallel channels to form replicas of periodic folding signals offset by one LSB. ZX detection of those replica folding signals resolves the relative location or the residue of the analog input x in the fold subrange. The fold (subrange) hit by x is resolved by a coarse channel with redundant ZXs dividing the FS range by the number of folds. Detection of the coarse channel ZX signals resolves the modulus symbol value, while the offset folding signals collectively forming a residue channel produce the residue symbol. The redundant ZXs in the coarse channel are "aligned" to the residue channel ZXs with "bit sync".

26 Parallel Residue Search: Residue Encoding

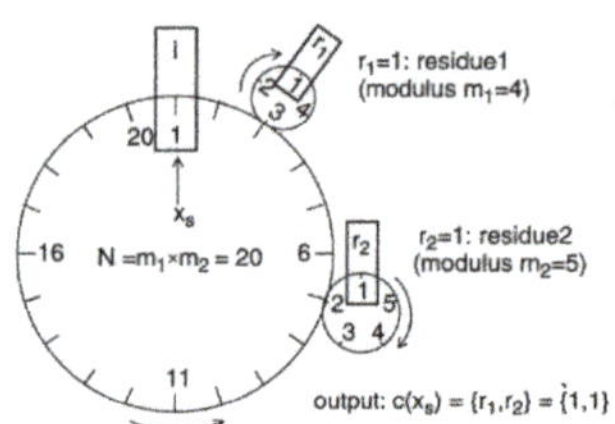

- Based on Chinese remainder theorem
- Enable carry-free addition
- Optical implementation
- Overwhelming bit-sync

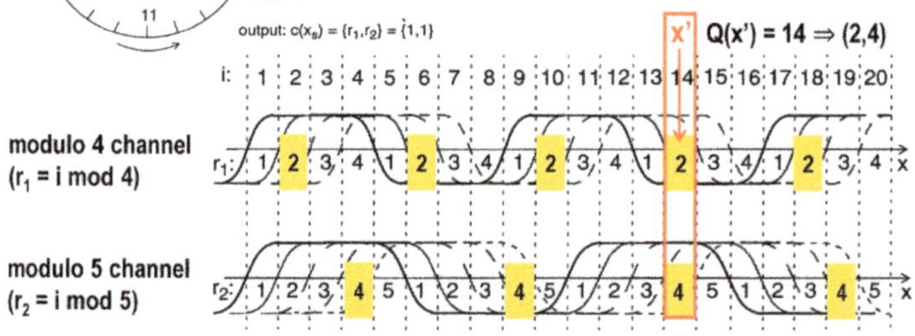

- Each symbol represents an independent partial search/quantization.
- Each encoding system represents a partition and search algorithm.

26 Parallel Residue Search: Residue Encoding

Instead of modulus-residue combination, multiple residue symbols also identify a unique subrange out of N ones as long as N is less than the product of the moduli Mj (the numbers of subranges) of the residue channels. In this example, residue channels 1 and 2 divide the FS into M1 = 4 and M2 = 5 subranges labeled from 1 to 4 and 5, respectively. The AND of the two subranges for residue r1 = 2 and r2 = 4 nails the unique fine subrange i = 14. The major advantage of the residue code is carry-free addition, which holds for operations within the dynamic range of N = M1*M2. For example, carry-free addition of (2, 4) and (4, 4) results in (2, 3), which corresponds to i = 14 + 4 = 18.

27 Pre-ZG Folding Reduces ZG Complexity

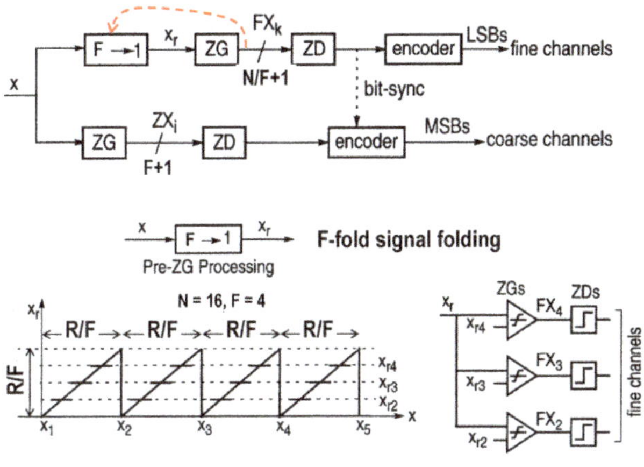

- F ➔ 1 mux (folding) moved to ZG input to reduce both ZGs and ZDs by F
- Overhead: analog folding and coarse channel to resolve 1 ➔ F ambiguity

The folding signals discussed so far are formed after the ZX generation by alternating adjacent ZXs and stitching them together. This reduces the number of ZX detectors (ZD) due to the effective operation of F è 1 multiplexing of ZX signals. To reduce the number of ZX generators (ZG), the F → 1 operation of the residue channels can be moved from the ZG outputs to the ZG inputs, leading to direct folding of the input signal x. This reduces the ZG by F times, but the pre-ZG analog signal folding is much more challenging because it must preserve the full signal accuracy required for the quantizer resolution. The accuracy tends to degrade around the tips of folding transitions.

28 Serial Folding without Bit-Sync Issue

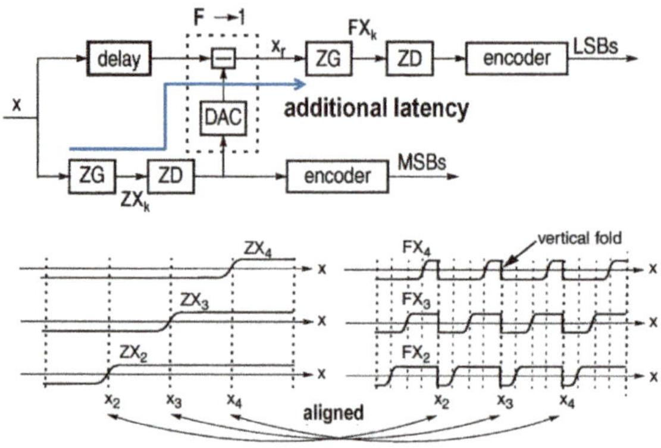

- Serial folding operation ensures bit sync at the cost of latency and matching.
- Analog delay can be avoided with a S/H driving both coarse and fine channels.
- Coarse threshold error does not matter if folding signal falls in fine quantizer FS.

We introduced the concept of signal folding as an automatic F → 1 multiplexing in an effort to parallelize quantization steps. But this does not prevent us from realizing signal folding in a serial fashion for use with serial quantization steps. In this example, the F → 1 analog pre-ZG folding or multiplexing is controlled by the MSB code through the DAC subtracting the delay-matched input signal. Since it is very hard to match the analog delay, a sample-and-hold (S/H) is used in this case to convert fast varying signals to DC samples. Given well-defined MSBs without metastability, the serial folding eliminates the need for bit-sync between the MSB and LSB channels, because the DAC subtraction realizes a vertical folding transition collapsing the redundant ZXs of the parallel LSB channels into one ZX regardless of channel mismatch.

29 Parallel Folding without Bit-Sync Issue

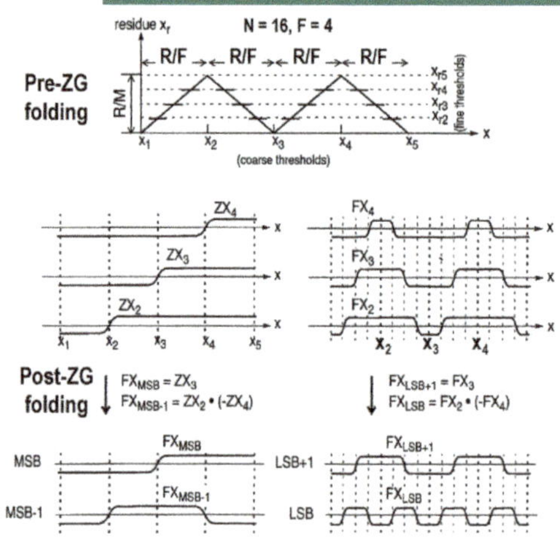

[van de Plassche, JSSC 1988]

- Parallel coarse-fine channels: low latency
- No vertical folding: robust quantization
- Gray code: no bit-sync and low complexity
- Rounded folding tip: linearity bottleneck

29 Parallel Folding without Bit-Sync Issue

For parallel pre-ZG folding, the bit sync between coarse and fine channels can be avoided using a triangular folding, where the coarse channel ZX points are located at the tip points of the fine channel triangular folding signal and no fine ZXs are defined for the tip points. This avoids the redundant fine ZXs and therefore eliminates the need for bit-sync. With some simple post-ZG folding operations, the triangular pre-ZG folding results in parallel analog encoding for Gray code. It makes sense that a signal folding without sharp or vertical transitions corresponds to the robust Gray code.

30 Ripple Through Folding for Gray Code Encoding

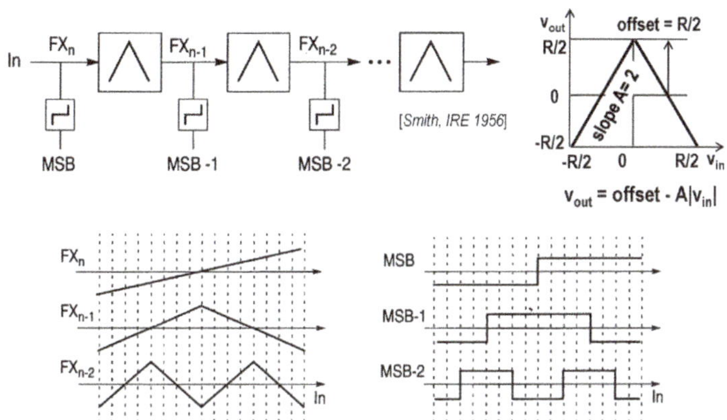

- Efficient discrete (rectifier based) ADCs for PCM, video and radar.
- $FS_{out}(i) = 2 * offset(i)$ should match $FS_{in}(i+1) = 4 * (offset(i+1) / A(i+1))$.
- $A = 2$ is not required as long as $offset(i+1) = (A(i+1) / 2) * offset(i)$.

Multiple (inverted) V-shape folding operations can be cascaded to generate the n-step folding signals and the corresponding Gray code directly without the need for bit-sync. The analog ripple-through action for signal folding without involving serial quantization decisions minimizes the latency and maximize the speed. As a pre-ZG folding, it also eliminates the need for extra ZGs. The minimum component count matters in the old days when quantizers were build on expensive discrete components. So this used to be the dominant quantizer architecture before the advent of VLSI technology.

31 Serial Folding for Binary Coding

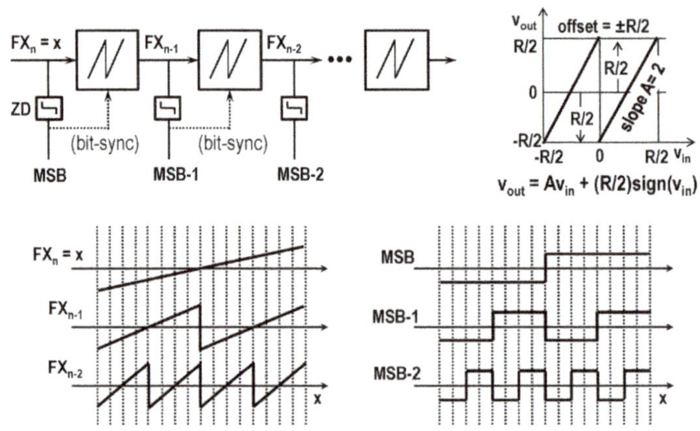

- Vertical folds requires the ZD-like decision and serial operation.

The ripple-through V-shape folding can be replaced with vertical folding as shown in the slide to generate the binary codes directly without the need for Gray code to binary code encoding. However, the vertical folding transition involves a crispy binary decision or ZX detection (ZD). The strict serial operation involving metastability-free ZDs slows down the quantization, which is usually overcome by pipelining each binary folding step as shown in next slide.

32 Pipelined Folding Using MDAC Subtraction

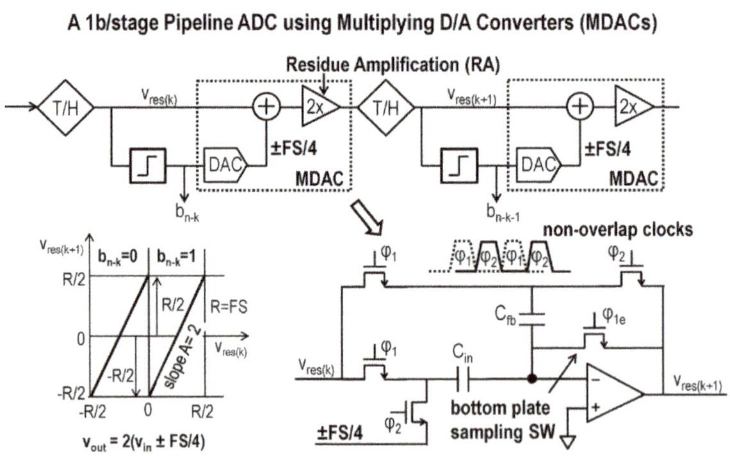

32 Pipelined Folding Using MDAC Subtraction

This is slide shows the 1b/stage pipeline ADC as an example of the aforementioned serial binary folding architecture. Started with the advent of CMOS technology that enabled switch-cap based track-and-hold (T/H) and MDAC (multiplying digital-to-analog converter) circuits to realize pipelined binary folding (or residue generation) and amplification, this ADC has evolved to the popular 1.5b/stage pipelined architecture with 0.5b over-ranging to overcome the ZD comparator offset error.

33 Discussions on Signal Folding

- **Signal folding reduces quantizer complexity with F→1 multiplexing on the analog signal (pre-ZG analog folding) or the ZX signals.**
- **The F→1 runs automatically involving no regeneration in parallel search based quantizers, preserving low latency and high speed.**
- **Automatic folding is popular in bipolar process where rectifiers are readily available to fold signals independent of any serial detection.**
- **Ripple-through cascaded V-shape folding was the major option for high speed A/D converters before the advent of IC.**
- **Folding involving serial ZD is popular in CMOS where switch cap circuits are readily available to pipeline the serial folding.**
- **Analog folding suffers full-accuracy requirement, while ZX folding suffers mutual loading & interference with limited speed benefit.**
- **Folding ADCs is going obsolete as DSPs become much cheaper than precision analog processing in advanced CMOS.**

CIRCUIT IMPLEMENTATIONS

34 Open-Loop Track/Hold (T/H)

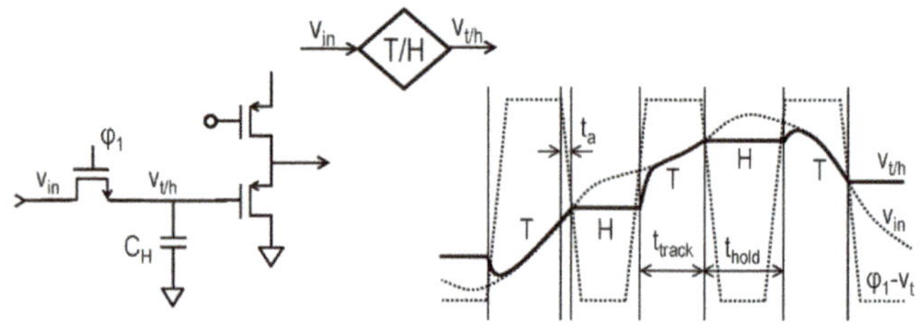

- Sampling aperture time t_a << $1/f_{in}$ limited by clock edge.
- Tracking BW_{track} > f_{in} limited by MOSFET f_T, power, etc.
- Linearity limited by switch on-resistance modulation, aperture time modulation, charge injection, nonlinear gate and junction parasitic capacitance, drooping/leakage, etc. [Iizuka, TCAS-I 2018]

This section covers some basic circuits used to build ADCs starting with the input track-and-hold (T/H). A simple T/H consists of a switch and a hold capacitor, along with a buffer driving the following ZG stage. CMOS technology is best fit for implementation of this switch cap circuit, for the switch is readily available as a MOSFET with its gate driven by the clock, and the hold capacitor can be realized with a MOM or MOS capacitor, or just the buffer gate parasitics. The T/H output linearity is limited by signal modulation of the switch on-resistance, the sampling aperture, charge injection, drooping, and so on. The distortion gets worse at higher frequencies as the switch impedance increases with respect to the hold capacitor. This T/H is usually used for ultra-fast low resolution (<8b) applications. A sample-and-hold (S/H) can be realized by cascading two such T/Hs.

35 Difference Amplifier (Preamp) for ZX Generation (ZG)

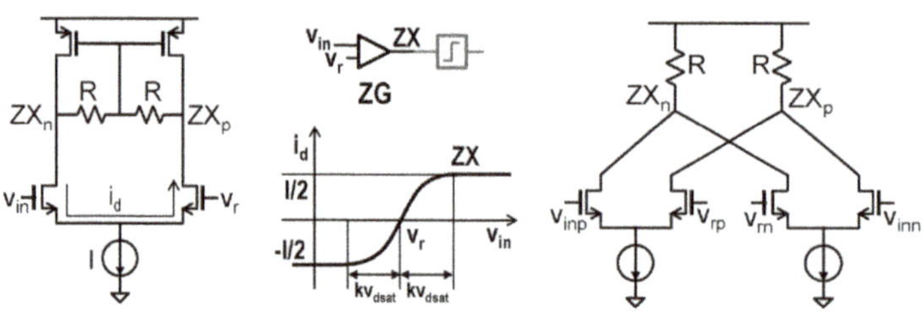

- Diff. output current i_d clips to ±I/2 for $|v_{in} - v_r| > kV_{dsat}$, $k \sim \sqrt{2}$.
- Diff. input and reference are applied to input diffpairs in p-p and n-n combinations to avoid ZX error from the current clipping.
- Preamp merges amplification with ZG to mitigate ZD errors by buffering ZD kickback and reducing input referred ZD offset.

35 Difference Amplifier (Preamp) for ZX Generation (ZG)

The ZX signal can be generated with a difference amplifier fed with signal and threshold voltages. For differential operations, two diffpairs are used with one for the positive side signal and threshold voltages and the other for the negative side voltages. At differential ZX, both diffpairs are at ZX without the need to maintain diffpair linearity over the FS range. Gain is usually included to overcome the dynamic offset and noise of the succeeding ZD. The threshold voltages v_{rk} are usually generated by tapping a resistor string with low enough resistance to absorb signal feedthrough and comparator kickback. ZX signal samples can be generated with switch-cap circuits to be discussed next.

36 ZX Generation using Switch Cap Array (CDAC)

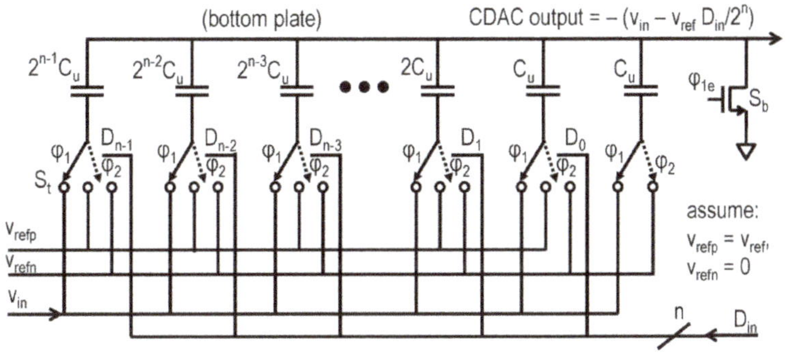

- ZX generation merged with T/H resulting in sampled ZX (residue)
- Serial generation of any of the 2^n ZXs corresponding to n-bit code D_{in}
- Used in SAR for ZX generation and MDAC for residue generation

When the T/H capacitor top-plate is switched from the input v_{in} to a threshold voltage v_r in the hold mode, the floating bottom-plate voltage becomes $-(v_{in} - v_r)$, which is a ZX signal at $v_{in} = v_r$. All the 2^n threshold voltages v_{rk} can be generated with a resistor string to cover the n-bit quantizer FS range. But a more efficient solution without burning static current through a resistor string is use a capacitor D/A converter (CDAC), where binary weighted capacitor top plates are switched from v_{in} to either ground or a fixed reference $v_r = v_{ref}$ to generate a ZX signal at any of the 2^n threshold voltages at the floating CDAC bottom-plate output. Applied in the MDAC of pipelined ADCs, the CDAC generates a residue (i.e., linear ZX) sample, while in SAR ADCs, the CDAC generates a ZX signal for each cycle or serial search step. It is noted that the capacitor plate connected to the input signal is referred to as top plate in this tutorial, though the physical plate is very likely the opposite.

37 Reference Generation with Ripple Cancellation

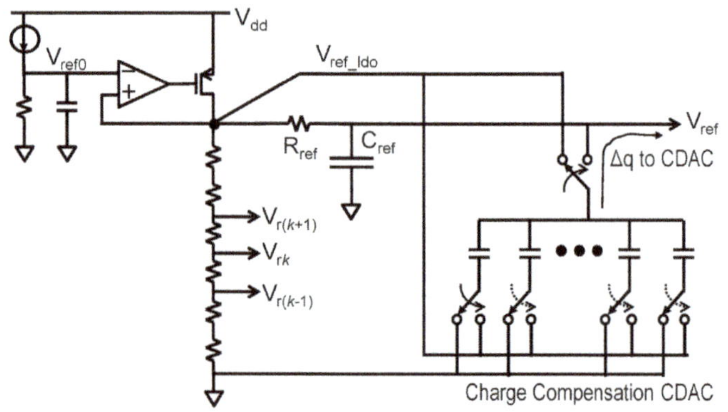

- Resistor string generates parallel quantization thresholds V_{rk}.
- Resistance is small enough to absorb comparator (ZD) kickback.
- V_{ref} ripple from charging a CDAC is cancelled with a compensation CDAC.

Signal dependent ripples appear on the reference V_{ref} charging a ZX generation CDAC, resulting in second and higher order harmonics in the CDAC and ADC output. With reference ripple cancellation (RRC), whenever the top-plate of the CDAC capacitor is switched from ground to V_{ref}, the bottom plate of the corresponding RRC capacitor is *simultaneously* switched from ground to $V_{ref\_ldo} \approx V_{ref}$ to provide the amount of charge Δq required for the CDAC. The RRC capacitors are pre-charged to V_{ref_ldo} in the CDAC tracking phase. As a result, the LDO voltage regulator and the decoupling C_{ref} can be substantially smaller saving reference power and area. The LDO provides power supply ripple rejection (PSRR) within its loop bandwidth, while the output RC filter (R_{ref} and C_{ref}) attenuates the supply noise beyond the loop bandwidth.

38 ZX Folding by Summation

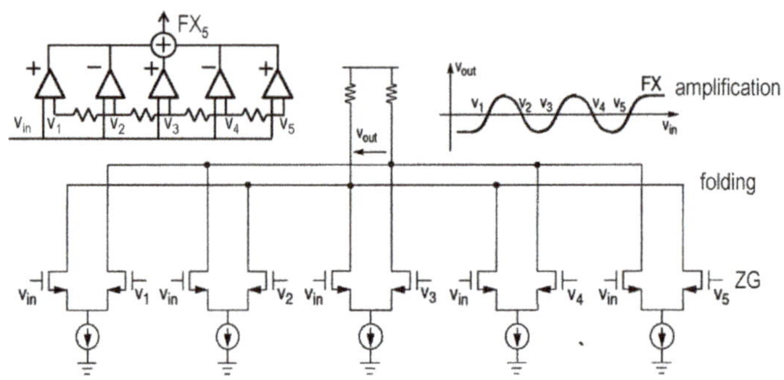

- ZX generation, amplification, and folding are merged.
- For each ZX (fold), remaining even no. ZXs ideally cancel.
- ZX folding is subject to all error sources and mutual loading.

38 ZX Folding by Summation

This slide shows generation of a post-ZG folding signal as an example of ZX signal processing. The ZX signals are alternated in polarity and "stitched" together by merging/summing F (an odd number) diffpair currents at the differential output nodes. This signal folding circuit has a few drawbacks. Since only one diffpair is active around a ZX point while the rest are clipped to the tail current with zero g_m, the loading to the active diffpair is increased F times. The inactive diffpairs cause different offset current if the even number of clipped currents from them do not differentially cancel at the output due to the tail current mismatch.

39 Additional Folding Signals from Interpolation

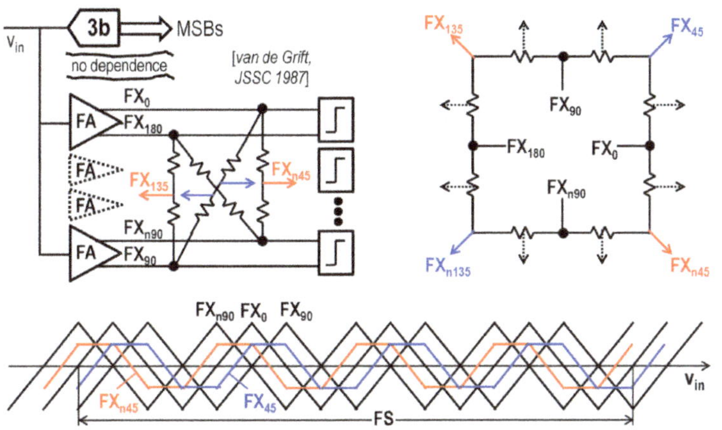

- No interaction between the folding signals
- No linearity requirement for 2x interpolation

Additional ZX or folding signals can be generated by interpolating between two existing ones. This slide shows a resistive interpolation using a resistor bridge with the two existing differential signals injected at the bridge null points so that they do not interact with each other. Interpolation yields better linearity than the existing ZX signals assuming negligible mismatch error in the interpolation network, which is usually the case. For example, interpolation of two odd symmetric ZXs generates the exact midpoint ZX. Away from the perfect midpoint, the interpolated ZXs are subject to some error from the nonlinearity in the existing ZXs, but the error is very limited because the closer and more linear ZX signal weighs more in the interpolation than the other less linear ZX signal. When the existing ZXs have random offsets, the interpolated ZXs are more accurate because they average out the two random offsets.

40 Multi-Step Folding by Summation or Multiplication

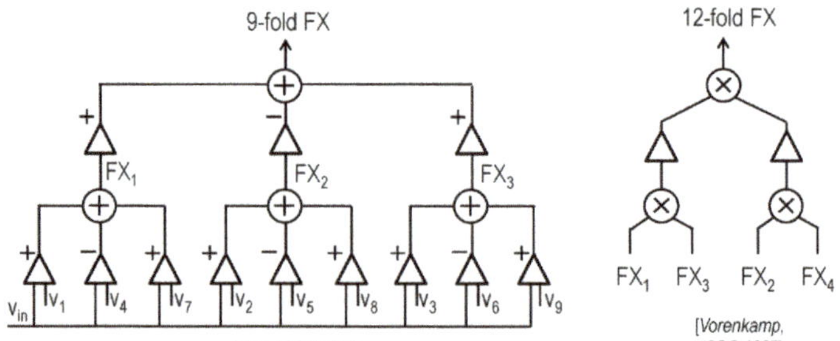

[Bult, JSSC 1997]

[Vorenkamp, JSSC 1997]

- ZXs are combined in multiple steps to avoid excessive output loading.
- Multistep folding is merged with multistep amplification for better BW.
- Pipelining is possible but is against the purpose of "flash" operation.

The post-ZG folding signal with a large number of folds can be generated in cascaded stages to reduce the number of ZXs that have to be merged at once. This alleviates the aforementioned issues associated with large number of folds formed by ZX current summation at the differential output nodes. In the case of merging multiple ZX signals into a folding signal by multiplication that preserves the ZXs, the folding or ZX signals have to be merged in pair using Gilbert cells in cascaded stages. The cascaded stages also allow gain distribution across multiple stages and possible pipelining of the signal folding and amplification.

41 Folding with Min/Max Follower(s)

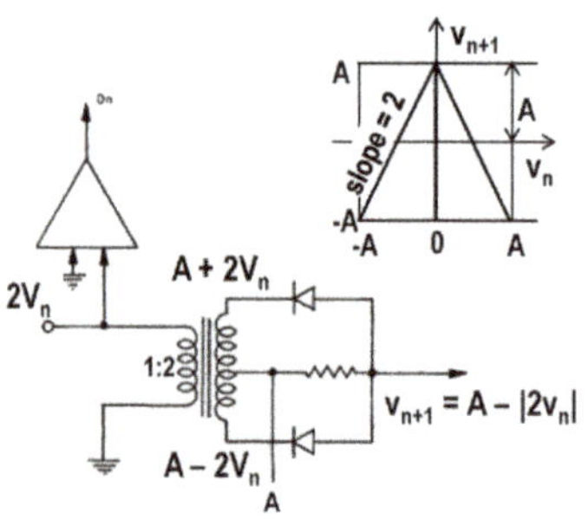

(a) [Smith, IRE 1956]

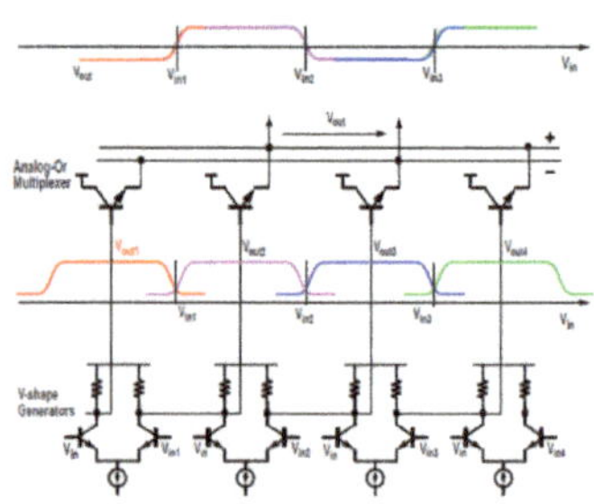

(b) [van de Grift, JSSC 1987]

- In (a), the diodes selects the minimum of the two voltages A ± 2v$_n$ to realize an inverting V-shape folding output: $v_{n+1} = A - |2v_n|$.
- In (b), the output emitter followers select the maximum among the input inverting U-shape folding signals and stitch them by alternating their polarity.

41 Folding with Min/Max Follower(s)

The V-shape folding was originally generated using two rectifying diodes in combination with a 1:2 balun. The diodes choose the lower side of the balun differential output voltages A +/- $2V_n$ as the output V_{n+1}. Assuming negligible diode forward biasing voltage as compared to the DC voltage A applied to the center tap, we have $V_{n+1} = A - 2|V_n|$, which appears as an inverted V-shape folding centered at the origin in a 2Ax2A box. The 2X slope required for the ripple-through inverted V-shape serial folding architecture is realized with 1:2 turn ratio of the balun. The noninverting V-shape folding $V_{n+1} = 2|V_n| - A$ can be realized by flipping the diode polarity and the center tap voltage polarity. The same principle of min or max follower can be applied to generate multiple U-shape folding signal and merge them into a single folding signal as shown in schematic (b).

42 Folding by Rectifier Self-Switching

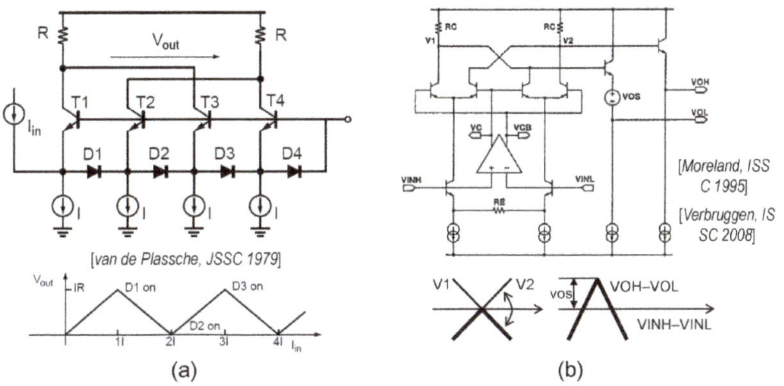

- At $I_{in} \geq kI$, Tk shuts off, Dk turns on, $I_{in}-kI$ switches side by T(k+1).
- At (b) output, input signal multiplies its polarity by self-switching.
- ZX is detected by an op-amp instead of a regenerative latch (ZD).

The aforementioned triangular folding signal can be generated using the self-switching property of diodes. The signal current I_{in} is fed to a string of diodes D1-D4 as shown schematic (a). The current is steered to the output node through T1, generating a differential output voltage ramp as the input current increases. At I_{in} = I, the T1 emitter current and the base-emitter voltage drop to zero causing D1 to automatically switch on and conduct the additional current from I_{in} to T2. Since T2 collector is connected to the negative side of differential output, the output starts to ramp down. In schematic (b), as the differential input VINH-VINL sweeps crossing zero, the op-amp detecting the input ZX automatically swap the differential output current polarity, causing signal folding at input zero. If the input is sampled, a regenerative ZD can be used to detect the ZX and drive switching of the differential polarity.

43 CML Latches for ZX Detection

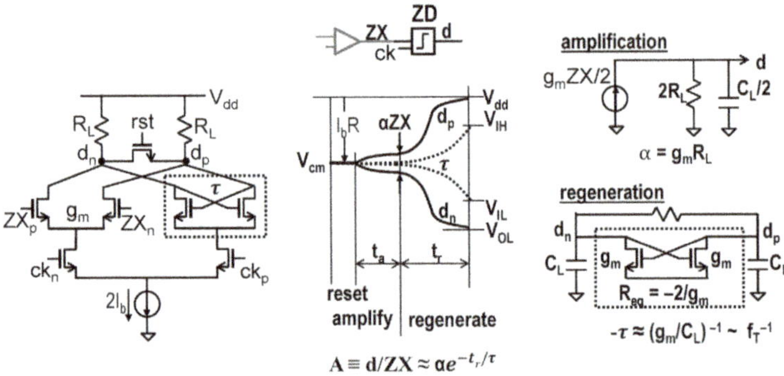

- ZD uses latch regeneration (positive feedback) to maximize the gain.
- Two-phase operation: tracking/amplification & sampling/regeneration.
- Regeneration speed or time constant (τ) is limited by technology (f_T).

ZX detectors (ZDs) are predominantly implemented with regenerative latches to attain the maximum possible gain through positive feedback. The cross-connected diffpair in the latch realizes a negative differential resistance R_{eq} » $-2/g_m$ (for $g_mR_L >> 1$) or right-hand side pole corresponding to a negative time constant $\tau = -(g_m/C_L)^{-1}$, which results in a gain A growing exponentially with the regeneration time t_r: $A = \alpha e^{-t_r/\tau}$ for $\tau < 0$. The static CML latches are compatible with the fast current mode logic (CML). The differential input is transferred to the latch output in the amplification phase and then strobed and regenerated to full CML level. The phase transition is realized by steering the constant bias current between the input diffpair and the regenerative diffpair with fixed output CM.

IMPERFECTIONS AND CORRECTIONS

44 Offset Auto-Zeroing and Correlated Double Sampling

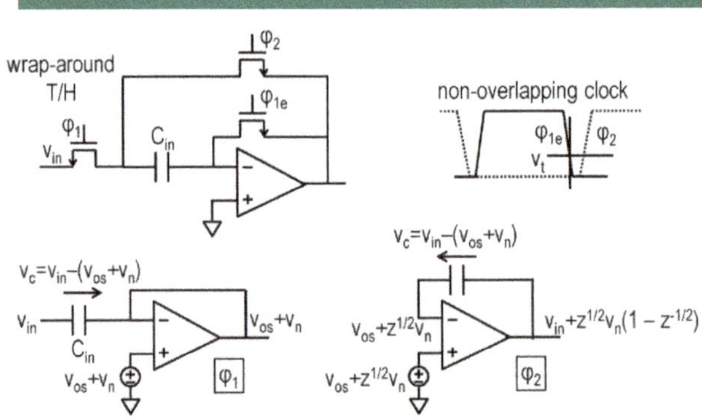

- Offset is sampled on C_{in} and cancelled in next phase.
- Op-amp input referred noise is filtered by $(1\text{-}z^{-1})$.
- Auto-zeroing is based on correlated double sampling (CDS) as slow varying op-amp offset is sampled twice in each cycle.

44 Offset Auto-Zeroing and Correlated Double Sampling

Auto-zeroing or correlated double sampling is commonly used in switch cap circuits such as T/Hs and comparator preamps to remove DC offsets or flicker noise, which is effectively slow varying offset. In the track phase ϕ_1, the op-amp offset is sampled on the S/H capacitor C_{in}. In the hold phase ϕ_2, the capacitor is wrapped around to the op-amp output, and the sampled offset cancels the op-amp offset at the virtual ground, leading to zero offset at the op-amp output. The offset is cancelled by sampling the same or correlated offset twice, namely, correlated double sampling.

45 T/H Bottom-Plate Sampling

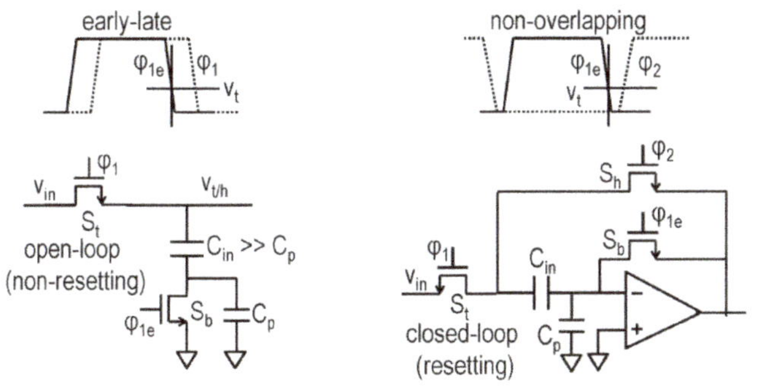

- Bottom plate switch S_b opens before top-plate switch S_t.
- Switch S_b determines the sampling time and aperture.
  - Imperfections are signal independent.
- Closed-loop virtual ground removes parasitic (C_p) effect.
  - Eliminate error from charge injection/feedthrough.

To overcome signal dependent sampling aperture and charge injection, a switch S_b is inserted between the bottom plate (the plate not connected to the input) of C_{in} and the AC ground. The input signal is sampled to C_{in} at the moment the bottom-plate switch S_b opens, because little charge can reach C_{in} once the charging path is open circuited for parasitic cap $C_p << C_{in}$. For the same reason, the top-plate switch S_t contributes no charge injection to C_{in} when it opens. Since S_b sees constant voltage, its sampling aperture and charge injection are signal independent. Due to finite C_p, a small amount of signal change and charge injection from S_t are sampled on C_{in} after S_b opens. In contrast, the closed loop bottom-plate sampling is parasitic insensitive. The error charge stored on C_p flows back to cancel the error charge on C_{in} when the bottom plate is restored to virtual ground in the hold mode.

46 Input Switch Gate Bootstrapping

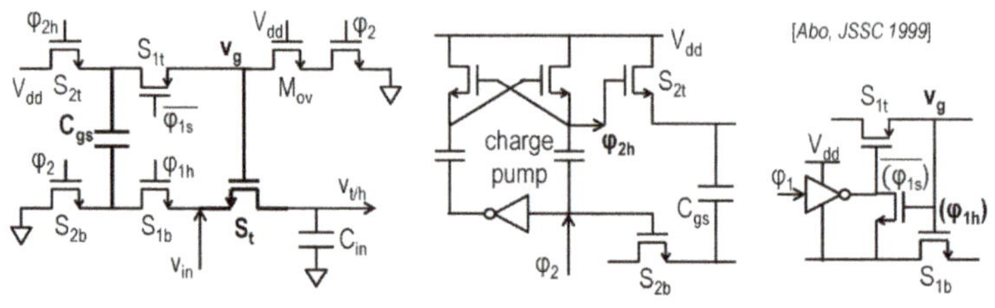

(a) S_t bootstrapped with C_{gs} (b) high-V clock φ_{2h} (c) high-V clock φ_{1h}

- Input switch gate is bootstrapped with $v_{gs} \approx V_{dd}$.
- S_t on-resistance R_{sw} is linearized except for body effect.
- No overvoltage between MOSFET terminals except for node voltages $\mathbf{v_g}$ and $\boldsymbol{\varphi_{2h}} > V_{dd}$.

To overcome the distortion from signal modulation of the T/H top-plate switch on-resistance, the switch gate is usually bootstrapped with the input voltage to maintain a constant overdrive up to V_{dd}. This not only linearizes the on-resistance but also minimizes it. In addition, the sampling aperture and charge injection dependence on the signal are also significantly reduced. A C_{gs} capacitor is charged to V_{dd} in the T/H hold phase Φ_2 with the charge pump providing an overdrive voltage about V_{dd} to turn on S_{2t}. The charged C_{gs} is connected between the switch MOSFET S_t gate and source nodes in the track phase Φ_1 through S_{1t} and S_{1b}, which use v_g and v_{in} as the rail voltages to avoid overvoltage. M_{ov} is added for overvoltage protection of the pull-down MOSFET.

47 Closed-Loop Switch Gate Bootstrapping

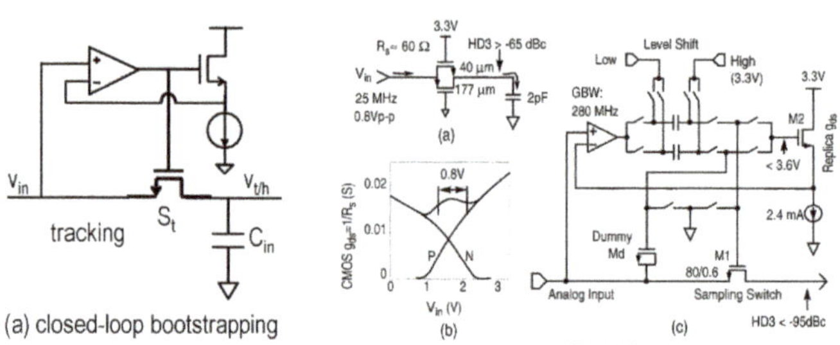

(a) closed-loop bootstrapping

[Pan, JSSC 2000]

- Super-linear switch with body effect compensated
- Less reliability concern with lower overvoltage
- Too much power for the op-amp driving the gate

47 Closed-Loop Switch Gate Bootstrapping

Due to the MOSFET body effect, the T/H switch threshold voltage V_t as a function of V_{sb} is modulated by the input voltage. Since the overdrive $V_{gs} - V_t$ determines the switch on-resistance, the bootstrapped $V_{gs} = V_{dd}$ does not completely remove the switch track-mode nonlinearity unless V_t or V_{sb} is also kept constant. A bootstrap technique that overcomes the body effect uses a high gain op-amp loop to copy the constant g_m of a source follower to the switch in the track mode. This active approach greatly enhances linearity but is power hungry.

48 Distributed Virtual Ground Switch (DVGS)

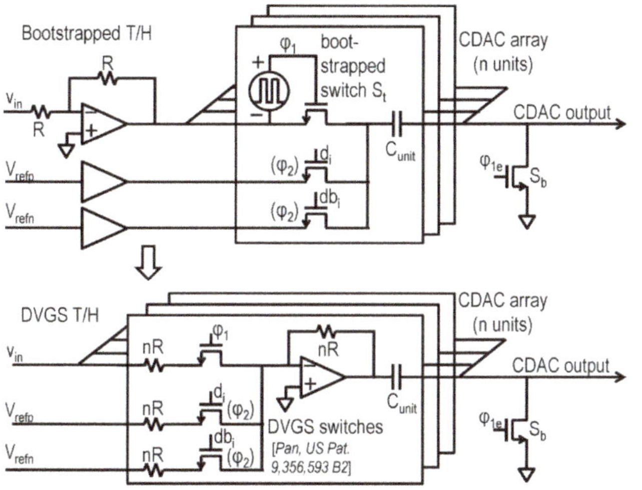

In embedded applications, it is common that an inverting buffer amenable to voltage scaling drives a CDAC through bootstrapped T/H switches (S_t). With the input buffer split uniformly to drive each unit capacitor C_{unit} individually, the T/H switch can be moved to the split buffer virtual ground node, and so can the reference switches. This eliminates the need for bootstrapping while keeping the total input buffer size unchanged. The offset mismatch averaged among the unit buffers in the tracking phase causes only a small constant DC output. The DVGS switch timing mismatch is not a concern either due to bottom-plate sampling by S_b. However, the gain mismatch degrades the CDAC linearity in the conversion phase, which is suppressed with the buffer op-amp loop gain.

49 KT/C Noise from Switch and Op-amp

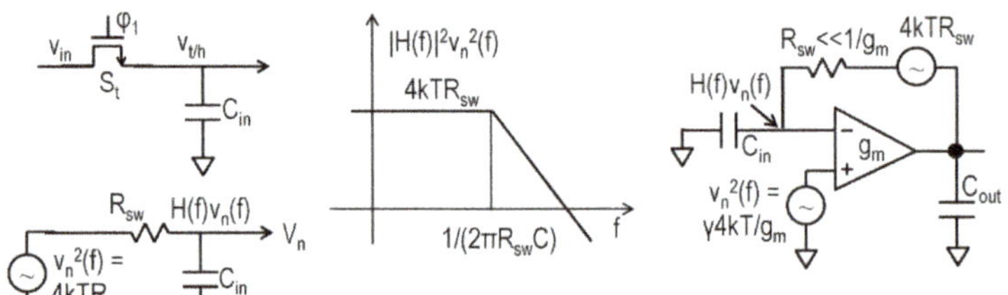

$$V_n^2 = \int_0^{\infty} |H(f)|^2(4kTR_{sw})df = kT/C$$

$\leftrightarrow$ energy equipartition: $CV_n^2/2 = kT/2$

- Effects of R_{sw} (or g_m) on noise BW and level cancel each other.
- Noise power inversely proportional to C and the power driving C.
- Large signal swing is preferred to reduce C and current quadratically.
- kT/C not valid for C → 0 due to quantum effect for f > kT/h ~ 6.2THz.

The kT/C noise can be directly derived from energy equipartition theorem or from the thermal power spectral density expression 4kTR, which are both based on classic statistical mechanics. However, it does not make sense that the noise goes to infinity as C reduces to zero. The energy equipartition and therefore the kT/C equations do not hold for very high frequencies (f) or very low temperatures (T) when the photon quantum energy (hf) gets close to or surpasses the thermal energy kT = 25.7meV at T = 25°C. Within the limited bandwidth (BW) of most practical T/H circuits, kT/C holds well, as long as BW << kT/h = 6.2THz. If it were not for the quantum mechanics that overcomes the ultraviolet catastrophe for thermal radiation, we would have a zero-capacitance catastrophe for the kT/C noise.

50 Sampling Jitter Effect on ADC Performance

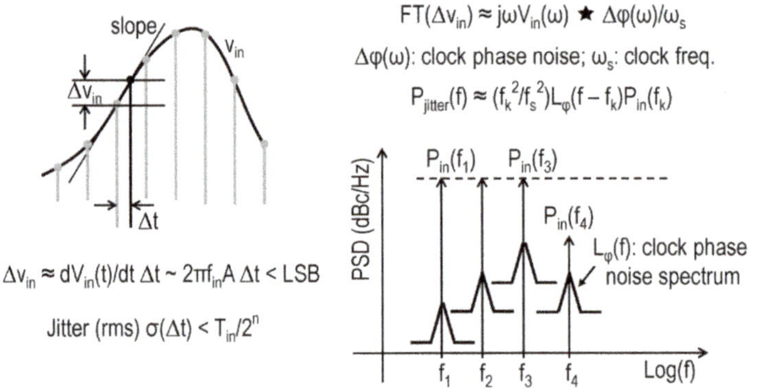

$FT(\Delta v_{in}) \approx j\omega V_{in}(\omega) \star \Delta\varphi(\omega)/\omega_s$

$\Delta\varphi(\omega)$: clock phase noise; ω_s: clock freq.

$P_{jitter}(f) \approx (f_k^2/f_s^2)L_\varphi(f - f_k)P_{in}(f_k)$

$\Delta v_{in} \approx dV_{in}(t)/dt\ \Delta t \sim 2\pi f_{in}A\ \Delta t < LSB$

Jitter (rms) $\sigma(\Delta t) < T_{in}/2^n$

- Noise due to clock jitter scales with sampled signal frequency and amplitude.
- Jitter spec can be considerably relaxed for wideband signals with large PAPR.
- Jitter slower than system tracking BW or observation time can be excluded.

50 Sampling Jitter Effect on ADC Performance

It is illustrative to present the effect of sampling jitter in the frequency domain. The spectrum of jitter induced sampling error is similar to the effect of reciprocal mixing, where the clock phase noise spectrum is proportionally passed to the signal. The similarity is understandable because sampling is mixing with infinite number of harmonics. The difference is also noticeable, that is, the sampling error spectrum does not only scale with the signal power but also with the power of the signal to sampling frequency ratio f_{in}/f_S. This reflects the time domain jitter effect, that is, the jitter induced sampling error is proportional to the signal slope. To avoid the jitter error amplification by $f_{in}/f_S > 1$ in subsampling cases, it makes sense to mix the high frequency signal to a low $f_{IF} << f_S$ before being sampled.

51 Preamp Input Referred Offset: ZX Error

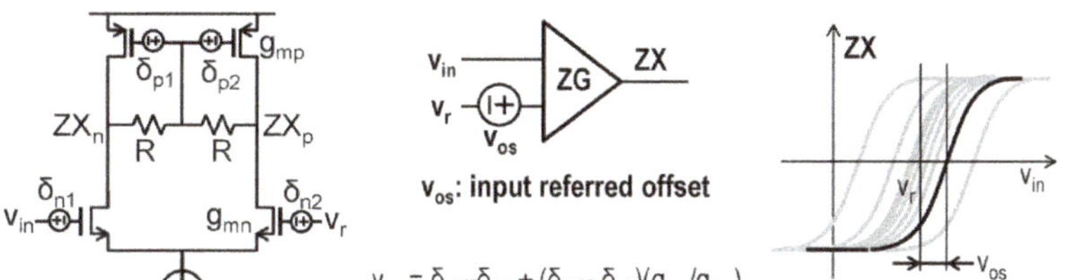

$V_{os} = \delta_{n1} - \delta_{n2} + (\delta_{p2} - \delta_{p1})(g_{mp}/g_{mn})$

Variance: $(\sigma_{os})^2 = 2(\sigma_{vt\_n})^2 + 2(\sigma_{vt\_p})^2 (g_{mp}/g_{mn})^2$

- Preamp offset arises from mismatch in MOSFET threshold V_t, carrier mobility, size, output load, etc.
- Static preamp offset is mitigated by auto-zeroing, sizing, averaging/spatial filtering, calibration, etc.
- Impact of coarse quantizer offsets is eliminated by over-ranging the fine quantizer to cover the increased residue.

Comparator preamps are usually used to suppress the dynamic kickback and the input referred dynamic offset from the succeeding regenerative latch, but they introduce offsets of their own due to random mismatch among the devices. The offset is minimized by reducing the load MOSFET g_m and scale up the size of all the devices. To maintain the same bandwidth, current biasing should also scale with the size, resulting in more power consumption. In flash ADCs, the random offsets of adjacent preamps of the comparator array can be suppressed by having them averaging each other.

52 MOSFET Mismatch Modeling

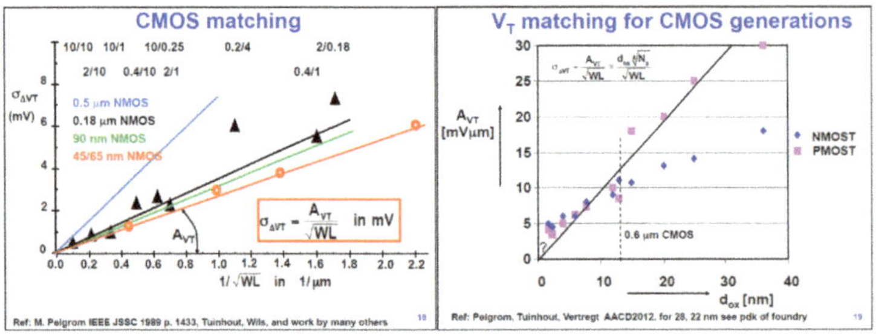

[https://indico.cern.ch/event/228972/session/8/contribution/222/material/slides/0.pdf]

- FET mismatch is inversely proportional to device area in power.
- Threshold mismatch scales with technology (d_{ox}) and dominates offset.

The MOSFET random mismatch is dominated by the threshold voltage variation caused by surface and space charge fluctuation at the thin oxide interface. This leads to two important observations. One is the threshold voltage mismatch (rms) is inversely proportional to the square root of the gate or thin oxide area; the other is that the proportion factor A_{VT} scales with the process geometry, specifically, the oxide thickness d_{OX}. As the area A = WL doubles, the thin oxide capacitance C_{OX} doubles, but the amount of random error charge ΔQ doubles in power. Therefore, we have the mismatch $\Delta V_t = \Delta Q/C_{OX} \sim \text{sqrt}(A)/A = 1/\text{sqrt}(A)$. If d_{OX} doubles, C_{OX} halves, leading to doubled ΔV_t and A_{VT}. Flicker noise is dominated by the slow time-varying error charge and is inversely proportional to sqrt(A) for the same reason. This fundamentally reflects the averaging effect of random quantities.

53 Interpolation and Offset Averaging

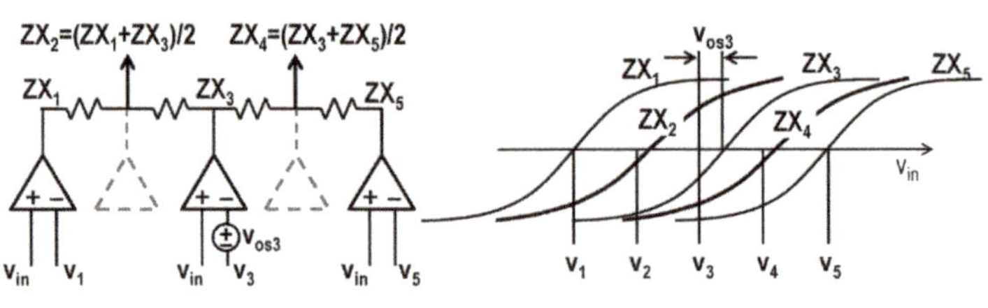

$$V_{os4} = (V_{os3} + V_{os5}) / 2 \rightarrow (\sigma_{os'})^2 = (\sigma_{os})^2 / 2$$

σ_{os}, $\sigma_{os'}$: **standard deviation before and after averaging**

- Interpolated ZXs ($ZX_{2,4}$) have less offset due to offset averaging
- But there is no averaging effect among the original ZXs ($ZX_{1,3,5}$)
- It is possible to extend the averaging effect to all ZXs ($ZX_{1\text{-}5}$)

53 Interpolation and Offset Averaging

The offset of the interpolated ZXs are improved by sqrt(2) due to the averaging effect between the adjacent ZXs. The ZX generator output impedance R0 is made much smaller than the interpolation impedance R1, or the aforementioned bridge-like interpolation network is used to prevent interaction between the ZXs. The small R0 consuming large power is unnecessary because the interactions cancel either other as long as translational symmetry is preserved across the ZX generator array. The finite array should be properly terminated with dummies to preserve the symmetry. The dummy overhead can be outweighed by the benefit of offset averaging among the ZXs that are allowed to interact. Matched spatial filtering theory is applied to optimize the net benefit. Interpolation effectively lump many small distributed ZXs to form large ZX generators with lumped avearging.

54 Spatial Filtering of Offsets in ZX Array

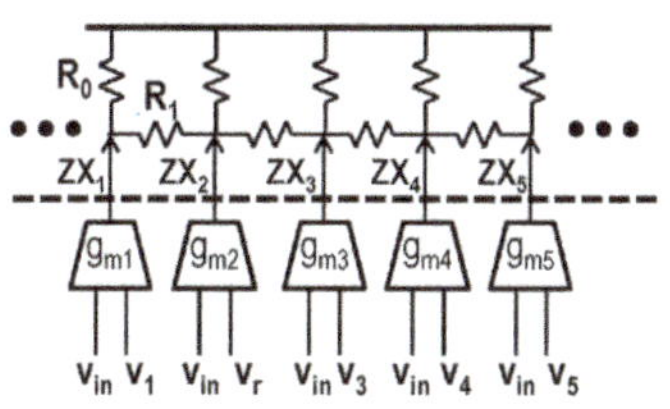

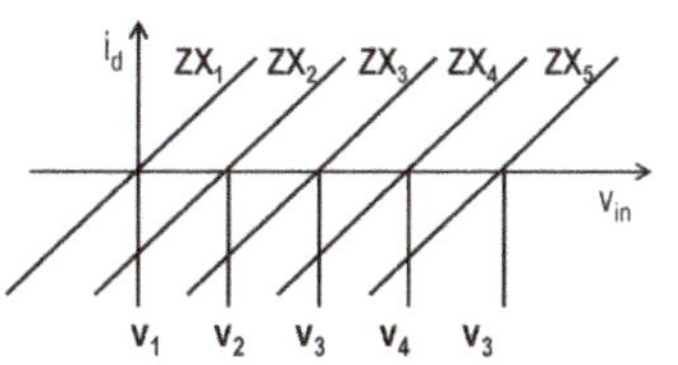

[Pan, TCAS-I 2003]

- Change small R_0 voltage-mode ZGs to Gm cells with large R_0 to allow interaction among the g_m array output currents in the R_0-R_1 network.
- Consider an infinite array of Gm cells that output current ZX's uniformly spaced with translational symmetry (as shown on the right above).
- Output error currents caused by input offsets $v_{os}(k)$ are considered as the "spatial noise" injected to the R_0-R_1 network - the "spatial filter."
- The input voltage increment needed to null the perturbation to the kth ZX_k caused by all $v_{os}(k)$ is a new offset $v'_{os}(k)$ for ZX_k.
- Error correction factor ECF = $\sigma(v_{os})/\sigma(v'_{os})$ is optimized under matched spatial filtering theory and boundary condition of translational symmetry.

The R0-R1 network appears as a spatial filtering to the g_{mi} output current array with a spatial impulse response decaying exponentially with spacing. The spatial stimulus current pulse width is determined by the g_m linear region of the ZX generator input diffpair, while the random offsets causes spatial white noise current. The ECF or the spatial filter SNR is optimized under the matched filter condition, that is, the spatial impulse width as a function of R0/R1 matches the number of ZX generators in the linear region. The optimal matched width to minimize the array size is 1/3 array for best tradeoff between offset reduction and dummy overhead. The bandwidth is preserved while the array area, input capacitance, and power are drastically reduced for the same offset (rms). The DNL improves even more because of offset correlation. Now the array size and power should scale by 4x instead of 8x for every bit increment in the flash ADC resolution.

55 Latch Dynamic Offsets and Simulations

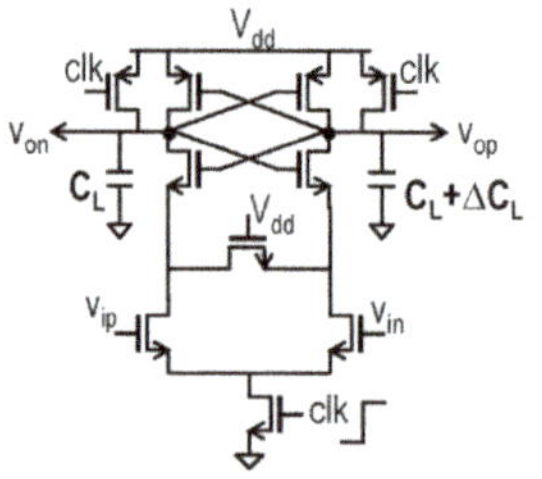

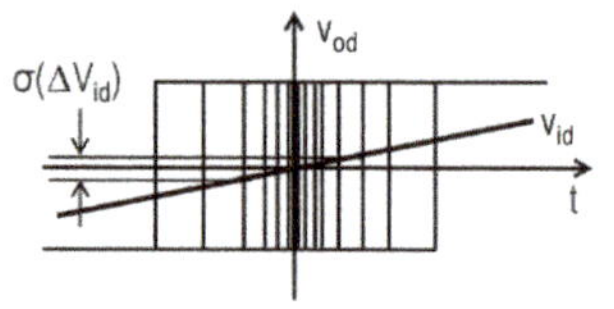

Transient Monte Carlo simulations of input referred dynamic offset $\sigma(\Delta V_{id})$

- Latch switching activities cause dynamic offsets [Kim, TCAS-I 2009]
  - Due to CM-to-DM conversion (e.g. $\Delta V_{out} = (\Delta C_L/C_L)\Delta V_{cm}$)
  - Due to kickback to the input stages such as the reference resistor string
  - Due to clock/signal feedthrough and coupling from supply/ground noise
  - Sensitive to variations in supply, clock/signal, temperature, parasitics
  - Mitigated with preamps, though corrections possible [Abidi, CICC 2014]
  - Simulated with random mismatch included and clock running

Dynamic latches do not consume steady current but exhibit large differential offset due to the CM jump associated latch strobing. At the clock rising edge, the common source switch shorts the input diffpair common source to ground while the switches connecting the output nodes to V_{dd} are opened. As the output CM is pulled down by the diffpair drain currents, differential voltage also develops across the output and the intermediate nodes depending on the differential input and mismatches. When the output CM drops by one PMOS threshold voltage V_{tp} for an input near ZX, the cross-coupled PMOS pair is turned on and starts to regenerate the voltage difference to V_{dd}. Adjustable capacitors can be added at the output nodes and/or the intermediate nodes to correct the offset.

56 Static CML Latch Improvements

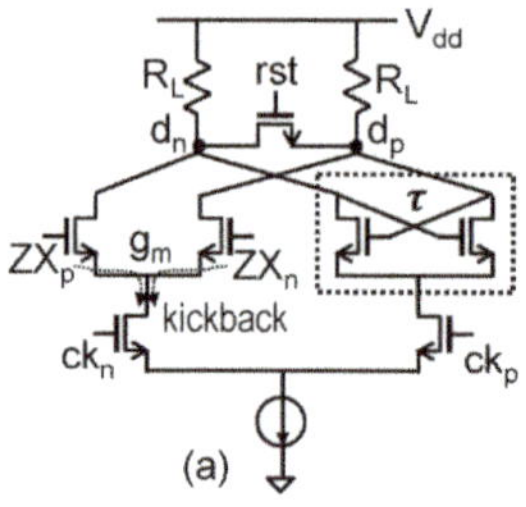

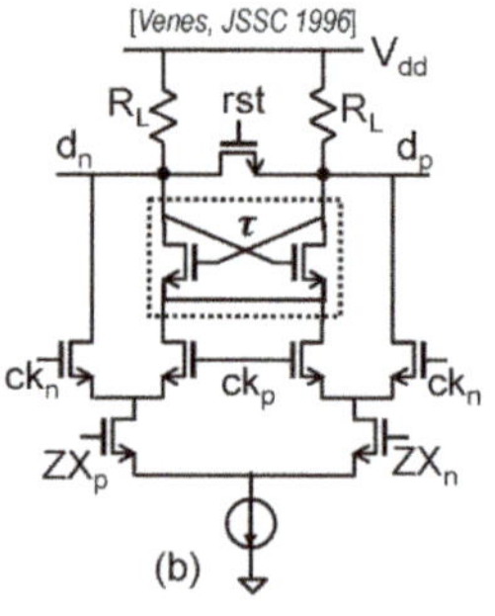

- Steady current bias and output CM minimize dynamic offset.
- Gilbert-cell topology (b) provides isolation between the input and output
  - Blocking the input feedthrough and output kickback with the stacked diffpairs
  - Minimizing clock feedthrough and clock kickback using the balanced clocks
- A reset pulse (rst) before amplification helps to erase the hysteresis.

56 Static CML Latch Improvements

The CM jump and the corresponding offset can be avoided using the CML latch biased with steady current as shown in schematic (a). However, current kickback to the input nodes occurs when the current bias of the input diffpair is steered on and off at transitions between amplification and regeneration modes. This is avoided by stacking current steering on top of the input diffpair so that the input diffpair is biased at constant current as shown in schematic (b). An additional benefit is that this Gilbert cell topology (b) also suppresses the input feedthrough that could over-write the output polarity even after regeneration.

57 Flash-ADC Bubble Error Correction

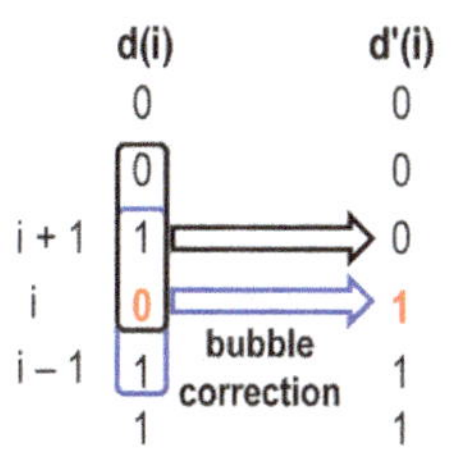

majority vote:
d'(i) = 1 when 2 out of 3
immediate neighbors are 1

$$d'(i) = d(i-1) * d(i) + d(i) * d(i+1) + d(i-1) * d(i+1)$$

[Mangelsdorf, JSSC 1990]

- Adjacent comparator thresholds may cross each other, causing bubbles in the thermometer code and sparkle error
- This may happen in high speed ADCs due to excessive comparator offset and clock skew among adjacent comparators.
- Bubbles are detected and corrected by looking at more than two adjacent bits as is done in the majority vote scheme.

When offset exceeds one LSB, it is possible that the adjacent quantization thresholds of flash ADCs cross each other causing bubbles in the output thermometer code. This phenomenon used to be common for high speed flash ADCs where the comparator latches strobe at different points of a fast vary signal that is not sampled and held. A majority vote logic was used to correct the bubbles that cause large encoding error. The amount of bubbles is reduced with a standalone T/H preceding the flash ADC.

58 SSCS Latch Regeneration and Metastability

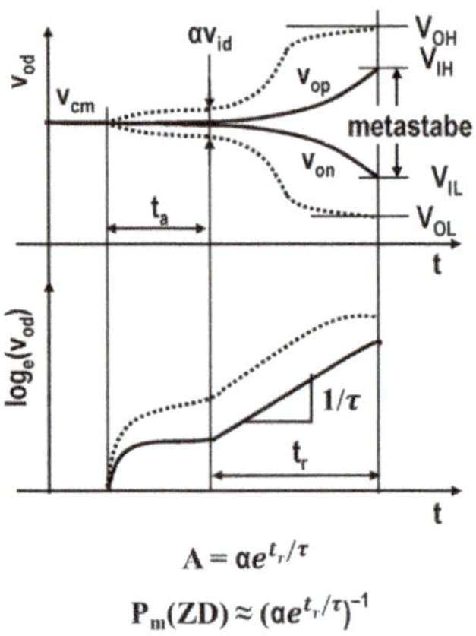

- Small signal regeneration is linear operation and can be modelled with a negative resistance and negative time-constant $-\tau$.
- Time constant τ can be obtained from the slope of $\ln(v_{od}(t))$.
- Metastability occurs when latch output v_{od} falls within $[V_{IL}, V_{IH}]$.
- Metastability Probability $P_m(ZD) = ((V_{IH}-V_{IL})/A)/FS \approx (\alpha e^{t/\tau})^{-1}$, not affected by noise. [Figueiredo, JSSC 2013]

The latch output should be regenerated to a logic level above V_{IH} or below V_{IL} before it is passed for succeeding (encoding) logic operations to avoid ADC output uncertainty. However, there is always a chance that the strobed input signal falls within a small region $(V_{IH} - V_{IL})/A$ around ZX causing the latch output to fall within the undefined metastability region. The probability of metastability is therefore $P_m = ((V_{IH} - V_{IL})/A)/FS \sim 1/A$, assuming the signal PDF is uniform across the input FS and the logic level $(V_{IH} - V_{IL})$ is comparable to FS. With the τ available from transient simulations, the max. clock rate can be calculated from the required P_m or vice versa: $f_{clk} = 1/(2t_r) < (1/(2\tau))/(-\ln(P_m)) < (f_t/2)/(-\ln(P_m))$. Given $\tau \sim 10$ps for the latest technologies, the latch should be clocked below $f_{clk} \sim 1.45$GHz for $P_m < 10^{-15}$.

59 SSCS ADC Metastability and Mitigations

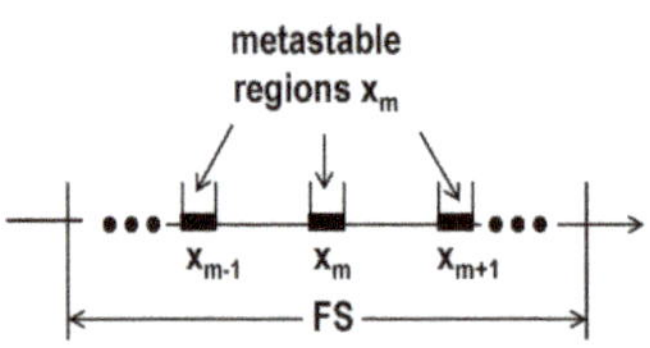

ADC Metastability Probability

$P_m(ADC) = (2^n x_m) / FS \approx 2^n P_m(ZD)$

Comparator Metastability Probability

$P_m(ZD) = e^{-t_r/\tau}/A$

- Metastability is a fundamental ADC issue that can not be eradicated.
- Noise may break a metastable event but does not alter the probability.
- Metastability causes large sparkle errors even at very high S/N levels.
- Metastability and its effect on ADC output error can be mitigated.
  - Increasing the effective regeneration time t_r
    - Cascading latches or DFFs before the encoding logic
    - Time interleaving the ADCs or the comparators
    - Asynchronous clocking [Chen, ISSC 2006] for need based regeneration time.
  - Reducing the metastability impact on encoding output [Portmann, JSSC 1996]
  - Reducing regeneration time constant in fast technologies

59 ADC Metastability and Migitations

The probability of ADC metastability Pm(ADC) is the sum of the Pm(ZD) associated with all the N ZXs. This is straightforward for flash ADCs where each ZX corresponds to a comparator. For the SAR ADC, the P_m(ADC) seems to be smaller as only one comparator is used for n << N times for each conversion. This is not true, because the FS range for each succeeding quantization step is halved and the P_m is doubled resulting the same P_m(ADC) given the same regeneration gain. However, for asynchronous SAR ADCs, P_m(ADC) can be substantially reduced, because the regeneration time is determined on a need basis instead of forced by a fixed clock frequency n times the conversion rate. The impact of ADC metastability is mitigated using a robust encoder immune to metastability or system level error correction such as FEC.

60 Error Correction by Digital Calibration

- Analog vs. Digital
  - Analog correction: errors are measured digitally but corrected in analog. *[Lee, JSSC 1984]*
  - Digital correction: errors are measured and corrected both digitally. *[Karanicolas, JSSC 1993]*
  - Digital overhead becomes affordable starting 65nm CMOS. *[Murmann, ESSCIRC 2013]*
- Foreground vs. Background
  - Foreground calibrations are done in dedicated time slots. *[Lin, JSSC 1991]*
  - Background calibration is transparent to the normal operations. *[Fu, JSSC 1998]*
- Statistic vs. Deterministic
  - Statistic measurements take millions of samples to converge. *[Galton, TCAS-II 2004; Murmann, JSSC 2003]*
  - Deterministic measurements against references need only thousands or less samples to converge. *[McNeil, JSSC 2005]*
- Design parameters to be calibrated
  - Flash: comparator threshold offsets, etc. *[Cao, JSSC 2010]* ...
  - Pipeline: inter-stage gain, reconstruction DAC linearity, etc. *[Chiu, TCAS-I 2004]* ...
  - Interleave: lane imbalance in offset, gain, skew, BW, etc. *[Razavi, JSSC 2013; Abidi, VLSI 2014]* ...

METRICS AND MEASUREMENTS

61 Integral & Differential Nonlinearity (INL & DNL)

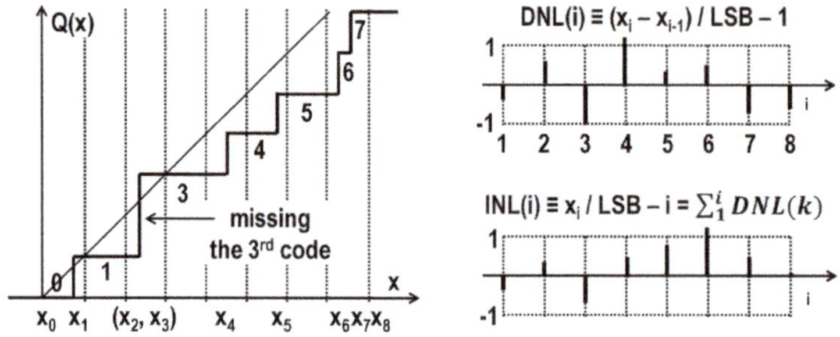

- Gain and offset errors are extracted by linear fitting and excluded from INL and DNL.
- INL and DNL both capture ADC nonlinearity as quantization threshold (x_i) errors in LSB.
- INL highlights low-order distortion from analog front-end such as input/ref. buffer, T/H, etc.
- DNL stresses high-order quantization errors from comparator offsets, RA inaccuracy, etc.
- INL and DNL do not capture signal dependent threshold errors (e.g. latch (ZD) hysteresis.)
- INL and DNL can be measured by static DC sweep or by dynamic code density test.

INL and DNL measure the deviation in quantization thresholds and intervals, respectively. It is common to take off the constant and linear components in the INL error because they are not inherent quantization error and can be easily compensated with DC and gain calibrations. The INL and DNL are meant to provide complete information about all the quantization thresholds. Even though they can be derived from each other, they stress different aspects of the quantization imperfections. The DNL that captures subtle localized errors such as missing codes corresponding to DNL = -1, which is important for applications with large signal dynamic range. The INL is relatively important for wideband communications applications that are concerned about spurs from quantizing large in-band blockers.

62 Output Code Density Test for DNL and INL

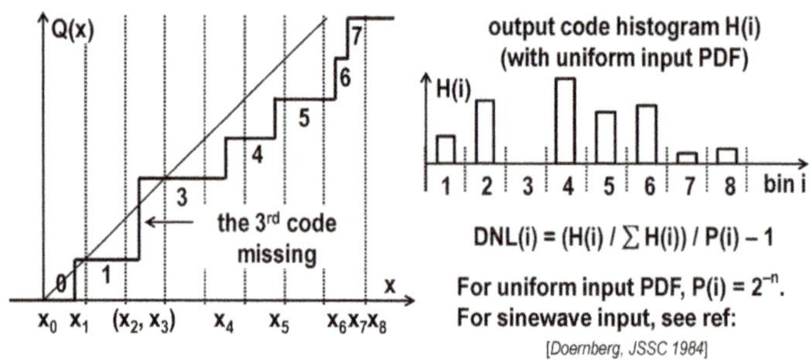

- Freq. of code occurrence is proportional to bin size R_i given uniform input PDF.
- FS sinewave asynchronous to f_s with well-defined P(i) is used as the test signal.
- H(i) averages out noise, resulting in accurate DNL/INL, gain, offset measurement.
- Frequency response and dependence on conversion rate are also captured.

62 Output Code Density Test for DNL and INL

Even though INL and DNL can be measured directly by sweeping the input signal and recording the trip points, they are commonly derived from output code density measurement using FS sinewave as input, based on the fact that the code density normalized with the signal PDF is proportional to the quantization interval. This statistical approach has many benefits. It averages out noise and signal dependent error, captures both DC and dynamic performance, reuses any measurement data available, and so on. For wireless applications, they are kind of redundant given many spectral characterizations available such as SNDR, SFDR, IMD, and NPR vs signal strength. In any case, the INL & DNL profiles are important debugging tools because they are directly related to the ADC architecture and the spectral signature.

63 Frequency Domain Characterization Using FFT

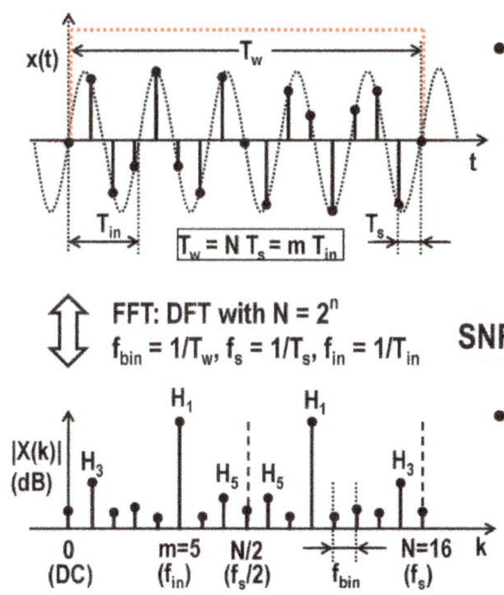

- Single-tone test: $|H_1|^2 \sim P_{FS}$

  SNDR $\equiv |H_1|^2 / (\sum_{k \neq m}^{N/2} |X(k)|^2)$

  SFDR $\equiv |H_1|^2 / \max|X(k)|^2_{k \neq m}$

  THD $\equiv \sum_2^I |H_i|^2 / |H_1|^2$ (e.g. $I = 15$)

  SNR $\equiv |H_1|^2 / (\sum_1^{N/2} |X(k)|^2 - \sum_2^I |H_i|^2)$

- Multi-tone test:

  $f_{inj} = m_j\, f_{bin}$; $j = 1, 2, \ldots, M$

ADC spectral characterization relies on FFT analysis. It is important to meet the coherent sampling requirement to have the test signals fall exactly on the FFT frequency bins, that is, $f_{in} = m^* f_{bin}$. For single-tone tests, all the harmonics higher than $F_S/2$ are folded back and forth between the $F_S/2$ and DC. A script is usually used to identify the location of all significant harmonics. To reduce the chance of harmonics falling in the same bin, m is preferred a prime number. For multitone test, uniformly spaced prime numbers are infeasible and unnecessary, but the FFT window T_W should be a integer number of the beat periods. The FFT points N should be large enough to ensure each quantization threshold is traversed, $T_W = N/F_S$ meets the required observation time, and/or $f_{bin} = F_S/N$ is finer than the min. channel bandwidth.

64 Effective Resolution Bandwidth (ERBW)

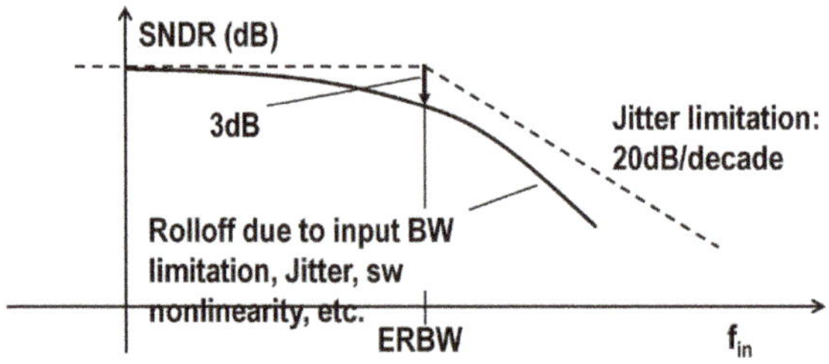

ERBW ≡ f_{in} @ ENOB = ENOB(dc) – 0.5

- An n-bit ADC is supposed to have ENOB(dc) > NOB – 0.5.
- An n-bit Nyquist ADC is supposed to have ERBW > f_s / 2.
- Metrics for FS single-tone test near Nyquist tend to over-specify ADCs.
  - Input signal freq. usually occupies the lower portion of the Nyquist band.
  - Input signal rms is usually much smaller than the full scale with peak-to-average power ratio (PAPR) < 13dB.

To be claimed a Nyquist ADC, the ENOB should not drop more than half bit over the entire Nyquist band, or ERBW > fs/2. Nowadays, it is common to see the ENOBs of many published ADCs drop by one bit or more at Nyquist. The relaxed specification can be justified for the following reasons. First, the ADC sample rate is intentionally chosen much higher than 2X signal BW to relax the antialiasing filtering, to lower the noise floor for processing gain, to ease timing recovery with oversampling, etc. Second, for high speed wireline applications, the low frequency portion dominates the baseband signal spectrum due to channel bandwidth limitations. Third, when the clock and data recovery (CDR) loop locks, the sampling happens at center of the eye with small slope.

65 Noise Spectral Density (NSD) and Noise Figure (NF)

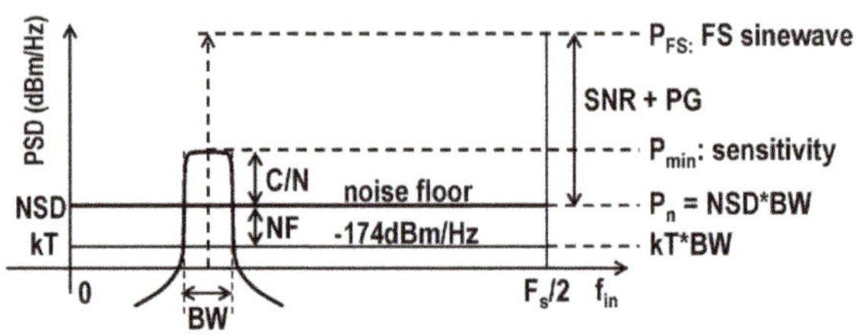

$$NSD \equiv P_{n(total)} / (F_s/2) = P_{FS} - SNR - 10\log(F_s/2) \quad (dBm/Hz)$$

$$\text{Noise Power } P_n \equiv NSD{*}BW = P_{FS} - SNR - 10\log(F_s/2BW) \quad (dBm)$$

$$\text{Processing Gain (PG)} \equiv 10\log(F_s/2BW) = 10\log(OSR) \quad (dB)$$

$$NF \equiv P_n / (kT{*}BW) = NSD\,[dBm/Hz] + 174 \quad (dB)$$

- ADC dominates receiver sensitivity unless wideband LNA precedes ADC.
  - For a 10b 1GS/s ADC with P_{FS} = 4dBm, NF = 4 – 62 – 10log(1G/2) + 174 = 29 (dB).
- ADC SNR spec can be relaxed by increasing conversion rate or PG.

65 Noise Spectral Density (NSD) and Noise Figure (NF)

For radio receiver applications, ADC spectral performance is characterized in terms of the noise floor and the dominant spurs under various test signal conditions. A common noise floor spec is the noise spectral density (NSD) in dBm/Hz , which is defined as the total Nyquist band noise (excluding the spurs) divided by the Nyquist bandwidth. The noise figure (NF) is readily available from the NSD. As the conversion rate doubles, the NSD drops and the in-band SNR increases by 3dB for a given SNR. The SNR improvement is a processing gain because it results from the succeeding digital processing (i.e. filtering) of the in-band signal. Given an NSD requirement and a fixed FoM, the ADC power is independent of the conversion rate, that is, oversampling does not necessarily burn more power. Therefore, oversampling is a free ADC design parameter that has been utilized in combination with analog and digital signal processing more often in advanced CMOS for better FoM.

66 Noise Power Ratio (NPR) for Broadband Applications

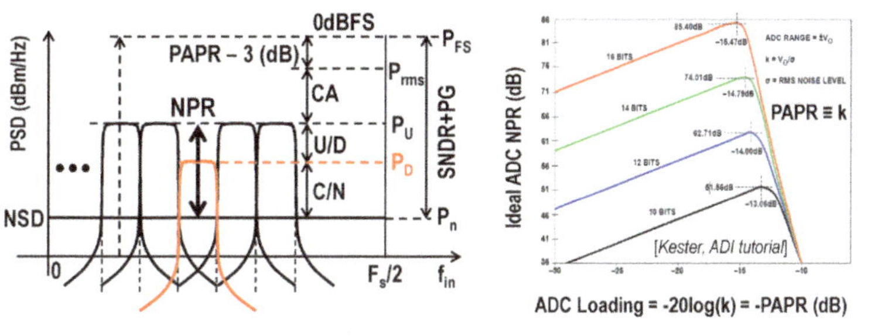

$$SNDR = C/N + U/D + (PAPR - 3) + (CA - PG) = NPR + PAPR - 3 \text{ (dB)}$$

(For CA ≡ 10*log(total # of carriers) = PG)

- Broadband signals look random and ADC output distortion appears as white noise.
- Clipping of the Gaussian PDF tails at ADC full scale leads to optimal PAPR & NPR.
- SNDR dominates sensitivity given signal independent spurs << –(SNDR+PG).

For multi-channel broadband communications, the noise floor is characterized with the noise power ratio (NPR), which measures the noise level of an unoccupied channel with respect to the signal level of the occupied channels. The test signal can be synthesized with a certain amount of white noise passing a notch filter or simply with multiple tones with one missing for a channel carrier. The noise floor is signal dependent due to distortions such as signal clipping to the ADC full scale. As the signal power grows, the clipping starts to dominate the noise floor causing a peak NPR and optimal ADC loading. For a 10b ADC, the optimal loading is around (-13+3) dBFS for Gaussian signals. Again, as the ADC sample rate doubles, the NPR increases by 3dB for the same number of carriers due to the processing gain.

67 Spurious Free Dynamic Range (SFDR)

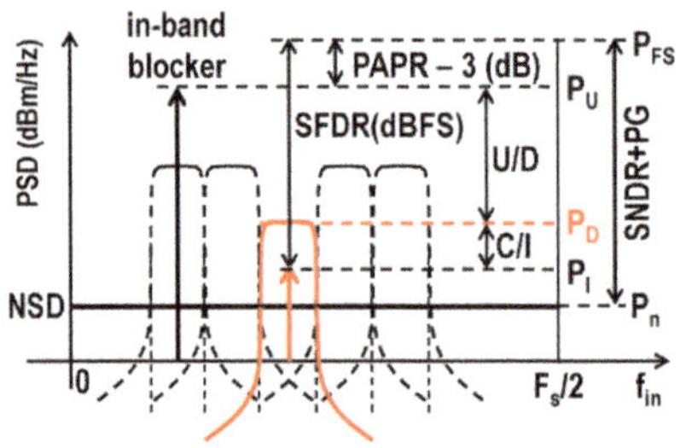

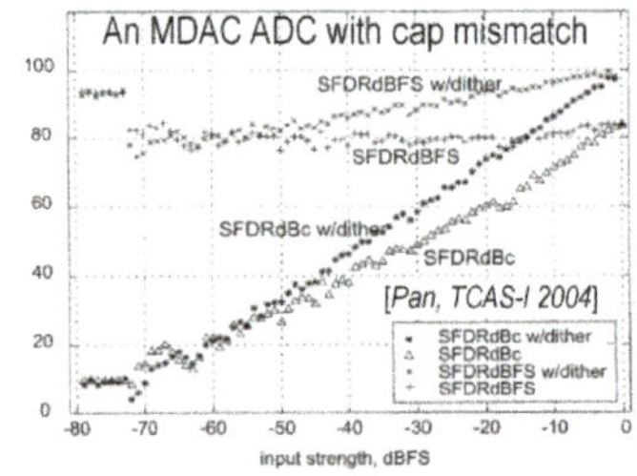

SFDR (dBFS) = C/I + U/D + PAPR – 3 >> SNDR

- SFDR usually determines mobile receiver sensitivity due to blocker interferences.
  - Given U/D~53dB, C/I~18dB, PAPR~12dBFS for a base station receiver, SFDR>80dBFS.
- ADC output spur level dependence on blocker strength could be complicated.
  - Low order spurs from T/H distortion of strong blockers usually dominate SFDR
  - High order quantization spurs grow with reduced blockers and could dominate SFDR

For wideband mobile receivers, in-band blockers impose a major challenge to the design of the ADCs that capture the entire RF or IF bands. The harmonic spurs of a CW blocker could land in a weak channel in the Nyquist band and swamp the signal. In this case, the SFDR should be high enough to accommodate a large range of signal power variations including fast fading that can not be compensated by the variable gain control (VGC) loop. At the SFDR~80dBFS level required for applications such as base stations, attention must be paid to high-order quantization spurs, which grow with reduced blocker power. To dither out those spurs with the inherent ADC noise, the ADC resolution is usually 2b or 3b higher than the required ENOB. It is common to use various dynamic and static scrambling techniques and architectures to spread out the spur energy.

68 ADC Figure of Merit (FoM)

- Walden (ISSCC) $FoM_W = P / (2^{ENOB} f_s)$
  - For low resolution (n<10b) full-flash ADCs where ENOB dominates the total number of comparators and total power.
- Schreier $FoM_S = (2^{ENOB*2} f_{nyq}) / P \approx SNDR + 10\log(f_s/2P)$ (dB)
  - For high resolution (n≥10b) sw-cap based ADCs, ENOB determines KT/C, and the capacitor C dominates the total power.
  - For $P=V_{FS}^2Cf_s=V_{FS}^2f_skT12/LSB^2=2^{ENOB*2}12kTf_s$, $FoM=(24kT)^{-1}\approx190$(dB).
- Many FoMs emphasize different aspects for different applications.
- Many factors affect the FoM
  - ADC definition (e.g. buffers/PGA/AAF), f_t, trimming, clock source jitter, digital calibration, decimation filter, post processing, etc.
- Many publications and surveys are there on FoMs
  - For example: http://converterpassion.wordpress.com/
  - B. Murmann, "ADC Performance Survey 1997-2016," [Online]. Available: http://web.stanford.edu/~murmann/adcsurvey.html.

68 ADC Figure of Merit (FoM)

For the purpose of ADC trend study and paper evaluation, it becomes popular to rank ADCs out of the application context using a single FoM that takes into account a couple of major ADC specs such as ENOB, conversion rate, and power. The Walden FoM normalizes ADC power with product of the conversion rate and the effective number of quantization intervals or conversion steps 2^{ENOB}. This ends up measuring the energy per comparison of an equivalent full-flash ADC. This is obviously unfair for ADCs of higher ENOB in which each comparison consumes more power for higher accuracy. This "unfairness" is addressed by the Schreier FoM, where the ADC power is normalized with the power of the quantization accuracy, namely, $(2^{ENOB})^2$, based on the fact that comparator power is inversely proportional to kT/C noise power. The number of comparisons is not used as part of the Schreier normalization because it grows only linearly with ENOB using serial binary search.

69 Survey Results: Schreier FoM vs. Fs

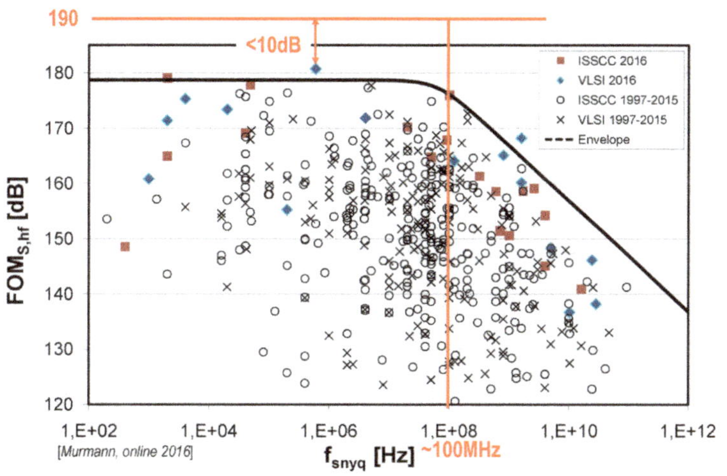

The survey of ADC Schreier FoM vs Nyquist rate F_{snyq} is usually characterized with an envelope or trend line with a subjective 10dB/decade roll-off for the following reasons. First, the T/H distortion at high Nyquist frequencies limits the ADC FS range and necessitates more power consumption to suppress the noise. Second, the clock generator and buffers must be scaled up in size and power to minimize the jitter that causes large sampling error at high Nyquist frequencies. Third, pipelining and time-interleaving techniques for high speed operation consume extra power. At low frequencies without the aforementioned power overheads, the envelop is surrounded by delta-sigma ADCs that attain high efficiency with signal processing. It seems theoretically possible to pass the 190dB KT/C limit using high-order noise-shaping that exponentially enhance accuracy or SNR at linear expense of power. The envelop is pushed by advancement in technology and efficient signal processing.

70 ADC Meta-Stability Test

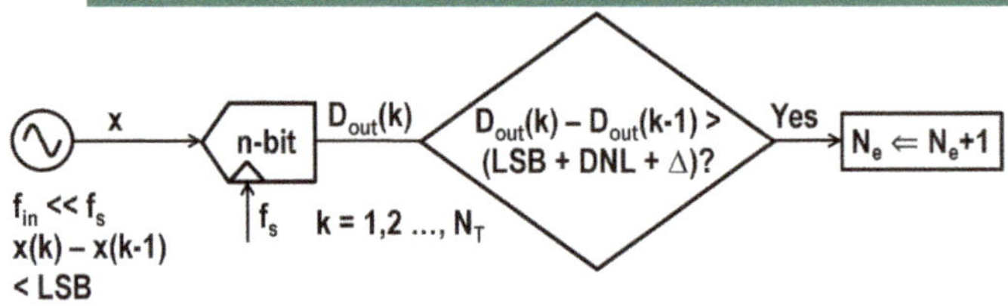

$$P_m(ADC) \approx N_e / N_T$$

- Meta-stability is hard to catch in transient simulations but can be easily measured on- or off-chip.
- Meta-stability probability P_m can be accurately estimated in simulation from the regeneration transient waveform.
- Comparator regeneration time constant can be estimated from measured metastability dependence on conversion rate.

The ADC metastability can be measured by counting the output "sparkle" errors in a high SNR environment with a very slow varying test signal. In this test setup, the difference between adjacent ADC output codes should be within the DNL limit plus one LSB if it were not for ADC metastability. For baud-rate communication systems it is important to characterize ADC metastability and make sure it is smaller than the required bit-error rate (BER), because each ADC sparkle could cause a symbol error. If the BER requirement is reduced from 10^{-16} to 10^{-8}, for example, the ADC can be clocked 2X as fast without noticeable SNR degradation from the sparkle error. The ADC metastability dependence on clock frequency can be measured to derive the comparator regeneration time constant.

EVOLUTIONS AND TRENDS

71 A 3.3V 50MS/s 12b ADC with 85dB SFDR in 0.6μm CMOS

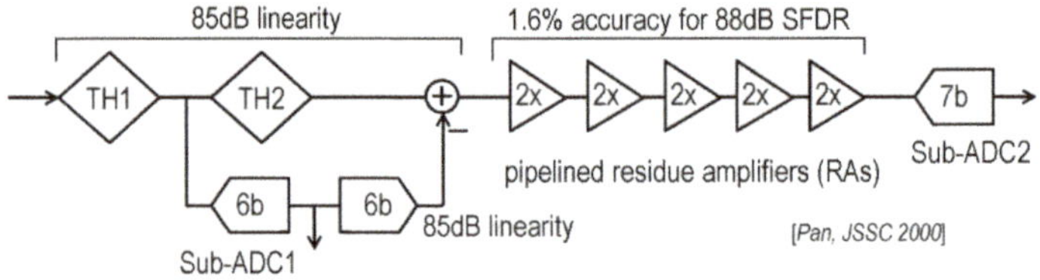

- Pipelined 2-step conversion achieves 50MS/s and 85dB SFDR in 0.6μm CMOS with V_{dd} lowered to 3.3V from standard 5V.
- 6b first step removes SFDR bottleneck imposed by the residue gain inaccuracy with the T/H and DAC linearity determine the SFDR.
- Pipelined RAs remove tradeoff between amplifier gain and BW.
- Signal Folding reduces the 6b sub-ADC and the DAC complexity.
- Optimal spatial filtering reduces input cap of each sub-ADC by 7X.

71 A 3.3V 50MS/s 12b ADC with 85dB SFDR in 0.6μm CMOS

To push the speed and linearity envelops without digital calibration in the then most advanced 0.6um CMOS technology, this ADC departs from the traditional 1.5b/stage pipeline architecture using 6b high-resolution first pipeline stage to desensitize the ADC SFDR to the inter-stage gain error. It relies heavily on pipelining technique to achieve the high throughput. A lot of power penalty is paid to get the premium performances. For example, TH1 burn a lot of power to drive a large load from the 6b flash ADC. TH2, necessary for pipelining the 6b coarse quantization, adds noise and a lot of power running at full speed and linearity. The high gain residue amplification is broken to pipelined stages for full speed operation with tightened accuracy for each stage causing a lot of power. The following slides show how this can evolve to the most efficient SAR ADC with technology advancement.

72 Evolution with Technology: Remove 1st T/H

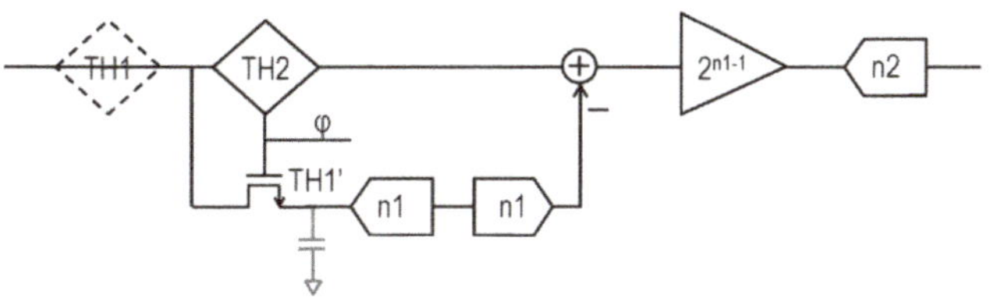

- A separate T/H driving the n1-bit sub-ADC needs only n1-bit accuracy and can use simple open-loop T/H.
- It is more feasible in advanced technology to meet the n1-bit accuracy requirement with simple T/Hs.
  - Better clock for timing alignment between TH2 and TH1'
  - Less error from sampling switch imperfections such as charge injection, bandwidth limitation, distortion, etc.
  - Possible digital calibration for the timing alignment and gain matching

It is too costly to drive the n1-bit coarse ADC with a full-linearity T/H TH1. Indeed, TH1 can be skipped using a separate low-cost T/H TH1′ to drive the n1-bit coarse ADC, as long as TH1′ samples the input at the same time as TH2 with n1-bit accuracy. The timing alignment is much easier to realize in a faster CMOS process that allows sharper clock edge and small aperture time for accurate sampling. Though TH2 should still run at full speed with full linearity, TH1′ may use simple open-loop topology to save power. It is noted that if the inter-stage gain can be digitally calibrated or trimmed to the required accuracy, there is no need to increase the first stage resolution. However, in the slow process before 65nm CMOS, the digital calibration was expensive compared to pure analog solutions.

73 Evolution with Technology: Remove 2nd T/H

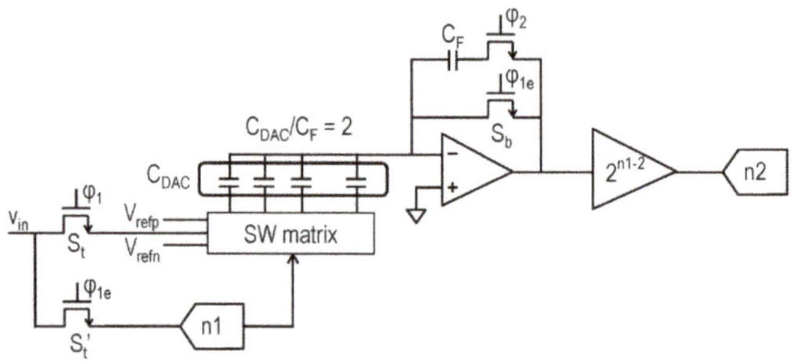

- The n1-bit sub-ADC latency is reduced with fast comparator decision in advanced technology, obviating the need for the 2nd T/H TH2.
- The reconstruction DAC capacitor top plates are directly driven by the input, eliminating the standalone input T/H and the associated KT/C noise, resulting in the so-called SHA-less ADC architecture. *[Lee, JSSC 2011]*

TH2 can also be skipped with the input sampled to the DAC capacitor directly using switch S_b that opens before S_t for bottom plate sampling. The sampling is aligned with the n1-bit coarse ADC top-plate sampling using the same clock ϕ_{1e}. Again, the timing alignment necessitates small aperture time that is feasible in fast processes. The pipelining effect of TH2 is now not needed because the n1-bit reconstruction DAC is not in the critical path, and the n1-bit coarse ADC latency becomes negligible given much smaller regeneration time constant in fast processes. Effectively, TH1 and TH2 are merged with the MDAC, resulting in the so-called SHA-less pipelined ADCs, where SHA stands for standalone sample and hold amplifier.

74 Evolution with Technology: Remove the RAs

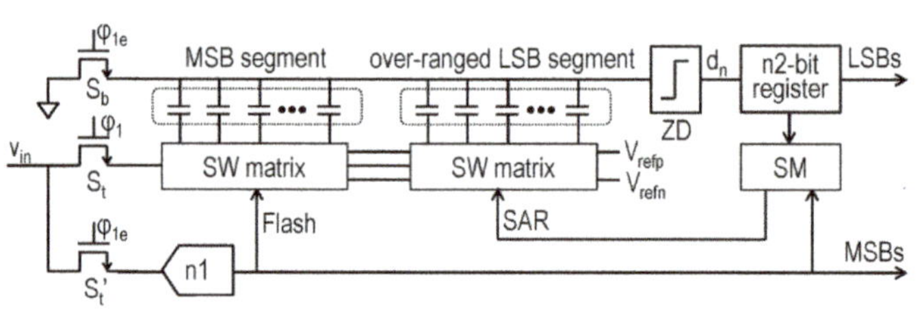

- The 2nd sub-ADC is implemented in a SAR architecture that does not need residue amplification for using only one comparator for the entire conversion.
- The SAR in an advanced process can achieve the required conversion rate.
- The 1st sub-ADC output directly control the MSB segment of the SAR DAC.
- The SAR LSB segment is over-ranged using radix < 2 for the DAC array.
- The 2-step pipeline ADC evolves to a sub-ranged SAR ADC *[Lin, VLSI 2010]* ...
- Technology has a drastic impact on ADC architecture evolutions and FoMs.

74 Evolution with Technology: Remove the RAs

The inter-stage residue amplification overcomes the comparator offsets of the succeeding flash ADC, but this becomes redundant if the fine ADC is implemented in the SAR architecture using only one comparator whose offset contributes only an output DC component. Again, the process is assumed fast enough that the SAR ADC speed can match the flash counterpart in the slow process. The SAR CDAC is combined with the MSB reconstruction DAC in the track phase to sample the input together using the same bottom plate switch S_b. After sampling, the MSBs are resolved in one step, followed by one LSB bit per step. The pipeline ADC evolves to the typical 1b/cycle SAR ADC with the MSBs reduced to only one bit. This demonstrates convergence to the most efficient serial binary search algorithm by removing all the auxiliary analog processing needed for low f_t.

75 Pipelined Sub-ranging SAR ADC

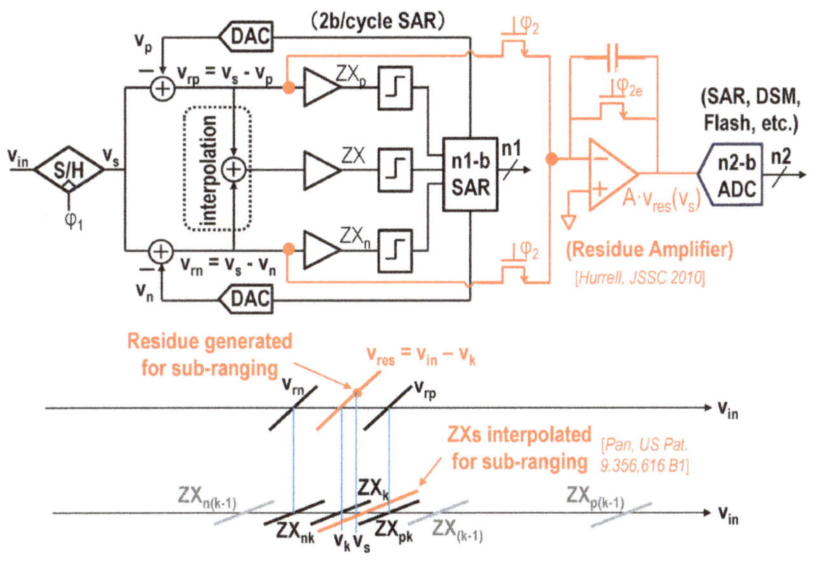

As technology advances, new wideband applications emerge, and time-interleaving is commonly used to overcome the SAR ADC speed bottleneck. Traditional MDAC based pipelining is also applied to one-bit or multi-bit/cycle SAR ADCs to enhance the lane speed. This reversal to the costly precision analog processing defeats the benefits of SAR ADCs involving only ZX signals. Residue generation and amplification with MDACs are usually based on closed-loop op-amps assisted with various gain calibrations. The required high DC gain for the opamp is not compatible with technology scaling. To preserve ZX only operation for pipelined SAR ADCs, the ZX signals readily available from the inputs or outputs of adjacent comparators of a multi-bit/cycle SAR ADC can be sampled and interpolated for fine quantization. Interpolation between two closely spaced ZX signals eliminates the need for accurate single residue generation and amplification.

76 Pipelined SAR ADC using Only ZX Signals

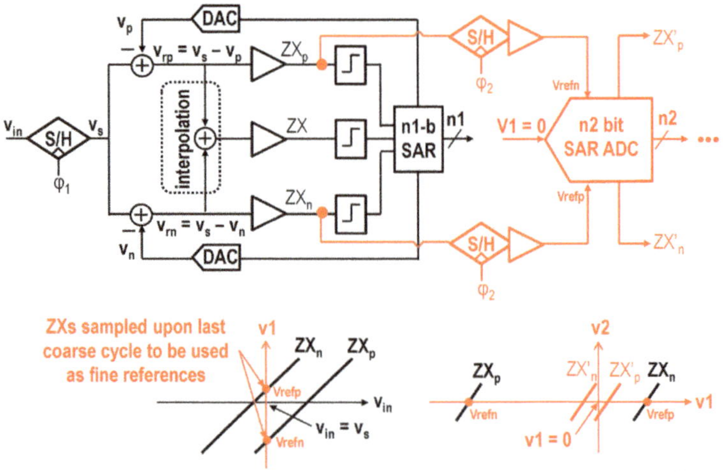

The SAR ADC is also used for the pipelined fine quantization to avoid reversal to the low efficiency interpolation based flash-like architecture. The two ZX signals from the last coarse conversion cycle are sampled and fed to the fine SAR ADC as references V_{refp} and V_{refn} while the input is fed with fixed zero. The fine SAR can be either one-bit or multi-bit/cycle, though the former is preferred for the last pipeline stage, as the single comparator offset contributes only a DC output component. The fine reference offsets can be corrected with the coarse DACs extended in resolution. The first SAR should resolve enough bits so that the inherent linearity of the sampled ZXs are sufficient, resulting in a true pipelined SAR architecture that enhances speed and resolution without reversing to precision analog signal processing. This is different from the traditional dual-residue ADCs where the residue magnitude can not be arbitrarily reduced due to exponential growth in the complexity of the flash-type coarse sub-ADC.

77 Quantizer Trend

- Quantization is a search process ranging from the most efficient serial binary search to the fastest brute-force parallel search.
- Parallel and serial-parallel search complexity is reduced with signal folding at cost of delicate analog signal processing (e.g. MDAC).
- Serial binary search combined with comparator reuse results in the most efficient SAR ADCs at the expense of conversion rate.
- SAR ADC speed bottleneck can be overcome by time-interleaving many SAR ADCs, with the mismatch errors digitally calibrated.
- Digital calibration and error correction overheads diminish in advanced CMOS processes, leading to SAR ADC popularity.
- SAR ADC resolution, speed, efficiency are enhanced with sub-ranging, pipelining, optimal switching, time-based operations, etc.

78 Closing Remarks

- ADC trends are driven by technology & application advancements
  - Low digital cost: more digital calibration, less precision analog
  - High f_t: more oversampling/signal processing, less pipelining/folding
  - High performance & low power: application specific system design
- ADC algorithms converge to the most efficient quantization
  - Serial binary search coupled with error correction and signal processing
  - Oversampling based statistic search coupled with noise shaping
- ADC architectures converge to simple quantizers coupled with sophisticated signal processing and error corrections
  - Time-interleaved asynchronous SAR with digital calibration
  - Delta-sigma ADCs including VCO based quantizers [Straayer, JSSC 2008]
- ADC circuits converge to simple (inverter based) ZX operations
  - Inverter ring as latch for ZX detection and as VCO for integration
  - Inverter chain as ZX amplifiers and time-domain ZX generators, etc.

79 References

1. Hui Pan, at al., "A 3.3-V 12-b 50-MS/s A/D Converter in 0.6-μm CMOS with over 80-dB SFDR," *IEEE J. Solid-State Circuits*, vol. 35, pp. 1769-1780, Dec. 2000.
2. R. A. Gomez. "Theoretical Comparison of Direct-Sampling vs. Heterodyne RF Receivers," *IEEE Trans. Circuits Syst. I*, vol. 63, pp. 1276-1282, Aug. 2016.
3. D. E. Crivelli, et al., "Architecture of a Single-Chip 50 Gb/s DP-QPSK/BPSK Transceiver With EDC for Coherent Optical Channels," *IEEE TCAS-I*, vol. 61, pp. 1012-1025, April 2014.
4. Hui Pan, Asad. A. Abidi, "Spectral Spurs due to Quantization in Nyquist ADCs," *IEEE Trans. Circuits Syst. I*, vol. 51, pp. 1422-1439, Aug. 2004.
5. Hui Pan, Asad. A. Abidi, "Signal Folding in A/D Converters," *IEEE Trans. Circuits Syst. I*, vol. 51, pp. 3-14, Jan. 2004.
6. Hui Pan, Asad. A. Abidi, "Spatial Filtering in Flash A/D Converters," *IEEE Trans. Circuits Syst. II*, vol. 50, pp. 424-436, Aug. 2003.
7. W. C. Black, Jr. and D. A. Hodges, "Time Interleaved Converter Arrays," *IEEE J. Solid-State Circuits*, vol. SC-15, No. 6, pp. 954–961, Dec. 1987.
8. S. H. Lewis and P. R. Gray, "A pipelined 5-MSample/s 9-bit analog-todigital converter," *IEEE J. Solid-State Circuits*, vol. 22, pp. 1022 - 1029, Dec. 1980.

9. Hui Pan, Ichiro Fujimori, "Hierarchical parallel pipelined operation of analog and digital circuits," US Patent no. 7012559 B1, Sept. 24, 2004.

10. Tetsuya Iizuka, Takaaki Ito, and Asad A. Abidi, "Comprehensive Analysis of Distortion in the Passive FET Sample-and-Hold Circuit," *IEEE TCAS-I,* vol. 65, pp. 1157-1173, April 2018.

11. B. D. Smith, "An unusual analog-digital conversion method," *IRE Trans. Instrum. Meas.*, pp. 155–160, June 1956.

12. R. J. van de Plassche and R. E. J. van de Grift, "A high-speed 7 bit A/D converter," *IEEE J. Solid-State Circuits*, vol. SC-14, pp. 938–943, Dec. 1979.

13. R. E. J. van de Grift, at al., "An 8-bit video ADC incorporating folding and interpolation techniques," *IEEE J. Solid-State Circuits*, vol. SC-22, pp. 944–953, Dec. 1987.

14. C. W. Moreland, "An 8b 150MSample/s serial ADC," in *ISSCC Dig. Tech. Papers*, San Francisco, CA, Feb. 1995, pp. 210–211.

15. B. Verbruggen, et al., "A 2.2mW 5b 1.75GS/s Folding Flash ADC in 90nm Digital CMOS", *ISSCC Dig. Tech. Papers*, pp. 252-253, Feb. 2008.

16. A. G. W. Venes, et al., "An 80-MHz, 80-mW, 8-b CMOS folding A/D converter with distributed track-and-hold preprocessing," *IEEE JSSC*, vol. 31, pp. 1846–1853, Dec. 1996.

17. K. Bult and A. Buchwald, "An embedded 240-mW 10-b 50 MS/S CMOS ADC in 1-mm ," *IEEE J. Solid-State Circuits*, vol. 32, pp. 1887–1895, Dec. 1997.

18. P. Vorenkamp and R. Roovers, "A 12-b, 60-MSamples/s cascaded folding and interpolating ADC," *IEEE J. Solid-State Circuits*, vol. 32, pp. 1876–1886, Dec. 1997.

19. S. H. Lewis, et al., "A 10b 20-Msample/s analog-to-digital converter," *IEEE J. Solid-State Circuits*, vol. 27, pp. 351–358, Mar. 1992.

20. M. J. M. Pelgrom, A. C. J. Duinmaijer, and A. P. G.Welbers, "Matching properties of MOS transistors," *IEEE J. Solid-State Circuits*, vol. 24, pp.1433–1440, Oct. 1989.

21. A. M. Abo and P. R. Gray, "A 1.5-V 10-bit 14.3-MS/s CMOS pipeline analog-to-digital converter," *IEEE J. Solid-State Circuits*, vol. 34, pp.599–606, May 1999.

22. C. W. Mangelsdorf, "A 400-MHz input flash converter with error correction," *IEEE J. Solid-State Circuits*, vol. 25, pp. 184–191, Feb. 1990.

23. B.-S. Song, et al., "A 12-bit 1-Msample/s capacitor error-averaging pipelined A/D converter," *IEEE J. Solid-State Circuits*, vol. 23, pp. 1324–1333, Dec. 1988.

24. J. Montanaro, et al., "160-MHz, 32-b, 0.5-W CMOS RISC Microprocessor," *IEEE J. Solid-State Circuits*, vol. 31, pp. 1703–1713, Nov. 1996.

25. A. A. Abidi and H. Xu, "Understanding the Regenerative Comparator Circuit (Invited)," in *Proc. CICC*, 2014.

26. D. Schinkel, et al., "A Double-Tail Latch-Type Voltage Sense Amplifier with 18ps Setup+Hold Time," *ISSCC Dig. Tech. Papers*, pp. 314-315, Feb. 2007.
27. Y-M Lin, B. Kim, and P. R. Gray, "A 13-b 2.5-MHz Self-Calibrated Pipelined A/D Converter in 3μm CMOS," *IEEE J. Solid-State Circuits*, vol. 26, pp. 628-635, April 1991
28. H-S Lee, D. A. Hodges, and P. R. Gray, "A Self-Calibrated 15 Bit CMOS A/D Coverter," *IEEE J. Solid-State Circuits*, Vol. SC-19, pp. 813 - 819, Dec. 1984
29. J. McNeil, et. al., ""Split ADC" Architecture for Deterministic Digital Background Calibration of a 16-bit 1-MS/s ADC," *IEEE JSSC*, vol. 40, pp. 2437-2445, Dec. 2005.
30. A. N. Karanicolas, H-S Lee, and K. L. Bacrania, "A 15-b 1-Msample/s Digitally Self-Calibrated Pipeline ADC," *IEEE J. Solid-State Circuits*, vol. 28, pp. 1207-1215, Dec. 1993.
31. B. Murmann, "Digitally Assisted Data Converter Design," 2013 *Proceedings of the ESSCIRC*, pp 24 – 31, 2013.
32. B. Murmann and B. E. Boser, "A 12-bit 75-MS/s Pipeline ADC Using Open-Loop Residue Amplification," *IEEE J. Solid-State Circuits*, vol. 38, pp. 2040 -2050, Dec. 2003
33. D. Fu, et. al., "A digital background calibration technique for time-interleaved analog-to-digital converters," *IEEE J. Solid-State Circuits*, vol. 33, pp. 1904–1911, Dec. 1998.
34. J. Cao, et al., "A 500mW ADC-Based CMOS AFE with Digital Calibration for 10Gb/s Serial Links Over KR-Backplane and Multimode Fiber," *IEEE J. Solid-State Circuits*, vol. 45, pp. 1172 - 1185, June 2010
35. Y. Chiu, et al., "Least Mean Square Adaptive Digital Background Calibration of Pipelined Analog-to-Digital Converters," *IEEE Trans. Circuits Syst.* I, vol. 51, pp. 38 - 46, Jan. 2004
36. B. Razavi, "Design Considerations for Interleaved ADCs," *IEEE J. Solid-State Circuits*, vol. 48, pp. 1806 - 1817, August 2013
37. A. Abidi, et al., "Adaptive Calibration of Time-Interleaved A/D Converters," in Proc. VLSI Circuits Short Course, Course 1: *Advanced Data Converter and Mixed-Signal Design*, 2014
38. I. Galton, "Digital Cancellation of D/A Converter Noise in Pipelined A/D Converters," *IEEE Trans. Circuits Syst.* II, vol. 47, pp. 185 - 196, March 2004.
39. J. Doernberg, H-S Lee, D. A. Hodges, "Full-Speed Testing of A/D Converters," *IEEE J. Solid-State Circuits*, vol. 19, pp. 820-827, Dec. 1984.
40. J. Kim, et. al., "Simulation and Analysis of Random Decision Errors in Clocked Comparators," *IEEE Trans. Circuits Syst.* I, vol. 56, pp. 1844-1857, August. 2009.
41. S-W. M. Chen and R. W. Brodersen, "A 6b 600MS/s 5.3mW Asynchronous ADC in 0.13μm CMOS ," *ISSCC Dig. Tech. Papers*, pp. 2350 - 2359, Feb. 2006
42. P. J. A. Harpe, et al., "A 26 W 8 bit 10 MS/s Asynchronous SAR ADC for Low Energy Radios," *IEEE J. Solid-State Circuits*, vol. 46, no. 7, pp. 1585 - 1594, July 2011.

43. M. Yoshioka, et al., "A 10b 50MS/s 820μW SAR ADC with On-Chip Digital Calibration," *ISSCC Dig. Tech. Papers*, pp. 384 - 385, Feb. 2010.

44. K. Gulati and H.-S. Lee, "A High-Swing CMOS Telescopic Operational Amplifier," *IEEE J. Solid-State Circuits*, vol. 33, no. 12, pp. 2010 - 2019, Dec. 1998.

45. E. Sackinger and W. Guggenbuhl, "A High-Swing, High-Impedance MOS Cascode Circuit," *IEEE J. Solid-State Circuits*, vol. 25, no. 1, pp. 289 - 298, Feb. 1990.

46. C. C. Lee, et al., "A SAR-Assisted Two-Stage Pipeline ADC," ," *IEEE J. Solid-State Circuits*, vol. 46, pp. 859-869, April. 2011.

47. Y-Z, Lin, et al., "A 9-bit 150-MS/s 1.53-mW Subranged SAR ADC in 90-nm CMOS," *IEEE Symp. on VLSI Circuits*, pp. 243-244, Jun. 2010.

48. C. P. Hurrell, C. Lyden, D. Laing, D. Hummerston, and M. Vickery, "An 18 b 12.5 MS/s ADC With 93 dB SNR," *IEEE J. Solid-State Circuits*, vol. 25, no. 12, pp. 2647 - 2654. Dec. 2010.

49. C. L. Portmann and T. H. Y. Meng, "Power-Efficient Metastability Error Reduction in CMOS Flash A/D Converters," *IEEE J. Solid-State Circuits*, vol. 31, no. 8, pp. 1132 - 1140. Aug. 1996.

50. P. M. Figueiredo, "Comparator Metastability in the Presence of Noise," *IEEE Trans. Circuits Syst.* I, vol. 60, pp. 1286 - 1299, May 2013.

CHAPTER 09

Measurement of High-Speed ADCs

Lukas Kull

Cisco Systems, Switzerland

Danny Luu

Cisco Systems, Switzerland

* The authors did the research of the chapter at IBM Research - Zurich but they are currently at Cisco Systems, Zurich, Switzerland.

This chapter focuses on the measurement of analog-to-digital converter (ADC) prototypes using needle probes. We explain a typical measurement setup, and the design and setup decisions that need to be considered for a successful measurement.

High-speed high-resolution ADCs are challenging in design and measurement. High accuracy on data and clock input have to be ensured. To minimize nonlinearity, it is important to avoid signal-dependent errors. With these goals in mind, we discuss design trade-offs that have to be considered for high resolution ADCs. Furthermore, a full measurement setup using needle probing for high-speed ADCs is described.

On-chip memory enables accurate analysis of the output samples without the need for a high-speed digital interface or data decimation. An efficient shift-register-based approach for an on-chip memory to handle the large aggregated output data of highly interleaved ADCs is presented. The shift-register-based custom memory is compared with a register-based synthesized memory in terms of area and energy efficiency.

1 Outline

- Motivation
- ADC Block Diagram
- Measurement Setup
- Shift-Register Memory
- Summary
- References

2 Motivation

- All papers talk about implementation and measurement results, but we will talk about ***how to measure an ADC***
- Full measurement setup is presented
- What problems occurred in our measurement setup?
- How to deal with high-speed output?

This chapter presents the full measurement setup used to characterize our 6-12b high-speed ADC prototypes. Potential tradeoffs and problems that occurred in this setup are discussed. Finally, we show an on-chip memory implementation to capture the high-speed ADC output at full speed.

3 Needle Probing

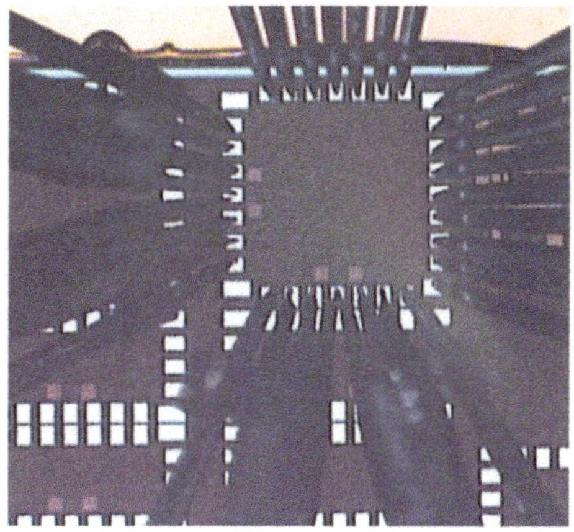

Note that all measurements were performed with needle probing. Some of the problems discussed here are more pronounced for needle probing compared to a PCB-based measurement setup.

4 ADC Block Diagram (I)

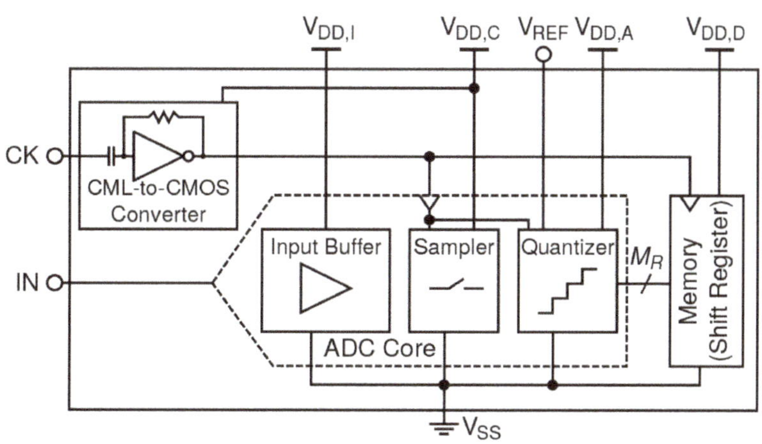

This slide shows a typical block diagram of an ADC testchip. The ADC core consists of an input buffer, a sample and hold circuit (sampler), and a quantizer. The CML-to-CMOS converter translates a sine-wave clock input to a CMOS square-wave clock. This CMOS clock is then used to drive the ADC and the output memory.

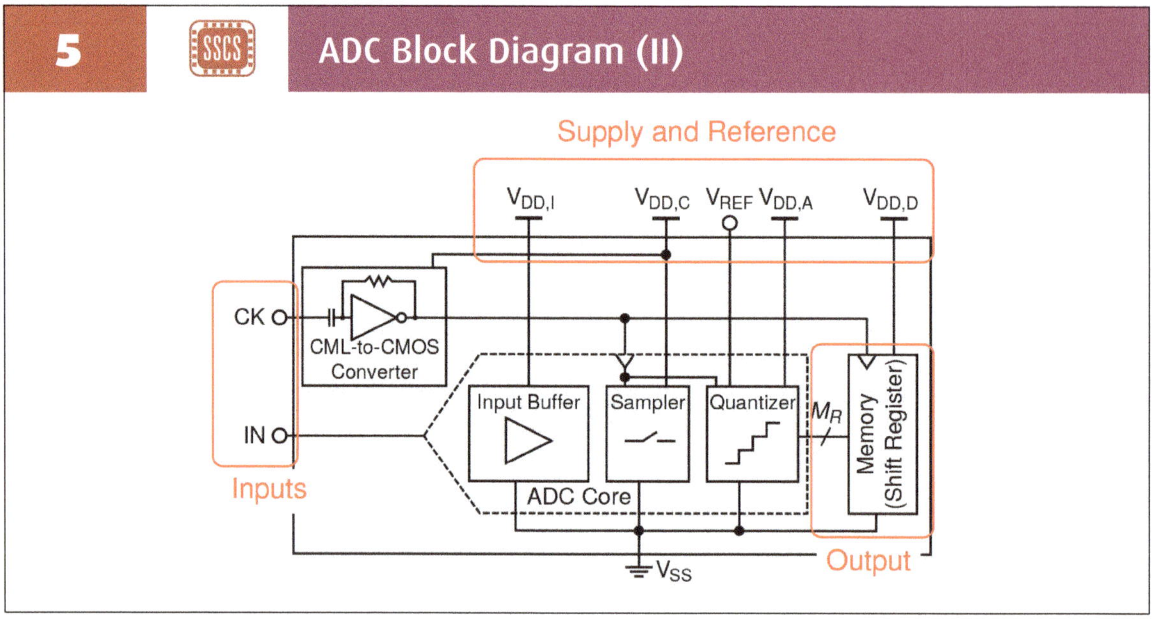

This chapter does not focus on the ADC core but on how to provide the input signals, the supply voltages, and on how to capture the output with an on-chip memory.

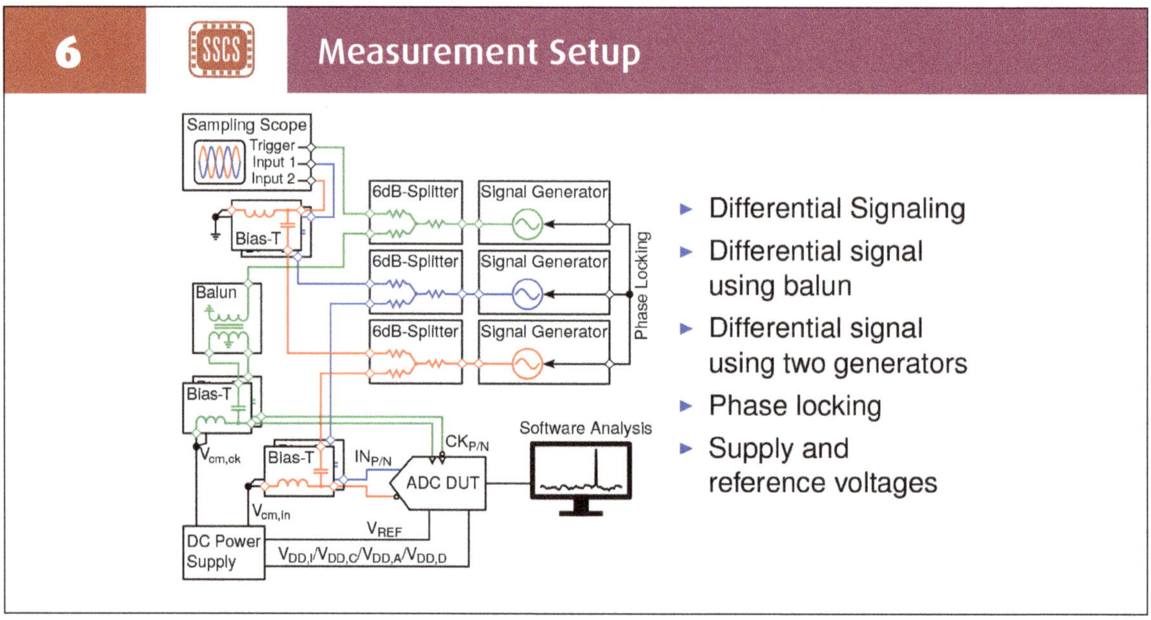

The full measurement setup is shown in this slide. Differential signaling is used for the input and clock signal.

7 Single-ended vs Differential Signaling

Differential signal	Single-ended signal
✓ Common-mode noise immunity	✓ Simple setup
✓ 100Ω termination not connected to supply	✗ Signal-dependent current drawn from 50Ω to ground
✓ Cancellation of even harmonics	
✓ Internal ADC blocks are differential	
✗ Signal generation with sufficient differential matching	

This table summarizes the advantages and disadvantages of differential versus single-ended signals. The main advantage of differential signals is the reduction of signal-dependent supply noise. Signal-dependent errors lead to non-linearity and can limit the overall performance of the ADC, especially for medium-to-high resolution converters. However, we need to generate differential signals with sufficient amplitude and phase matching. Two methods to generate differential signals will be discussed next.

8 Measurement Setup (I)

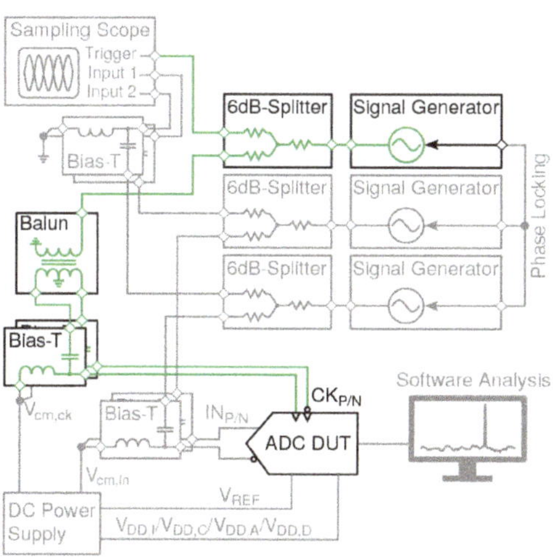

- Differential Signaling
- Differential signal using balun
- Differential signal using two generators
- Phase locking
- Supply and reference voltages

The first method to generate a differential signal is applied for the ADC clock and uses a single signal generator and a balun single-ended-to-differential converter. The common mode is set with bias-Ts. The clock signal is split and one output triggers the sampling scope and the other output connects to the ADC differential clock input. The sampling scope is required for the second method. Alternatively, an additional signal generator can be used to trigger the sampling scope.

9 Measurement Setup (II)

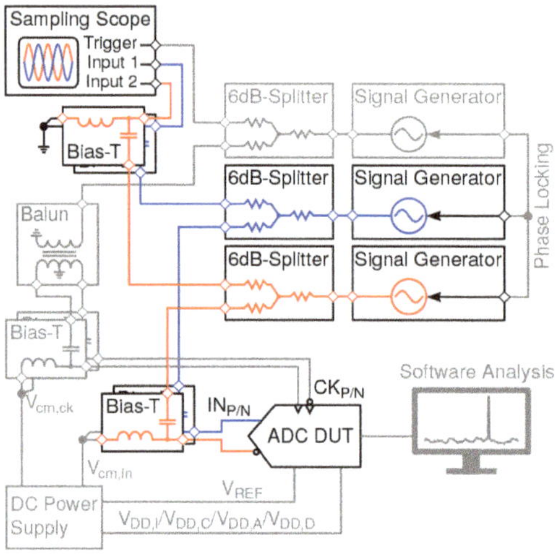

- Differential Signaling
- Differential signal using balun
- ► Differential signal using two generators
- Phase locking
- Supply and reference voltages

The second method requires two signal generators to create a differential signal. The output of each signal generator is split to drive the ADC input and the sampling scope simultaneously. The sampling scope is used to monitor and calibrate the differential amplitude and phase. Dummy bias-Ts are inserted towards the sampling scope to match the loss of the signal path towards the ADC.

10 Differential Signal Generation (I)

Balun	Two generators
✓ Simple setup	✗ Two generators and sampling scope
✗ Frequency-dependent amplitude and phase matching	✓ Any amplitude and phase matching possible
✗ Sampling scope can only measure amplitude	✓ Amplitude and phase exactly known through sampling scope
✓ No phase drift	✗ Two generators must be phase-locked

This table summarizes the advantages and disadvantages of the two presented methods to generate differential signals. The balun-method is used for the clock signal to simplify the setup because the differential matching of the clock is not as important to the ADC performance as the matching of the differential input signal.

11 Differential Signal Generation (II)

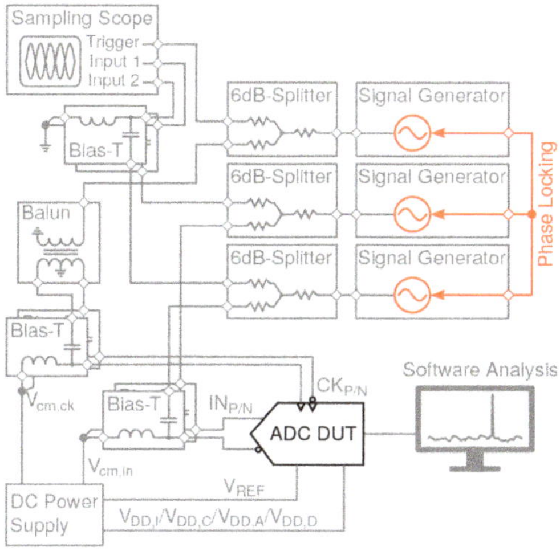

- Differential Signaling
- Differential signal using balun
- Differential signal using two generators
- **Phase locking**
- Supply and reference voltages

The two-generator method used for the input signal requires long-term phase locking to avoid phase drift between the two differential signals during the measurement.

12 Phase Locking Options (I)

- Standard 10 MHz reference locking
- High frequency reference locking (100 MHz, 1 GHz)
- Full coherency by running from a master PLL

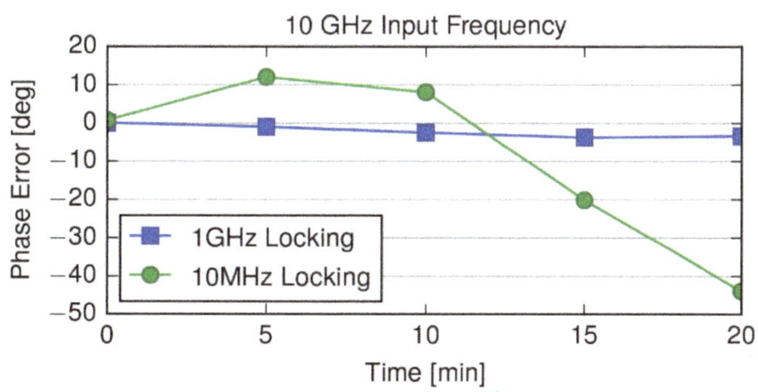

All signal generators offer 10 MHz locking of the internal reference. Depending on the signal generator model, there are different options for enhanced phase locking. Higher locking frequencies reduce phase drift significantly, as shown in the plot. The plot shows the phase drift with a standard 10MHz locking versus a high frequency 1GHz locking. Full coherency, when all generators run from a master PLL, would be the optimum solution, but this is useful for our application only if the phase between the two output signals can still be adjusted. This configuration was not available in our setup.

13 Phase Locking Options (II)

- 1 GHz locking after 20 minutes

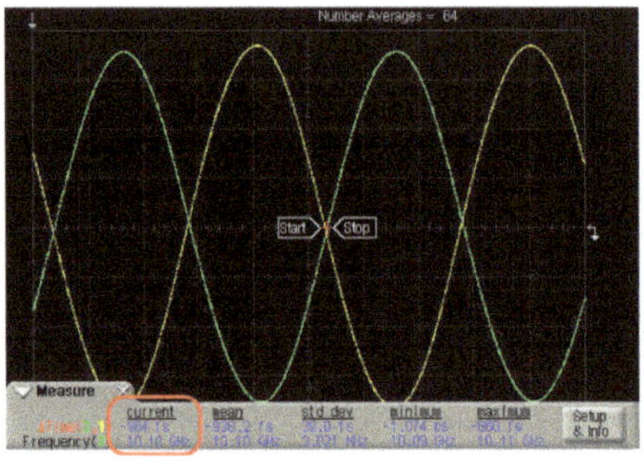

964 fs = 3.5° @ 10 GHz

- 10 MHz locking after 20 minutes

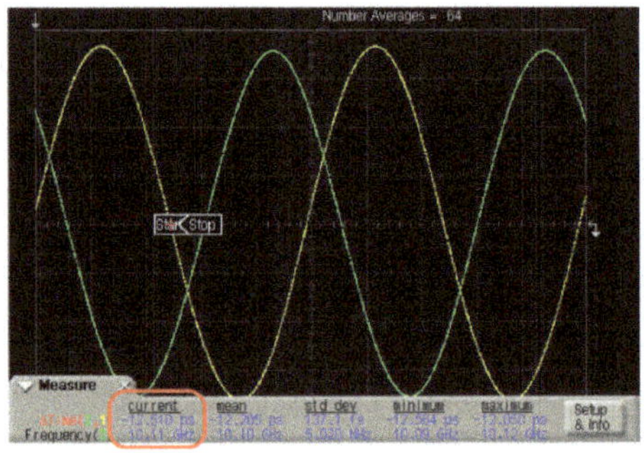

12.5 ps = 45° @ 10 GHz

This slide compares the phase error 20 minutes after calibration for 10 MHz and 1GHz locking frequency. The resulting phase error is 3.5° for 1 GHz and 45° for 10 MHz locking. Including foreground calibration of the ADC, measurements at a single frequency can take several minutes. If 10MHz locking is used, phase calibration is required about every minute to ensure no performance degradation. This results in much longer measurement times.

14 Phase Locking Options (III)

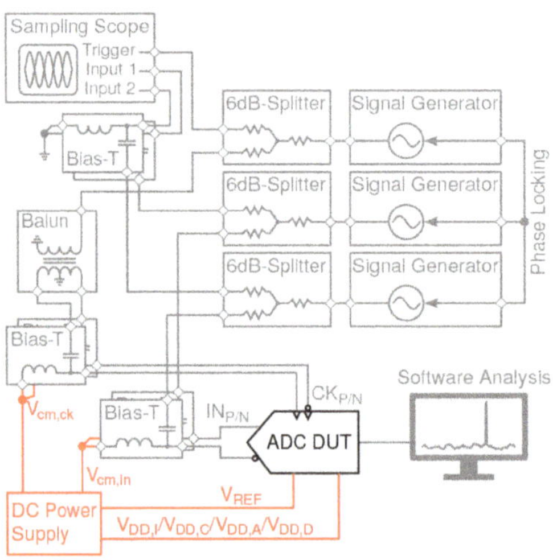

- Differential Signaling
- Differential signal using balun
- Differential signal using two generators
- Phase locking
- Supply and reference voltages

Several DC power supplies are required to generate the supply voltages, reference voltages and input common mode voltages.

15 Supply and Reference Voltages (I)

- $V_{DD,I}$: Input Buffer
  $V_{DD,C}$: Clock generation and sampling
  $V_{DD,A}$: ADC quantizer
  $V_{DD,D}$: Memory and digital interface
- Separate jitter-critical supply ($V_{DD,C}$), signal-dependent supplies ($V_{DD,I}$, $V_{DD,A}$, $V_{DD,D}$) and noisy supplies ($V_{DD,A}$, $V_{DD,D}$)

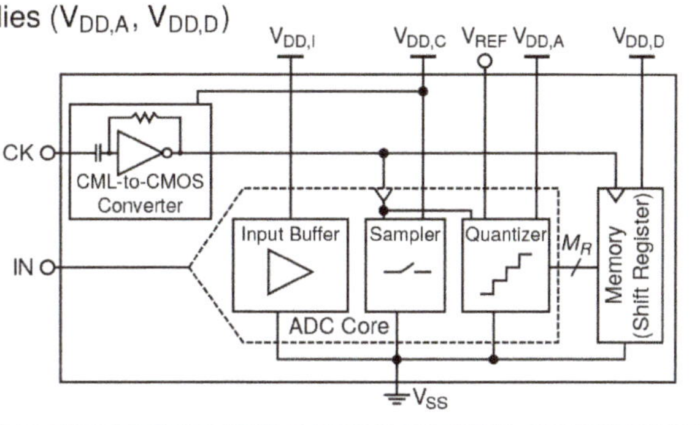

Our designs have up to four power domains: input buffer, clocking & sampling, ADC, and digital circuits. The idea is to separate jitter-critical supplies, signal-dependent supplies and noisy/switching supplies. Furthermore, the supply voltages can be set individually for each block to achieve best power efficiency. For example, the input buffer supply can be raised for more headroom and linearity, whereas the other supplies are set depending on the sampling speed.

16 Supply and Reference Voltages (II)

- Reference voltage V_{REF} sets the conversion gain
- Can be generated externally or on-chip
- Any noise on V_{REF} affects SNDR directly
- Separate V_{REF} for each sub-ADC
- Used for gain calibration in TI ADCs

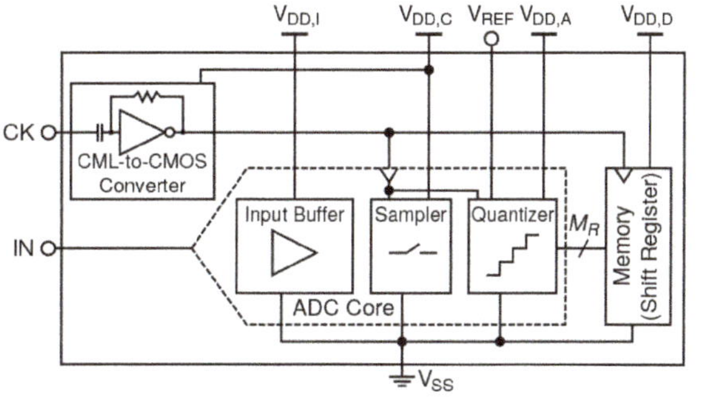

The reference voltage of the ADC that sets the conversion gain is the most sensitive supply. Any noise or ripple on the reference voltage directly changes the conversion gain and can affect SNDR. For time-interleaved ADCs, separate reference voltages per sub-ADC allow for gain calibration and avoid crosstalk between channels.

17 Comparison of External Reference with and without Decoupling (I)

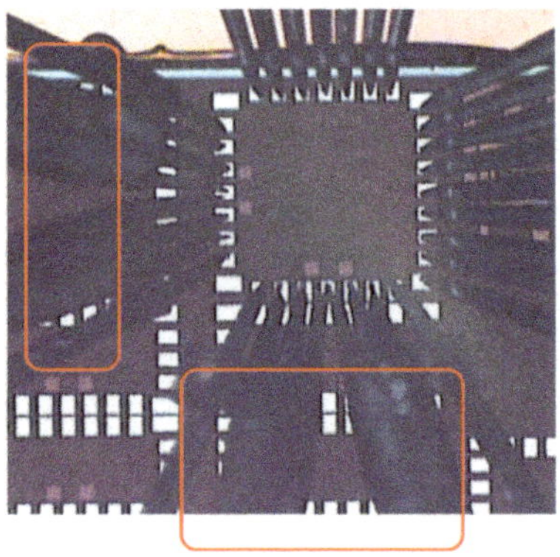

For a first prototype, the reference voltage was supplied externally. Two sets of measurements have been performed: first, using standard needle probes without decoupling capacitors, and second, using custom needle probes with decoupling capacitors at the tip of the needle that supplied the reference voltage.

18 Comparison of External Reference with and without Decoupling (II)

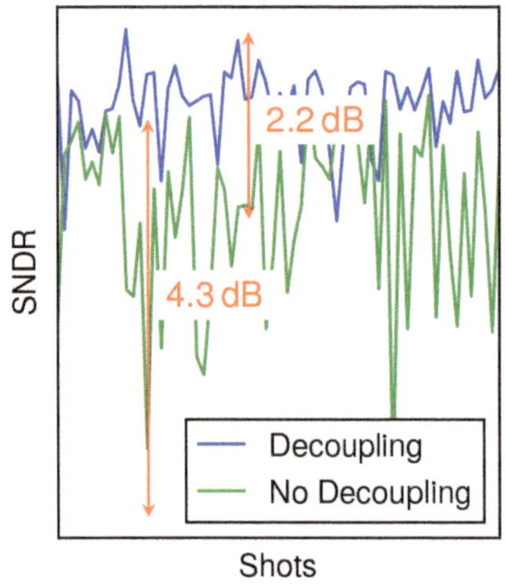

External electromagnetic interferences at a few megahertz were observed in our measurement lab. This interference coupled into the reference voltage and resulted in large variations in the measured SNDR, when no decoupling was used. Multiple snapshots at approximately one second intervals were taken for the figure on the left. To minimize SNDR degradation due to external interferences for future prototypes we implemented an on-chip reference generation.

19 Comparison of External Reference with and without Decoupling (III)

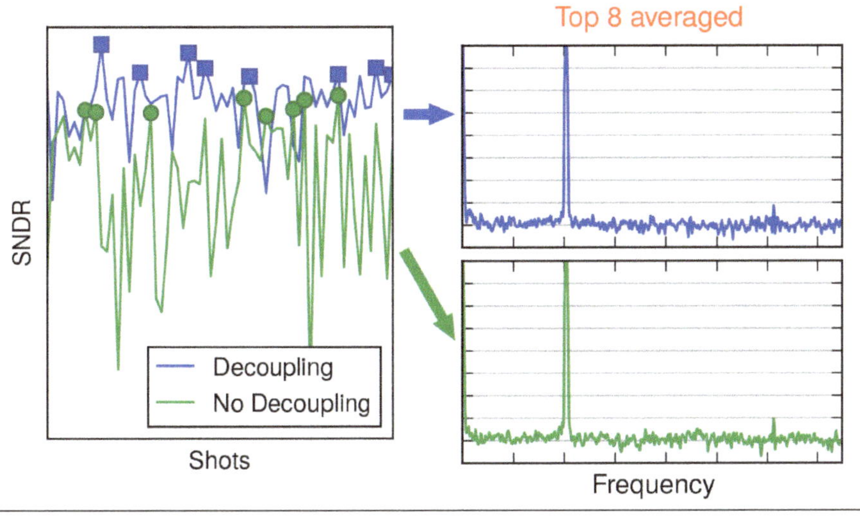

The power spectrum by averaging the top-8 shots for each measurement set shows similar performance. With decoupling, the noise floor is slightly reduced, which results in better SNDR.

20 Comparison of External Reference with and without Decoupling (IV)

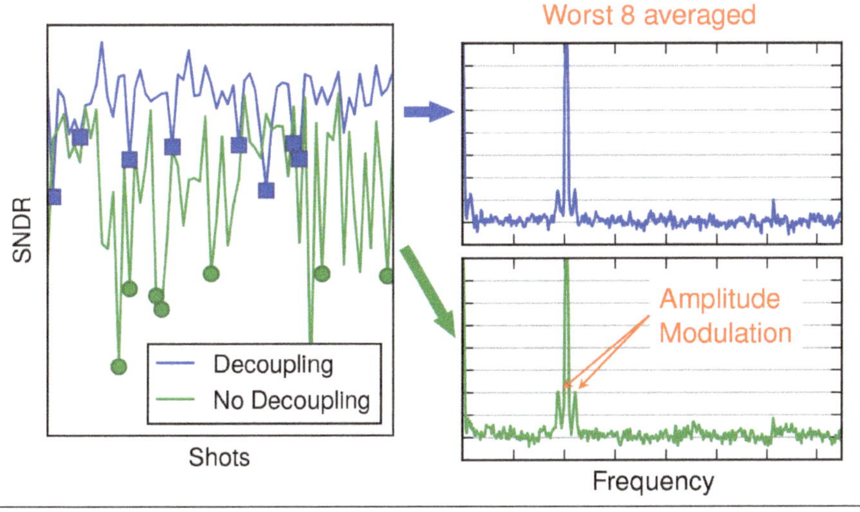

The power spectrum by averaging the worst-8 shots for each measurement set shows a clear difference when decoupling capacitors were used. The interference results in an amplitude modulation of the converted signal, which can be seen as spurs adjacent to the input signal. Not only is the interference greatly reduced by decoupling capacitors, but the chances of capturing a shot that is not corrupted be the interference is improved. Thus, measurement time can be reduced.

21 Measurement Setup

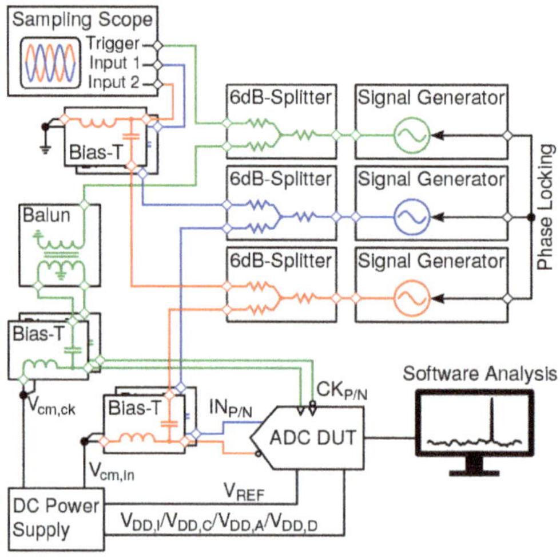

- Differential Signaling
- Differential signal using balun
- Differential signal using two generators
- Phase locking
- Supply and reference voltages

The full measurement setup is remotely controlled by software and enables an automated setup to calibrate the input signals, DC voltages and differential matching. Thus, the measurements are easily reproducible.

22 Shift-Register Memory (I)

- High-speed digital interface
- Data decimation
- On-chip memory

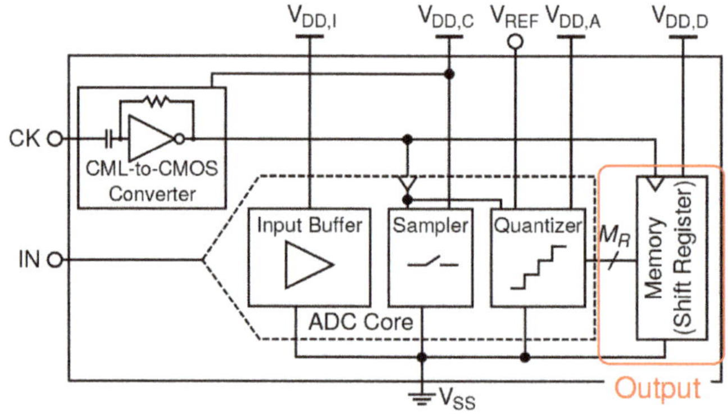

One of the challenges in high-speed ADC prototype measurements is to deal with the high data rate of the ADC output. The data can be sent off-chip at full rate. However, this requires several high-speed data interfaces, which are a challenging design by themselves and can occupy more area than the ADC. The processing of this data would likely require an ASIC or FPGA to capture and/or process data in real time. Data decimation sends only every N^{th} sample resulting in a much lower data rate but at the cost of information loss. Our prototypes implement an on-chip memory that stores the ADC output at full rate and is then read by a low data-rate digital interface.

23 Shift-Register Memory (II)

Requirements

- Sequential read and write
- High data-rates (100Gb/s – 1Tb/s)
- Area efficiency
- Low switching noise

Memory Options

- Digital synthesis with SRAM macro
- Digital synthesis with Registers
- Shift-register Memory

ADC outputs are sequential, thus there is no need for a random access memory. A first-in-first-out shift-register memory is sufficient to store and process the ADC output. The next slides present an easily customizable shift-register memory architecture that can reach high data rates at low area and power consumption. The presented memory can be implemented without access to SRAM macros or a digital synthesis flow.

24 Building Blocks (Flip-Flop) (I)

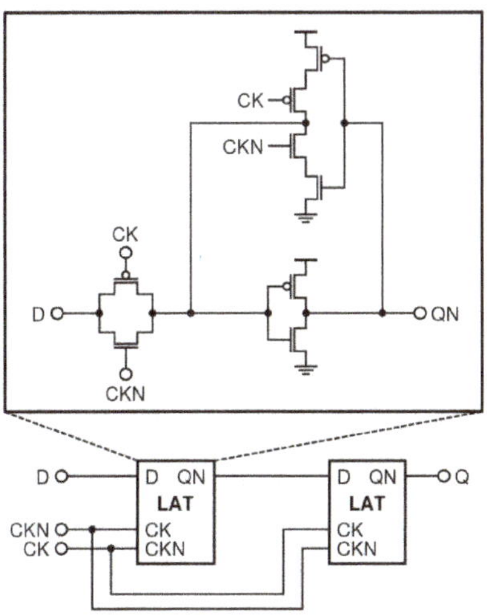

Flip-Flop (FF) using two 8-transistor inverting latch

- No special SRAM design rules are used
- very compact layout
- all drains and sources can be shared
- output shared with following input
- 2× smaller than a standard cell FF

The most important building block for this memory is the flip-flop (FF). The largest contributor to the overall area are the combined area of all FFs. Thus, the FF has been optimized for low area without the use of special SRAM DRC rules.

25 Building Blocks (Flip-Flop) (II)

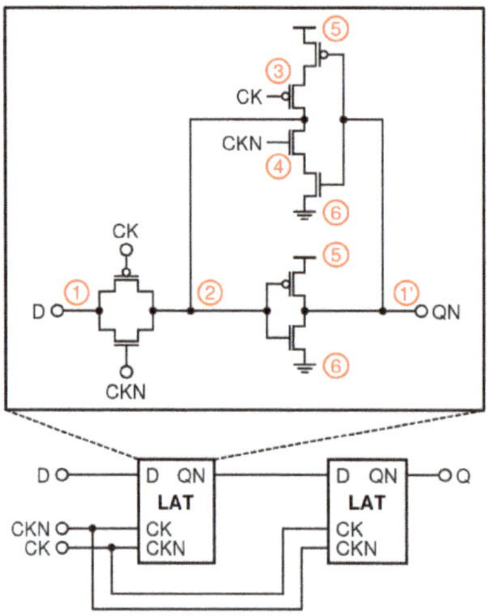

Flip-Flop (FF) using two 8-transistor inverting latch

- No special SRAM design rules are used
- very compact layout
- all drains and sources can be shared
- output shared with following input
- 2× smaller than a standard cell FF

The FF consists of two 8-transistor inverting latches. The latch structure was chosen such that every transistor shares drain and source with another transistor. This results in a very compact layout. The output of a latch overlaps with the input of the following latch. Thus, the resulting FF is array-placeable for a simple shift-register layout.

26 Building Blocks (Shift Register)

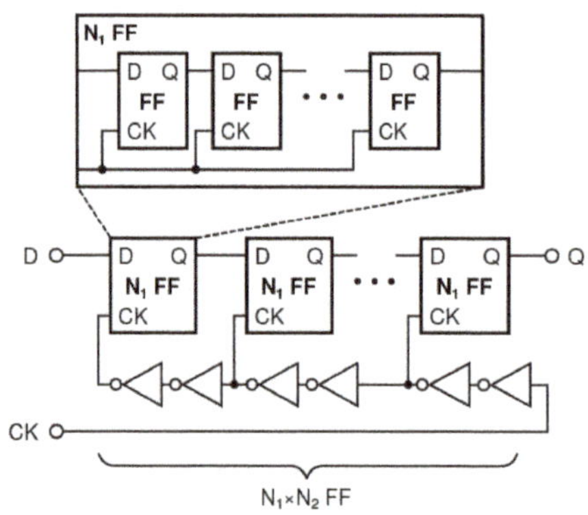

Shift Register (SR)

- N_1 FF clocked simultaneously
- Last FFs clocked first to avoid hold violations
- Split long SR into several clock-gated SR to avoid long inverter chain and reduce power

$N_1 \times N_2$ FFs are connected in series to form a shift-register. N_1 FFs are clocked simultaneously. The FFs at the end of the shift-register are clocked first to avoid hold violations because of clock skew. To achieve long SRs, multiple clock-gated $N_1 \times N_2$-SRs are combined.

27 Building Blocks (Clock-Gated Shift Register) (I)

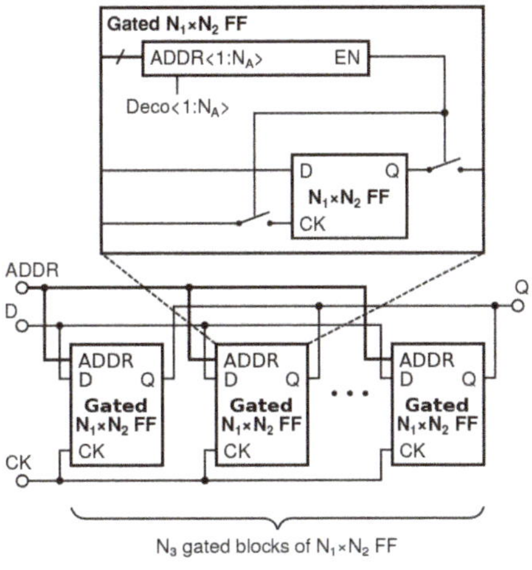

Clock-gated SR

- N_3 SR with gated clock and gated output
- Deco<1:N_A> is used to set the address

N_3 SRs with gated clocks and gated outputs are connected in parallel to increase the memory size. Deco<1:N_A> is used to configure a different address for each block, where $N_A = \log_2 N_3$. Thus, only 1 out of N_3 SRs is clocked to save power and reduce switching noise.

28 Building Blocks (Clock-Gated Shift Register) (II)

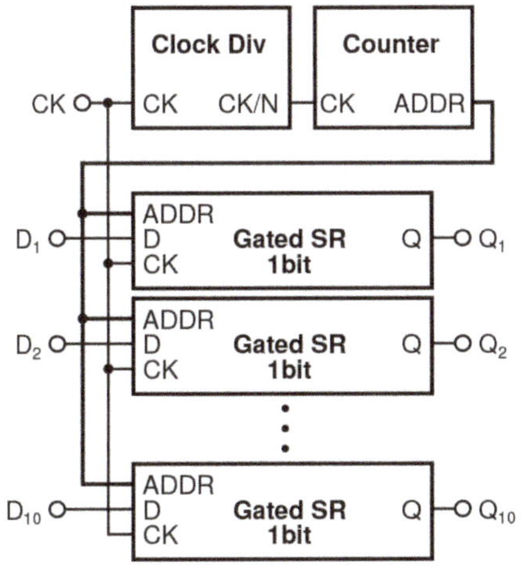

Clock-gated SR

- N_3 SR with gated clock and gated output
- Deco<1:N_A> is used to set the address
- ADDR generated by a counter
- ADDR shared among all bits
- Equivalent to an 10 bit SR (Example)

The address for the clock-gated SRs is generated by an N_A-bit counter, which is shared among all bits. The clock divider defines the number of words stored consecutively in one clock-gated SR before switching to the next one. This reduces the number of bit transitions on the address line. The shown structure is equivalent to a SR with a width of 10 bit and a depth of $N_1 \times N_2 \times N_3$ words.

29 Shift-Register Memory Interface

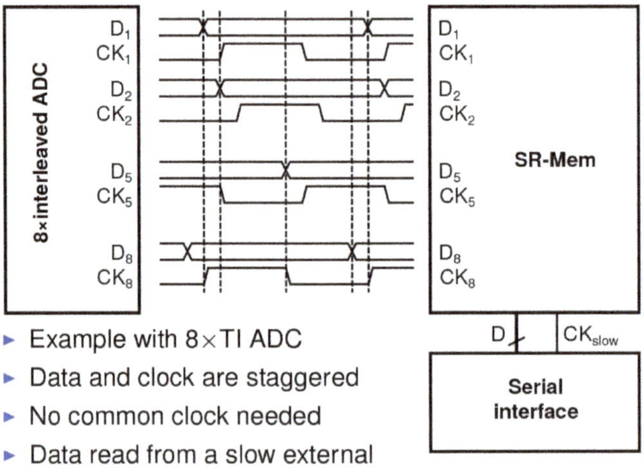

- Example with 8×TI ADC
- Data and clock are staggered
- No common clock needed
- Data read from a slow external interface

Multiple instances of the presented SR can be used to store the output of a time-interleaved (TI) ADC at full rate. Each sub-ADC provides the data and the corresponding clock staggered with respect to the other sub-ADCs. There is no need to synchronize all the data to a common clock, because each SR can operate independently. The stored data is then read by a slow serial interface.

30 Enabling and Disabling the Shift Register (I)

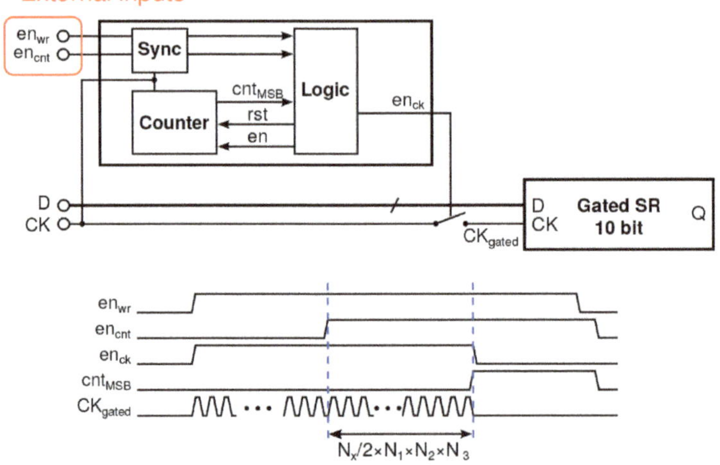

Two external signals are required for correct operation of a single SR, write enable en_{wr} and counter enable en_{cnt}. These signals are set by the slow serial interface and are then synchronized to the fast ADC clock. A $\log_2(N_x{\times}N_1{\times}N_2{\times}N_3)$-bit counter is used to disable the clock to the SR.

31 Enabling and Disabling the Shift Register (II)

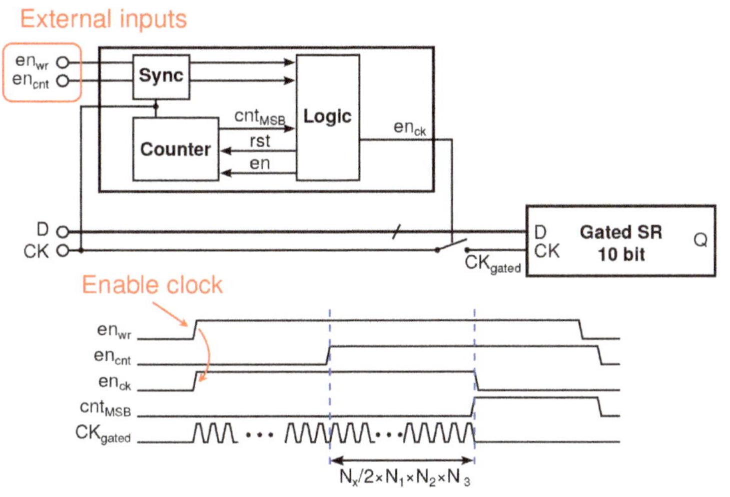

Write enable en_{wr} enables the clock to the SR, and the memory starts to store data. Any transient in the power grid due to enabling the memory settles before the counter is enabled. As long as the counter is not started, the continuous operation of the memory is not interrupted, and the power consumption of the memory can be measured.

32 Enabling and Disabling the Shift Register (III)

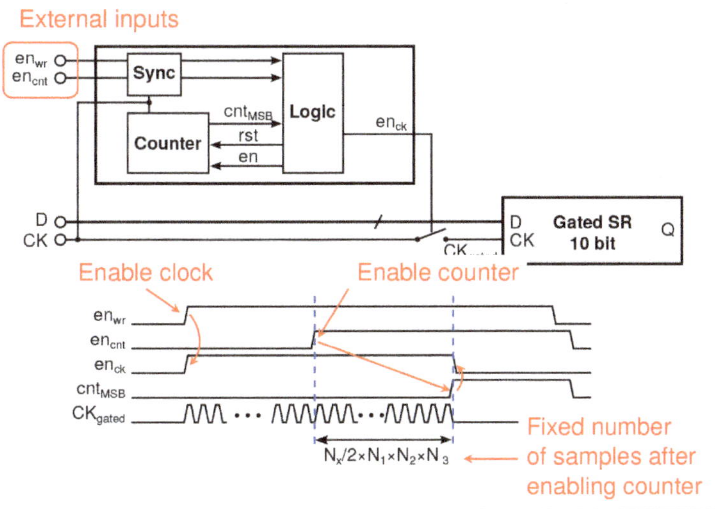

Counter enable en_{cnt} starts the counter. The MSB of this counter is used to disable the clock to the SR. In this way, we ensure that the clock of the SR is disabled after exactly $N_x/2\times N_1\times N_2\times N_3$. The last $N_1\times N_2\times N_3$ samples are retained in the SR and can be read by the external interface. N_x is chosen to be >2, such that any errors due to transients while enabling the counter are not captured in memory.

33 Enabling and Disabling the Shift Register (IV)

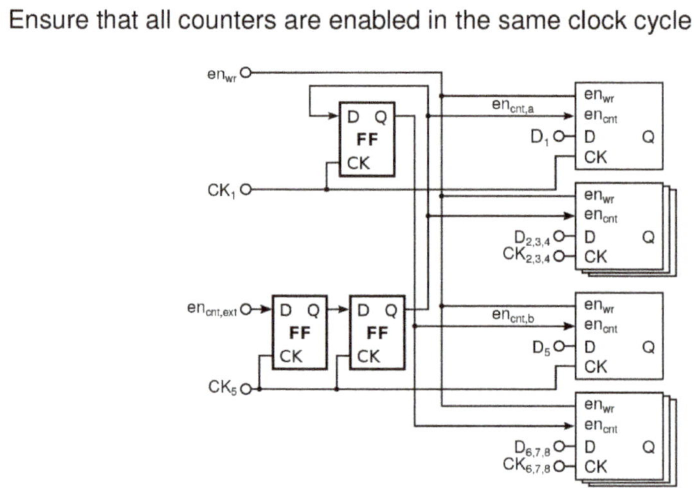

For correct operation in a TI ADC, all SRs must be enabled in a single sub-ADC clock cycle. This figure shows an 8×TI example that synchronizes the external counter enable signal $en_{cnt,ext}$, such that all counters are enable in a single cycle without any timing violations.

34 Enabling and Disabling the Shift Register (V)

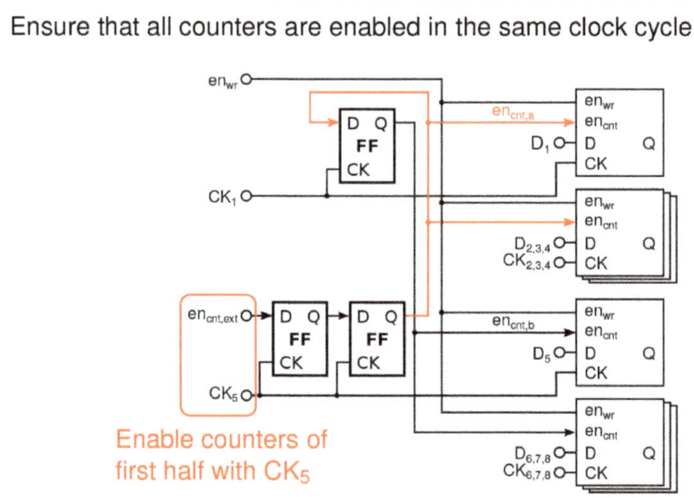

The first half of the SR memories is enable with $en_{cnt,a}$, which is synchronized with CK_5. This ensures that the available settling time for $en_{cnt,a}$ is half of the sub-ADC cycle time.

35 Enabling and Disabling the Shift Register (VI)

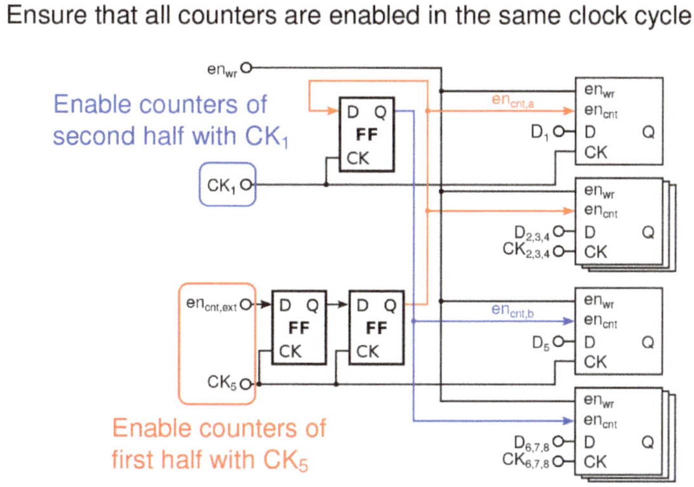

The second half of memories is enabled by $en_{cnt,b}$, which is synchronized by CK_1. Also here, $en_{cnt,b}$ has half the sub-ADC cycle time to settle.

36 Memory Comparison (I)

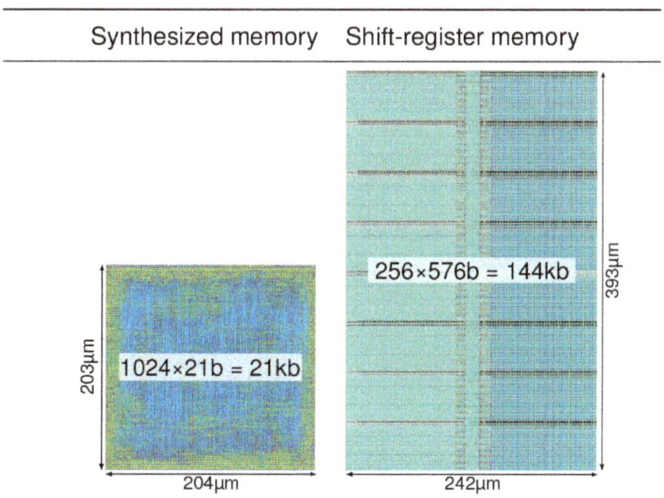

Two memory prototypes have been implemented in a 14nm CMOS FinFET process. The first is a register-based synthesized memory with random read access of 1024 21b-words. The second memory is the shift-register memory presented above for a 48×TI ADC. It implements $N_1 = 16$, $N_2=2$, $N_3=8$, $N_x=8$ and a width of 48×12b.

37 Memory Comparison (II)

	Synthesized memory	Shift-register memory
Nominal speed	1 GHz	1 GHz
Max. speed	2 GHz	2 GHz
Input width	21 bits	576 bits
Storage length	1024 samples	256 samples
Area	$203\times204\ \mu m^2$	$393\times242\ \mu m^2$
Power	20 mW at 0.9 V	95 mW at 0.7 V
Power/Gbps	0.95 mW/Gbps	0.16 mW/Gbps
Area/bit	$1.9\ \mu m^2$	$0.6\ \mu m^2$

- 6× improvement in power-efficiency
- 3× improvement in area-efficiency

This table compares the two memory implementations. The presented shift-register memory achieves 6× improvement in power-efficiency and 3× improvement in area-efficiency compared with the synthesized memory.

38 Summary (I)

Measurement of High-Speed ADCs

- Complex measurement setup is required
- Setup tradeoffs were highlighted

We presented a full measurement setup for high-speed ADCs and highlighted some tradeoffs and problems of the setup.

39 Summary (II)

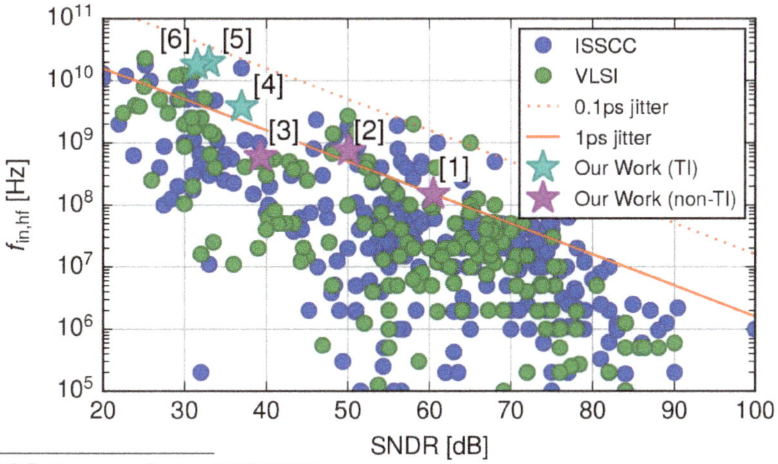

B. Murmann, "ADC Performance Survey 1997-2016"

This setup (or variants with small modifications) was used to characterize our ADC prototypes that show state-of-the-art performance in terms of speed and resolution.

40 Summary (III)

Measurement of High-Speed ADCs

- Complex measurement setup is required
- Setup tradeoffs were highlighted

Shift-Register Memory

- No digital design flow needed
- Area and power efficient
- Very modular
- Successfully measured in a 48×TI 10b ADC

In the second part, we presented a shift-register memory to aggregate the ADC output at full speed. Compared to a synthesized memory, the architecture is sufficiently simple and modular such that custom layout is feasible.

41 References

1. D. Luu, L. Kull, T. Toifl, C. Menolfi, M. Braendli, P. A. Francese, T. Morf, M. Kossel, H. Yueksel, A. Cevrero, I. Ozkaya, and Q. Huang, "A 12b 61dB SNDR 300MS/s SAR ADC with inverter-based preamplifier and common-mode-regulation DAC in 14nm CMOS FinFET," in 2017 *Symposium on VLSI Circuits*, Jun. 2017, pp. C276–C277.

2. L. Kull, D. Luu, C. Menolfi, M. Braendli, P. A. Francese, T. Morf, M. Kossel, H. Yueksel, A. Cevrero, I. Ozkaya, and T. Toifl, "A 10b 1.5GS/s pipelined-SAR ADC with background second-stage common-mode regulation and offset calibration in 14nm CMOS FinFET," in *ISSCC Dig. Tech. Papers*, Feb. 2017, pp. 474–475.

3. L. Kull, T. Toifl, M. Schmatz, P. A. Francese, C. Menolfi, M. Brandli, M. Kossel, T. Morf, T. M. Andersen, and Y. Leblebici, "A 3.1 mW 8b 1.2 GS/s single-channel asynchronous SAR ADC with alternate comparators for enhanced speed in 32 nm digital SOI CMOS," *IEEE Journal of Solid-State Circuits*, vol. 48, no. 12, pp. 3049–3058, 2013.

4. L. Kull, T. Toifl, M. Schmatz, P. A. Francese, C. Menolfi, M. Braendli, M. Kossel, T. Morf, T. Andersen, and Y. Leblebici, "A 35mW 8b 8.8GS/s SAR ADC with low-power capacitive reference buffers in 32nm digital SOI CMOS," in 2013 *Symposium on VLSI Circuits*, Jun. 2013, pp. C260–C261.

5. L. Kull, T. Toifl, M. Schmatz, P. A. Francese, C. Menolfi, M. Braendli, M. Kossel, T. Morf, T. M. Andersen, and Y. Leblebici, "A 90GS/s 8b 667mW 64 interleaved SAR ADC in 32nm digital SOI CMOS," in *ISSCC Dig. Tech. Papers*, Feb. 2014, pp. 378–379.

6. L. Kull, J. Pliva, T. Toifl, M. Schmatz, P. A. Francese, C. Menolfi, M. Braendli, M. Kossel, T. Morf, T. M. Andersen, and Y. Leblebici, "A 110 mW 6 bit 36 GS/s interleaved SAR ADC for 100 GBE occupying 0.048 mm2 in 32 nm SOI CMOS," in 2014 *IEEE Asian Solid-State Circuits Conference (A-SSCC)*, Nov. 2014, pp. 89–92.

CHAPTER 10

Practical Dynamic Element Matching Techniques for 3-level Unit Elements

Khiem Nguyen

Analog Devices Inc., USA

This chapter deals with the design of Dynamic Element Matching (DEM) techniques that are found in oversampling Sigma-Delta (SD) with 3-level unit elements. A brief review of the known 2-level DEM techniques will be provided first, followed by the detail analysis of 3-level counterparts. This chapter does not intend to burden the readers with any intensive mathematical analysis of the techniques but rather it provides the intuitive insights and practical solutions to the problems from an engineering point of view. Examples are shown to clarify non-trivial concepts, and they serve as the starting point for the readers to develop new and more effective solutions. The chapter concludes with a few application-dependent concepts of DEM that are very useful to reduce power consumption, silicon area, and enhance the design performance.

1 Outline

- **Brief background**
  - Motivations and Applications
  - 3-level vs. 2-level unit element: advantages and challenges
  - What problems DEM does and does not solve
- **Practical dynamic element matching (DEM) techniques**
  - 2-level DEM review
  - 3-level DEM
  - Limitations of each technique
  - Solution spaces
- **Higher-order DEM**
  - Issues in 2nd-order DEM
  - Solutions
- **Application specific concept of DEM**
- **Q and A**

This chapter was written with attentions focused on the practical design aspects of 3-level dynamic-element-matching (DEM) techniques. The analytical analysis offers the appropriate level detail for the readers to appreciate the math behind each technique so that the intuitive understanding, and actual implementation tricks and tips can be easily picked up.

The chapter begins with review of 2-level DEM concepts, terminologies, and observations as these serve as the foundation for 3-level DEM algorithms. Examples and schematics are provided to help clarifying non-trivial concepts. A few future works are presented in conclusion to offer the readers with directions for DEM development that are more a part of the overall application use case.

2 Applications of 3-level unit elements

2 Applications of 3-level unit elements

Oversampling SD data converters with 3-level unit elements are used in a wide variety of applications. They range from 16-bit resolution, 500kHz bandwidth in the ADC in the receiving path of automotive 24GHz radar systems, to 18b-bit high performance audio ADC and DAC in automotive infotainment applications such as cabin noise cancellation, active sound design (ADS), to portable low power audio codec application such as premium acoustic noise cancellation headphone.

3 What is Dynamic Element Matching?

- A technique to achieve perfect matching of analog elements ***over time***.

The key idea of dynamic element matching is the averaging effect resulted from using different inherently mismatched analog elements over time to represent the analog output, hence achieve the desirable linearity in the long term sense.

4 Euclid's 5th definition for plane geometry

"Through a pair of points in space, there exists one and only one straight line"

An interesting property that we learnt in basic middle-school math class back then.

5 What it means in the converter world...

• Can achieve "perfect linearity" if the circuit output has only 2 states

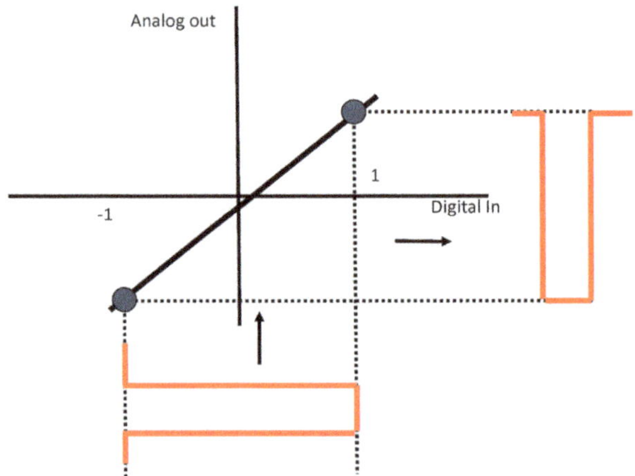

And what it actually means in the converter world. A perfectly straight-line transfer function created by only 2 possible output states is the key to reproduce an analog output with the highest level of linearity.

6 Year 1995...

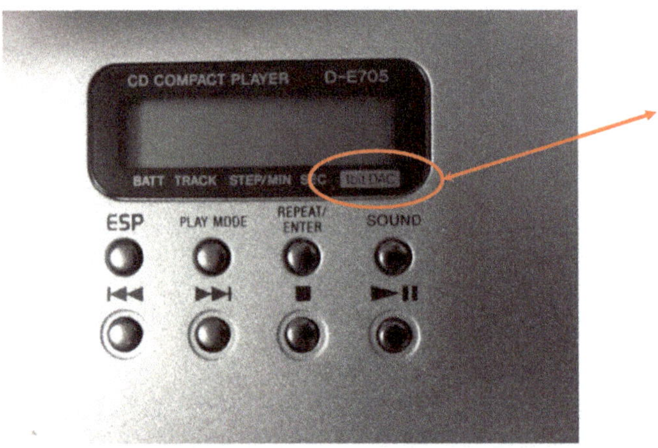

The "Perfect DAC" !

The 1990's was an era of 1-bit digital-to-analog (DAC) converters, popularized by portable CD players such as the one in the picture above.

7

Why multi-bit ?

- Drawback of single-bit
  - High out-of-band noise → Requires stringent post-analog filter → costly, bulky
  - Limited usable input range with high-order noise shapers → Challenging to improve dynamic range
  - Stringent bandwidth and slew rate requirements for reconstruction amplifier → Challenging to reduce power consumption
- Multi-bit
  - Solves all the above but introduces linearity issue due to element mismatch → Mismatch shaping is the remedy
    - Each element is 2-level
- *Why 3-level element ?*

The excellent linearity of the 1-bit converter comes with a cost of more stringent requirements on the post analog filter performance. The high out-of-band shaped noise energy, a result of the 1-bit architecture, if not adequately filtered out, can quickly cause severe distortion to the immediate following stage in the signal chain.

Furthermore, 1-bit output modulators need a higher-order noise transfer function (NTF) to achieve the desired SNR. Such aggressive NTF limits the usable input range to not cause modulator instability, and require stringent bandwidth and slew rate performance for the reconstruction amplifier.

Multi-bit modulators with 2-level unit elements offer a much more efficient solution to achieve the desired high SNR performance. Dynamic element matching (DEM) techniques offer an elegant solution to the non-linearities caused the mismatch of the analog elements in this type of converters.

This raises the question of why pursuing the use of 3-level unit element ?

8

Current steering DAC cell comparison

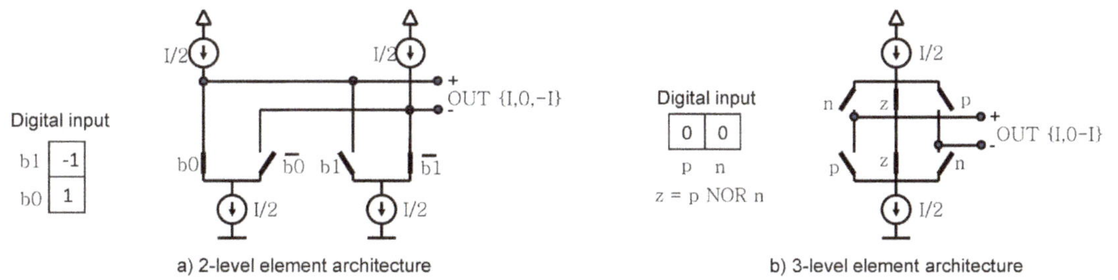

a) 2-level element architecture

b) 3-level element architecture

- **3-level element properties**
  + Power efficient
  + Best noise performance: No noise contribution in the "zero" state.
  - ***Needs a new dynamic element matching logic***
  - Needs a low power solution for ISI problem
  - Drain voltage modulation degrades THD

8 Current steering DAC cell comparison

Let's take a look at a simple comparison between a 2-level and a 3-level current steering DACs in Figure a) and b), both produce the same output {I,0,-I}.

For a zero digital input, the current sources in Figure b) are completely disconnected from the output, hence contributes no noise. It follows that at a low level signal condition (-60dB full scale), where the modulator outputs only 1 LSB, the 3-level architecture has a superior noise performance, at a lower power consumption level.

The use of 3-level unit element does require a new DEM design to handle the zero state, a new switch driver design to achieve low inter-symbol interference, and new remedies for the drain voltage modulation effect.

The scope of this chapter is limited to the solution space for 3-level DEM techniques, while other mentioned challenges have been addressed by several different techniques as found in the references of this chapter.

9 Example 1: High performance ΣΔ audio DAC

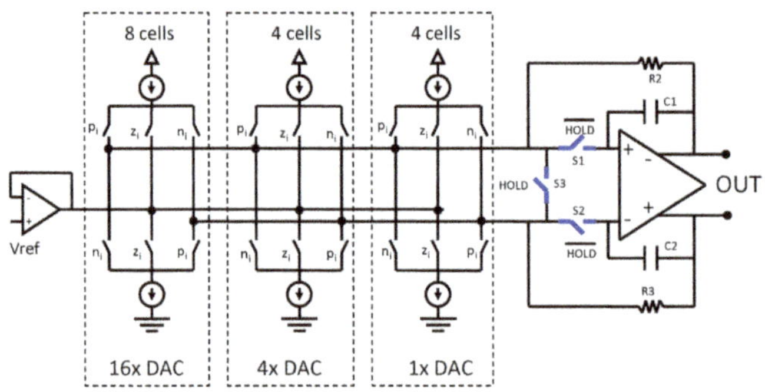

Process	0.18μm CMOS
Supply	1.8 V
Full scale differential output	0.9 Vrms
Digital power consumption	0.4 mW / channel
Analog power consumption	0.7 mW / channel
OSR	64
Clock frequency	3.072 MHz
SNR (A-weighted)	108 dB
THD+N	-97 dB

The above simplified block diagram provides the reader with a feel for how 3-level unit elements are used in a high performance low-power audio DAC. The 3 sub DACs' 16x, 4x and 1x reconstruct a 6-bit analog current output while the I-to-V output stage then converts the DAC current to the voltage, and provides the necessary drive capability needed for the external load.

The switches at the summing junction of the amplifier are used to mitigate the ISI problem, and a small, low-power amplifier is used to maintain the unused DAC cell drains at a known voltage. This amplifier helps to minimize the transients of the DAC cell drain voltage when being switched from and to the I-V amplifier summing junctions.

Three separate DEM logics control the selection of the DAC elements while a noise-shaped splitter provides a first-order high-pass shaping to the gain errors between the DAC segments.

10 SSCS

Example 2: High performance ΣΔ audio DAC

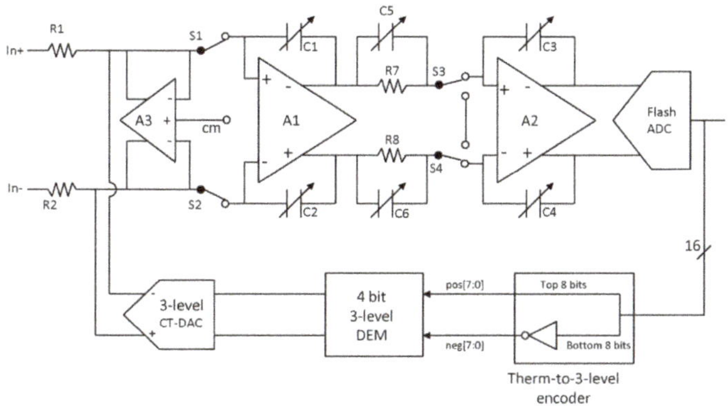

Process	0.18μm CMOS
Supply	3.3 V
Digital power consumption	0.4 mW / channel
Analog power consumption	2 mW / channel
OSR	128
Clock frequency	6.144 MHz
SNR (A-weighted)	111 dB
THD+N	-97 dB

Another example is a high performance audio ADC whose block diagram is shown above. The 4-bit, 3-level unit element feedback DAC is linearized by the a 1st-order DEM rotational barrel shifter.

11 SSCS

3-level element in ΣΔ ADC

- Additional advantages:
  - Minimum noise contribution from current steering DAC at low level signal (below -60dB full scale)
  - Equivalent to a current noise from a large resistor R, $i_N = \sqrt{4kt/R}$
    - I-DAC current noise is not the dominant source anymore
  - Dominant noise sources : input R and amplifier
    - For the same SNR target : Larger allowable R → Smaller integration capacitor size → Silicon area saving
  - Less switching activity at low level → Less spectral leakage due to ISI
- Cautions
  - Signal dependent switching activity → A source of harmonic distortions
  - Slight noise modulation due to signal amplitude

ADC with 3-level unit element feedback DAC also sees benefit in smaller integrating capacitor size when compared with 2-level unit architecture with the same SNR. The reason is in the fact that at low input signal level, the noise contribution of I-DAC is much less than that of the 2-level, hence the drain current of each I-DAC cell can be reduced further, equivalent to a larger input resistor. It follows that for the same integrator time constant (tau), the integrating capacitors can be made smaller, resulting a good amount of silicon area saving.

Other benefits include less switching activity at low input level, which results in better ISI immunity.

It is noted that the switching activity at the summing junctions is signal dependent, and thus, is a main cause of harmonic distortions in the ADC. Additionally, the output noise floor is slightly modulated by the signal amplitude, which may be undesirable in some applications.

12 DEM: What it does and does not solve

- Solves
  - Cell-to-cell static mismatch (born or due to aging)
    - Between the +1s' and the -1s'
  - 1/f noise of cells
- Does not solve
  - Magnitude error between "+1" and "-1" of a single cell
    - Circuit design technique exists as remedy for this issue
  - Cell-to-cell mismatch due to switching dynamics

In the particular case of 3-level unit element, it can be seen that a DEM technique can solve the matching issue between any pair of elements, any pairs of 1s' or -1s'. Additionally, another beneficial effect of DEM is to modulate the 1/f noise of the DAC cells to Fmod/2 when the input signal is relatively low (-60dB full scale).

What a 3-level unit element DEM cannot solve is the matching of the "+1" and "-1" within each cell. It will require circuit techniques to achieve the necessary level of matching between those 2 output states. The scope of this chapter does not cover the discussions on the circuit techniques.

13 Recap

- Brief background
  - Motivations and Applications
  - 3-level vs. 2-level unit element: advantages and challenges
  - What problems DEM does and does not solve
- **Practical dynamic element matching (DEM) techniques**
  - **2-level DEM review**
  - 3-level DEM
  - Limitations of each technique
  - Solution spaces
- Higher-order DEM
  - Issues in 2nd-order DEM
  - Solutions
- Application specific concept of DEM
- Q and A

14 Review: Noise shaping concept

- High precision multi-level analog quantity is "difficult" to reconstruct
- Noise shaping: Trade off frequency band usage for word-width
  - 16b @ 44.1kHz → 1b at 5.6488MHz

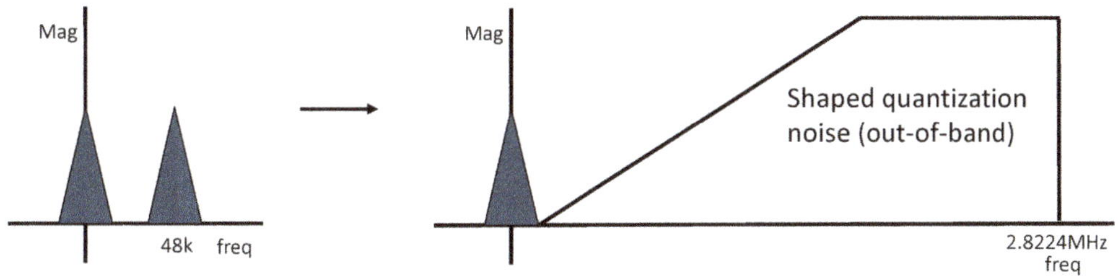

The noise shaping concept in SD converters is a cost-effective solution widely used to achieve the high performance level in data converters. The concept relies on a feedback loop to push the large quantization noise of the 1-bit converter into the frequency region beyond the band of interest. The out-of-band noise can then be filtered out by the subsequent stage.

15 1-bit versus multi-bit DAC

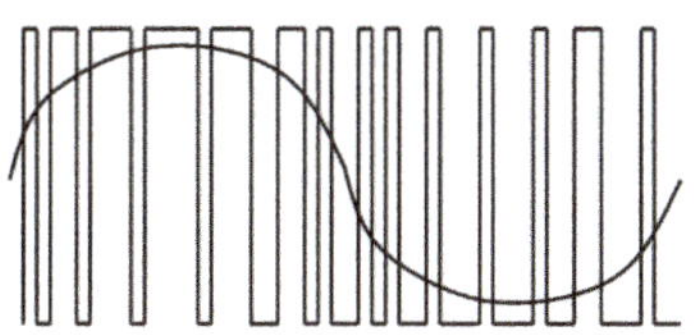

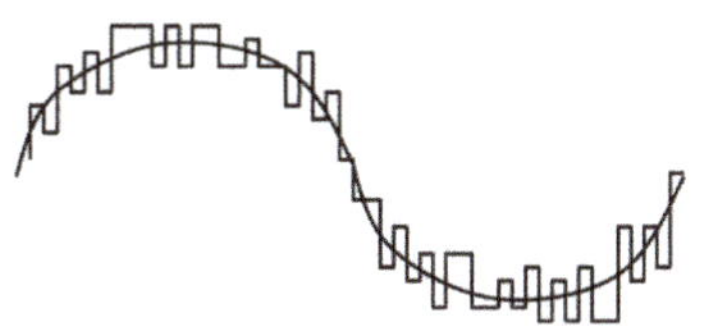

- **1-bit Advantages**
  - **Linear – no matching issue**
- **1-bit problems**
  - **Large step size → jitter sensitivity**
  - **Tonal quantization noise can cause "idle tones" (quantizer can't be dithered properly)**
  - **High-order loops become unstable with large inputs**

- **Multi-bit advantages**
  - **Tone-free quantization noise (can be dithered)**
  - **Lower-order loops can often be used (easier stability)**
  - **Small steps (low jitter sensitivity, less filtering required)**
- **Multi-bit problems**
  - **Matching; DAC element errors cause distortion + noise**

As mentioned earlier, 1-bit architecture offers the best linearity but has several severe drawbacks such as high sensitivity to clock jitter, idle tones and limited input range. Multi-bit architecture, on the other hand, is less stringent on the circuit requirements, less prone to having idle tones, less sensitive to clock jitter but has an inherent linearity problem due to the mismatch between the elements.

16 Multi-bit with element mismatch

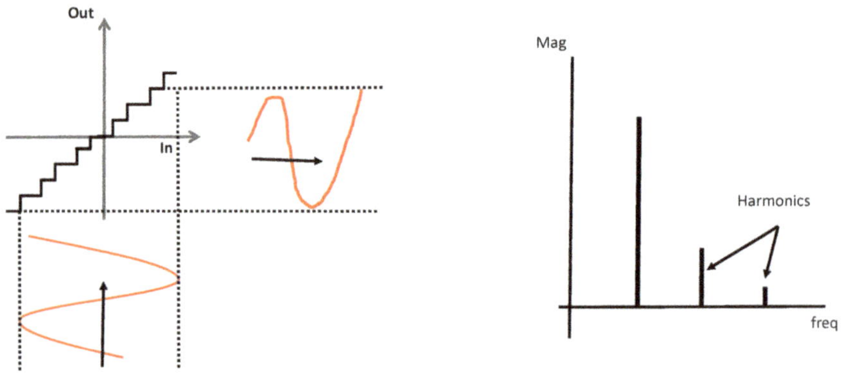

The figure above illustrates how element mismatch can cause linearity problem in a data converter. Mismatch in the elements causes the transfer function to have unequal step sizes. When the analog signal is reconstructed, it will have harmonic distortion components created as a result.

17 Example of multi-bit audio DAC

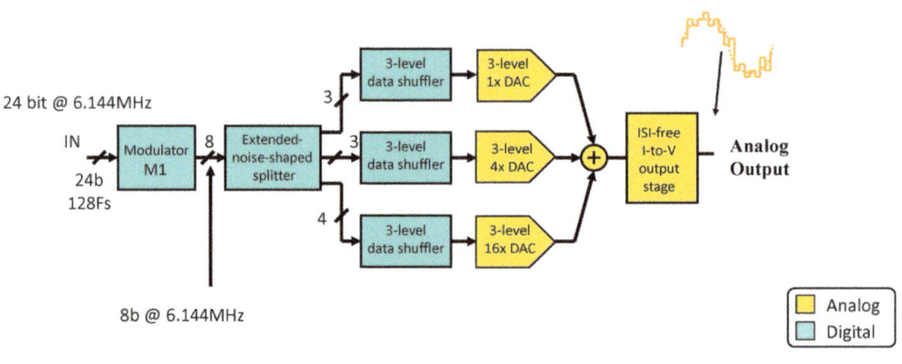

The block diagram above illustrates how the dynamic element matching technique is used to achieve high linearity in a high performance audio DAC [3].

The digital input signal is first interpolated to a 24b resolution at 6.144Mhz before being modulated to a 2nd-order, 8b-output noise shaper. Existing designs of DEM logics cannot handle 8b-resolution due to their impractical large silicon area. To avoid this problem, the 8b word is then split into 3 sub-words in such a way that their gain errors are also shaped by a 1st-order high-pass function.

Three separate 3-level DEM logics then receive their respective inputs and produce the linearized outputs which are then summed and converted to voltage output.

18 The essence of DEM

- Linearizing the transfer function
  - Making each element appears as average value → Reconstructed values are "perfectly" linear in the band of interest
- Requirements
  - Use thermometer coded data (uniformly weighted elements)
  - Must have multiple ways (redundancy) to reconstruct any value
  - Have an "out-of-band" region to push shaped mismatch errors to. i.e., oversampling ratio larger than 1 (practical OSR >= 16)

For a dynamic element matching algorithm to work effectively, there are three sufficient conditions. First, uniformly weighted elements are used in the design. Meeting this condition will ease the design of DEM logics. Second, the system needs to be able to produce an analog value in different ways. Third, the system clock has an appropriate over-sampling ratio (OSR) such that the residual in-band shaped mismatch noise does not degrade the SNR.

19 Multi-bit transfer function

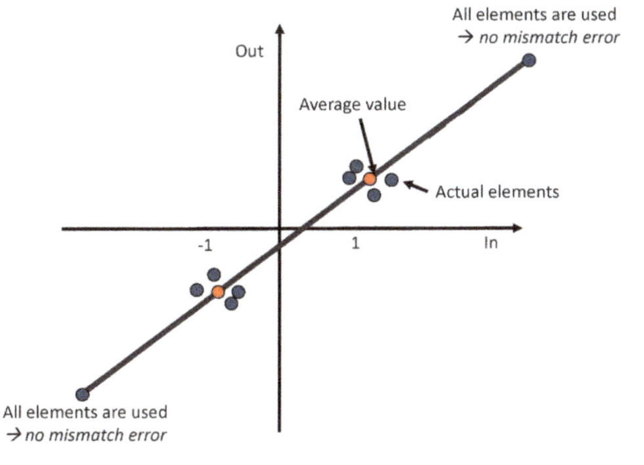

A simplified drawing of a 3-bit transfer function spanning from -4 to +3 is show above to illustrate how DEM works with the sufficient conditions.

At the 2 end points, all elements are used, and hence there is no mismatch error. At any other analog value on the transfer function, there are more than one way to produce the desired value. With a properly designed DEM algorithm, each time an analog output is created from a different set of elements from the previous each cycles. Over time, the analog outputs will approach their average values, and thus, gives rise to a perfectly linear transfer function.

20 Spectral contents with and without DEM

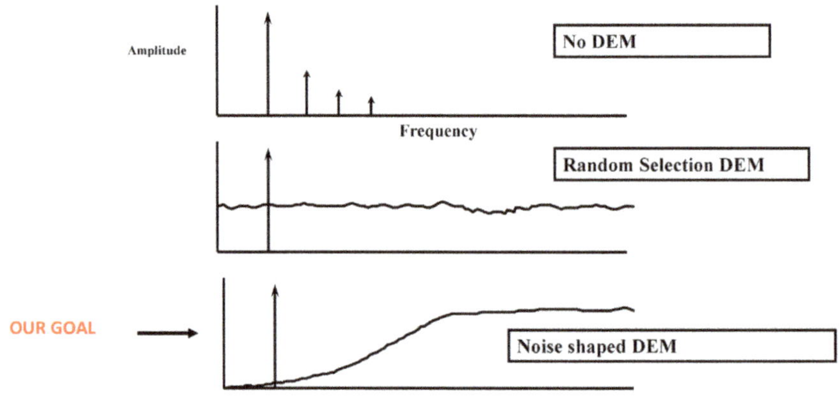

A DEM algorithm can either spectrally shape the mismatch error by a high-pass transfer function or just convert the mismatch error into white noise. Depending on the end application that the designer would select the appropriate approach.

This slide illustrates in the frequency domain how DEM works. Figure a) shows a simplified FFT plot of the output of a converter with the effect of element mismatch exhibited as harmonics in the baseband.

Figure b) show a simplified plot of the output of a converter with non-shaped mismatch error. In this case, the mismatch error is converted to white noise spreading from DC to Fs/2 where Fs is the system clock frequency.

Figure c) show a simplified plot of the output of a converter with shaped mismatch error. In many case, the shaping function is a 1st-order high-pass which can be realized efficiently as will be shown later.

For Nyquist converters, particularly SARs, mismatch error is typically converted to white noise. For oversampling converters, mismatch error is pushed to the higher frequency region and to be filtered out in the subsequent signal processing step.

21 Approaches to 2-level unit element DEM

1. Data directed
2. Rotational
3. Tree structure
4. Vector quantization
5. Real-time

There are essentially five approaches to DEM as listed above. Let's review each of them in the context of 2-level architecture which will serve as the platform to extend into the 3-level architecture.

22 SSCS 1. Data directed shuffling

- Consider a 2-bit thermometer code DAC
  - Cell A = 1 + e1
  - Cell B = 1 + e2

AB	Out	Instantaneous error (ei)
00	-(2+e1+e2)	0
01	-e1 + e2	-e1+e2
10	e1 – e2	e1-e2
11	2+e1+e2	0

- Goal: Make the cumulative error approach zero.
  - *Done by alternating use of OUT1 and OUT2 when A ≠ B*

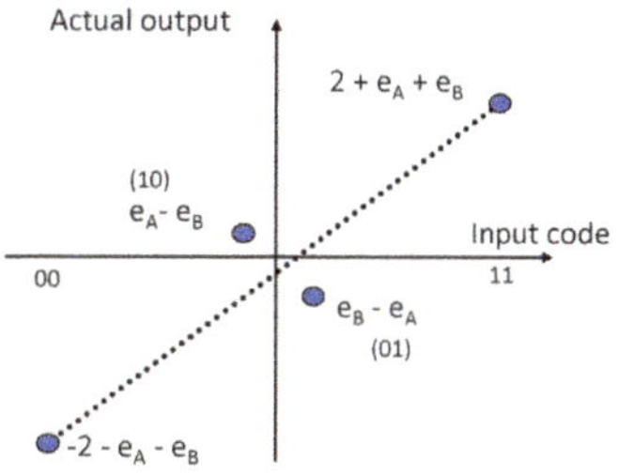

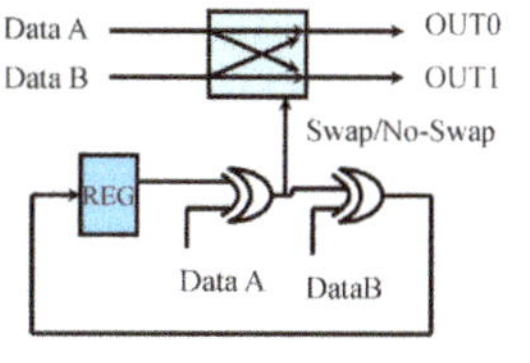

Swapper cell

Data directed shuffling (or scrambling) is best described in an example with 2 units elements as above. One assumption which holds true in most high OSR cases is the high input data activity.

The DEM cell comprises of the swapper cell and a decision logic which decides whether to route Data A to OUT0 or OUT1, and similarly for Data B.

The true table above shows the 4 possible cases of the instantaneous error for the possible input combinations. These errors are derived from the transfer function on the right hand side. As mentioned before, at the 2 end points, the instantaneous error is zero, and for the points in between, the instantaneous error can take different values depending on which unit element is selected to produce the output.

It follows that if the decision logic decides to route the input to the output such that it forces the cumulative error (the first integral of the instantaneous error over time) to be zero, then the mismatch error transfer function will be a high-pass shape.

This goal is achieved by the simple logic circuit shown above.

23 SSCS How to make a multi-bit shuffler

To create a multi-bit data-directed shuffler, the shuffler cells are connected by a butterfly network. Each input bit can reach any output and the decision whether it passes through or gets swapped depends on the data seen at each individual cell locally. Hence, as each input bit traverses through the network , it will be compared to all other inputs, which allows the network to keep track of the cumulative errors of all output bits.

The network thus serves as an "usage" correlator such that over time, all the output bits will be used equally.

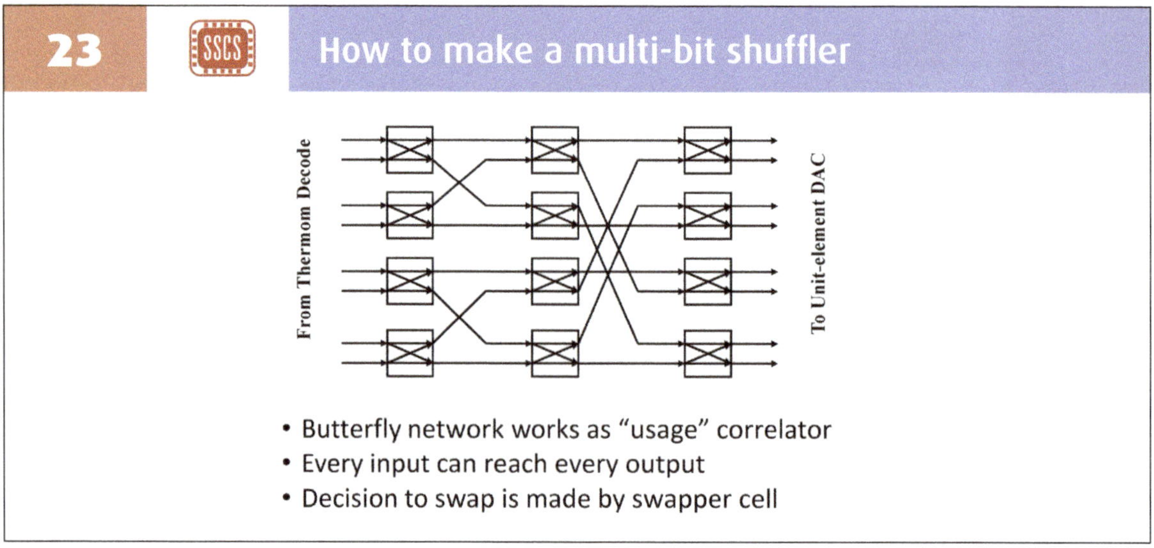

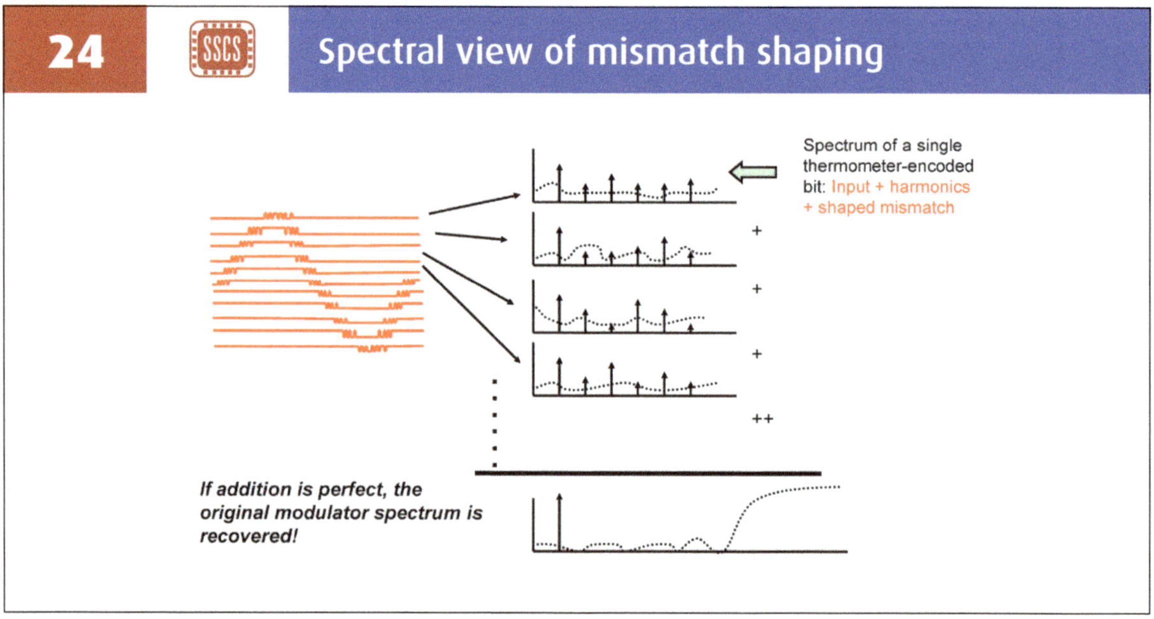

The output of a DEM logic thus consists of a group of 1st-order modulated 1-bit data streams. Each of those has the original baseband information and idle tone components created by the 1st-order usage shapers.

At a first glance, there seems to be severe problem in reconstructing the original signal. However, there are two factors that greatly help to reduce the effects of these tones. First, the summation of all the DEM outputs contains zero idle tone components in the digital sense. It follows that the analog mismatch (or gain error) between the elements would be the only mechanism to produce any residual idle components.

As mentioned earlier, since the difference between each pair of elements has a high-pass shape, it follows that the gain errors will then be shaped by this high-pass transfer function as well.

It is worth mentioning that large enough idle tone amplitude coupled with a large born mismatch error, tones can still be present in the reconstructed analog output. There a several remedies for this problem and will be discussed in a bit later.

25 SSCS 2. Rotational DEM

- Example: Digital in = 3, element selection in 4 clock cycles

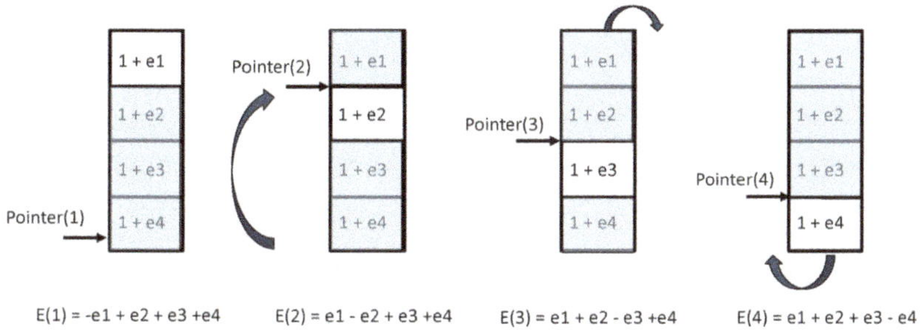

- **Cumulative error after 4 cycles = 2(e1 + e2 + e3 + e4)**
- **(e1+e2+e3+e4) : all elements are used equally → cumulative error = zero!**

The 2nd approach to DEM is the popular rotational scheme described in the example above.

The converter is assumed to have 5 possible analog output levels (0,1,2,3,4) generated by the summation of 4 different analog elements. Each of these elements has an deviation from the ideal value labelled as e_n. A single pointer is used to indicate the current position for selecting the elements.

To see how this DEM works, assume the input is a static 3. In the first clock cycle, element 2, 3 and 4 are selected, and the pointer advances as shown. In this cycle, the output is E(1). As the system clock ticks, output E(2), E(3) and E(4) are then generated. When all elements are used equally, i.e., after the 4th clock cycle, the cumulative mismatch error is zero.

In real use cases, the input data is likely an active signal due to the OSR requirement in the main SD modulator, hence, the cumulative error versus frequency will traverse the high-pass shaping function as described earlier.

26 SSCS 3. Tree structure DEM

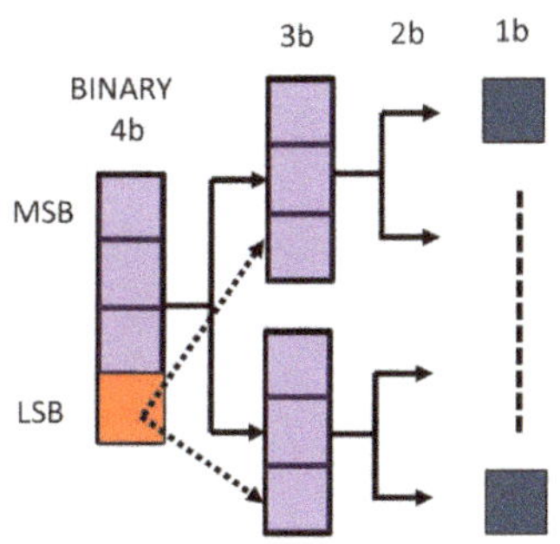

- **DEM algorithm**
  - **Start with N-bit binary word**
  - **Divide by 2 (shift left by 1)**
  - **If LSB = 1, use ΣΔ modulator to decide which sub-word to receive the LSB**
  - **Update modulator**
- **Observation**
  - **Difference between 2 words is a noise shaped sequence → Gain mismatch is high-passed**
- **Note:**
  - **Needs headroom to not overflow**
  - **Works fine for first–order shaping, what happens if input is EVEN for a while!**

26 3. Tree structure DEM

Another popular approach is the tree structure proposed by Galton [7]. The algorithm as described above also assumes that the LSB at each node is active so that the difference between sub-words will be effectively shaped by the modulator. In real use cases, that condition may not be met with low OSR design which can lead to idle tone behaviors. Dithering can be used to mitigate idle tones but will be limited by the headroom available.

A higher-order DEM logic in theory can be achieved with this approach, where at each node, a higher-order modulator is used. In practice, a higher-order tree-structure does not performance well due to the fact that with low input data activity, the higher-order integrators in the modulators will quick grow due to insufficient feedbacks. The modulators, while still performs as a higher-order modulator, become ineffective in shaping the mismatch error [1].

27 4. Vector quantization

- Generalized version of all DEM algorithms → Excellent model to study DEMs
- Hardware intensive

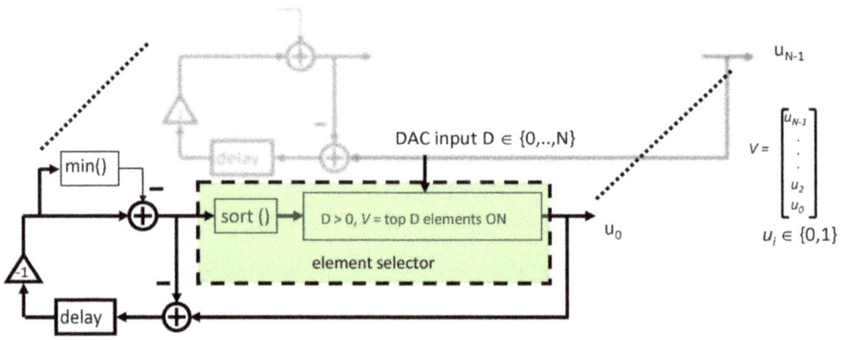

Example of a 1st-order DEM (using the error feedback ΣΔ structure)

The example above shows a generalized way to analyze 1st-order DEM algorithms [1]. A closer look at the diagram reveals that the model behaves as an error-feedback sigma-delta modulator, which gives rise to the 1st-order mismatch shaping function. Each unit element has a separate integrator to keep track of its usage, the integrators will then be sorted and the least used elements will be selected at the earliest chance in order to balance the usage of all elements. Any common mode values between the integrators will be subtracted out to not cause any over-range.

The need of an element selection logic makes the implementation of this approach hardware intensive for designs with larger than 2-bit data. Nevertheless, the model presents a very useful way to analyze higher-order DEM algorithms, and thus is a great starting point for higher-order DEM algorithm study.

28 Element selection logic

- Rank N input vectors with least used first
- Turn on (assert "1") D output elements in output vector V
- Feedback the elements of V to the corresponding integrators (or loops)
- Can apply the use of a dithering vector to the input of the sorting routine to cure idle tone behavior

The function of the element selection logic (ESL) includes ranking the content of the integrators, selecting a number of least used elements, and producing a feedback signal for each individual modulator. Thus, the ESL is acting as a quantizer in a traditional SD modulator design. It follow that to dithering signal can also be applied to the ESL input vector to mitigate any idle tone problem.

29 5. Real-time DEM

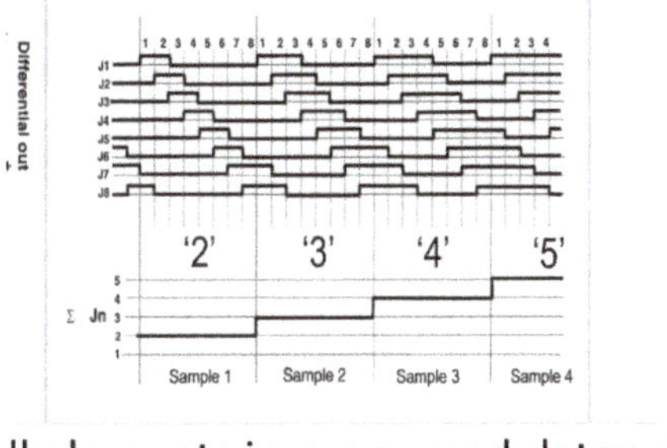

- Use all elements in every modulator clock cycle
  - Advantage:
    - Very simple logic
  - Drawbacks:
    - Need a fast clock (2^N times) for an N-bit DAC
    - Cannot go beyond 1st-order shaping

An effective DEM approach is to use all elements equally in every modulator clock cycle [14]. Thus, the cumulative error is always zero at each clock cycle which results in a very effective 1st-order mismatch shaping. Non-linearity such as inter-symbol-interference (for current-steering elements) is also minimized since each element effectively is a return-to-zero current-steering DAC element.

This technique has several drawbacks since it cannot be extended to a higher-order shaping, and it requires a 2^N-times faster clock, where N is the number of elements. For this reason, the approach is not suitable for low power designs.

30 Recap

- Brief background
  - Motivations and Applications
  - 3-level vs. 2-level unit element: advantages and challenges
  - What problems DEM does and does not solve
- **Practical dynamic element matching (DEM) techniques**
  - 2-level DEM review
  - 3-level DEM
  - Limitations of each technique
  - Solution spaces
- Higher-order DEM
  - Issues in 2nd-order DEM
  - Solutions
- Application specific concept of DEM
- Q and A

31 Approaches to 3-level unit-element DEM

1. **Data directed shuffling**
2. Rotational
3. Tree structure
4. Vector quantization

32 Encoding style for 3-level data

- Best to treat data as **positive and negative thermometer codes** separately
- Results in simplest logic manipulation and decoding logic in the current steering cells
- Truth table

Positive thermometer code	Negative thermometer code	Analog value
1	0	+1
0	1	-1
0	0	0
1	1	Unused

One fundamental difference between 2 and 3-level DEM logics is how the data is encoded. Since the cell can create both push or pull currents, theoretically, there can be 2 states that produce zero output signal current, one is with both current sources turned off, the other is with both current sources turned on. The latter is an not used since it contributes noise to the output as in the case of 2-level scheme.

The best way in terms of logic encoding and decoding complexity is to use signed thermometer codes. This scheme is described in the truth table above where each cell will take in a positive and a negative thermometer bit which are mutually exclusive. This scheme will result in a very low decoding complexity as will be shown later.

33 Approach #1: Data directed shuffling

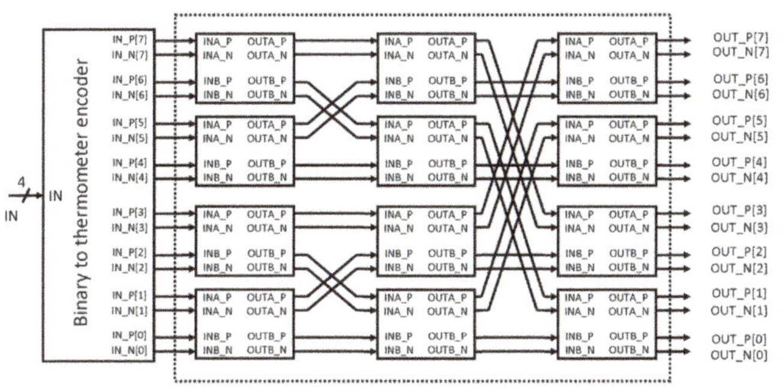

A 4-bit example

- Butterfly network serves as the usage correlator

$$\#\ of\ cells = \frac{N}{4}\log\left(\frac{N}{2}\right),\ for\ N+1\ output\ levels$$

33 Approach #1: Data directed shuffling

In principle, the data-directed 3-level DEM approach follows the same design philosophy of the 2-level counterpart. It leverages the similar butterfly network previously shown as the usage correlator. The block diagram above shows an example of a 4-bit 3-level data-directed shuffler. The 4-bit binary input is first encoded to signed-thermometer codes arranged into 8 pairs.

Since each shuffling cell now works with 2 pairs, the number of shuffling cells is now N/4log(N/2) which is ½ of what the 2-level DEM design would require. However, each 3-level cell complexity is higher as will be shown in the next slide. Thus, in terms of hardware usage, the two designs have similar silicon area requirement.

The significant difference is that the dynamic power consumption of the 3-level design for small signal levels is lower due to a much lower data activities in the design.

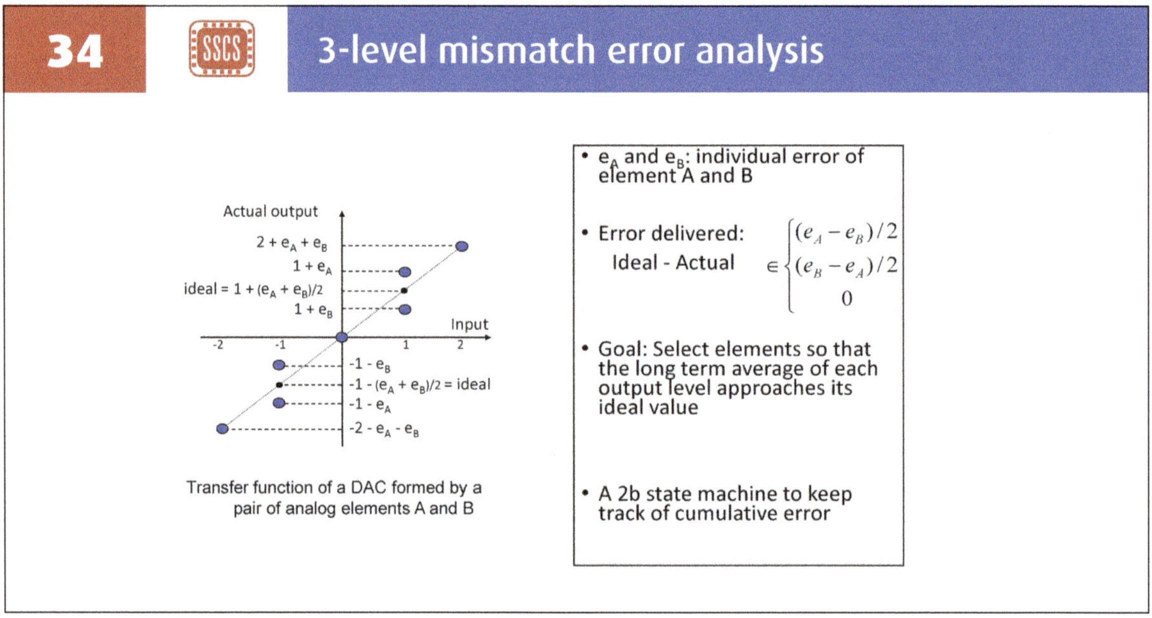

To understand how 3-level mismatch error shaping works, let's look at the analysis of a DAC cell consisting of 2 elements A, and B, each is capable of delivering analog quantities of +1, -1, and 0.

Let *eA* and *eB* be the error of each element respectively from the ideal value (the average of the 2 elements), the transfer function on the left can then be derived. It follows that at the end points and the origin, there is no mismatch error contributed from either elements. Hence, the goal of the mismatch logic is to monitor the usage, then select the elements such that the long term average of the +1 and -1 outputs converge to their ideal values. This principle is not different from that of the 2-level element system.

35 Cell implementation

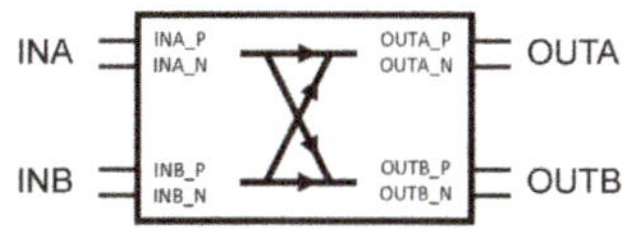

INA, INB, OUTA, OUTB are 2b quantities	
Value	Code (PN)
+1	10
-1	01
0	00
unused	11

- Output value = OUTA + OUTB, $\in\{-2,..0..,2\}$
- Simple modification of thermometer code
  - A and B always have the same sign
- Input can be rerouted to either output
- Needs a state machine

To simplify the logic implementation, signed thermometer coding scheme is used. Each input is now a 2-bit number designated as *p* and *n* which can take the value shown in the table above. The zero state is encoded as "00" as it implies both current sources are disconnected from the output.

Later on when the output the shuffler cell is used in the analog DAC cell, there is one small decoding circuit needed to create the zero state to drive the DAC cell.

The function of the shuffler above is to re-route the inputs to the outputs while keeping the cumulative error zero over time.

36 State diagram

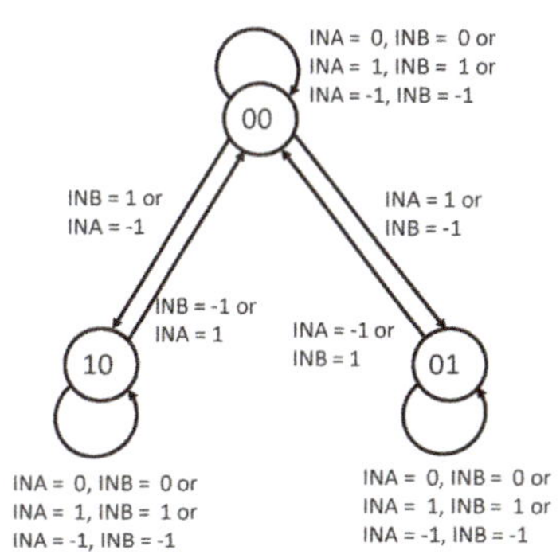

- Cumulative error:
  - "00" : *zero*
  - "01" : $(e_B - e_A)/2$
  - "10" : $(e_A - e_B)/2$
- Apply the DEM rules to reroute data and keep cumulative error at zero
- The shuffler cell operates similarly to the generalized DEM model

37 State diagram

The finite-state machine for the shuffler cell has the state diagram showed above.

The operation of the state machine can be described as follow. At the "00" state, the cumulative error is zero, hence the input data INA and INB will pass through directly to OUTA and OUTB. Depending on the input data, that the state will then be either"00", "10" or "01".

At the states "01" and "10", the cumulative error will be *(eB-eA)/2* and *(eA-eB)/2* respectively, depending on the input conditions shown in the state diagram that the inputs will either be swapped or not before being routed to the outputs to return the cumulative error back to zero.

It is apparent that this shuffler cell is performing the dynamic element matching on a pair of 3-level input data.

37 Idle tones in first-order data-direct shuffling

- Reasons:
  - Each shuffling cell is a 1st-order noise shaper (can create large inband tones)
    - With gain mismatch between cells, these result in idle tones
  - At low level (below -60dBFS), input has a tendency to travel through only a certain path due to the construct of the butterfly network
  - At low level input, some cells in the network become "starve" in activity
  - Can be detected by sweeping the input amplitude with small DC range
- Remedy
  - Precede the shuffler network with a randomizer
  - Ensure that each cell will have equal amount of data activity over time
  - Mismatch shaping effectiveness is slightly degraded but still can perform very well with high OSR (>=64)

One known issue with data-direct shuffling is the idle tone problem. Since each cell output is a sequence of 1st-order noise shaped signal, the input data activity has a strong effect on the behavior of the idle tones in the output sequence. At low level signal, -60dBFS or below, only the few cells at the bottom of the network can receive the data. This condition leads to very large idle tone present in each output. With a inherent large gain error between each analog element, the cancelation of these tomes are incomplete which results in in-band spectral residues that severely affects the SNR and low level harmonic distortion performance.

An effective remedy for this problem is to precede the butterfly network with a randomizer. This randomizer allows the cells in the first column of the butterfly network to have equal probability of receiving input data at any clock cycle, and hence, effective dithering the data activity to the cells. The end result is that any pseudo-periodic behavior, or idle tones, in the outputs will be completely destroyed.

38 Measured result ADAU1361 at -0.5dbFS

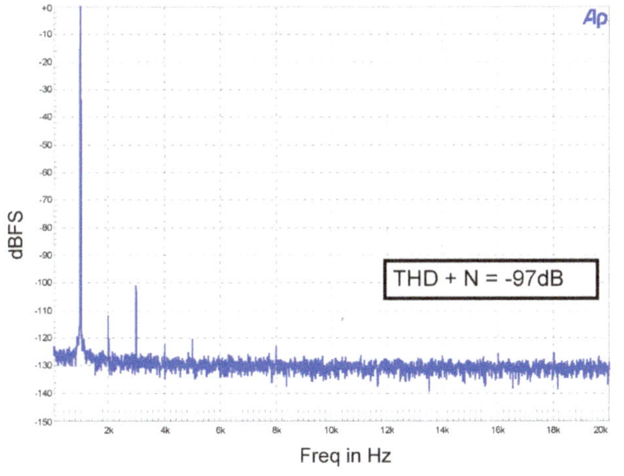

A measured spectrum at -0.5dBF of a high performance audio DAC output using 3-level DEM technique. This DAC uses the data-directed 3-level DEM technique, with a pre-randomizer as a remedy for the idle tone issue. The THD is dominated by the third harmonic as a result of the amplifier power consumption optimization.

39 Measured result ADAU1361 at -60dbFS

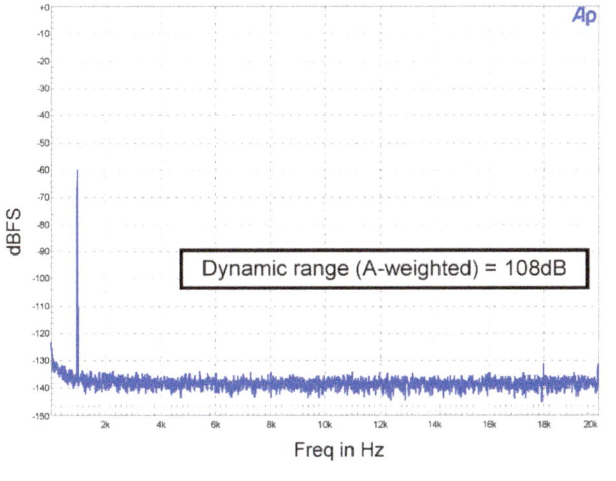

At low input level, DEM techniques would be most vulnerable to pseudo-periodic patterns present in the individual bit streams. These are generated due to lack of enough activities in the individual shuffler cells as mentioned earlier. This audio DAC uses a position randomizer in front of the butterfly network to ensure that the all the cells receive adequate data activity.

The measured spectrum at -60dBFS above of the same audio DAC shows the effectiveness of the idle tone remedy.

40

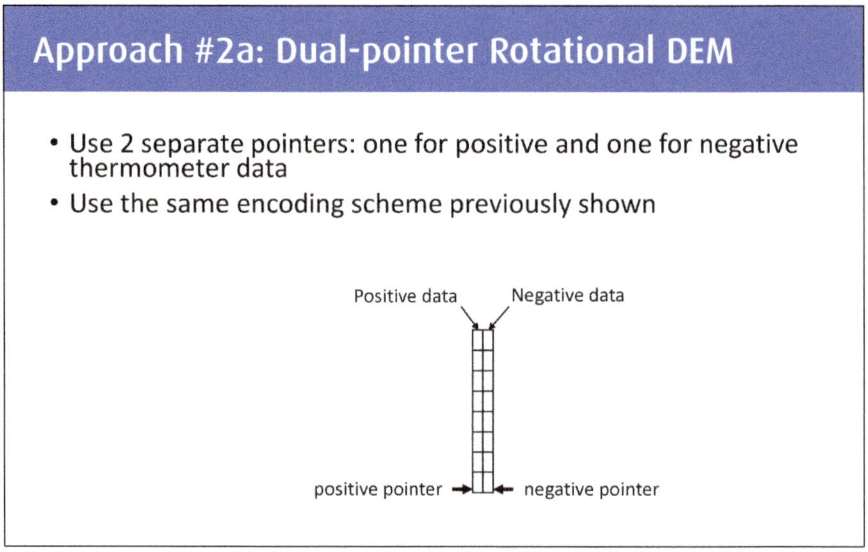

The second approach to 3-level DEM is based on the well-understood 2-level rotational DEM technique. In the 3-level case, the technique presented here uses 2 different pointers, a positive and a negative, to indicate the current position of the cells to be used in the reconstruction of the positive and negative analog output.

The hardware implementation of this technique is quite simple. The binary data is first encoded into signed thermometer codes. At the same time, the binary data is fed to 2 separate accumulators, each responds to either a positive or negative input. The accumulators are N-bit counters, where N is the binary input word width. The counter outputs then directly control the $2N^{-1}$-position barrel shifters which receive and shift the signed thermometer codes accordingly.

41

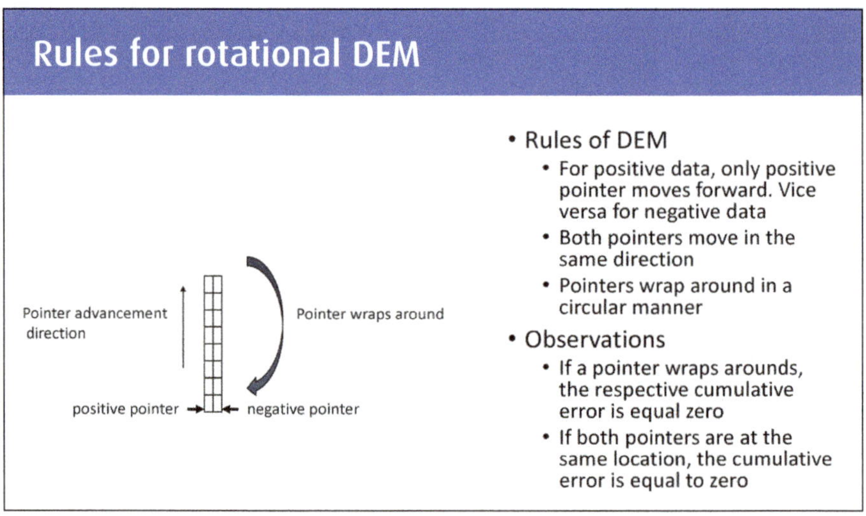

The DEM algorithm applied here bears some resemblance to its 2-level counterpart.

As shown in the drawing above, for positive data, only the positive pointer would be updated and advanced to the new location in the next clock cycle, and vice versa with negative data.

Both pointers are independent of each other and are free to wrap around as needed. One interesting observation is that when the 2 pointers are at the same location, the cumulative mismatch error in each output at that particular time is zero. This observation is important because it will be used later to develop a remedy for the idle issue in this DEM approach.

42 Example

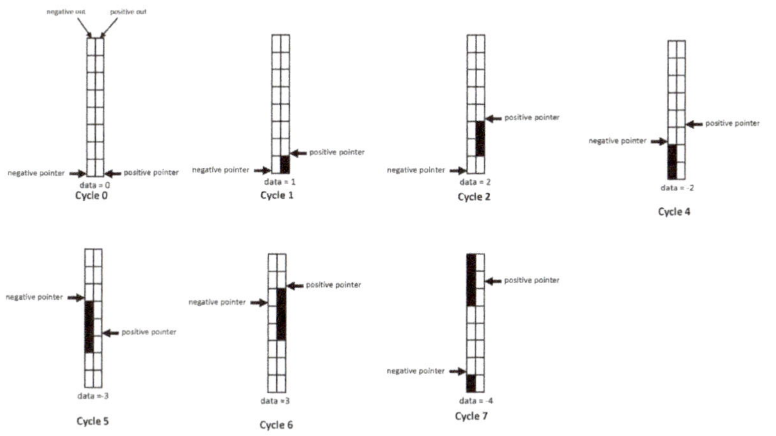

To help the reader visualizing this DEM approach, the example above is provided. In this case, a DAC with 8 elements spanning between +8 and -8 is used. The advancing of the pointers versus the input data at each clock cycle is illustrated in the above drawing.

43 Implementation of dual pointer DEM

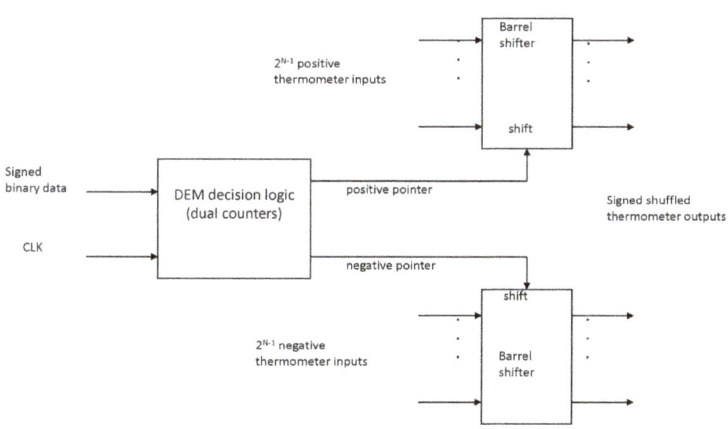

As mentioned earlier, the implementation of this DEM approach is quite simple and illustrated in the block diagram above. Depending on the polarity of the input data that only 1 of the 2 barrel shifter gets updated and passes the data to the output. The inactive barrel shifter will output all zeros.

44

Implementation of dual pointer DEM with sign-magnitude data

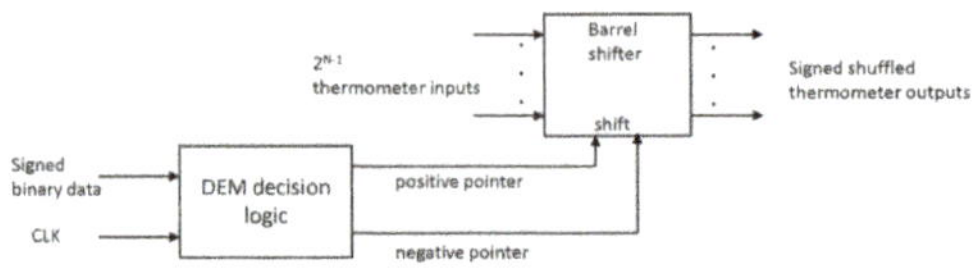

- Area efficient single barrel implementation
- Only 1 pointer is active and used at a time
- Barrel shift control is dependent on sign
- Barrel shifter only takes in magnitude
- Sign bit is used as a decoding

A hardware optimized implementation of the rotation DEM approach is shown above. The barrel shifter now handle only the magnitude of the input data in thermometer code form. The sign bit is simple provided to the analog side for proper decoding of the controls for individual current cells.

45

Idle tones in dual pointer DEM

- Reasons:
  - Similar to those found in 2-level single pointer DEM
  - Static born-mismatch pattern shows up at low input level
- Remedy
  - Observe that when both pointers are at the same location, the cumulative error is zero
    - **Both pointers can be repositioned to a randomly chosen location**
    - Static mismatch error pattern will be broken up
  - A slight degradation to the SNR due to the introduction of the randomness into the first-order single-bit noise-shaped sequences
  - Mismatch shaping effectiveness is still very good with high OSR (>=64)

At low input signal level (typically -60dBFS), the static mismatch error patterns in the array of analog cells can convolve with the low activity digital input to create in-band tones as mentioned before. In this case, an effective yet simple solution exists and can be implemented with very low hardware addition.

The idea is to break up the idle tones at the appropriate time while generating minimal disturbances to the high-passed behavior of each output. This goal can be achieved by re-seeding both pointers to a random location in the next clock cycle if they are at the same location in the current clock cycle. When both pointers are at the same location, the cumulative mismatch error in each output stream is at zero, which is the optimal instant for re-seeding the pointer location.

Simulations with different audio signal profiles, and appropriately high oversampling ratio (OSR > 32) were used to verified the effectiveness of this solution.

46 Example: With idle tone remedy

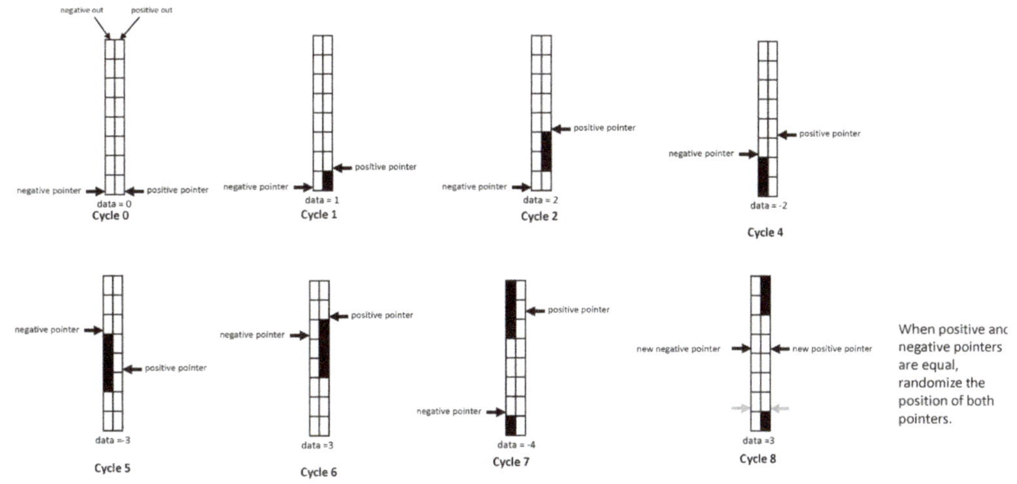

The example above helps to visualize how this idle tone remedy works. The input data and the pointer locations are shown as they change from the clock cycle 0 to 8. At the 8th clock cycle, both pointers are at the same location which trigger a small state machine to re-seed both pointers to a new location as described previously.

47 Measured result ADAU1966 at -0.5dbFS

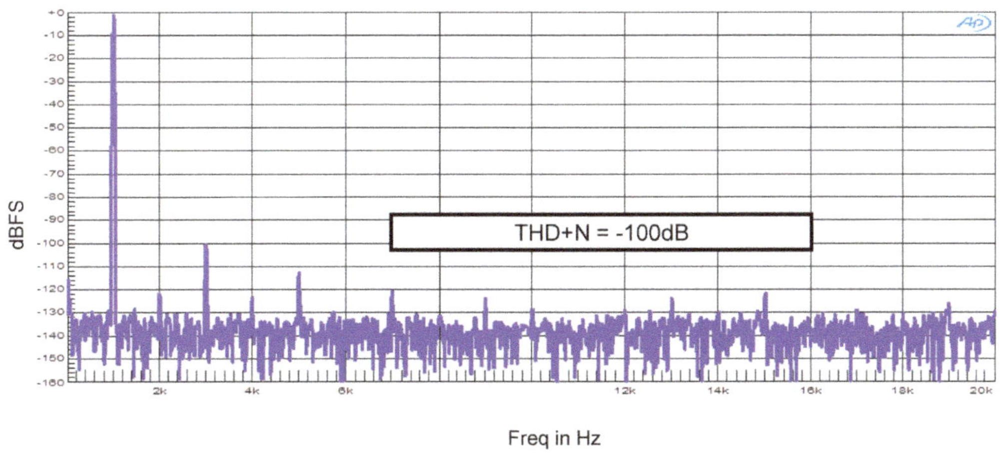

The rotational DEM technique was successfully used in a commercial 120dB audio DAC product. A measured spectral plot is shown above.

48

Measured result ADAU1966 at -60dBFS

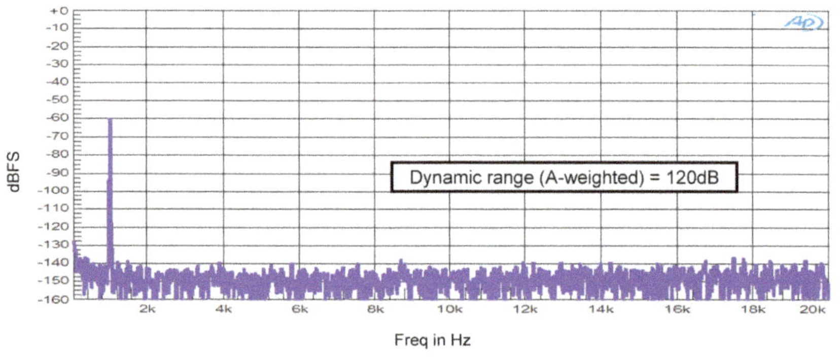

The solution to the idle tone problem was implemented and measured result at -60dBFS is shown above. The noise floor is quite clean of low frequency behavior. The input was also swept in small DC steps around the zero data to ensure the effectiveness of the solution.

49

Approach #2b: Single-pointer rotational DEM

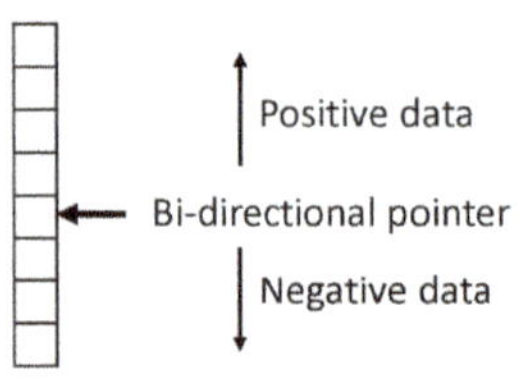

- One pointer that moves forward when data is positive and backward when data is negative
- Possible remedy for idle tone
  - Set up a reference point
  - Keep track of wrapping around the reference point
  - After wrapping around (single or multiple times) and ending at the reference point, reposition pointer to a new randomly chosen reference point.
  - Repeat the entire procedure

Another approach to rotational DEM using a single pointer that travels in both directions was presented in [8]. The pointer would move in one direction for positive data, and the opposite direction for negative data.

To get an intuitive understanding of how this approach works, consider an example with 2 elements, the analysis for such case shows that the element selection logic is identical to that of a data-directed shuffler cell.

It follows that if the input data amplitude is small (below -60dBFS), then the pointer will likely move back and forth from the original starting point, and slowly advance in one direction depending on the amplitude of the input. As a result, idle tones will be generated if no provision is included to break up any pseudo-period behavior.

One modification to the previously proposed idle tone solution is to designate a starting point so that when the pointer returns to that position, its new location will be randomized. The frequency of such action can be moderated accordingly to ensure that the it does not generate more mismatch error.

50 Approach #3: Tree structure DEM

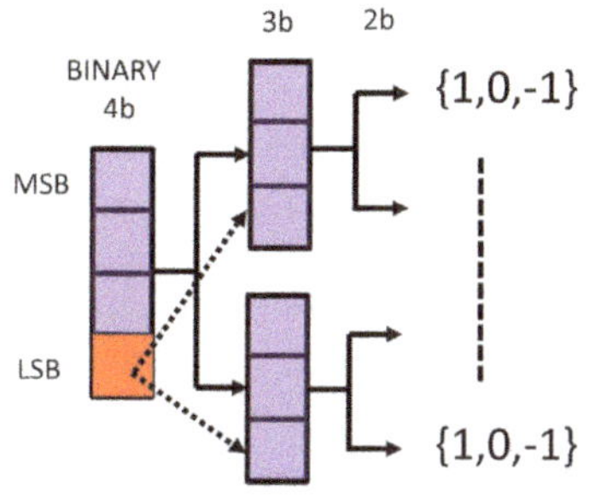

Example of a 4b implementation

- Use the tree structure to split the word until the last 2MSBs
- Decode the leaf cells to get the {+1,0,-1}
- Remedy for idle tone
  - If data is even for a while, consider add 1 to one branch and subtract 1 in the other branch. This increases the data activity.
  - Watch out for headroom issue in the digital words
- **Note:** that each bit is still a 1st-order noise shaped sequence which has the fundamental component and shaped noise.

The tree structure approach [7] in 2-level is readily applicable to 3-level architecture. The example above shows how such approach is implemented. The input word is continuously split until the last 2 MSBs remain which serve as the leaf cell. At every stage of splitting, the LSB can be added to either output words based on the decision of a simple error feedback sigma-delta loop. Dither can be added to this loop to decorrelate any pseudo-periodic behavior.

A potential problem with this technique is that if the splitting results the LSB of a stage being zero over some period of time consistently (due to the nature of the input signal), idle tones can still be produced due to this lack of activity. This condition can be monitored and data dynamics can be enhanced by injecting a +1 to one output word and -1 to the other, which helps to break up the idle patterns. However, as the words progressively become smaller as due to splitting, the headroom will limit such addition activity, and hence limit the effectiveness of this remedy.

51 Other possible solutions to tree-structure DEM

- Applicable when input word width is large for any straight implementations of DEM
- Consider noise shaped segmentation principle
  - Only one component in the sub-level has the signal fundamental. Easier to derive solution for idle tones.
  - May be done recursively
  - A and B can then be followed by a butterfly, rotational or tree-structure shuffler

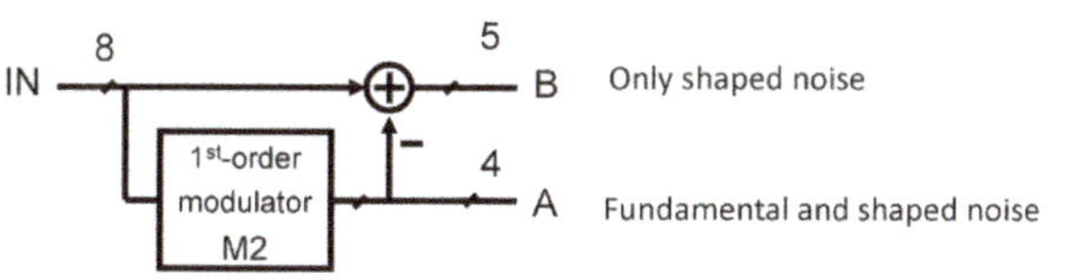

51 Other possible solutions to tree-structure DEM

Another approach to split a large input word to a manageable size for DEM logics is illustrated above. The fundamental difference between this approach and the tree structure is that the signal content remains in only 1, the A in this case, of the 2 output words. This splitting results in 2 smaller sub-words suitable for direct application of DEM or further spitting if necessary. The modulator M2 can be of 1st or higher-order which the designer has full control to make trade offs in the design process.

52 Approach #4: Vector quantization

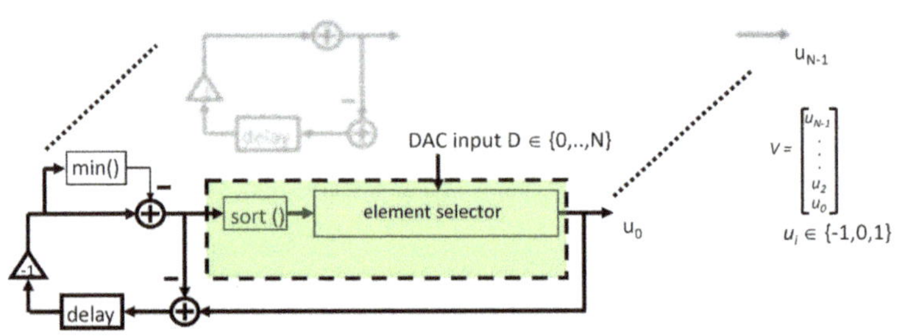

- A slight change in the ESL to accommodate 3-level data
  - Rank N input vectors with least used first
  - ***Assert {-1,0,1} in D output elements in output vector V***
  - Feedback the elements of V to the corresponding integrators (or loops)
  - Can apply the use of a dithering vector to the input of the sorting routine to cure idle tone behavior

The general model for DEM algorithm [1] can be used as a platform to develop higher-order 3-level DEM techniques. For simplicity, the 1st-order error-feedback DEM model is illustrated above. For an N-element design, there are N slices of the error feedback modulators, each responds to a single element. The quantizer comprises of a sorting logic, and an element selection logic. The sorting logic takes in the N multi-bit words and sorting them in descending order denoting the least used element at the top. The result is then passed to the selection logic which selects the top D elements to use in the reconstructing the analog output. The difference between the input to the quantizer and the vector D is then fed back to the loop filter in the same fashion as found in a error feed-back noise shaper. To avoid overflow, any common denominator in the inputs to the quantizer is subtracted out as shown.

Dither can also be used to break up any idle tone in this case.

53 Approach #5: Real-time DEM

- Direct concept can be re-used
  - Requires fast clock
  - Switching loss due to logics may become significant to not be considered for low power implementation

Real-time DEM concept can also be applied to the 3-level architecture with the same pros and cons as previously stated.

54 Recap

- Brief background
  - Motivations and Applications
  - 3-level vs. 2-level unit element: advantages and challenges
  - What problems DEM does and does not solve
- Practical dynamic element matching (DEM) techniques
  - 2-level DEM review
  - 3-level DEM
  - Limitations of each technique
  - Solution spaces
- **Higher-order DEM**
  - Issues in 2nd-order DEM
  - Solutions
- Application specific concept of DEM
- Q and A

55 Why need 2nd-order DEM

In designs with lower OSR, 1st-order DEM might not be able to sufficiently shape the mismatch error due the limited out-of-band frequency region. Over-clocking the DEM logic and associated analog elements may yield an incremental improvement at the cost of higher power consumption, higher clock jitter sensitivity, and potential SNR degradation due to dynamic switching noise.

Higher-order DEM offers an alternative solution but it has a few serious drawbacks such as having worse performance than 1storder design at low input level, and silicon area intensive usage with number of elements larger than 8.

55 Why need 2nd-order DEM

- 1st-order DEM
  - Effective when OSR >= 64x
  - At lower OSR
    - Shaped mismatch noise degrades SNR.
    - DEM tones appear inband.
    - Over-clocking DEM (2x) may offer small improvement (not enough). Switching dynamics become a source of noise.

- **Solution**: 2nd-order shaping

- Issues with 2nd-order:
  - Ineffective for low input amplitude.
  - Area intensive especially with number levels >8.

56 Approaches to 2nd-order DEM

- Data directed : Cannot produce second-order mismatch shaping effectively
  - When inputs are both 11 or 00, mismatch shaper cells get "overloaded"

- Rotational: Can produce a second-order, but extremely hardware intensive, tends to revert to first-order shaping

- Tree-structure: Can produce second-order shaping but tends to revert back to first-order in real life use case (signal conditions)

- Vector quantization: Very hardware intensive due to realization of sorter and element selector
  - ***Hardware implementation at size = 2-bit (4 elements) is affordable***

In an effort to design a 2nd-order DEM logic, known DEM architectures were examined in various previously reported publications which reveal the weakness in each approach.

The fundamental problem with the first three DEMs is the tendency to degrade and revert back to a 1st-order type of performance under low data activity while the last approach becomes impractical due to its large silicon area. Nonetheless, this technique is the only one that yields robustness and performance under different signal conditions, and hence worth the time and effort to search for a practical implementation.

57 Design challenges and solutions

- Full implementation of vector quantization is extremely hardware intensive
  - Solution: Segmentation (or tree structure)
    - Break it down to smaller chunks where realization is affordable

One immediate observation in the vector quantization approach is that the vector quantizer takes a significantly large silicon area. This intensive area is due to the hardware based sorting and ranking logics which largely depend on the number of elements in the design. When the number of elements is limited to 4 or below, the required area becomes greatly reduced. Thus, a higher order DEM design of combined tree-structure and vector quantization would make use of the advantages in both techniques.

58 2nd-order DEM for 3-level unit elements

- Each leaf cell needs 2 bits: $\begin{cases} 00 \text{ for } 0 \\ 01 \text{ for } +1 \\ 11 \text{ for } -1 \end{cases}$
- At one level higher in the tree, minimum size = 3 bits
  - Will need 4 cells to represent {+3,..., -4}

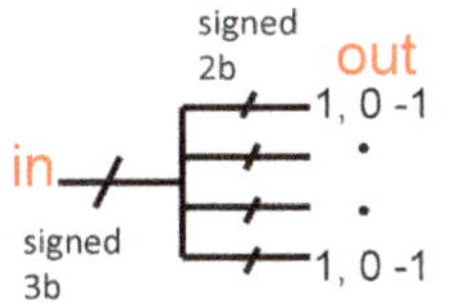

Operations:

1) OUT[1:0] = IN >> 2
2) RESIDUE = 2 LSBs of IN
3) Use a 2nd-order NS loop to decide how to distribute the residue to the outputs

One such example is described here. As previously mentioned, the 3-level DEM leaf cell is a 2-bit quantity. To create an effective 2nd-order DEM, the elements can be grouped in size of 2 or 4. In the immediate layer higher than the leaf cells, the word width is 3, which means it can be split into 4 elements as shown above. A 2nd-order vector quantization based mismatch shaping loop is then used to ensure the mismatch error between these 4 elements is shaped accordingly.

59 Block diagram of a 3-bit, 2nd-order DEM

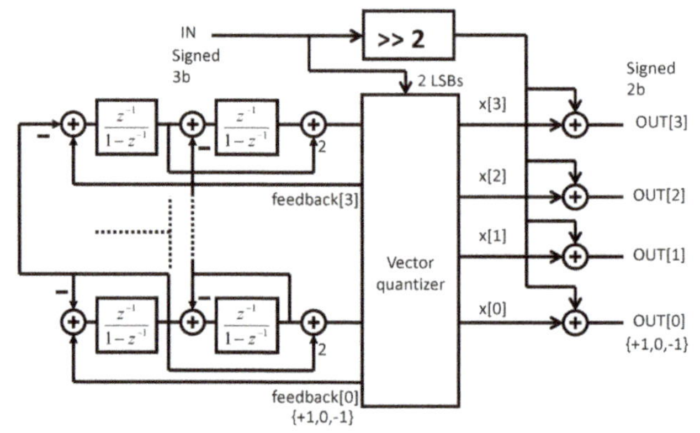

Feedback is ONLY produced when there is an instantaneous error made

The block diagram of a 2nd-order, 3-level DEM logic is shown above with an input of 3-bit signed binary data. It comprises of 4 error feedback modulators working on the errors generated between from the vector quantizer output and the input 2 LSBs. An important observation is that the error is zero when the input is zero. More discussions on this observations will be provided in the later slides.

60 Block diagram of a 5-bit, 2nd-order DEM

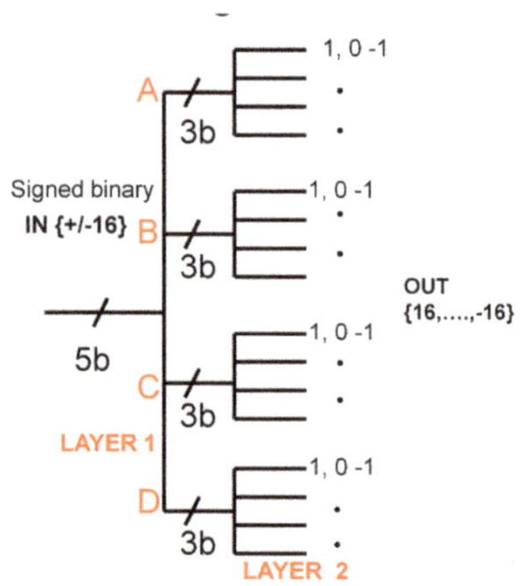

1. **A + B + C + D = IN.**
2. **The spectral difference between every pair in layer 1 must show a 2nd-order highpass function**
   - DAC-to-DAC gain error will be 2nd-order shaped.
3. **The spectral difference between any pair of unit elements in layer 2 must also show a 2nd-order highpass function**
   - Element mismatch will be 2nd-order shaped.

An example of a 5-bit 2nd order DEM is shown above. Although the word splitting is done from left to right, it is easier to understand the concept if the tree is traversed from right to left.

Starting out with the right-most groups of 4 elements, going to the higher level, a split by 2 or by 4 is. In this example, a split-by-4 is used to get to the 5-bit input using one layer. The gain errors between the groups in layer 1 is shaped by a 2nd-order mismatch shaper similar to the one used in the layer 2.

61 Simulation result

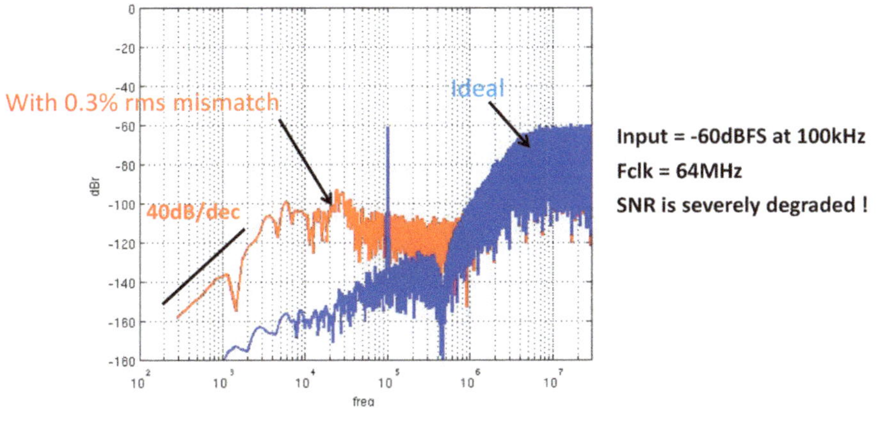

A 3rd-order, 5-bit analog modulator using the 2nd-order DEM logic was designed and simulated. At a signal condition of -60dB, the simulation result reveals a severe SNR degradation due to the effect of the DEM logic. Although the mismatch shaping still maintains a 40dB/dec slope, but the effective 3-dB frequency becomes very low. With a 0.3% rms mismatch, the SNR is basically dominated by the inband mismatch errors.

62 Closer look at the integrator outputs at low input level

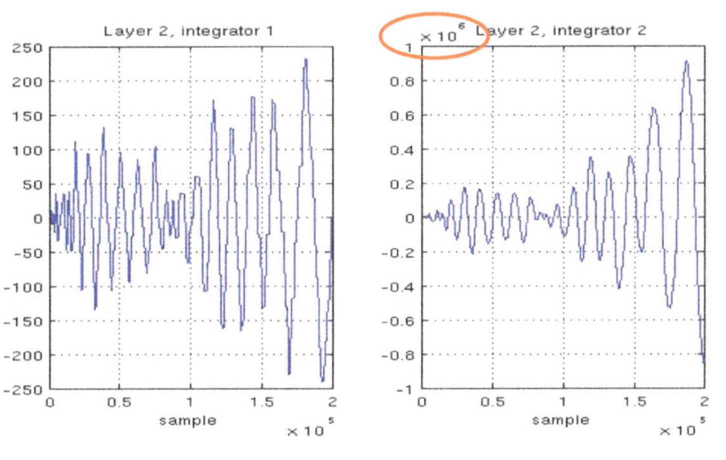

A closer examination of the design reveals that the dynamic range of the 2nd integrators in each layer is very high which implies the required large hardware implementation.

63 The cause: low data activity

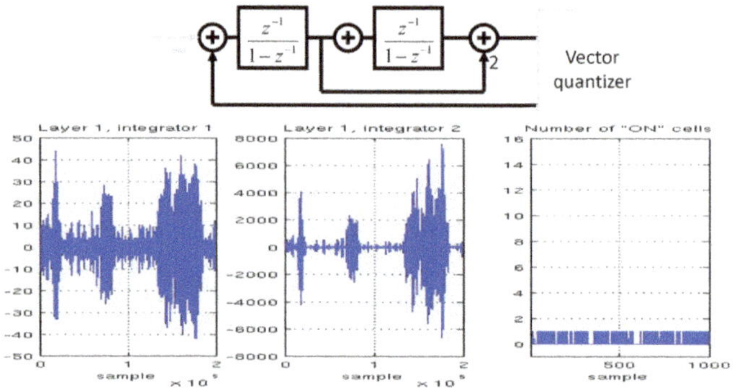

- Very large dynamic range integrator required
- Clipping makes the matter worse

Such large dynamic range is the result of the low data activity explained as followed. With low input amplitude, the data is mostly zero, which as previously mentioned, means the fed back error is zero. The 1st integrator will then be idling at its previous value while the 2nd integrator keeps on increasing every clock cycle. This effect causes the mismatch error logic to try to correct for it and hence creates this large fluctuation in the 2nd integrator, and results in the low 3dB frequency in the mismatch shaping high-pass function.

Attempt to clip the integrator output will make the problem worse since it is equivalent to injecting a large correlated noise into the vector quantizer input.

64 Design challenges and solutions

- Full implementation of vector quantization is extremely hardware intensive
  - Solution: Segmentation (or tree structure)
    - Break it down to smaller chunks where realization is affordable
- Inadequate data activity: cause large growth in integrator word widths
  - Solution: Trade off a small thermal noise performance: Dynamics enhancement

One solution is to enhance the data activity when needed.

65 Dynamics enhancement (DE) (I)

- Increases the data activities while maintaining the digital **value** unchanged.
  - Add +1 and -1 to digital input when possible
- Trades off thermal noise for mismatch
  - Goal: keep # of ON cells as low as possible
- Reduces the integrator output swing tremendously. Very hardware efficient.

The concept of dynamics enhancement is described as above.

66 Dynamics enhancement (DE) (II)

- If 2LSB = "00", start counting
- If count exceeds a designated number "wait_time"
  - Check if signal(s) are in safe range
    - Add +1 and -1 to the branches. Reset wait_time.
- Parameters to adjust
  - Wait_time: affects integrators word width
  - Amount of +1/-1 added, restricted by thermal noise performance
  - Choice of where to direct +1 and -1 to.

Dynamics enhancement works based on observing the data activity and injecting +1 and -1 with a net of zero into the output words such that they generate errors to be fed back to the mismatch shaper. Design parameters for the designer include the time between injections, the amplitude of injection. The main trade off in this solution is the increased thermal noise the output due to the added activities.

67 Combining 2nd-order DEM and dynamics enhancement

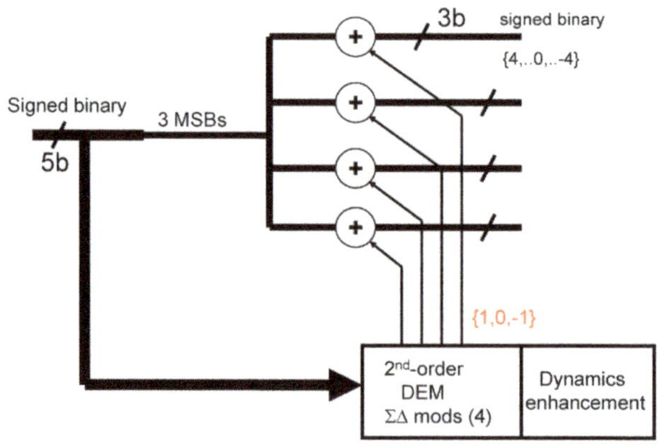

The above block diagram shows an implementation the first splitting layer in a 2nd order DEM design with dynamics enhancement.

68 Simulation: With dynamics enhancement

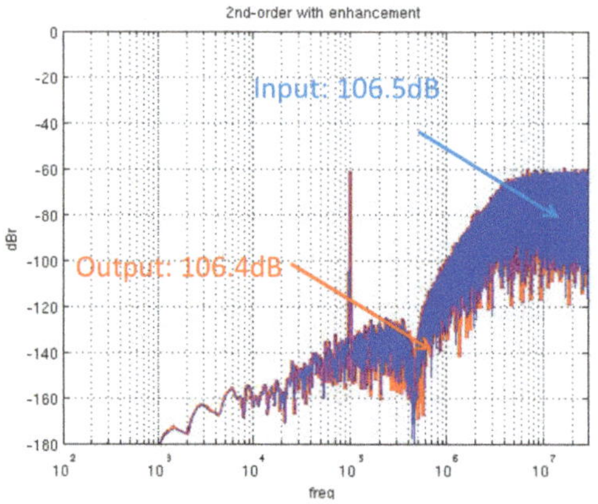

Input = -60dBFS at 100kHz

Fclk = 64MHz

Mismatch: 0.3% rms

With dynamics enhancements properly added, the design shows that it now can achieve a 2nd order mismatch shaping function, and does not degrade the SNR as before.

69 Integrator outputs with DE

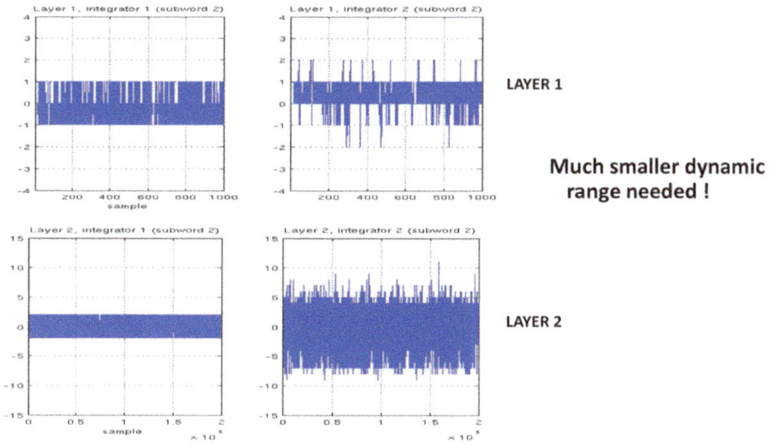

Further probing into the integrator outputs at both layers shows that the digital dynamic range is greatly limited to a very practical level that can be implemented by a 4-bit data width.

70 Number of DAC cells used at -60-dB input

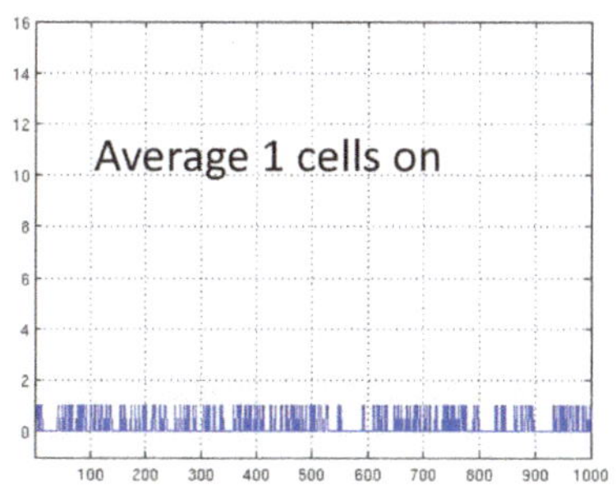

WITHOUT dynamics enhancement

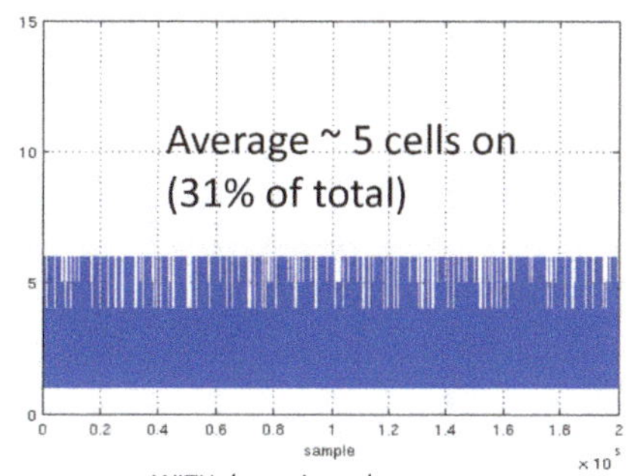

WITH dynamics enhancement

The level of thermal noise penalty depends on the design parameters mentioned previously. In this case, dynamic enhancement increase the cells usage by an average of 31% when compared in the exact the same input condition.

71 Layout of a Multibit Continuous Time ΣΔ ADC"

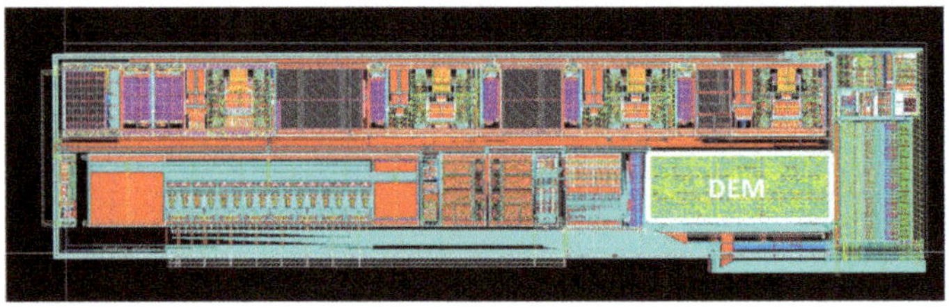

A CT sigma delta 3rd-order modulator with 2nd-order DEM

The design was used in a 3rd order modulator in a 0.18u CMOS process.

72 Recap

- **Brief background**
  - Motivations and Applications
  - 3-level vs. 2-level unit element: advantages and challenges
  - What problems DEM does and does not solve
- **Practical dynamic element matching (DEM) techniques**
  - 2-level DEM review
  - 3-level DEM
  - Limitations of each technique
  - Solution spaces
- **Higher-order DEM**
  - Issues in 2nd-order DEM
  - Solutions
- **Application specific concept of DEM**
- **Q and A**

The following section introduces several concepts to reduce the power consumption in 3-level DEM designs.

73 Problem definition

- **Reduce power consumption in the 3-level unit element DAC**
- In particular, when the input is small (-60dB) , only 1 or 2 cells are used, the rest of the cells are not used but still burn power.
- **Goal**: to turn off those unused cells without degrading performance.
- Note: *Dynamically turning off cells will cause degradation due to power consumption constraints and other dynamic effects.*

74 Variable-length DEM (I)

- Group size changes according to the envelop of the input signal
- Depends on the input data amplitude, the full group, half-group or quarter-group of the elements is used to create the signal
- The switch over to smaller group sizes is done at the time when the CUMMULATIVE ERROR is zero → DOES NOT interfere with the DEM activities
- Once the smaller group is selected, the unused elements (outside of the smaller group) will be turn off to save power
- With small group of elements, there is LESS mismatch error → Smaller shaped mismatch error in output.
- With less elements, the dynamics of the DEM is more → the DEM becomes more effective!

Variable-length DEM concept is about increasing or decreasing the number of elements needed to reconstruct the output signal such that it results in smaller mismatch error, and reduces the total power consumption.

75 Variable-length DEM (II)

- Example of a 16 units 3-level thermometer-code DAC.

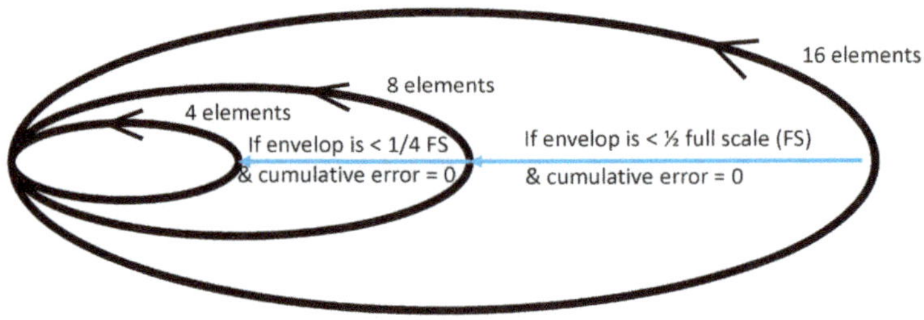

Rotation rings of DEM

The example above shows that from a group of 16 elements, the group can shrink to 8, 4 based on monitoring the input signal envelop.

76 Example: Application in audio DAC

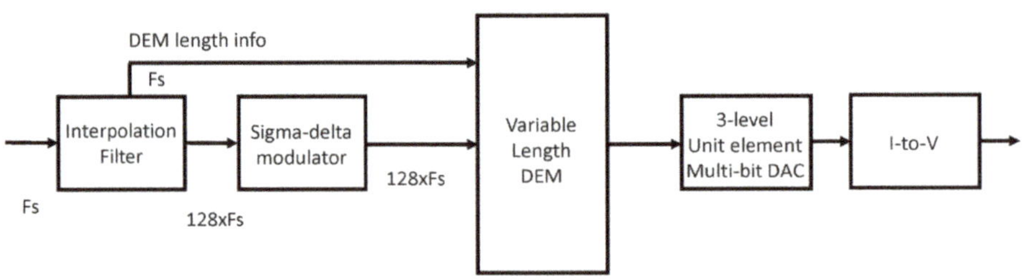

- Peak detector is used at the interpolation filter input
- Hysteresis is needed to avoid chatting (in and out of variable length mode)

The block diagram above shows how variable-length DEM is used in the design of an audio DAC. The envelop information is provided to the DEM logic so that it can perform the active group size reduction.

77 Simulation case

- 2nd order sigma delta , 128x OSR, 6b output
- Input frequency = 1kHz
- Analog elements have 0.2% rms mismatch error
- Amplitude changes from full scale to -60dB FS
  - Demonstrate the switching from full-size group to quarter-size group
  - Demonstrate the improvements in SNR
  - Demonstrate the ability to turn off the unused elements to save power

A simulation was done to verify the effectiveness of the variable-length DEM concept.

78 Simulation result: No Variable-length

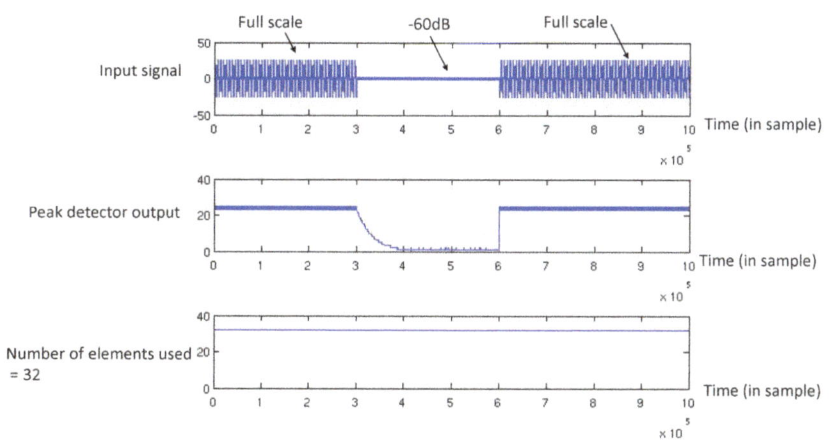

Without variable length DEM, the full 32 elements are used to in the signal reconstruction at all time.

79 No Variable-length: FFT of the -60dB input

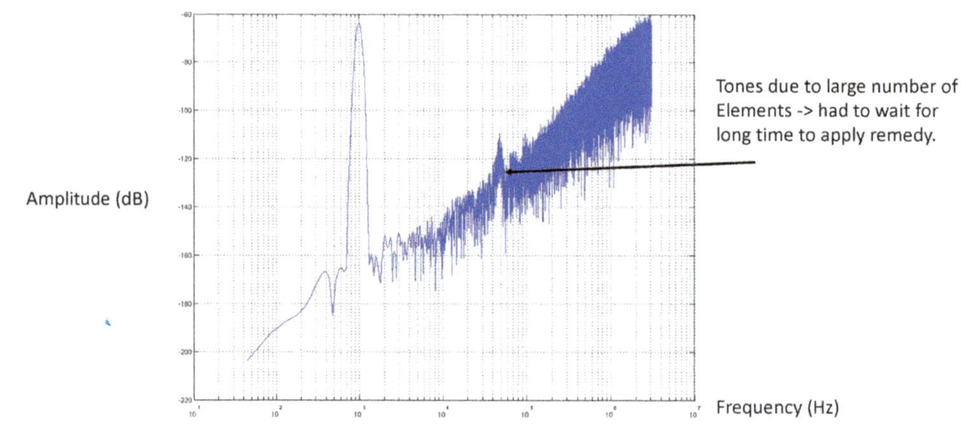

Idle tones were observed in the spectral plot due to some low data activity.

80 Simulation: With Variable-length DEM

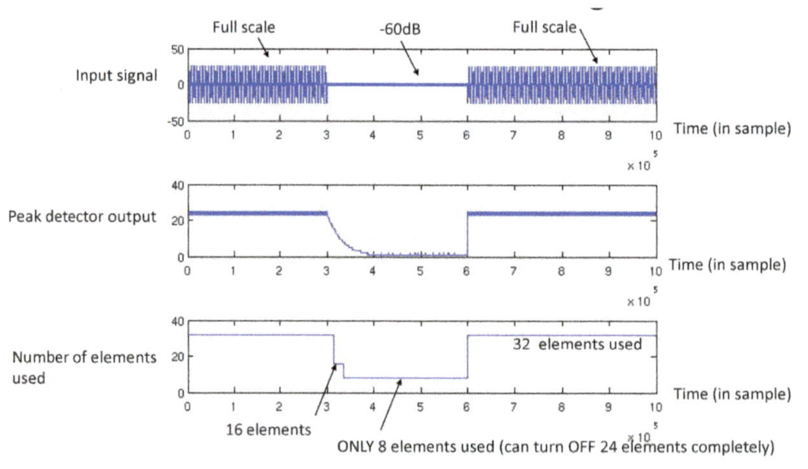

This plot shows the VL-DEM concept in action. When the signal level is below the designated threshold, the group size is dynamically reduced to 16 , then 8. When the signal amplitude jumps back to full scale, all elements are immediately put back to the active group. Harmonic distortion is a not an issue at the signal amplitude boundary.

81 With Variable-length: FFT of the -60dB input

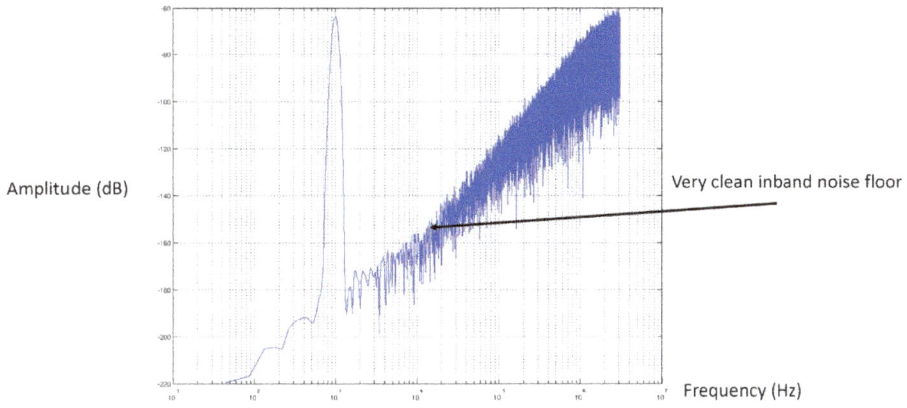

Idle tones were also eliminated due to a much more dynamic DEM system when the group size is reduced.

82 Performance compared to 2ndorder DEM @ -60dBFS, 0.25% rms mismatch

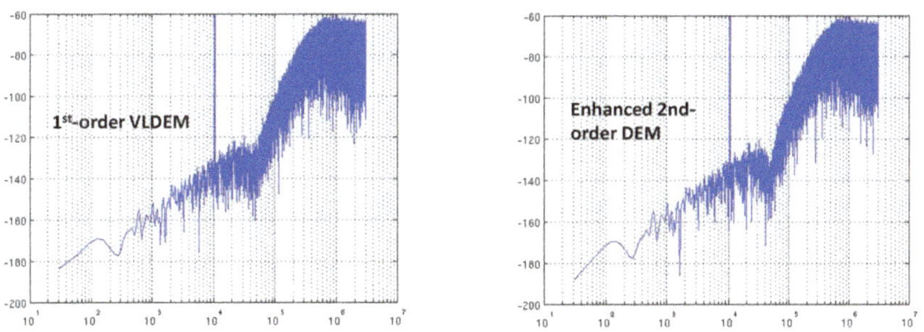

- Performs comparably to a **properly done** 2nd-order DEM (Both @ 103dB SNR)
- Much lower hardware requirement

This concept is used in a comparison between a 1st -order VL-DEM and a dynamics enhanced 2nd-order DEM design. It shows that VL-DEM performs just as well as a 2nd order DEM in steady state, and under the same input signal condition with much lower hardware requirement.This concept is used in a comparison between a 1st -order VL-DEM and a dynamics enhanced 2nd-order DEM design. It shows that VL-DEM performs just as well as a 2nd order DEM in steady state, and under the same input signal condition with much lower hardware requirement.

83 Notes

- Optimized power consumption for DAC at low input level
- Very effective first order DEM for the sub segment(s)
- Very effective for DEM idle tone remedy
- Trade off the gain accuracy across signal amplitude range (there is NO distortion because there is no switching back and forth between smaller DACs and the large DAC , on the fly).
- Suitable for applications that require high AC linearity and not DC accuracy

84 Recap

- **Brief background**
  - Motivations and Applications
  - 3-level vs. 2-level unit element: advantages and challenges
  - What problems DEM does and does not solve
- **Practical dynamic element matching (DEM) techniques**
  - 2-level DEM review
  - 3-level DEM
  - Limitations of each technique
  - Solution spaces
- **Higher-order DEM**
  - Issues in 2nd-order DEM
  - Solutions
- **Application specific concept of DEM**
- **Q and A**

85 References

1. Richard Schreier et al., "Delta-Sigma Data Converter", IEEE press 1977
2. Tom Kwan and Bob Adams, " A stereo multibit DAC with asynchronous master clock interface", JSSC vol 31, no. 12, pp. 1881-1887, Dec 1996.
3. Khiem Nguyen et al., "A 108dB SNR, 1.1mW Oversampling DAC with Three-level DEM technique", ISSCC 2008
4. A. Bandyopadhyad , K Nguyen, "A 120dB SNR 21.5mW CT SD DAC", ISSCC 2011
5. I. Fujimori et al., "A Multibit Delta–Sigma Audio DAC with 120-dB Dynamic Range", JSSC vol. 38, no. 8, Aug 2000
6. I. Galton et al., "Why Dynamic-Element-Matching DACs Work", Trans on Circuits and Systems II, vol. 57, no. 2, pg. 69-74, Feb 2010.
7. I. Galton et al., "Simplified Logic for First-Order and Second-Order Mismatch-Shaping Digital-to-Analog Converters", Trans on Circuits and Systems II, vol. 48, no. 11, pg. 1014-1027, Nov 2001
8. P. Romboutset al., "A Study of Dynamic Element-Matching Techniques for 3-Level Unit Elements", Trans on Circuits and Systems II, vol. 47, no. 11, pg. 1077-1087, Nov 2000
9. A. Bandyopadhyad , K Nguyen, "A 96dB SNR 600kHz CT SD ADC", Symposium on VLSI Circuits Digest of Technical Papers 2014
10. E. Fogleman et al., "A Dynamic Element Matching Technique for Reduced-Distortion Multibit Quantization in Delta–Sigma ADCs", Trans on Circuits and Systems II, vol. 48, no. 2, pg. 158-170, Feb 2001
11. R. Wang et al., "Split-set data weighted averaging", Electronic Letters, 16th February 2006 Vol. 42 No. 4.
12. Khiem Nguyen et al., "A 2.4mW, 111dB SNR Continuous-time SD ADC With A Three-level DEM Technique", CICC 2017.
13. Anas Hamoui and K. Martin, "Linearity enhancement of multi-bit SD modulators using pseudo data weighted averaging", ISCAS 2002, pg. 285-288.

CHAPTER 11

Time-Interleaved ADCs: A Supposedly Clever Thing I'll Never Do Again

Aaron Buchwald

Beechwood Analog, USA

Only a decade ago, terms like digitally-assisted analog were coined to acknowledge the increasingly prominent role of DSP in analog design. Unfortunately, this modifier conjures up an image whereby analog still does all the heavy lifting while the DSP remains a mere assistant. As design work is underway in 7nm CMOS, and with the introduction of 3nm on the horizon, the capacity for DSP has increased by two orders-of-magnitude just in the short time since 2008 when the 90nm ADC described herein was designed. Such an enormous increase in signal processing potential demands of analog designers that they consider, and reconsider, the amount of digital content planned for every design at every processing node. Doing so is no longer simply good practice, but essential.

That's not to say that simply throwing millions of digital gates at a problem automatically makes it better. Often times the simplest solution is the best, but the designer needs to be aware of all available options. As analog circuits get more complicated, it's easy to lose sight of what is important. A block-diagram with arrows and labels showing sophisticated signal-path monitoring and parameter manipulation may look impressive, but can easily mask multiple sources of crosstalk not included in the model. These eventually limit performance as has been my experience, in particular with one such supposedly clever idea: using system-identification techniques via a slow reference ADC for calibration of time-interleaved ADCs. After uncovering all its pitfalls, I'm quite certain this is something I'll never do again.

1

A Runner...

I'm a runner — or at least I like to think I am. I suppose if I want to call myself a "real" runner, despite having arrived at my hotel at 3am the night before my CICC 2017 presentation in Austin, Texas, I would've woken up early the next morning and took a run along the river, or the "lake" as the locals like to call it. Real runners actually do what I only wish I could. That's one of the reasons I admire elite runners so much.

2

Jordan Hasay: Elite Runner

Especially Jordan Hasay, who, just a few days prior to my arrival in Austin, ran the fastest debut marathon by any American woman by nearly three minutes. I was cheering her on from my living room with familiar excitement: I'd followed her career from the time I first saw her as a prep runner — a ninth-grader, coming to the Arcadia Invitational to take on the best high-school runners in the country, the tiniest girl on the track with a waist-length flowing blond ponytail, taking the lead from the gun in the 3200m and beating everybody a foot taller than she, including the ones from Corona del Mar (Newport Beach) and Woodbridge (Irvine) High Schools who I had come to cheer for.

3 Coach Alberto Salazar

It was particularly emotional for Jordan and her coach, Alberto Salazar, who had become like a second parent to her after the death of her mother just six months earlier. Salazar knew all about running the Boston Marathon and how important that is.

4 Salazar vs Beardsley

In 1982, Alberto Salazar had an epic battle with Dick Beardsley in one of the most exciting marathon finishes of all time. Salazar was the premier distance runner in the country. He was the reigning New York Marathon Champion, the US 10,000m champion and had a stellar career at Oregon, whereas, people outside of running circles didn't know who Dick Beardsley was, despite his continued PRs (Personal Records) in successive races (13 in a row) and having won the inaugural London Marathon the year before, but in a time still three minutes slower than Salazar's win in New York.

5 Pushing beyond Normal

Salazar had grown up in the Boston area. This was "his" race to win. He was not about to get beat by someone in a goofy hat without Salazar's exceptional track resume. So Salazar did everything he had to do to win, including bypassing every water stop on a brutally hot day — as those from the Boston area can attest, temperatures on Patriot's Day have a huge variance — and in1982 it was sunny and hot, not at all ideal for distance running. The two ran side-by-side the entire race and broke away together running the last nine miles with only the two of them. Salazar suffered mightily, dug deep, pushing beyond normal pain thresholds to just barely out kick Beardsley to the finish line to take the win. This kind of winning-at-all-costs — especially in the false-bravado world of engineering where the measure of ones' value as a person is invariably linked to the amount of hours worked past 8pm, but somehow never on the efficiency of those hours nor the hours before 10am — Salazar's achievement is what we admire — what we aspire to — or is it? Should it be?

6 When Do You Persevere: When Do You Stop

Salazar won the race but collapsed at the finish line before getting medical care where he was given six liters of saline solution intravenously because he had not drunk during the race.

"Salazar tried to convince himself that he hadn't blown it at Boston; that despite drinking so little, and running so furiously, he hadn't done himself lasting damage, but throughout 1983, he suffered one heavy cold after another: deep, racking, bronchitis-style colds: one a month. At the Marathon Trials in 1984, he struggled to a 2:12, second-place performance. It earned him a berth in the Games, but for Salazar, finishing second—especially in a race restricted to other Americans—was like finishing last. He was

6

When Do You Persevere: When Do You Stop

never the same again." (*Dual in the Sun*, by John Brant).

By now you might be wondering what, if anything, this has to do with electrical engineering or data-converters in particular. The answer is simple: Everything.

One of the biggest challenges we face as engineers is maintaining awareness — knowing when to persevere with an idea and when to stop. Knowing what is best for the application — not just best for the engineer. Much more can be gained by mastering awareness skills, knowing what can be thrown out and when. The biggest gains in engineering are in the architecture — not trying to squeeze the last dB out of a well-established approach. Knowing when to stop — when to switch — when to save your energy for another day — becomes the most important engineering skill we have.

7

No Solid Basis

As engineers, we might feel perfectly suited to ponder just these sort of questions. After all, we deal with data constantly. We pride ourselves on objectivity. We arrogantly look down upon the pseudo-sciences for lacking what we believe we've already mastered. For example, we can look at a survey ranking the Ten Most Livable Cities in Texas and immediately see such a survey as a quasi-analytical ranking with no solid basis in the type of rigor we deal with on a daily basis.

8 Subjectivity in Engineering vs. Social Science

- Cultural and ethnic diversity
- Income per capita
- Crime statistics
- Housing costs
- School quality
- Hospital access
- University proximity
- Cultural events
- Music
- Food
- Seat of state government
- Minimal sized belt buckles and hats

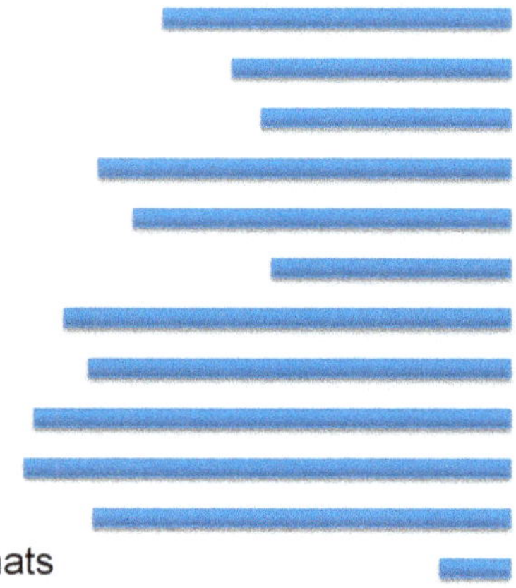

Even if each of the metrics chosen to quantize the study have very accurate and unambiguous data accumulated for years, it still begs the questions: why these metrics? Who's to say that these alone can measure something as subjective as the livability of a city. Besides, even if these were the proper set, what's the weighting. How do I rank them. We engineers see a survey such as this and immediately recognize the ambiguity and subjectivity that permeates every step. We see it because we are not invested in the process.

9 Subjectivity in ADCs

- INL
- DNL
- SNR
- SFDR
- THD
- SNDR
- NPR
- MTPR
- IM3
- Aperture uncertainty
- Power
- FoM

9 Subjectivity in ADCs

But, somehow, when we move away from the social sciences and apply measured metrics to circuits in engineering, we suddenly lose all objectivity. We've become so invested in the belief that we are objective scientists, that we can't accept — can't even see — that engineering is highly subjective. Holding to the belief that there is no subjectivity in engineering is our biggest downfall.

Why is THD, as measured with a full-scale sine wave, even a relevant metric to judge the quality of an ADC designed for broadband capture? Why is Figure-of-Merit for one block even relevant when it's the total power of the system that matters, not just sub-divisions of blocks partitioned in such a way to make them look more efficient than the system actually is. There are trade-offs made for all parts of the system: accounting for only one line-item at a time while ignoring the rest leads to sub-optimal products.

Can one judge the quality of a painting by looking at the RGB ASCII codes? You can't see the picture when focused on the technical details. This is similar to judging ADCs by metrics, divorced from application, measured using inputs that the ADC will never see. We need a clearer view, to step back, with awareness, and think about the fact that nearly everything we do in engineering is every bit as subjective as ranking the best cities in Texas.

10 Biases

Even when we admit to certain biases, we still want to believe that data is the true currency we work with: data has higher weight than anything else. But, this isn't always the case, nor is it the smart thing to do in the context of the real world. I'm not saying that anyone falsifies data or tries to suppress the truth, the problem I'm referring to is more subtle. When your career, your stock price, your patent portfolio and your future contracts all hang in the balance, all of which encompasses your family's well-being and not just your own, the best decision is not always the most objectively obvious one.

Let me give you an example to make the problem a little more concrete — and I'm not saying, one way or the other, whether I'm guilty of this, despite the fact that I know the dynamics of this scenario all too well. Assume that a particular ADC architecture has suddenly become en vogue; there is a lot of money

10 Biases

available for research, contracts and investments. You find yourself as one of the leading experts in this area. Because you know more about the limitations and because you've already worked out the second-order issues of which most who are coming to this area fresh are still unaware, your opinion is that there is nothing special about this architecture — no breakthrough — no advantage — it's just different, that's all, but suffers from the same fundamental issues like gain and speed as do other architectures. You also don't particularly believe it's the best approach for the applications for which its being touted. However, you realize that your company is now in a leading position to secure a lucrative contract to further develop this approach: Do you A: tell the program mangers: "No thank you." Their money would be better spent elsewhere, or B: give a jaw-dropping presentation highlighting your knowledge and expertise in these matters in hopes of being awarded a lucrative contract to keep your business afloat for a couple more years. You might not even do this knowingly, as sometimes your own explanations are persuasive enough to where you start believing your own hype — on a sugar-high from your own Kool-Aid.

11 Personal Battle

Before finding myself in a situation like that, it's good to get some distance and be aware of my own biases. Then it's easier to realize when I am being falsely enticed. The first thing I ever wanted to be was an Olympic Champion in the 1500m. When I learned early that I didn't have world-class speed, I then wanted to be a Rock Star. Lacking more than rudimentary musical ability, my next choice was to be a novelist. It turns out that what I was best at was engineering. But I didn't go into this profession because of a love of technology and gadgetry nor did I have a desire to create and design devices that keep people apart while they're standing together in line at Starbucks nor make it ever easier to manipulate minds for the profit of advertisers or perhaps in order to steal an election. Given my circumstances, I saw engineering as a creative and competitive outlet for my other passions where my talent was insufficient — a substitute — a search for Truth buried in the physics and the data.

So my default approach for any design or project is that it won't interest me unless it's hard, otherwise anyone could do it. Therefore, I only want to work on new, difficult, non-standard, non-commodity designs, (does this sound like a familiar nightmare to readers who happen to also be managers). And, I especially was drawn to the kind of applications where my creativity and insight would lead to new and better designs.

That's my default. That's how I eventually get my gold medal, my platinum album or my Booker prize. Understanding this about myself is vital in considering what happened to Alberto Salazar and objectively looking at every situation and deciding to take my finger off the scale, set aside biases and let the data speak for itself.

Because most of what we do in this business is "ordinary" and comprises just-getting-the-job-done, we have to find ways to find the special in the ordinary. For me, that comes down to finding the purest and best fit of all possible designs for a given application. That's the trick.

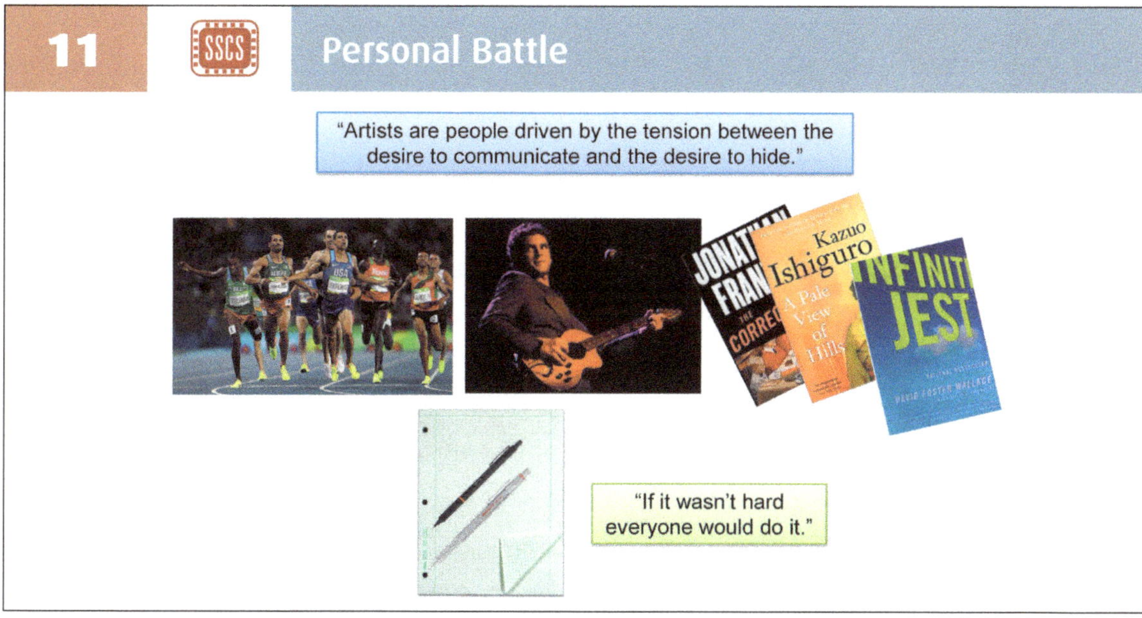

This is best explained by an example. A Standard Toilet represents a simple ADC as shown on the left while a background calibrated ADC is represented on the right by a system which contains all of the state-of-the-art features listed above and a few highlighted below:

- Tornado/Siphon Jet Flushing System
- Oscillating and Pulsing Comfort Washing
- Auto Lid open/close, Auto Flush
- Automatic Air Purfying System
- Adjustable Heated Seat
- ewater+® - mists the bowl with Electrolyzed Water, reducing the need for harsh cleaning chemicals
- Back-up Manual Flush

Assume further that you find yourself in an ideal

12 Creative Outlet: Design Example

position to be a world leader in fancy toilet design from having gone to the right school at the right time and worked on mixed-signal products that have put you in a position to be an innovator and pioneer in this field. Doing so could bring publications, patents, contracts, potential press coverage and notoriety, whereas designing a standard ADC relegates you to obscurity inside a gray, fluorescent-lit cubical.

But you realize that the key requirement in all the specifications is that the unit needs to operate when the power is out, meaning that both units pictured above behave virtually identically in that case — as simple manual-flush toilets — except one of them carries with it a lot of unnecessary overhead, risk and cost.

The question we all need to ask ourselves, if we are put in this situation is this: will we bypass personal goals and biases and pass up an opportunity to work on something new and exciting, or will we advocate for the Standard Toilet knowing it's the lowest risk, lowest cost, best fit for the application?

13 The Plan

Time Interleaved ADCs:
The Plan

The preceding provides background for what I intended to do in starting a company, Mobius Semiconductor, to develop and advance the state-of-the-art in calibrated time-interleaved ADCs. To be a viable company we needed to make something *good*. To provide differentiation, our solution had to be unique, otherwise, where's the value?

The goal was to make a unified architecture that could be used for both high-resolution and high-speed. Calibration was essential, both linear and non-linear, to allow minimally designed signal-paths. The calibration needed to run in the background and be independent of signal statistics. The architecture would be scalable and modular so that a small team could make a variety of products without redesigning from scratch at every speed and resolution.

14 Time-Interleaved ADCs Using a Slow Reference ADC

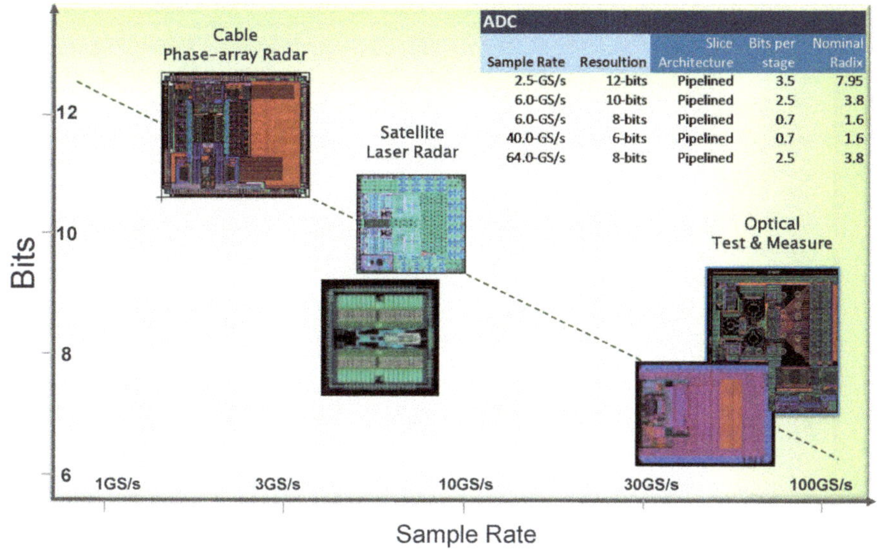

Sample Rate	Resoultion	Slice Architecture	Bits per stage	Nominal Radix
2.5-GS/s	12-bits	Pipelined	3.5	7.95
6.0-GS/s	10-bits	Pipelined	2.5	3.8
6.0-GS/s	8-bits	Pipelined	0.7	1.6
40.0-GS/s	6-bits	Pipelined	0.7	1.6
64.0-GS/s	8-bits	Pipelined	2.5	3.8

This approach was used to design five calibrated time-interleaved ADCs ranging from 12-bits at 2.7-GS/s to 8-bits at 64-GS/s, all using the same basic architecture with interleaved pipelined slices.

15 Linear Compensation: 2003-2004

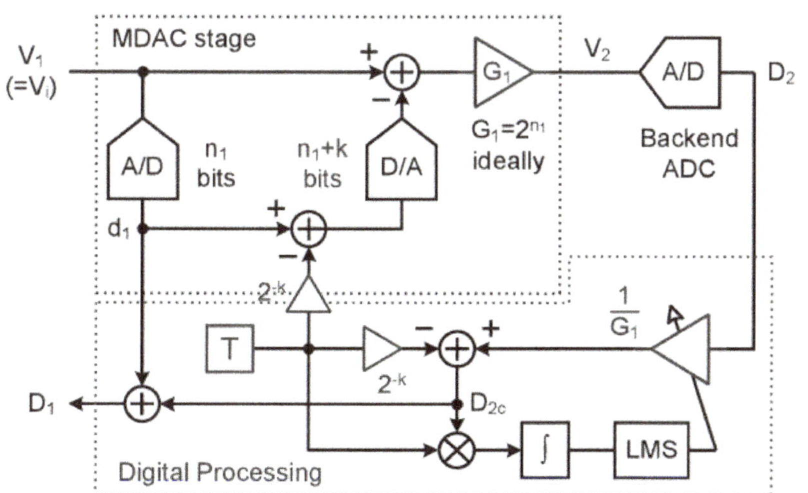

- Pseudo-Random Modulation

Ian Galton, Eric Siragusa, 2003-2004

15 Linear Compensation: 2003-2004

Previous work had been done to provide continual background calibration of pipelined ADC without disrupting the normal operation. The most common, and most often used is the approach by Ian Galton and Eric Siragusa: a pseudo-random dither is injected at the MDAC where the signal is already sampled and the DAC function is natural. Signal-path parameters of the ADC are estimated in such a way to minimize any remaining correlation. After subtracting the dither from the quantized signal and correlating against the pseudo-random sequence, ideally, if the signal-path gains are known, the dither should be subtracted exactly and the correlation should approach zero, provided the pseudo-random sequence is long enough to insure statistical independence and convergence under a slowly varying environment.

There are few drawbacks to this approach, other than sacrificing a small amount of useable signal range for the dither and that the convergence can be slow since it requires de-correlation, or separation, of a long sequence from the signal. Attempts to split the ADC (Li and Moon) and apply the dither in a common-mode fashion can help separate the dither from the signal and speed convergence.

16 Non-Linear Compensation: 2003

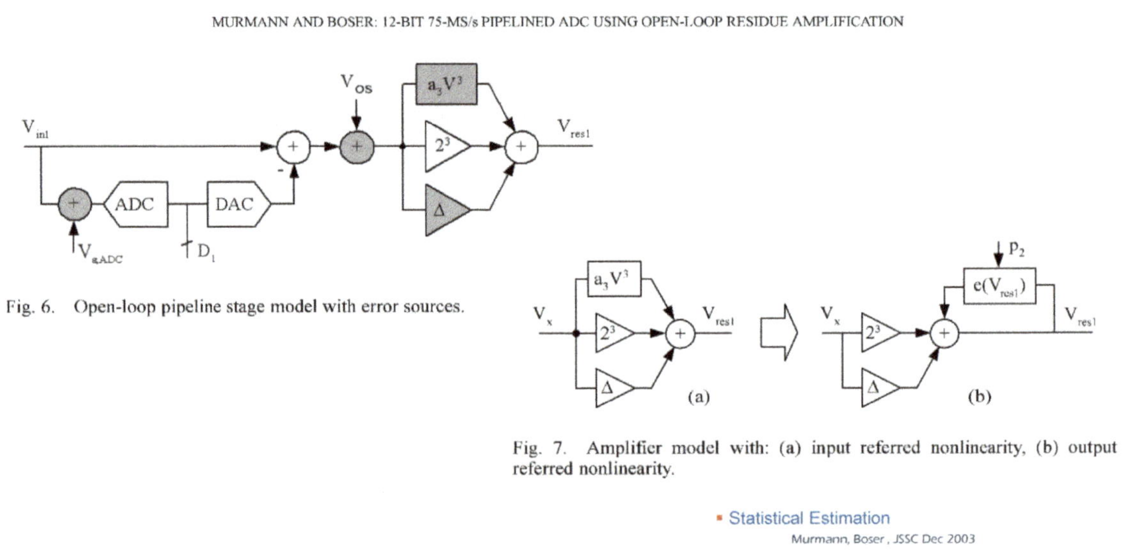

At the time we contemplated other options. Murmann and Boser had described a method for achieving a significant improvement in accuracy by using both linear and non-linear calibration. This allowed the use of simple — even open-loop — amplifiers where the native inaccuracies are regained via calibration. For background operation the approach used statistical methods.

17 Non-Linear Compensation: 2006

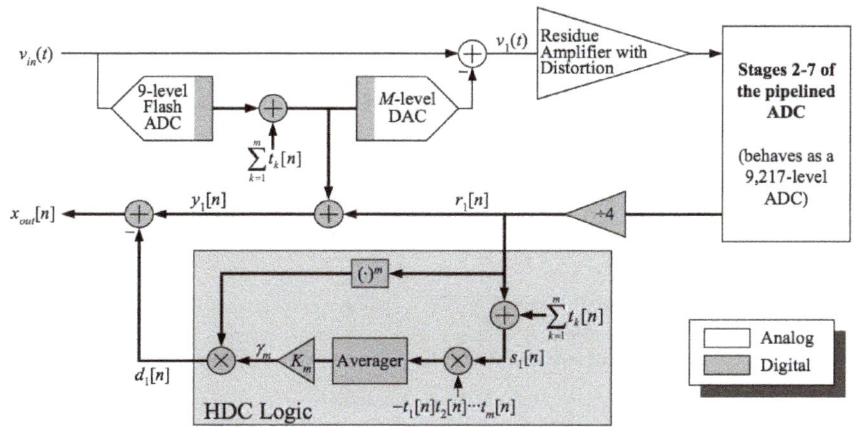

Fig. 3: An example of the HDC technique for correction of *m*th-order residue amplifier distortion.

- Pseudo-Random Modulation

Ian Galton, Andrea Panigada, 2006

The pseudo-random dither-based method was extended by Galton and Panagada to include the effects of nonlinear distortion. This was achieved by using additional sequences and estimating information about nonlinearity by correlation of cross-products: essentially extending a scalar correlation calculation to an autocorrelation matrix of cross-correlation. The problem here was that the individual sequences for linear calibration were already long so the accuracy requirement for estimating smaller nonlinear errors with the same method required very long filters for high enough processing gain such that convergence was slow — maybe too slow in practice.

I should point before moving on that, to my knowledge, the *best* approach for background calibration of a non-interleaved ADC, today, is the linear pseudo-random approach of Galton and Siragusa. By adding memory effects to correlate against past samples one can cancel kickback. By applying additional random dither to smooth residual errors (Ahmed Ali) much of the artifacts of calibration and sub-ranging can't be seen in the final spectrum. The reason that the nonlinear extension shown above is not as widely used is also an artifact of Murden's Rule, which is explained in more detail in the next two figures.

18 Causal Feedback Block Diagram

Frank Murden of Analog Devices points out that negative feedback in amplifiers is not there only to stabilize closed-loop gain: lest we forget, it also desensitizes the circuit to a myriad of error source including, power supply noise and cross-talk in addition to providing common-mode rejection. Errors coming from these sources are hard to estimate and therefore difficult to calibrate or eliminate, so there often needs to be a minimal amount of negative feedback to keep these error sources under control.

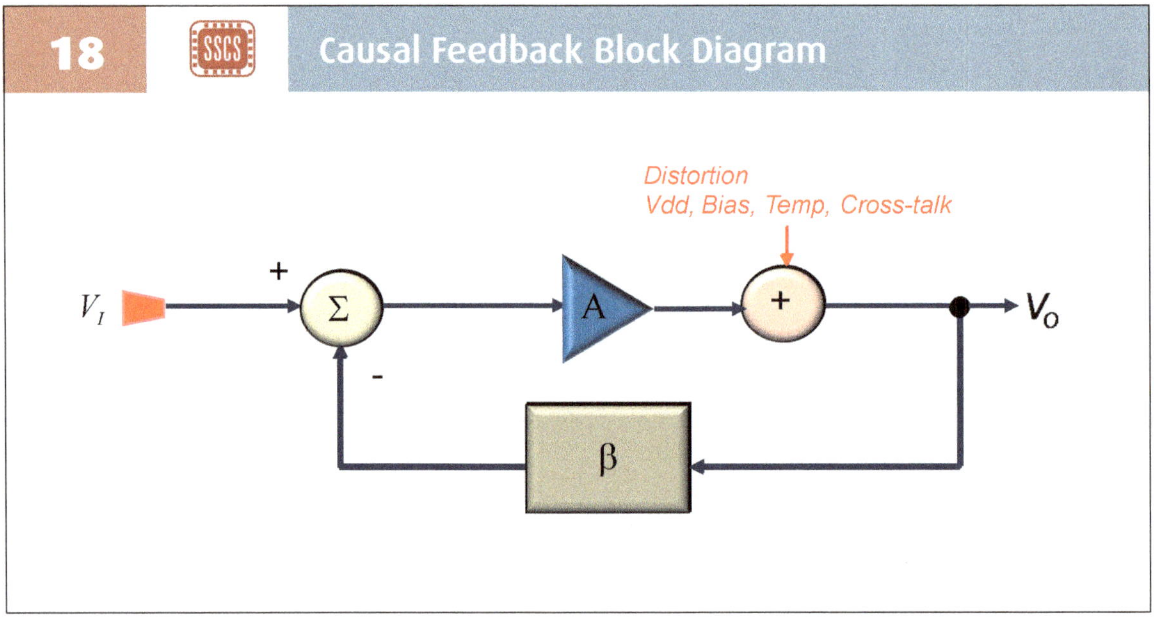

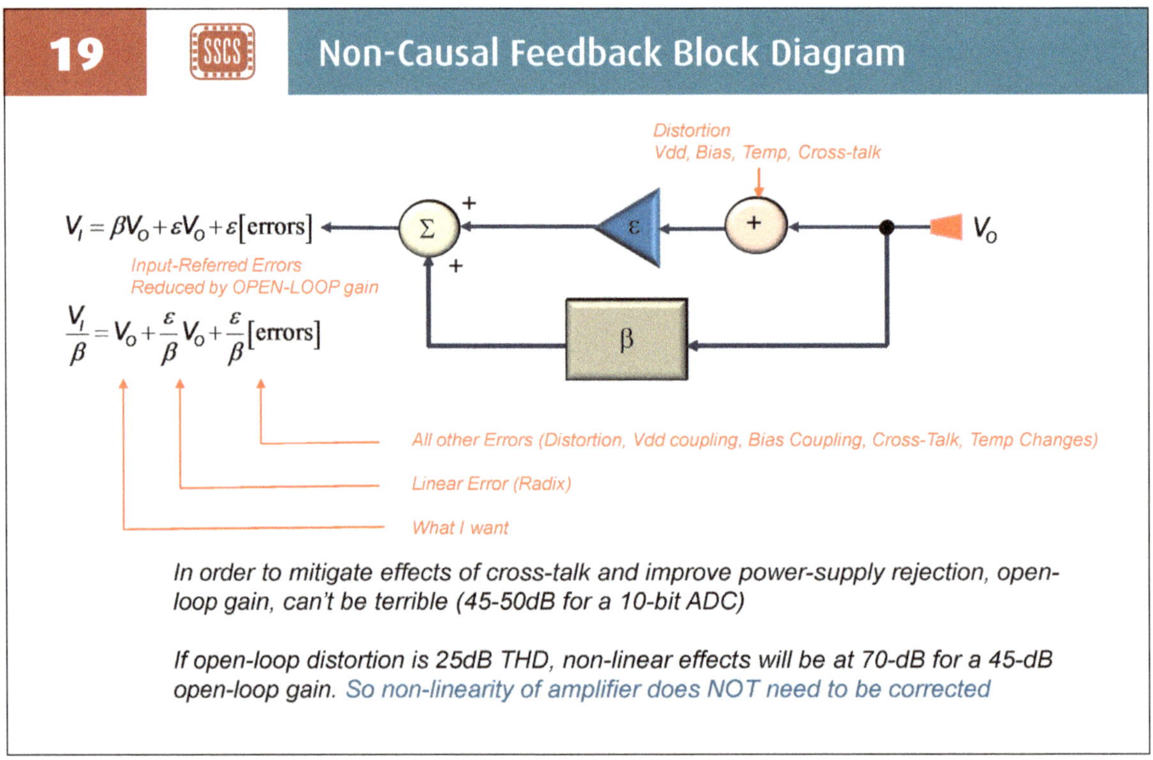

Although the loop-gain is often insufficient to achieve the desired absolute closed-loop gain accuracy so that linear calibration is needed, once there is a minimal amount of negative feedback to mitigate all other error sources, the amplifier is usually sufficiently linearized to where nonlinear calibration is not necessary.

20 Reference ADC Solution: No De-correlation Required

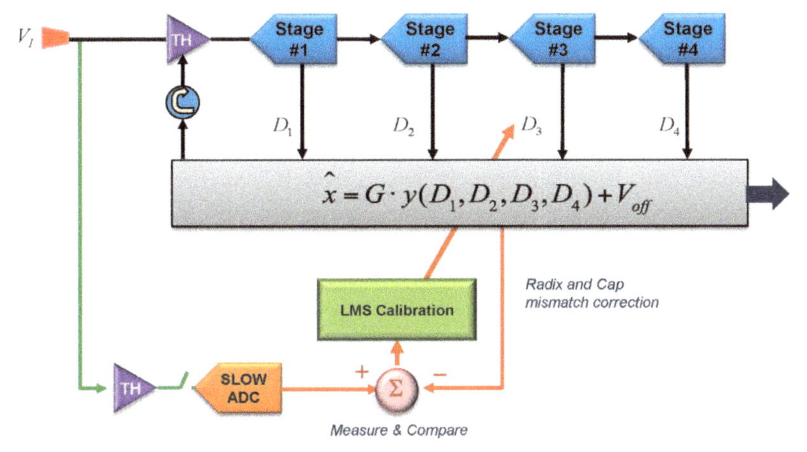

There are three main reasons for departing from pseudo-random dither in hopes of a better and more elegant solution for a calibrated ADC. The first was to extend the approach to include nonlinear effects of the amplifier in a simple way. The second was to reduce the convergence time and the third was to extend the approach to time-interleaving which I'll discuss in a bit.

The approach adopted is a standard system-identification (SI) approach using a slow, but accurate, ADC to obtain a sub-sampled ideal measurement with which to create an error signal to drive an LMS loop to minimize errors between the reference and the main ADC. Because such an approach is common in control systems and adaptive filters, the convergence properties are well-known and attractive. In addition, the approach is fairly independent of signal statistics, requiring only sufficient signal activity and a signal that is not synchronous with the sub-sampled rate.

21 Reference ADC Solution: With Dummy Reference

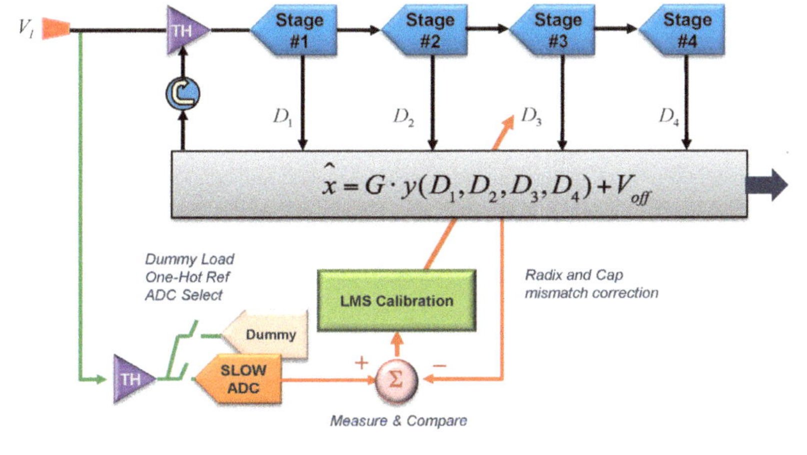

21 Reference ADC Solution: With Dummy Reference

This technique was not new. The first I had seen it applied to background calibrated ADCs was in the work at UC Davis by Dyer and Lewis. My team was well aware of the potential cross-talk hazards of using a slow reference ADC. I had personally dealt with a nightmare of mutual coupling and phase-pulling between multiple LC VCOs on the same die, so I was cautious of cross-talk potential with any architecture.

The essential problem is similar to the Heisenberg Uncertainty Principle and the measurement dilemma. The act of measuring the circuit disturbs it in such a way that the measurement becomes a biased reflection of circuit performance at other times. This can be suppressed by using a dummy measurement device. Although the reference ADC only samples rarely, a dummy sampler is inserted to sample the signal as often as the main ADC to keep the conditions the same in normal operation as they are during the rare measurement. Of course there will be slight differences, but it should be possible to keep the cross-talk, power-supply droop, kickback and other error sources near enough the same for all samples within the desired degree of accuracy of the ADC, and for a single-slice ADC that is fairly true.

22 Unified: Non-Linear Radix & Time Interleaving (I)

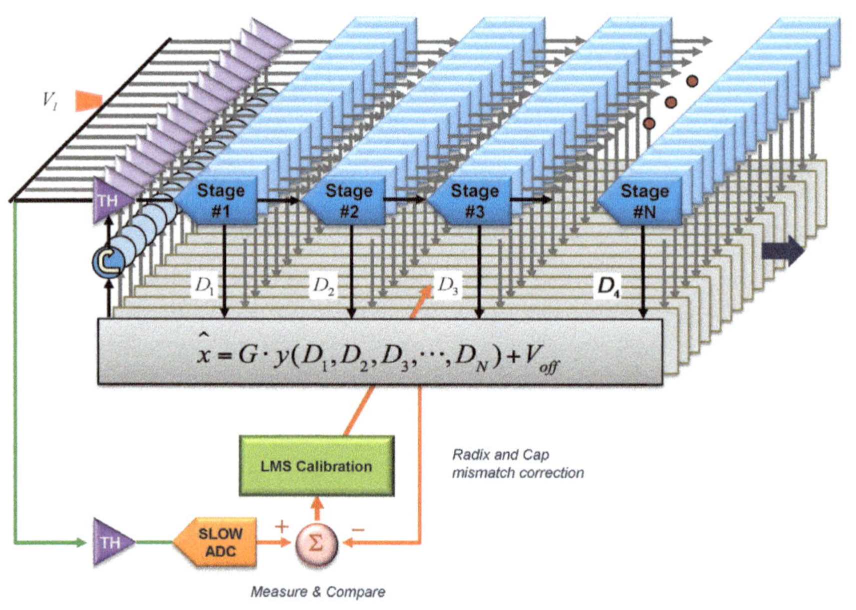

However, the primary reason for using a reference ADC in the first place was for time-interleaving as it allowed a simple means to calibrate all time-interleaving artifacts, especially time-skew, in the background. Since each slice of the main ADC is calibrated against the same reference, all slices are implicitly calibrated against each other. This provides the straightforward, modular and scalable design that we were seeking. The trick is to make sure that the operation of the slow ADC does not create more cross-talk than the baseline resolution of the circuit. Whether it does or not will determine the final verdict on this architecture.

23 Unified: Non-Linear Radix & Time Interleaving (II)

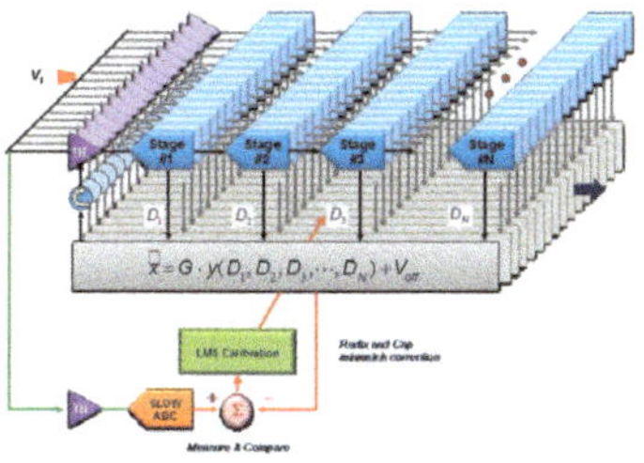

- Calibration of Linear and Nonlinear errors in Pipelined ADCs
  - Amplifier Gains, Capacitor Mismatches, Offsets
  - Nonlinearities
- Time Interleaving of ADCs
  - Mitigate effects of Gain, Offset and Clock Skew mismatches
- Background Continuous Operation
  - Track and compensate for variations caused by power supply, temperature and process
- Benefits of Digitally-Assisted Design
  - Significantly lower power than state-of-art
  - DSP-based algorithms result in best-in-class analog performance
  - Designs are scalable in speed and resolution
  - Portable across processing nodes
  - Enable next generation of commercial products

To summarize, the approach allowed for calibration of linear and nonlinear errors, was easily extendable to time-interleaving, operated in the background, was mostly independent of signal statistics, had fast and reliable convergence and was highly scalable. What more could one want.

24 A New Name

Liger

The first opportunity for a real application was for Liger, which was an 8-bit 6-GS/s ADC in 90nm CMOS intended for laser radar.

25 My Favorite Animal

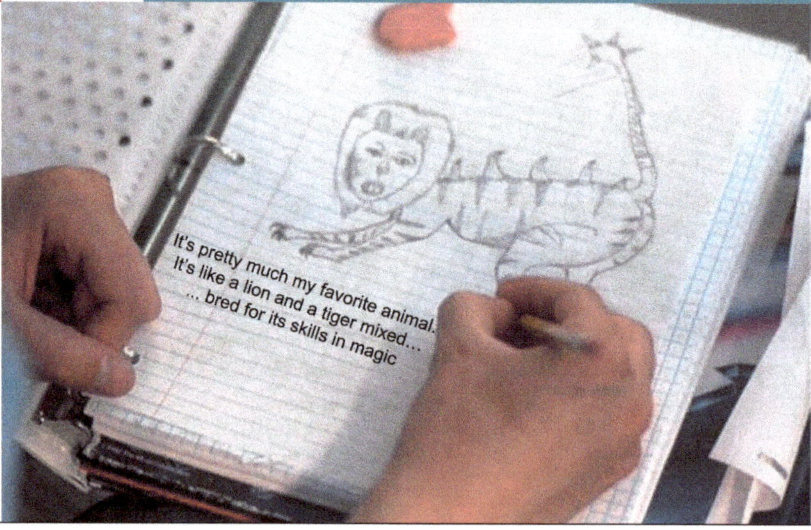

The name was chosen after watching the movie, *Napoleon Dynamite* and then looking up *Liger* on Wikipedia. It's pretty much my favorite name for a mixed-signal ADC.... bred for its skills in magic.

26 On Genomics

I highly recommend that anyone waiting for a simulation to finish, spend a few minutes researching the topic of genomic imprinting which is the unequal expression of genes depending on the parent of origin. Ligers are HUGE. They are a hybrid offspring of a male lion and a female tiger.

"Lions are known to reproduce well in captivity, with a male mating bout lasting 3-4 days while copulating 20-40 times a day. This high frequency of mating likely increases the chances of fertilization. Male and female lions have different goals for their offspring: Males want their offspring to survive, and so the paternal genes promote size to ensure that his offspring can out-compete siblings that were, perhaps, fathered by other males. Lionesses, on the other hand, want survival and quality in their offspring, so maternal genes inhibit fetal growth to increase offspring survival. Comparatively, tigers are solitary animals, and females usually only encounter and mate with one male during the breeding season. Therefore, males are not in competition and females have not acquired any mechanism to minimize the growth of their fetuses. When a male lion mates with a female tiger, however, the male lion contributes its growth-promoting genes while the female tiger has no means to inhibit growth. Thus, a liger grows to be significantly larger than either of its parents." (Zoe McKinnell and Gary Wessel, Brown University).

26 On Genomics

27 Liger: Dual 8-bit 6-GS/s ADC in 90nm

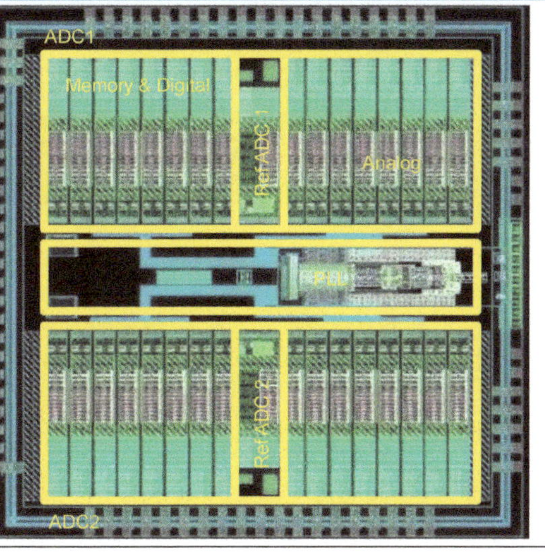

Because the application was for Laser Radar the input signal only appears in short bursts. There was no guaranteed signal activity in the absence of an echo. In the context of this system, the ADC calibration is continually updated during the dead-time between Laser pulse emissions, thus mitigating against environmental changes. A DAC was required that was muxed to the input of the ADC to provide a signal and a known reference to perform radix calibration of the ADC, while the reference ADC still performed all the background calibration for time interleaving.

The chip has two ADCs, serviced by a single PLL. Each ADC comprised 16 pipelined slices running at a slice rate of 6-GS/s-over-16 or 375-MHz. A reference ADC for each instantiation is placed in the middle. The reference ADC also runs at the slice rate and is calibrated with a low-speed DAC at start-up and in between Laser pules. The top and bottom of the chip shows the DSP calibration logic and a 4-ksamples capture memory per ADC.

28 10-Stage Radix=1.6 Pipelined Slice

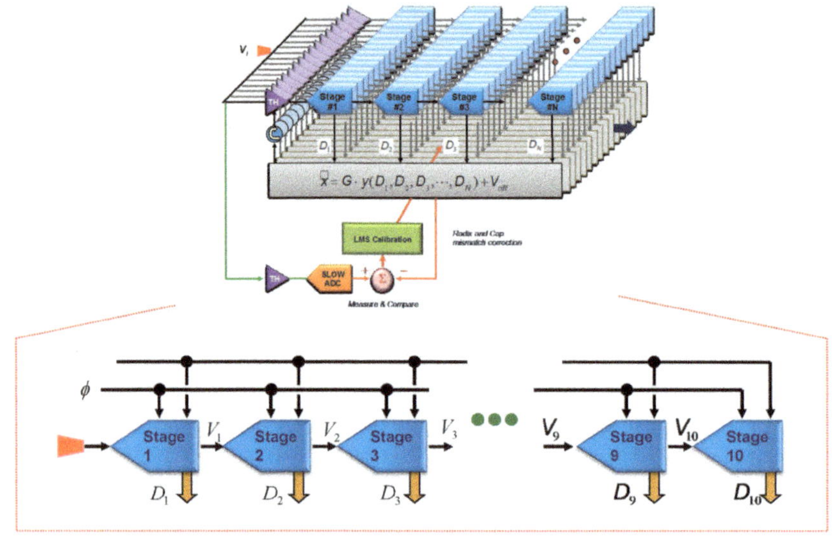

A block diagram of the 6-GS/s ADC is shown above where all 16 slices are calibrated against a single reference ADC. The 8-bit pipelined ADC actually consists of 10 stages with a nominal radix of 1.6 to achieve redundancy with the one-comparator-per-stage architecture.

The reference ADC has 12-stages instead of the 10-stages for each slice which reduces it's quantization error and helps in achieving better calibration accuracy.

29 Radix < 2 MDAC Stage (I)

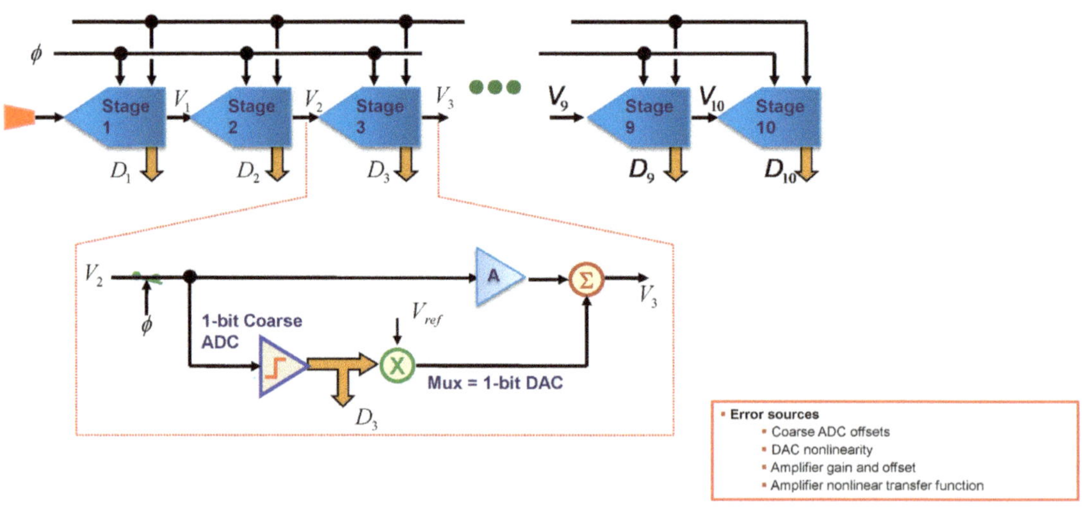

The MDAC section is represented here. An advantage of a radix-less-than-two sub-range is that only one comparator per stage is required: because it's decision is purely based on the polarity of the input, no explicit reference generation is needed for the comparator.

30 Radix < 2 MDAC Stage (II)

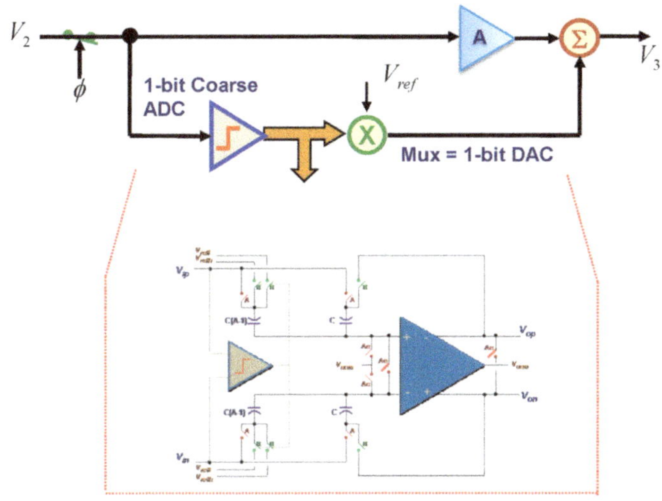

- Switched Capacitor MDAC Stage
  - Single zero-crossing comparator
  - Flip-around switched capacitor structure

The circuit diagram of the MDAC shows the switched capacitor implementation. The signal is sampled onto both capacitors: in the hold-phase one of the capacitors is flipped across the amplifier to achieve a nominal closed-loop gain of 1.6.

31 Radix < 2 MDAC Stage (III)

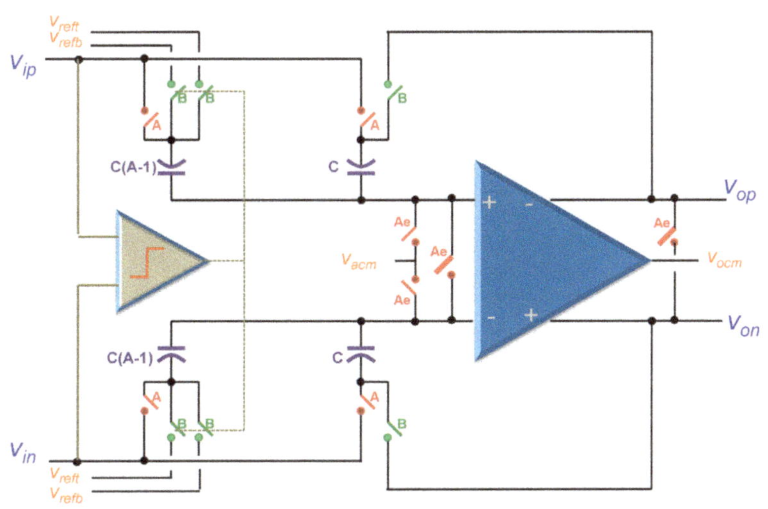

The operation of the switching is shown in the following diagrams.

32 Radix < 2 MDAC Stage (IV)

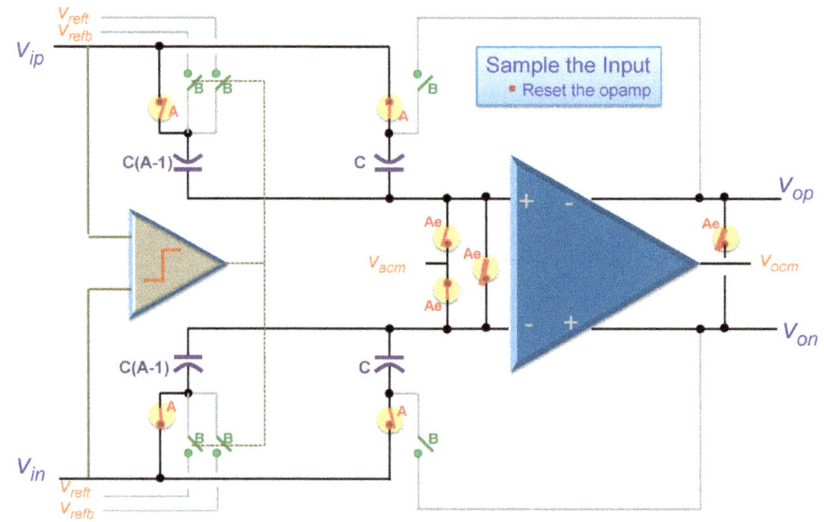

First the input is sampled onto the capacitors relative to the common-mode at the amplifier input, which has a main shorting switch and two smaller switches to keep this node anchored to the desired common-mode while the opamp is also reset.

33 Radix < 2 MDAC Stage (V)

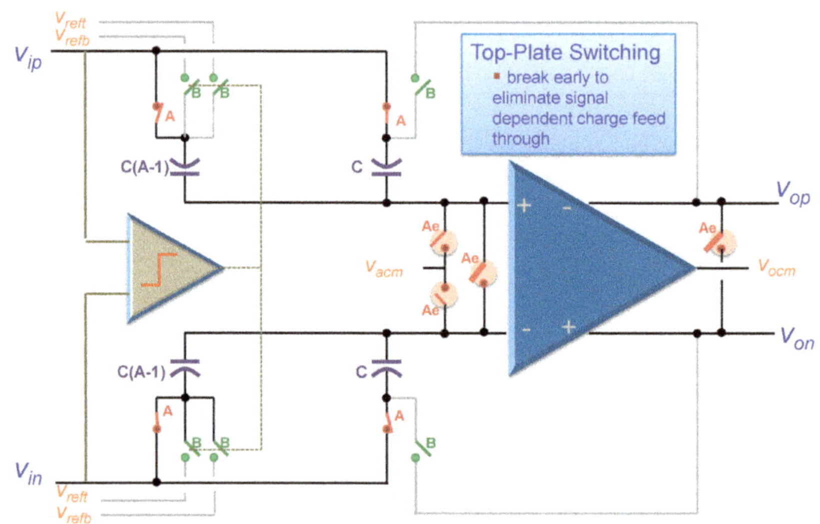

The top-plate switches are released (denoted Ae, for phase-A-early) before the input switch is released which minimizes signal-dependent charge injection.

34 Radix < 2 MDAC Stage (VI)

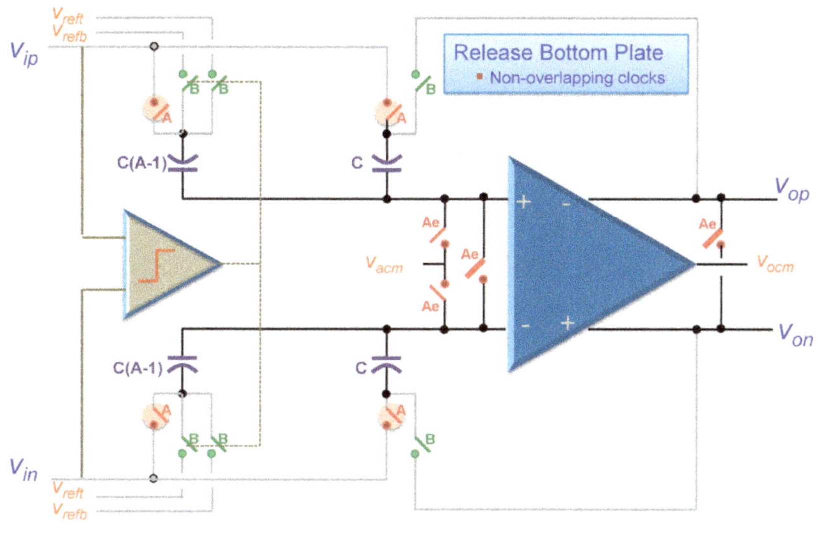

Then the input switches are released

35 Radix < 2 MDAC Stage (VII)

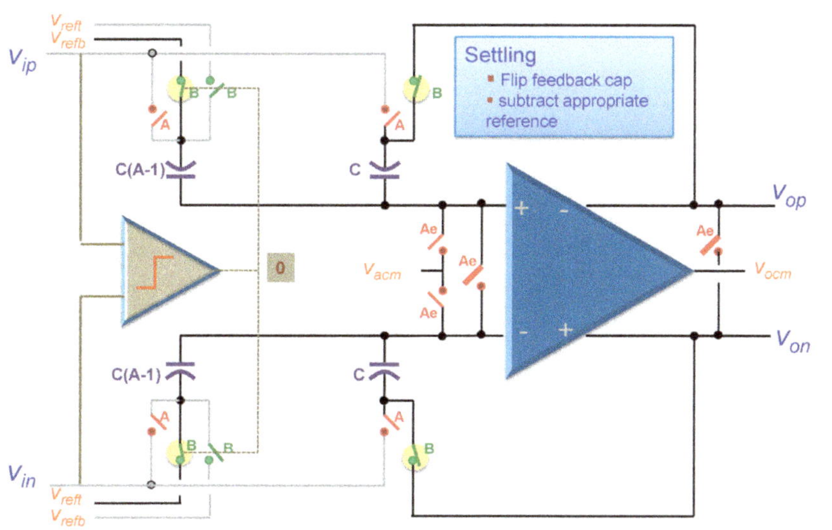

On phase-B, the feedback capacitor is flipped around the amplifier while the input cap is connected to either the positive- or negative-reference as directed by the comparator decision. The case for a zero result from the comparator is shown.

36 Radix < 2 MDAC Stage (VIII)

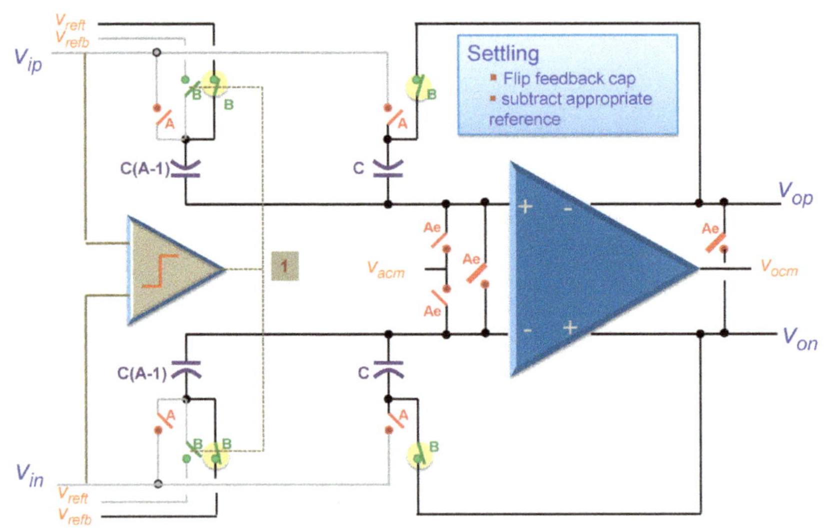

When the comparator output is "one" the reference connections to the input capacitor are reversed.

37 Simple Amplifier

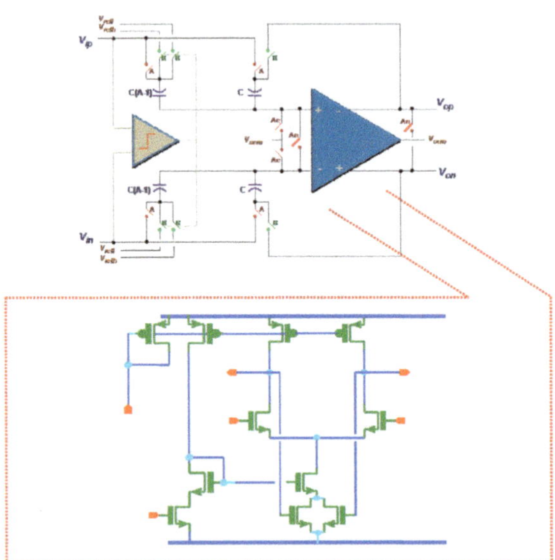

- Simple Amplifier
  - Local triode-degeneration common-mode feedback
  - DAC nonlinearity
  - Amplifier gain of only about 14-16 (24dB)
  - Amplifier nonlinear transfer function

37 Simple Amplifier

Calibration relaxes the absolute accuracy requirements of the amplifier. Therefore, a very simple, single-stage differential amplifier was used with embedded common-mode feedback using two triode-resistors in parallel which creates a degeneration resistance modulated by the common-mode level, thus providing feedback to keep it at the desired level to within the matching of the bias circuit and main amplifier.

As a side note: I was not aware of Murden's Rule at the time I opted for this amplifier but learned quickly. The gain is only about 14-16 (23-24dB). As the design progressed, it became clear that there were many other sources of errors to deal with which are often neglected when the gain is high and the input nodes appear like a nice virtual ground. Simple, low-gain amplifiers are viable, and should be the subject of continued investigation but methods to mitigate the effect of not having the rejection that feedback provides will be necessary — Violate Murden's Rule at your own peril.

38 Foreground Calibration

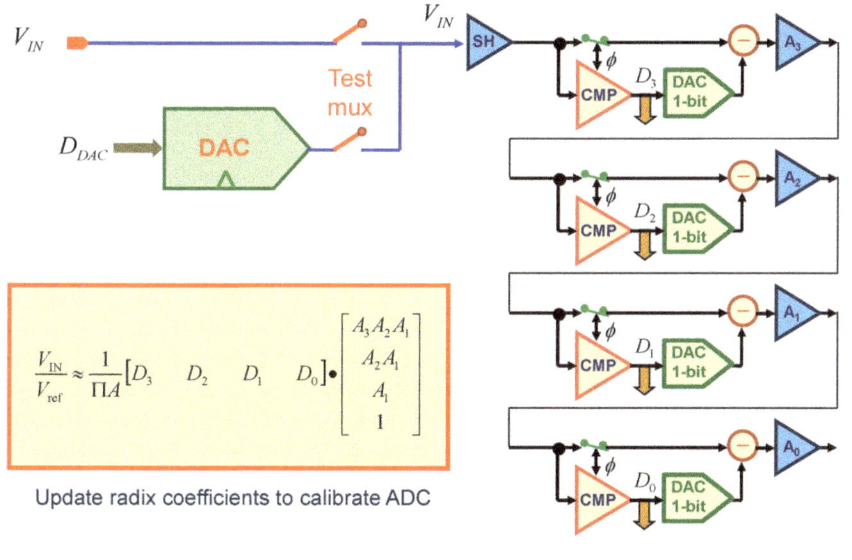

$$\frac{V_{IN}}{V_{ref}} \approx \frac{1}{\Pi A}[D_3 \quad D_2 \quad D_1 \quad D_0] \bullet \begin{bmatrix} A_3A_2A_1 \\ A_2A_1 \\ A_1 \\ 1 \end{bmatrix}$$

Update radix coefficients to calibrate ADC

Foreground calibration is implemented using a current-mode DAC with shuffled rows and columns which provides a very linear (on average) input signal. The transfer function is dependent on all of the gains in the signal path. An LMS calibration loop estimates these gains to produce the reconstructed radix vector.

39 Radix Calibration

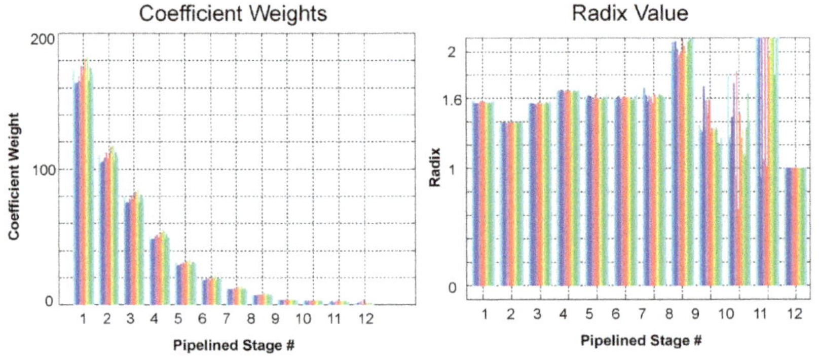

The plot on the left shows the radices for all 12-stages of the reference ADC. These radix coefficients are then de-embedded in the plot in the right to show the estimated gain of each stage. The different colors show the gain estimate for each of the 16 converter slices. The gains for all MDAC stages are "close" to 1.6, although the second-stage gain is lower. This was expected: in order to increase speed (reduce settling time) the capacitance was aggressively scaled in the second stage, so some of the signal is lost to parasitics. Also the gain of stage nine is closer to 2. This is because minimal size stages were used at the end of the pipeline and the transition from a larger stage driving a smaller stage resulted in this discontinuity in gain. As the ADC is only designed for 8-bits of noise, the radix accuracy for the low resolution stages is poor. By definition, the weight for the final stage is unity.

40 Radix Evaluation and Settling Behavior

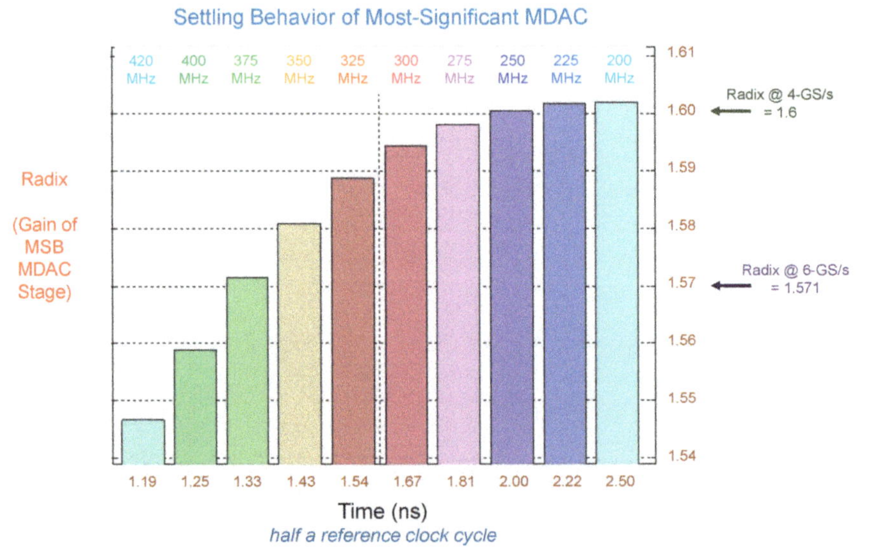

40 Radix Evaluation and Settling Behavior

Radix calibration can be used to determine the settling time of each MDAC stage. The gain values across all 16 slices for the first MDAC stage were calibrated as a function of the sampling rate, averaged and then plotted here as a function of settling time (inverse of sampling rate). For low frequencies the MDAC has enough time to settle to its DC value and achieve a gain of 1.6. As the sample frequency is increased, the amplifier doesn't have enough time to settle and the LMS algorithm estimates a smaller gain, corresponding to incomplete settling.

In order to save power, the amplifier was purposely designed to not fully settle such that the remaining error would be removed by the radix estimation. This is illustrated by the dashed line at the intended sample rate of 375-MHz per slice, the MDAC has not fully settled resulting in a radix of slightly less than 1.6.

41 Did Anything Change After the Calibration?

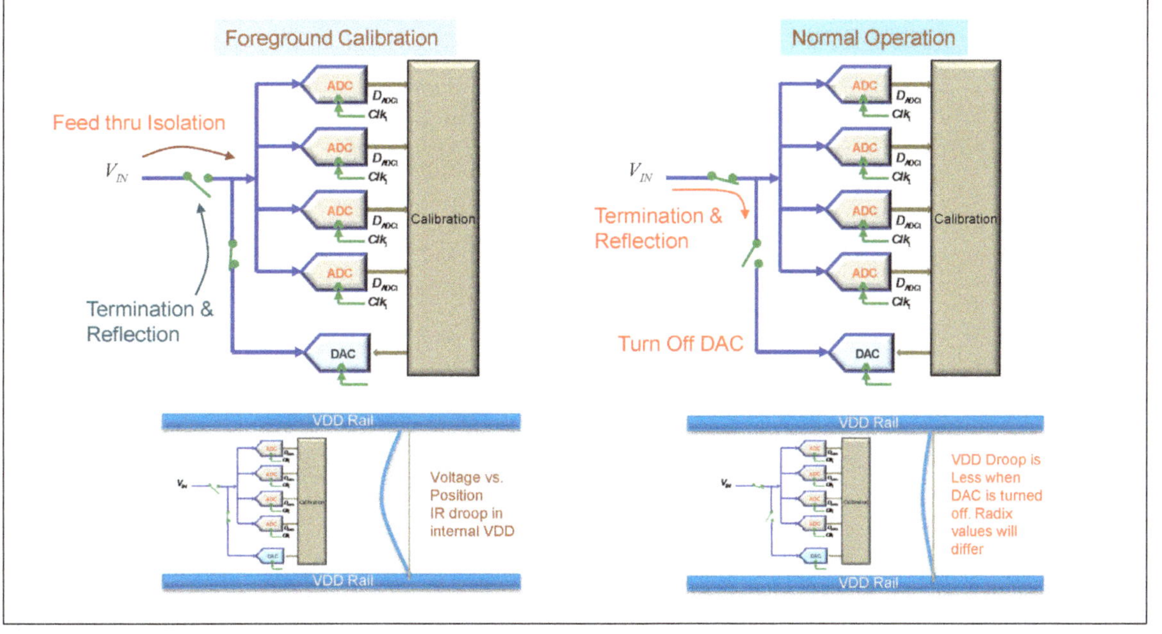

An important consideration is to make sure that the environment for the ADC is the same during calibration as it is in normal operation. Whenever there is a foreground calibration, muxes are involved, so care must be taken to match the multi-path conditions as well as necessary.

The diagram above on the left shows the DAC connected to the input during foreground calibration. Not only is the termination and the kickback path different, than in normal operation but the DAC is also connected to the same power supply as the ADC causing a slightly different voltage droop than exists in normal operation when the DAC is shut down. Since the MDAC amplifier gain is a function of the supply voltage, small systematic errors in radix calculations were found because of this phenomena. Plotting the radix errors for ADC slices across the die showed a second-order error indicative of a droop in VDD.

42 Reference ADC Synchronization: 4-Slice Example

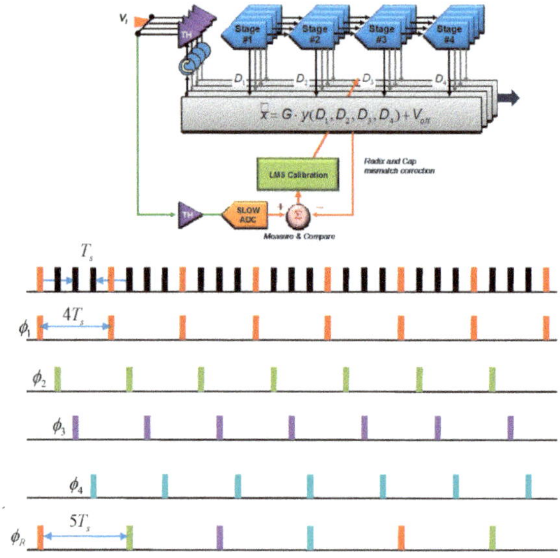

A divide-by-16 block generates clocks for the ADC slices and a divide-by-17 generates a clock to the reference ADC. Every 17th sample the reference will align with an ADC slice. Then 17 samples later, the reference ADC will align with the next slice, cycle-slipping through until 16 x17 samples later when the pattern repeats.

This principle is illustrated above for the case of a 4-slice ADC with a reference ADC sampling at Fs/5. Because of a relative-prime set of frequencies, the reference ADC will cycle through and align to each slice, repeating every 20 samples. A single LMS loop adjusts the radix, offset, gain and time skew so all slices match the reference and all timing is based on a full-speed clock.

43 Reference Clock Crosstalk: $2f_s/17 \pm f_{in}$

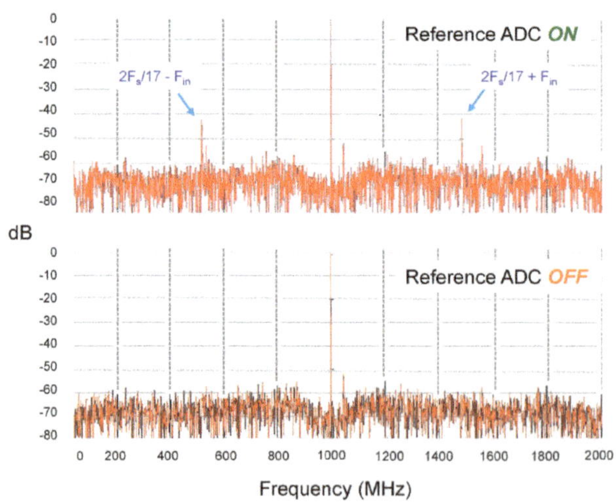

43 Reference Clock Crosstalk: $2f_s/17 \pm f_{in}$

Having multiple frequencies in the same circuit can be problematic. Despite attempts to keep the cross-talk to a minimum, the walls are thin, so to speak, and information at the reference rate unintentionally modulates the signal. This is shown clearly in the spectra above. The top plot shows Liger after calibration, but with the reference ADC still running. Spurious tones at 2Fs/17 ± Fin are visible at about 42-dBFS. The bottom plot shows the same conditions after the divide-by-17 is held in reset, effectively killing the clock to the reference ADC. By carefully examining the distribution of the parasitic coupling on each of the 16 ADC slices it was observed that the coupling appeared random and not systematic with respect to the physical location of the slice. It was surmised that the cause was an inter-modulation of the phases of the Fs/16 and Fs/17 clocks in the clock generation circuitry caused by crosstalk in the bias and through the power supply.

44 f_{in}= 251-MHz: f_s= 4.096-GHz

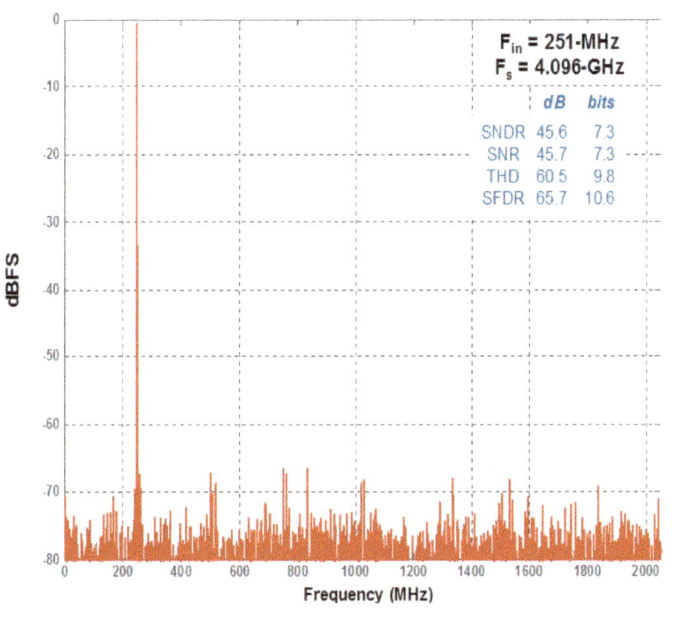

For many applications full background calibration is not needed as was the case here for the intended application of laser radar and laser range finding where the ADC is quiet most of time and can be calibrated offline. In such use-cases the ADC calibration is updated so when the system receives an echo it accurately quantizes the return signal.

Spectra of the circuit using the reference ADC for calibration and then turning it off in normal operation while waiting for an echo are shown above and in the following plot.

45 f_{in}= 1051-MHz: f_s= 4.096-GHz

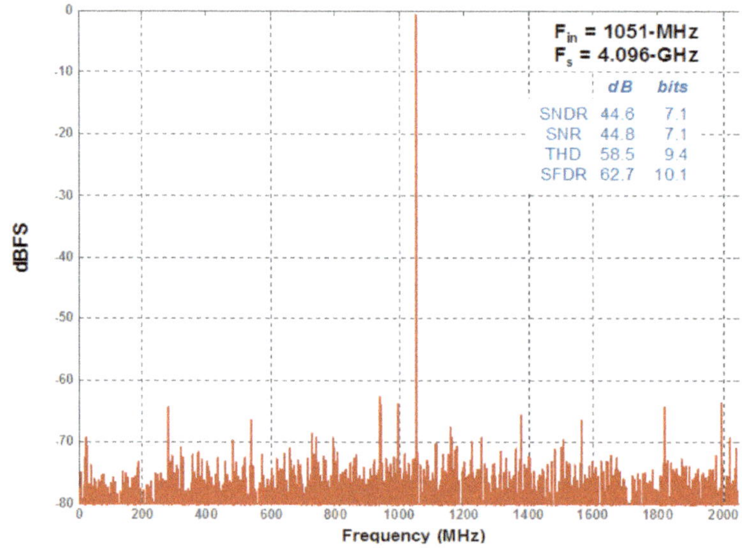

The ADC achieves better than 7.3-effective bits with a spurious-free dynamic range (SFDR) of 65.7-dB (10.6-bits) for an input of 251-MHz. For an input over 1-GHz, performance is maintained at 7.1-effective bits where (SFDR) is 62.7-dB (10.1-bits). Results were promising, but we knew we would have to take better care to isolate the reference ADC in any further implementation.

46 The Next ADC Designed

Chelsea

The next ADC designed was Chelsea, a 2.7-GS/s 12-bit eight-way time interleaved pipelined architecture implemented in a 130nm BiCMOS process. Several changes were planned for the reference ADC to mitigate inter-modulation observed in Liger. Because Chelsea is a 12-bit design, the reference ADC was made more accurate. The sample rate was reduced to approximately 1-MS/s and a recirculating architecture was used that relied on dynamic element matching for the capacitors. In order to improve isolation, separate power supplies and bias circuits where used. The reference ADC was also physically separated from the main.

47 3-GS/s 12-bit ADC, PLL & Memory

The layout of a 2.7GS/s 12-bit ADC with 75-dB SFDR through the first Nyquist zone and 62-dB through the 3rd Nyquist zone is shown above. All digital circuitry is included to perform background calibration. An 8k-sample memory is used to aid in data capture for bench-testing. The chip measures 4.5mm x 4.7mm and is fabricated in IBMs 8HP 130nm BiCMOS process. The reference track-and-holds and quantizers are on the leftmost middle of the chip, while the main ADC branches to the right: the 8 slices are approximately at the chip center.

48 8-Way Interleaved: 4-Stage Pipelined ADC

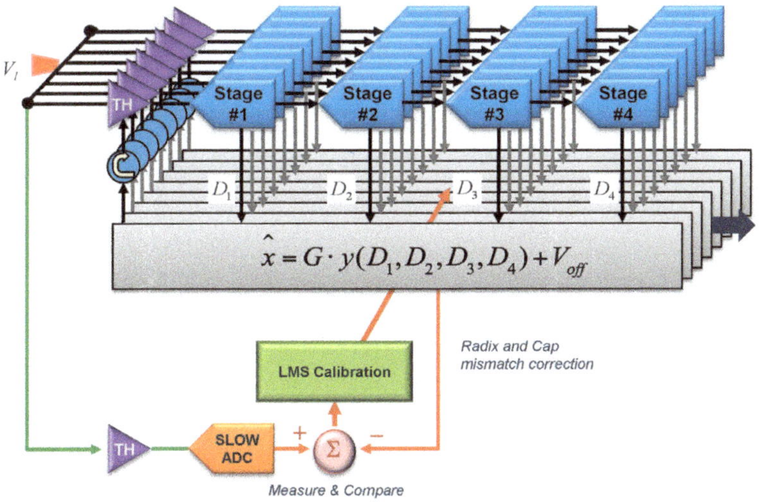

48 8-Way Interleaved: 4-Stage Pipelined ADC

The above shows a block diagram of an eight-slice time-interleaved pipelined ADC using a SI (System Identification) slow ADC for a reference for full background calibration. The gain, time-skew and offsets are also corrected so that all slices appear as identical and uniformly spaced in time.

To align each of the eight slices to the reference ADC, a full-speed 2.7-GS/s clock is run through a delay line ranging from one-to-eight unit samples. A clock selection mux chooses one of eight clocks to align with any of the eight slices with the exact precision of one unit-interval step, but in a user selectable random order. All calibration circuitry resides on chip as does an 8k-sample memory for use in test. Error correction is performed exclusively in the digital domain, directed only by the magnitude of the sub-sampled error between the main ADC slices and the reference ADC, with the exception of the timing skew correction which requires both magnitude and direction in order to converge. It is not sufficient to have only the sample from the reference ADC. Therefore, a slope estimator is also included which when multiplied by the absolute error from the reference produces the sign of the update direction. The LMS engine uses this information to drive a digital code, which in turn adjusts the edge position of eight capacitive clock-delay-DACs, thus closing time-skew correction in the analog domain. Randomization of the reference clock phases eliminated signature spurs from the clocking of the reference ADC.

49 Input Buffer Stage

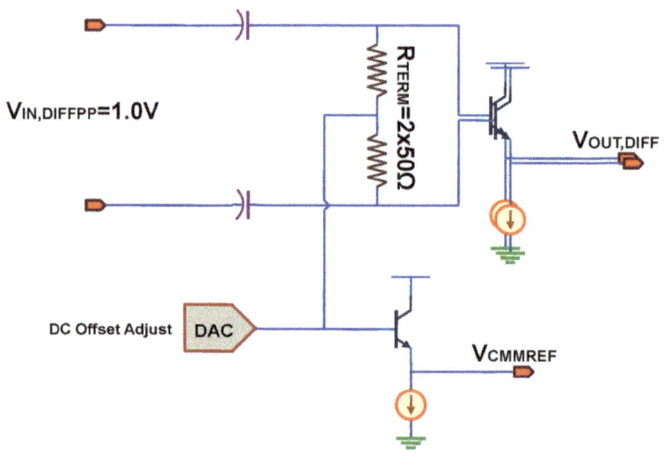

Simple AC coupling is used to connect the input. Bipolar transistors in the BiCMOS process were utilized as emitter follower buffers. A DAC is used to set the common-mode and a replica emitter-follower buffer tracks the common-mode level-shift with respect to the signal input.

50 Sample & Hold

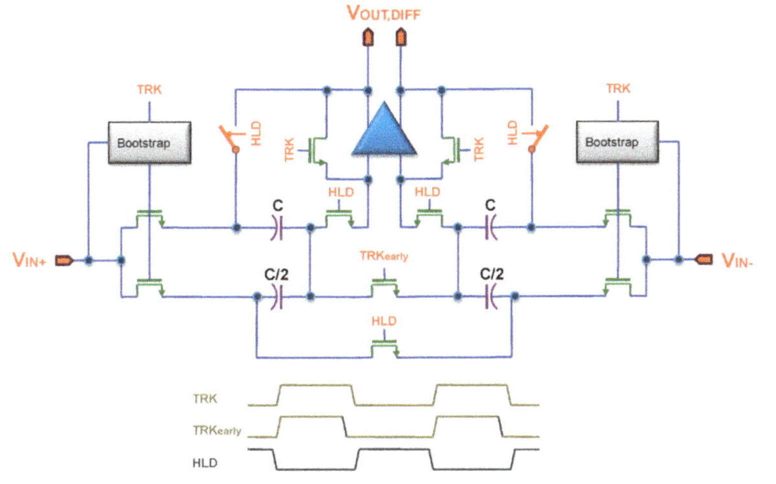

The sample-and-hold circuit is a flip-around type with a fast amplifier. Input switches are bootstrapped to improve linearity.

51 Track Mode

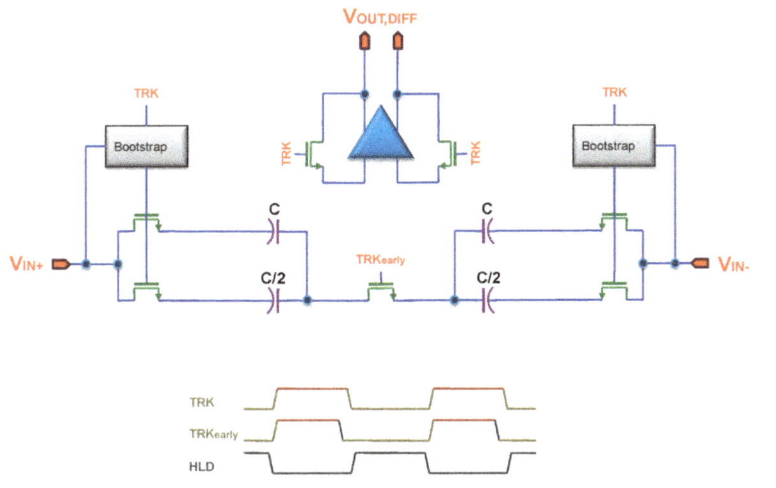

In sample-mode the input is applied to the capacitors whose top plates are shorted together by the track-early switch while the amplifier is held in reset by connecting output to input.

52 Hold Mode

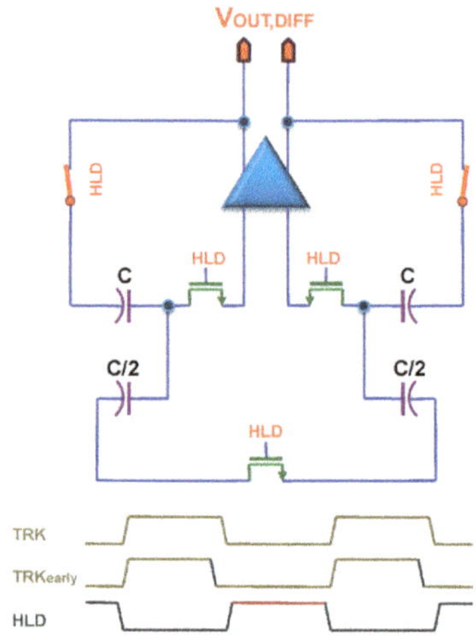

In hold mode, the larger cap is flipped across the amplifier and the smaller caps are shorted together. This provides a gain of 1.5 which eased the KT/C requirements for the capacitors of the following MDAC stages.

53 Sample & Hold Amplifier

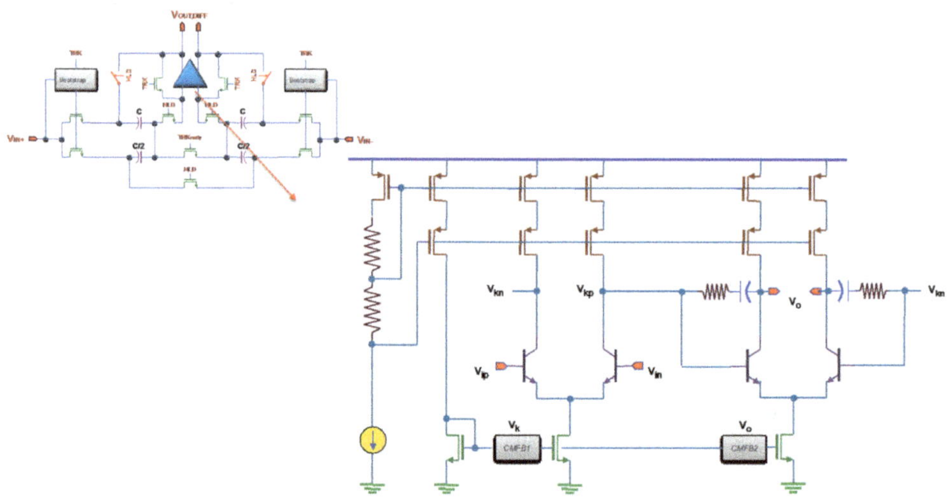

The amplifier uses two gain stages with Miller compensation and common-mode feedback around each stage. Bipolar devices are used in both gain stages. The base-current in the first stage causes droop on the input capacitors. This would not be desirable for an MDAC as this would result in a radix error, which would convert to a nonlinearity due to the sub-ranging action of the MDAC. For the track-and-hold, where no sub-ranging occurs, this only results in a gain error but not a nonlinearity which is tolerable.

54 3.5-bit-per-Stage Pipelined Slice

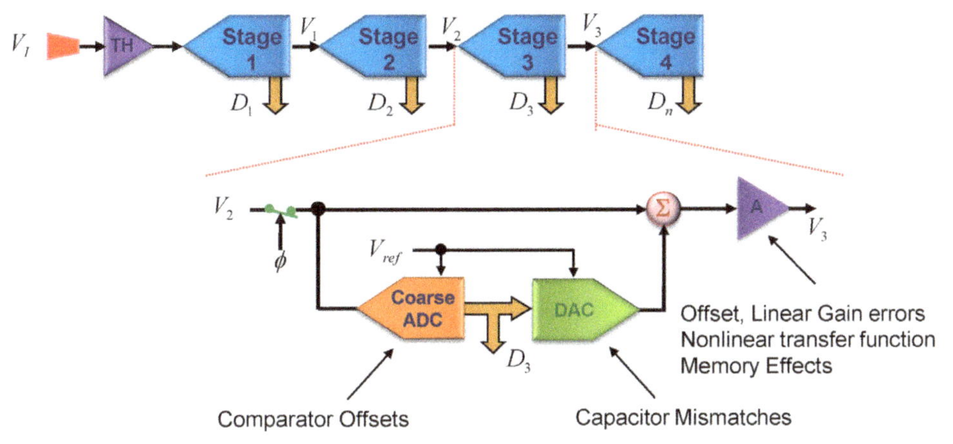

The pipelined slice used four stages of 3.5-bits per stage, using the traditional 14 comparators in the course ADC.

55 MDAC 3-bits Resolved per Stage (I)

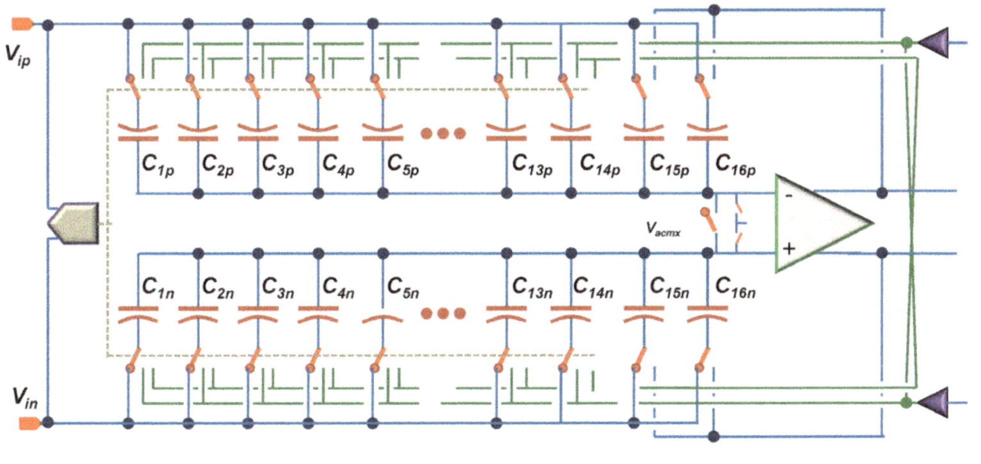

There are a total of 16 differential capacitors for the MDAC. The 14 comparators control the polarity of the reference switches to the first 14 pairs of capacitors while the last two are flipped across the amplifier to achieve a nominal closed-loop gain of eight.

56 SSCS MDAC 3-bits Resolved per Stage (II)

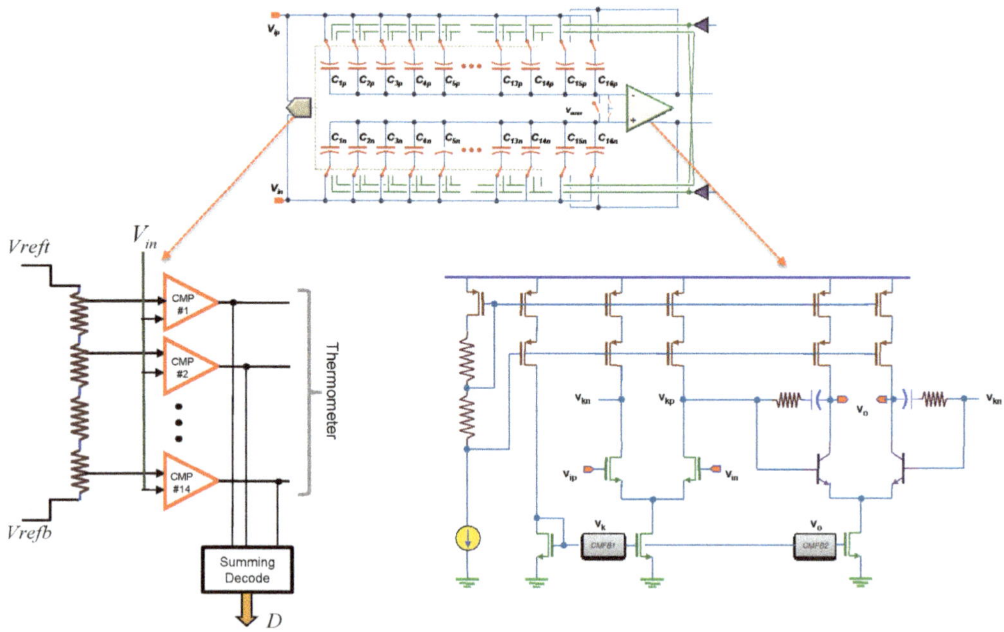

The amplifier for the MDAC is almost identical to the track-and-hold amplifier except that the input stage is an NMOS pair to eliminate any input current that would result in signal loss. The 14 reference values for the coarse ADC are obtained from a resistive ladder.

57 SSCS The ADC Performance at 2.5-GS/s

Performance vs. F_{in}

The following plots show the ADC performance at 2.5-GS/s as a function of input frequency, starting at DC.

58 SSCS f_{in}= 0: Input Terminated to 50-Ω

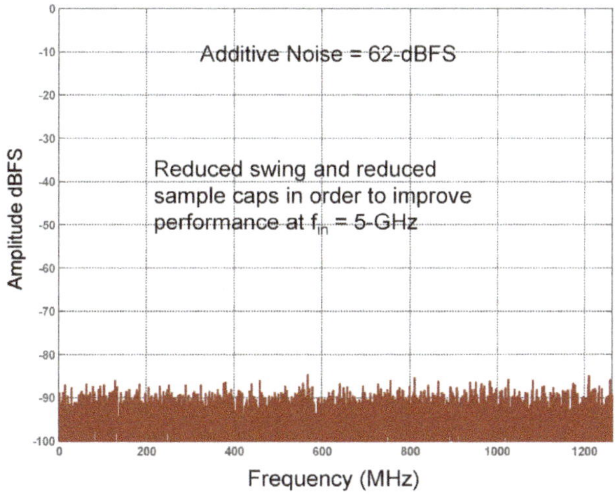

A "zero-input" test is a good place to start when testing any ADC. There should be no impact of nonlinearity or jitter. What is measured is the baseline additive noise. If the noise looks non-white, then something is feeding through and needs to be addressed — probably before any other testing is done.

59 SSCS Operation in 1st Nyquist Zone: F_s=2525-MHz

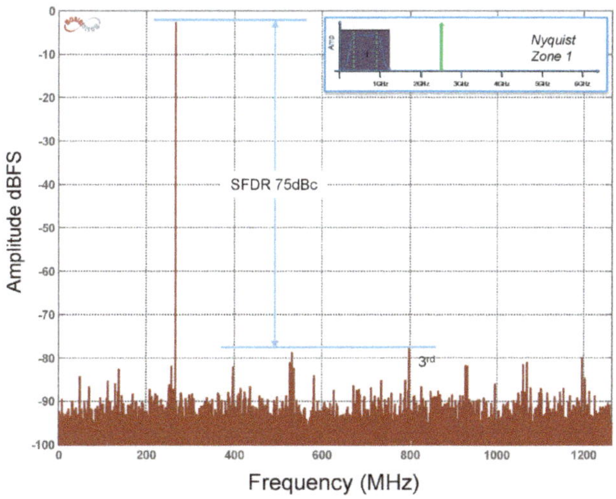

Measured frequency response with an 8k-sample FFT for an input in the first Nyquist zone, fin= 252-MHz and fs=2.525GS/s.

60 F_s=2500-MHz: F_{in}=1052-MHz

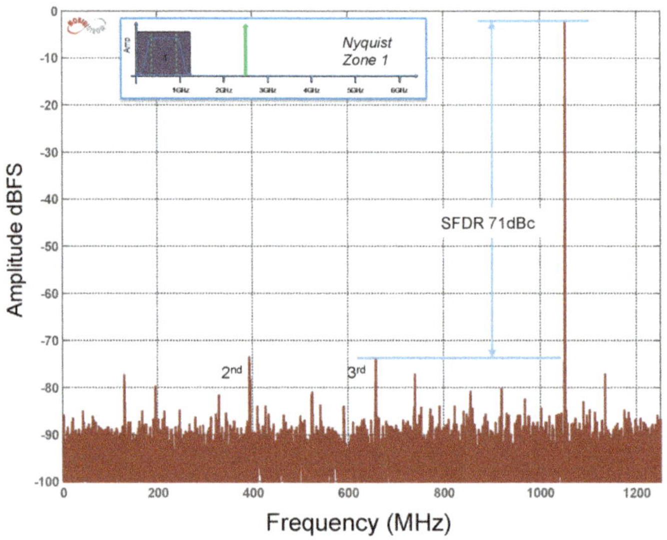

First Nyquist zone, fin= 1052-MHz and fs=2.525GS/s: 8k-sample FFT.

61 F_s=2525-MHz: F_{in}=2259-MHz

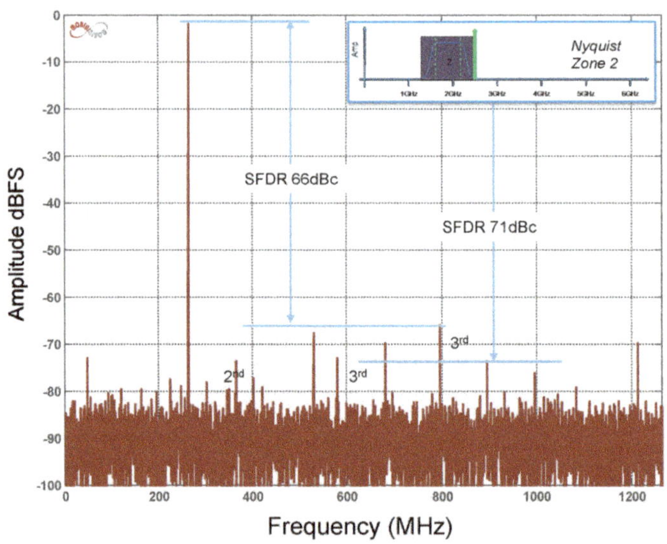

Second Nyquist zone, fin= 2259-MHz and fs=2.525GS/s: 8k-sample FFT.

62 F_s=2525-MHz: F_{in}=3225-MHz

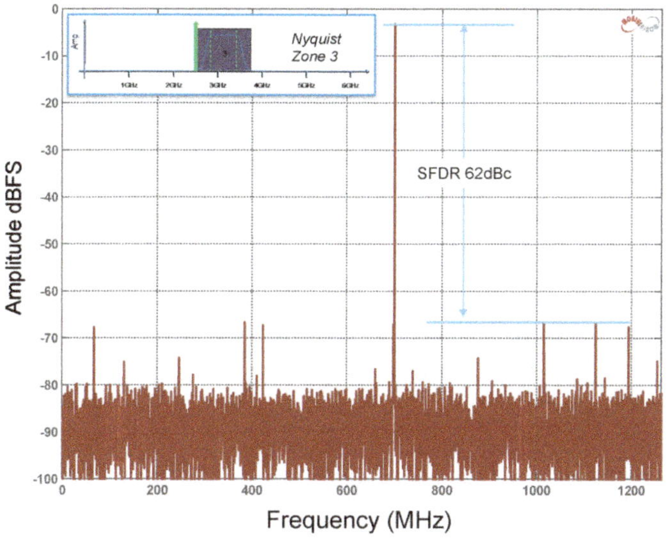

Third Nyquist zone, fin= 3225-MHz and fs=2.525GS/s: 8k-sample FFT.

63 F_s=2525-MHz: F_{in}=4783-MHz

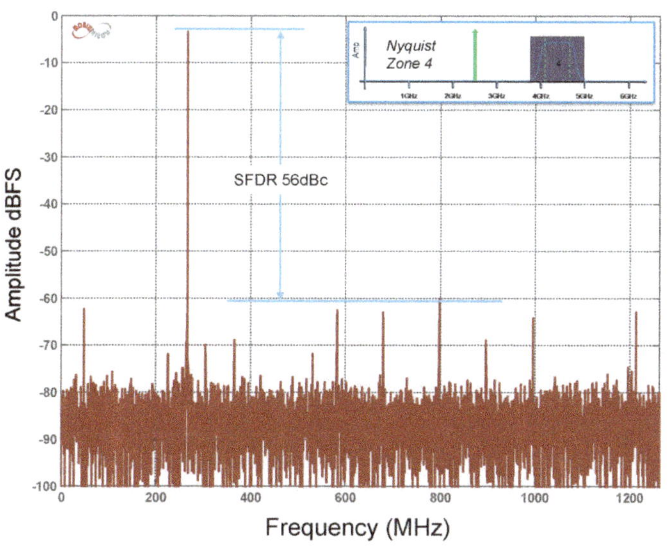

Fourth Nyquist zone, fin= 4783-MHz and fs=2.525GS/s: 8k-sample FFT.

64 F_s=2525-MHz: F_{in}=5000-MHz

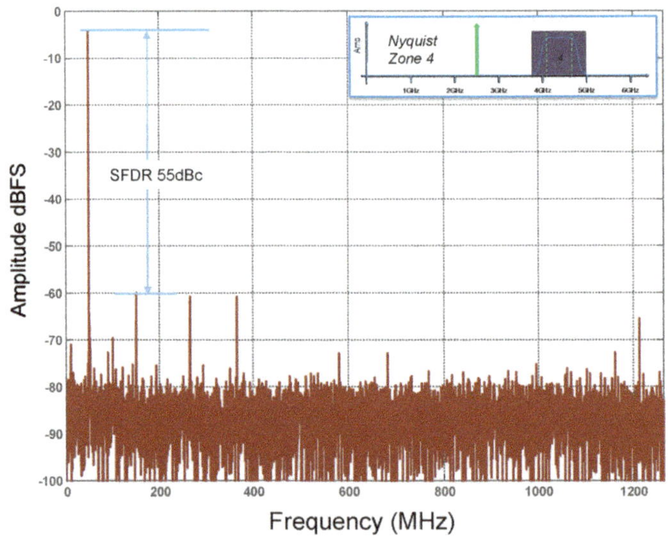

Fourth Nyquist zone, fin= 5000-MHz and fs=2.525GS/s: 8k-sample FFT.

65 F_s=2525-MHz: F_{in}=5150-MHz

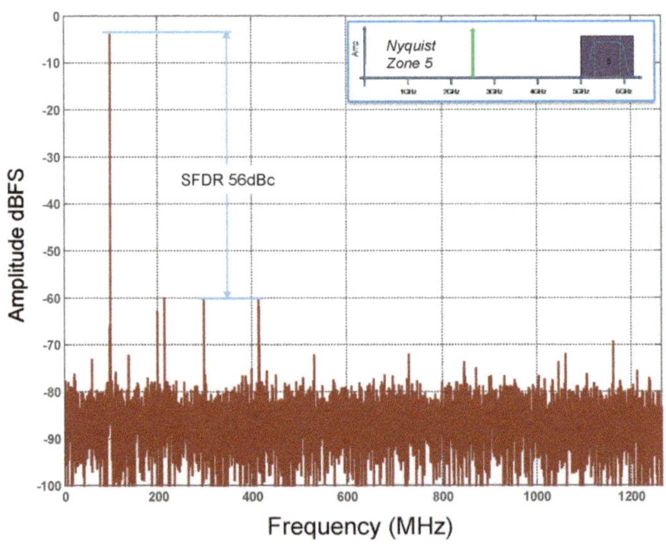

Fifth Nyquist zone, fin= 5150-MHz and fs=2.525GS/s: 8k-sample FFT.

66 SFDR vs. Fin for F_s=2.525-GS/s

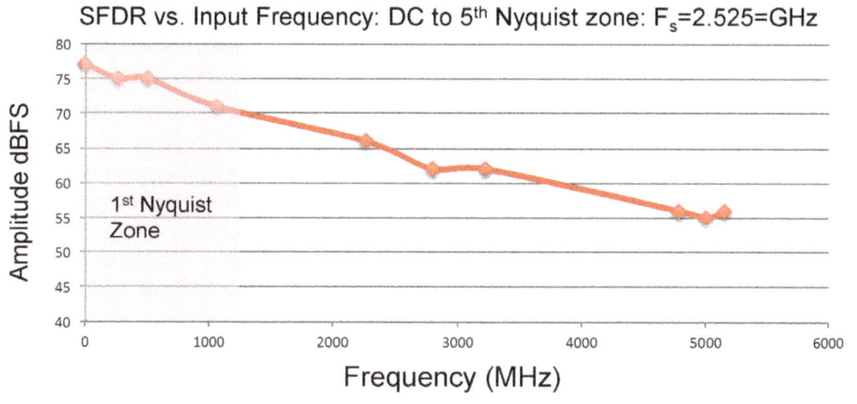

- **SFDR Results Preliminary**
  - Limited by HD3
  - Have not optimized bias / common-mode
  - Static linearity correction of buffer implemented in next-gen ADC

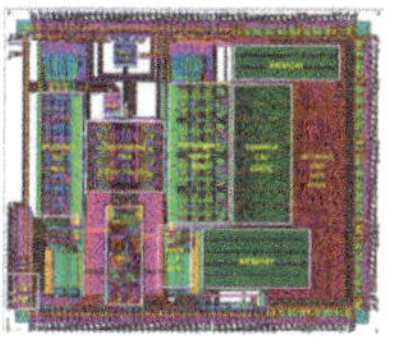

The SFDR as a function of input frequency from Nyquist zones 1-5.

67 F_s=2525-MHz: F_{in}=2790-MHz: Before Calibration

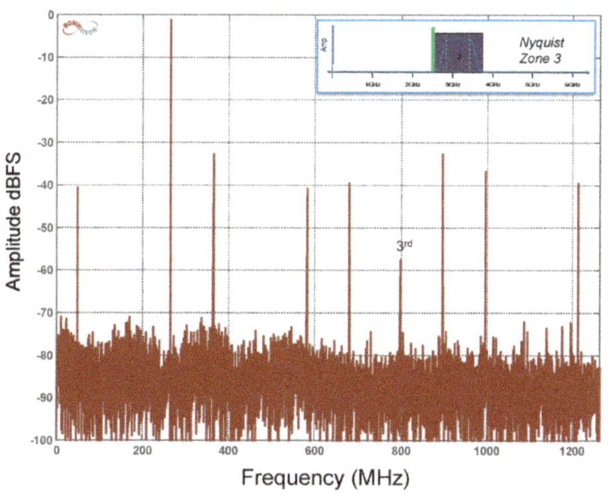

Above is the obligatory "before calibration" plot for a signal in the 3rd Nyquist zone. Time-interleaving spurs are very large because the slope of the signal in the 3rd Nyquist zone is so large causing several tens of LSB errors for a skew mismatch.

68 SSCS F_s=2525-MHz: F_{in}=2790-MHz

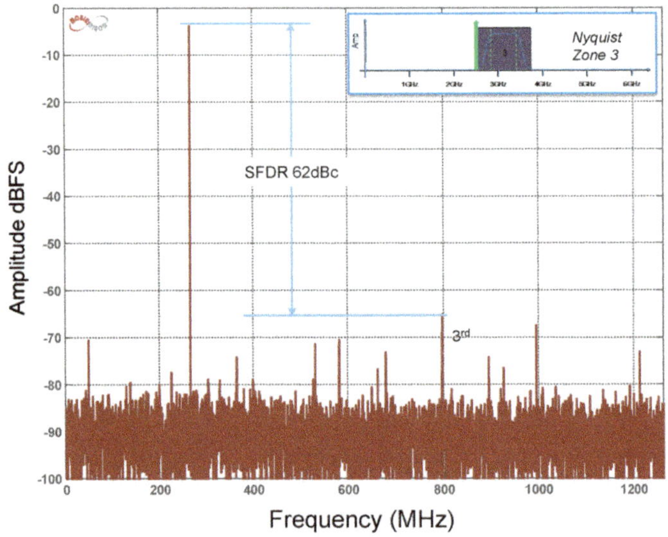

After calibration in the third Nyquist zone, fin=2790-MHz and fs=2.525GS/s: 8k-sample FFT, the spectrum looks pretty "normal." SFDR is limited by the 3rd harmonic and smaller artifacts of time interleaving are still visible.

69 SSCS F_s=2525-MHz: F_{in}=498-MHz

The SFDR of 75dBc exceeded the requirements for the circuit in the first Nyquist zone, where fin=498-MHz and fs=2.525GS/s: shown above with an 8k-sample FFT. Despite calibration, artifacts of time interleaving remain in the output spectrum. Some adjustments to the original plan for autonomous background calibration was required to obtain the best performance that is shown above. The limited resolution of the time skew correction circuit (50-fs step size for skew adjust) is often thought to account for residual time-interleaving errors. However, freezing the background calibration and manually manipulating the time skew controls reduced extraneous spurs, thus providing evidence to suggest that larger error sources than quantized time skew steps existed. Despite a reference ADC that showed much less crosstalk and intermodulation than was seen in Liger, there was still evidence of residual errors when the reference ADC was turned on: these went away when the reference ADC clock was held in reset. It was evident that the background calibration settings were not providing the best solution.

Convergence of a single LMS algorithm is never ideal when error sources are non-orthogonal. Such interaction between errors sources can occur in the presence of crosstalk, intermodulation and nonlinearities. So we sought to determine the possible source of errors that improved when the reference ADC was turned off that were limiting our performance when calibration was running.

69 F_s=2525-MHz: F_{in}=498-MHz

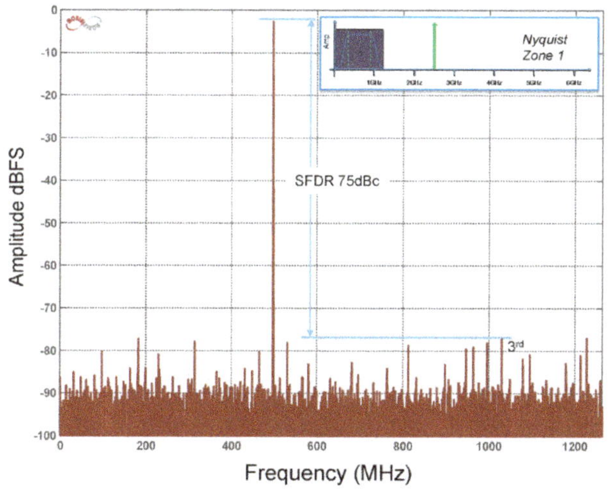

70 Kickback Pulse Response (Nonlinear) (I)

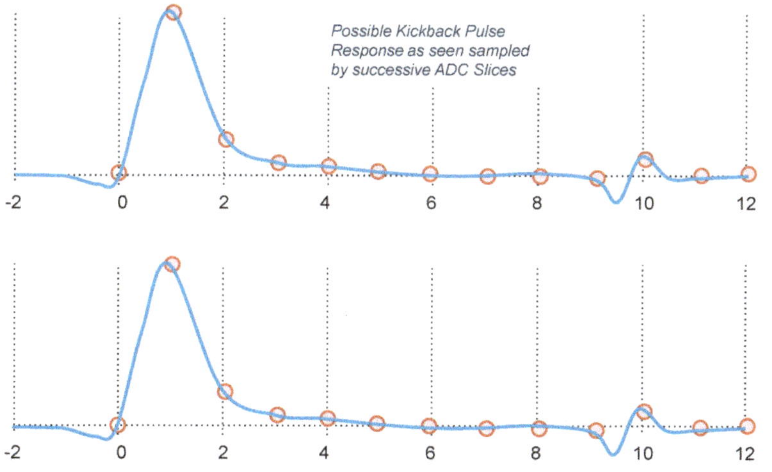

Kickback was identified as the most likely cause of the nonidealities and interaction between errors (Jurgen van Engelen). An effective way to think about kickback in a time interleaved ADC is to view it as a MIMO system (Multi-Input, Multi-Output). If it were a purely linear system then a MIMO equalizer at the output of the ADC slices could correct for the errors. However, the kickback is highly nonlinear. To better visualize this type of interaction, one could imagine the kickback to have a kick-response as shown above. I hesitate to use the term pulse-response because it becomes too likely to think of this as a linear system, whereas this pulse is nonlinear and signal dependent. If the reference ADC aligns with slice zero, there is not enough time for the kickback to impact the sample taken on slice zero. However,

70 Kickback Pulse Response (Nonlinear) (I)

the energy from the reference ADC is transmitted across the input distribution network where it gets sampled by the subsequent slices. If this happened identically in every sample the problem is no different than for a non-interleaved ADC. The portion of the kickback that is signal independent would be mixed to DC and appear as an offset. The portion that is linearly related to the signal will appear as a gain error. The portion that is nonlinearly related will cause compression or expansion in the transfer function. As energy stays on the line for multiple samples it acts as a nonlinear filter causing all aforementioned errors to be frequency dependent.

71 Kickback Pulse Response (Nonlinear) (II)

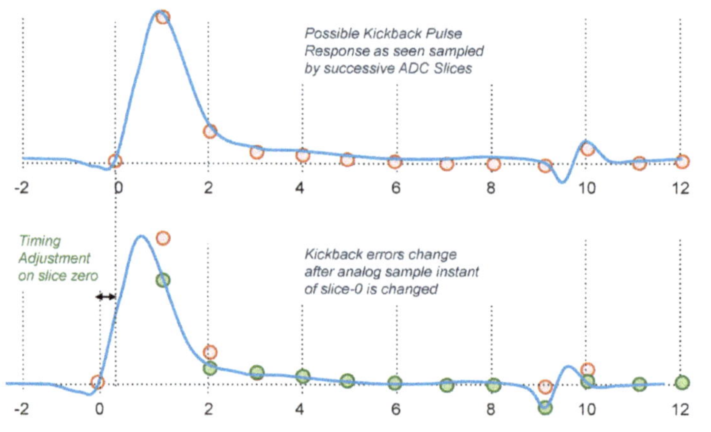

None of these artifacts are desirable but they can all be accommodated easily if the error was the same for every sample as it would be for a non-interleaved ADC. However, when the ADC is time-interleaved and the interleaved array uses a single reference ADC, even with dummy switching, the kickback becomes synchronous with the slice rate and not the full ADC sample rate. Therefore, the kickback is not the same on each sample. The only way to try to make it equal is to provide a reference ADC in every slice but even with the excessive overhead of extra power for dummy track-and-holds this does not solve the problem that one reference is needed to provide timing alignment.

To illustrate the non-orthogonal interaction of errors, consider a case where timing adjustments are made in the analog domain as shown above. When the timing instant moves, all of the samples of the kick-response move as they are sampled at a different position. Thus a change in timing for a single slice, impacts the offset that is mixed to DC, impacts the gain error, changes the nonlinearity and also modulates the bandwidth and group-delay of the circuit in a frequency dependent way as the sampled kick-response acts like a filter which also changes the timing instant. Therefore a single timing update causes everything to change, making the LMS algorithm have unpredictable convergence properties. It may either limit-cycle, trying to converge all errors simultaneously, or contain a systematic bias, or both. Interactions of error sources will always prevent the total system error from converging to its ideal value.

72 Time-Interleaved ADC with Digital Calibration

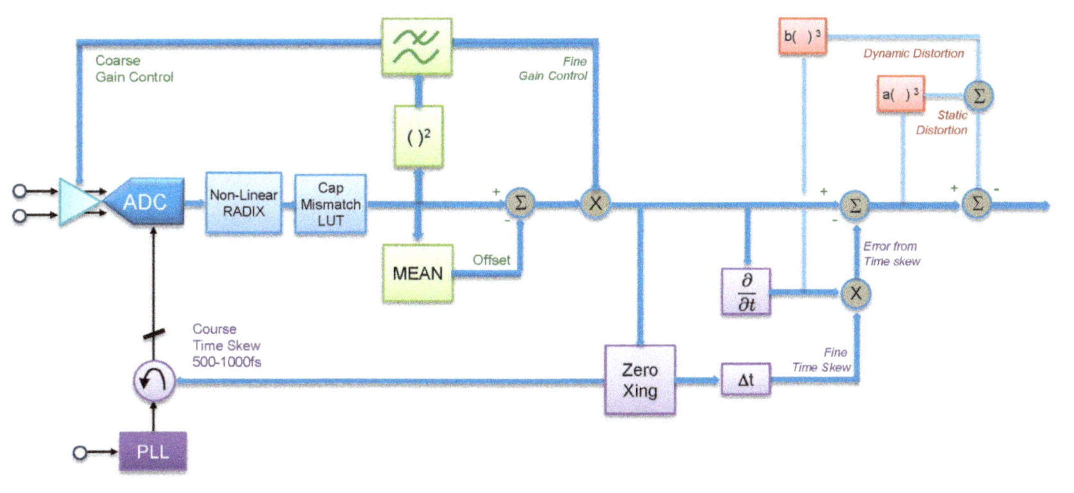

Because there is no perfect isolation between slices, error sources interact when adjustments are made in the analog domain. Kickback and instantaneous Vdd droop modulates the signal as does crosstalk through any others means such as via bias lines. Although good design practice can minimize the interaction, when aiming for high resolution, especially above 10-bits it becomes difficult to squelch crosstalk to a 70-80dB level especially as frequencies increase and parasitics become more important. In order to minimize this interaction a hybrid approach can be adopted. Course analog adjustments are used to get the circuit close to ideal performance followed by fine adjustments in the digital domain. A slope filter can be implemented relatively cheaply, especially when it is known that the time skew errors are small and the slope filter need not be complex as a few taps are adequate for the required accuracy of the estimation.

Implementing timing correction with a slope estimator is economical when it can serve a dual purpose. As the nonlinearity of a track-and-hold is related to both the signal value and slope, this information is useful in a Volterra filter to compensate for front-end distortion An efficient slope estimation filter above can correct for both time skew and be used as an input to a Volterra filter which corrects for front-end nonlinearity.

Digital fine-correction breaks the interaction between control loops making error sources more orthogonal which leads to more predictable convergence. This architecture has the ability to achieve high linearity of the front-end via calibration and very low residual time-interleaving artifacts.

73 Proud of These Chips... But

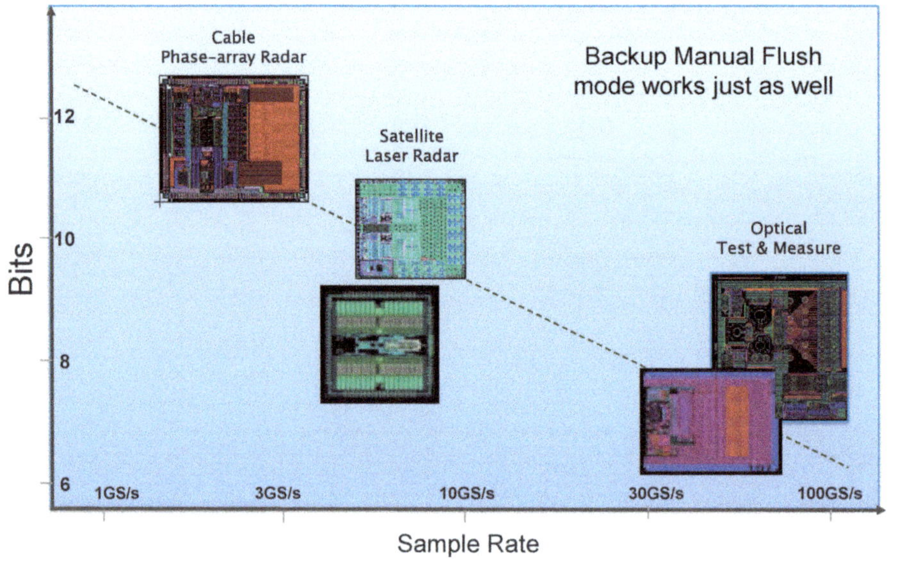

I'm satisfied with the results of all the time-interleaved ADCs above. They all met the requirements. Several innovations were needed to get the performance to the level that was measured. However, none of the ADCs converged to the best possible performance for a variety of reasons. Our highest resolution ADC was 12-bits and showed time-interleaving artifacts at the 75-80dB level. These were shown to be a result of kickback from the reference ADC, which by nature of the architecture, must sample at the slice rate (dummies could be inserted to sample on every slice, but this is highly wasteful).

So if I step back and look at my own circuits objectively, what I had was a really nice toilet with lots of fancy features. The key question to ask was whether or not the same performance could be achieved without anything fancy. Could the ADC be foreground calibrated and would the error drift due to temperature and environmental changes be any larger than the kickback artifacts of our calibration. Honestly — they would have been about the same, but have been able to achieve in a much easier and lower risk way.

That's hard to admit — especially when so much effort and energy was put into the solutions and implementations. So, because of its limitations, I'm fairly certain this is something I will never do again — but, with ever shrinking DSP and further advances in time-varying MIMO Voterra crosstalk and echo cancelers in 7nm or even 3nm, — one should never say never. In the words of T.S. Elliot: "We shall not cease from exploration. And the end of all our exploring will be to arrive where we started and know the place for the first time."

So one never knows, now does one, now does one, now does one.

PART IV

DC-DC CONVERTERS AND VOLTAGE REGULATORS

CHAPTER 12

DC-DC Power Converter Designs

Ramesh Harjani

University of Minnesota, USA

This tutorial aims to provide an overview of DC-DC converters architectures and circuit implementations. In particular, the tutorial will provide an introduction to linear converters, inductive converters and capacitive converters. Due to the gradual reduction of the digital power supply digital LDOs have recently been introduced. Fully integrated inductive converters using integrated inductors show promise and can be designed to exploit the close proximity of the integrated environment. For capacitive converters, a unified framework for circuit design, recent programmable prototypes and some novel converters including flying domain and resonant mode converters are introduced.

1

Talk Outline

- **Introduction to DC-DC converters**
  - Motivation, uses
  - Types of converters
- **Linear converters**
  - Series/shunt
  - Classification, design, examples
- **Inductive converters**
  - Passives + integration
  - Control, design methodology
- **Capacitive converters**
  - Framework
  - Design methodology
  - Circuit examples
  - Special capacitive converters

We will start our discussion with the introduction to the growing needs and design issues of modern day power management. Next, we will move to the analysis and details of linear regulators. Both series and shunt regulators will be taken up during the discussion. Next will be inductive DC-DC converters. We will then walk through the integrated inductor design, control mechanisms and various design methodologies for inductive DC-DC converters. Our last topic of discussion will be the capacitive converters. We will start with design framework for capacitive converters. Before wrapping up the discussion, we will show some circuit examples and two special categories of capacitive switching converters.

MOTIVATION

2

Power Is Precious

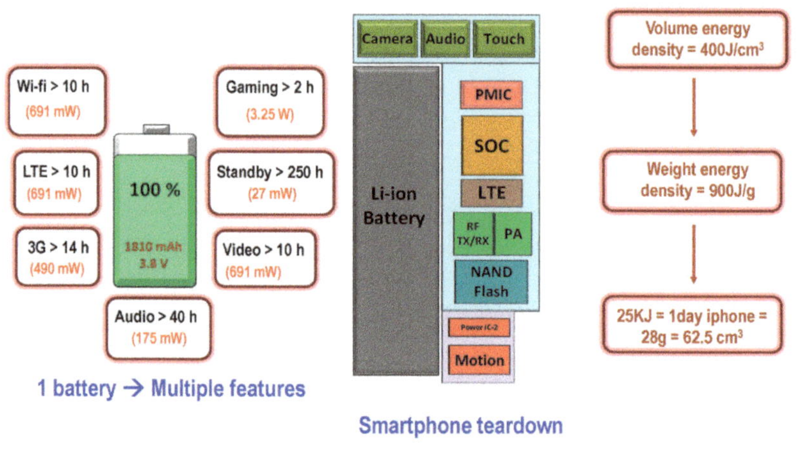

In the era of smart devices with ever-growing applications, efficient power management is important. And, in handheld devices, a battery is the only source of power. The center of the slide shows a teardown of a typical smart phone. Loaded with features such as camera modules, audio codecs, applications SOCs, memories, RF modules, motions sensors etc, the battery here occupies 30~40% by volume and weight. The thing to note is that, we have a single battery to meet all applications' requirements. Be it high power gaming or low power audio or even sleep mode, a single battery has to cater to a vast range of power and voltages. Therefore the DC-DC converter, which transfers power from battery to circuits, needs to be versatile and very power efficient.

3

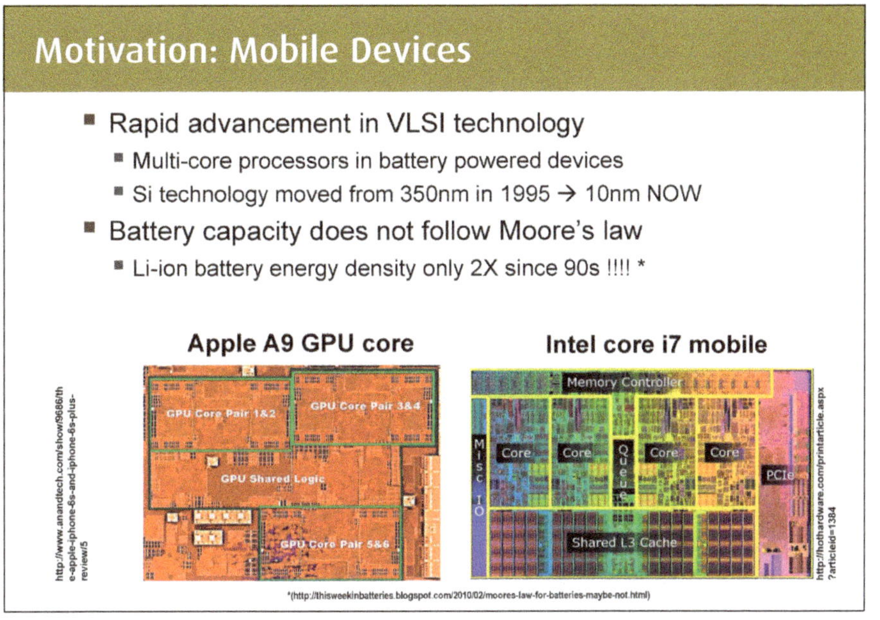

Take the case of mobile devices further, VLSI technology-scaling has been far more aggressive than battery technology. While silicon technology scaled 35X, the battery energy density has improved only 2X since 90s. So circuits techniques have to be used to compensate for this gap to ensure effective power management. Multicore designs have been the most successful method used to cut down redundant power consumption and to streamline the power utilization in applications. Apple A9 GPU and intel i7 are the two important commercial examples of this technique.

4

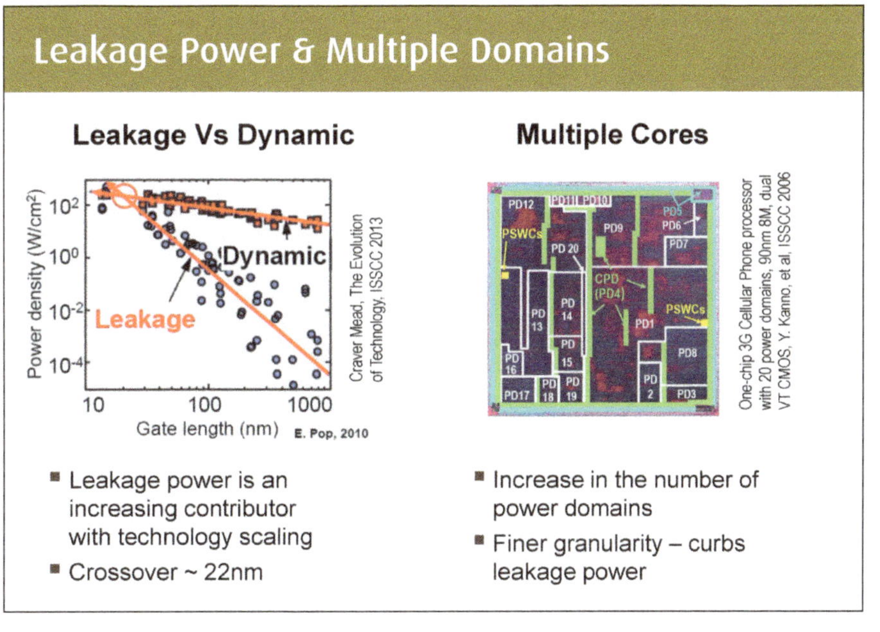

One of the principle drawbacks of VLSI technology-scaling is the increase of leakage power. As left figure shows, when we go to lower nodes, leakage power becomes a substantial portion of the total power. At the 22nm node, leakage power can become equal to the dynamic power. As the previous slide described, multiple core architectures are becoming popular as they streamline the power consumption among applications. The increased granularity and flexibility in active silicon area help to mitigate the increased leakage problem due to technological scaling.

5

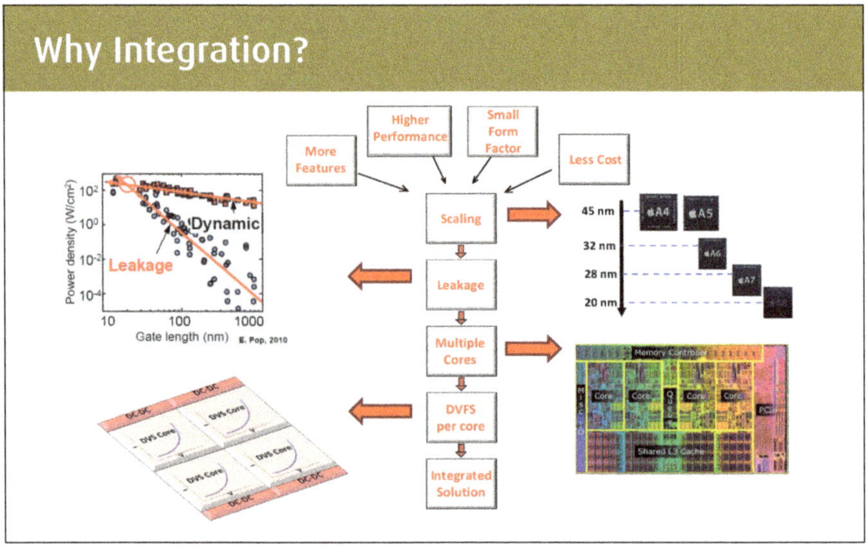

Lower cost, higher performance, small form-factor, all these factors push the need for VLSI scaling. With scaling comes the problem of leakage. Circuit techniques like multiple cores can handle increase leakage overhead. To further enhance the power handling in each core, a popular digital technique called dynamic voltage and frequency scaling (DVFS) has been used extensively. DVFS stems from the fact that for dynamic power is roughly proportional to P~Vdd3. So, if in a digital system if we lower the supply voltage, we can bring down the power consumption significantly. Thus this gives an additional requirement to the DC-DC converter. The DC-DC converter should be an integrated solution so as to provide DVFS to each core independently and efficiently.

TYPES OF DC-DC CONVERTORS

6

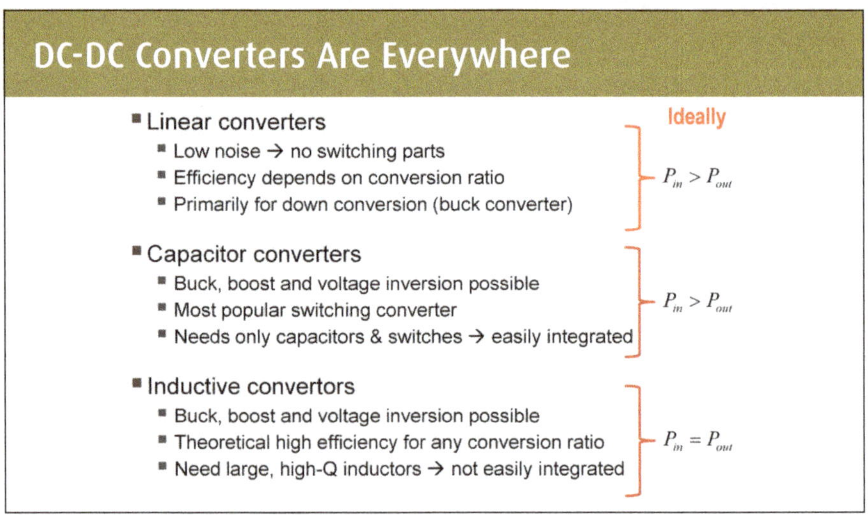

DC-DC converters are of three types- linear regulators, capacitive converters and inductive converters. In principle both linear and capacitive converters will be less efficient compared to inductive converters. But due to their ease of implementation in CMOS, they are popular choice for DC-DC conversion.

Linear regulators have less output voltage noise. The maximum attainable efficiency is Vo/Vin. Also these types of converters will always be step down.

Capacitive converters are the most popular types of DC-DC converters in integrated solutions. They are the easiest to implement in VLSI because they just require switches and capacitors. Capacitive converters can be buck, boost or negative types.

Inductive converters can be buck, boost and negative types. In principle, inductive converters are more efficient than their capacitive and linear counterparts owing the fact that inductors can drop a voltage across with zero power dissipation.

7 DC-DC Converter Modelling (Ideal)

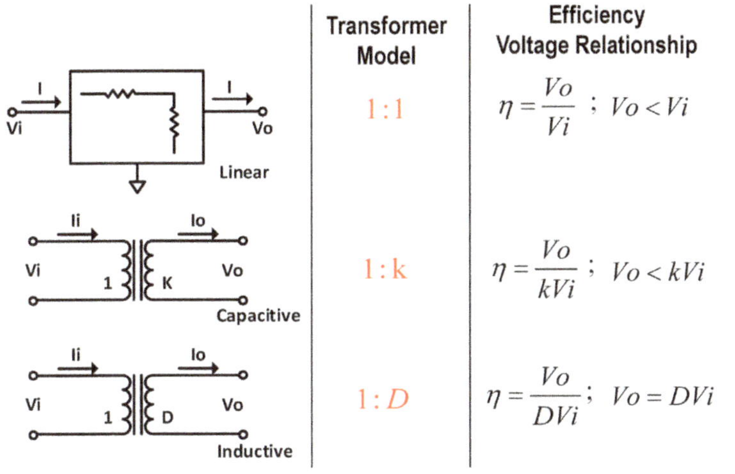

This slide models the three types of DC-DC converters into conceptual transformer models. This kind of modelling is beneficial for design and comparison of integrated DC-DC converters. Linear regulators can be modeled as 1:1 transformers, capacitive converters as 1:k transformers (where k is Vo/Vi) and inductive converters as 1:D transformers (where D is the duty cycle).

LINEAR CONVERTERS

8 Linear Regulators

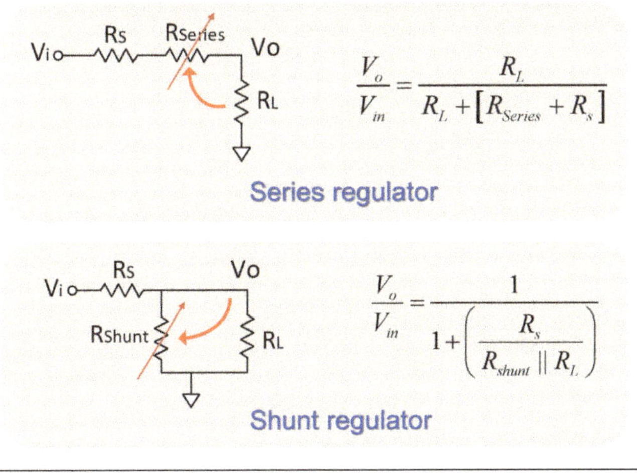

Linear converters are of two types- series and shunt. Top figure shows the working principle of the series regulator. Here the load circuit is in series with source and the sense resistor. This type of DC-DC converter works on a voltage division principle. Another kind regulator is the shunt regulator as shown in the bottom figure. This is a less popular variant due to its low power efficiency. We will focus more on the series type of voltage regulators in the subsequent discussions.

9 Series Regulator

- The output voltage Vo is kept constant by varying R_R
- Efficiency = power in load /power from source
- If same current is flowing in R_R and R_L then $\eta = {}^{Vo}\!/_{Vin}$

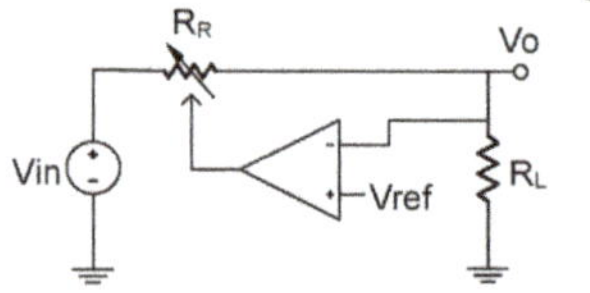

Linear Converter Block Diagram

$$\eta = \frac{Vo^2}{R_L}\frac{R_R+R_L}{Vin^2} = \left(\frac{Vo}{Vin}\right)^2 \frac{R_R+R_L}{R_L}$$

$$= \left(\frac{R_L}{R_R+R_L}\right)^2 \frac{R_R+R_L}{R_L} = \frac{Vo}{Vin}$$

$$\eta = \frac{I_L \cdot Vo}{I_L \cdot Vin} = \frac{Vo}{Vin}$$

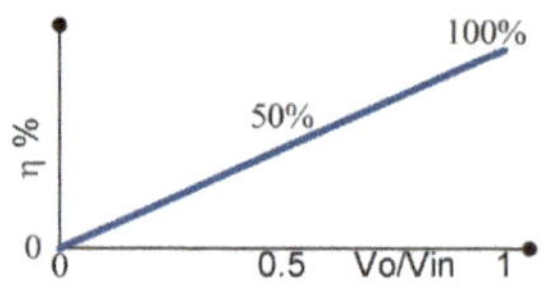

Conversion Efficiency vs Vo/Vin

Voltage divider → efficiency is ratio of voltages

In this slide we will talk about the efficiency of series type linear converters. Power efficiency is given by output power divided by input power. Putting in the values of power terms in the efficiency expression, we get η = Vo/Vin, provided the input and output currents are the same. Since the input and output currents are same, so the output to input voltage ratio also represents the respective power ratios. Interestingly the relation between efficiency and Vo/Vi is a straight line with a slope of 1.

10 Model For Linear Converter

- Linear regulator = linear converter = aka LDO
- Linear regulator model ←→ transformer model
- Transformer has a turns ratio of 1
- Conversion efficiencies

$$\eta = \frac{Vo}{1 \times Vin} = \frac{Vo}{Vin}$$

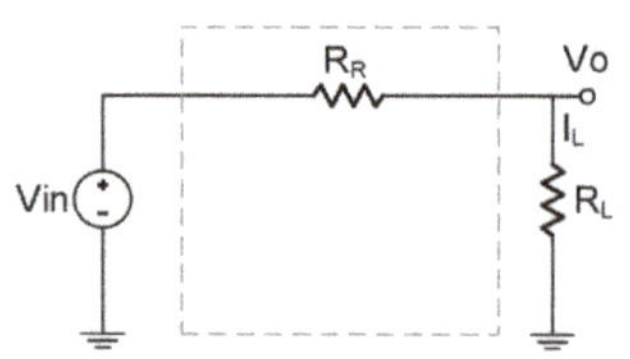

Linear Converter Model

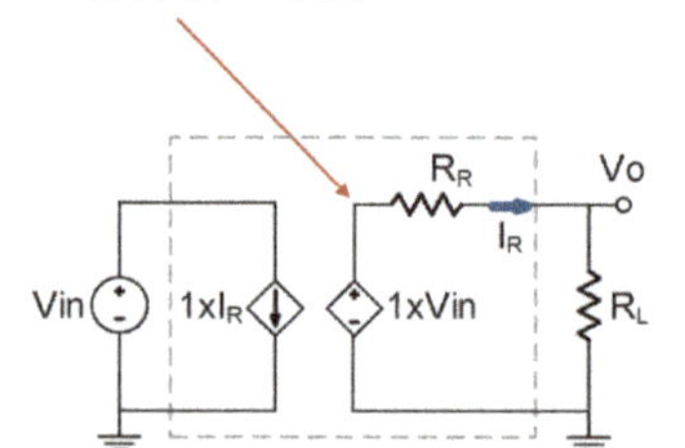

Linear Converter Transformer Model

This slide repeats the analysis discussed on the previous slides using the transformer model. Linear converter turns out to be a 1:1 converter with series resistor being the output resistor of the model.

11 DC & AC Characteristics

- DC characteristics
  - Line regulation

$$\frac{\Delta V_o}{\Delta V_{in}} = \frac{1}{1+LG}$$

  - Load regulation

$$\frac{\Delta V_o}{\Delta I_o} = \frac{R_{o,open}}{1+LG}$$

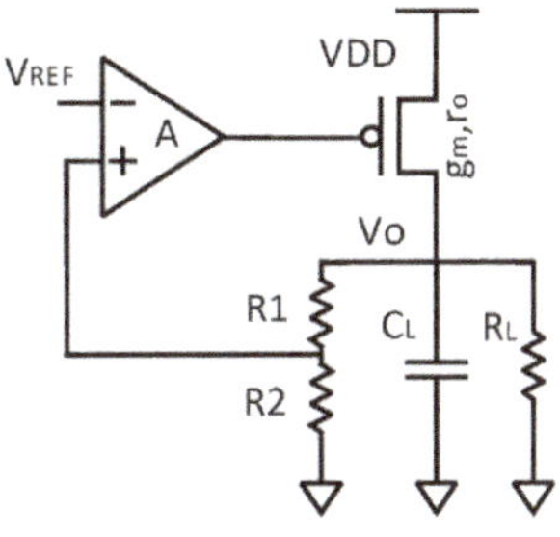

$$\beta = \frac{R_2}{R_1+R_2}$$

$$R_{o,open} = r_o \,||\, (R_1+R_2) \,||\, R_L$$

$$LG = \beta A.g_m(R_{o,open})$$

- AC characteristics
  - Input ripple rejection (PSR) → AC equivalent of line regulation
  - Output ripple rejection → AC equivalent of load regulation

Now we look at the DC/AC characters and implementation of the series type linear DC-DC converter. This slide shows a typical PMOS implementation of the converter. It consists of a PMOS power switch, opamp, feedback resistors and output filter capacitor. The aim is to maintain the output voltage equal to a pre-decided reference voltage, Vref. The feedback resistors scales the output voltage to be compared to the Vref. The opamp establishes the negative feedback and thus maintains the output voltage directly related to Vref as described in the slide.

Once the negative feedback is established, DC characteristics like line and load regulations can be derived directly from basic feedback theory.

12 Low Drop Out Regulator

- Low Drop Out (LDO) regulator
  - (Vi-Vo) is desired as low as possible
  - Very popular research topic

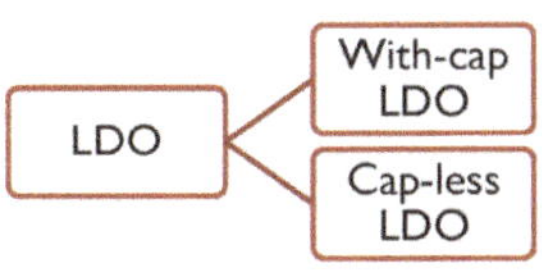

- With Cap LDOs
  - Off-chip output capacitor, higher power levers, discrete
- Cap-Less LDOs
  - Integrated, fast/higher bandwidth, large load transient droop

Series type LDOs have been a popular topic in integrated DC-DC converters. Integrated series LDOs are of two types – with output filter capacitor (often referred as cap LDO) or without output filter capacitor (capless LDO). LDOs with output filter capacitors are used for higher power levels and mostly in discrete power solutions. Capacitor less LDOs are fast and have higher bandwidth. The drawback with them is the large voltage droop at large load transients due to the response time of the loop.

13 With-Cap LDOs

- With Cap LDOs
  - Output capacitor, C_L, is the dominant pole capacitor
  - Small bandwidth, less droop
  - Easy compensation
- Design constraints
  - V_E cannot be lower than a ΔV
  - $V_{o,max} = VDD - \Delta V$
  - A_E, A_P must be constant during the operation of regulation
  - Width of the pass transistor goes up rapidly for lower VDD

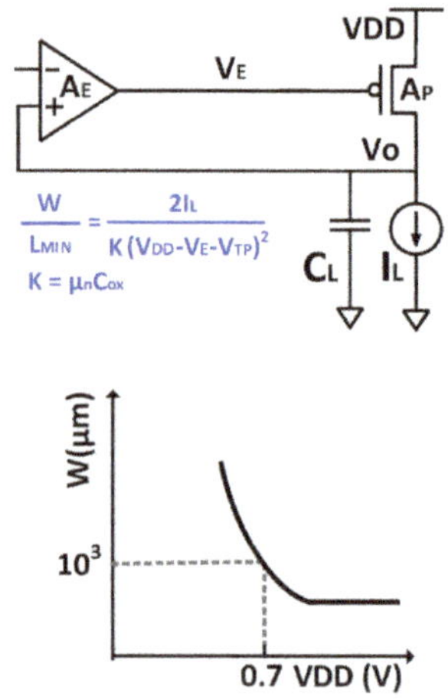

Let us discuss about the with-cap LDOs. Since CL is quite large as compared to internal parasitic capacitors, it forms the dominant pole at the output. This proves easy for compensation and incurs lower output voltage droop at the transients. But this makes the LDO suffer from a smaller loop bandwidth.

Coming to the design aspects of the capacitor based LDOs, the constraints that we have are that we need to have a voltage drop equal to an overdrive across the gm power transistor in order to be in saturation. Also, there is a minimum output voltage that the error amplifier can hold. Consequently, for a fixed output voltage, if we reduce the supply voltage of the LDO drops below say 0.7V, the switch size will blow up.

14 Compensation

- Dominant pole
  - Output capacitor, CL, is dominant pole capacitor
  - Small bandwidth, easy compensation
- Non dominant poles
  - Error amplifier internal poles
  - Gate of the pass transistor
- Impact of load
  - More instability at lighter loads

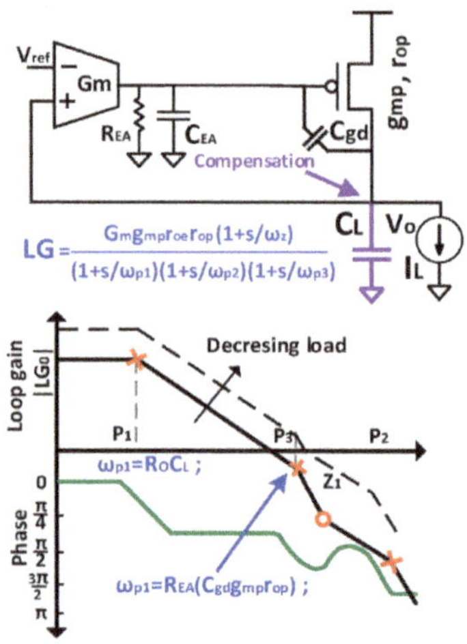

Now we focus on the compensation of the LDO. The output node is our dominant pole here and we end up with a secondary pole at the gate of the power FET. We also have a zero caused by the Cgd. In this scenario we can safely use the common compensation techniques for two pole systems.

In this type of compensation, instability increases at lighter loads as shown in the bode plots at the bottom of the slide.

15 CAP-Less LDOs

- Dominant pole
  - Pass transistor is larger
  - Dominant pole of the EA
  - Large bandwidth, difficult compensation
- Non dominant poles
  - Output pole
  - Error amp internal poles
- Impact of load
  - More instability at lighter loads as gain increases

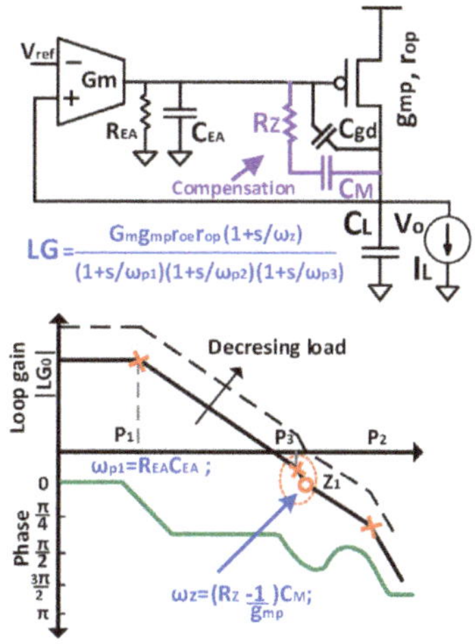

Now we see at the case of cap-less LDOs. These are also two pole systems. But the output capacitor is either not present or comparable to the internal parasitic capacitor. Due to this, the dominant pole is the internal parasitic pole at the output of the error amplifier and the output node pole of the LDO is the non-dominant pole. To compensate this design, we need to a Miller capacitor – resistor pair as shown. In this case also, as we decrease the load, the instability increases mainly due to the increased gain of the overall system.

16 Classification: Digital LDO vs A-LDO

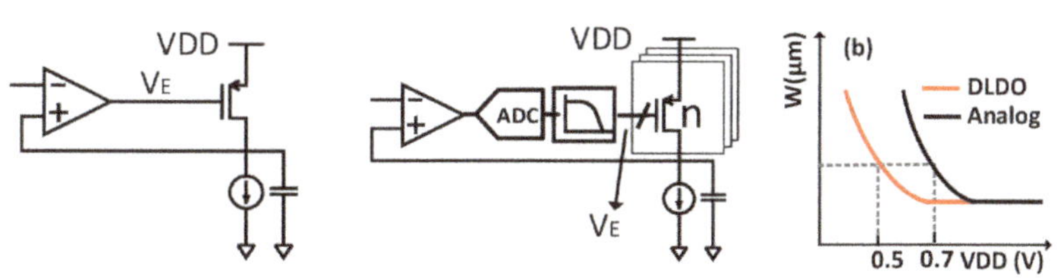

- Analog amplitude
  - No steady state output ripple
  - Analog compensation
- Digital amplitude
  - Steady state ripple
  - Digital compensation

Conventional LDOs

Digital LDOs

Talking about integrated LDOs, there are two types of implementations – digital and analog. Analog LDOs are the convectional LDOs which can be either cap LDOs or cap less LDOs. But with ever decreasing supply voltages (which is the input for the LDOs), the digital implementation is growing more popular.

Roughly put, digital LDOs are basically DAC based current supplying sources, which try to match the load current demand with current supply capability of the power switches by switching part of them on/off. Due to this, they have an inherent output voltage ripple even at steady-state with a constant load at the output.

17 Digital LDOs

- Motivation - analog vs digital
  - As VDD lowers, Ve is bottleneck
  - Make Ve zero
- DLDOs
  - Pass transistor either OFF/ON
  - Mimic LDO operation
  - More complex design
  - Easy compensation
  - Intrinsic ripple due to quantization

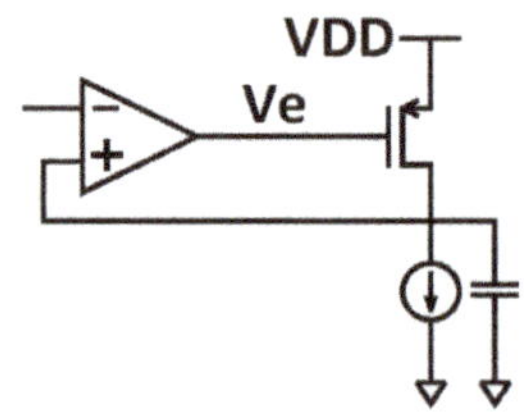

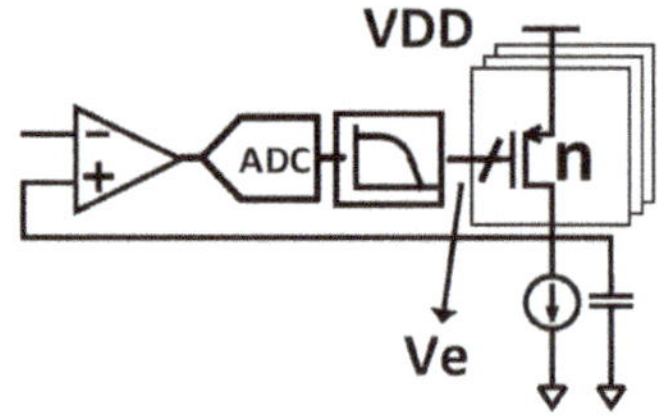

Looking at the digital LDOs, we note that it consists of two parts – a DAC based current supplying switch bank and digital control loop. The difference between output voltage and the reference voltage is sampled digitally and either via PI or PID fashion, the control loop is closed using ADCs and digital filters.

18 An Example of Digital LDO

A 100nA-to-2mA SAR DLDO With PD Compensation And Sub-LSB Duty Control

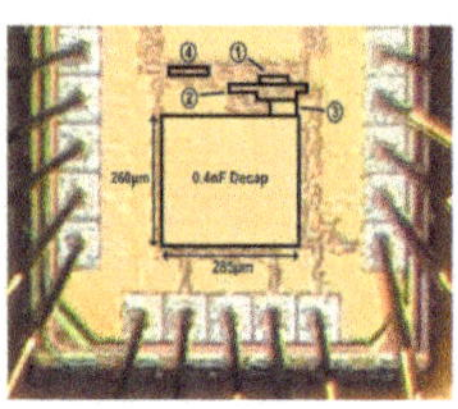

Process: IBM 130nm CMOS
Loai G Salem, Patrick Mercier
ISSCC 2017

Let us take up an example of digital LDO from Prof Patrick Mercier's group published in ISSCC 2017 in order to understand the mechanisms of digital LDOs.

19 Example - SAR DLDO

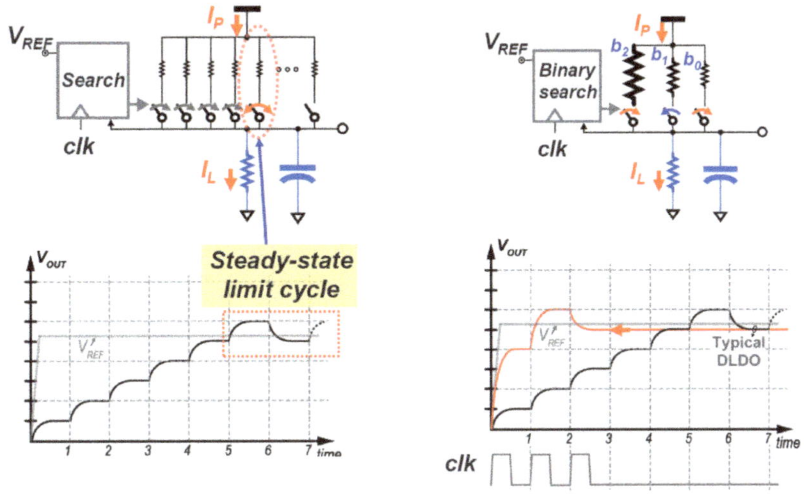

In this slide we study the transient response in the digital LDOs. The control mechanism of this digital LDO is based on successive approximation. In a conventional digital LDO, the transient recovery will happen step by step (linear search) as shown in bottom left. But in this paper, they apply a SAR logic. It is like using the binary search algorithm to find the right current supply range to support the load requirement as shown in bottom right. It has a much faster response time.

20 Measurement Results

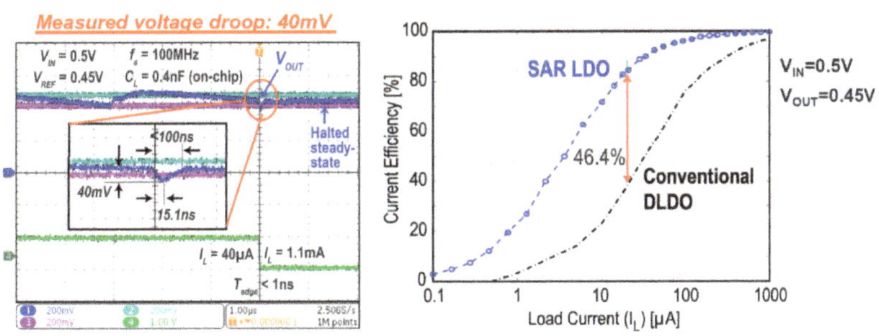

	Proposed LDO	Prior LDO
Response time (T_R)	15.1ns	377.5ns
Settling time	<100ns	1.37μs
FOM	199.4ps ↔	638ns

$>2^{2N}/N$

These are the measurement results of the work. Due to SAR logic, there is a significant improvement in response and settling time for this type of LDOs. This work also reports an improvement in the current efficiency owing to fewer number of switching transitions during the search process.

21 Shunt Regulator

- **Regulation**
  - Lowest operating point (minimum Vin-Vo)
  - Shunt resistor regulates
  - Unequal input and output currents

    $I_{In} \neq I_{Load}$
  - Current divider → efficiency is ratio of currents

$$\eta = \frac{I_{Load}}{I_{in}} = \frac{1}{1+\frac{I_{Shunt}}{I_{Load}}}$$

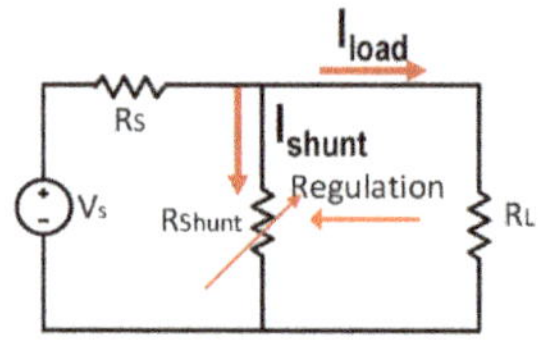

- **Clamp/Zener***
  - Used for clamping the input voltage
    - E.g., use a Zener
  - Zener need to accommodate large wasted currents for small load currents
  - Less popular (bulky)

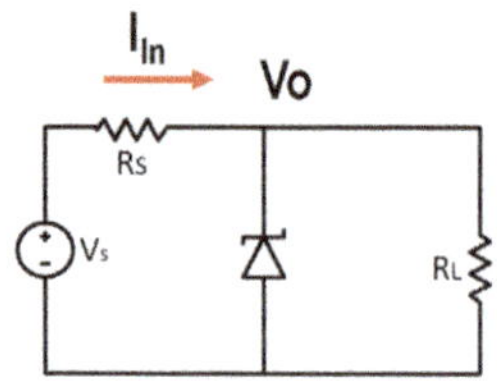

Before we conclude our discussion about the linear regulators, let us talk about the less popular but extremely useful shunt linear regulators for the sake of completion.

These types of regulators are best suited for clamping based applications. Also as is evident from the circuit diagram, these types of regulators are less efficient at higher voltage levels.

INDUCTIVE CONVERTERS

22 Inductive Power Converter

- Switch
  - Chops up the input voltage
  - Can be implemented using a Diode-FET, NMOS-PMOS, or NMOS-NMOS combinations

$$V_{out} = D \cdot V_{in}$$

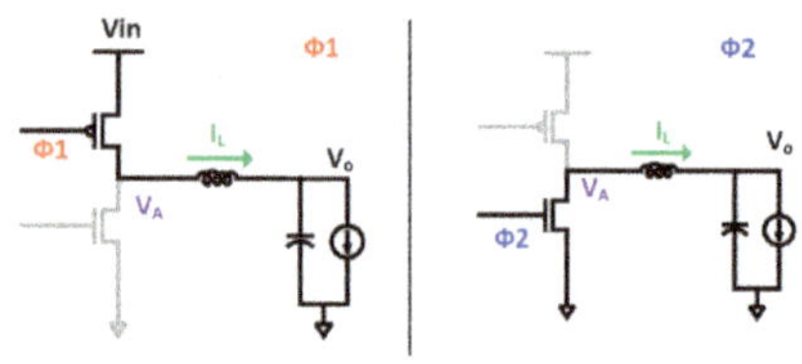

- Filter
  - Ripple current is proportional to switching frequency and inductance
  - Ripple voltage is proportional to ripple current and output capacitance

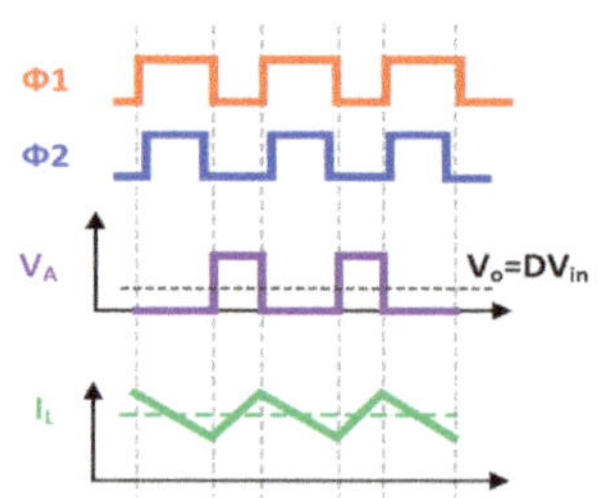

We take the popular buck (step-down) converter topology as a vehicle to discuss the inductive converters. Non overlapping clocks phi1 and phi2 create a square wave voltage at the drain junction of the two switches. The inductor and capacitor network forms a second order filter which filters the square wave into a near DC output. In phase phi1, the inductor sees a voltage drop of Vin-Vo, the current builds up in the inductor and in phase phi2, the voltage across inductor is –Vo, so the current ramps down. The average of this inductor current is the load current.

23 Operating Parameters

$$V_{o,avg} = DV_{in}; \quad D = \text{Duty-cycle}$$

$$L\frac{dI_L}{dt} = V_{in} - V_o \Rightarrow \Delta I_L = \frac{(V_{in} - V_o)\Delta t}{L}; \phi 1$$

$$L\frac{dI_L}{dt} = -V_o \Rightarrow \Delta I_L = -\frac{V_0 \Delta t}{L}; \phi 2$$

$$\Delta I_{L,pp} = \frac{(V_{in} - V_o)D}{Lf_s}$$

$$\Delta V_{ripple} = \frac{(V_{in} - V_o)D}{LCf^2}$$

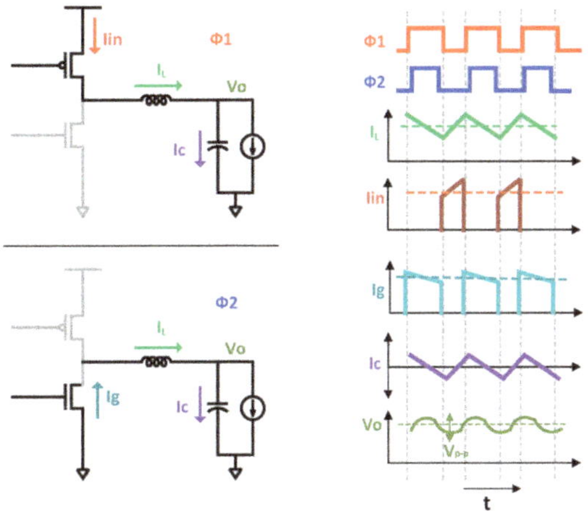

This slide shows the derivation for the inductor current ripple and output voltage ripple under the assumption that the voltage ripple at the output is very small.

24 Integration - Inductors

- **Output filters require high inductance (1µH to 100µH)**
- **High inductance results in**
  - Large area
  - High series resistance resulting in low efficiency
- **1-3nH integrated inductors have reasonable series resistances**
- **Low inductance generates large current ripple resulting in**
  - Large output voltage ripple
  - Low efficiency
- **Strategy for integration**
  - Interleaving: output less dependent on inductor current ripple
  - Magnetic coupling: reduces current ripple magnitude at inductor
    - On-chips are close and so have coupling

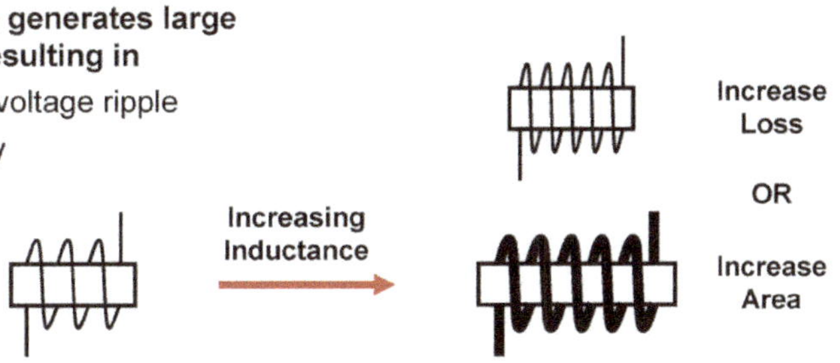

Inductors are the main passive component of the inductive DC-DC converters. This slides explains the differences between off-chip and on-chip inductive converters. Some of the techniques we often use in inductor integration are interleaving and magnetic coupling.

25 Why Fully Integrated Converters ?

- Passives off-chip
  - ✓ Large L and C possible
  - ✓ Good quality of L and C
  - x Higher BOM
  - x Large pin-count
  - x Power domains → limited #
- Fully integrated converter
  - ✓ Small pin count
  - ✓ Lowest BoM
  - ✓ Power domains → large #
  - x Small L and C , Quality of L low
  - x Consumes silicon area
- Goal: reduce on-chip overhead

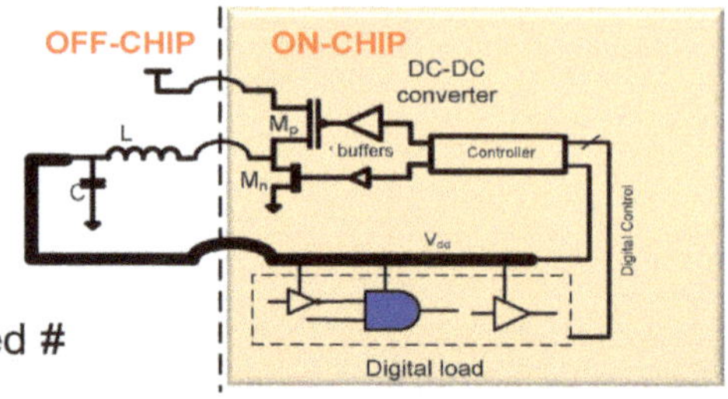

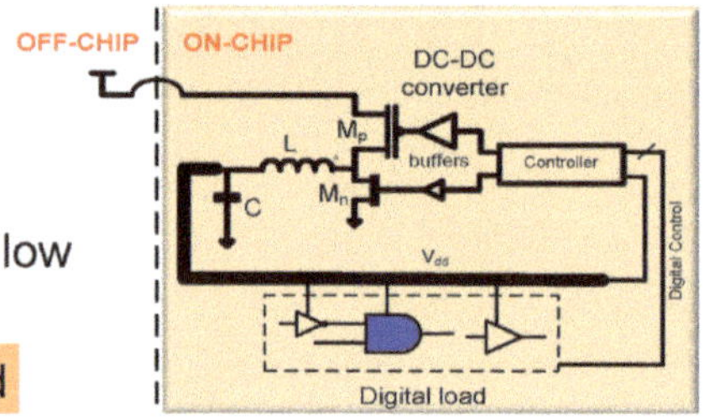

The primary benefit for integrated inductive DC-DC converters are smaller pin count, low bill of materials, and support for multiple power domains. Off-chip passives in non-integrated converters tend to be bulky.

26 Inductors

- Inductors
  - Area bulky (**1nH/0.2mm2**)
  - High quality factor – a problem (typically **1nH/150mOhm**)
- Examples

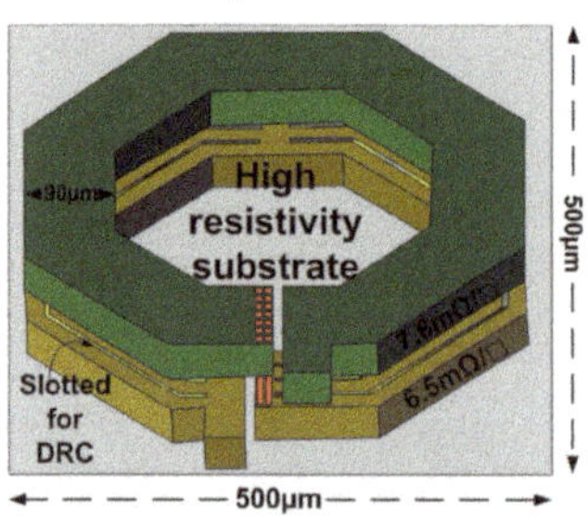

- Diameter: 600 μm
- Metal Width: 75 μm
- Turns: 1.75
- Spacing: 4.5 μm

Thanks to RF IC design, the design of integrated inductors has matured a lot. The primary property to look for an inductor in inductive DC-DC converter design is the DC resistance. Typical integrated inductors have approximately a 1nH/150mOhm inductance to resistance ratio.

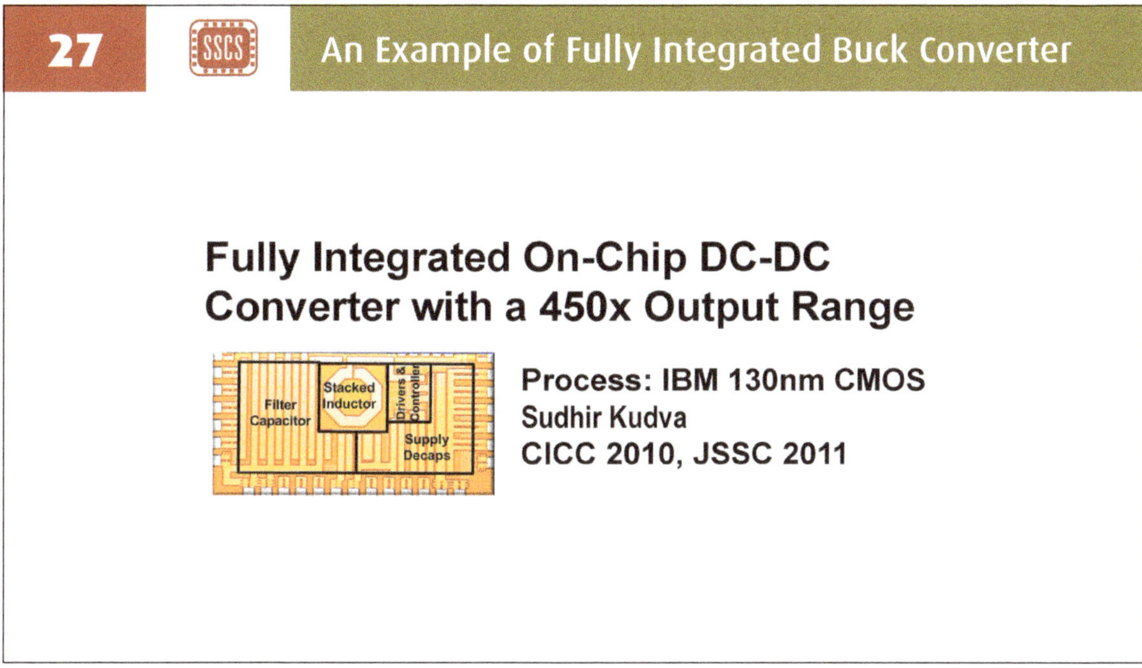

To gain more insight into inductive converter designs, we take up an example of fully integrated buck converter in 130nm CMOS.

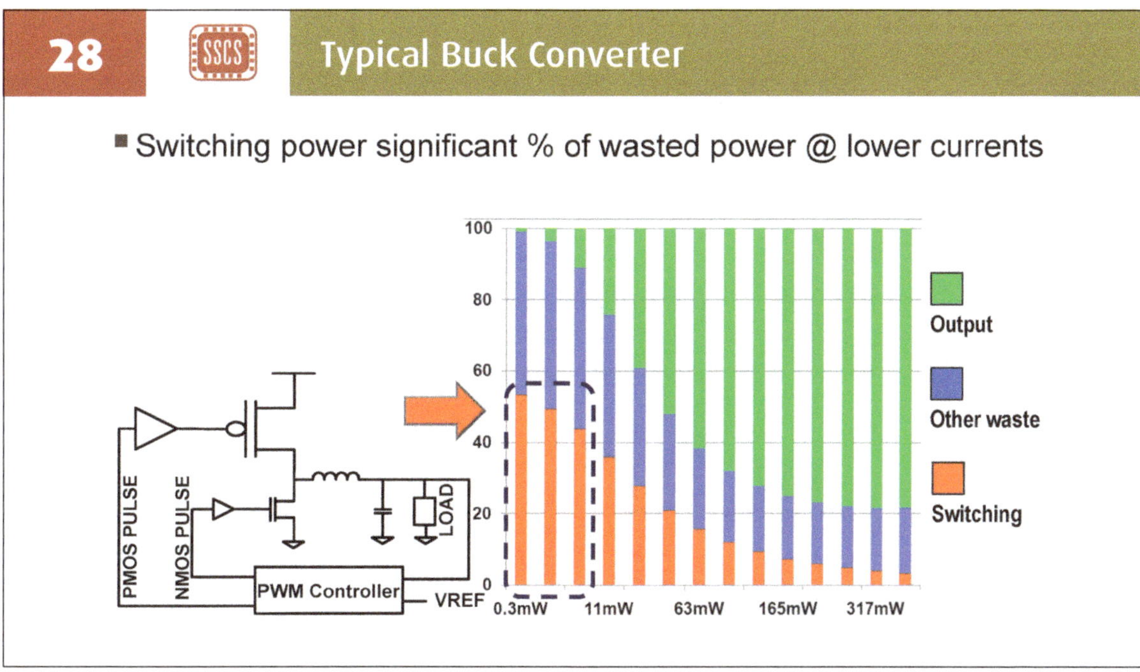

This slide shows a typical power breakup profile for an integrated DC-DC converter. We observe that at lower powers, the switching loss dominates thus play a primary role in determining the overall efficiency.

29 Modified Circuit

- Split PMOS switch and the drivers
- Size of the nail decides the hammer
- Only PMOS variable → larger PMOS to handle larger current

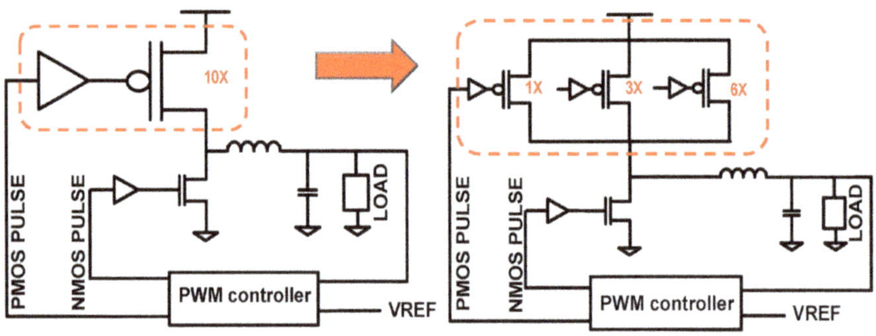

Since the PMOS switch in a buck converter has to handle a larger current (in usual applications, duty cycle > 0.5), it has to be carefully designed. One of the important technique discussed in this work is of splitting the PMOS switch and corresponding drivers so as to add flexibility in cutting down switch losses at lighter loads.

30 Hysteresis Controller

- Improved efficiency at high output loads
- However, still low efficiency at low loads
- Switch pulse frequency depend on the load current
- Effectively variable switching frequency
- PMOS and NMOS both OFF for some time

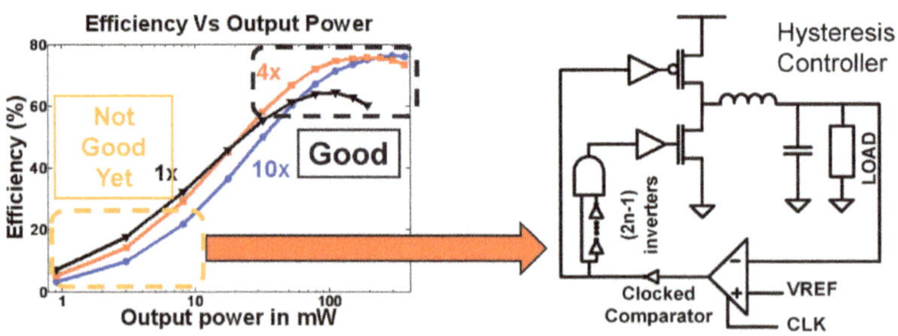

This slide explains the hysteretic control mechanism of the inductive converter.

31 Integrated Converter

- Variable PMOS switches → 10x,4x,1x
- High load ➔ PWM, Low load ➔ Hysteresis

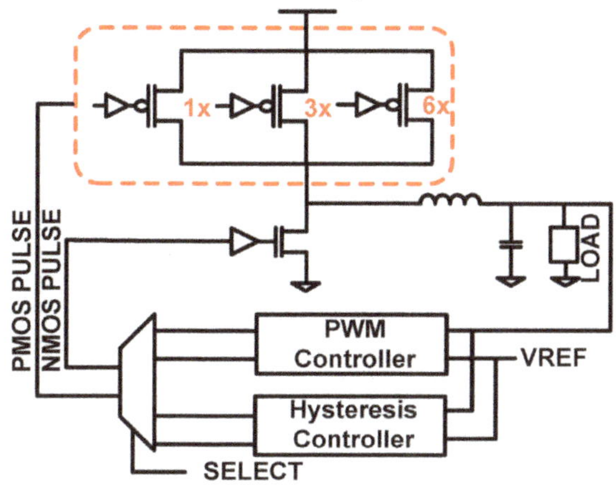

The inductive converter has two parallel control loop circuits. First one is the PWM controller which is for the higher current loads. Second one is the hysteretic controller for the lighter current load.

32 Chip Photograph

- IBM 130nm CMOS process
- Converter core area – 1.13mm²
- Total area with supply decaps – 1.592mm²
- Capacitance – 5nF   Inductance - 2nH

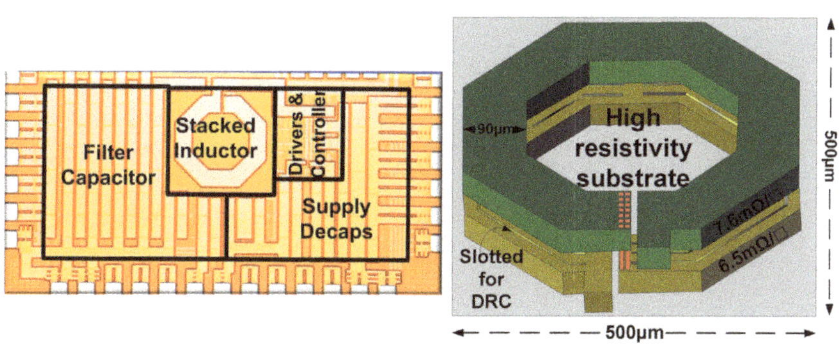

This slide shows the chip micrograph of the converter. It uses a 5nF filter capacitor and a 2nH switching inductor. The inductor is realized by the top two metal layers.

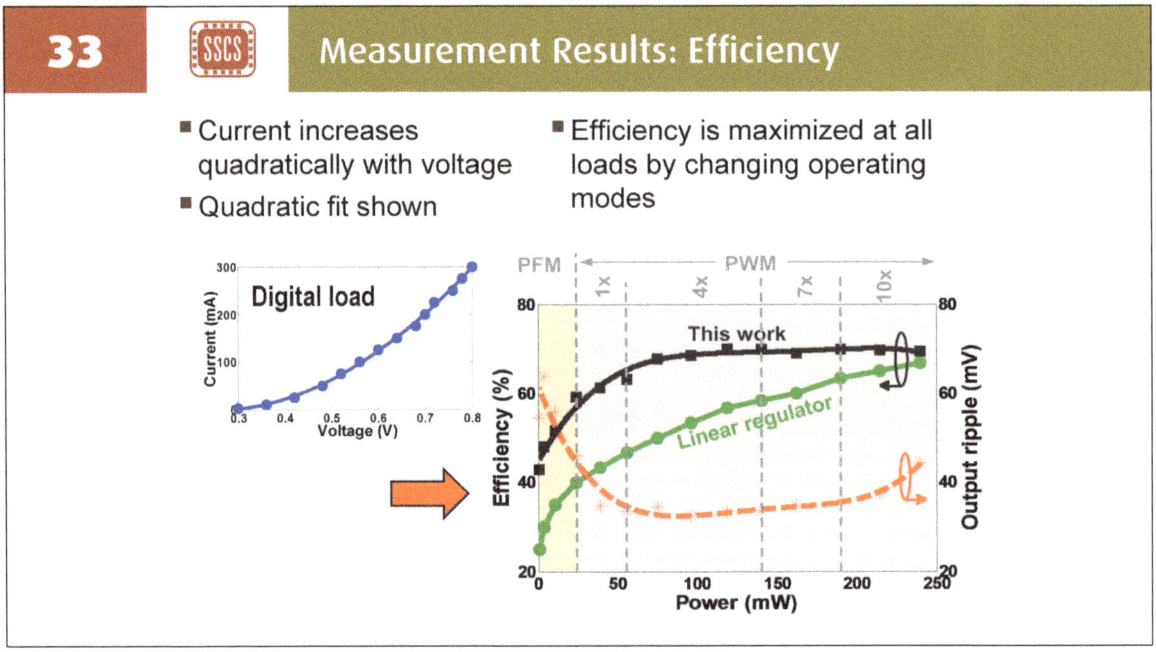

This inductive converter was designed for a quadratic DVS load. The slide shows the efficiency and output voltage ripple verses output power relation. As we discussed before, PWM techniques are used for higher loads and PFM (hysteretic control) are used for lighter loads.

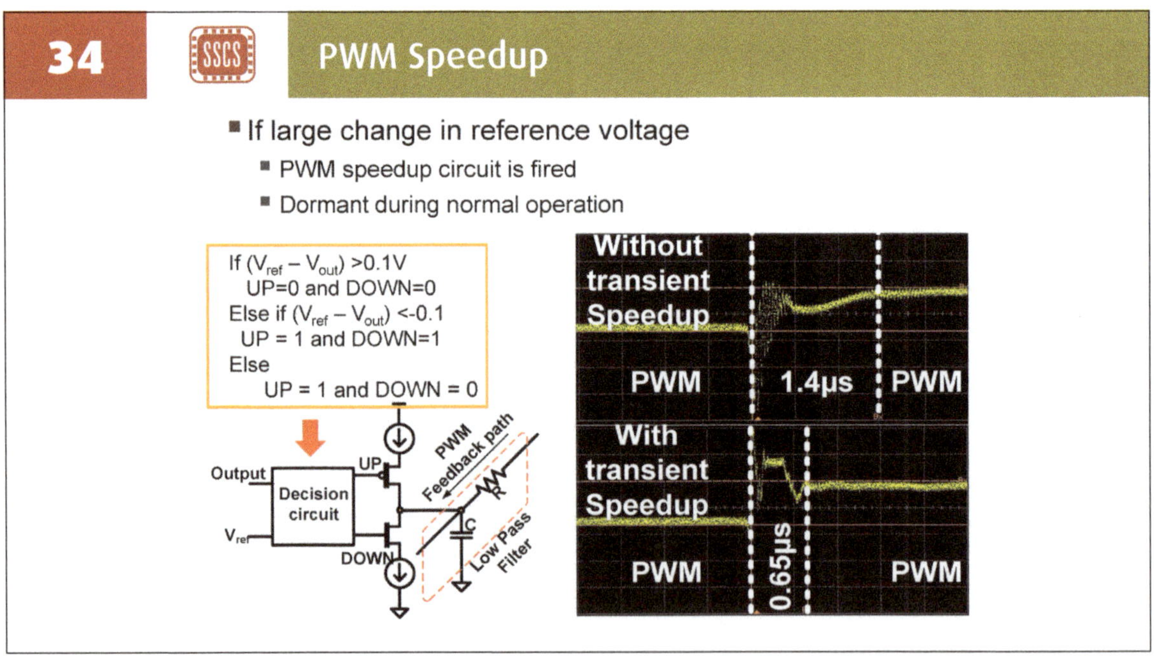

This slides shows the details of the PWM speedup mechanism for the inductive DC-DC converter. The figure on the right shows the corresponding transients.

35 Mode Selection (I)

- Additional resistance → efficiency ↓
- We reuse inductor series resistance
- DC drop across the R_s ∝ load current
- Process, temperature variation
  - Use programmable offset

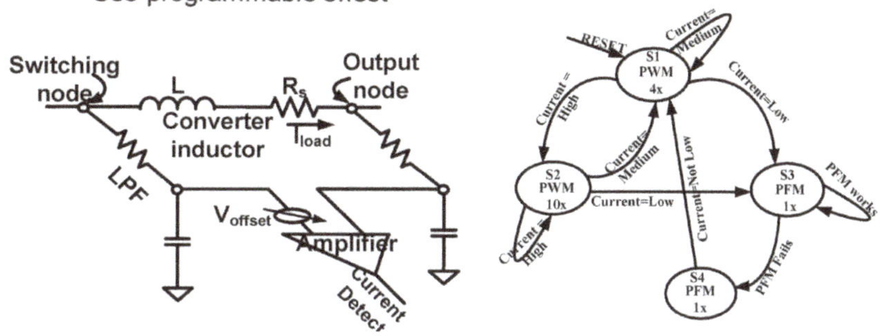

This slide illustrates the algorithm for the mode selection. The state diagram on the right explains the overall procedure.

36 Mode Selection (II)

- State machine decides operation mode
- Transition time from M1 → M2 → M1

Load Voltage	Load Current	Mode
374mV	30mA	PFM 1X (M1)
717mV	153mA	PWM 10X (M2)

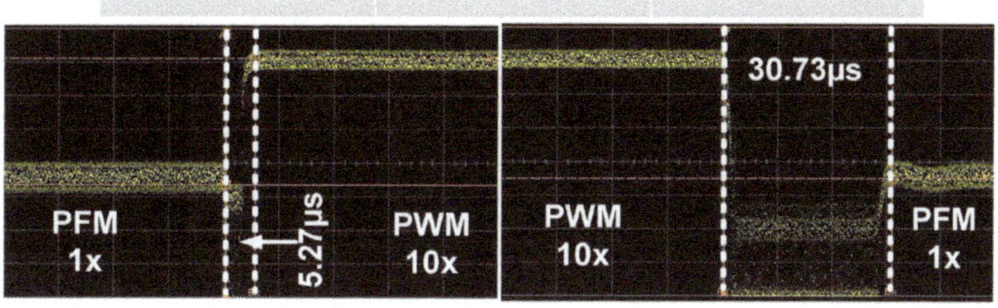

This slide continues on the mode selection algorithm and shows the transient response details in PFM and PWM.

37 Inductive: Comparison

	O/P Power (mW)	O/P Power range	Efficiency (%)	Inductor/ Capacitor	Passive Components
Musunuri PESC 2005	50 – 200	~4X	40 - 60	0.1μH/30nF	MEMS inductor
Xiao JSSC 2004	0.15-600	~4000X	70 - 92	10μH/47μF	Off-chip inductor and capacitor
Wibben JSSC 2008	3 – 315	~100X	10 - 78	2x2nH/5nF	On-chip inductor and capacitor
Abedinpour T PE 2007	95 – 400	~4X	35 - 64	2x11nH/6nF	On-chip inductor and capacitor
This work	0.6–266	~450X	42.8 - 74.5	2nH/5nF	On-chip Inductor and capacitor

This work provides the best on-chip performance

This slide shows the overall design comparison of this inductive converter with other designs. This is a fully integrated converter implementation and achieved a wide range of output power (~450X).

CAPACITIVE CONVERTERS

38 Capacitive Conversion Efficiency Trends

- Conduction loss : Fundamental – sets upper limit
- Parasitic loss: sets peak value
- DC control losses: almost constant with load – dominate at lighter loads
- Partial charging loss: dominates at higher loads

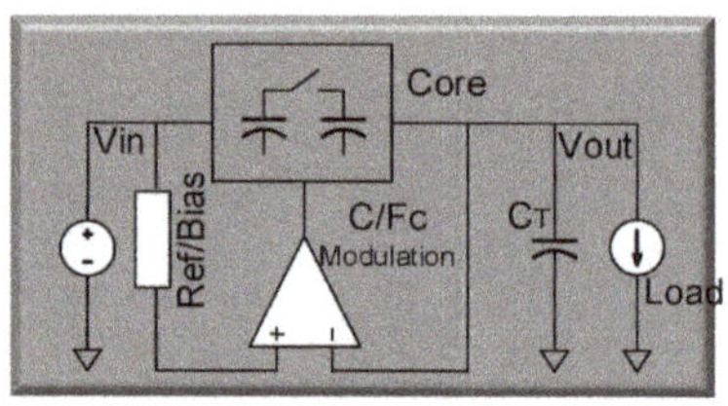

Capacitive DC-DC converter

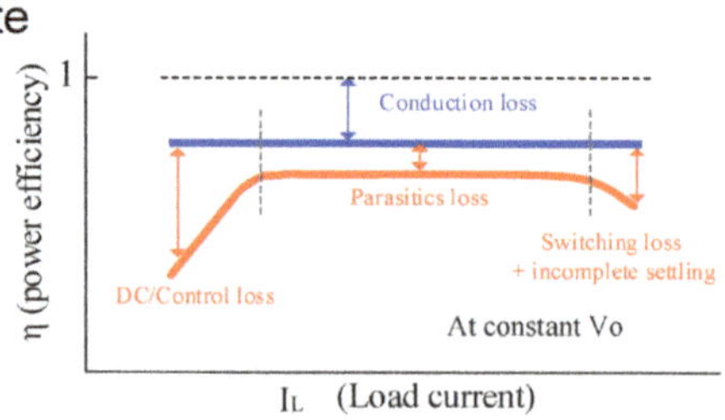

The efficiency of the capacitive converter is shaped by different loss mechanisms as shown for a constant Vo in the graph in the right bottom. In this figure, the converter is assumed to have voltage regulation mechanisms that maintain a constant output voltage for different current loads. The upper efficiency limit, identified by the solid blue line is topology dependent and is a function only of the output voltage. Actual measured efficiencies will be lower than this upper limit, due to parasitics and is further limited at low currents by DC/control losses and limited at higher currents by incomplete settling and switching losses

39 Present Day Integrated Capacitors

- Power density → capacitive density
- Desired qualities of integrated capacitor
  - Low bottom plate capacitance
  - High density, low leakage, low ESR
- MIM (Metal Insulated Metal)
  - Low density, low bottom plate parasitic, expensive
- MOM (Metal Oxide Metal)
  - High parasitic, low density, cheap
- MOS (Inversion)
  - High density, high leakage, medium bottom plate
- Accumulation Floating Junction (AFJ) Capacitor (CICC:16-8)
  - Low Leakage, higher density

Integrated capacitors are the important design component of the capacitive converters. Available integrated capacitors are – MIM (metal insulator metal), MOM (metal oxide metal) and MOS capacitors. Some recent examples also investigate the use of MOS capacitors in accumulation mode as well.

40 Series Parallel Capacitive DC-DC Converters

Unified Series Parallel Capacitive DC-DC Converter Framework

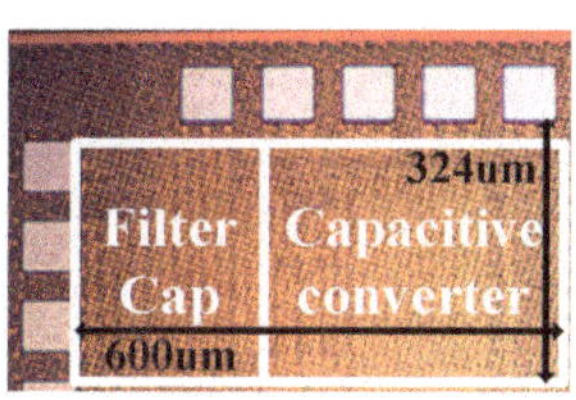

Ramesh Harjani, Saurabh Chaubey, CICC 2014

Let us discuss the design methodology and modelling of series parallel capacitive DC-DC converters.

41

We list out all practical series-parallel converter topologies. To make the problem tractable, we use a set of rules to prune out the less effective topologies.

(a) All circuits have only 3 terminals. Vi for input, Vo and a common ground, used by both input and output. Multiple input and output systems are not considered.

Design Rules

- Motivation
  - Currently, ad hoc single point solutions
  - No generalized set of topologies & equations
- Heuristic rules
  - Ripple voltage is small
  - Only three terminal designs
  - All bucket caps same size
  - All bucket caps move as one group
  - The bucket caps are in series or parallel
  - A cycle is composed of two phases
  - All three terminals used during each cycle

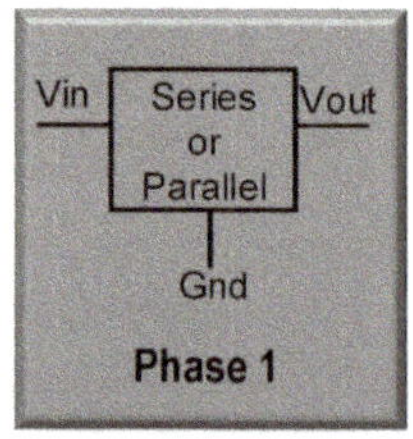

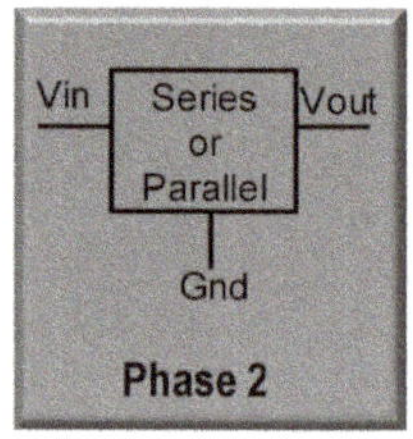

Ref: Ron Balczewski and Ramesh Harjani, IEEE International Symposium on Circuits and Systems, 2001

(b) All bucket capacitors are of the same size to ensure maximum area efficiency.

(c) In a clock phase, all capacitors are either connected in series or in parallel, but never a combination of both.

(d) Only two phase circuits are considered. Each phase uses a particular configuration. A converter is defined by the configurations used in the two phases.

(e) All three terminals are used in each cycle. This is a required condition for the capacitive converter to function. If the input is not connected, no charge enters the capacitive converter. If the output is not connected, no charge leaves the capacitive converter and if the ground terminal is not used, the converter does not function.

42

We use a three letter naming convention to represent all the possible configurations. The naming convention is explained here. The first letter in the naming convention represents the first terminal that the capacitors are connected to, with possible values of I(input), O(output) and G(ground). The second letter represents the configuration that the capacitors are connected in, i.e., series or parallel. And the third letter represents the second terminal that the capacitors are connected to, i.e., I, O, or G.

Nomenclature

- Phase notation we will use

Position	Code	Description
1st	I, G, O	1st connection (Input, Output, Gnd)
2nd	P, S	Series or Parallel
3rd	I, G, O	2nd connection (Input, Output, Gnd)

$3 \times 2 \times 2 = 12$ phases

IPG	IPO	GPI	GPO	OPI	OPG
ISG	ISO	GSI	GSO	OSI	OSG

43 12 Distinct Configurations

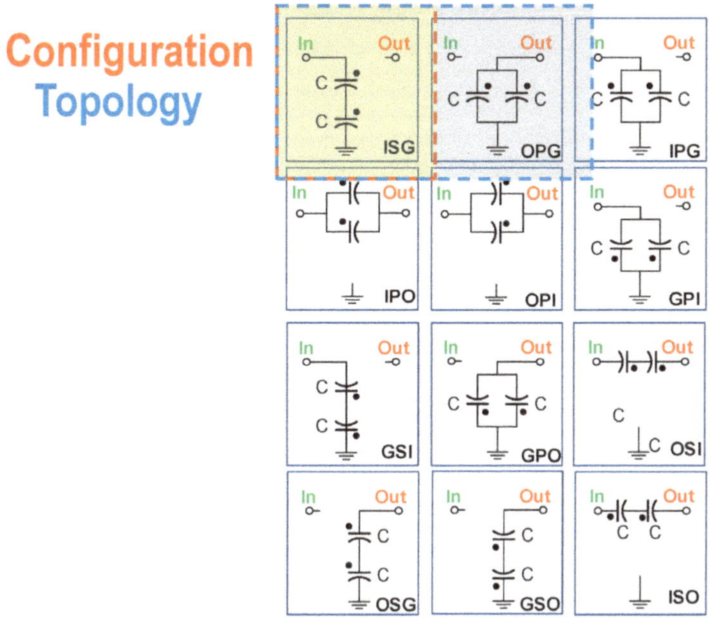

Using the heuristic rules (discussed earlier) all the 12 non-redundant configurations that are possible using the rules are listed is this slide.

44 Configuration to Topology

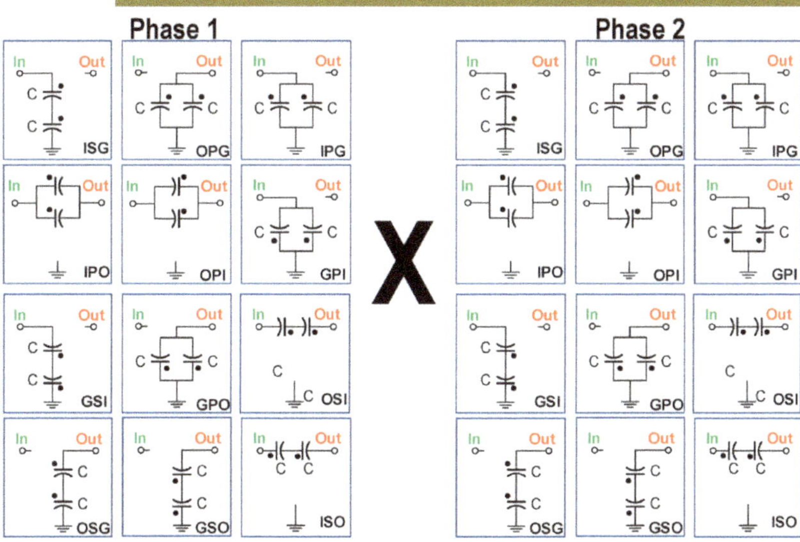

We can use any of the 12 configurations from table during phase 1. During phase 2 we only have 11 configurations possibilities as one has been selected in phase 1, giving us a total of 121 different converter topologies.

45

Enumeration

- 12 configurations for 1st phase
- For two phases (12 x 11=132) combinations
- Apply the heuristic rules: reduces to 96
- More heuristic rules
  - Same 2 phases but different order: ISG-IPO ; IPO-ISG
  - Series or parallel in both phases: ISG-ISO = IPG-IPO
  - Flipping 1st and 3rd letter (IPG-IPO=GPI-OPI)

18 structurally unique topologies

IPG-IPO	IPG-GPO	IPG-OPI	IPG-OPG	IPG-ISO	IPG-GSO
IPO-GPO	IPO-OPG	IPO-ISG	IPO-GSI	IPO-GSO	GPO-ISG
GPO-OSI	IPG-OSI	IPO-OSG	IPG-OSG	GPO-ISO	GPO-GSI

However, three of these 11 configurations in phase 2 do not use all three terminals, i.e., do not provide power, such that we finally end up with (12×8 = 96) converter topologies that provide power. Of these 96 topologies, several topologies are redundant or less efficient and can be eliminated. For example, the topologies IPO-OSG is exactly the same as OSG-IPO with the phases swapped. Removing these redundant topologies reduces the number of topologies to 48. Topologies in which the capacitors are connected in series in both phases can also be implemented using topologies in which the capacitors are connected in parallel in both phases. Since the charge transferred is higher with parallel bucket capacitors, we can remove the topologies with the capacitors in series in both cycles. This reduces the number of topologies to 36. Since, the first and second terminals are interchangeable, e.g., IPG-IPO is the same as GPI-OPI. We can now reduce the total number of topologies to the 18 practical ones.

46

Derivation of Performance

- Generic performance equations

$$Vmax = K \times Vin$$ → No load voltage

$$Iout = (F_{sw} \cdot C) \times f(Vin, Vout, N)$$ → Output current

$$Iratio = \frac{Iout}{Iin} = K$$ → Conversion ratio

$$Peff = \frac{Pout}{Pin} = \frac{Vout}{Vmax}$$ → Power efficiency

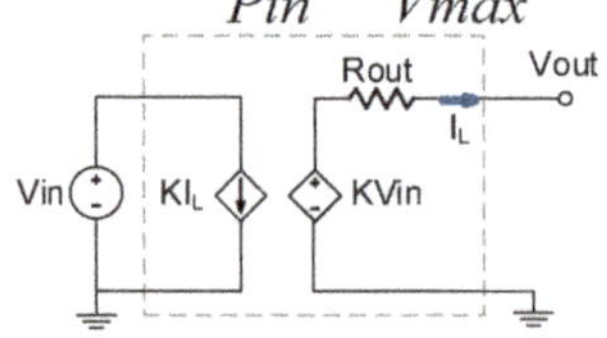

We develop the converter performance equations that will allow us to evaluate conduction losses that sets the maximum efficiency for a topology at a given output voltage, i.e., the maximum efficiency. The different topologies are evaluated based on three parameters, Io the output current, Vmax the maximum output voltage and Iratio the ratio of output current Io to input current Ii. The equations for these parameters are developed in terms of, f the frequency of operation, C the bucket capacitance size, N the number of capacitors, Qin the charge input to the capacitive converter and Qout the charge output from the capacitive converter.

47 1:1 Converter (IPG-OPG)

- Figure shows a 1: 1 capacitor converter
  - I.e., the maximum voltage at the output is Vin, $(Vo \le Vin)$
- Bucket capacitor, C_b, charged to Vin during input Φ_1
- C_b discharge to Vo during Φ_2
- The tank capacitor is much larger $C_T >> C_b$
  - I.e., Vo can be assumed to be fairly constant

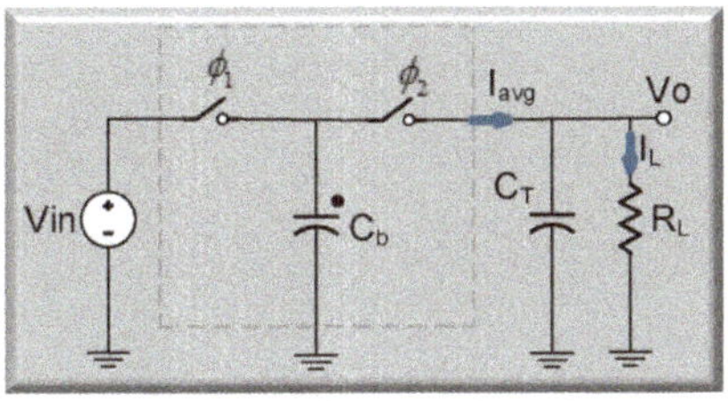

IPG-OPG Capacitive Converter

$$\Delta Q = (Vin - Vo)Cb \text{ per cycle}$$

$$I_{avg} = \Delta Q \cdot F_c = (Vin - Vo)Cb \cdot F_c = I_L$$

$$R_{avg} = \frac{\Delta V}{I_{avg}} = \frac{(Vin - Vo)}{(Vin - Vo)Cb \cdot F_c} = \frac{1}{Cb \cdot F_c}$$

This slide shows the working of a 1:1 capacitive DC-DC converter using the transformer model.

48 Transformer Model for IPG-OPG

- Transformer model for 1:1 capacitive converter
- Load line for IPG-OPG converter
- Maximum current when Vo approaches zero
- When Vo=Vin → no charge transferred to output

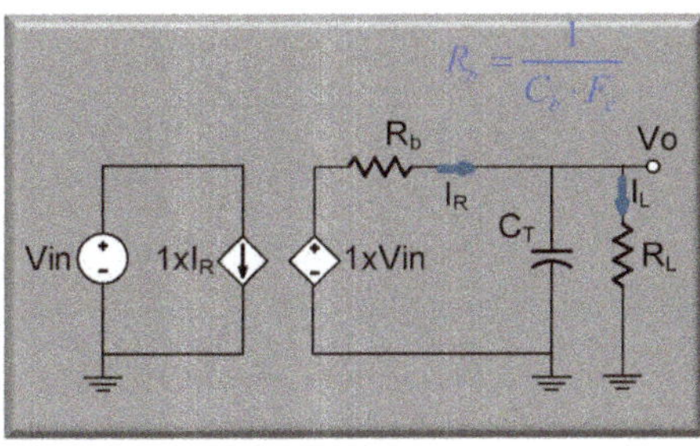

IPG-OPG Converter Transformer Model

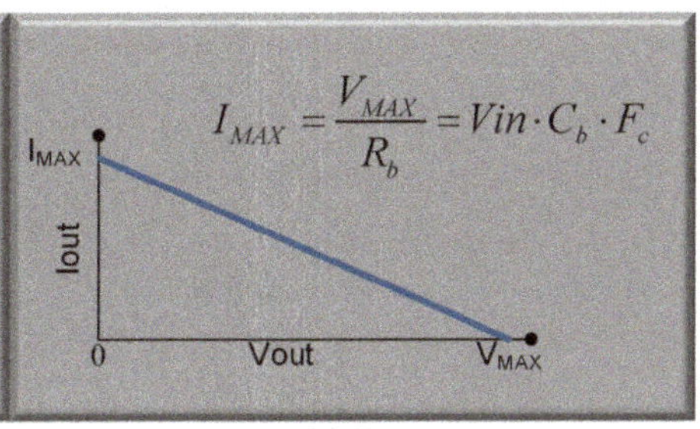

IPG-OPG Load Line

Rout of this 1:1 topology is 1/(C.F), where C is charge transfer bucket capacitor and F is the switching frequency. We note that as we depart more and more from the no load voltage of the topology (Vmax), the current capability increases as illustrated in the bottom right graph.

49 Buck Converter #1 (ISG-OPG) (1 : ½)

- Φ_1 each cap charged to Vi/2
- Φ_2 both caps discharged to Vo

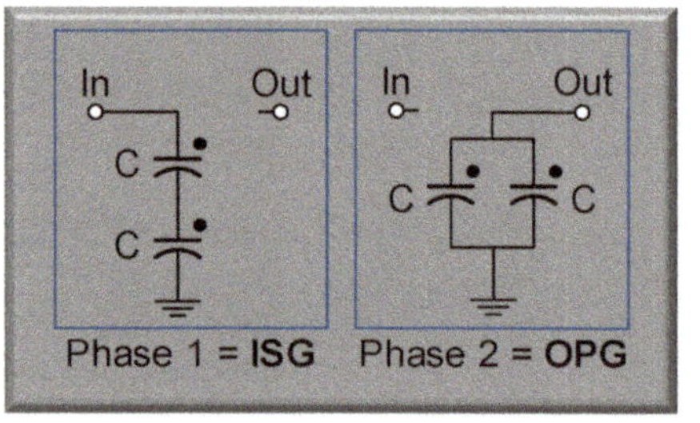

$$\Delta Q_1 = 0$$

$$\Delta Q_2 = 2\left(\frac{Vi}{2} - Vo\right)C$$

$$Io = \left(2\left(\frac{Vi}{2} - Vo\right)C + 0\right)F_c$$

$$Ro = \left(\frac{Vmax - \text{Vo}}{Io}\right) = \frac{1}{2 \cdot C \cdot Fc}$$

$$Vmax = \frac{Vi}{2}$$

$$Ro = \frac{1}{N \cdot C \cdot Fc}$$

$$Area = N \cdot Area_c$$

This slide goes through analysis of a 2:1 (or 1:1/2) topology.

50 Buck Converter #2 (IPO-OPG) (1 : ½)

- Φ_2 caps parallel to Vout
  - Charged to Vout
- Φ_1 caps in parallel between input and output
  - Caps charged to Vin-Vo
- Charge transferred to output during each phase
- Calculating Vmax

$$(Vin - Vo) = Vo \rightarrow V_{\max} = \frac{Vin}{2}$$

$$Area = N \cdot Area_c$$

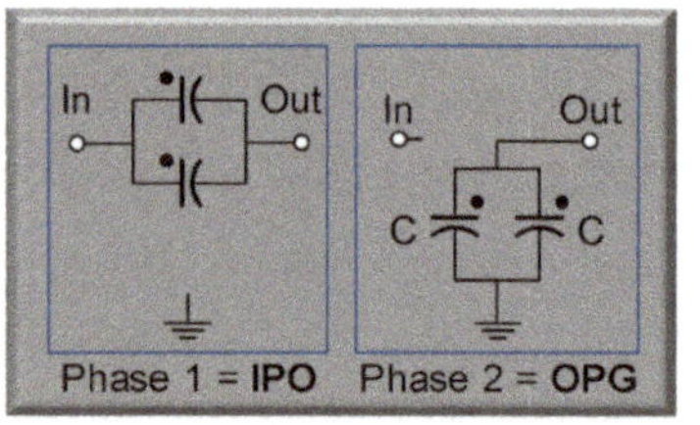

$$\Delta Q_1 = \left[(Vi - Vo) - Vo\right]2C$$

$$\Delta Q_2 = \left[(Vi - Vo) - Vo\right]2C$$

$$Io = (\Delta Q_1 + \Delta Q_2)Fc$$

$$= 2\left(\frac{Vi}{2} - Vo\right)4C \cdot Fc$$

$$Ro = \frac{1}{4N \cdot C \cdot Fc}$$

This slide calculates the output impedance of this topology (1:1/2).

51 Quick Summary: Topology Selection

- Same total capacitance used to compare designs
  - N=2, C=1
- Some topologies are just better than others
  - Lower Ro ➔ can provide larger current
  - **IPO-OPG:** charges during Φ1 and Φ2 ➔ lower ripple

$$Ro = \frac{1}{4N \cdot C \cdot Fc}$$

$$Ro = \frac{1}{N \cdot C \cdot Fc}$$

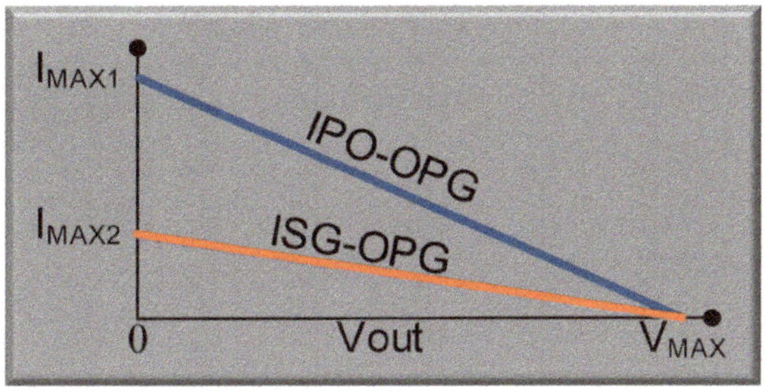

Some of topologies aren't as efficient as others at a given voltage conversion ratio. For example, the IPO-OPG topology always performs better than ISG-OPG for a given output voltage. Thus, the less efficient topologies can be eliminated.

52 Boost Converter: (IPG-OPI) (1 : 2)

- Φ_1 parallel capacitors connected to Vi
- Φ_2 parallel capacitors between Vo and Vi
  - Note phase of capacitors
- Calculating Vmax

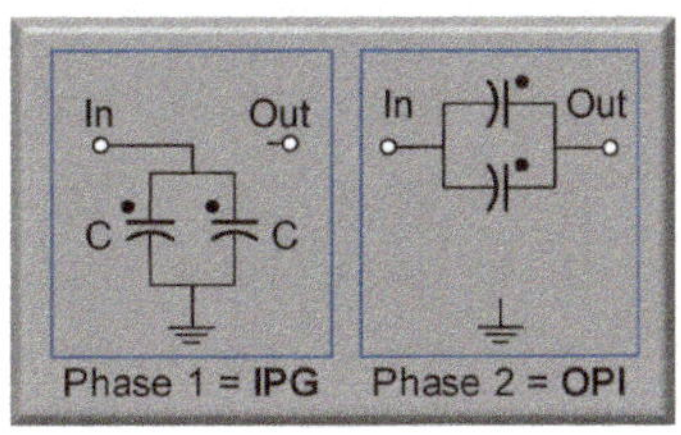

$$\Delta Q_1 = 0$$

$$\Delta Q_2 = [(Vi - Vo) + Vi]2C$$

$$Io = [0 + (2Vi - Vo)2C]Fc$$

$$Vi = -(Vo - Vi)$$

$$Vmax = 2Vi$$

$$Ro = \frac{1}{N \cdot C \cdot Fc}$$

$$Area = N \cdot Area_c$$

This slide deals with a 1:2 boost converter topology.

53 Performance Equations: Final Set

MODE	K	Rout	Nmin
IPG-OSI	N+1	$\frac{N}{fC}$	2
IPG-OPI	2	$\frac{1}{fCN}$	1
ISG-OPI	$\frac{N+1}{N}$	$\frac{1}{fCN}$	2

Boost performance equations

MODE	K	Rout	Nmin
ISG-GPO	$-\frac{1}{N}$	$\frac{1}{fCN}$	2
IPG-GPO	-1	$\frac{N}{fCN}$	1
IPG-GSO	-N	$\frac{N}{fC}$	2

Negative performance equations

MODE	K	Rout	Nmin
IPG-OPG	1	$\frac{1}{fCN}$	2
IPO-OSG	$\frac{N}{N+1}$	$\frac{N}{fC(N+1)^2}$	1
IPO-OPG	$\frac{1}{2}$	$\frac{1}{4fCN}$	2
ISO-OPG	$\frac{1}{N+1}$	$\frac{N}{fC(N+1)^2}$	2

Buck performance equations

IPG-OSG	OPI-OSG	IPO-ISG	ISG-OPG
ISO-GPO	IPG-ISO	IPG-IPO	OPG-OPI

Only 10 Optimum Topologies

Following the analysis in our previous discussions, we propose the 10 most optimum topologies which are classified as either buck, boost and negative topologies.

54 Modeling Capacitor Parasitics

- Top / bottom parasitics modeled as conduction loss
- Capacitor parasitics
  - Parasitic input load
  - Parasitic output load
  - Parasitic parallel converter

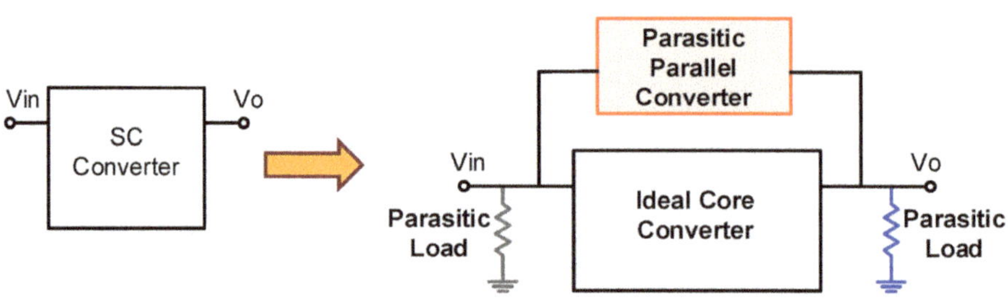

Once the efficiency's upper bound is set by conduction losses, the parasitic losses set the maximum attainable efficiency. These losses are determined by the top and bottom plate parasitics associated with the capacitor and switches.

55 Effect Of Parasitics

- Example: IPO-OPG
  - Top plate acts as 1:1 parallel converter
    - Takes charge from Vin in Φ_1 dumps on load in Φ_2
  - Bottom plate acts as a passive load
    - Takes charge from Vout in Φ_1 and dumps to ground Φ_2

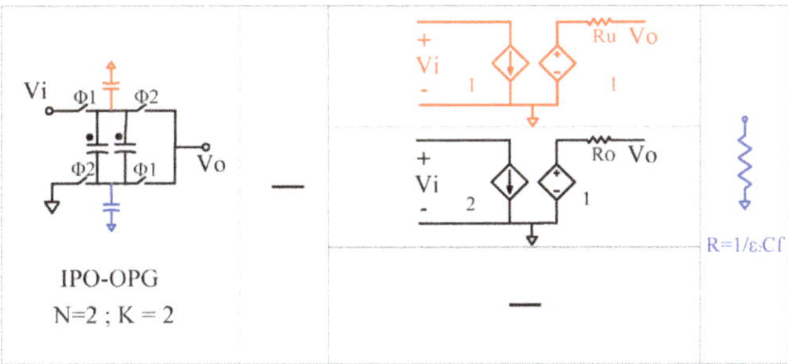

We illustrate the impact of these parasitics for a single capacitor based IPO-OPG converter topology shown in figure. The desired converter is modeled in black in the center row. The top plate is modelled as the parasitic 1:1 converter (in red). Also the bottom plate capacitor is modelled as a parasitic resistive load at the output. (shown in blue in the table)

56 Parasitic Modeling

- Bottom plate further reduces efficiency

$$\eta = \frac{\left[4\left(\frac{Vi}{2}-Vo\right)\cancel{C_b f}+\varepsilon_1\left(Vi-Vo\right)\cancel{C_b f}-\varepsilon_2 Vo\cancel{C_b f}\right]}{\left[2\left(\frac{Vi}{2}-Vo\right)\cancel{C_b f}+\varepsilon_1\left(Vi-Vo\right)\cancel{C_b f}\right]}\left(\frac{Vo}{Vi}\right)$$

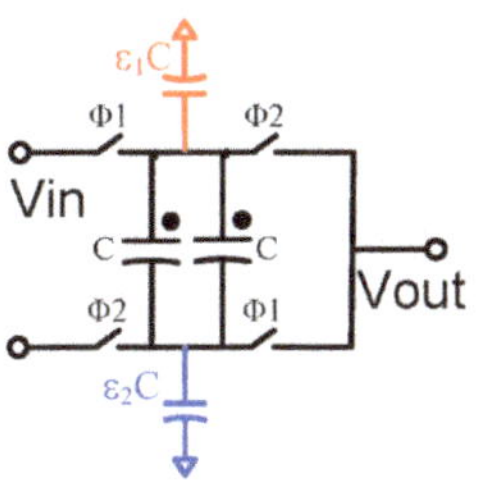

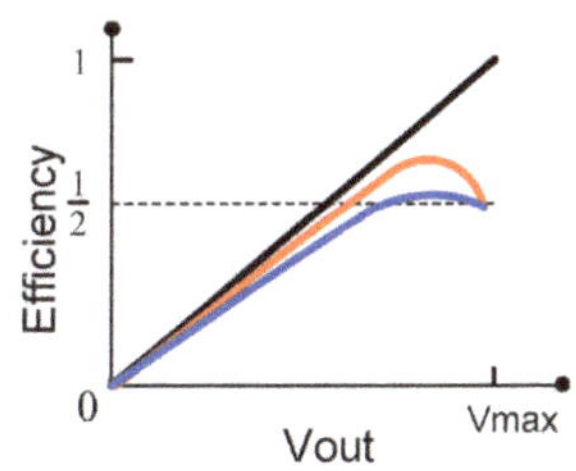

The top plate parasitic, shown as red ε_1C, is connected to the input during ϕ1 and to the output during ϕ2, i.e., it behaves as a 1:1 IPG-OPG converter. This parasitic converter transfers charge from the input to the output in parallel with the IPO-OPG converter and is shown in the top row of the "core" column of the figure. The bottom plate capacitance, shown as blue ε_2C is connected to the output during ϕ1 and to and during ϕ2, i.e., it behaves as additional load resistance equal to 1/ ε_2Cf on the output node. It affects only the output side and is shown in the "output" column. For this particular converter topology there is no parasitic impact on the input side. Therefore, the "input" row is blank. The overall converter, including parasitics, now needs to consider two converters in parallel (IPO-OPG and the parasitic IPG-OPG) and the parasitic load at the output node.

57 Parasitics

- All 10 topologies
- For each topology
  - Primary converter
  - Parasitic input load
  - Parasitic output load
  - Parasitic converter

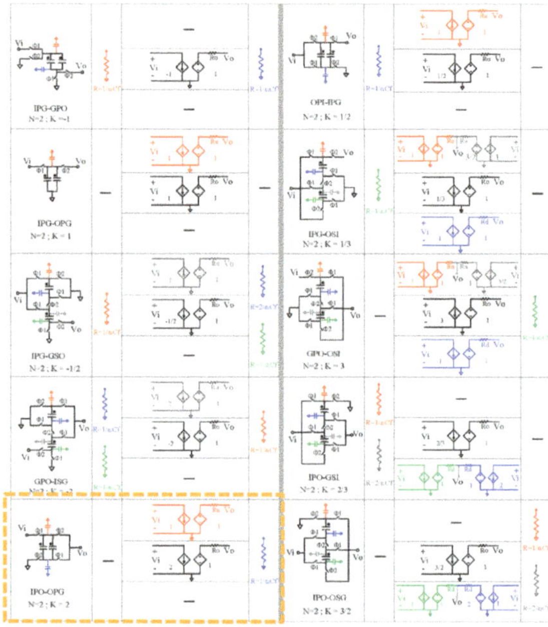

Following a similar analysis we extend our model to incorporate parasitic loss models for the 10 optimal topologies in the figure.

58 An Implementation Example

Software Defined 3000X Output Current Range SC DC-DC Converter

Process: TSMC 65 GP
Saurabh Chaubey
CICC 2017 (Session 16-8)

Let us walk through an implementation example. This paper presents a fully integrated, software defined capacitive DC-DC converter. The converter implements K-F-C tuning (K = conversion ratio, F = frequency and C =capacitance) in real time so as to accommodate any output load. It has a 4X tunable output voltage, supports a 3269X output load current range while achieving a peak efficiency of 82.1%.

59 Efficiency: Open Loop

- Open-loop efficiency
  - Maximum at a given output voltage
- Conduction losses at lower output voltages
- Bottom plate losses at higher output voltages
- K-tuning: different conversion ratios

$$K = \frac{V_i}{V_o}$$

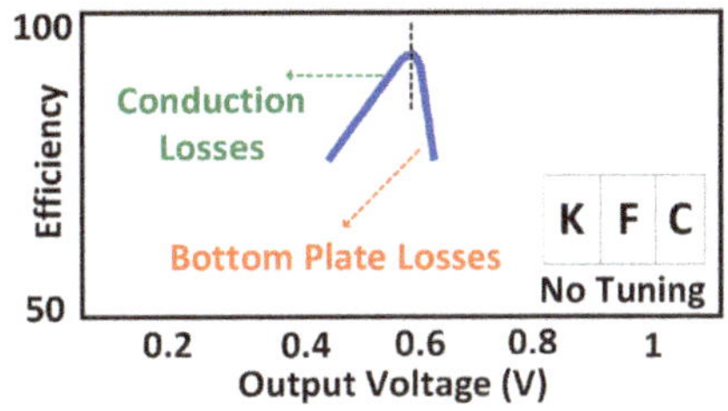

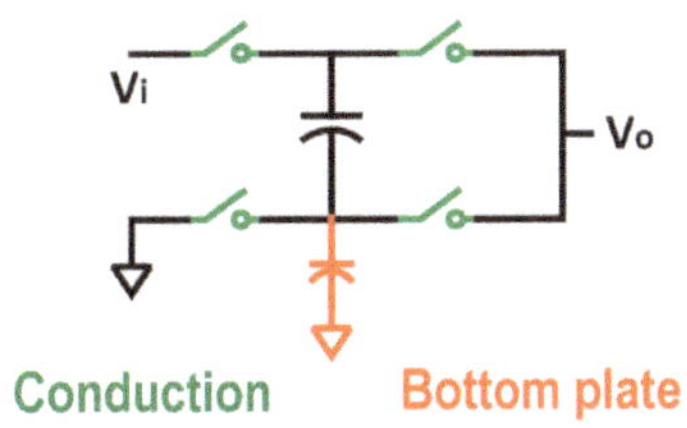

This slide shows the voltage-efficiency profile of a single capacitive converter. We observe that at voltages near the no-load voltage, the efficiency loss is dominated by bottom plate parasitics while at lower voltages the dominant loss mechanism is conduction losses. The peak efficiency is obtained close to the no-load voltage. Variable output voltages are best handled via multiple voltage conversion ratios.

60 Efficiency: F-CTuning

- F-Tuning: frequency tuning
  - Accommodate I_{load} changes
  - Constant output voltage
  - Gate drive losses and incomplete settling at higher load currents
- C tuning cap-modulation – on/off bucket capacitors
- F-C tuning – PFM and capacitance modulation simultaneously

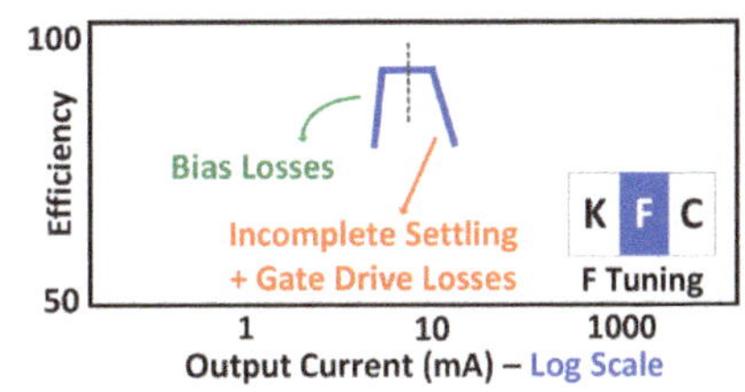

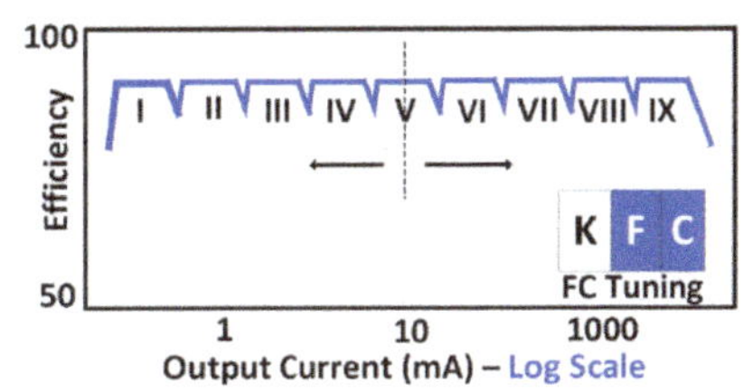

We observe that by changing the switching frequency (F), we can regulate the output of the converter for a finite range. At higher currents the dominant loss mechanism is incomplete settling and at lighter loads constant losses dominate. If we need to increase the load current range appreciably, we need to increase the size of the bucket capacitors being switched, i.e., capacitive tuning (C Tuning). As shown in the slide in bottom-right, nine identical capacitor cores are used to increase the load drive capability. Each of the nine cores can have their frequency tuned independently. When both F and C are tuned, we term that as F-C tuning.

61 Software Defined DC-DC: KFCTuning

- K-F-C – spans wide output voltage and loads
- K tuning - change in voltage conversion ratio
- F tuning – fine load regulation(plateau)
- C tuning – coarse load regulation(many plateaus)
- K-F-C- software controlled

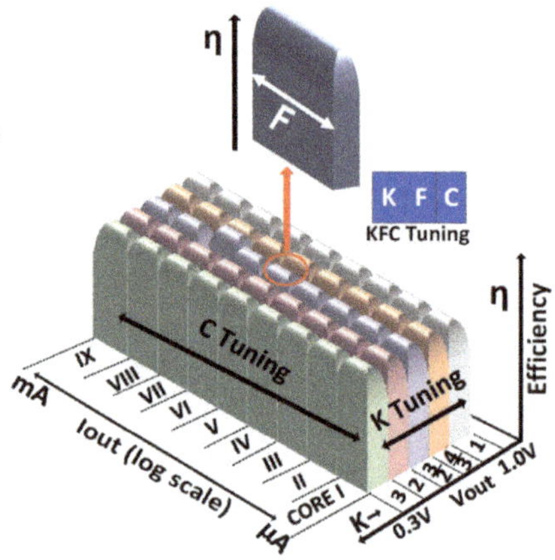

By using all three forms of tuning, i.e., K, F and C tuning, allows the system to support the largest load range while maintaining high efficiency as shown by the 3-D surface plot in the slide. This scheme forms the core of the proposed software defined capacitive DC-DC converter.

62 Overall Software Defined DC-DC

- Modular: standard cell based
  - 2:1 cell is standard cell - 6 cells / core (1 cell = 180pF)
  - Conversion ratio → tile (one or more cells make a tile)
  - 9 bucket capacitor cores (1 core = 1.08nF)
- Dual loop modulation
  - Capacitive modulation (nine levels) - Outer
  - Frequency modulation (RO based VCO) - Inner

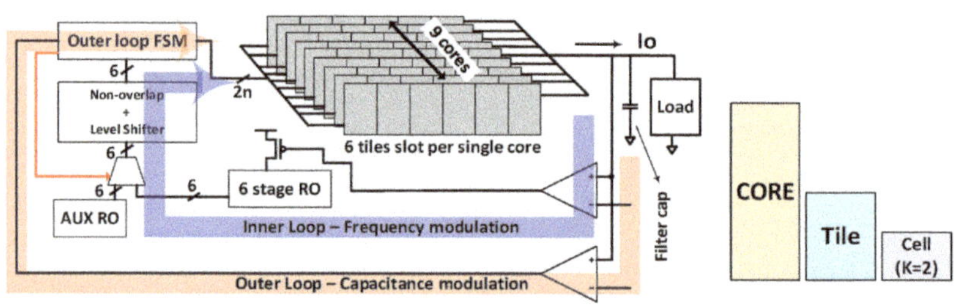

The proposed fully programmable converter can be digitally tuned to meet the electrical needs of any load (non-DVS and/or DVS). The proposed design enables the converter to support five step down (K:1) voltage ratios (K=1, 1.33, 1.5 2, 3) for a wide range of load currents. This architecture mimics a "digital standard cell" approach and is made to be easily scalable. We propose the use of both capacitive modulation and frequency modulation for load regulation in order to achieve a 3269X current load range. Capacitive modulation (or C-tuning) is realized by an asynchronous digital FSM based (marked in orange) system while frequency modulation (or F-tuning) is a VCO based analog-PID implementation (marked in purple). C-tuning implements coarse voltage regulation by turning off/on bucket capacitors. The design interpolates between two states of C-tuning using F-tuning as is discussed later

63 Open Loop Core Details (K=2)

- ☐ Cell: IPO-OPG
- ☐ Tile: a 2:1 cell
- ☐ Core:
  - Six 2:1 tiles
  - No extra switch
  - Six interleaving
- ☐ IPO-OPG can be repeated for other K values

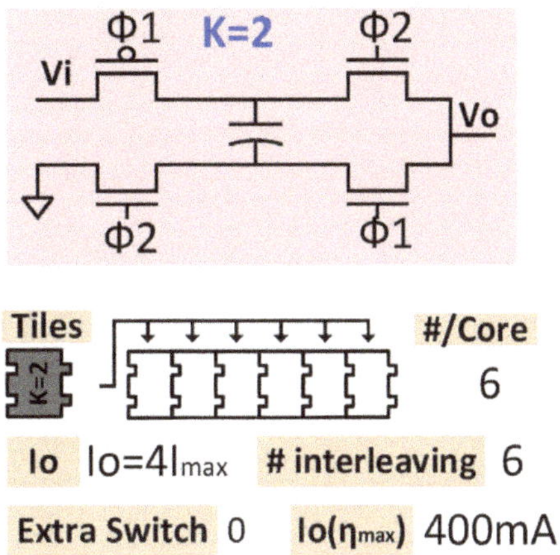

This slide shows the basic cell tile for the proposed programmable converter. The tile is flexible IPG-OPG (K=1) / IPO-OPG converter (K=2) depending on switch conditions. Six such tiles make up a single conversion core. This single core can serve as any combination of six 1:1 (or six 2:1) or three 3:2 (or three 3:1) or two 4:3 converters. The overall capacitive bank is made up of nine such cores.

64 Open Loop Core Details (K=3)

- ☐ Cell: IPO-OPG
- ☐ Tile: two 2:1 cells
- ☐ Core:
  - Three K=3 tiles
  - One extra switch
  - Three interleaving
- ☐ Extra switch connects both cells

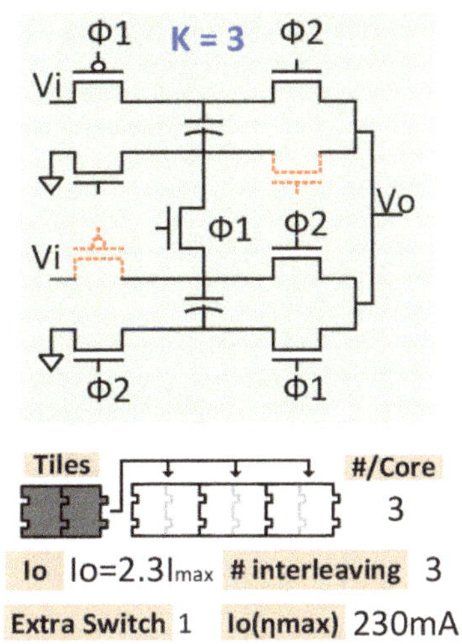

This slides tells how two IPO-OPG (K:1, k=2) tiles are fused together to realize 3:1 converter topology.

65 Open Loop Core Details

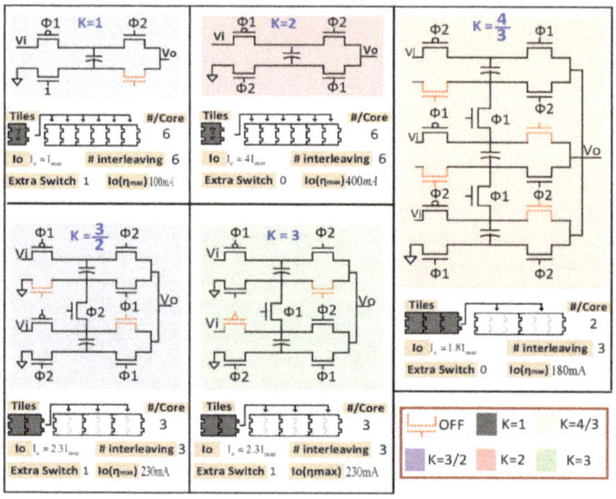

As shown in the slide, different conversion ratios are generated by a combination of individual tiles. For example, the modules of K=3/2 and K=3 can each be formed by stitching two K=2 modules in series. In this table the orange switch is always OFF, switch with a "1" on gate is always ON. Similarly, modules of K=4/3 can be formed by stitching three K=2 modules in series. Modules for K=1 and K=2 are the standard tile itself. We note that only K=1 and K=3/2 uses one additional switch in series in comparison to the minimal design. For this reason we have not scaled the switches for these two modes accepting the lowered fmax instead of lowered efficiency. As all five ratios utilize the same core tile we are able to use all the available capacitance for conversion at all times.

66 Dual Control Loops

- Dual control loops
  - Out digital loop - # of cores
    - Coarse regulation
    - On/freeze scheme
  - Inner analog loop
    - F tuning
    - Fine regulation per core
    - No degeneracy
- KFC algorithm
  - First choose K
  - Next adjust C to I_L
  - Next tune F for I_L

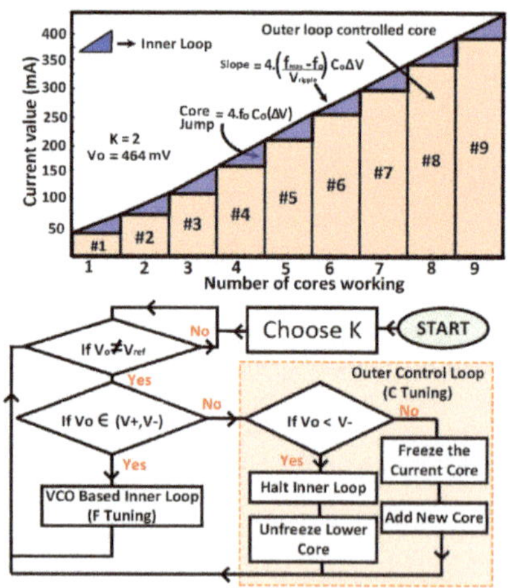

The operation of the two loops are best illustrated via the orange staircase (C-tuning) and the purple interpolation ramp (F-tuning). Once the number of cores is selected by the FSM the 6 phase RO starts at an fmin=55MHz and can increase to an fmax=520MHz depending on the bias control voltage. We observe that when an additional core is switched-on, there is a current step and within a given core the internal loop ramps-up the current capability with a fixed. As an example, when we are traversing (F-tuning) core #6, all the cores #1 - #5 are operating at fmax to maximize their current drives and RO only changes the frequency of core #6. Figure at the bottom right shows the control flow for KFC-tuning. The algorithm starts with selecting the K value (manually implemented in this prototype). Once the K is set the control moves to the interplay between the inner and outer control loops (F-C tuning). V+ and V- are the upper and lower thresholds for outer control loop, which are programmable on outer loop (C-tuning) state via a string DAC.

67

Efficiency

- Peak efficiency across different load currents
  - Resembles the constant flat efficiency plateau
  - K-F-C enabled high efficiency

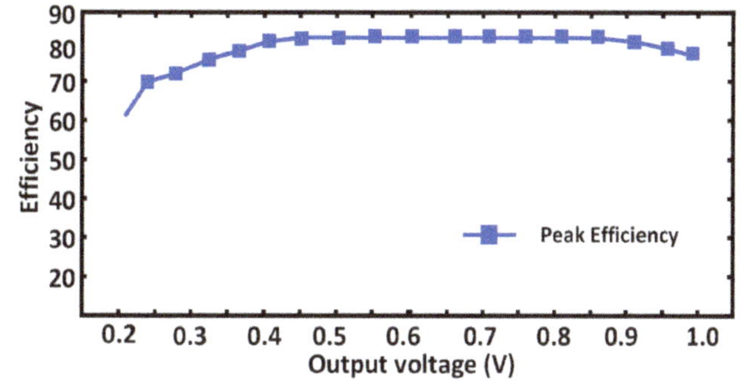

This slide shows the peak efficiency verses output voltage for different loads. For output voltages below 0:4V, (i.e., lower currents) the peak efficiency drops due to control and biasing losses. For higher voltages the peak efficiency starts to drop slightly due to switching losses and partial charging. During the voltage change from 0.25V to 0.95V the current can change from 0.13mA-425mA (3,269X). As an addition example, we test a DVS based load using a 9 stage ring oscillator.

68

Summary for Software Defined DC-DC

- Software defined DC-DC converter
  - Standard cell (reconfigurable)
  - Programmable # of phases
- Dual control loop
  - Capacitive & Frequency
- Maintains high efficiency plateau using KFC
- 3000X output current range support
- 40X leakage improved in standard CMOS
- Power density → 1.05W/mm2

We have implemented a fully programmable capacitive DC-DC converter which supports a wide range of output loads for every given output voltage between 0.25V to 0.95V. In order to achieve high power density we used a novel high density composite capacitor. This new PMOS-based capacitor has the lowest measured leakage (40X) of any structure available in standard CMOS. Further, simulations suggest that the accumulation mode MOS-cap inherits it low leakage properties for 45nm and smaller processes. The methodology can be applied to NMOS devices in twin-well processes as well, showing its versatility. A dual control loop achieves zero dark silicon for all five voltage modes.

69 Fully Integrated Capacitive DC-DC Converter

Fully Integrated Capacitive Converter with All Digital Ripple Mitigation

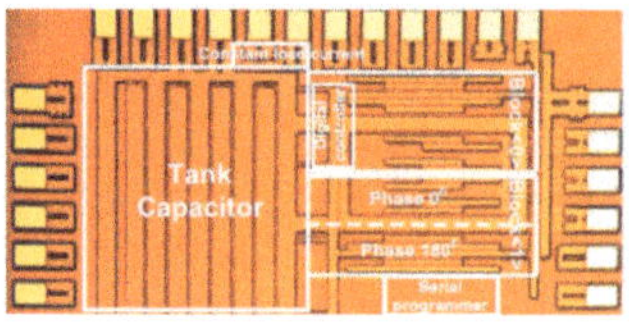

Process: IBM 130nm CMOS
Sudhir Kudva
IEEE CICC 2012, JSSC 2013

This paper presents an adaptive all-digital ripple mitigation technique for fully integrated capacitive DC-DC converters. Ripple control is achieved using a two-pronged approach where coarse ripple control is achieved by varying the size of the bucket capacitance, and fine control is achieved by charge/discharge time modulation of the bucket capacitors used to transfer the charge between the input and output, both of which are completely digital techniques. A dual-loop control was used to achieve regulation and ripple control. The primary single-bound hysteretic control loop achieves voltage regulation and the secondary loop is responsible for ripple control.

70 Ripple Control Techniques (I)

- Overshoot above nominal voltage
  - Results in wasteful power
- Ripple ↓ overall system efficiency ↑
- Conventional techniques
  - Increase switching frequency
  - Increase the interleaved stages
- Losses↑ or area↑ or need high frequency component
- Bucket capacitors designed for maximum load current
- Regulate the amount of charge transferred

This design focuses on reducing the ripple as lowering the ripple will reduce the overall system efficiency. The conventional ways to do so are either increase the switching frequency or increase the number of interleaving phases.

71 Ripple Control Techniques (II)

- Regulate the charge dumped into decap
- Unconventional ways to control charge transfer
- Three possible methods
- Switch resistance (method #1)
  - Gate voltage to vary resistance
  - Requires analog components

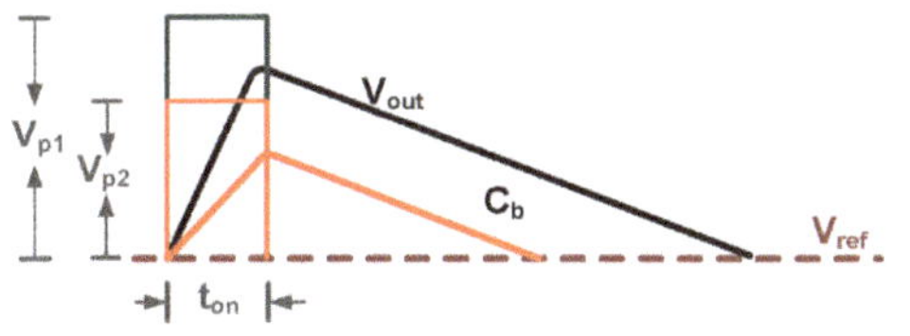

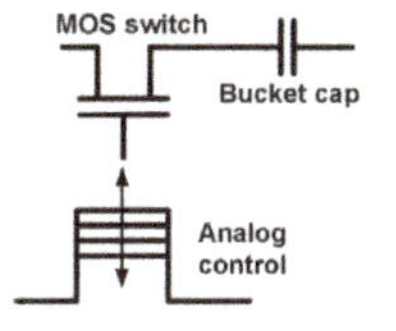

First method of ripple control is switch resistance control. Analog modulation control the gate voltage to increase/decrease the switch resistance.

72 Ripple Control Techniques (III)

- Capacitance modulation (method #2)
  - Vary bucket cap size
  - Hard to attain fine resolution

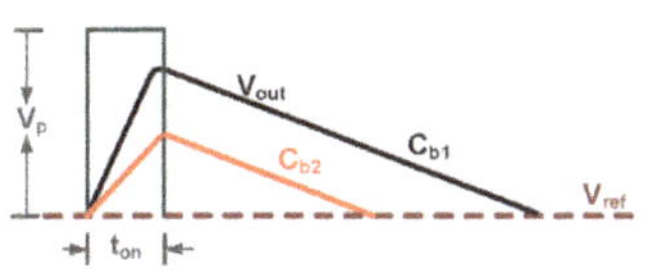

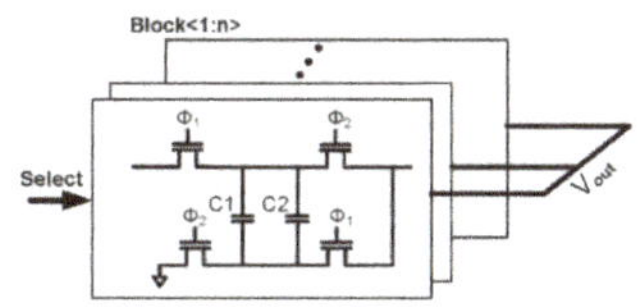

- Pulse width modulation (method #3)
  - Vary time of charge/discharge
  - Limited range of control

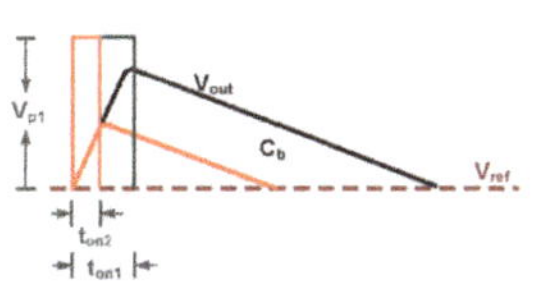

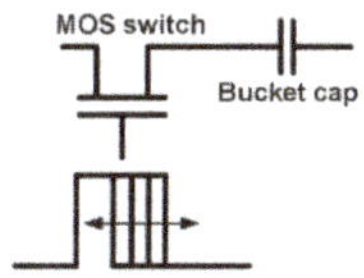

Second method to control the ripple is to vary the bucket capacitor size. The problem with this approach is to attain fine resolution. Third method is to vary the pulse width of clock pulses. But with this approach there is limited range of control.

73 Cap + Pulse Width Modulation (I)

- Both cap + pulse width modulation digital in nature
  - → No analog components
- Easily scalable with technology
- Primary loop for regulation
  - Changes switching frequency

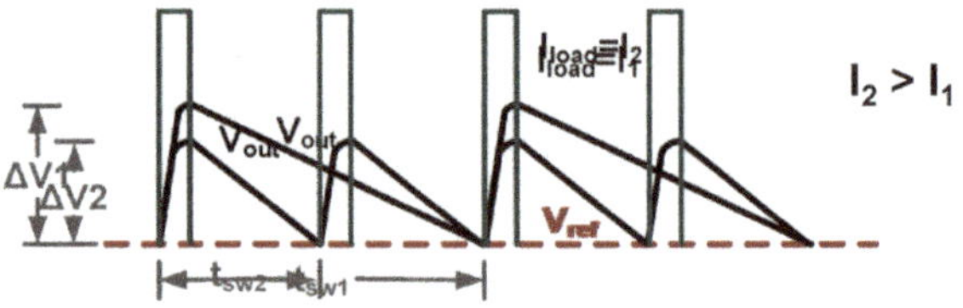

- Freq used as an indirect measure of ripple

If we use both bucket capacitor and pulse width modulation together, we can mitigate ripple in a technology portable way.

74 Cap + Pulse Width Modulation (II)

- Two step control
  - Course control → capacitance modulation
  - Fine control → pulse width modulation

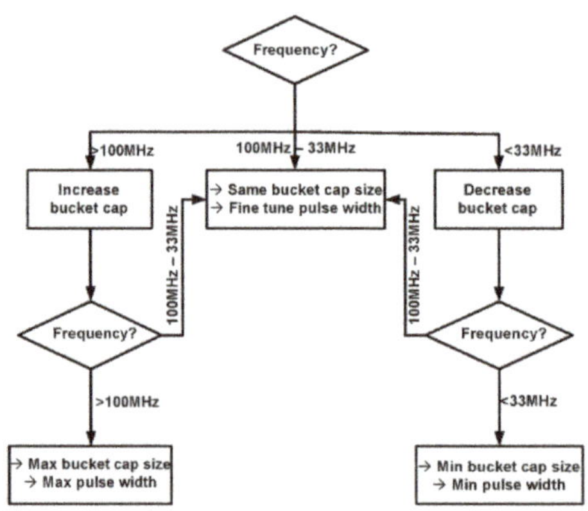

All digital detection mechanism

Easily scalable with technology

This slide describes the flowchart of the control mechanism for switching regulator.

75 Partial Charge/Dischage → Efficiency

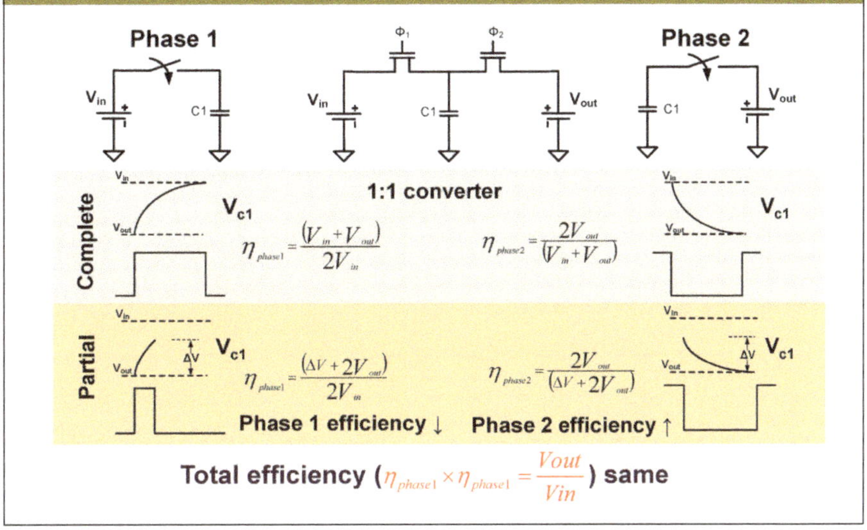

In partial charging, the efficiency of the charging phase is less than that for complete charging. This is because the capacitor is charged when the voltage drop across the switch is large, i.e., larger losses. However, the efficiency in the discharging phase for partially charging is higher than that for complete charging as now capacitor has to discharge from a lower value and the drop across the switch during the discharge process is smaller. The same explanation holds true if the capacitors are completely charged and partially discharged. Hence, by utilizing partial charging/discharging the overall efficiency of the converter does not change but the output ripple voltage is reduced.

76 Regulation

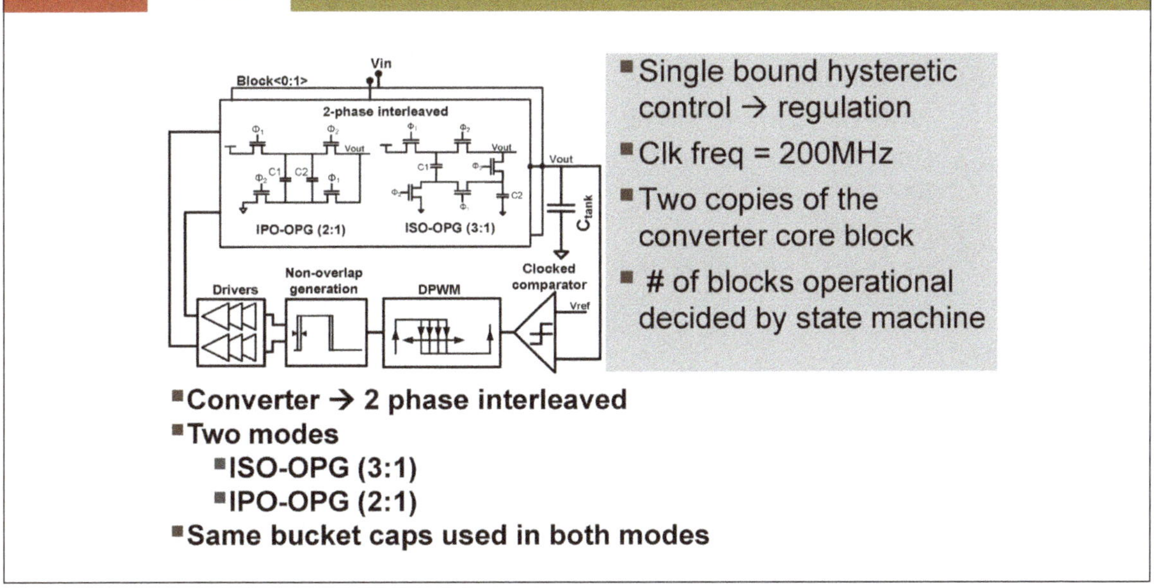

This slide shows the PWM control mechanism details in this DC-DC converter.

77 Fully Integrated Converter

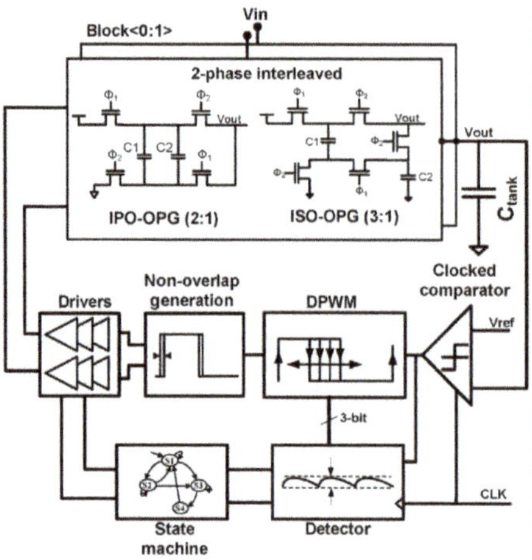

- Input to secondary loop
  - Primary switching frequency
- Capacitance modulation
  - All caps switching
  - Half of caps switching
- Pulse width → 3 bits
- Interaction between primary and secondary loop

The dual-loop control modulates the charge/discharge pulse width in a hysteretic variable-frequency environment using a simple digital pulse width modulator.

78 Measurement Results (I)

- Max efficiency mode → all bucket caps switching
- State machine control
  - Based on ripple selects mode
  - Lower efficiency → higher frequency
  - Less ripple (next slide)
- Experiment
  - Load current varied @ different Vout
  - Manual control (no S/M control)
  - With state machine control

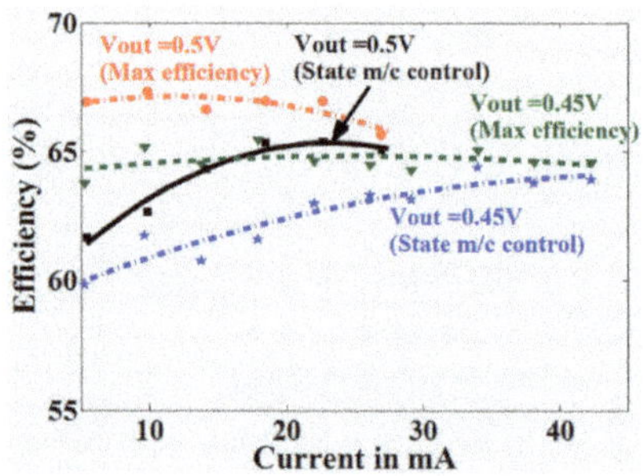

The efficiency of the converter at fixed output voltages and different load currents is shown in slide. The maximum efficiency curves represent the efficiency of the converter with the secondary loop disabled.

79 Comparison With Prior Work

- Achieves efficiency comparable to designs in lower tech. nodes
- Efficiency expected to increase with technology scaling

Work	Technology	Design technique	Max efficiency	Comments
[Lee07]	600nm	Resistance modulation	87%	Off-chip capacitors
[Le11]	32nm	32-phase interleaved	80%	Fully integrated
[Ramadass10]	45nm	Capacitance modulation	69%	Fully integrated
This work	**130nm**	**Cap + time modulation**	**70%**	**Fully integrated**

This prototype design implemented in 130 nm process shows efficiency which is comparable to the fully integrated designs in lower technology nodes with excellent ripple performance. Additionally, the efficiency is expected to improve further with device scaling.

SPECIAL CAPACITIVE CONVERTERS

80 The Flying Domain DC-DC Switch Capacitor Converter

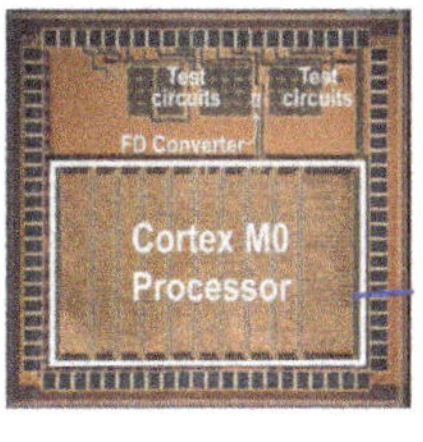

Flying Domain SC DC-DC
Process: 180nm SOI
Loai G. Salem, ISSCC 2016

We present the flying domain DC-DC switch capacitor converter.

81 Flying Domain DC-DC Converter

- Flying domain DC-DC converter
- Interchange the bucket capacitor with load
- 100X area reduction, no parasitic loss
- Requires SOI

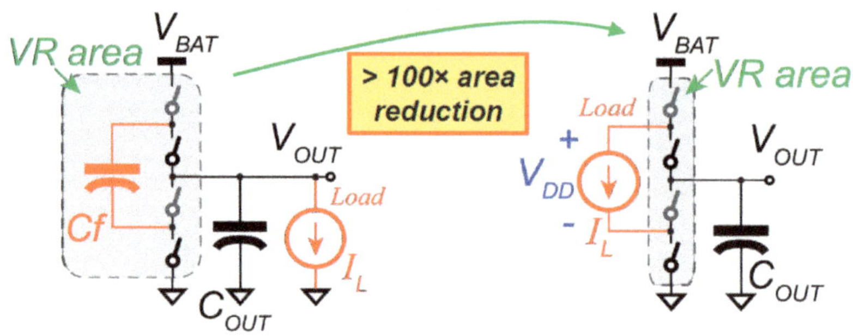

This paper presents a new class of DC-DC converters that perform input-to-output voltage conversion not by switching (or flying) an integrated passive (i.e., inductor or capacitor), but by instead flying the voltage domain of the load itself. The proposed flying-domain (FD) converter, directly replaces an SC flying capacitor with a load circuit, thereby creating an implicit DC-DC conversion network that achieves nonsingular loop and cut-set matrices.

82 Efficiency Measurement

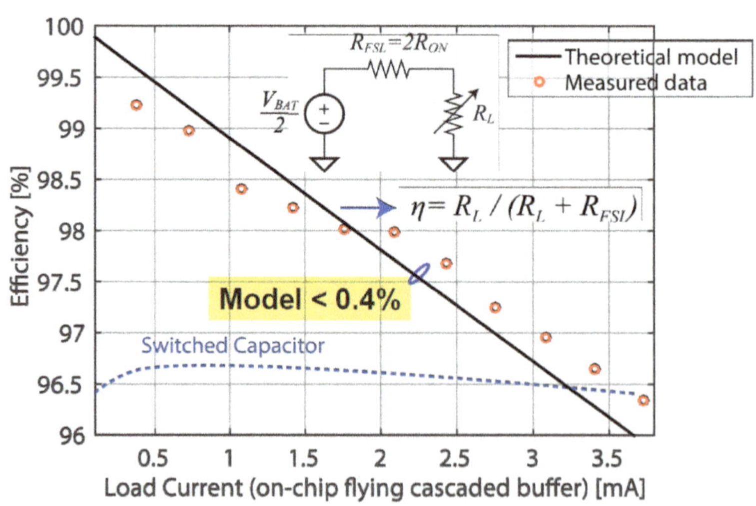

This slide shows the efficiency profile of this flying domain DC-DC converter.

83 Intuition of Why It Works

- For digital load
  - Every digital load has intrinsic decoupling capacitor
  - During transition from one phase to another, decoupling and trace capacitors maintain voltage across the flying load
- For purely resistive load
  - During dead time the trace and parasitic capacitance hold the voltage across the load

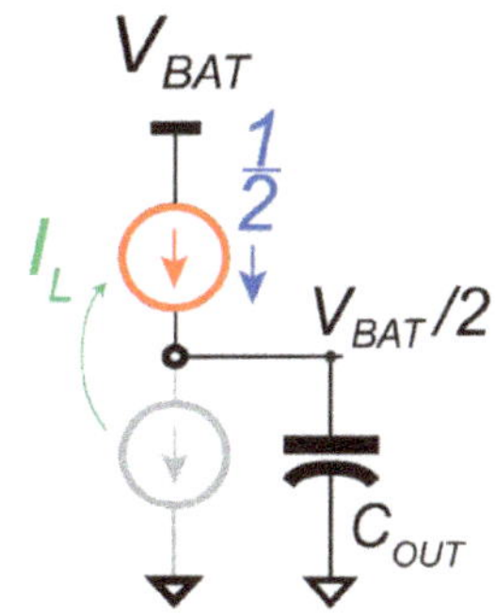

2-phase switched network KVL equations enforces $V_{LOAD} = V_{BAT}/2$

This slide describes the intuition behind the flying domain DC-DC converter.

84 Resonant Switching Converters

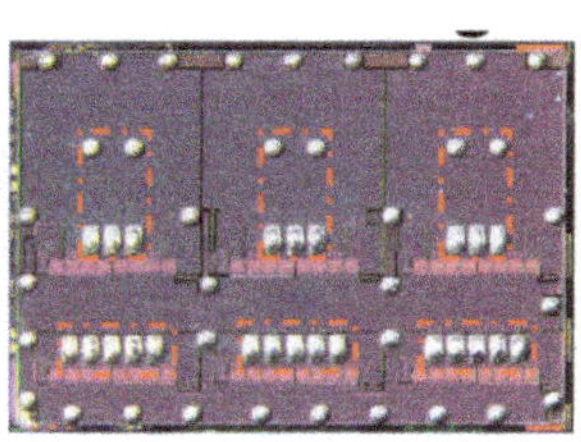

Resonant SC DC-DC
C. Schaef, ISSCC 2015

Now we look at the resonant switching converters which are becoming popular.

85 Resonant Switched Capacitive (ReSC)

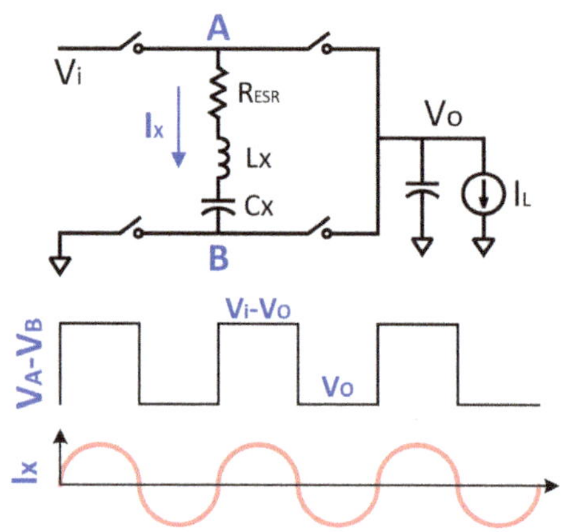

- Similar to SC converter
  - Uses similar series-parallel architectures to SC converters
  - Resonate capacitor Cx with Lx
  - Output impedance
    - RESR + switch resistance
- Advantages of ReSC
  - Improves trade-offs between switching and conduction loss
  - Narrow band energy transfer
  - Zero current switching
  - Helps to achieve lower output impedance at lower frequency

Resonant switched capacitor (ReSC) converters can be considered an extension of the SC approach in that similar hierarchical and cascaded topologies are used, which provide comparable device utilization and scaling advantages. However, a small amount of magnetic energy storage is used to implement soft switching with zero-current transitions. This can eliminate charge-sharing loss, significantly reduce bottom-plate loss and device stress, and enable operation at lower frequencies, providing high efficiency even with lower-cost bulk-CMOS processes. While the disadvantage of the ReSC approach compared to SC topologies is the need for high-Q magnetic components, the amount of magnetic energy storage is typically orders-of-magnitude lower than is required for buck or boost converters operating at comparable efficiencies and switching frequencies. With modern capacitor technologies supporting the bulk of passive energy storage, magnetic components can reasonably approach monolithic integration, or in this case, small air-core geometries embedded in the interposer, or circuit board.

86 Resonant Capacitive Converters

This slide shows a comparison of the resonant switch capacitor (ReSC) to normal switched capacitor (SC) architecture in a 2:1 configuration where converters operate with the same flying capacitance and series resistance, RESR. A key FoM, effective output resistance (REFF) captures the resistive loadline and resulting converter conduction loss. In the slide, REFF is normalized to RESR and frequency is normalized to the SC 'fast switching limit' (FSL) boundary. The ReSC approach outperforms comparable SC converters in the frequency-REFF space by operating at fundamental and subharmonic resonant minima. By tuning out the reactive impedance of the flying capacitor(s), the ReSC approach achieves the best-case limiting conduction loss of an SC converter, but with much lower switching loss.

86 Resonant Capacitive Converters

- REFF –effective output impedance
- SC: In FSL energy transfer limited by ESR + switch R
- ReSC achieves similar resistance but at lower frequency
- Opportunities for fundamental/ sub-harmonic operation

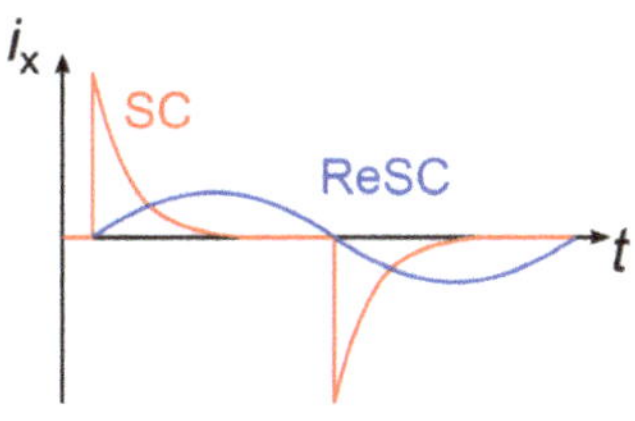

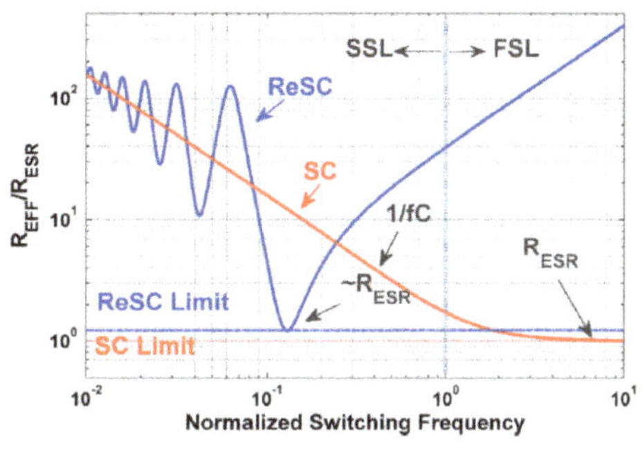

87 Measurement of ReSC

- 5.5 nH inductors give 90% efficiency
- Can provide load line compensation (on-time modulation)
- Significantly higher efficiency than Reff-modulation

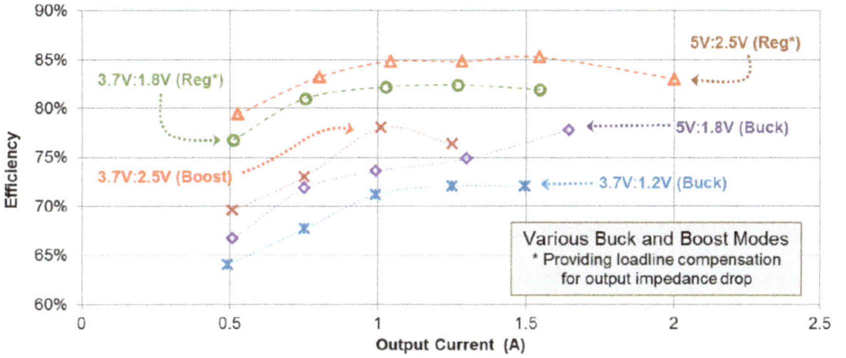

This slide shows the measured data trends for efficiency verses output current.

88 Overall Conclusions

- Linear converters
  - 1:1 conversion ratio, series/shunt
  - Easy to implement
  - LDOs vs DLDOs (for lower supply voltages)
- Inductive converters
  - Theoretical 100% efficiency
  - Integrated inductors are problem. Low quality factor
- Capacitive converters
  - Series parallel design framework
  - Capacitive and frequency based output regulation
- Special capacitive converters
  - Flying domain
  - Resonant converters

We started our discussion with linear regulators covering analog and digital LDOs. Next we discussed inductive converters and their design issues. Then we looked into design methodology of capacitive converter design and few examples of ripple control in capacitive DC-DC converter design.

89 Acknowledgements

- The author thank the funding agency that has supported this work for a number of years, i.e., the Semiconductor Research Corporation (SRC).
- The author thanks his current and former students, Josh Wibben, Sudhir Kudva, and Saurabh Chaubey for research contributions.
- In particular, the author thanks Saurabh Chaubey for transcribing the text from the video for this talk.

CHAPTER 13

Circuit Design Techniques for Fully Integrated Voltage Regulator Using Switched Capacitors

Hanh-Phuc Le

University of Colorado Boulder, USA

Efficient power delivery and management are current daunting challenges to secure performance and technology advancements in virtually all electronic systems, particularly in mobile applications, data centers, and high-performance IT systems. Although demanding different power levels, these applications are tightly related and share similar needs in power delivery. Particularly, data centers and cloud computing process power-hungry intensive computations, allowing mobile/wearable devices to handle local low-power computations and thus be small, light and enjoyable. With exponential increase in mobile devices and their data, challenges on performance and energy efficiency with stringent space constraints in these applications have never been more stressed simultaneously before. Among many efforts to deal with these challenges, a direct solution is to employ distributed fully-integrated voltage regulators in sub-systems to improve overall power delivery and management efficiency. This chapter feature switched-capacitor (SC) DC-DC converter as a good candidate for the solution.

In this chapter, the speaker will first discuss the needs and trade-offs between different types of converters to motivate the use of switched capacitor DC-DC converter for full integration. The main part of the chapter will cover circuit techniques that can be used in designing this type of converter in order to improve its power density, efficiency, transient response and input/output voltage range. Basic operation, design analysis and optimization of SC DC-DC converter, as well as specific circuit techniques, including clock level shifter, power switch driver, auxiliary voltage generator, MOS capacitors construction and biasing, parasitic capacitor reduction, and closed-loop control will be discussed. The speaker will also provide his perspective on a crucial system-level method that co-optimizes power supply with its load to optimally achieve system goals. Design examples in literature and from the speaker's research will be used to for the design techniques.

MOTIVATION FOR FULLY INTEGRATED VOLTAGE REGULATOR (FIVR)

1 SSCS The Internet of Everything

- **Needs power supply solution that is compact, efficient, and fast transient response.**

Icons from GraphicStock.com

In this era of the Internet of Everything, connected devices becomes ubiquitous with a fast-increasing number. At the same time, we want these devices to seamlessly integrate with the environment, requiring them as well as their power supply solution to be very compact.

2 SSCS Power Supply in Smart Phones

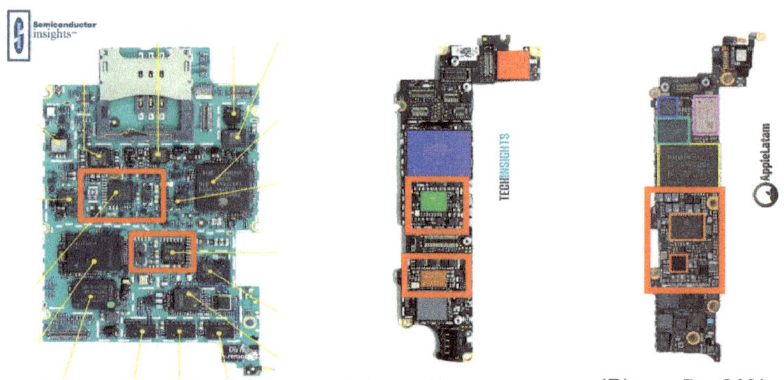

iPhone 3GS: ~10% iPhone 4S: 15% iPhone 5: ~20%

- **Space for power supply is increasing**

Taking example of power supply in smartphones, while other components, digital and analog, are shrinking in size with technology scaling, power management does not seem to follow this trend. Space portion for power supply is actually increasing.

3 Conventional Power Management Consumes Significant Board Space

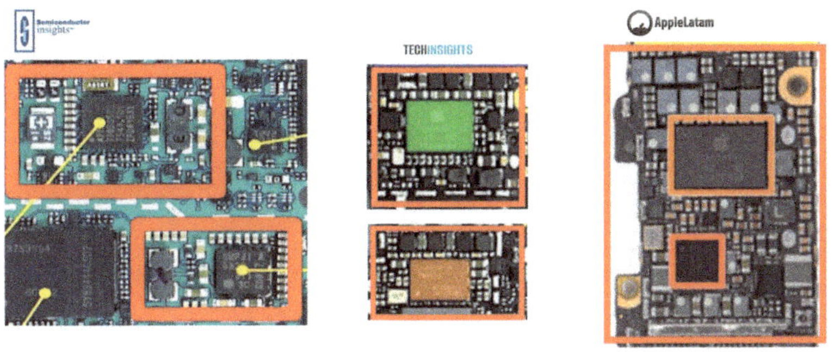

iPhone 3GS: ~10% iPhone 4S: 15% iPhone 5: ~20%

- **Space dominated by discrete passives**
- **Discrete passive components do not scale**

Looking closer into the implementation of power management unit, it can be clear that although the chip areas are relatively constant or increase slightly, the dominating space for passive components increased quickly with a lot more in numbers. This increase reflects more complex requirements in later system generations. Note that discrete passive components do not scale anywhere close to the on-chip active counterparts.

4 Power Waste with Single Supply

Apple A8

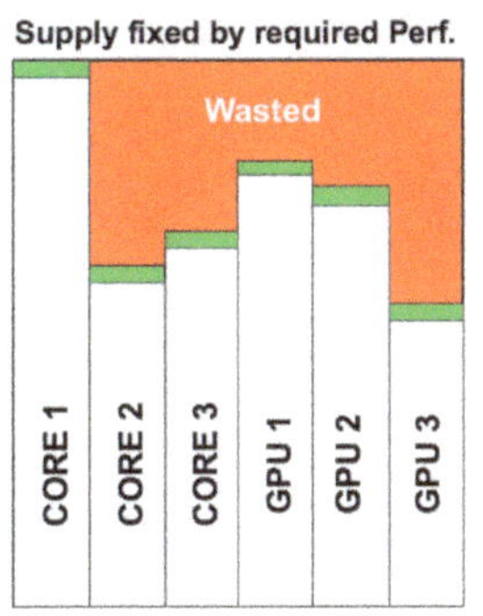

- **Clear need for multiple independent supplies**
  - Per-block power management, dynamic supplies, ...
- **But, using many off-chip supplies is costly**

The reason why the system needs a lot more power rails and complexity in power management is because it is desirable to have independent power supplies for different cores and function blocks to efficiently manage the power of the whole system. To meet this requirement, only having power supply using off-chip devices will result in an fast increase of discrete components and their space.

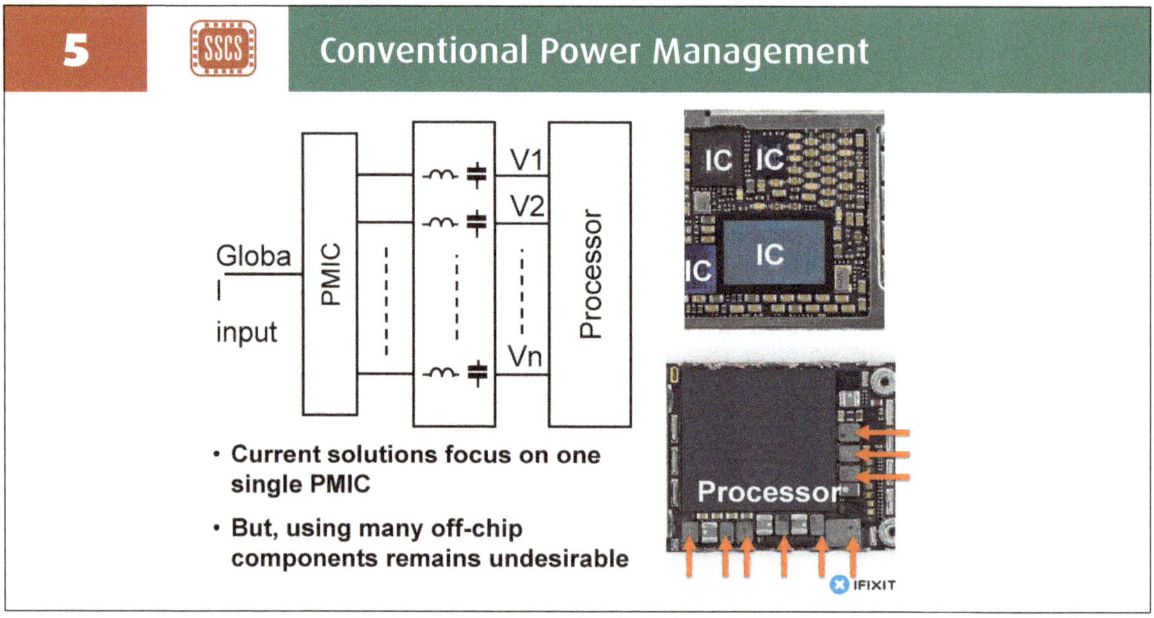

A clear example of current implementation for power management in these mobile systems can be seen in iPhone 6's board. It uses only one power management IC (PMIC) but a lot of capacitors and inductors, marked with red arrows.

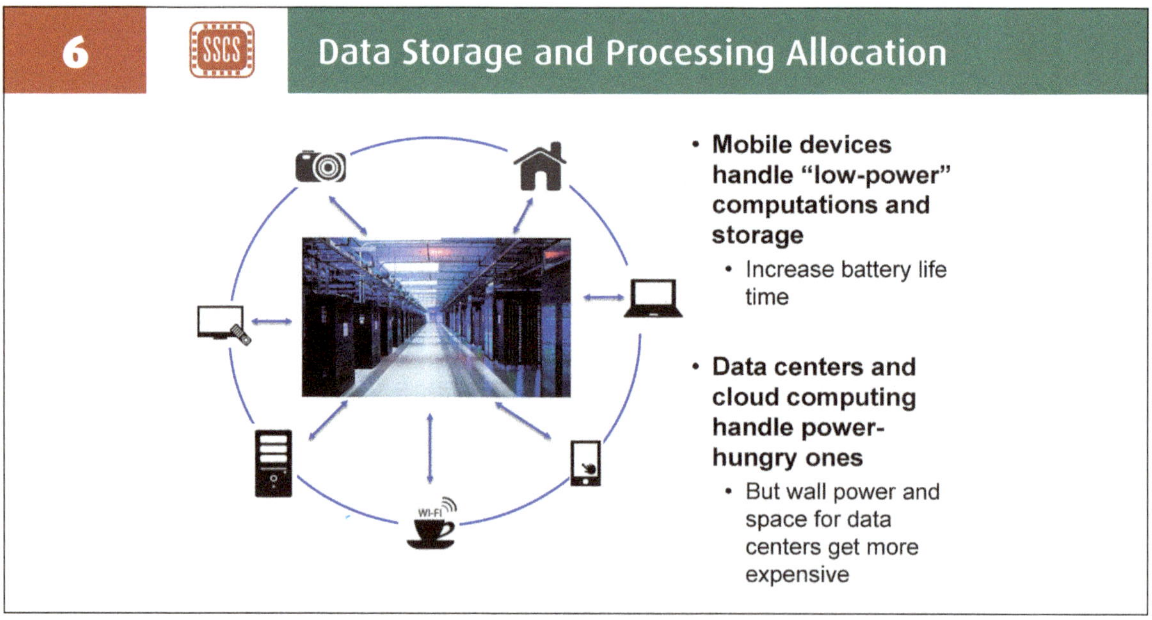

Another example for the need of more compact and scalable power supply can be seen in data center. While mobile devices only need to handle "low-power" computations, data center needs to handle all power-hungry tasks and at a massive scale.

7 Costs of Data Centers

- **Power**

Year	End-use Energy (B kWh)	Elec. Bills (US, $B)	Power plants (500 MW)	CO2 (US) (million MT)
2013	91	$9.0	34	97
2020	139	$13.7	51	147
2013-2020 increase	47	$4.7	17	50

Ref: NRDC, Data Center Efficiency Assessment, 2014

- **Space constraints to satisfy fast data booming rate**
  - Global data center traffic to triple from 3.1 ZB/year (2013) to 8.6 ZB/year (2018)

Data center indeed consumes a lot of power, i.e. 139B kWh expected in 2020, to satisfy the booming data traffic for cloud computing of 8.6 ZB per year in 2018, that is triple the number in 2013.

8 How Will Integrated Power Electronics Help?

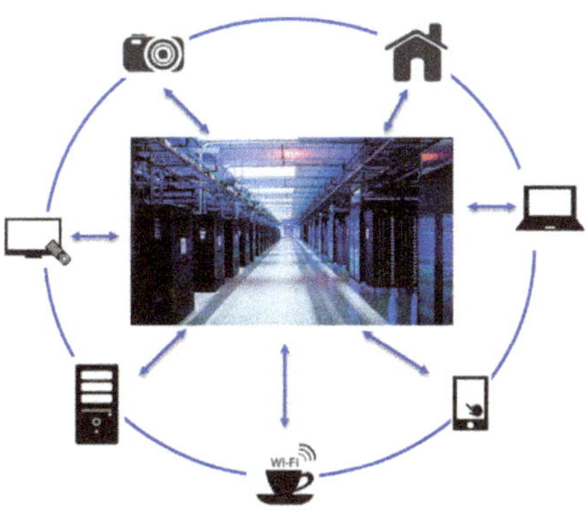

- **Power Efficiency**
- **Area/Volume Efficiency**
- **Scalability**

Integrated power electrics can help deal with this problem with higher power- and space-efficiency, and more granular scalability.

9 Fully-Integrated Voltage Regulators

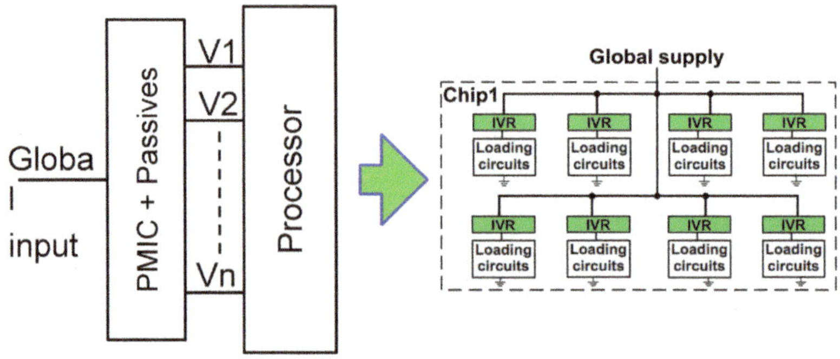

- **No external components → "zero mass" solution**
- **Free board space for other features, e.g. more battery**

Fully integrated voltage regulator (IVR) can be used where all passive components are integrated on the same chip with active parts. This can finally lead to a solution where IVR can be in the ultimate point of load (POL) position - on the same chip with the load. This can help free board space for other features, e.g. more battery for mobile applications.

10 High Power Density is the Key!

$$\textbf{Power Density} = \frac{\textbf{Output Power}}{\textbf{Total Implementation Area}}$$

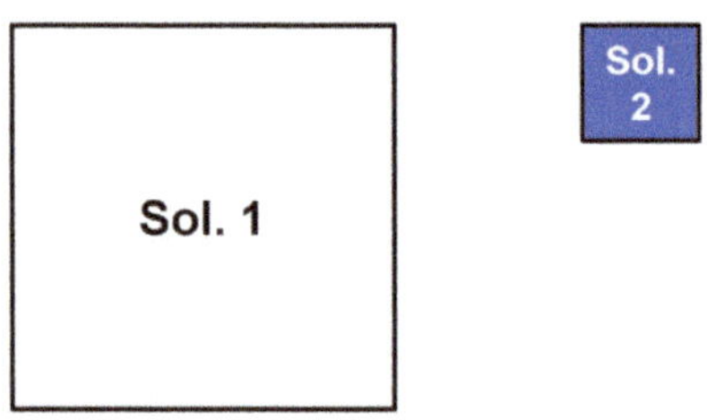

- **Same capabilities, the option with less resources will most likely win.**
  - Lower Cost
  - Higher Scalability

To meet all these requirements, high power density is the key. Power density is defined by the ratio between output power to the total implementation area. Using this metric, for the same output power a solution two with smaller implementation area will be a clear winner associated with lower cost and higher scalability.

WHY SWITCHED-CAPACITOR FOR FIVR?

11 Content

- Motivation for fully integrated voltage regulator (FIVR)
- **Why switched-capacitor for FIVR?**
  - How about linear or switched-inductor regulators?
- Switched-capacitor DC-DC converter: the basics
  - Design analysis and optimization
- Circuit techniques in SC converter design to improve
  - Power density
  - Efficiency
  - Transient response
- Other circuit techniques
- Conclusions

Now to implement fully integrated voltage regulator, why would switched-capacitor circuit be a compelling option? Let's discuss that in the next section.

12 DC-DC Converter Types

Linear Regulator

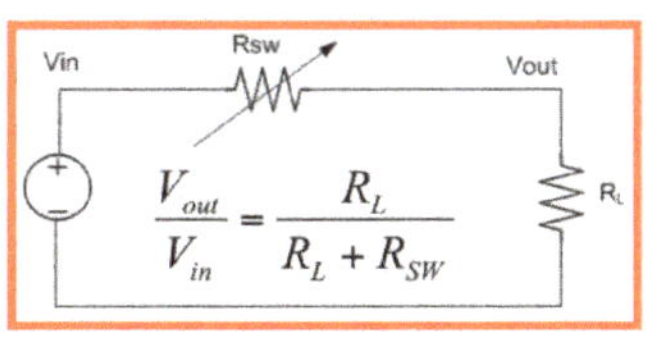

Switching Regulator

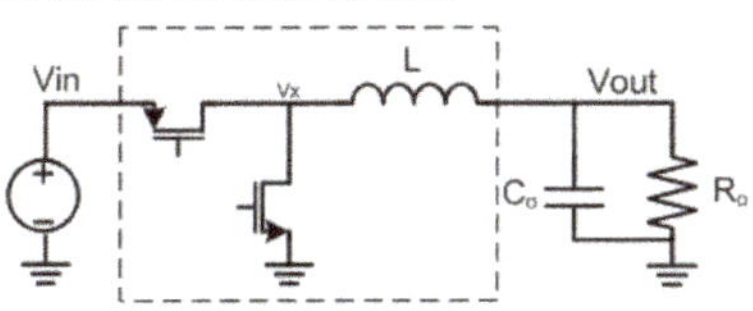

Switched-inductor (buck)

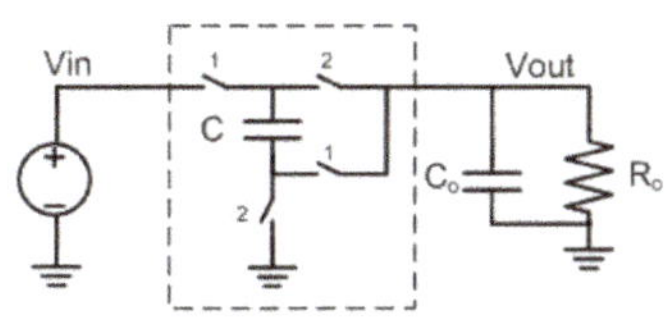

Switched-capacitor

There are two options to implement a DC-DC regulator, a linear regulator and a switching regulator. In switching regulator, switched-inductor and switched-capacitor are the most popular types.

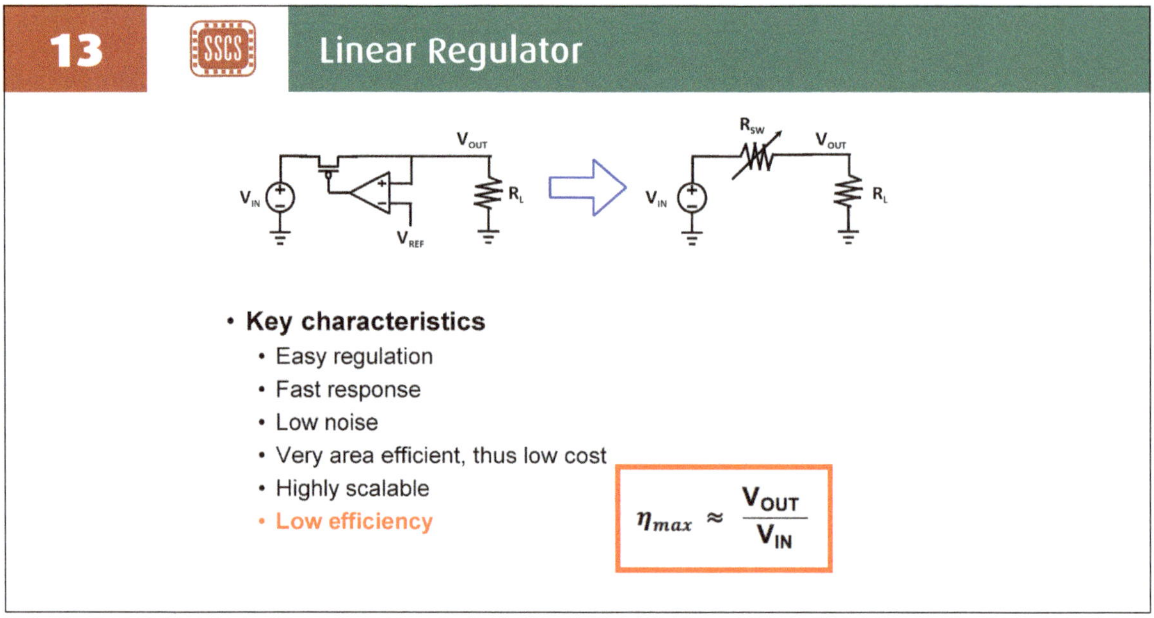

Linear regulator can be implemented simply by a series power switch whose on-resistance is controlled to regulate the output voltage. It is easy to implement, has fast response, low noise and can be very area-efficient, highly scalable and low cost. The fundamental drawback for linear regulator is that its efficiency is limited by the ratio of output voltage and input voltage.

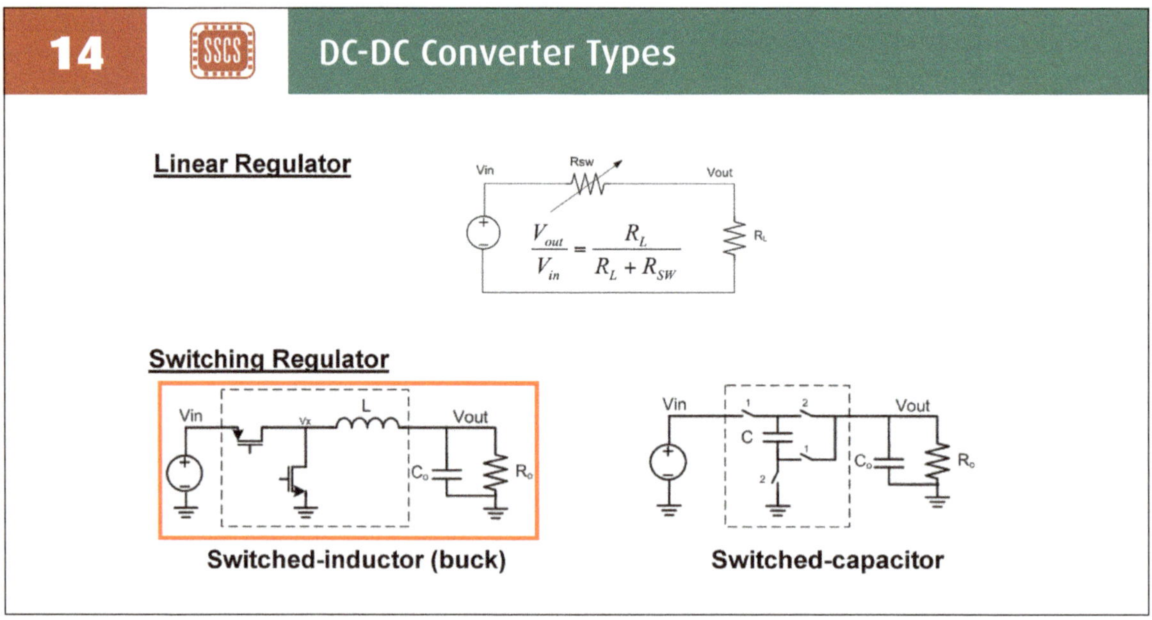

To avoid this undesirable and fundamental deficiency in linear regulator, we can use switching converter. First let's look at the most popular Buck converter.

15 Switched-Inductor Regulator

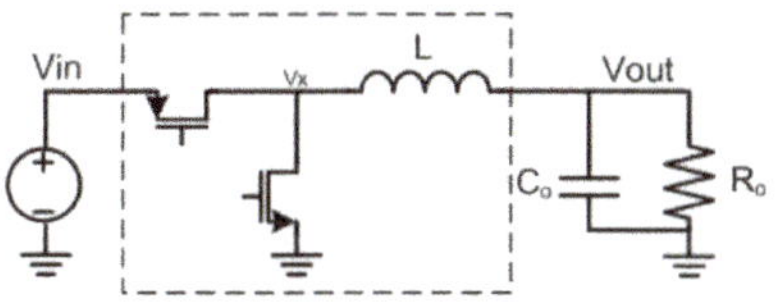

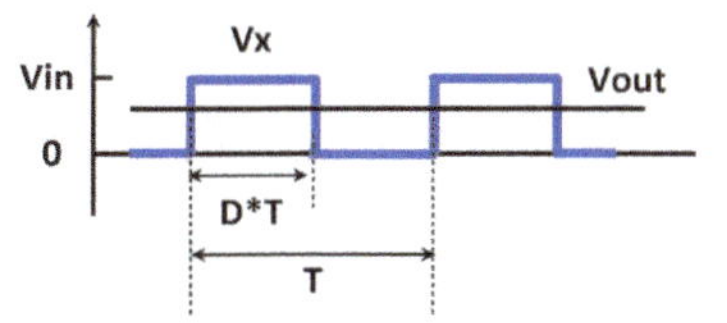

Vout = D * Vin

- **Key characteristics**
  - Transfer energy in form of current in inductor
  - High efficiency
  - Good fine regulation
  - Switching noise present
  - "Slow" response: mainly limited by switching frequency
  - Very popular for off-chip implementation
  - **Difficult/expensive for full integration because of inductor**
  - **Not easy to scale**

Buck converter transfer the energy from input to output in form of current in the inductor that enable efficient fine output voltage regulation. It is arguably the most popular off-chip converter. The key drawbacks of buck converter are the difficulty and cost in integrating the inductor on-chip. It is also very costly to do so. Many active researches are on inductor integration.

16 Industry and Academic Efforts on Integrated Power Conversion

TI uSiP, 2013:
2.3x2.9x1mm, 0.6A

Micrel, 2014:
2.5x3x1mm, 1A

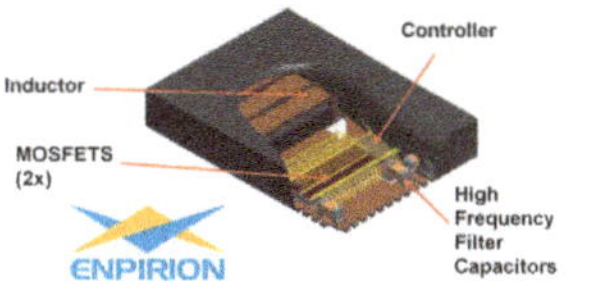

Enpirion, 2014: 8x11x3mm, 4A

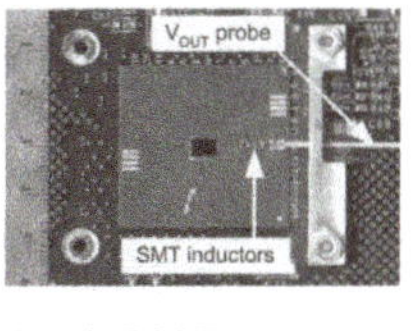

Intel, 2005
0.33A/mm^2, 83.2%,
233MHz

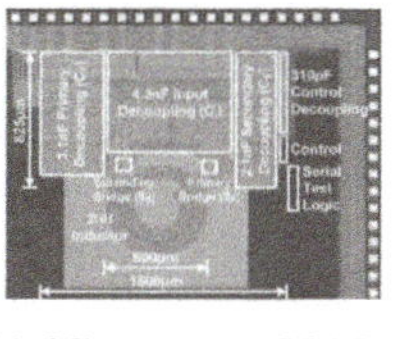

U. Minnesota, 2008
0.2A/mm^2, <80%,
170MHz

U. Columbia, 2012-2013
3.21A/mm^2, low 70s%
125-200MHz

Examples of these efforts are in both industry, i.e. Micrel, Enpirion, Intel, University of Minnesota, and particularly University of Columbia. These implementations still suffers from either low current density or low efficiency.

17 Available On-chip SI Products

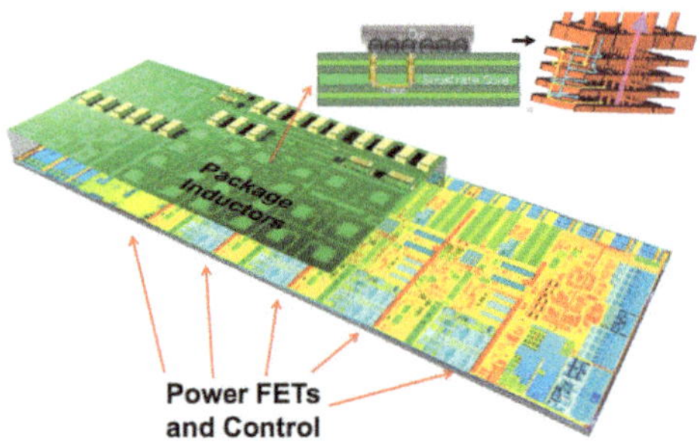

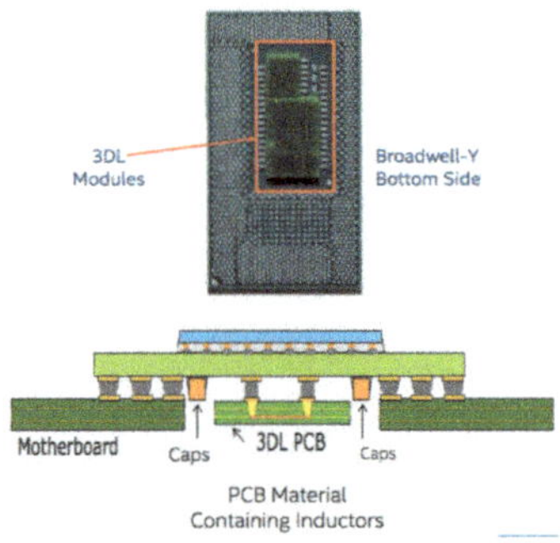

Intel Haswell, 2013
- Co-integrated switches
- Package inductors

Intel Broadwell, 2014
- Inductors on 3DL PCB

Remarkably, Intel started a serious commercialization of FIVR with integrated inductor in Haswell and Broadwell chips. These are successful products that create a boost for FIVR in both research and industry interest.

18 DC-DC Converter Types

Linear Regulator

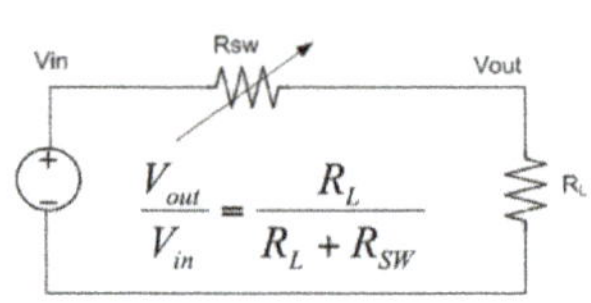

$$\frac{V_{out}}{V_{in}} = \frac{R_L}{R_L + R_{SW}}$$

Switching Regulator

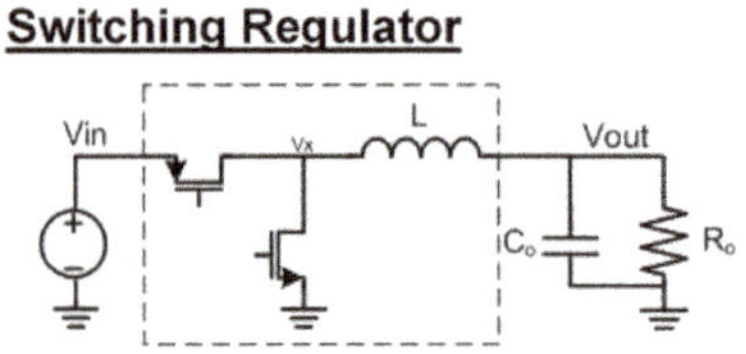

Switched-inductor (buck)

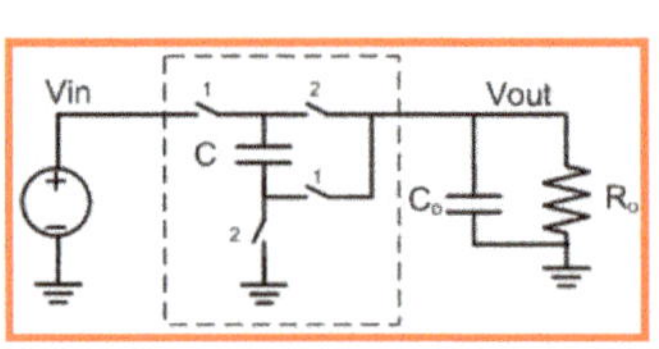

Switched-capacitor

Now, as an alternative solution to the integration issues in switched inductor converter, switched-capacitor (SC) converters recently draw a lot of attentions.

19 Switched Capacitor Operation and Topology

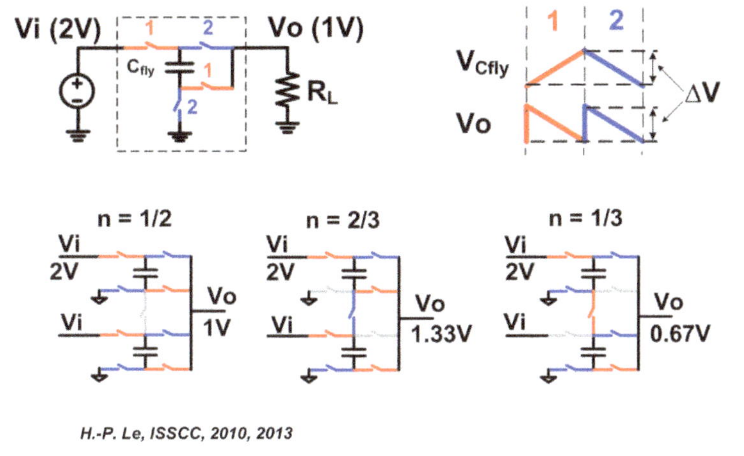

An example of a 2-to-1 SC converter is shown. It works in two phases. The output gets the charge by charging and discharging the capacitor. Two SC converter units can be combined into a 2-capacitor 9-switch network that is capable of providing three reconfigurable topologies.

20 On-chip Sw-capacitor Developments

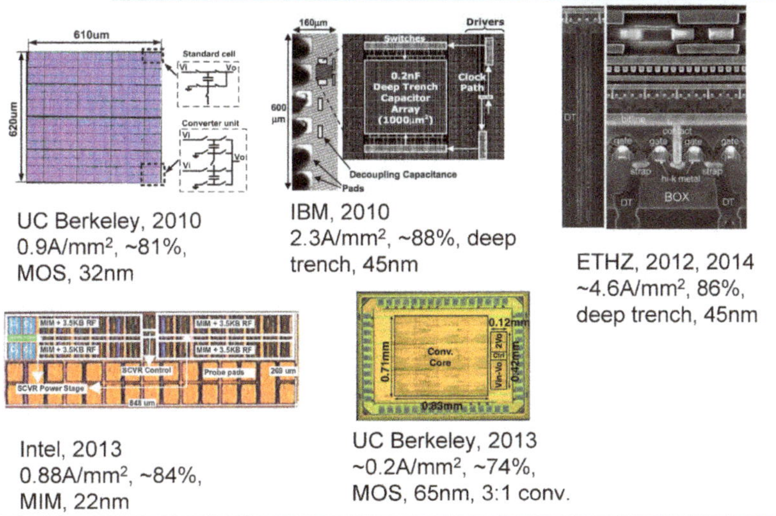

Both academic and industry research efforts have been put into SC converters that can achieve both high efficiency and high output current density. Notably are the work from UC Berkeley in 2010 using a 32nm commercial CMOS process, and IBM work using a special deep trench capacitor technology.

21 Fully Integrated Implementation

	Linear	Sw-Inductor	Sw-Cap
	V_{IN} V_O	V_{IN} V_O	V_{IN} V_O
Efficiency	✗ ✗	✓	✓
Power Density	✓	✗	✓
Ripple/Regulation	✓	✓	✓*
Scalability and Cost	✓	✗	✓

**** subject to conversion ratios (will discuss more)***

Since capacitors can be found readily available on-die and can be very scalable for different loads, SC converters have clear advantage over switched-inductor counterpart in the integration context.

SWITCHED-CAPACITOR DC-DC CONVERTER: THE BASICS

22 Content

- Motivation for fully integrated voltage regulator (FIVR)
- Why switched-capacitor for FIVR?
  - How about linear or switched-inductor regulators?
- **Switched-capacitor DC-DC converter: the basics**
  - Design analysis and optimization
- Circuit techniques in SC converter design to improve
  - Power density
  - Efficiency
  - Transient response
- Other circuit techniques
- Conclusions

Now, in the next section we will review how to design an optimal SC converter.

23 Switched-Capacitor Loss

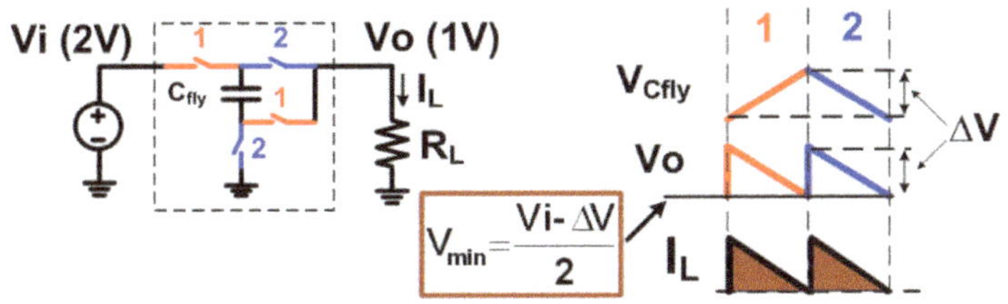

- **Converter supplies digital circuits**
  - Performance (f_{CPU}) set by V_{min}
- **Intrinsic switched-capacitor loss:**

$$P_{C_{fly}} = 2x \frac{I_L^2}{M_{conv,cap} f_{sw} C_{fly}}$$

Note that the load of interest is a high-performance digital circuits, e.g. processors. The performance of this load is defined by a minimum voltageVmin, making all ripple over this Vmin turn into loss. In this case, SC loss will double because it does not contribute to the load performance.

24 Multi-Phase Interleaving

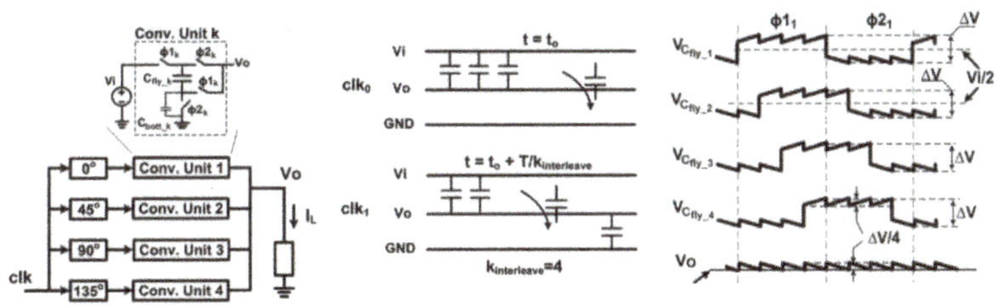

- **Good news:**
  - Interleaving can reduce ripple without changing V_{min}

$$P_{C_{fly}} = \left(1 + \frac{1}{k_{interleave}}\right) \frac{I_L^2}{M_{conv,cap} f_{sw} C_{fly}}$$

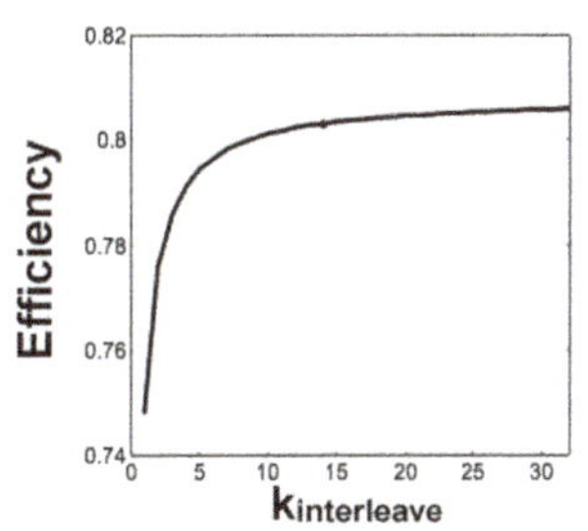

Fortunately, multi-phase interleaving operation can be used to minimize output voltage ripple, reduce this SC loss and thus increase SC converter efficiency. Our analysis shows that an interleaving factor over 8 would be sufficient to eliminate most of this SC loss.

25 Layout for Multi-Phase Interleaving

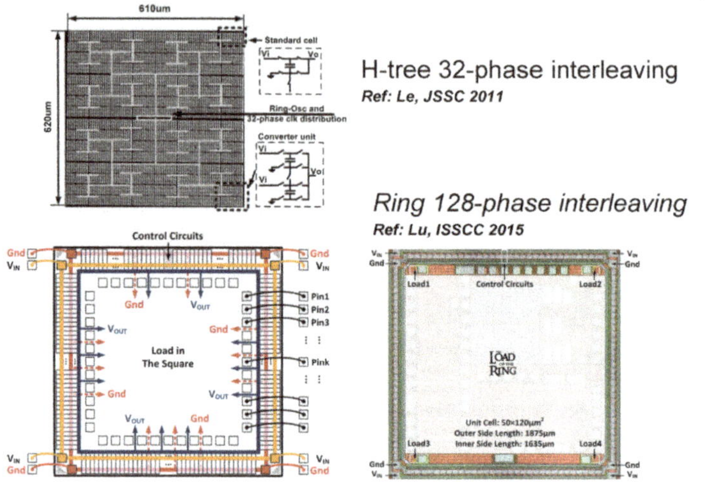

The implementation for multi-phase interleaving can be done using H-tree distribution for a localized converter, or in a ring for a distributed converter, for examples.

26 Loss Optimizationand Regulation

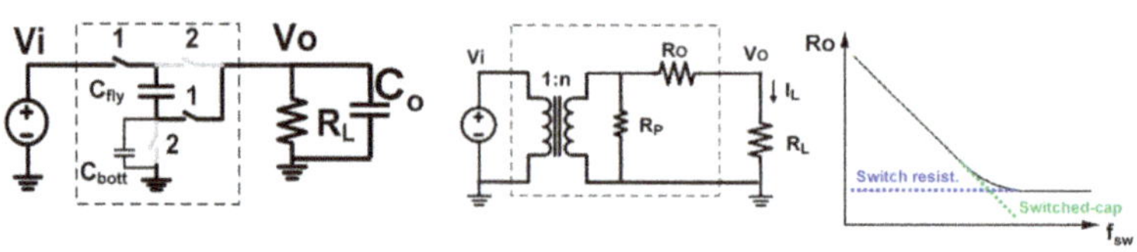

- **4 main loss components**

$$P_{loss} = P_{Rsw} + P_{C_{fly}} + P_{bott.cap} + P_{gate.cap}$$

$\alpha\ I_L^2 \frac{R_{on}}{W_{sw}}$ Switch resist. | $\alpha\ \frac{I_L^2}{f_{sw} C_{fly}}$ Switched cap | $\alpha\ C_{bott} f_{sw}$ Bottom Plate | $\alpha\ W_{sw} C_{gate} f_{sw}$ Gate loss

Series Losses (Ro) | Parasitic Losses ($R_{//}$)

The loss components of a typical SC converter includes switch resistance loss and switched capacitor loss that contribute to the output impedance, i.e. series loss, as well as bottom-plate and gate parasitic capacitance switching losses.

27 Optimization for Efficiency

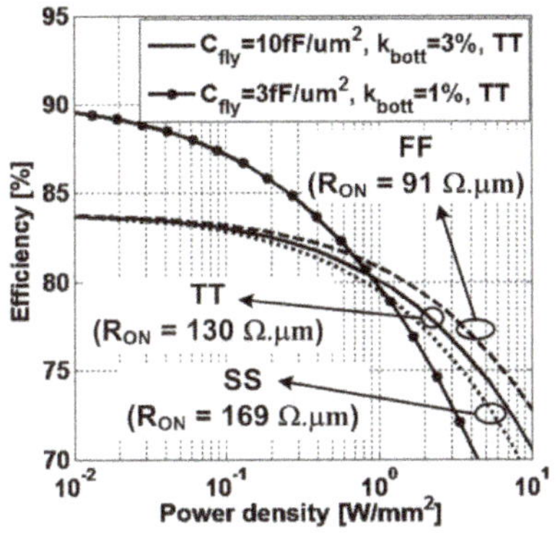

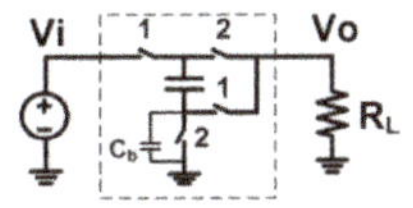

- Efficiency trades off with power density
- Low power density:
  - Bottom plate critical
    - Topology and operational phases
- High power density:
  - Flying cap critical
    - Better cap: deep-trench, DRAM

$$\frac{P_{loss}}{P_{Load}} \approx \sqrt{M_{conv}k_{bott}} + \sqrt{\frac{1}{\sqrt{M_{conv}k_{bott}}}\frac{V_{sw}^2 R_{on} C_{sw}}{V_o^2 R_L C_{fly}}}$$

Optimizing these losses can lead to very promising efficiency for SC converters. The optimization also shows that capacitors with high capacitance density is preferred at heavier load even if they have high bottom-plate parasitic capacitance,while low bottom-plate parasitic capacitance is the key for high efficiency at light load.

KEY CIRCUIT TECHNIQUES IN SC CONVERTER DESIGN

28 Content

- Motivation for fully integrated voltage regulator (FIVR)
- Why switched-capacitor for FIVR?
  - How about linear or switched-inductor regulators?
- Switched-capacitor DC-DC converter: the basics
  - Design analysis and optimization
- **Circuit techniques in SC converter design to improve**
  - Power density
  - Efficiency
  - Transient response
- Other circuit techniques
- Conclusions

In the next part, we will discuss some circuit techniques in SC converter design to improve power density, efficiency, and transient response.

29 Circuit Techniques for Better SC Converter Design

$$P_{loss} = P_{Rsw} + P_{C_{fly}} + P_{bott.cap} + P_{gate.cap}$$

$P_{Rsw} \propto I_L^2 \frac{R_{on}}{W_{sw}}$, $P_{C_{fly}} \propto \frac{I_L^2}{f_{sw} C_{fly}}$ → Ro

$P_{bott.cap} \propto C_{bott} f_{sw}$, $P_{gate.cap} \propto W_{sw} C_{gate} f_{sw}$ → R_P

- Output voltage ripple
  - Multiphase interleaving
- **Efficiency vs. regulation trade-off (Ro)**
  - More topologies, multiple conversion ratio
- **Switch driver design (R_{on}, W_{sw}, Q_{gate})**
  - Use of thin-ox devices
  - Utilization of main and auxiliary voltages
  - Depends on converter topology
    - Will be discussed in specific converter examples
- Clock level shifter (Rp)
- Integrated capacitor design (Ro, C_{fly})
  - Layout and well bias
  - Parasitic capacitor reduction
- Feedback control and regulation (Ro, C_{fly}, R_{on})

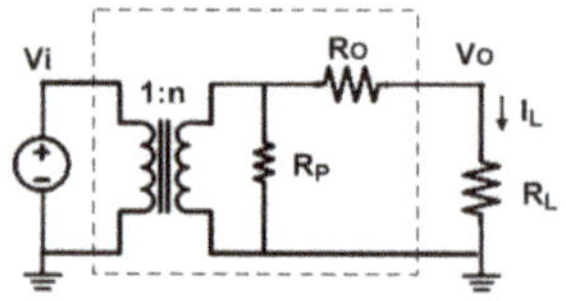

We have talked about using multiphase interleaving operation can reduce output voltage ripple and the SC loss. Now let's discuss some circuit techniques to deal with the fundamental trade-off of efficiency and fine regulation and methods to design good switch drivers to ensure optimal losses.

30 Regulation vs. Efficiency

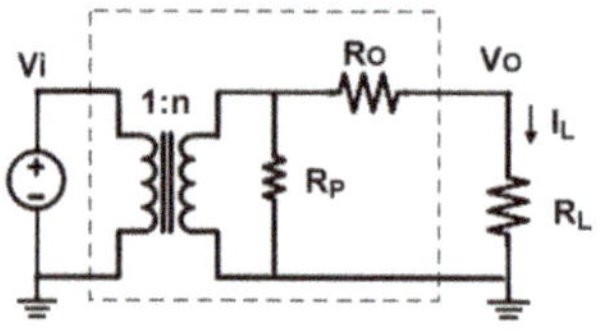

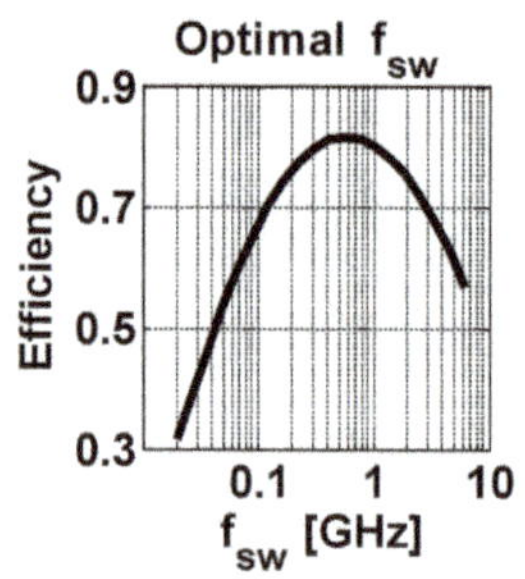

- **SC converter efficiency is high at approximately ideal conversion ratio**
  - Efficiency linearly goes down with higher conversion ratio
  - Similar to linear regulator
- **Need to have more topologies**
  - Increase optimal points

SC converter has a fundamental characteristic that it achieves high efficiency at near its ideal conversion ratio for a topology, but efficiency goes down quickly with larger conversion ratios. More reconfigurable topologies, and thus ideal conversion ratios, can be added to mitigate this issue.

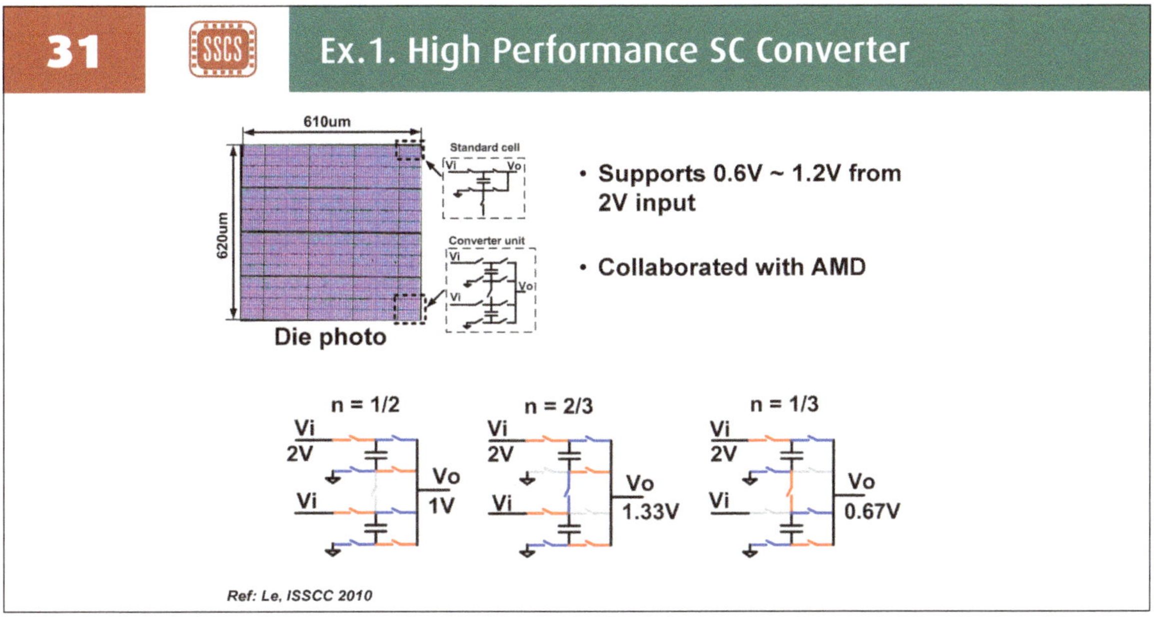

In the work published in 2010, a reconfigurable 3-topology converter was implemented on a 32nm technology that can support a 0.6V-1.2V output from 2V input.

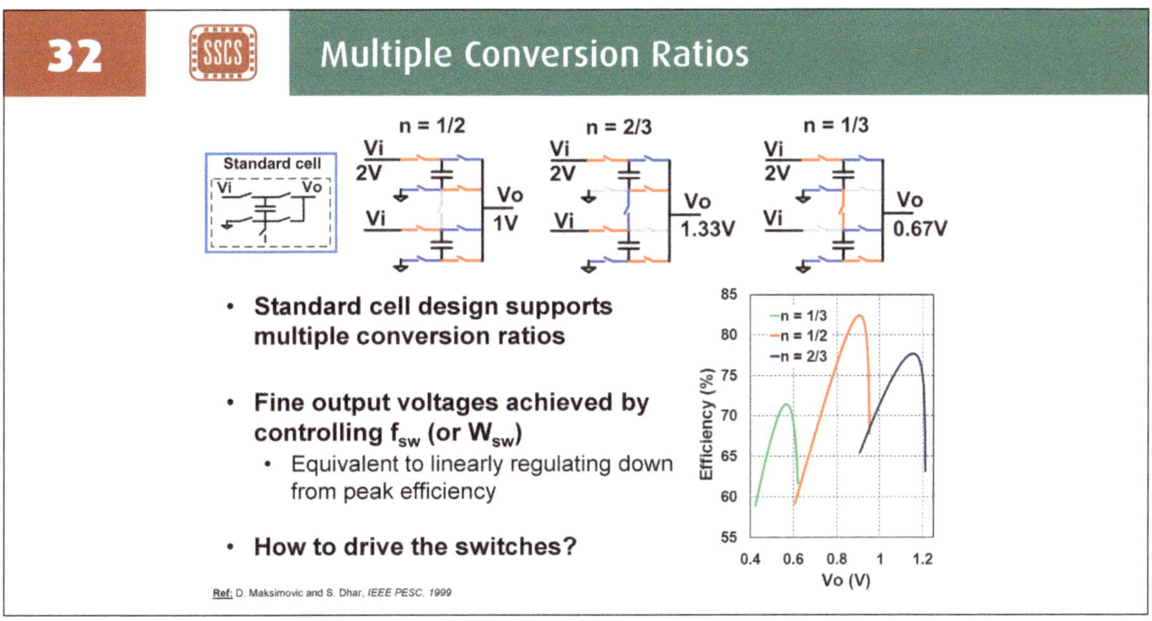

The 2-capcacitor 9-switch topology can support three reconfigurable topologies to achieve three optimal conversion ratios and thus optimal efficiency points. Intermediate output voltages can be linearly regulated by controlling switching frequency or switch strength, i.e. regulating the output impedance of the converter.

33 Switch Drivers

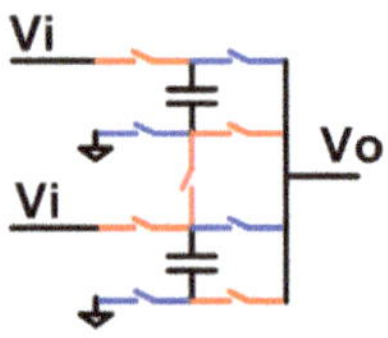

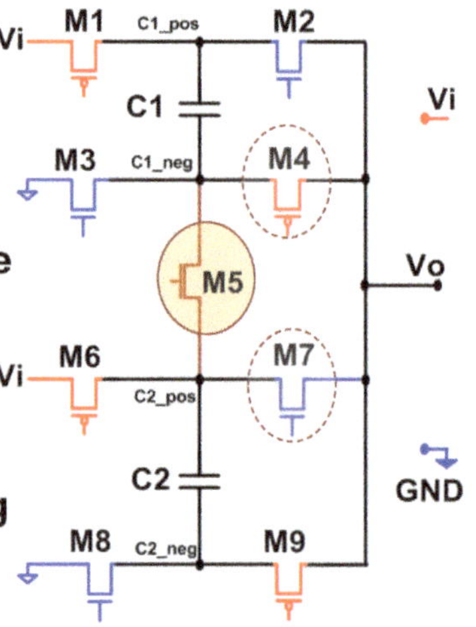

- **Most switches easy to drive**
  - 2 voltage domains:
    - (Vi – Vo)
    - (Vo – GND)
- **M4, M5, and M7 challenging**
  - Experience voltages between the two domains

In the strategy for switch drivers in this circuit, switches M1-M3, M6, and M8-M9 are relatively simple to drive since they only interact with either Vi-Vo or Vo-GND voltage domains. Switches M4, M5, and M7 are more challenging since they experience both of these voltage domains in their operation. Let's focus on M5 driver in this discussion. Drivers for M4 and M7 can be found in the paper by Le, et al, in JSSC 2011.

34 Switch Driver -M5

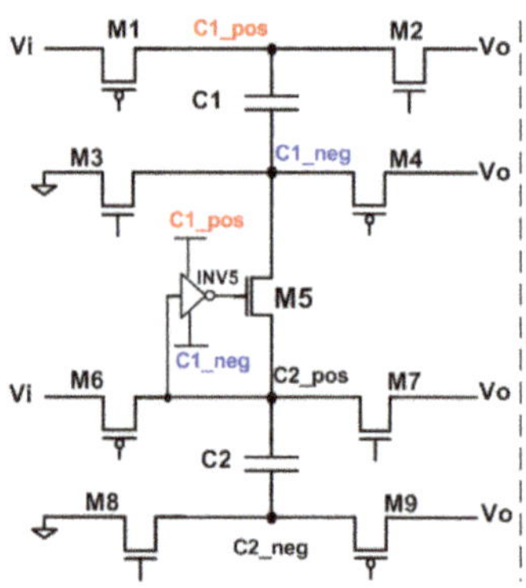

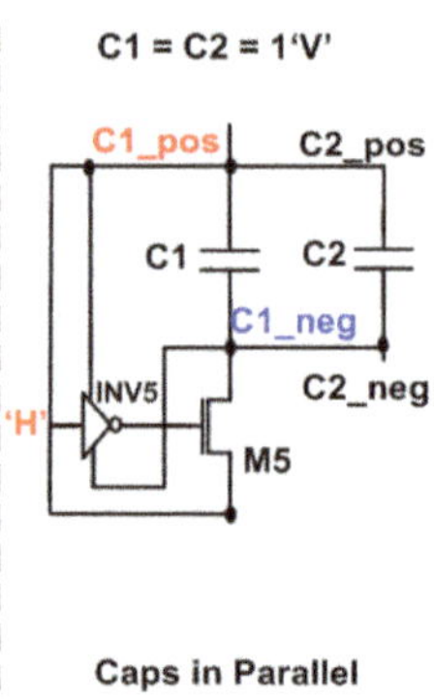

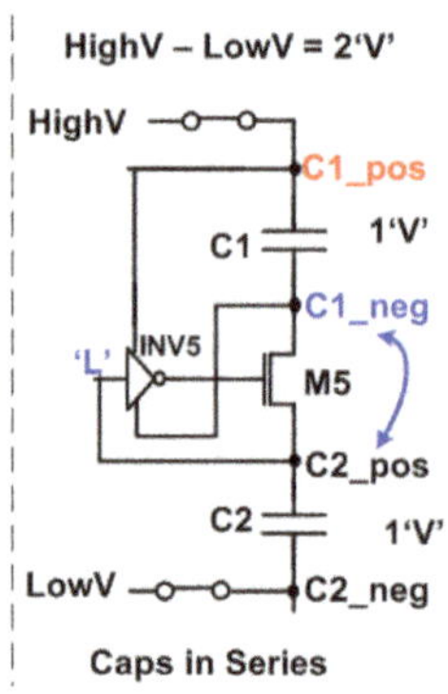

- **"Flying" inverter INV5 powered off of C1**
  - Controlled by top-plate of C2
- **Automatically synchronized by operation of other switches**

M5 is driven by a "flying inverter" INV5 that is powered off of the voltage over capacitor C1 and controlled by the voltage at the top plate of C2. While other switches are controlling voltage levels of C1 and C2, operations of INV5 and M5 are automatically synchronized to turn OFF M5 when C1 and C2 in parallel or turn ON M5 to stack C1 and C2.

35 Ex.1. Measured Eff. vs. Power Density

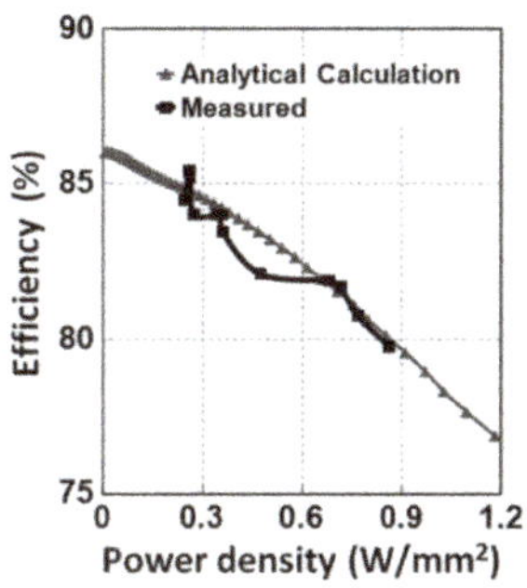

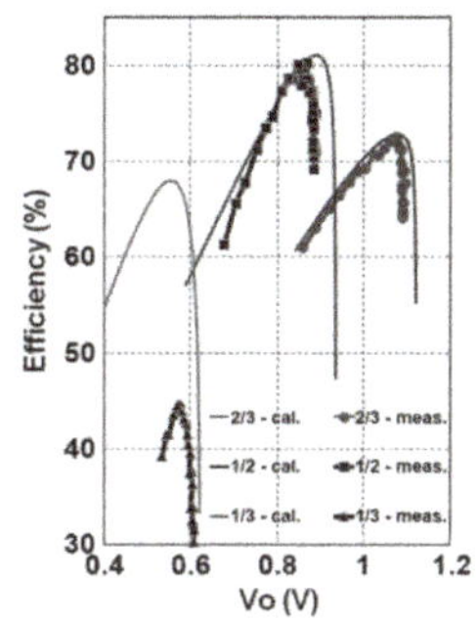

- **Measured in 1/2 mode (Vi = 2V, Vo ≈ 0.88V)**
- **Achieved ~80% efficiency @ 0.86 W/mm²**

Ref: Le, JSSC 2011

In measurement, this SC converter achieved ~80% efficiency at 0.86 W/mm², which is still a record performance for an SC converter implementation on a regular commercial process and MOS capacitors.

36 Ex.2. SC Converter w/ Deep-trench Caps

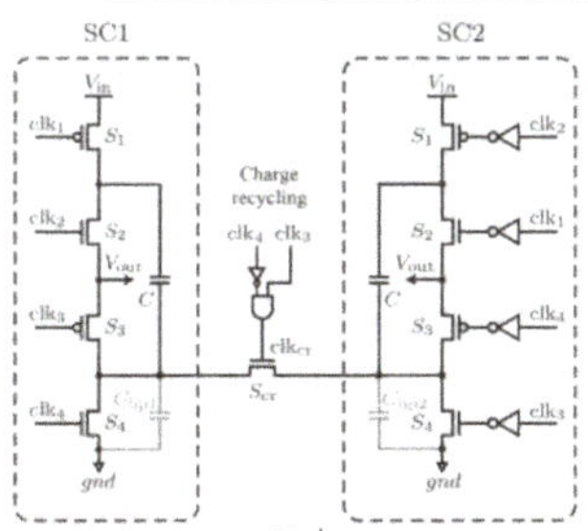

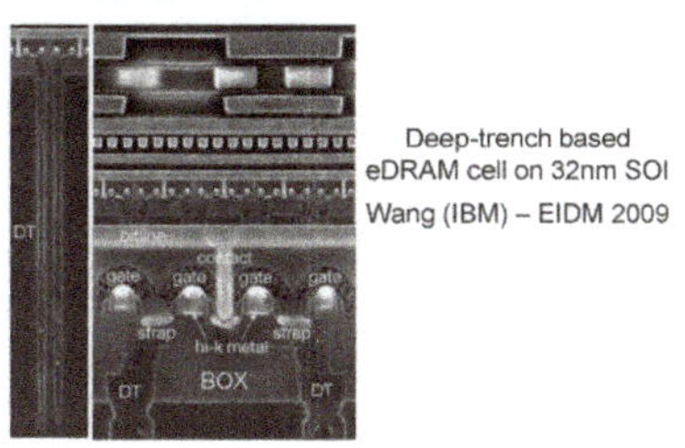

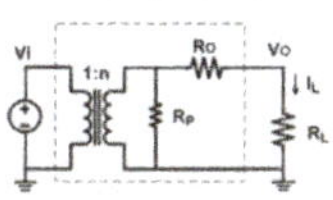

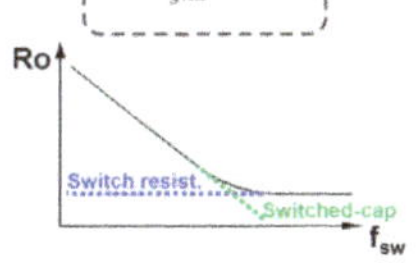

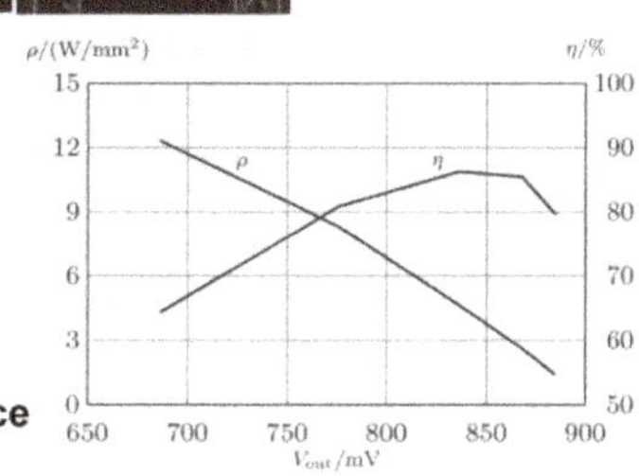

- **Use deep-trench capacitor**
  - >200 fF/um² (MOSFET: <10fF/um²)
- **Significantly improve conv. performance**

Ref: Andersen, PwrSoC 2012

In another implementation example, a similar circuit has similar active components but deep-trench capacitors on an IBM process that has ~20X higher capacitance density compared with MOS capacitors. This superior passive technology enables this SC converter to achieve 88% peak efficiency at ~4.5 W/mm². This number is significantly higher than any available switched-inductor counterparts.

37 Ex.3. Recursive SC Ratio Converter

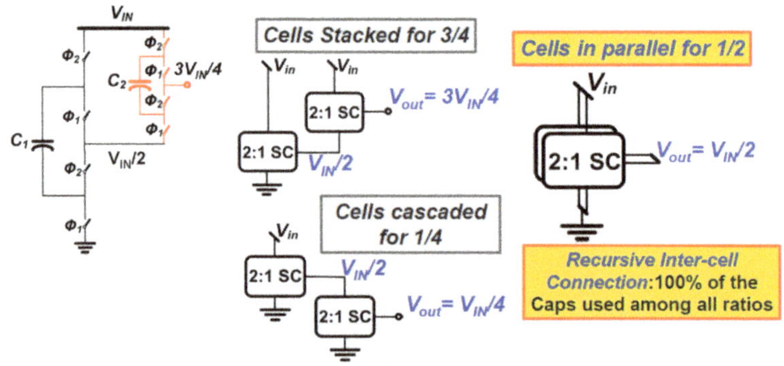

- **Generate more voltage levels**
  - Similar to a DAC
- **Need to utilize all available capacitors**

Ref: Bang, ISSCC 2013; Salem, ISSCC 2014

In the next design example, SC converter can have a more complex structure and operations to support many more reconfigurable topologies. It essentially works similar to a digital-to-analog converter (DAC) with multiple divide-by-2 units.

38 Ex.3. Switch Configuration – 2-bit Conversion

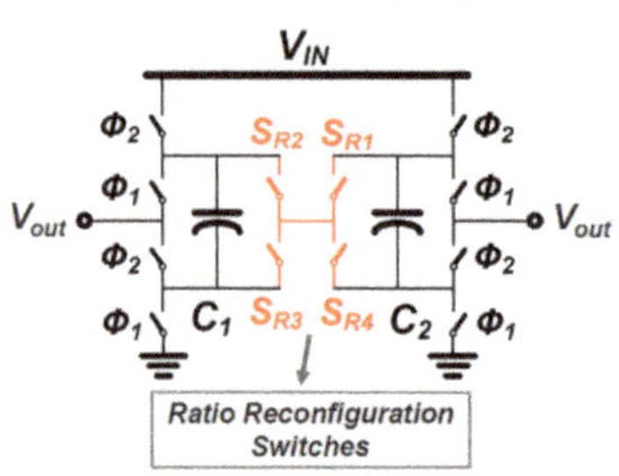

- **H-Bride to configure cell stacks**
- **Switches experience Vin**
  - Need to use thick-ox devices
  - Low power/current density

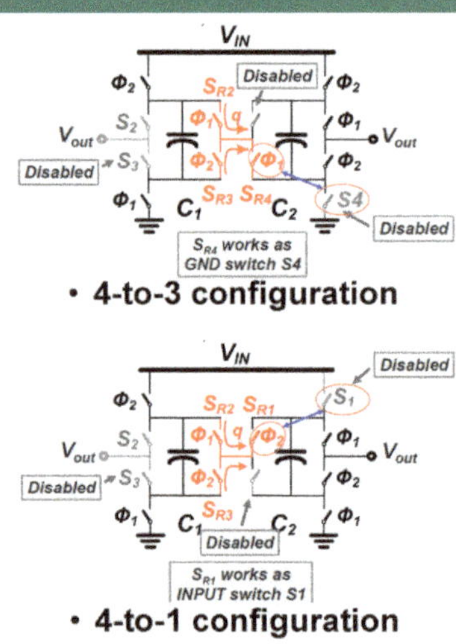

Ref: Salem, ISSCC 2014

The key contribution in this work is to add an H-Bridge from 4 switches to enable the configurability. The switches in this circuit are all thick-oxide devices for simple gate drivers.

39 Ex.3. More Configuration – 3-bit Conversions

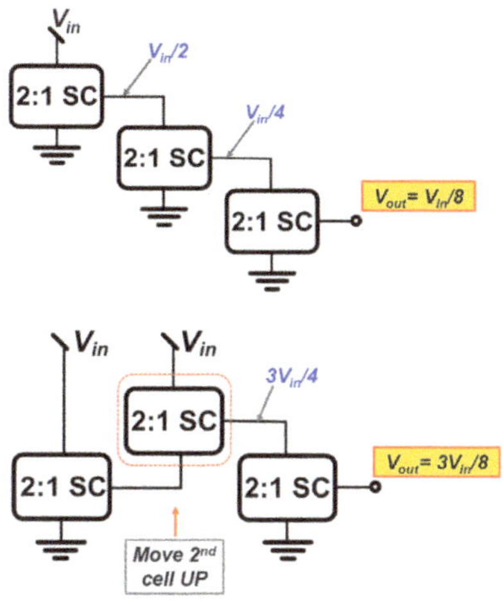

Ref: Salem, ISSCC 2014

- **Put more 2:1 SC cells in the stack**
- **Need a strategy to fully utilize all SC**
  - Put them more in parallel and avoid series connections if possible
  - Distribute capacitor cells aiming at the same charge density
- **More similar topologies:**
  - Jung, ISSCC 2016, 12.1
  - Lutz, ISSCC 2016, 12.4

More 2-to-1 SC cells can be added to achieve more conversion ratios. A significant effort in this work focuses on increasing capacitance utilization for SC operations.

40 Ex.3. Prototype

- **Prototype: 4-bit SC**
- **Use thick-ox 0.25 um 2.5V CMOS**
- **2.37 x 1.96 mm^2**
- **Current density: 0.15 mA/mm^2 @ 2:1**

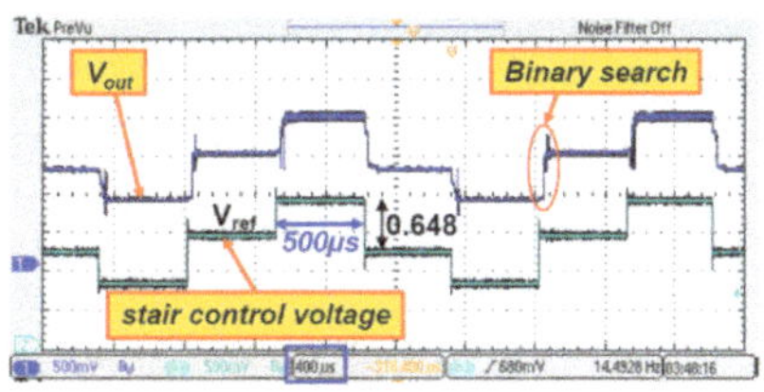

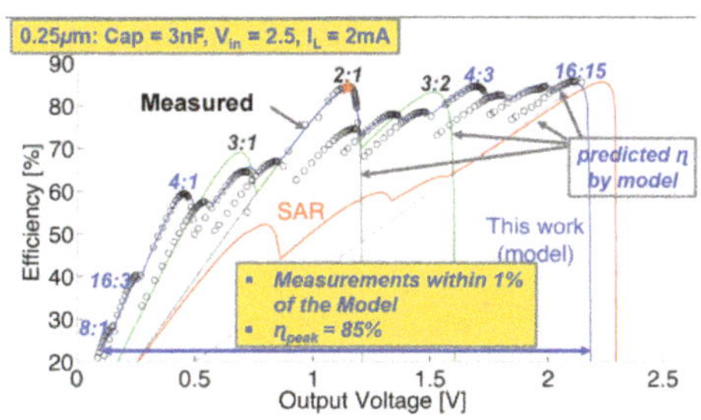

The converter prototype was implemented in a 0.25 um technology. It only achieves 0.15 mA/mm^{2}output current density, but allows 16 reconfigurable topologies to keep efficiency relatively flat over a wide range of output voltage from a fixed input voltage.

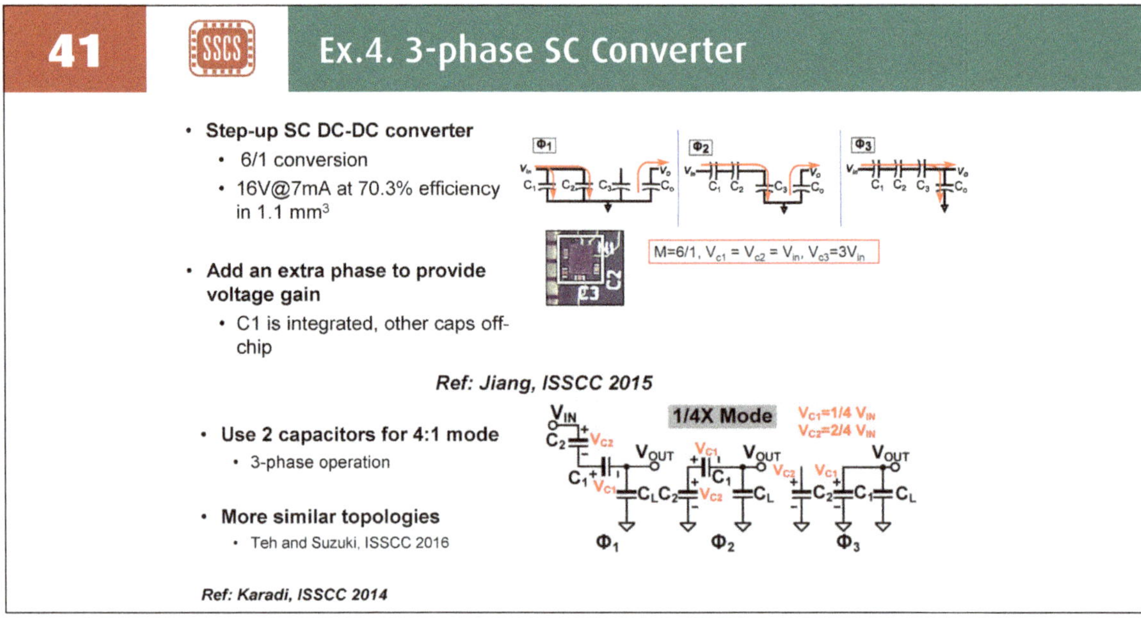

In the design example 4, rather than a regular 2-phase operation SC converters can use a 3-phase operation to generate different topologies and thus different conversation ratios.

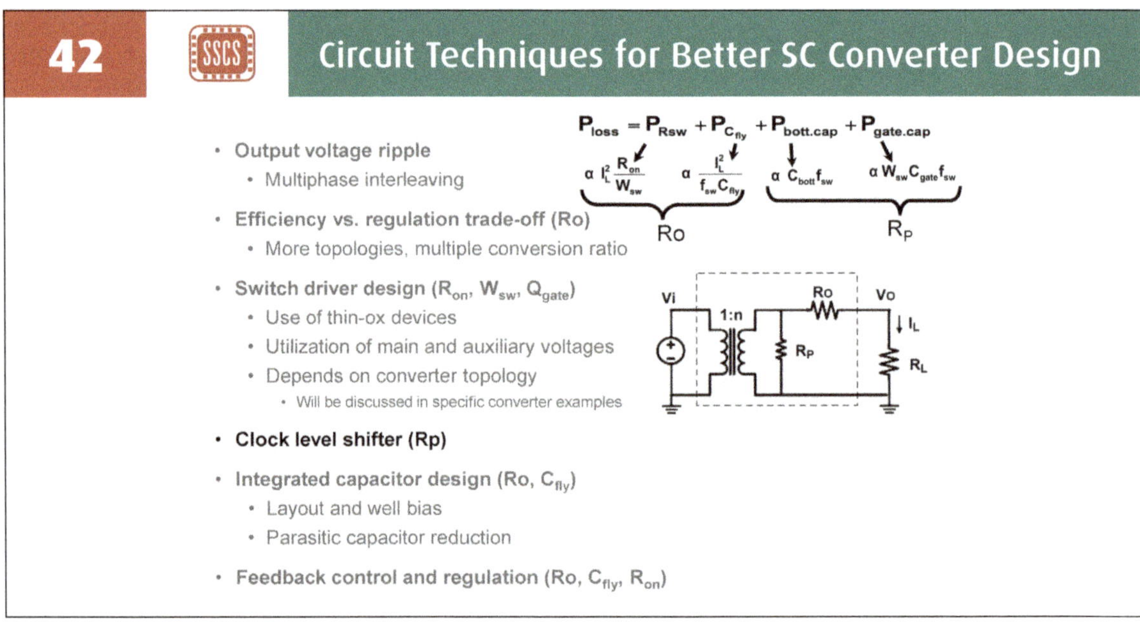

In the SC converter design, we have talked about different techniques to minimize series loss, i.e. the output impedance, let's review about other circuit techniques that can improve a SC converter design, particularly those for clock level shifter for gate drivers and integrated capacitor design to reduce parasitic switching losses.

43 Drv.1. Clock Level Shifter 1

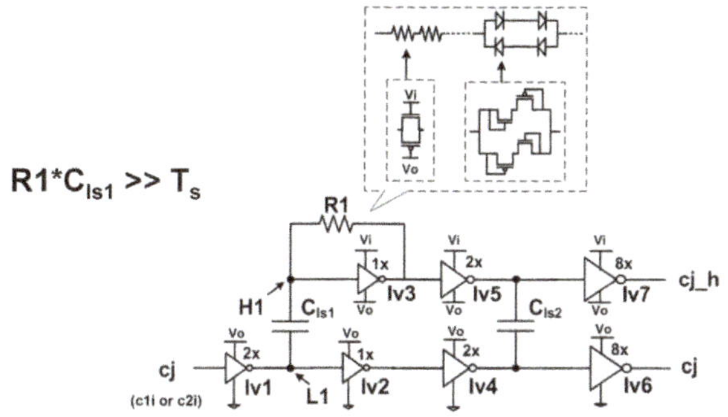

- **More suitable for high switching frequency range**
- **Can level shift to different voltage domain**
  - Original swing (lower domain) need to larger for reliability

Ref: Le, JSSC 2011

For clock level shifter, this design example 1 shows a circuit that uses an capacitor C_{ls1} to couple clock or driving signals at different voltage domains, e.g. Vin-Vo and Vo-GND, and a weak feedback resistor R1 to bias the receiving inverter Iv3 at its trip point. This circuit operation relies on the time constant of R1* C_{ls1} much larger than the switching period.

44 Drv.2. Clock Level Shifter 2

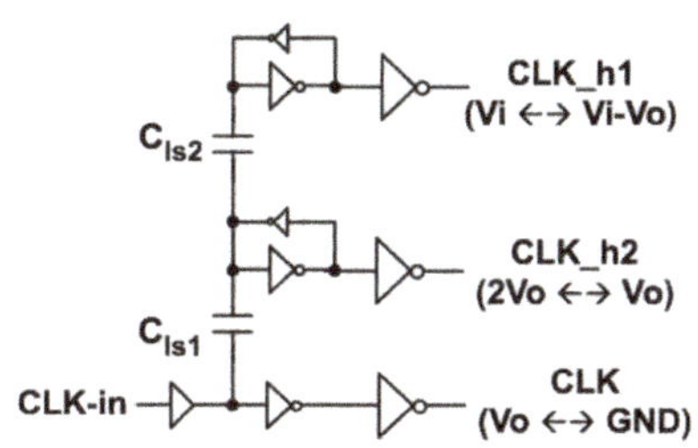

- **Use feedback inverter instead of resistor**
  - Can work at very low switching frequency
- **Can level shift to different voltage domain**
  - Original swing (lower domain) need to larger for reliability

Ref: Le, JSSC 2011

To avoid the dependence on the R1* C_{ls1} time constant and allow much smaller switching frequency, a small feedback inverter can be used in place of R1 as shown in clock level shifter example 2. It is preferable that the original signal swing, i.e. Vo-GND in this case, is larger than the level-shift domains for circuit reliability.

45 Drv.3. Auxiliary Converters

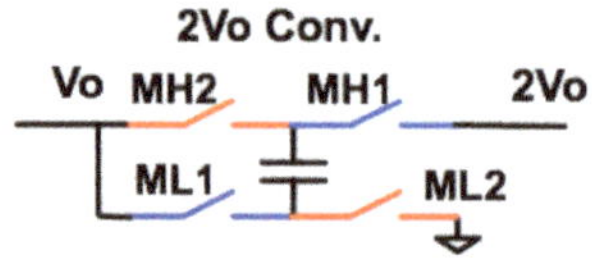

$V_{cap} = Vo$

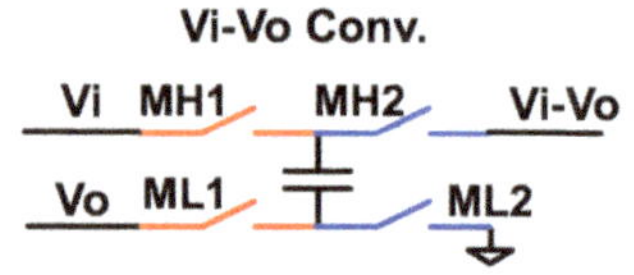

$V_{cap} = Vi - Vo$

- **Similar circuit used for 2 auxiliary converters**
  - Different in switch operation and rail connections
- **Capacitor in the Vi-Vo converter handles Vi-Vo**
  - Uses thick-ox MOS-based capacitor

As we can see in previous design examples SC converters requires different voltage levels for its power switch operations. The two converter topologies on this slide can generate 2*Vo and (Vi-Vo) auxiliary levels.

46 Drv.3. Auxiliary Converters –Driver Circuit

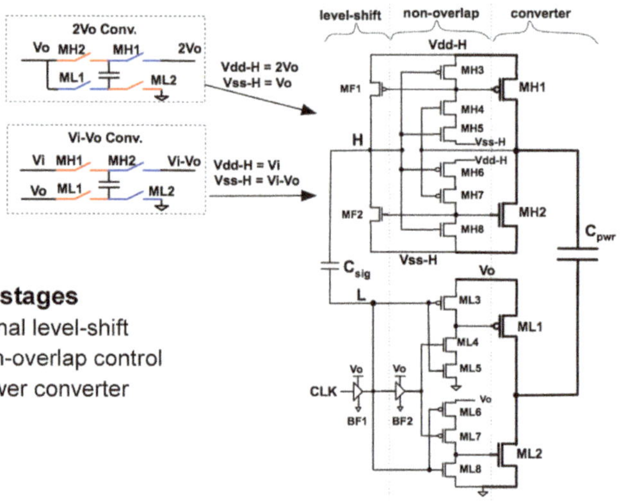

- **Three stages**
  - Signal level-shift
  - Non-overlap control
  - Power converter

Both of these auxiliary converters can use the same 1-capcitor 4-switch circuit and its driver as shown on the right. This converter has three stages: a power converter stage with the 4 power switches and power capacitorC_{pwr}, a non-overlap driver circuit, and a level-shifting stage with a signal-coupling capacitor.

47 Drv.3. Auxiliary Converters Operation (I)

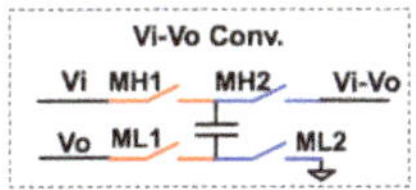

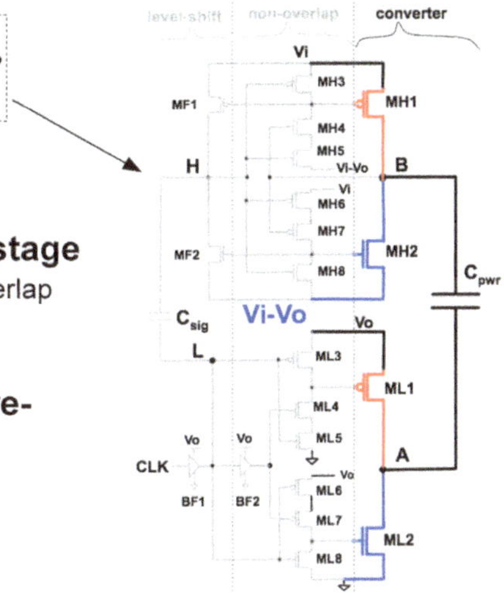

- **Power conversion stage**
  - Need to ensure non-overlap operation
- **Utilize "break-before-make" operation**

The operation of this converter utilize "break-before-make" operations for the power switches to ensure non-overlaps and shoot-through currents.

48 Drv.3. Auxiliary Converters Operation (II)

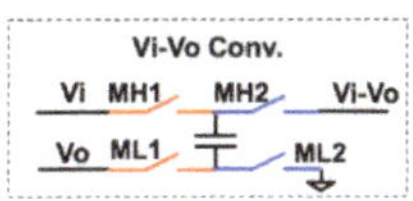

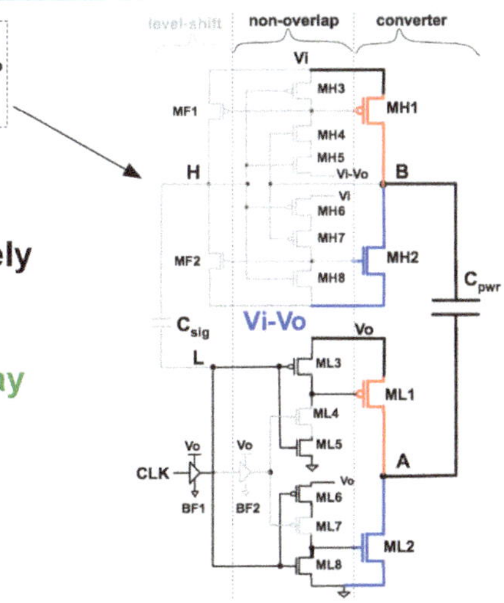

- **"Break" immediately**
  - Switches Mx3, Mx8
- **"Make" after a delay**
  - Switches Mx4-7

The non-overlap driver circuit is implemented so that turning off power switches, i.e. "break" operation, happens immediately while turning them on, i.e. "make" operation, only happens after a delay.

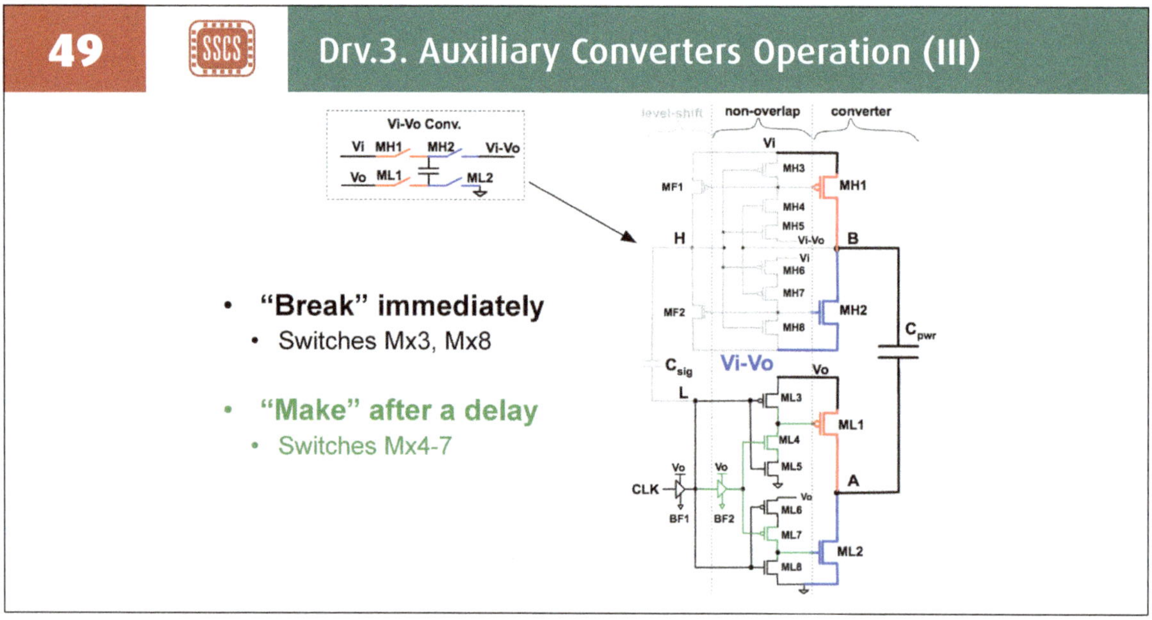

This delay is to make sure that a "break" operation is completed before a "make" operation.

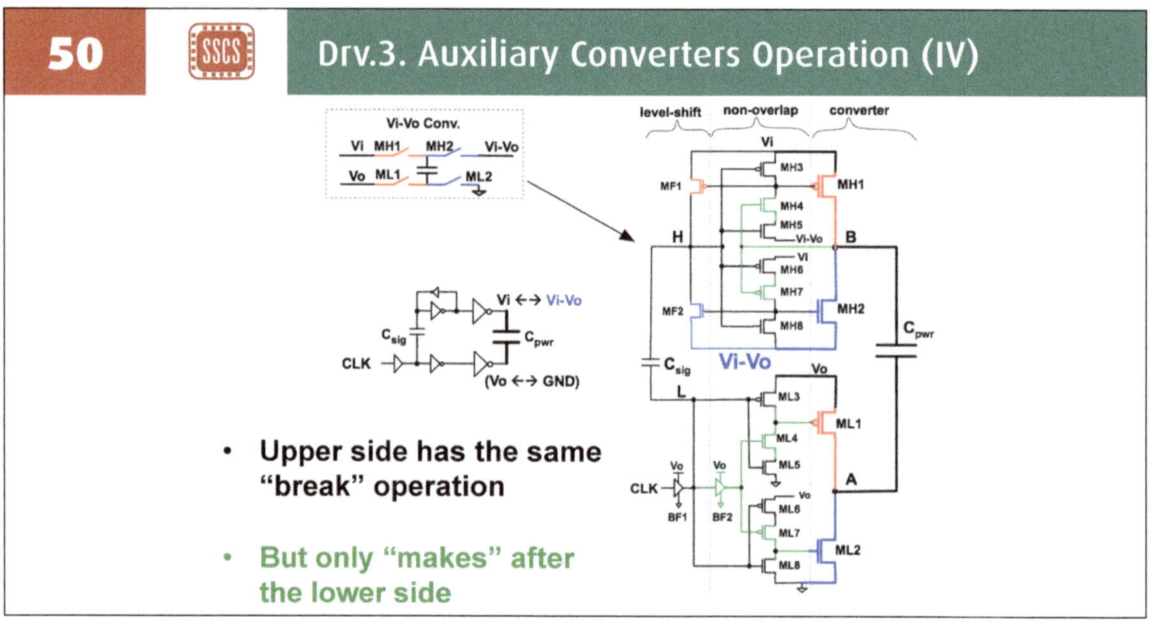

The high-side circuitry share the same "break" operation with the low-side to ensure "break" first. However, it has additional delay for "make" after the low-side "make" operation. This delay is introduced through the power converter stage to eliminate shoot-through current across high-side and low-side power switches via C_{pwr}.

51 Drv.3. Auxiliary Converters Operation (V)

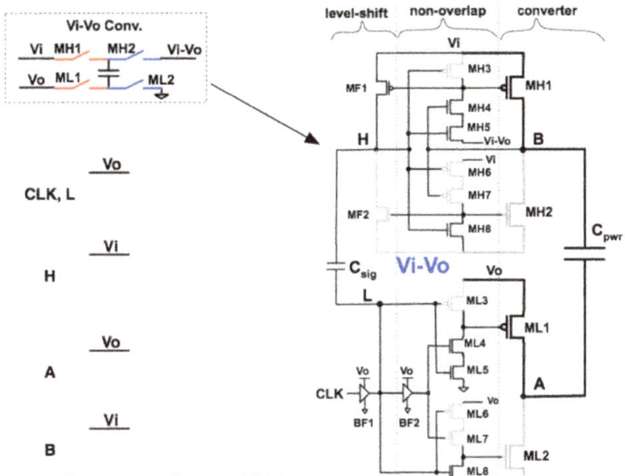

State: charge the flying cap C_{pwr} to Vi-Vo

In a detailed operation of this circuit, when node L, H, A, and B starts high at Vo, Vi, Vo, and Vi, respectively. MH1 and ML1 are both ON. This makes the voltage over capacitors C_{sig} and C_{pwr} charged to the difference between Vi and Vo, i.e. Vi-Vo.

52 Drv.3. Auxiliary Converters Operation (VI)

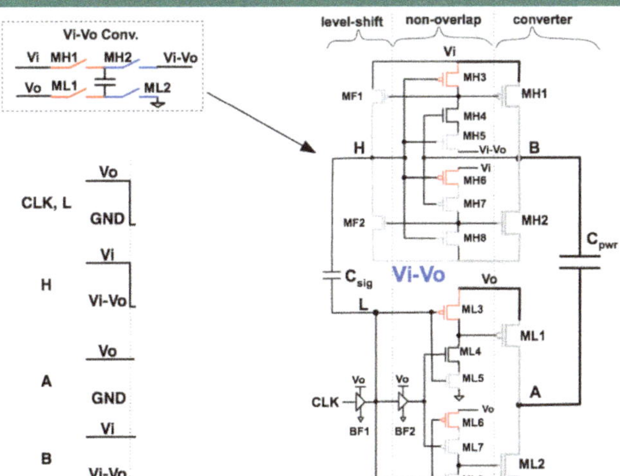

State: transfer signal and "break" for non-overlap

As soon as CLK and L transition to low at GND, the coupled H goes to Vi-Vo, immediately turning on transistors Mx3 and Mx6 and thus turning off both MH1 and ML1. As expected, at this transition the "break" happens immediately.

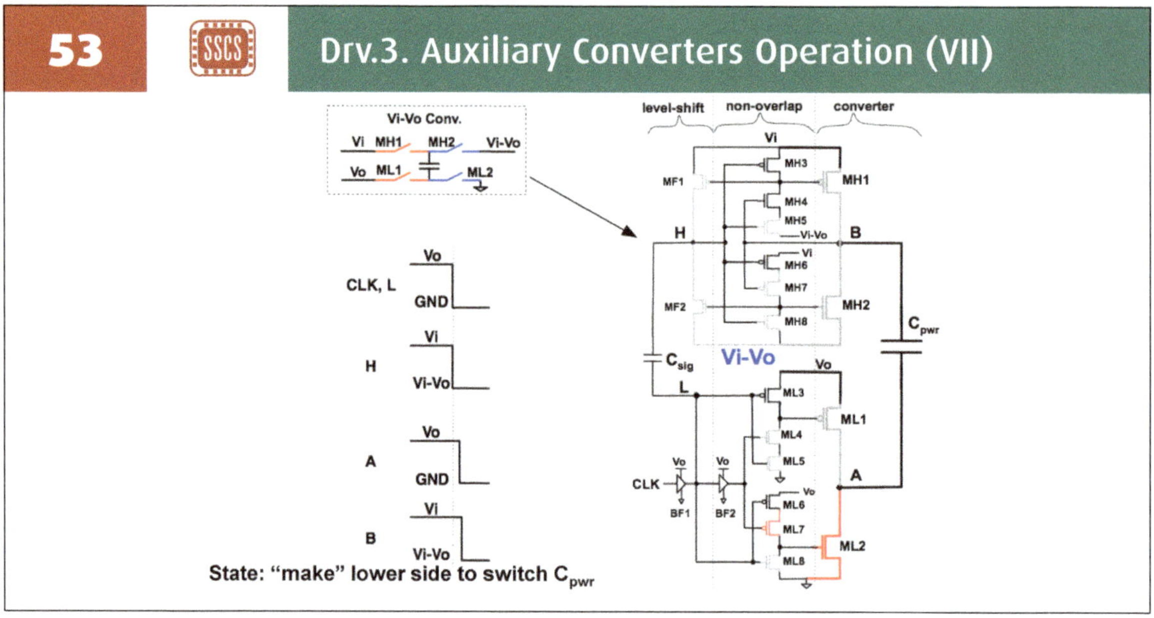

After a delay over BF2, ML2 is turned ON that is the "make" operation for the low-side circuit.

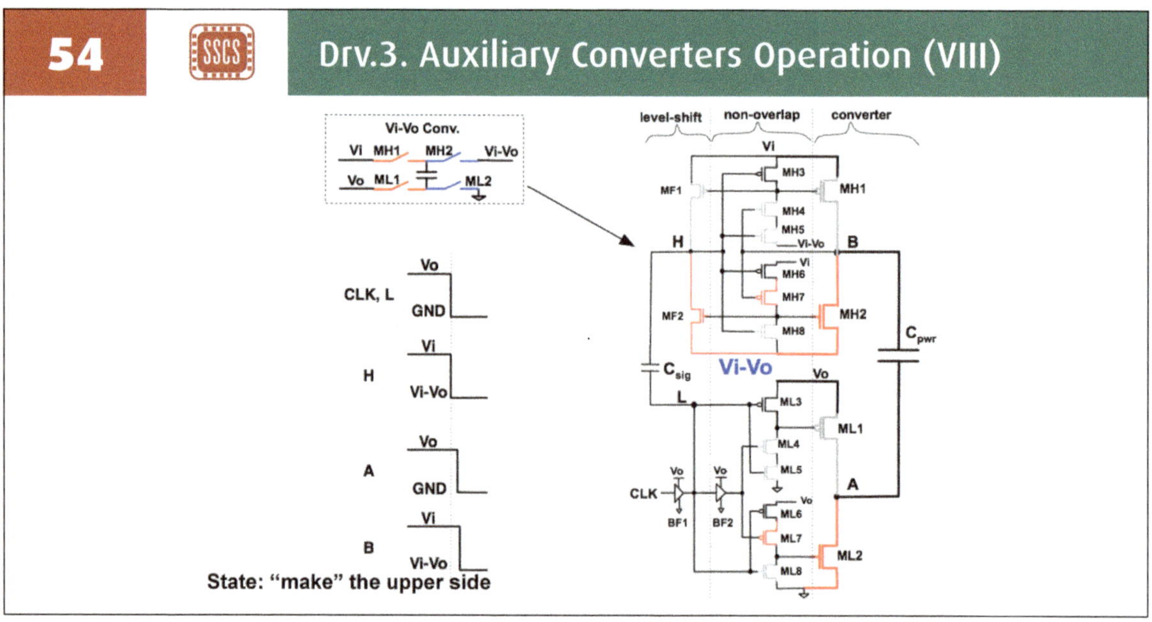

After "make" for low-side, C_{pwr} and thus node B are pulled low, turning on MH7. This action turns on MH2 and MF2 and complete the "make" operation for the high-side circuit.

55 Drv.3. Auxiliary Converters Operation (IX)

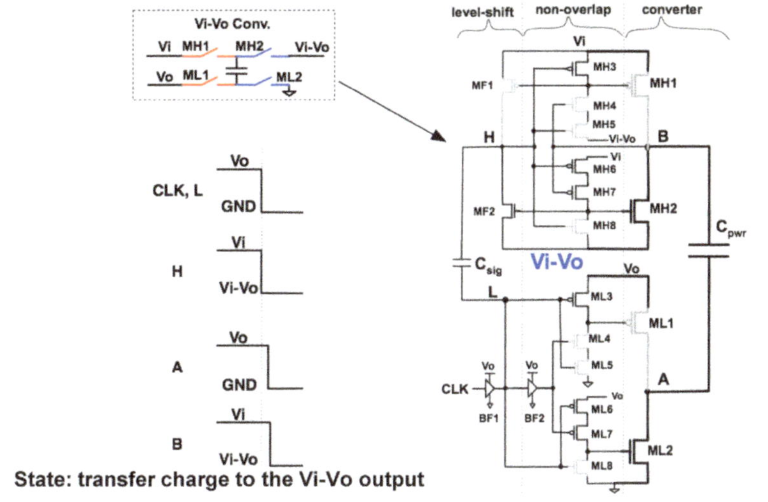

In this state, C_{pwr} is discharged to the Vi-Vo output.

56 Drv.3. Auxiliary Converters Operation (X)

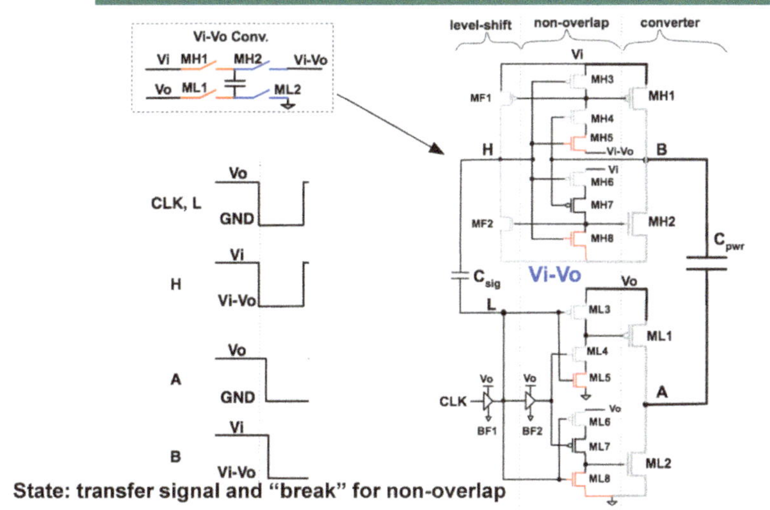

When CLK, nodes L, and node H transition back to High, immediately turning on Mx5 and Mx8 and turning off Mx3 and Mx6. This action turns off ML2 and MH2. Again, the "break" operation takes place first.

57 Drv.3. Auxiliary Converters Operation (XI)

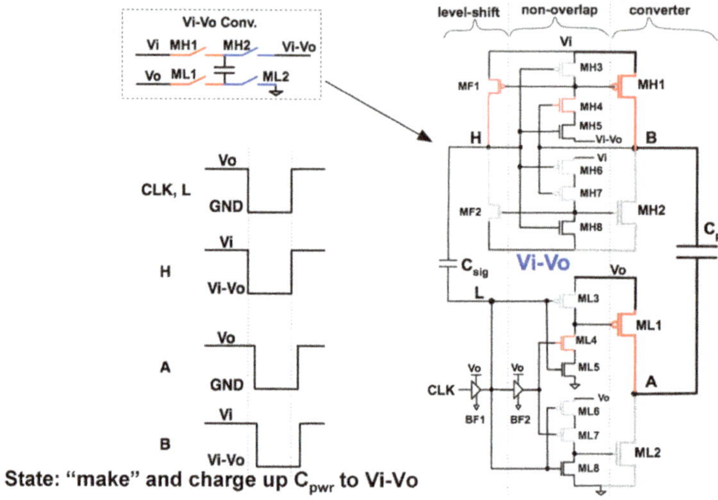

State: "make" and charge up C_{pwr} to Vi-Vo

Then similar to the previous transition, the make operations to turn on ML1 and then MH1 follows after a delay. In this state, C_{pwr} is again charged to Vi-Vo. This completes one operation cycle of the converter.

58 Circuit Techniques for Better SC Converter Design

$$P_{loss} = P_{Rsw} + P_{C_{fly}} + P_{bott.cap} + P_{gate.cap}$$

$$P_{Rsw} \propto I_L^2 \frac{R_{on}}{W_{sw}}, \quad P_{C_{fly}} \propto \frac{I_L^2}{f_{sw} C_{fly}}, \quad P_{bott.cap} \propto C_{bott} f_{sw}, \quad P_{gate.cap} \propto W_{sw} C_{gate} f_{sw}$$

(first two terms: Ro; last two terms: R_P)

- Output voltage ripple
  - Multiphase interleaving
- Efficiency vs. regulation trade-off (Ro)
  - More topologies, multiple conversion ratio
- Switch driver design (R_{on}, W_{sw}, Q_{gate})
  - Use of thin-ox devices
  - Utilization of main and auxiliary voltages
  - Depends on converter topology
    - Will be discussed in specific converter examples
- Clock level shifter (Rp)
- **Integrated capacitor design (Ro, C_{fly})**
  - Layout and well bias
  - Parasitic capacitor reduction
- Feedback control and regulation (Ro, C_{fly}, R_{on})

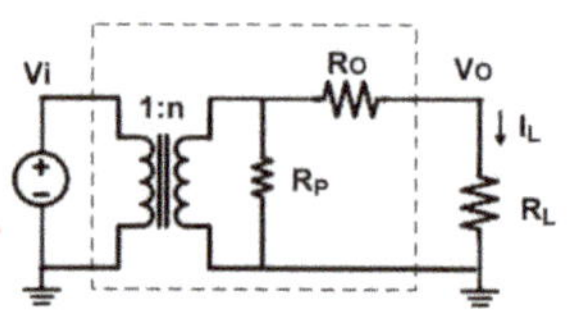

The next section will discuss a method to bias flying capacitors to reduce parasitic capacitance.

59 Par.1. MOS Capacitor Well Bias 1

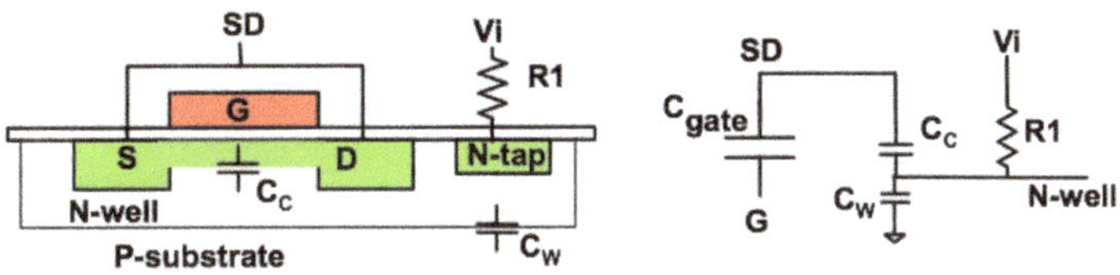

- **Use a large resistor R1 to bias the N-well**
  - Puts the channel cap C_c and Nwell-substrate cap C_w in series
  - Parasitic cap is dominated by C_w ($<<C_c$)
- **Reduces the total parasitic capacitor to 1.2% C_{gate}**

Ref: Le, ISSCC 2013

In this first design example, PMOS gate capacitance can be utilized for flying capacitors. To avoid a significant channel-to-body capacitance C_C, the PMOS's N-well can be biased with a large resistor R1 to essentially putting C_C in series with the well-to-substrate capacitance C_W. The effective parasitic capacitance is reduced to C_W that is much smaller than C_C.

60 Par.1. MOS Capacitor Well Bias 2

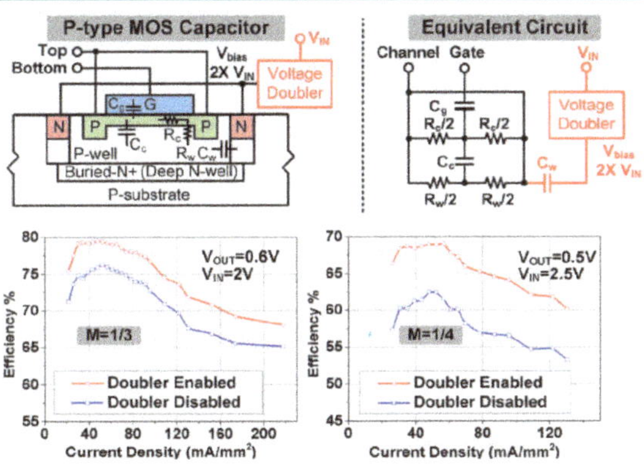

- **Use P-type MOS capacitor**
- **Bias DNW at higher voltage**

Ref: Jiang, ISSCC 2015

As shown in this parasitic reduction example 2, triple-well can be used to put more smaller parasitic capacitance in series to further reduce effective bottom-plate capacitance. In this case, a P-type MOS (NMOS gate) capacitor is put inside a P-well and a deep N-well.

61 SSCS Par.3. Parasitic Capacitor Switching Strategy

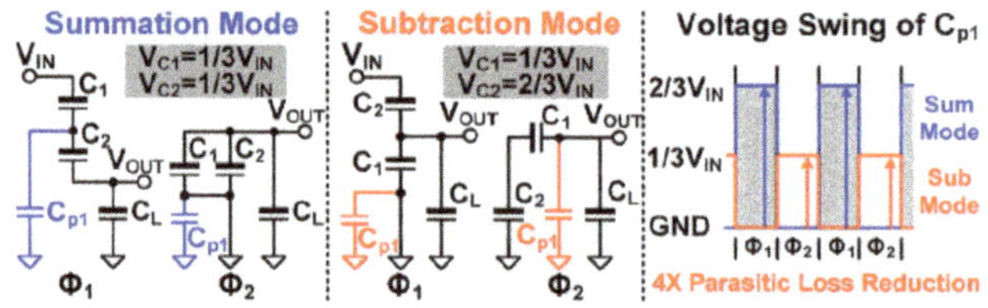

- **Modify topology operation**
  - Reduce voltage swing for parasitic cap

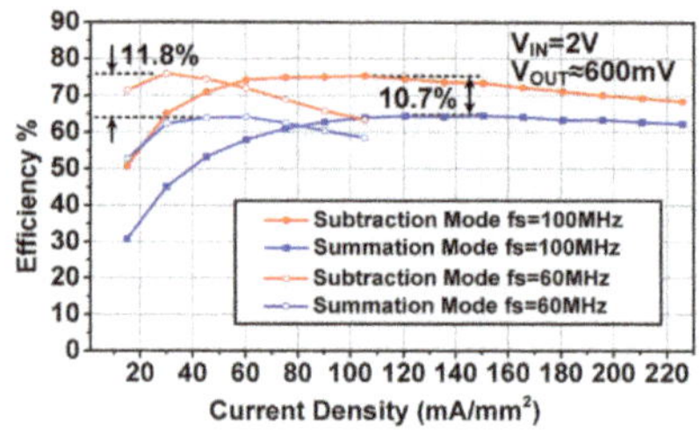

Ref: Jiang, ISSCC 2015

Operation of the main SC converter can also be altered to enable reduction of bottom-plate capacitor switching loss as shown in this example 3.

62 SSCS Par.4. Bottom-Plate Cap Charge Recycling 1

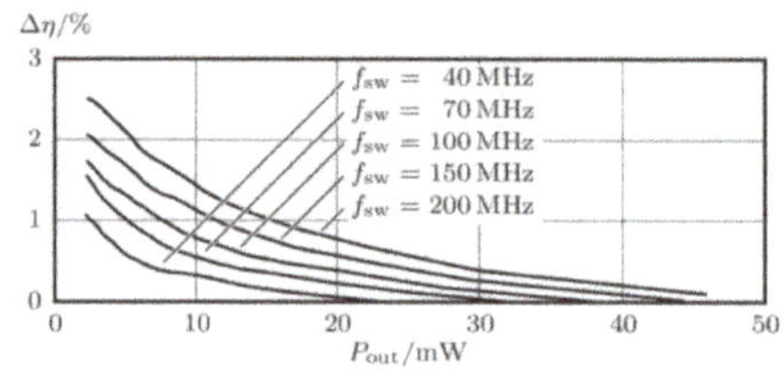

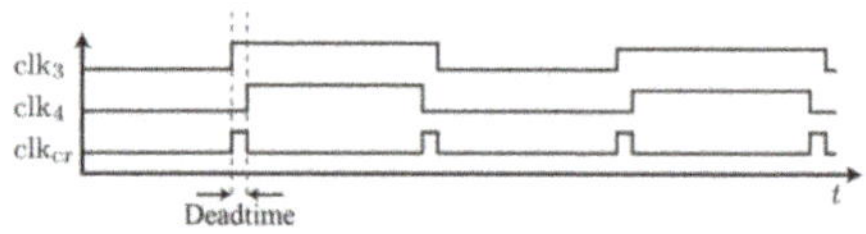

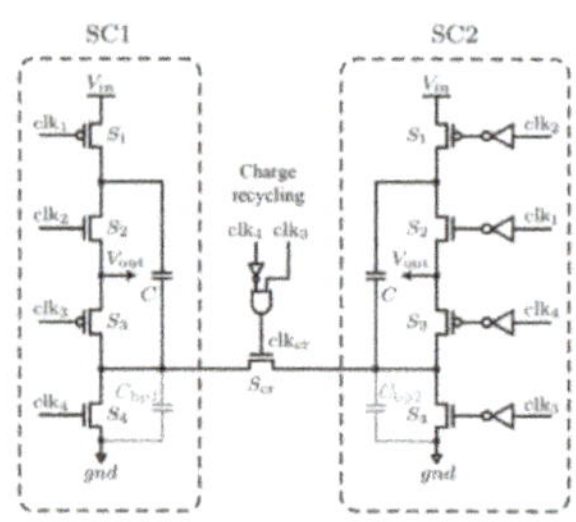

- **Recycle bottom-plate cap charge with additional phase**
  - Effective and necessary for large capacitance
- **Half $C_{bp}V^2f$ loss: $C_{bp}V^2f \rightarrow 2C_{bp}(V/2)^2f = (½)C_{bp}V^2f$**

Ref: Andersen, PwrSoC 2012;
Tong, CICC 2013

The charge in bottom-plate capacitors can be recycled using an additional phase. In example 4, shorting one charging bottom-plate capacitor and one discharging one can cut this parasitic switching loss in half.

63 Par.4. Bottom-Plate Cap Charge Recycling 2

- **Use more charge redistribution steps (CRS)**
  - More SC cells connected via switch matrix
  - Charge redistribution bus (CRB)
  - 16 cells, 9 CRS, 72 connections

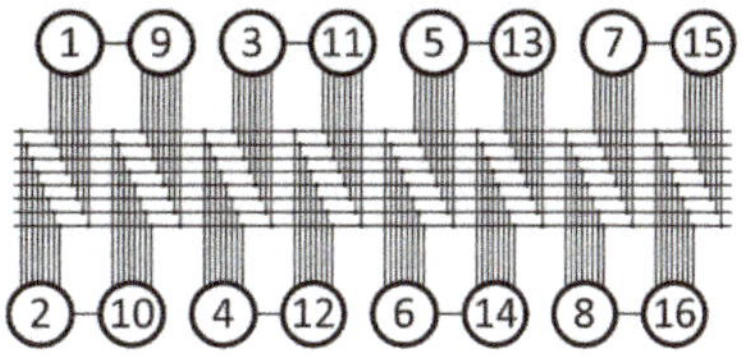

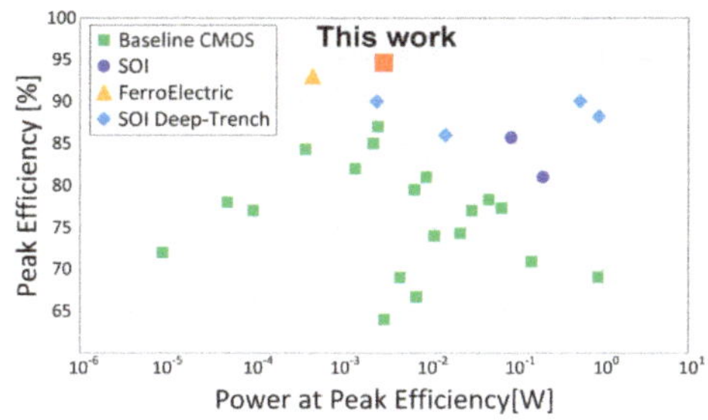

Ref: Butzen, ISSCC 2016

In example 5, the charge recycling can be done in multiple steps using more phased in a SC converter to effectively eliminate its bottom-plate switching loss.

64 Circuit Techniques for Better SC Converter Design

- Output voltage ripple
  - Multiphase interleaving
- Efficiency vs. regulation trade-off (Ro)
  - More topologies, multiple conversion ratio
- Switch driver design (R_{on}, W_{sw}, Q_{gate})
  - Use of thin-ox devices
  - Utilization of main and auxiliary voltages
  - Depends on converter topology
    - Will be discussed in specific converter examples
- Clock level shifter (Rp)
- Integrated capacitor design (Ro, C_{fly})
  - Layout and well bias
  - Parasitic capacitor reduction
- **Feedback control and regulation (Ro, C_{fly}, R_{on})**

$$P_{loss} = P_{Rsw} + P_{C_{fly}} + P_{bott.cap} + P_{gate.cap}$$

$$P_{Rsw} \propto I_L^2 \frac{R_{on}}{W_{sw}}, \quad P_{C_{fly}} \propto \frac{I_L^2}{f_{sw} C_{fly}} \quad (\text{Ro}); \qquad P_{bott.cap} \propto C_{bott} f_{sw}, \quad P_{gate.cap} \propto W_{sw} C_{gate} f_{sw} \quad (R_P)$$

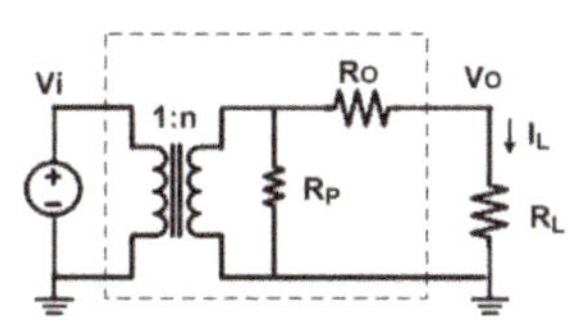

We have discussed circuit techniques to reduce loss and improve SC converter efficiency and power density. In the next section, we will review feedback control and regulation.

65 Ctrl.1. Simple Closed-Loop Control

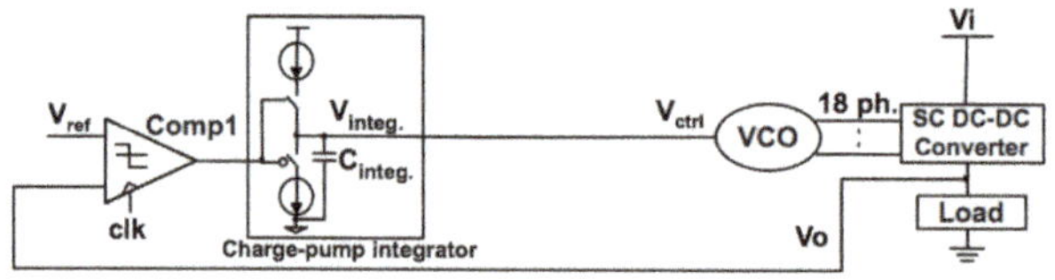

- **Output impedance of the regulator set by f_{sw}**
  - Rout α $1/(f_{sw}C_{fly})$
- **Switching frequency set by (slow) integral control loop**
- **Key challenge: response to 0 → I_{max} load step**

Ref: Le, ISSCC 2013

In this control example 1, a SC converter regulates the output voltage via its switching frequency using a feedback loop with a comparator (clocked at the same converter switching frequency), a charge-pump integrator and a voltage-controlled oscillator (VCO). With the integrator in the loop, it is difficult to have fast response to an extreme load step from 0 to max output current.

66 Ctrl.2. Control Loop with Fast Load Response

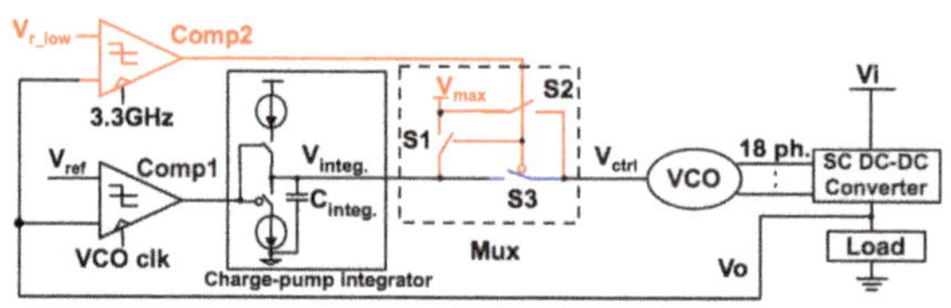

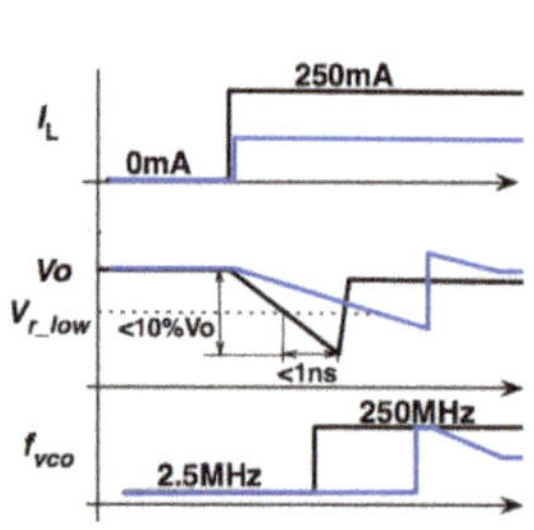

- **Additional comparator "jumps" f_{sw}**
- **Need sub-ns response time for <10% droop**
  - Comparator must sample Vo at high frequency

To handle extreme load current transient, an additional non-linear feedback path is added to the loop in the control example 2. This path is activated at 3.3GHz and allows the SC converter frequency to jump to its max switching frequency, i.e. lowest output impedance, when an extreme droop is detected at the output.

67 Ctrl.2. Load Step Measurement

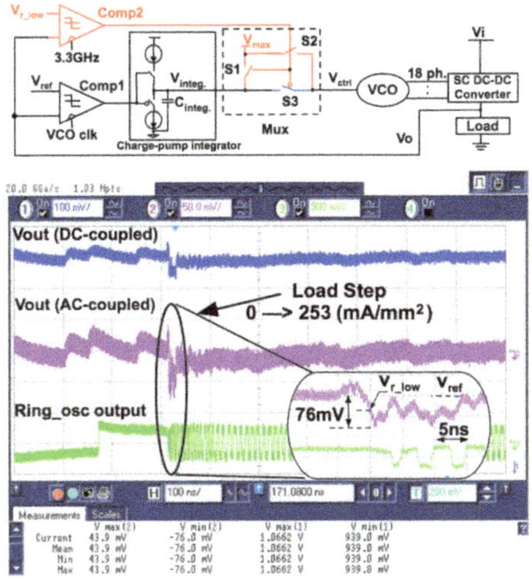

- **Achieves 7.6% droop under a full load step of 253mA/mm²**
  - Sub-ns response
- **Load step generated by on-die load circuitry**
  - 50ps rise time

Measurement results showed that this control allows the converter to have a sub-ns response to limit droops to 7.6% at full load transients.

68 Ctrl.3. Lower-Bound Control

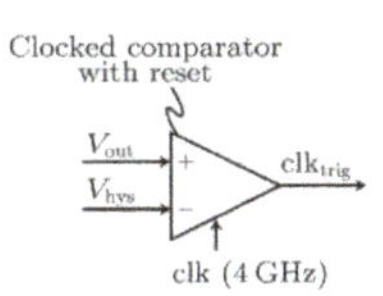

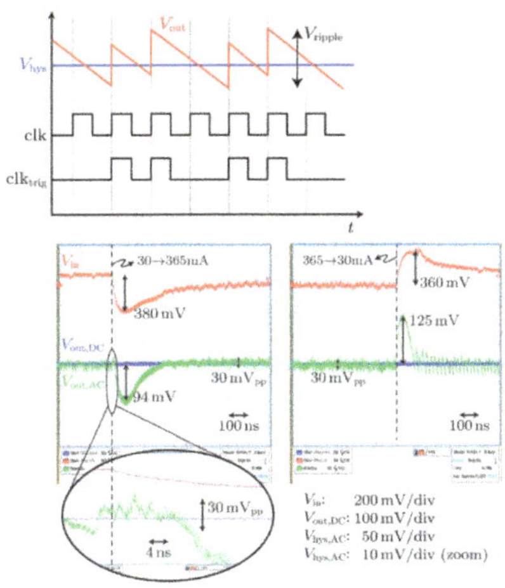

- **Simple and fast load transient**
  - Approach sub-ns response
- **But have higher output ripple**
  - due to limit cycle switching
  - Undefined ripple frequency

Ref: Andersen, ISSCC 2014

The control example 3 usesa lower-bound control with a comparator clocked at high frequency. The fast clocked comparator enable nanosecond responses. However, the converter suffers from high output voltage ripple at undefined frequency as the result of limit-cycle switching.

69 Ctrl.4. Series Linear Regulation

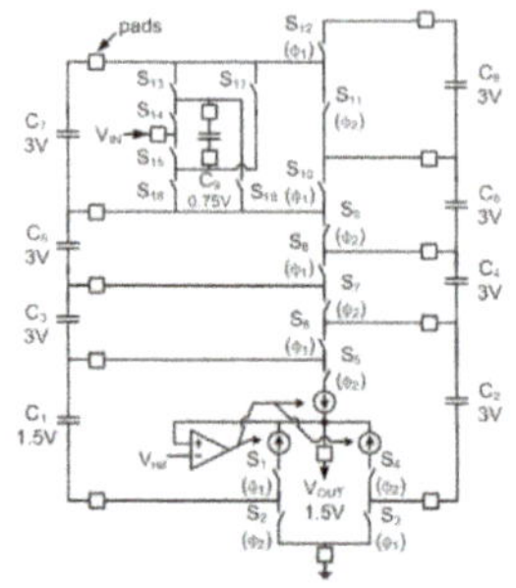

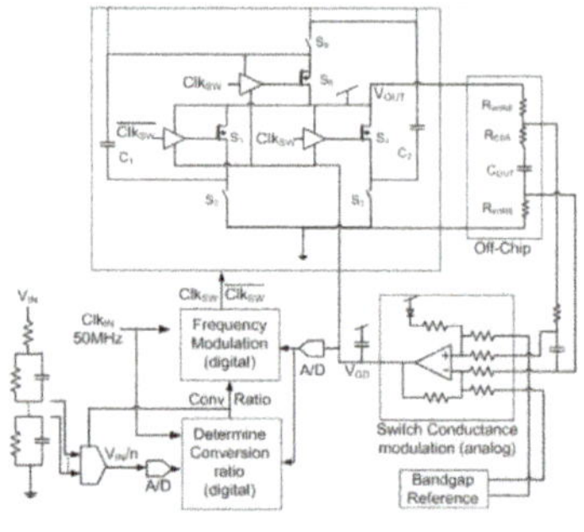

- **Regulate output switch on-resistance**
  - Similar to LDO

Ref: V. Ng, ISSCC 2012

Output of SC converter can also be fine-regulated by controlling the strength, i.e. on-resistance, of the power switches as shown in the control example 4. This is similar to adding an LDO following the SC converter.

OTHER CIRCUIT TECHNIQUES

70 System Techniques

- **Put SC converter in a system and in context**
  - Have system view of configuration and operation
- **SC converter has issue with efficiency for fine regulation**
  - Seeking for fine regulation in SC alone is not desirable
  - The issue also lies in the traditional method of using a regulator
- **SC converter operation and its discrete optimal efficiency can be leveraged**
  - May not require a lot of reconfigurable topologies

In order to improve SC efficiency even further, techniques at the system level can be used. The key is to understand that SC would often not need to be stand-alone. If it is put in the right context and operation, it could give a boost to the system efficiency.

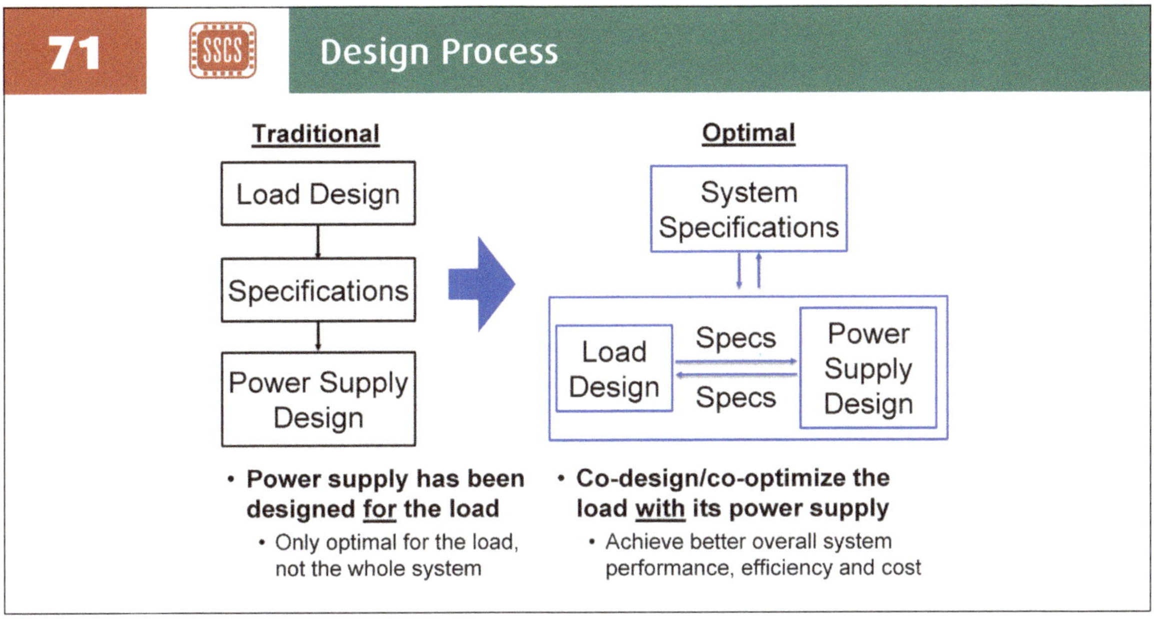

The first system technique is to recognize the issue that in conventional design process, power supply only receives requirements from the load. If power supply can also give a strong feedback to the load design to enable a co-design and co-optimization with the load, better overall system performance can be achievable.

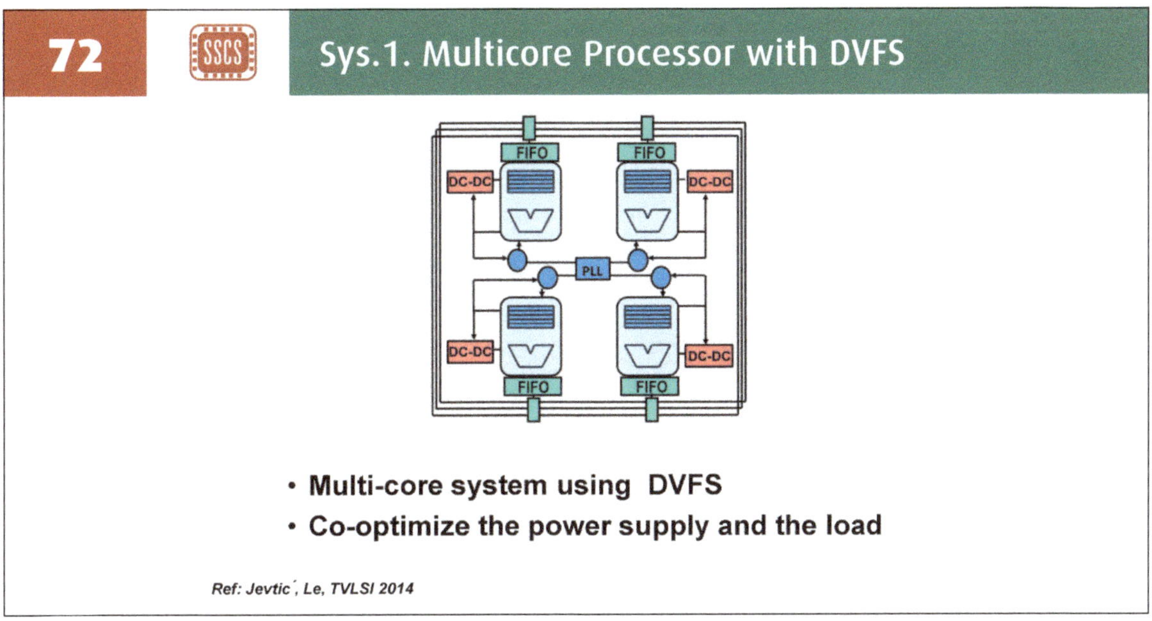

This technique can be seen in the system example 1 design where a multi-core system using dynamic voltage and frequency scaling (DVFS) is co-optimized with its power supply.

73 Switched-Capacitor Loss

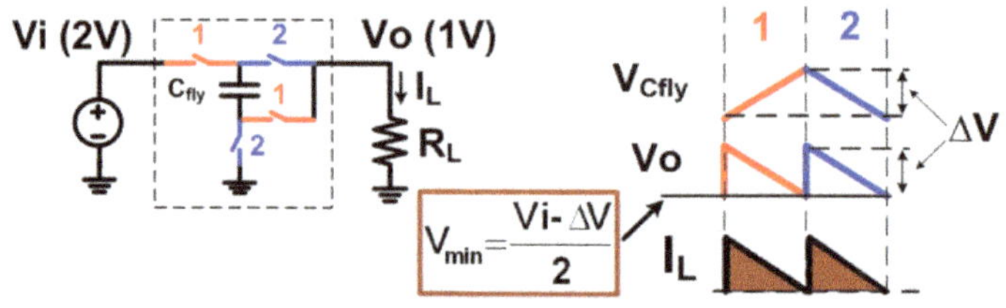

- **Conventional digital circuits**
  - Performance (f_{CPU}) set by V_{min}
- **Can we utilize the ripple power for computation?**

Remember that SC has significant SC loss because of the Vin requirement in processors. The question is can we utilize the ripple power for computation and essentially remove the fixed Vmin requirement.

74 Sys.1. Optimize the Load for System Efficiency

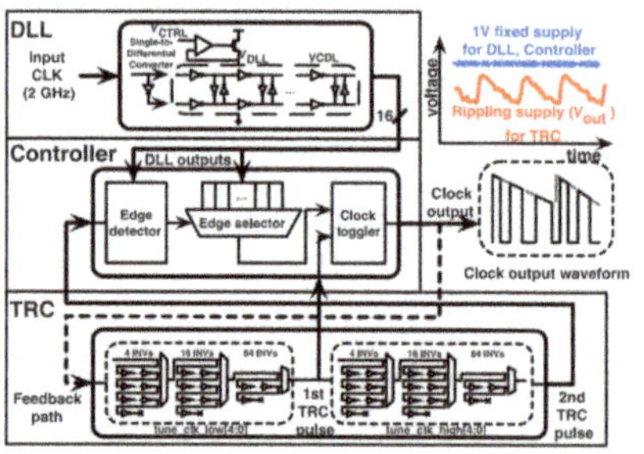

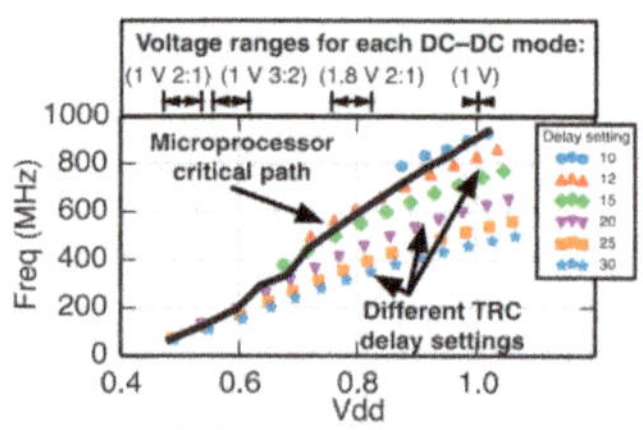

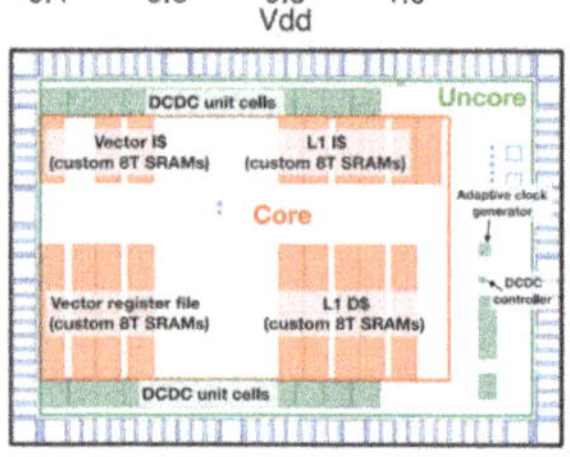

- **Multi-core System using DVFS**
- **Utilize ripple power**
  - Eliminate the charge-sharing loss

Ref: Jevtić, Le, TVLSI 2014;
Zimmer, VLSI 2014 and JSSC 2016

We know that processors already have the DVFS capability. In this work, SC converter is allowed to have large output voltage ripple at its switching frequency. The processor tunes its clock frequency along with this ripple to efficiently utilize the charge supplied by the SC converter.

75 Sys.1. More Optimal System Efficiency

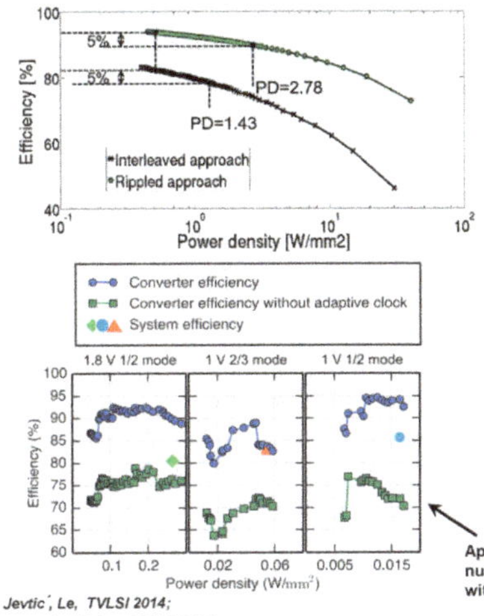

Approximate converter efficiency measured by numerically integrating voltage waveforms with pre-characterized load current.

- **High SC efficiency with minimized switched-capacitor loss**
- **Energy saving of 10%-20% for the whole system**
- **Booted Linux**
  - Zimmer, VLSI 2015, JSSC 2016

Ref: Jevtic', Le, TVLSI 2014; Zimmer, VLSI 2014 and JSSC 2016

As the SC loss, i.e. also called charge-sharing loss, is effectly minimized, the system achieved 10%-20% energy saving for the same required operation. Linux was installed and started using this system.

76 Sys.2. Flying Load Instead of Flying Capacitor

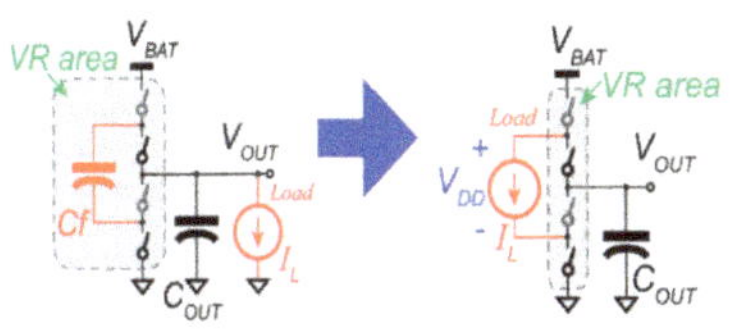

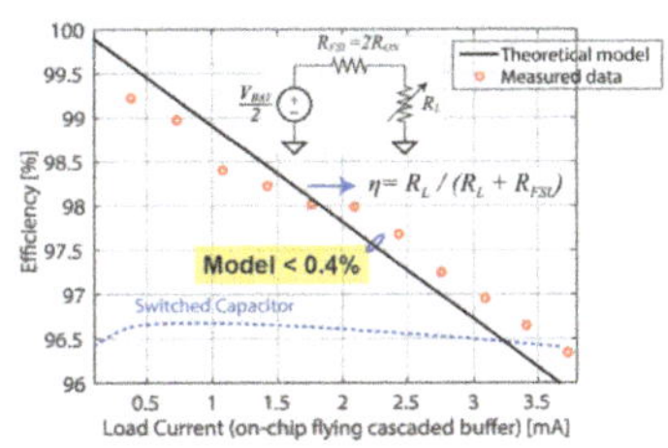

- **Originated from the Sys.1. and stack-domain converter**
  - Switched load and flying capacitor operation
- **Allow load voltage domain to be flying**
  - Eliminate large flying capacitor
- **High "power density" and efficiency**
  - But complicated load ground reference

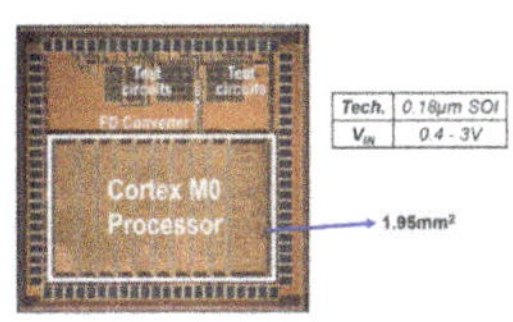

Ref: Salem, ISSCC 2016

In the system technique example 2, the author group also wants to eliminate the charge sharing loss by removing the output filtering capacitor. In this design, instead of switching the flying capacitor the load is switched. Although charge sharing is effectively eliminated and high efficiency is achieved, this implementation has a complicated load ground reference if used in a larger system.

CONCLUSIONS

77 The Enablers for Efficient iPower

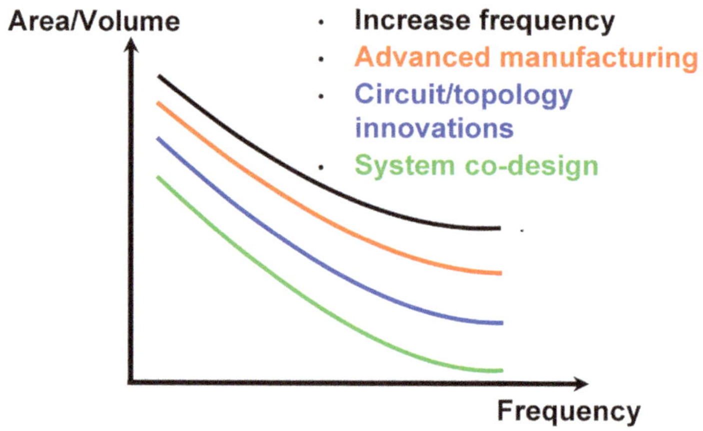

In summary, the techniques to achieve integrated power supply that is both power- and space-efficient can be characterized in four categories: (1) increased switching frequency for smaller passives, (2) advanced manufacturing for better passives and integration, (3) circuit and topology innovations, and (4) system co-design. While (3) is most appealing and manageable in designers' territory, (4) could be difficult since it requires collaborations at a higher level.

78 Conclusions

- **Clear need to have efficient integrated power supply**
  - SC DC-DC converter is a good candidate for both high power density and high efficiency
- **SC Converter design techniques**
  - Multiphase interleaving for smaller output ripple
  - More converter topologies
  - Efficient switch drivers
  - Parasitics reduction for integrated capacitors
  - Feedback and regulation
  - System techniques
- **More power converter design innovations coming**

In conclusion, SC is a good candidate for to address the need for efficient integrated power supply. In this chapter, we have discussed various design techniques for good SC converter designs, including circuit, topology, control and system techniques. I am positive that more converter design innovations are coming and look forward to it.

79 Thank you for your attention!

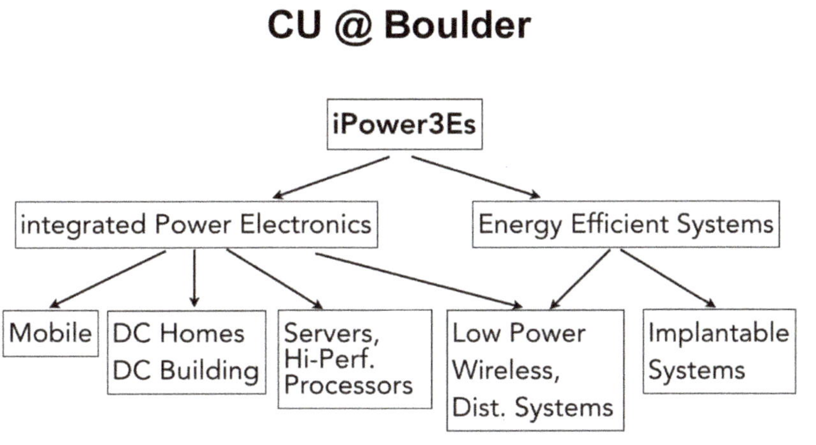

Thank you from the integrated Power Electronics and Energy Efficienty systems (iPower3Es) group at the University of Colorado Boulder.

80 References

1. D. Maksimovicand S. Dhar, "Switched-capacitor DC-DC converters for low-power on-chip applications," *30th Annual IEEE Power Electronics Specialists Conference*. Record. (Cat. No.99CH36321), Charleston, SC, USA, 1999, pp. 54-59 vol.1.

2. J. Wibbenand R. Harjani, "A high-efficiency DC-DC converter using 2 nHintegratedinductors," in *IEEE Journal of Solid-State Circuits*, vol. 43, no. 4, pp. 844–854, Apr. 2008.

3. H. Leet al., "A 32nm fully integrated reconfigurable switched-capacitor DC-DC converter delivering 0.55W/mm² at 81% efficiency,"*2010 IEEE International Solid-State Circuits Conference - (ISSCC)*, San Francisco, CA, 2010, pp. 210-211.

4. L. Chang, R. Montoye, B. Ji, A. Weger, K. Stawiasz, and R. Dennard, "A fully integrated switched-capacitor 2:1 voltage converter with regulation capability and 90% efficiency at 2.3 A/mm ," in *IEEE Symp. VLSI Circuits Dig.*, Jun. 2010, pp. 55–56.

5. H-P. Le, S. Sanders and E. Alon, "Design Techniques for Fully Integrated Switched-Capacitor DC-DC Converters," in *IEEE Journal of Solid-State Circuits*, pp. 2120 -2131, Vol. 46, Iss. 9, Sep. 2011.

6. V. Ng and S. Sanders, "A 92%-efficiency wide-input-voltage-range switched-capacitor DC-DC converter,"*2012 IEEE International Solid-State Circuits Conference*, San Francisco, CA, 2012, pp. 282-284.

7. N. Sturckenet al., "A 2.5D Integrated Voltage Regulator Using Coupled-Magnetic-Core Inductors on Silicon Interposer," in *IEEE Journal of Solid-State Circuits*, vol. 48, no. 1, pp. 244-254, Jan. 2013.

8. T. Tong, X. Zhang, W. Kim, D. Brooks and G. Wei, "A fully integrated battery-connected switched-capacitor 4:1 voltage regulator with 70% peak efficiency using bottom-plate charge recycling,"*Proceedings of the IEEE 2013 Custom Integrated Circuits Conference*, San Jose, CA, 2013, pp. 1-4.

9. H-P. Le, J. Crossley, S. Sanders and E. Alon, "A Sub-ns Response Fully-Integrated Battery-Connected Switched-Capacitor Voltage Regulator Delivering 0.19W/mm2at 73% Efficiency," *IEEE International Solid-State Circuits Conference (ISSCC) Dig. Tech. Papers*, pp. 372-373, Feb. 2013.

10. T. M. Andersen et al., "A 4.6W/mm2power density 86% efficiency on-chip switched capacitor DC-DC converter in 32 nm SOI CMOS," in *IEEE Applied Power Electronics Conference and Exposition (APEC)*, Long Beach, CA, 2013, pp. 692-699.

11. T. M. Andersenet al., "4.7 A sub-ns response on-chip switched-capacitor DC-DC voltage regulator delivering 3.7W/mm2at 90% efficiency using deep-trench capacitors in 32nm SOI CMOS," 2014 *IEEE International Solid-State Circuits Conference Digest of Technical Papers (ISSCC)*, San Francisco, CA, 2014, pp. 90-91.

12. L. G. Salem and P. P. Mercier, "4.6 An 85%-efficiency fully integrated 15-ratio recursive switched-capacitor DC-DC converter with 0.1-to-2.2V output voltage range," 2014 *IEEE International Solid-State Circuits Conference Digest of Technical Papers (ISSCC)*, San Francisco, CA, 2014, pp. 88-89.

13. R. Karadiand G. V. Pique, "4.8 3-phase 6/1 switched-capacitor DC-DC boost converter providing 16V at 7mA and 70.3% efficiency in 1.1mm3," 2014 *IEEE International Solid-State Circuits Conference Digest of Technical Papers (ISSCC)*, San Francisco, CA, 2014, pp. 92-93.

14. E. A. Burton et al., "FIVR—Fully integrated voltage regulators on 4th generation Intel Core SoCs," in *Proc. Appl. Power Electron. Conf. (APEC)*, Mar. 2014, pp. 432–439.

15. J. Jiang et al., "20.5 A 2-/3-phase fully integrated switched-capacitor DC-DC converter in bulk CMOS for energy-efficient digital circuits with 14% efficiency improvement," 2015 *IEEE International Solid-State Circuits Conference - (ISSCC) Digest of Technical Papers*, San Francisco, CA, 2015, pp. 1-3.

16. Y. Luet al., "20.4 A 123-phase DC-DC converter-ring with fast-DVS for microprocessors," 2015 *IEEE International Solid-State Circuits Conference -(ISSCC) Digest of Technical Papers*, San Francisco, CA, 2015, pp. 1-3.

17. R. Jevtiĉ et al., "Per-Core DVFS With Switched-Capacitor Converters for Energy Efficiency in ManycoreProcessors," in *IEEE Transactions on Very Large Scale Integration (VLSI) Systems*, vol. 23, no. 4, pp. 723-730, April 2015.

18. H-P. Le, "Fully integrated power conversion and the enablers," 2015 *9th International Conference on Power Electronics and ECCE Asia (ICPE-ECCE Asia)*, Seoul, 2015, pp. 1778-1783.

19. B. Zimmeret al., "A RISC-V Vector Processor With Simultaneous-Switching Switched-Capacitor DC–DC Converters in 28 nm FDSOI," in *IEEE Journal of Solid-State Circuits*, vol. 51, no. 4, pp. 930-942, April 2016.

20. N. Butzen and M. Steyaert, "12.2 A 94.6%-efficiency fully integrated switched-capacitor DC-DC converter in baseline 40nm CMOS using scalable parasitic charge redistribution," 2016 *IEEE International Solid-State Circuits Conference (ISSCC)*, San Francisco, CA, 2016, pp. 220-221.

PART V

IoT SECURITY CIRCUITS AND TECHNIQUES

CHAPTER 14

Energy Efficient and Ultra Low Voltage Security Circuits for Nanoscale CMOS Technologies

Ram K. Krishnamurthy
Sanu Mathew
Sudhir Satpathy
Vikram Suresh
Intel Corporation, USA

Low-area energy-efficient security primitives are key building blocks for enabling end-to-end content protection, user authentication, and consumer confidentiality in the IoT world that is estimated to surpass 50billion smart and connected devices by 2020. This chapter describes design approaches that blend energy-efficient circuit techniques with optimal accelerator micro-architecture datapath, and hardware friendly arithmetic to achieve ultra-low energy consumption in security platforms for seamless adoption in area/battery constrained and self-powered systems. Industry leading energy-efficiency is demonstrated with three designs, fabricated and measured in advanced process technologies : 1) A 2040-gate arithmetically optimized composite-field Sbox based AES accelerator achieves 289Gbps/W peak energy-efficiency while offering 432Mbps throughput in 22nm tri-gate CMOS, 2) Hybrid Physically Unclonable Function (PUF) circuit leverages burn-in induced aging to reduce bit-error, followed by temporal-majority-voting, dark-bit masking, and error-correction conditioning techniques to generate a 100% stable full-entropy key with 190fJ/bit energy consumption in 22nm tri-gate CMOS. 3) A light-weight all digital TRNG uses in-line correlation suppressor and entropy-extractor circuits to achieve >0.99 min-entropy with 3pJ/bit measured energy-efficiency while operating down to 300mV in 14nm tri-gate CMOS.

The advent of IoT era has led to a dramatic rise in miniaturized computing platforms encompassing hand-helds, wearables, sensor nodes, drones, robotics, implantable devices, health and environmental monitoring aids etc. Increasing compute capability of these devices and seamless connectivity in a ubiquitous compute ecosystem has led to massive rise in generation, consumption and transfer of user data. Information security and privacy protection have hence become integral to most day-to-day applications running on modern SoCs that power these IoT devices. Poor energy-efficiency and diminishing performance resulting from implementing these security protocols in general purpose processors necessitate hardware acceleration to meet real-time throughput demands under a limited power budget. Advanced Encryption System (AES), True Random Number Generator (TRNG), Physically Unclonable Functions (PUFs) are some of the key primitives that enable hardware acceleration of a significant number of security applications targeted for advanced content protection, signature generation and verification protocols.

Conventional implementations of security accelerators achieve high throughput at the expense of large die-area and power dissipation, rendering them unsuitable for use in area/power constrained mobile and wearable systems. The energy and thermal constraints of such systems motivate lightweight low-cost hardware implementations of these security primitives with compact layout footprint and ultra-low leakage energy consumption. This paper describes circuit design techniques including arithmetic and micro-architectural optimizations that enable significant improvement in energy-efficiency and die-area reduction in these edge IoT devices. The remainder of this chapter is organized as follows: We describe an on-die lightweight nanoAES hardware accelerator built around a single 8-bit composite-field Sbox circuit that is arithmetically optimized for minimum area to achieve a compact layout footprint spanning 2090 gates; Later, we present a process-voltage-temperature (PVT) variation-tolerant all-digital PUF array targeted for on-die generation of 100% stable, device-specific, high-entropy keys that features a hybrid PUF circuit to leverage burn-in induced device aging to improve stability; an all-digital TRNG circuit is then discussed that uses three independent self-calibrating entropy sources, coupled with correlation suppressors and a real-time BIW extractor to enable ultra-low energy consumption of 3pJ/bit, while generating cryptographic-quality keys over a wide dynamic supply voltage down to 300mV.

1 Security in the Internet of Things

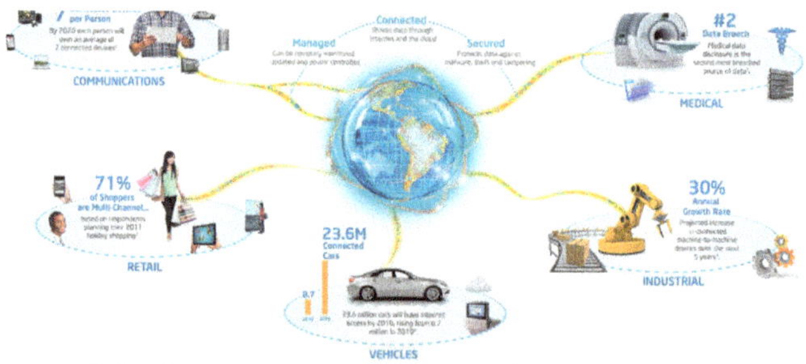

- **Billions of devices generate and exchange data:**
  - **Makes end-to-end security and privacy critical.**
- **Edge devices require ultra-lightweight and energy-efficient security features:**
  - **Private Key and Public Key Cryptography**
  - **Random Number Generation**
  - **Secure ID/Key Generation**

The advent of IoT era has led to a dramatic rise in miniaturized computing platforms encompassing hand-helds, wearables, sensor nodes, drones, robotics, implantable devices, health and environmental monitoring aids etc. Increasing compute capability of these devices and seamless connectivity in a ubiquitous compute ecosystem has led to massive rise in generation, consumption and transfer of user data.

2 Motivation: Secure IoTPlatforms

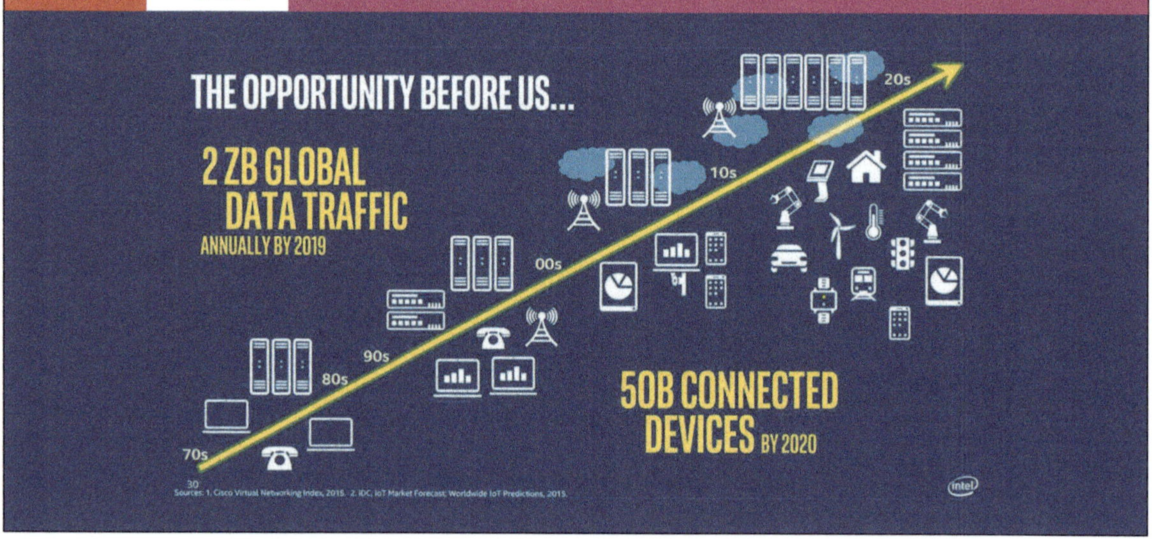

Information security and privacy protection have hence become integral to most day-to-day applications running on modern SoCs that power these IoT devices. Poor energy-efficiency and diminishing performance resulting from implementing these security protocols in general purpose processors necessitate hardware acceleration to meet real-time throughput demands under a limited power budget.

Advanced Encryption System (AES), True Random Number Generator (TRNG), Physically Unclonable Functions (PUFs) are some of the key primitives that enable hardware acceleration of a significant number of security applications targeted for advanced content protection, signature generation and verification protocols.

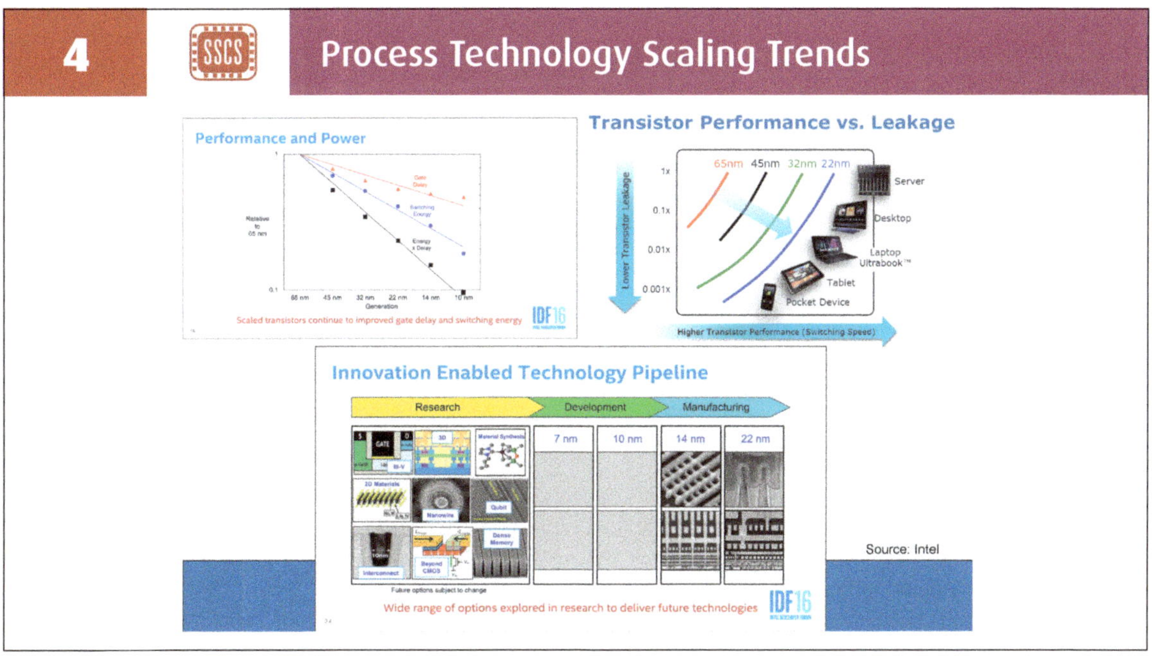

Conventional implementations of security accelerators achieve high throughput at the expense of large die-area and power dissipation, rendering them unsuitable for use in area/power constrained mobile and wearable systems. The energy and thermal constraints of such systems motivate lightweight low-cost hardware implementations of these security primitives with compact layout footprint and ultra-low leakage energy consumption.

5

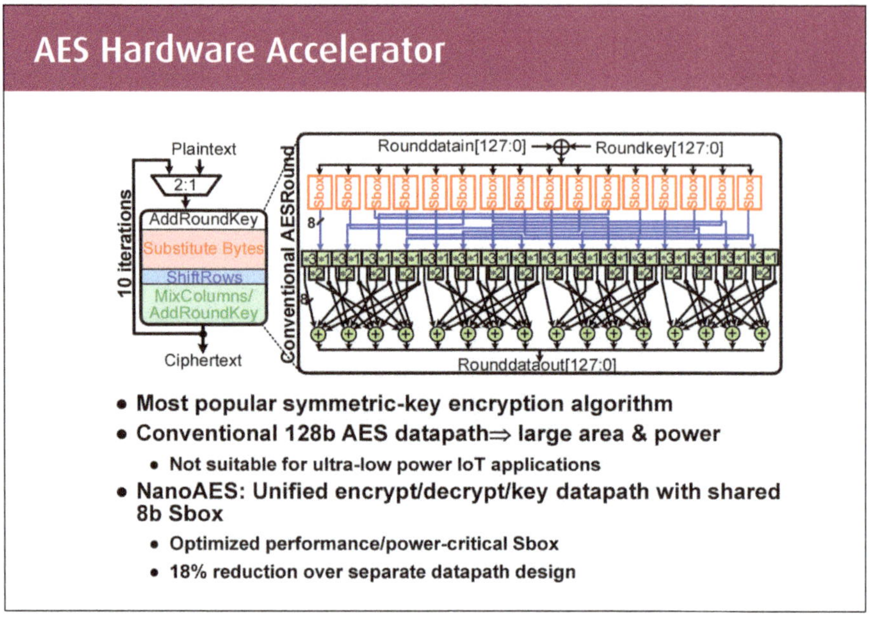

The advanced encryption standard (AES) is the most popular symmetric key block-cipher in use today, underpinning the security foundation of media content protection and memory encryption. AES uses a shared key to encrypt/decrypt information that is transmitted on an insecure channel. Data is iteratively processed for 10 rounds, 128bits at a time using a series of addition, permutation, substitution and mixing operations.

6

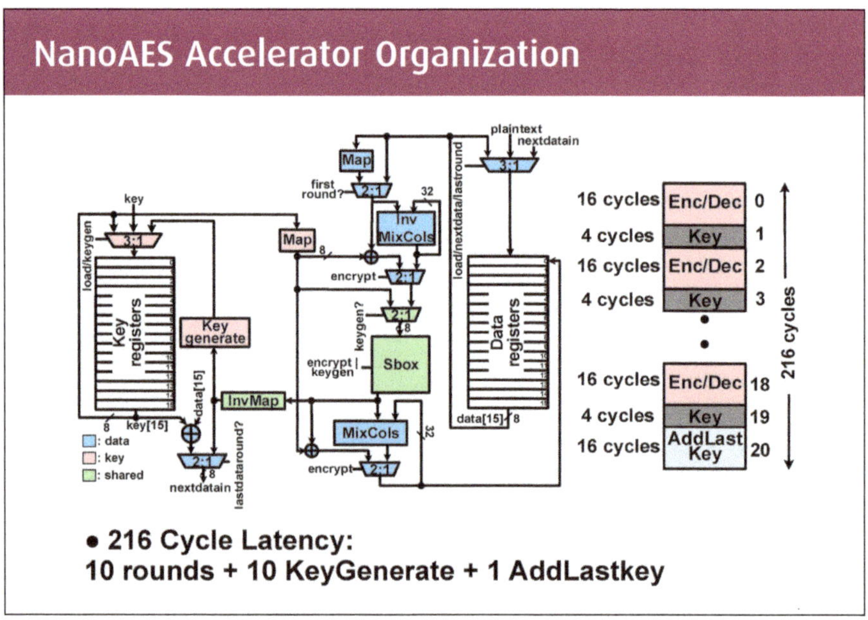

Conventional hardware accelerators use 128bit datapath that is organized as 16 byte-sized slices with 64 wiring tracks for shift-row permutation to achieve single cycle round latency at the cost of high area and energy consumption of parallel round units, rendering them unsuitable for IoT systems. In contrast to conventional 16 byte-slice datapath designs, the nanoAES accelerator is organized around a 1byte Sbox circuit that computes the performance critical Galois-field inversion along with affine and inverse affine transformations. Plain-text and key bytes are mapped from the standard-specified prime-field of $GF(2^8)$ to a composite-field of $GF(2^4)^2$ prior to first round iteration, allowing all downstream Sbox and Mixcolumns computations occur in the composite-field, thus amortizing the cost of field-transformation over 10 rounds.

7 SSCS AES Polynomial vs. Area Trade-off

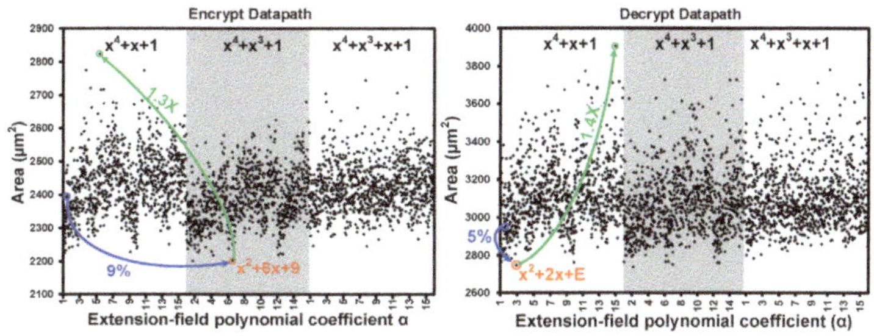

- **Choice of polynomial among 2880 potential combinations:**
  - **Impacts all GF(2^4) arithmetic blocks in Sbox**
  - **Impacts Map and Inv.Map operations**
  - **Impacts scaling factors in MixColumn operation**
- **Impact on NanoAES area:**
  - **30% spread between smallest and largest area design**
  - **9% area reduction over prior-art polynomials**

The circuit processes 1byte of plain/cipher text every cycle, generating 128bit output after 16 cycles that is stored in shift-row permuted format in a shift register. Following round processing, the accelerator switches to key generation mode that uses existing Sbox and other datapath logic to compute the key for the subsequent round. This organization enables reuse of the dominant combinational circuits in the accelerator, resulting in 18% area reduction over a separate datapath design.

8 SSCS NanoAES Accelerator Die Micrograph

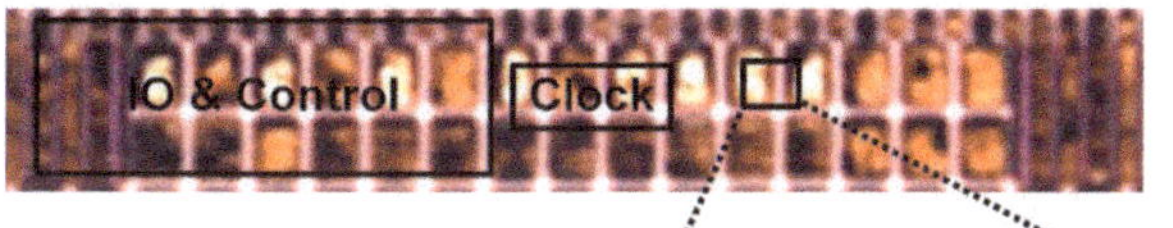

	Encrypt	Decrypt
Process	22nm tri-gate high-K metal gate CMOS	
Die area	0.19mm²	
Gate Count	1947	2090
Ground-field poly	x^4+x^3+1	x^4+x+1
Extension-field poly	x^2+6x+9	x^2+2x+E
Area (μm²)	2200	2736

Smallest AES accelerator reported-to-date

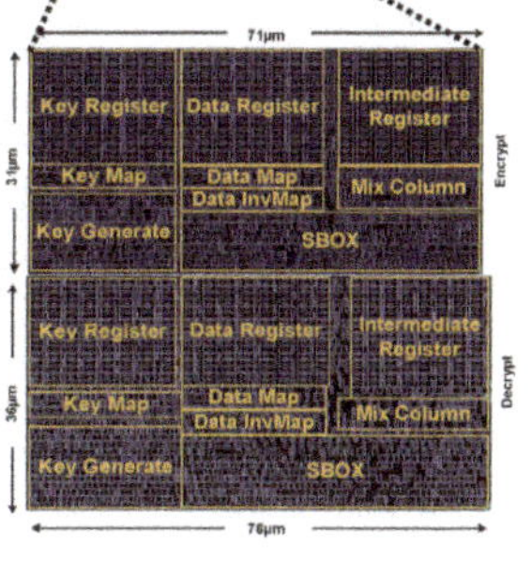

The performance critical Sbox circuit is optimized by avoiding direct computation of inverse in GF(2^8), that requires 8x8bit multiplication and 16bit reduction circuits. Instead, plain-text data and keys are mapped to a composite-field of GF$(2^4)^2$ using a mapping transformation in the first round. All subsequent iterations are computed in this composite-field where elementary operations like squaring, additions and inverse operations involve simpler 4-bit logic. The native Sbox design removes the mapping and inverse-mapping transformations from the performance critical AES round logic, this reducing critical-path delay by 12%. Although operations in the prime-field are defined by the Rjindael polynomial that is described in the AES standard, the arithmetic complexity of the composite-field accelerator datapath is determined by a pair of polynomials (ground-field and extension-field) selected at designer's discretion. These polynomials not only determine the mapping/inverse-mapping transforms but also the logic complexity of the Sbox, Mix-columns and scaling circuits. A framework to exhaustively evaluate the entire design space of all 2880 valid polynomial pairs was developed to further optimize the accelerator datapath.

9

22nm Performance/Power Measurements

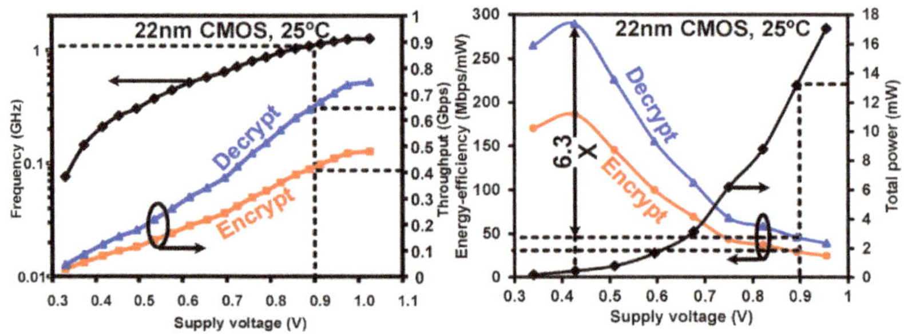

- **Performance:**
  - **1.1GHz, 432/671Mbps throughput at 0.9V, 25˚C**
  - **29-788Mbps throughput across 340mV-1.1V supply**
- **Power/Energy:**
  - **13mW total power consumption at nominal 0.9V**
  - **Peak energy efficiency of 289Gbps/W at 430mV**

Results of the polynomial-based area optimization using 22nm tri-gate high-k/metal-gate CMOS standard-cell library shows accelerator layout area grouped into three regions representing the 3 valid ground-field polynomials. Within each region, the extension-field coefficients are swept from 0x0 to 0xF, with a 1.3´ spread between the largest and smallest area polynomial-pair. The lowest area design occupying 2200µm² was obtained for a ground-field polynomial of x^4+x^3+1 and extension-field of x^2+6x+9, representing a 9% area reduction over previously-reported polynomial choices. Decrypt datapath optimization results show a 1.4´ spread in area, with 5% area reduction for the lowest-area design occupying 2736µm² obtained for a polynomial-pair of x4+x+1 and x2+2x+E. The arithmetically optimized composite-field Sbox circuit based AES accelerator with unified round compute and key expansion datapath was fabricated in a 22nm tri-gate high-K/metal-gate CMOS technology with encrypt/decrypt designs occupying 2200/2736µm2 (1940/2090 equivalent gate count) respectively. At nominal supply of 0.9V, the design operates at 1.133GHz resulting in 432/631Mbps AES-128 encrypt/decrypt throughput. Ultra-low voltage circuit optimizations enable reliable operation over a wide dynamic supply voltage down to 340mV, with peak energy-efficiency of 289Gpps/W measured at near-threshold supply of 430mV.

10

Full-Entropy TRNGs

True Random Number Generators are the bedrock of secure platforms, providing uniformly distributed full-entropy keys for data encryption, secure communications and media content protection. A full-entropy output implies a bitstream where each n-bit pattern has uniform occurrence probability approaching 2-n over a large infinite dataset. While thermal-noise based entropy sources produce bitstreams with high-levels of entropy, they may still contain non-uniformities and serial-correlations due to extreme process corners, intermittent supply or coupling noise or control loop artifacts that render them sub-optimal for direct use are full-entropy cryptographic keys. Conventional designs overcome such non-idealities by coupling these entropy sources with block-cipher based entropy extractors such as AES in CBC-MAC mode, where the raw bitstream is encrypted with an arbitrary key over multiple iterations. These iterations are repeated till 2n bits of min-entropy are accumulated at the input, at which point the output n-bit word is guaranteed to contain 'n' bits of entropy. This approach, while being effective, consumes large amounts of area and energy, making them unsuitable for use in energy-constrained IoT and wearable platforms.

10 SSCS Full-Entropy TRNGs

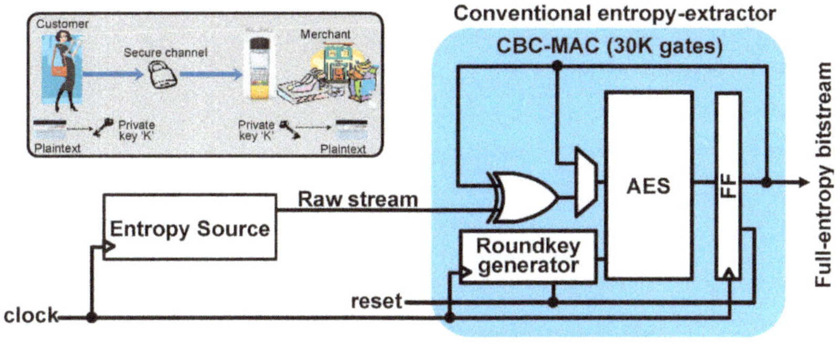

- Full-entropy keys are foundation of data security
- Conv. TRNGs: 1 entropy source with heavyweight extractor
  - Extracts n-bit key from raw 2n-bit min-entropy stream
- Large area and energy consumption ⇒ unsuitable for IoT

11 SSCS All-digital Entropy Source

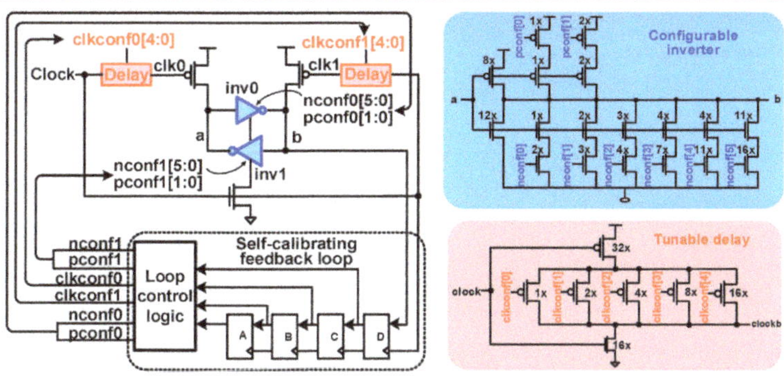

- Pre-charged cross-coupled inverter
  - clk=0: unstable state
  - clk=1: resolves based on magnitude of diff. thermal noise
- Systematic mismatches can disrupt random behavior
- Self-calibrating digital mechanisms to tune out mismatches

Three digital mechanisms are provided to tune out imbalances in the circuit. The first involves modulating the strengths of the cross-coupled inverter devices using digital configuration bits nconf0, pconf0, nconf1 and pconf1. The second tuning mechanism introduces a relative skew between the precharge clocks that control nodes a and b. The common clock signal is split into clock0 and clock1 that are generated using a pair of tunable delay buffers shown here. The absolute delays of these buffers are controlled using a pair of 5 digital configuration bits that can delay clk1 relative to clk0 or vice-versa. A delayed clock will hold the corresponding inverter diffusion node in precharge for a fraction of time longer than the complementary node, thereby increasing its bias to '1'. Finally a self-calibrating feedback loop examines a window of 4 consecutive output bits and updates one of these 6 configurations to move the overall system towards a state of higher entropy.

12

Full-Entropy µRNG Organization

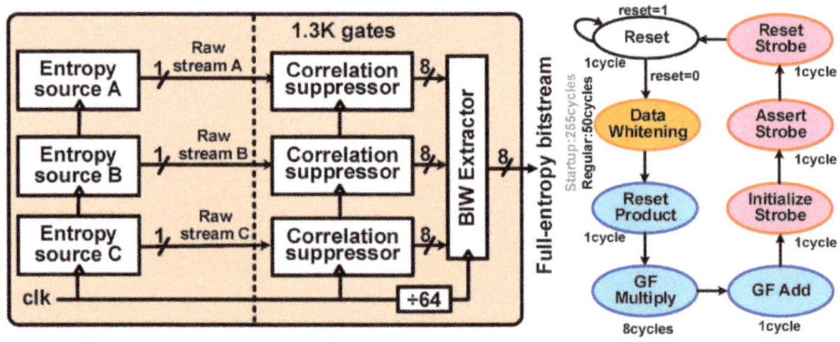

- **Multiple sources, each with uniform min-entropy ≥ δ.n**
  - **BIW circuit extracts n-bit entropy from 1/δ samples**
- **De-correlators ensure independence of sources**
- **Lightweight design requires a compact entropy source**

In contrast, the uRNG introduced here produces a full-entropy output from multiple parallel copies of the same entropy source. If each source produces min-entropy of delta-n, the BIW extractor uses simple Galois-field arithmetic operations to produce a full-entropy extracted output using 1-over-delta sources. In this design, we use 3 entropy sources A, B and C, with each source containing at least 0.33n bits of entropy. The BIW extractor also requires that input streams be serially and mutually uncorrelated. This is achieved by inserting 3 correlation suppressors that perform data whitening of the raw bitstreams. This approach of using multiple entropy sources coupled with a lightweight entropy extractor results in a more compact energy efficient full-entropy TRNG, resulting[13] in over 30x reduction in total gate count and 60% lower energy consumption compared to conventional SHA or AES based entropy extractors. However, the use of multiple entropy sources motivates the need for a compact implementation that is tolerant to process-voltage-temperature variations typically encountered in high-volume-manufacturing scenarios.

13

Die Micrograph & Entropy Results

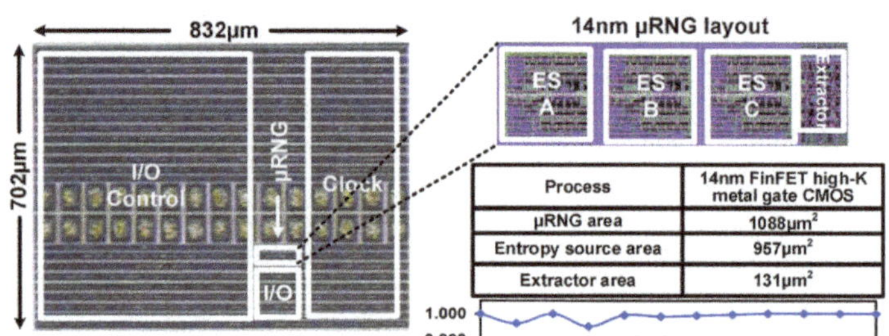

Process	14nm FinFET high-K metal gate CMOS
µRNG area	1088µm²
Entropy source area	957µm²
Extractor area	131µm²

- **Shannon entropy up to 0.99999999995**
- **Passes all NIST tests with p-value>0.09 (9x higher than threshold)**

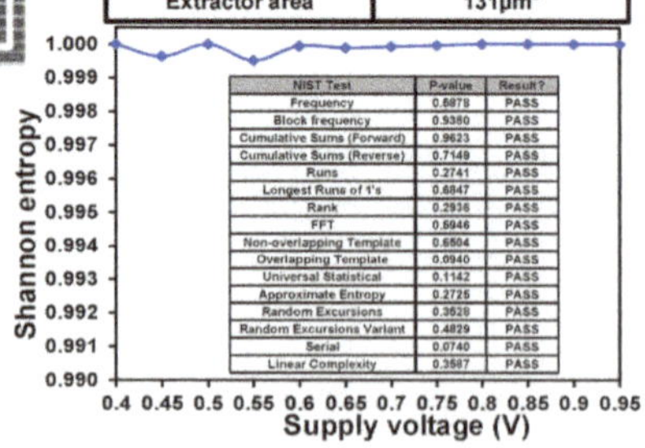

NIST Test	P-value	Result?
Frequency	0.5878	PASS
Block frequency	0.9380	PASS
Cumulative Sums (Forward)	0.9623	PASS
Cumulative Sums (Reverse)	0.7149	PASS
Runs	0.2741	PASS
Longest Runs of 1's	0.6847	PASS
Rank	0.2936	PASS
FFT	0.5946	PASS
Non-overlapping Template	0.6504	PASS
Overlapping Template	0.0940	PASS
Universal Statistical	0.1142	PASS
Approximate Entropy	0.2725	PASS
Random Excursions	0.3528	PASS
Random Excursions Variant	0.4829	PASS
Serial	0.0740	PASS
Linear Complexity	0.3587	PASS

The all-digital full-entropy uRNG is fabricated in a 14nm high-k metal-gate finfet CMOS technology, with the figure here showing the die-micrograph and layout details. Total die area of the design is 0.58mm² with the uRNG occupying 1088um². 88% of the layout area is occupied by the 3 entropy sources with the remaining 12% area occupied by the lightweight BIW extractor.

14 Frequency & Power Measurements

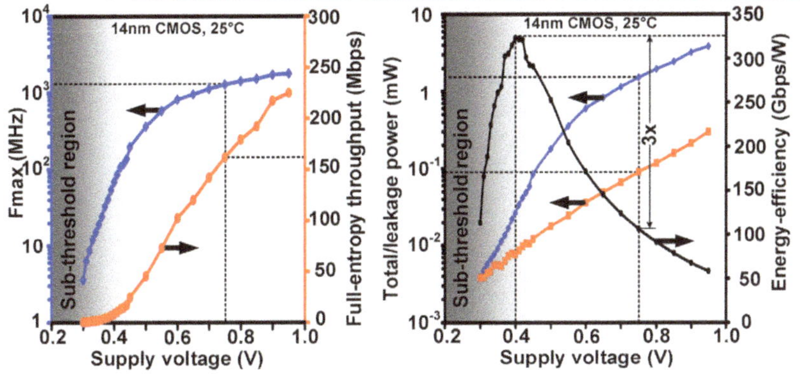

- **1.3GHz, 163Mbps throughput at 0.75V, 25°C**
- **0.4-225Mbps throughput, 4μW-4mW total power across 300-950mV**
- **323Gbps/W peak energy-efficiency at 400mV(3x higher than nom.)**

The graphs here show frequency, throughput and power measurements of the uRNG obtained by sweeping the supply voltage in a temperature-stabilized environment of 25C. At the nominal supply voltage of 0.75V, the uRNG operates at a maximum frequency of 1.3GHz, resulting in full-entropy throughput of 163Mbps. Scalability of the all-digital approach is demonstrated over a wide operating range with throughput scaling from 225Mbps at 0.95V, down to 400Kbps measured at 300mV. Total power consumption of the uRNG at nominal voltage of 0.75V is 1.5mW, with leakage component of 90uW. Total power scales down to 4uW at 300mV. Energy-efficiency measurements show that efficiency increases as supply voltage is scaled down as total power reduces at a faster rate than performance degradation. Energy-efficiency peaks at 323Gbps/W at near-threshold voltage of 400mV, with full-entropy throughput of 8.6Mbps, total power consumption of 27uW and leakage power of 7uW. At the peak-efficiency point, the uRNG has a total energy consumption of 3pJ to produce 1 full-entropy bit.

15 Physically Unclonable Function

Paradigm shift in physical security

- **Secure generation of device-specific keys/IDs**
  - **Generate ID by exploiting random physical phenomenon**
- **Raises security-bar vs. fuses**
  - **Created without manual intervention**
  - **ID not stored physically on-die**
  - **Unclonable: Immune to chip-decapsulation & visual inspection**

15 Physically Unclonable Function

Physically Unclonable Functions are low-cost cryptographic primitives used for generation of a stable, repeatable device-specific secure key. They represent a paradigm shift in physical security by moving from explicitly programmed digital IDs to an ID that is generated by exploiting some intrinsic characteristic of devices on the die. The volatile nature of PUFs provide a high-level of security and tamper resistance against invasive probing attacks compared to conventional one-time programmable fuses. They also raise the security bar vs. fuses since they are created without any manual intervention and hence avoid the insider attack problem with fuses. Furthermore, unlike fuses, they are immune to visual inspection after chip decapsulation.

16 Hybrid PUF Circuit

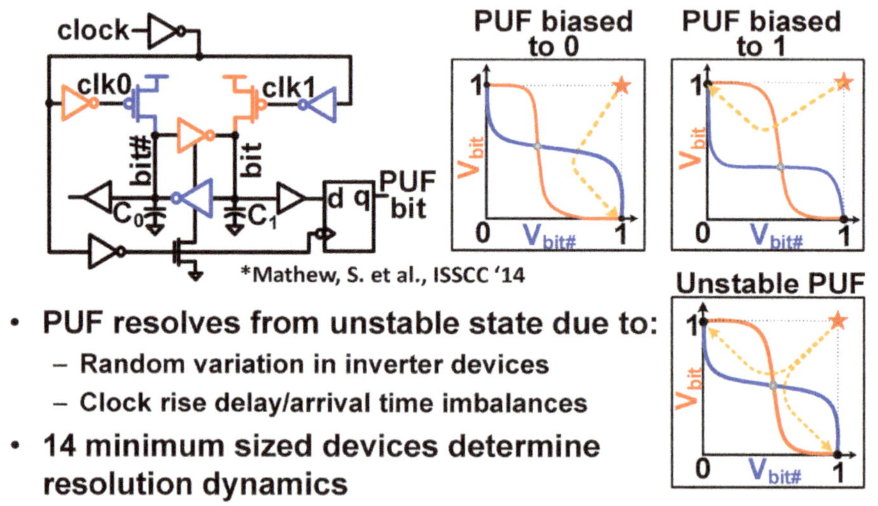

The cells used in the PUF array are based on the hybrid PUF circuit shown here. It consists of a cross-coupled inverter pair that is pushed into an unstable state by pre-charging nodes bit and bit# to Vcc using a pair of pre-charge transistors. When the clock goes high during evaluation, the circuit snaps to one of 2 stable states, represented by the butterfly curves shown on the top right. Which direction it goes to is determined by the relative strengths of variation-impacted minimum size inverter devices. Additionally, random variations in precharge and clock delay devices produce mismatches in clock rise and arrival times that introduce a transient dimension of uncertainty into PUF resolution dynamics. This hybrid approach combines the stability and compactness of SRAM-based PUFs, with the resistance to invasive probing attacks that we get with delay-based PUFs. PUF cells that have insufficient net random variation generate unstable bits that resolve to 0 or 1 based on thermal noise or voltage and temperature conditions.

17 PUF Key Generator Organization

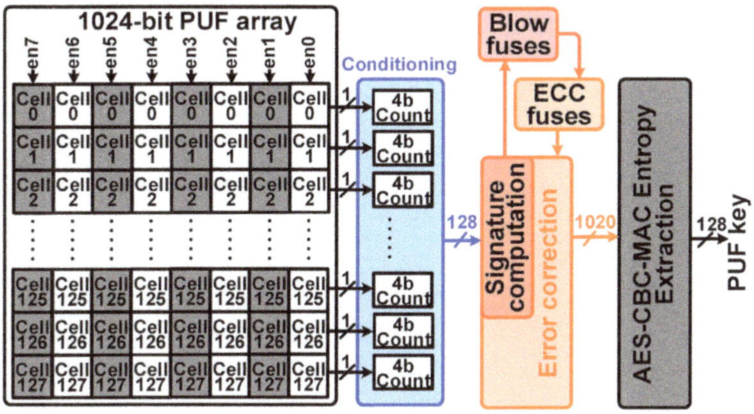

- Goal: Generate die-specific full-entropy 128b key
- PUF bits stabilized by conditioning, error correction

In this presentation, we will look at how a stable 128-bit PUF key is generated from an 1024-bit array of PUF cells some of which may be inherently unstable in their raw state. In the first evaluation of this array during tester-time operation, a golden key is generated from the 1st PUF value after some conditioning. An ECC signature computed from this golden key is then blown into fuses on the die. During every subsequent evaluation in regular operation, the raw PUF value, combined with the stored ECC signature is used to recreate the golden key with 100% accuracy. This golden key then goes through an entropy extraction function like AES-CBC-MAC that finally produces a full-entropy 128-bit PUF key. To minimize ECC overheads, we require a PUF circuit that produces an output bit with a high level of stability.

18 PUF Conditioning Circuits

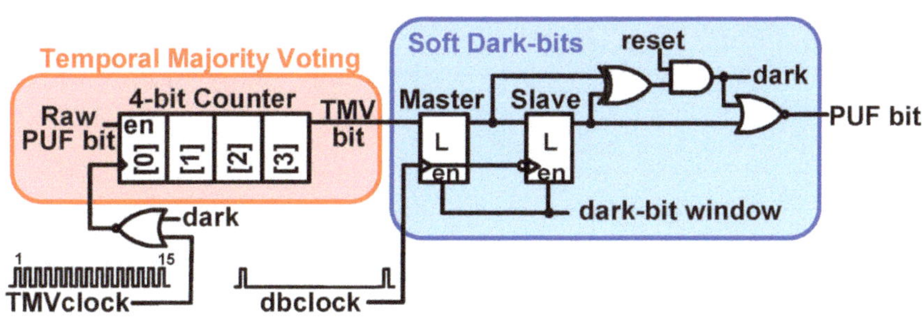

- **Temporal Majority Voting (TMV) stabilizes mildly unstable bits**
- **Dark-bit mask: detect, suppress highly unstable bits at runtime**
- **Volatile soft mask generated by comparing master/slave latches during 100 cycle dark-bit window**

18 PUF Conditioning Circuits

As described earlier, the raw PUF values from the array may contain unstable bits that produce noisy responses. Two types of conditioning circuits are used to stabilize the noisy bits to reduce overall bit-error-rate. The first conditioning circuit, called the temporal majority voter computes the quantized mean of PUF responses within a voting window. This is implemented using a 4bit counter that keeps tracks of the number of times, the cell evaluates to '1' during a voting window of 15 cycles. Stable cells that result in count values above 8 are considered to be 1's and those with counts 7 or below are counted as 0's. Mildly unstable bits with error probabilities <8% use TMV-15 to reduce their effect error rate to 10-6. Cells with counts around the threshold of 6-9 represent highly unstable bits that are unaffected by TMV. These bits as identified as those that change value at least once within a darkbit window of 100 cycles. The change in value is detected by comparing master and slave latch values at the end of each TMV cycle. The presence of complementary values in these latches indicates a bit that was unstable even after TMV. This bit is marked as a darkbit and masked off from the final PUF key. Note that the darkbit mask is a volatile softmask that is not explicitly stored anywhere on die and has to be recreated at each power-up.

19 Die Micrograph and 14nm CMOS Layout

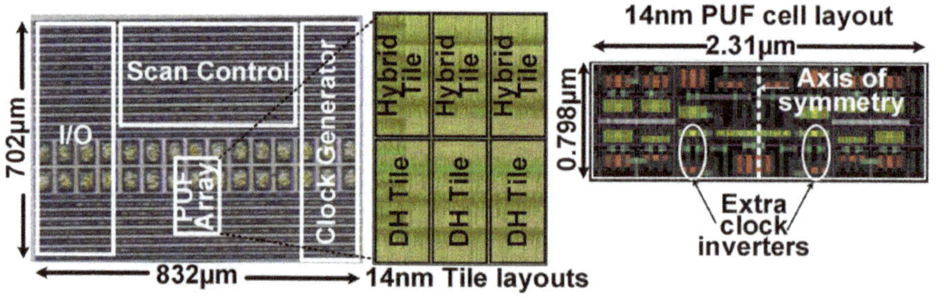

Process	14nm tri-gate CMOS
Die area	$0.58mm^2$
PUF cell area	$1.84µm^2$
PUF array organization	64x16

To characterize the behavior of the PUF cells under various operating conditions, we fabricated 3 tiles each using the basic hybrid PUF array as well as the delay hardened cell in a 14nm trigate CMOS technology. Each tile contains 1024b to satisfy the minimum entropy requirements to generate a 128b full-entropy PUF key. The 14nm PUF cell layout occupies 1.84um2 and uses a common centroid layout with the axis of symmetry going through the center of the cell. The extra clock buffers of the delay hardened cell are placed on either side of the cross-coupled inverter in a mirrored layout topology.

20 Summary

- **Reliability of IoT platforms depends on robust security features.**
- **NanoAES: Ultra-lightweight 2090-gate composite-field AES accelerator in 22nm CMOS**
  - **Wide operating voltage range of 340mV-1.1V**
  - **Peak energy efficiency of 289Gbps/W at 430mV**
- **µRNG: All-digital full-entropy RNG with BIW extractor.**
  - **Throughput of 163Mbps at 1.5mW total power**
  - **Ultra-low energy consumption of 3pJ/bit**
- **Physically Unclonable Function: Hybrid PUF cell**
  - **4fJ/bit energy consumption, BER < 1.3%**
  - **Immune to power-up probing attacks**

CHAPTER 15

System-on-Chip Security Assurance for IoT Devices: Co-operations and Conflicts

Sandip Ray

University of Florida at Gainesville, USA

Security is a critical component for computing devices targeted towards Internet-of-Things applications. Unfortunately, IoT security assurance is a challenging enterprise, involving cooperation and conflicts among a variety of stake-holders working in concert with a variety of architecture and design collateral generated across various points in a complex design life-cycle. Furthermore, the long life and aggressive energy/performance needs of IoT applications bring in new challenges to security designs. In this paper we discuss some of the challenges, the current industrial practice to address them, some gaping holes in the state of the practice, and potential research directions to address them.

1 The Internet of Things Regime

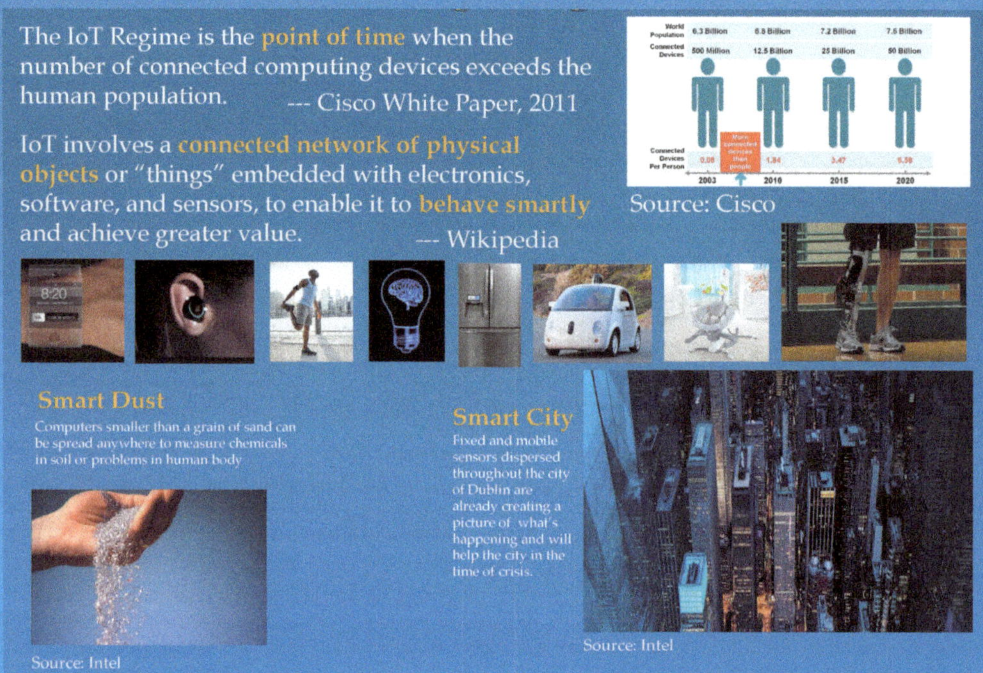

The Internet-of-Things (IoT) has become a buzzword in recent years. Unfortunately, the topic is not very well-defined. Indeed, depending on the organization and its research, technical, or business goals, the definitions might be very different. So it is important in any work on Internet-of-Things to explain at the outset the scope of the problem being considered. One definition that has received some traction at least in various industrial contexts is from Cisco, who propounded it in a white paper in 2011. We will use this notion for the rest of the document. The idea is to view IoT not as a collection of devices, or their networks, or the physical objects they are associated with, or the cloud that performs data analysis. Instead, we will think of IoT as a "regime". In particular, the IoT regime denotes that point of time from which the number of connected electronic and computing devices exceeded the human population. By that measure, the IoT regime started in middle of 2008, and we are deep in that regime. This regime has been characterized by an unprecedented growth in the number of electronic devices, perhaps the largest growth among any sector in any point of the human civilization. To understand what these devices are doing, we turn to a more operational definition from Wikipedia. The devices are attached to physical objects and they are connected to the Internet, to make the physical devices "smart". The term "smart" here is operative: we have everything smart including light-bulbs, watches, cars, ear-buds, even prosthetics and implants. If we look at the application range, then that is about as diverse as can be potentially imaginable. On the one hand, we have computing devices of the size of a grain of sand, forming what are called "smart dust". On the other hand, we have applications of the scale of a smart city.

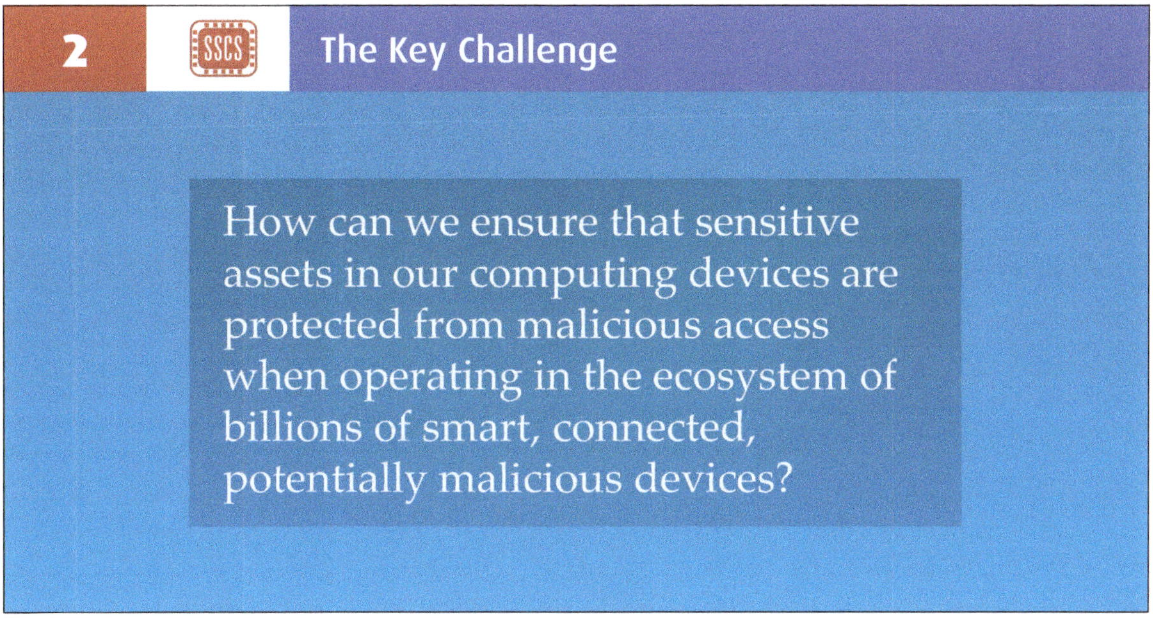

The key challenge now is to determine how we can develop technologies to ensure that all these devices, operating in the environment in which there are billions of other devices (many potentially malicious) executing, act reliably, in a trustworthy and secure manner.

The basic story of security here is of course pretty obvious. We are generating too much data, much of which is private, sensitive, and in need of protection. We are also communicating this massive amount of sensitive data. So, we are increasing exposure to threats of malicious access. The problem is often cheekily called "Beckstrom's Law of Cybersecurity. It says, "Everything connected to the Internet can be hacked" and now "Everything is connected to the Internet", so by elementary deduction law, we can get "Everything can be hacked".

SoC SECURITY vs TRADITIONAL COMPUTER SECURITY

Before discussing SoC security, it makes sense to first discuss the difference between these and traditional computer security. In particular, computer security has been with us for a long time. So what changed? And what changed in the SoC ecosystem.

The first key issue is changing computing landscape. Till about early 2000, it was easy to categorize a computing system as either (1) general-purpose system, i.e., desktop, laptop, server, or (2) an embedded system, e.g., data organizer, iPod etc. The first category had versatility and programmability (as well as connectivity), but with a demarcation between hardware and software components at least for most part. The second category had unique use-case constraints, tight hardware/software integration, and customized designs, but for very narrow use cases. This demarcation started to blur with the advent of smartphones, tablets, and later smart wearables. These are still embedded systems in that they still have to satisfy the embedded system constraints. But they are also versatile, programmable, and as complex as general-purpose systems. Even the definition of embedded systems has changed significantly from two decades. In particular, embedded systems are now encompassing everything other than desktop, laptop, or server.

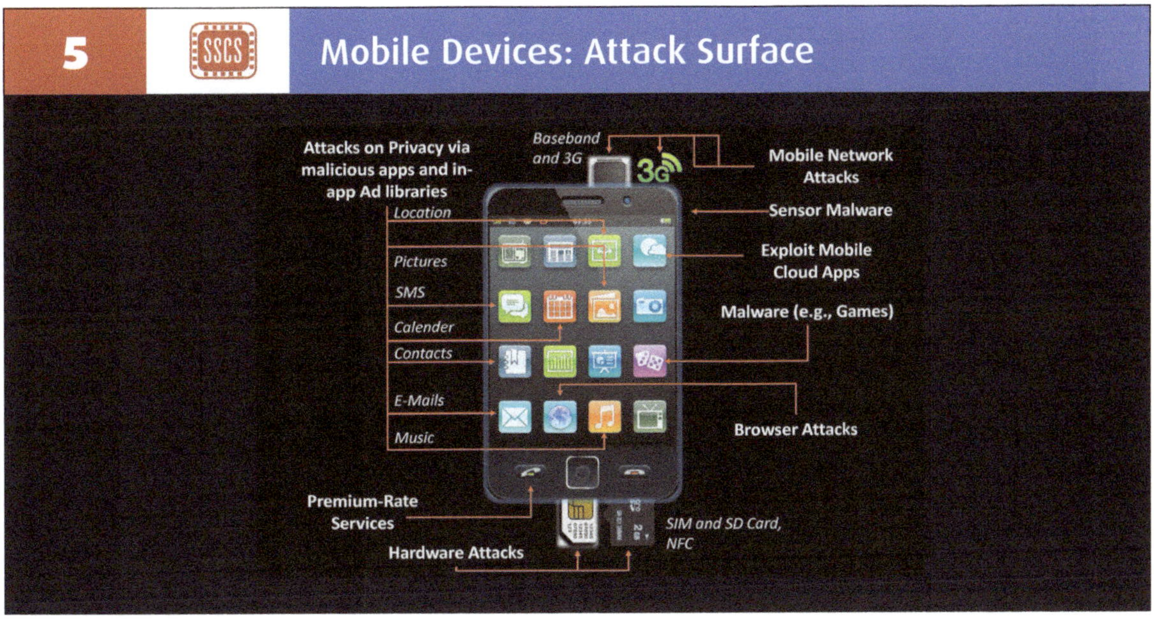

It is instructive to look at how many ways a smartphone can be attacked, just to understand the complexity. Almost any component of the device can be targeted, including apps, premium rate services, network, hardware, and sensors. With an attack surface this broad it is almost surprising to find one that has not been hacked yet.

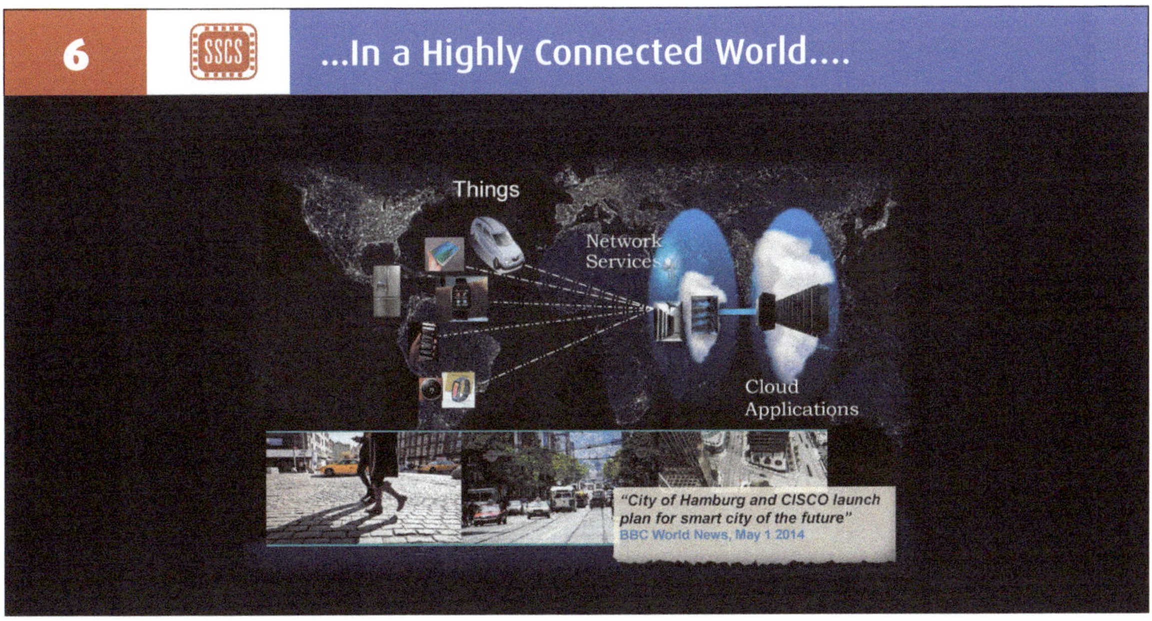

Furthermore, they are all connected and produce complicated applications like a smart city. However, from security standpoint this means that an attack on one can be used to infect others. If there are billions of devices, there are possibly millions that are malicious. So every device must work in an ecosystem in which there are millions of malicious devices trying to attack them.

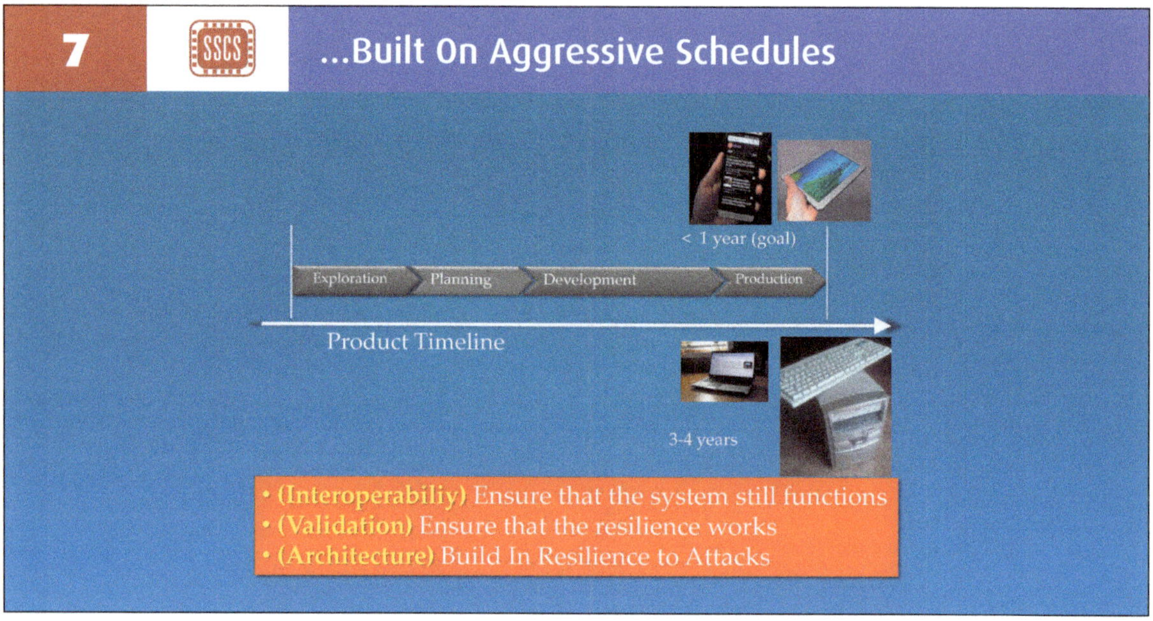

Finally, time is not on our side. The traditional life cycle from exploration to production used to be 3-4 years for desktops and laptops. Now for mobile and wearable devices they are projected to be less than a year. This means very little time for designing resilience and validating that designs indeed are resilient. To address this challenge, we must consider the three parts, architecture, validation, and interoperability together, exploiting their synergies.

ELEMENTS OF SoC SECURITY

We now look at SoC platform security and what it is about.

8

What is platform security assurance? It includes reviewing the entire platform (which includes software, firmware, hardware, potentially services, as well as the board) for potential exploits (which includes investigation of all potential attacker motivations as well as entry points), that undermine security objectives (which can come from feature, operating system, architecture, or even the user's expectation), in ways not covered by other validation (to minimize duplication of efforts) requiring the security designer and validator to understand other validation methodologies and gaps in them that can affect security.

ASSETS AND POLICIES

Let's dive a little deeper into what needs protection, i.e., the assets, and the protection requirements, i.e., policies.

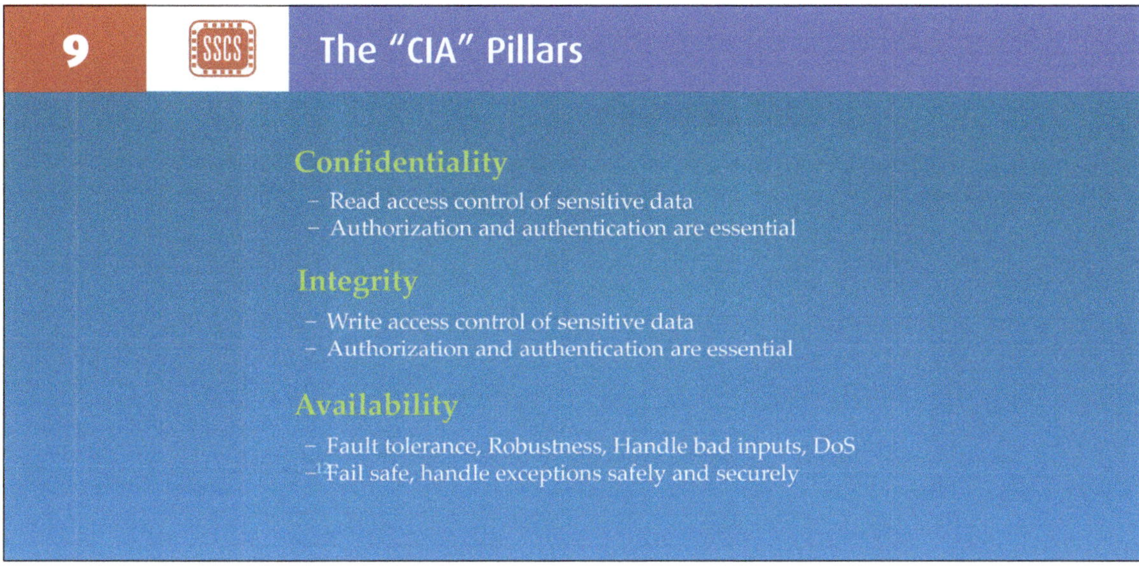

Traditionally, security has been defined by three components, confidentiality, integrity, and availability.

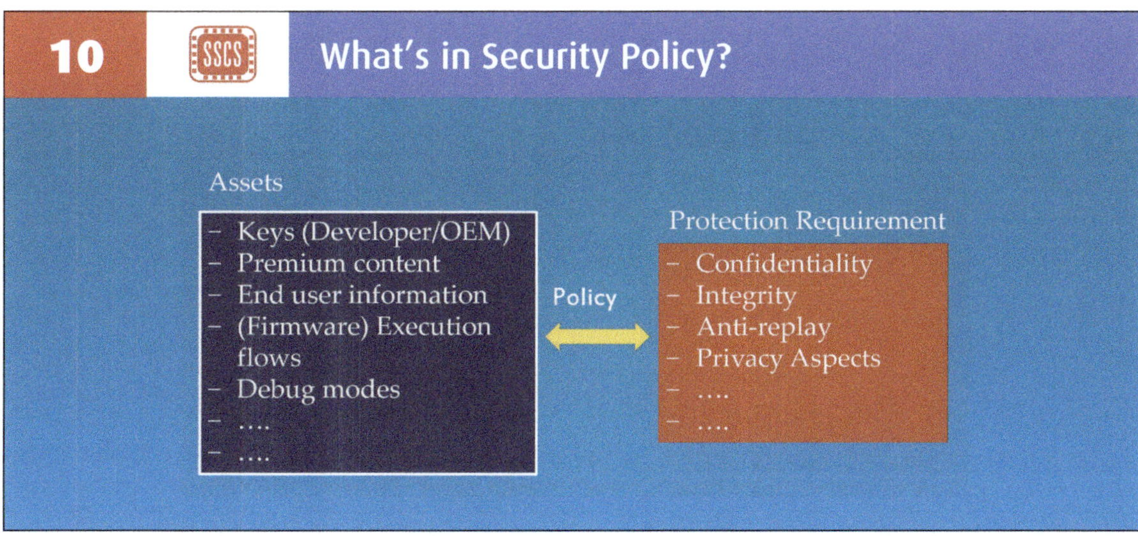

However, for each asset, it's not generally clear what kind of protection will be necessary just from the CIA requirements. This is addressed by security policies. For each asset in the system, a policy specifies what protection the asset must have. Note policies do not specify how to *implement* the protection, merely what the protection should be.

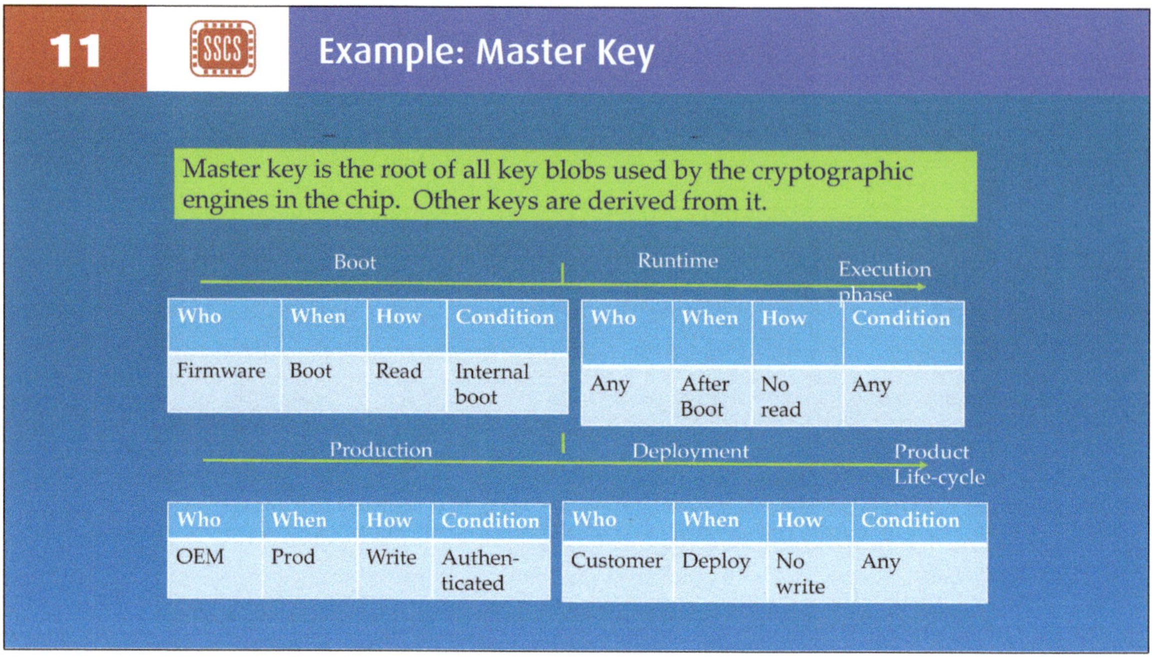

Let us look at a simplified example of security policy to illustrate the complexity of security policies

The asset we look at is the master key of the SoC. The master key can be used to derive a set of keys to be used for crypto engines on the chip. The master key can be read by the authenticated firmware during boot time in an internal boot. It cannot be accessed through other boot modes and should be blocked from access after boot. This master key can only be programmed by OEM during the production of the system when the OEM pass the authentication. The customer of the OEM cannot update the master key under any circumstances during the product deployment.

We can see that the same asset might or might not be accessible depending on the execution phase and chip life cycle. And this example is a largely simplified version of the security policy.

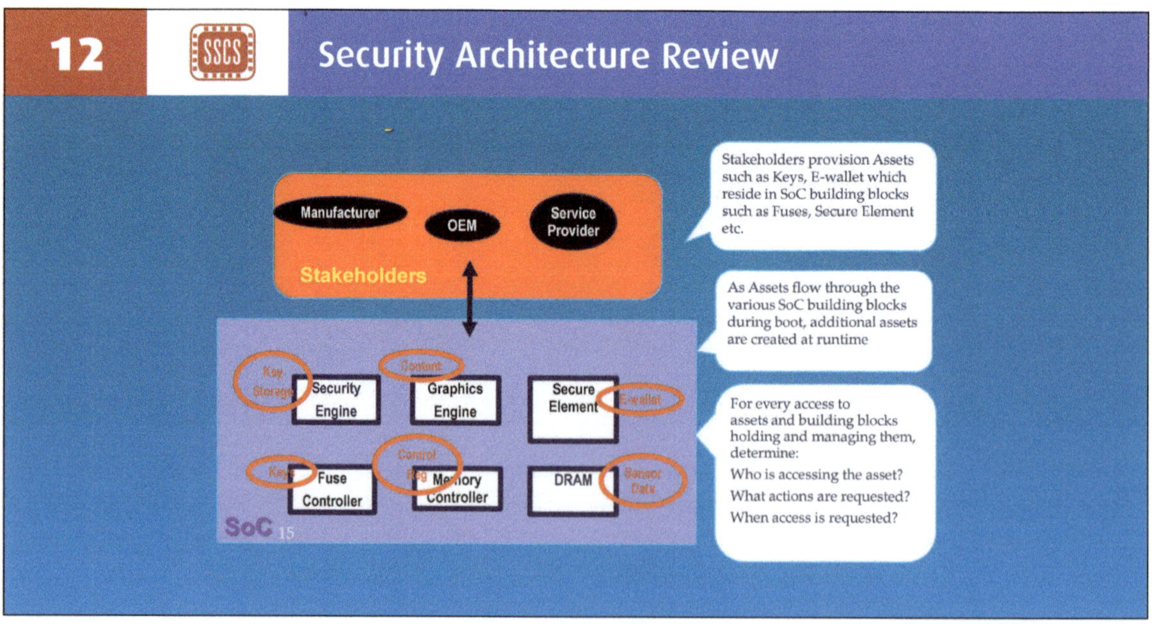

12 Security Architecture Review

How can we design an architecture that implements security policies? This is unfortunately pretty manual. Note that assets are provisioned by numerous stake-holders. Also, some are inserted statically (e.g., keys in a fuse controller) and others created during system execution (e.g., key storage in security engine). So the architect must consider for every potential access request at every stage of execution how to handle it.

13 Industrial Primitives

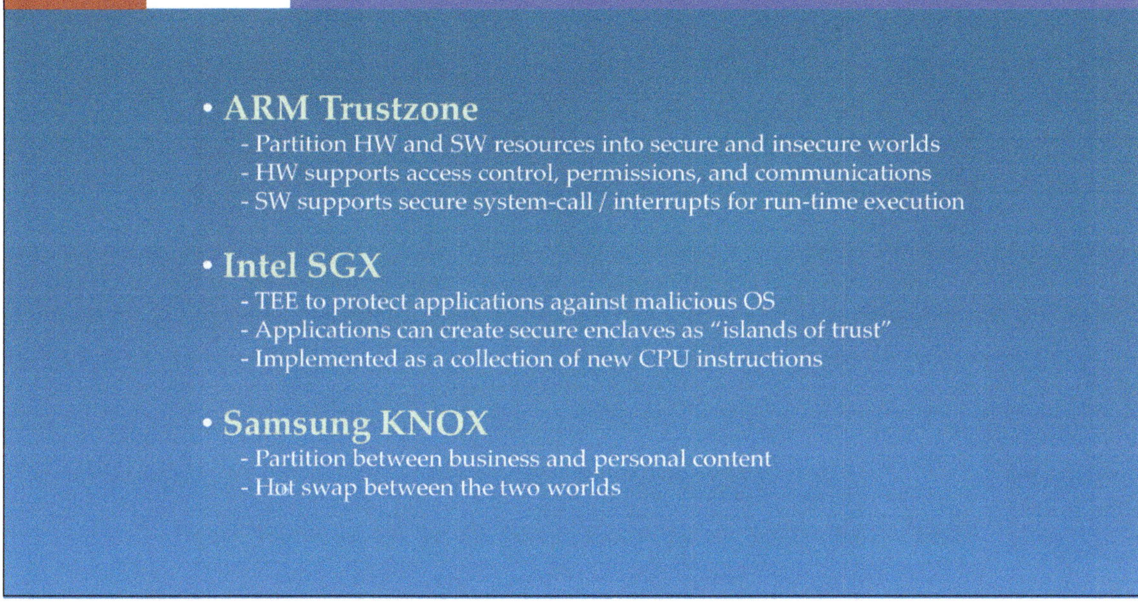

That is not to mean that there is no automation or architectural support. Almost each company has, for example, a trusted execution environment (TEE) that provides some isolation of various system computation. But of course that is only a primitive. It is up to the architect to use these TEEs effectively to implement security policies.

14 Enforcement Requirements

Some of the requirements to enable smooth policy implementations are shown here. They involve techniques for specification, architecture, and validation of policies. None of these exist in industrially usable form at this point.

14

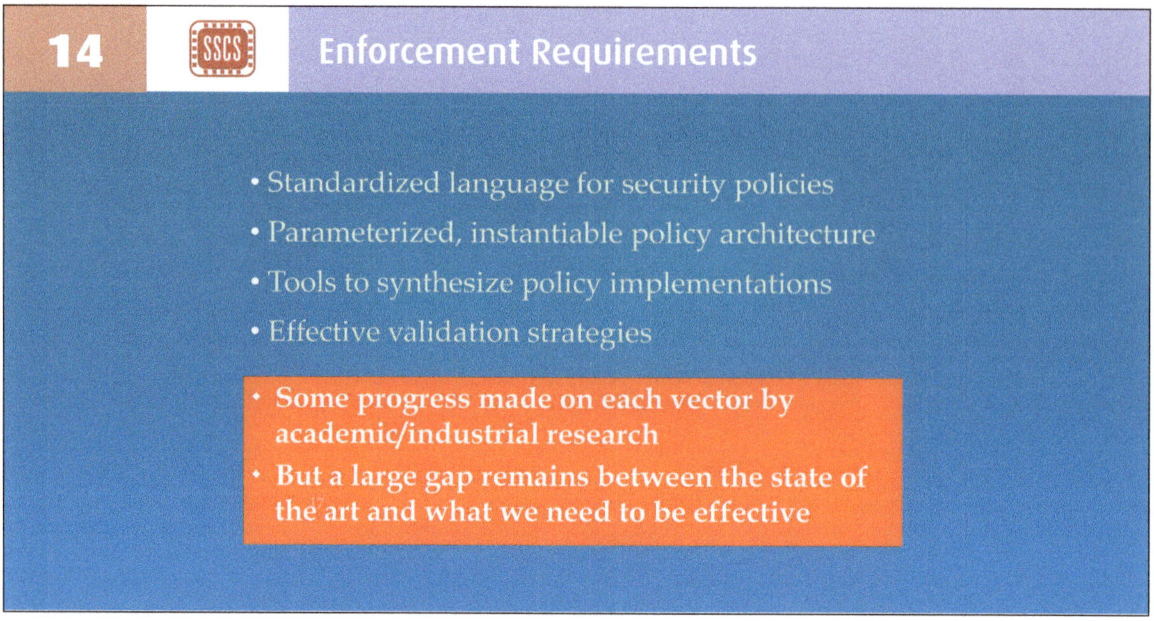

SECURITY VALIDATION

We will now consider security planning and validation activities.

15

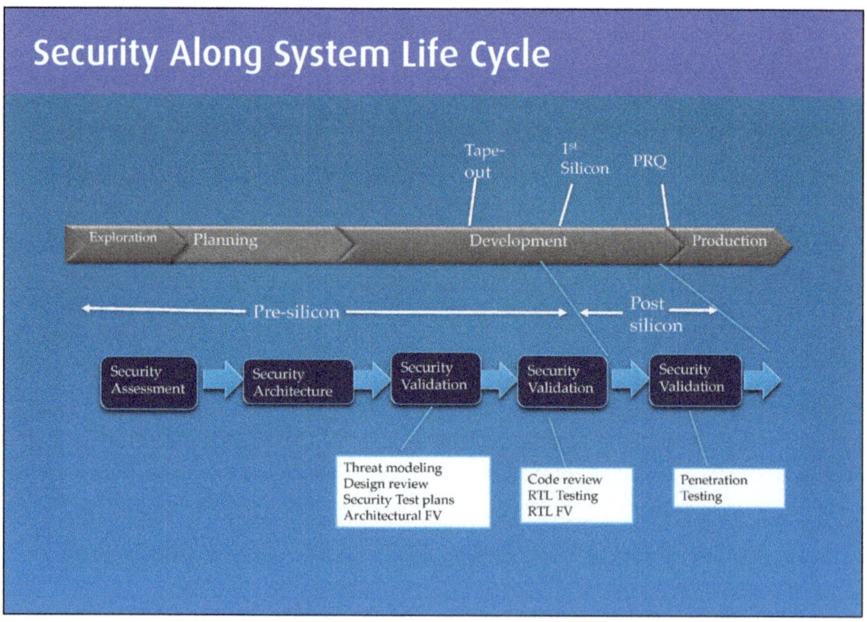

We show the different stages of a system development from exploration to production. At some point the design is ready to go to fabrication, a process known as tape-out. After that a pre-production silicon implementation is produced, which is used for validation. Finally, there is a process called "PRQ" through which the decision is taken whether the design would go to production (or be cancelled). Security activities pervade across the life-cycle, and in particular security validation goes across from pre-silicon to post-silicon.

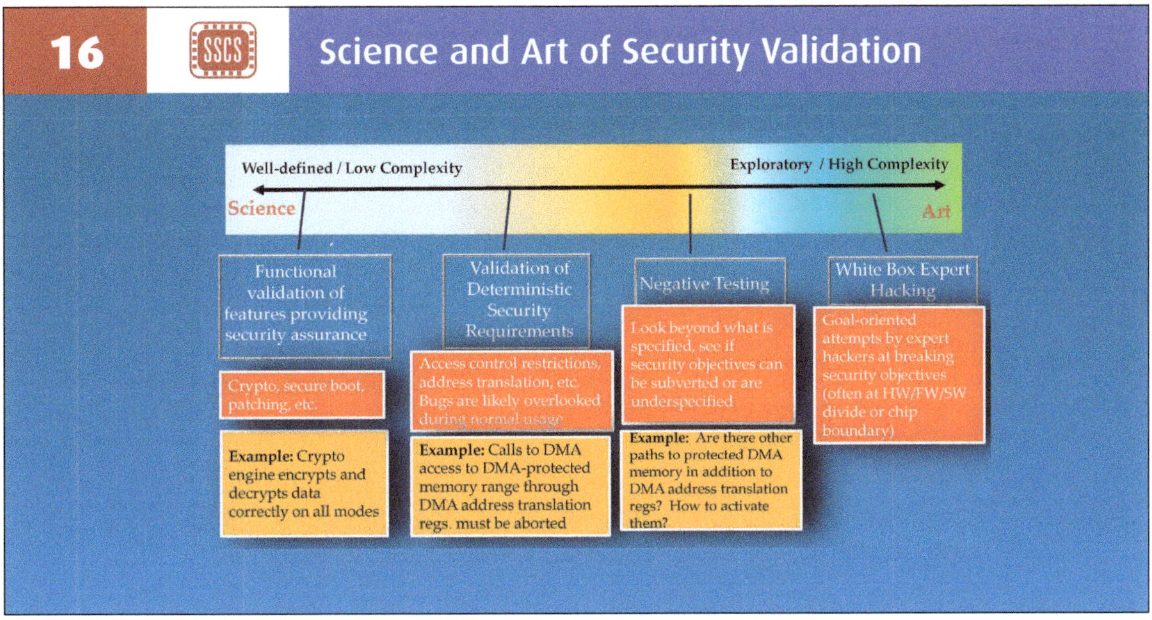

We discuss the diversity of things that need to fall under the overall umbrella of security validation. They include some activities that are essentially extension of functional validation (e.g., more thorough validation of crypto modules), but then can move quickly to art and even black magic where the human finds ways to hack into a system to expose vulnerabilities.

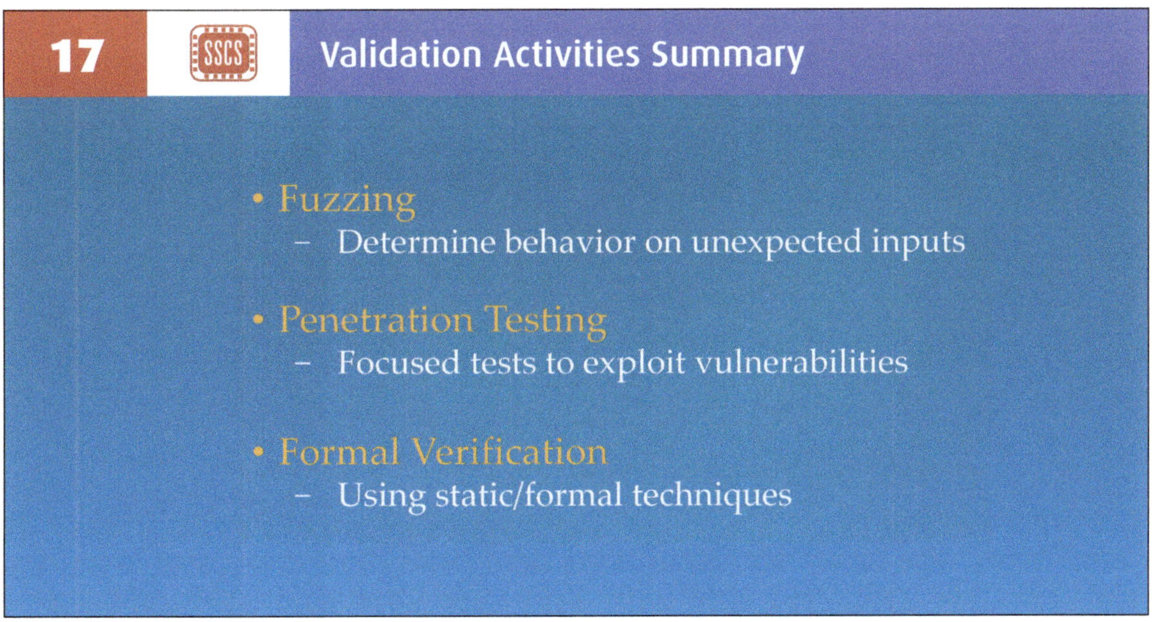

We can divide validation activities into three categories:

- **Fuzzing**, i.e, determining behavior of the system on unexpected inputs.
- **Penetration testing**, i.e., focused tests to exploit vulnerabilities
- **Formal verification**, i.e., the use of mathematical methods to prove the design free of vulnerabilities (rather than the use of tests).

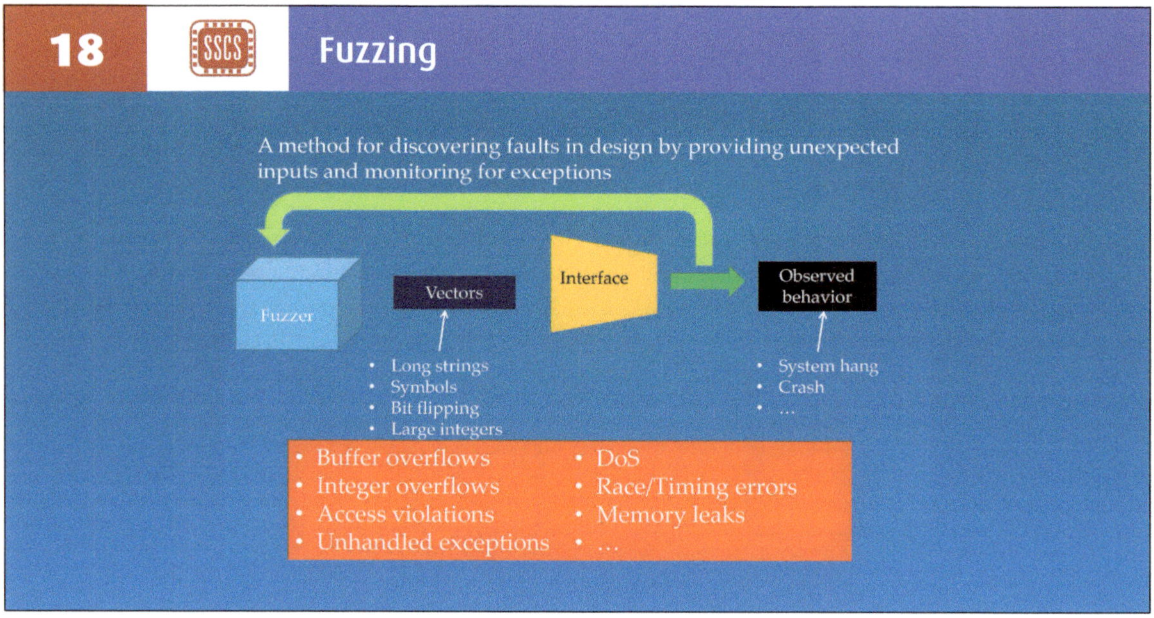

Fuzzing is conceptually very simple. Instead of giving the design expected inputs, one provides unexpected ones, often random inputs or random mutation of valid inputs. In spite of simplicity, this method is surprisingly effective. One reason for that is that it can be automated and it is inexpensive to apply. It is effective for a number of different security violations, e.g., overflows, races, unhandled exceptions etc. But it cannot find "deep" vulnerabilities that requires a very carefully constructed input stimulus to exhibit.

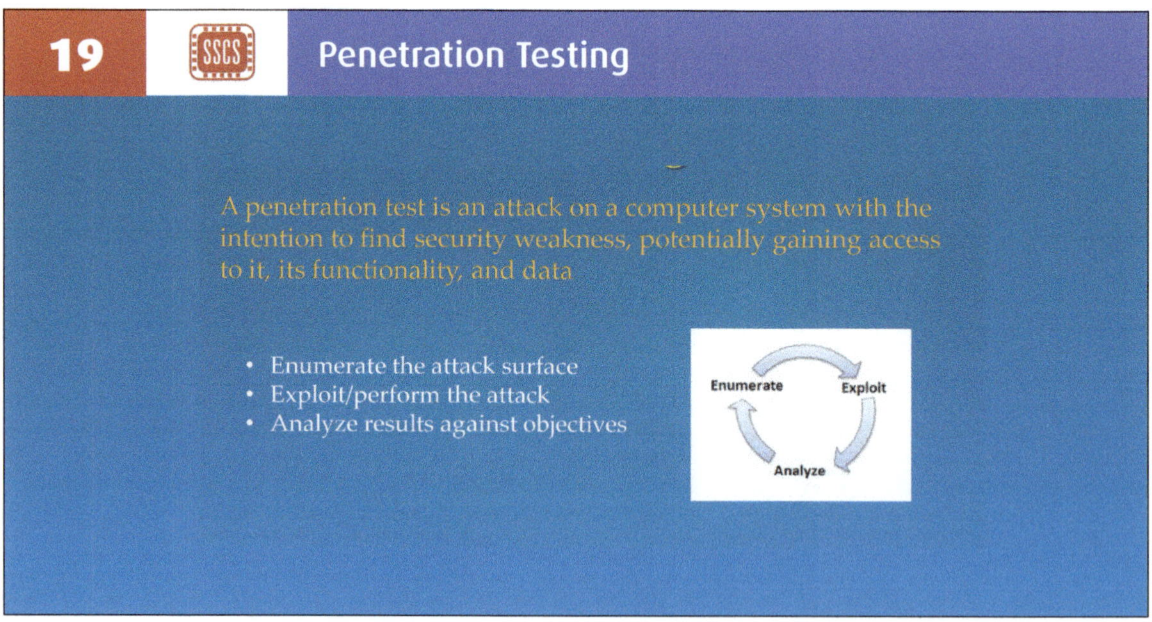

19 Penetration Testing

Penetration testing is the other extreme of security validation, where the human finds a way to hack the system. It includes three phases.

Phase 1. Enumerate attack surface
Referred to as fingerprinting, scanning, discovery, or identification, the overall goal of the enumeration phase is to figure out what aspects of the target are accessible/available for attacking. Activities in this phase typically include network port/service scanning, OS and technology identification, documentation review, and fuzzing/brute forcing to discover entry points.

Phase 2. Exploit/perform an attack
Once what is accessible to attack is determined, one can attempt applicable attacks and exploits against select target areas. This may require research into known vulnerabilities, looking up applicable vulnerability class attacks, engaging in vulnerability research specific to the target, and writing/creating the necessary exploits.

Phase 3. Analyze your results against objectives
If the attack was successful, then the resulting state of the target should be compared to the security objectives to see if an objective was compromised. If an objective was indeed compromised, then one has identified and demonstrated a security vulnerability and should perform the necessary steps to handling an identified security bug. Even if a security objective wasn't directly compromised, it is possible an attack could create additional attack surface. So we go back to phase 1, analyze the newly acquired attack surface, and continue to penetrate deeper into the system.

20

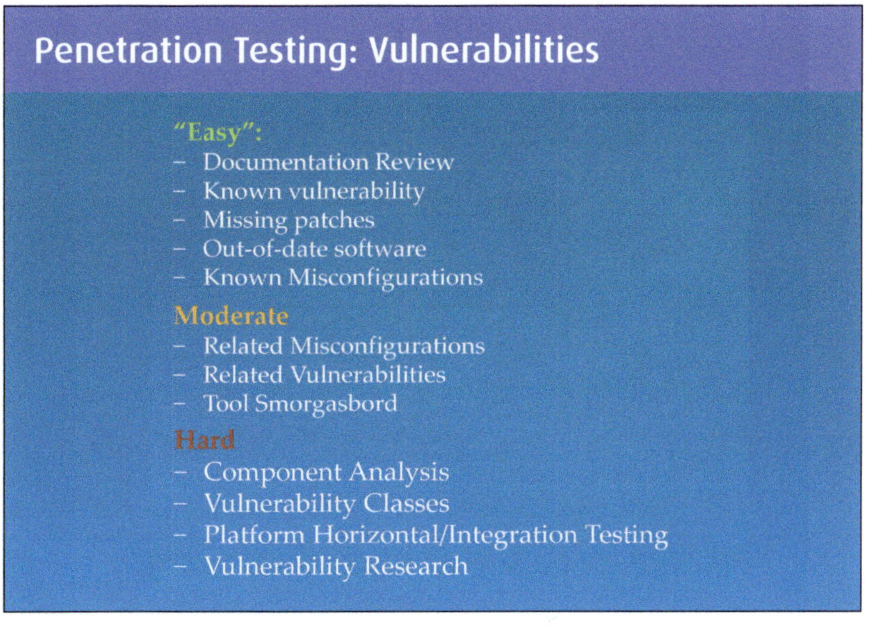

Penetration testing is manual. However, there are guidelines on what a human can do, categorized into "Easy", "Intermediate", and "Hard". Roughly, "easy" means that some instance of the vulnerability itself can be found somewhere, e.g., through documentation review or by noticing a missing patch that was supposed to fix a vulnerability, etc. "Related" means that an analogous vulnerability is known, and one has to project from there and adapt it to identify how to exercise a vulnerability in the current target. "Hard" means that there is no such hint and one essentially has to rely on creativity.

21 Formal Verification

Use of mathematical and symbolic methods to formally prove a desired property of the system

Heavy-weight

Full verification of critical modules, e.g., cryptographic core

Light-weight

Using static analysis methods to exercise different system paths and expose vulnerabilities

Formal verification is another approach for validating security. Here, instead of testing one employs mathematical methods to *prove* that the design satisfies desired properties. When applicable it gives a high assurance. Even when not completely done, some techniques from this method can be borrowed to effectively explore the design space. However, in general, formal methods suffer from scalability issues and cannot be directly applied to large designs without significant abstractions.

SECURITY IN IoT

We now consider some unique challenges for security in IoT, some of which arise from factors that one does not think about typically when thinking about security.

22 Some Unique Features of IoT

- Long, complex life cycle
- Mass produced in same configuration
- Equipment never intended to be connected
- Machine-to-machine
- Normal C-I-A often reversed
- Requires holistic view of device to gateway to cloud and the communications between them
- Many traditional protection mechanisms not applicable due to form factor, deployment, power constraints

22 Some Unique Features of IoT

One critical factor is that these devices may have a very long life, e.g., a car can remain in field for a decade. Security requirements change in that time-frame. Another problem is that we are connecting things that were never intended to be connected, e.g., refrigerators and light bulbs. It's unclear today what security issues can be created by a hacked light-bulb that can communicate with a refrigerator.

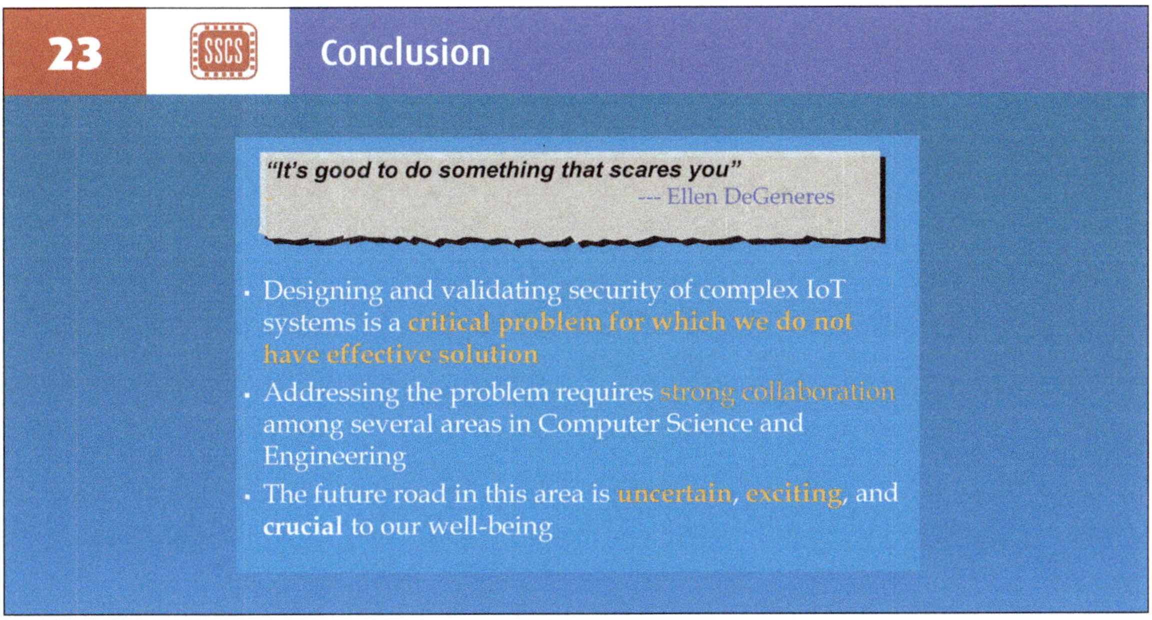

System security is a critical problem with no effective solution in sight. This suggests that it provides a fertile ground for exciting, collaborative research.

About the Editors

Ali Sheikholeslami

Ali Sheikholeslami (S'98-M'99-SM'02) received the B.Sc. degree from Shiraz University, Iran, in 1990 and the M.A.Sc. and Ph.D. degrees from the University of Toronto, Canada, in 1994 and 1999, respectively, all in electrical engineering.

In 1999, he joined the Department of Electrical and Computer Engineering at the University of Toronto where he is currently Professor. He was on research sabbatical with Fujitsu Labs in 2005-2006, and with Analog Devices in 2012-2013. His research interests are in analog and digital integrated circuits, high-speed signaling, VLSI memory design, and CMOS annealing. He has coauthored over 70 journal and conference papers, 10 patents, and a graduate-level textbook entitled "Understanding Jitter and Phase Noise". He was a co-author of the CICC2017 Outstanding Student Paper Award.

Dr. Sheikholeslami served on the Memory, Technology Directions, and Wireline Subcommittees of the International Solid-State Circuits Conference (SSCC) in 2001-2004, 2002-2005, and 2007-2013, respectively. He currently serves as the Education Chair for both ISSCC and the Solid-State Circuits Society (SSCS). He is a Distinguished Lecturer (DL) of the Society and oversees its DL and webinar programs. He is an Associate Editor for the Solid-State Circuits Magazine, in which he has a regular column entitled "Circuit Intuitions". He was an Associate Editor for the IEEE TCAS-I for 2010-2012, and the program chair for the 2004 IEEE ISMVL.

Dr. Sheikholeslami has received numerous teaching awards including the 2005-2006 Early Career Teaching Award and the 2010 Faculty Teaching Award, both from the Faculty of Applied Science and Engineering at the University of Toronto. He is a registered professional engineer in Ontario, Canada.

Jan Van der Spiegel

Dr. Jan Van der Spiegel is a Professor of Electrical and Systems Engineering at the University of Pennsylvania. He is former Department Chair of Electrical Engineering, and Associate Dean of Education and Professional Programs of the School of Engineering. He was a Senior Visiting Professor at the Institute of Microelectronics and the Department of Electronics Engineering at Tsinghua University (Sept. 2017 till Feb. 2018).

Dr. Van der Spiegel received his Master's degree in Electro-Mechanical Engineering and his Ph.D. degree in Electrical Engineering from the University of Leuven, Belgium. His primary research interests are in mixed-mode VLSI design, CMOS vision sensors for polarization imaging, bio-inspired image sensors and brain-machine interfaces. He is the author of over 250 journal and conference papers and holds 8 patents.

He is a Life Fellow of the IEEE, a Distinguished Lecturer of the Solid-State Circuit Society, and is the recipient of the IEEE Major Educational Innovation Award, the IEEE Third Millennium Medal, the UPS Foundation Distinguished Education Chair and the Bicentennial Class of 1940 Term Chair. He received the IBM Young Faculty Development Award and the Presidential Young Investigator Award. He has served on several IEEE program committees (IEDM, ICCD, ISCAS, and ISSCC) and was the technical program chair of the 2007 International Solid-State Circuit Conference (IEEE ISSCC).

He is an Associate Editor of the Transaction of BioCAS, a member of the Editorial Board of the IEEE Proceedings and Section Editor of the Journal of Engineering of Institute of Engineering and Technology (IET). He was the president of the IEEE Solid-State-Circuits Society (2016-2017). He is currently the Conference Chair of the IEEE International Solid-State Circuits Conference (ISSCC).

Yanjie Wang

Yanjie Wang (S'97–M'08–SM'16) received the M.A.Sc. degree from Carleton University, Ottawa, ON, Canada, in 2002, and the Ph.D. degree from the University of Alberta, Edmonton, AB, Canada, in 2009.

He was a ASIC Design Engineer with Nortel Networks, Ottawa, and AMCC, Ottawa, working on high-speed wireless and optical products from 1999 to 2003. In 2007, he was a Ph.D. Graduate Student Researcher at the Berkeley Wireless Research Center, University of California, Berkeley, CA, USA. Since 2008, he has been with Intel Corporation, Hillsboro, OR, USA, where he is currently a Senior Staff Research Scientist focusing on the next-generation 5G mm-wave and energy-efficient broadband RF transceiver circuits and systems.

Dr. Wang was a recipient of the Best Invited Paper Award of the IEEE CICC2015, the Best Paper Award (2nd place) of the IEEE CICC2015, the Best Paper Final List of the IEEE RFIC 2008, the Queen Elizabeth II Doctoral Scholarship from 2006 to 2009, the Doctoral Entree Award in 2005, and the Faculty of Graduate Study Research Abroad Award in 2007, University of Alberta. He has been served as a Technical Program Committee Member of the IEEE Radio Frequency Integrated Circuits Symposium and Organization Committee Member of the IEEE Custom Integrated Circuits Conference since 2012.

About the Authors

Aaron Buchwald

Aaron Buchwald is Founder of Beechwood Analog Inc. and Adjunct Professor at UC Irvine and sometimes Adjunct Professor at the Hong Kong University of Science and Technology. He received his Ph.D. in electrical engineering from UCLA in 1993. He joined Broadcom in 1994 as the first member of the Analog Group (later Central Engineering). In 2003 he founded Mobius Semiconductor to develop time-interleaved calibrated ADC, where he served as CEO.

His research interests are in data converters and mixed signal circuits for communication. He currently serves as an Associate Editor for *IEEE Journal of Solid State Circuits* and previously served on the Data Converters Subcommittee for the *International Solid-State Circuits Conference (ISSCC)*.

[Chapter 11]

Tolga Dinc

Tolga Dinc received the B.S. and M.S. degrees in electrical engineering from Sabanci University, Istanbul, Turkey, in 2010 and 2012, respectively. In 2018, he received the Ph.D. degree in electrical engineering from Columbia University, New York, NY.

He is currently with Kilby Labs of Texas Instruments, Dallas, as an RF/mm-wave engineer, working on high-speed integrated circuit and transceiver design. Previously, he held internship positions with Texas Instruments, Dallas, and IBM Research, Yorktown Heights, in 2015 and 2016, respectively.

Tolga is the recipient of several honors and awards, including the Columbia EE Jury Award in 2017, the IEEE MTT-S Graduate Fellowship in 2016, the IEEE RFIC Symposium Best Student Paper Award (1st Place) in 2015, the 2012 Sabanci University Gursel Sonmez Research Award and the IEEE MTT-S Undergraduate/Pre-Graduate Scholarship Award in 2010. He has authored/co-authored more than 30 peer reviewed papers in the area of RF/mm-Wave integrated circuits, antennas and systems.

[Chapter 2]

Amr Fahim

Amr Fahim received his BASc, MASc and Ph.D. degrees from the University of Waterloo in Canada in Electrical Engineering in 1996, 1997 and 2000 respectively. His Ph.D. thesis title was *Agile frequency synthesizers for wireless applications*. He has over two decades of experience in the design and modeling of RF and mixed-signal integrated circuits for system-on-chip (SoCs) for both wireless and wireline applications. He authored and co-authored several papers and patents. He is the author of two textbooks titled *Radio Frequency Integrated Circuit Design for Cognitive Radio Systems* and *Clock Generators for SoC Processors*. He has also given a number of talks in advanced wireless communication systems and analog circuit design. He has served as a TPC member for several IEEE conferences. His current research interests include millimeter wave frequency synthesis and wireless transceiver design.

[Chapter 4]

Fa Foster Dai

Fa Foster Dai received Ph.D. degree in electrical engineering from The Pennsylvania State University in 1998. From 1997 to 2002, he was with Hughes Network Systems, YAFO Networks, and Cognio Inc. In August 2002, he joined Auburn University, USA, where he is currently an Ed and Peggy Reynolds Family Endowed Professor in electrical and computer engineering. His research interests include analog and mixed-signal circuit designs, RFIC and MMIC designs, and high performance frequency synthesis. He co-authored six books and book chapters such as *Integrated Circuit Design for High-Speed Frequency Synthesis* (*Artech House Publishers*, Feb., 2006).

Dr. Dai served as Guest Editor for *IEEE Journal on Solid State Circuits* in 2012 and 2013 and Guest Editor for *IEEE Transactions on Industrial Electronics* in 2001, 2009 and 2010. He served on the technical program committees (TPC) of the *IEEE Symposium on VLSI Circuits* from 2005 to 2008. He currently serves on the executive committee of *IEEE Bipolar / BiCMOS Circuits and Technology Meeting (BCTM)* and the steering committee of *IEEE Custom Integrated Circuits Conference (CICC)*. He was the TPC Chair of 2016 BCTM conference and the General Chair of 2017 BCTM conference. He serves as the TPC Chair 2019 CICC conference. He was elected *IEEE Fellow* in 2009 "for contributions to high-speed frequency synthesis and radio frequency integrated circuits".

[Chapter 5]

Ramesh Harjani

Ramesh Harjani received the B.S. degree in electrical engineering from the Birla Institute of Technology and Science, Pilani, India, in 1982, the M.S. degree in electrical engineering from the Indian Institute of Technology, New Delhi, India, in 1984, and the Ph.D. degree in electrical engineering from Carnegie Mellon University, Pittsburgh, PA, USA, in 1989. He is currently the Edgar F. Johnson Professor with the Department of Electrical and Computer Engineering, University of Minnesota, Minneapolis, MN, USA. Prior to joining the University of Minnesota, he was with the Mentor Graphics Corporation, San Jose, CA, USA. In 2001, he cofounded Bermai Inc., a startup company that develops CMOS chips for wireless multimedia applications. He has been a Visiting Professor with Lucent Bell Labs, Allentown, PA, USA, and the Army Research Laboratories, Adelphi, MD, USA. His research interests include analog/RF circuits for communications.

[Chapter 12]

Hossein Hashemi

Hossein Hashemi is a Professor of Electrical Engineering, Ming Hsieh Faculty Fellow, and the co-director of the Ming Hsieh Institute at the University of Southern California. His research interests include electronic and photonic integrated circuits. He received the B.S. and M.S. degrees in Electronics Engineering from the Sharif University of Technology, Tehran, Iran, in 1997 and 1999, respectively, and the M.S. and Ph.D. degrees in Electrical Engineering from the California Institute of Technology, Pasadena, in 2001 and 2003, respectively. Hossein is an Associate Editor for the IEEE Journal of Solid state Circuits (2013 – present). He was a Distinguished Lecturer for the IEEE Solid-State Circuits Society (2013 – 2014); member of the Technical Program Committee of IEEE International Solid-State Circuits Conference (ISSCC) (2011 – 2015), IEEE Radio Frequency Integrated Circuits (RFIC) Symposium (2011 – present), and the IEEE Compound Semiconductor Integrated Circuits Symposium (CSICS) (2010 – 2014); an Associate Editor for the IEEE Transactions on Circuits and Systems—Part I: Regular Papers (2006–2007) and the IEEE Transactions on Circuits and Systems—Part II: Express Briefs (2004–2005); and Guest Editor for the IEEE Journal of Solid state Circuits (Oct 2013 & Dec 2013).

Hossein was the recipient of the 2016 Nokia Bell Labs Prize, 2015 IEEE Microwave Theory and Techniques Society (MTT-S) Outstanding Young Engineer Award, 2008 Defense Advanced Research Projects Agency (DARPA) Young Faculty Award, and a National Science Foundation (NSF) CAREER Award. He received the USC Viterbi School of Engineering Junior Faculty Research Award in 2008, and was recognized as a Distinguished Scholar for the Outstanding Achievement in Advancement of Engineering by the Association of Professors and Scholars of Iranian Heritage in 2011. He was a co-recipient of the 2004 IEEE Journal of Solid-State Circuits Best Paper Award for "A Fully-Integrated 24 GHz 8-Element Phased-Array Receiver in Silicon" and the 2007 IEEE International Solid-State Circuits Conference (ISSCC) Lewis Winner Award for Outstanding Paper for "A Fully Integrated 24 GHz 4-Channel Phased-Array Transceiver in 0.13um CMOS based on a Variable Phase Ring Oscillator and PLL Architecture".

Hossein is the co-editor of the books "Millimeter-Wave Silicon Technology: 60 GHz and Beyond" published by Springer in 2008, and "mm-Wave Silicon Power Amplifiers and Transmitters" published by the Cambridge University Press in 2016.

[Chapter 1]

Eric A. M. Klumperink

Eric A. M. Klumperink (IEEE Member '98, Senior Member '06) was born on April 4th, 1960, in Lichtenvoorde, The Netherlands. He received the B.Sc. degree from HTS, Enschede (1982), worked in industry on digital hardware and software, and then joined the University of Twente, Enschede, in 1984, shifting focus to analog CMOS circuit research. This resulted in several publications and his Ph.D. thesis "Transconductance Based CMOS Circuits: Circuit Generation, Classification and Analysis" (1997). In 1998, Eric started as Assistant Professor at the IC-Design Laboratory in Twente and shifted research focus to RF CMOS circuits (e.g. sabbatical at the Ruhr Universitaet in Bochum, Germany). Since 2006, he is an Associate Professor, teaching Analog & RF IC Electronics and guiding PhD and MSc projects related to RF CMOS circuit design with focus on Software Defined Radio, Cognitive Radio and Beamforming. He served as an Associate Editor for the IEEE TCAS-II (2006-2007), IEEE TCAS-I (2008-2009) and the IEEE JSSC (2010-2014), as IEEE SSC Distinguished Lecturer (2014/2015), and as member of the technical program committees of ISSCC (2011-2016) and the IEEE RFIC Symposium (2011-..). He holds several patents, authored and co-authored 175+ internationally refereed journal and conference papers, and was recognized as 20+ ISSCC paper contributor over 1954-2013. He is a co-recipient of the ISSCC 2002 and the ISSCC 2009 "Van Vessem Outstanding Paper Award".

[Chapter 6]

Ram K. Krishnamurthy

Ram K. Krishnamurthy (M'98-SM'03-F'11) is Senior Research Director and Senior Principal Engineer at Circuits Research Lab, Intel Labs, Hillsboro, Oregon. He heads the high performance and low voltage circuits research group. In this role, he leads research in high performance, energy efficient, and low voltage circuits for microprocessors and SoCs, and has made contributions to the circuit design of various generations of Intel products, including Intel® Itanium®, Pentium4®, Core®, Atom® and Xeon® line of microprocessors and SoCs. He has been at Intel Corporation since 1997. Krishnamurthy has filed over 175 patents (121 issued), and has published 200 papers and 3 book chapters on high performance energy efficient circuits. He serves as chair of the Semiconductor Research Corporation (SRC) technical advisory board for circuits, has been a guest editor of IEEE Journal of Solid-State Circuits, associate editor of IEEE transactions on VLSI systems, and on the technical program committees of ISSCC, CICC, and SOCC conferences. He served as Technical Program Chair/ General Chair for the IEEE International Systems-on-Chip Conference and presently serves on the conference's steering committee. Krishnamurthy serves as an adjunct faculty of the Electrical and Computer Engineering department at Oregon State University, where he taught advanced VLSI design. He is a board member of the industry advisory board for State University of New York. Krishnamurthy has received the IEEE International Solid State Circuits Conference distinguished technical paper award, IEEE European Solid State Circuits Conference best paper award, outstanding industry mentor award from SRC, Intel awards for most patents filed and most patents issued, Intel Labs Gordon Moore award, Alumni recognition award from Carnegie Mellon University, distinguished alumni award from State University of New York, MIT Technology Review's TR35 innovator award, and recognized as a top ISSCC paper contributor. He has received two Intel Achievement Awards for pioneering the first 64b sparse-tree ALU technology and the first Advanced Encryption Standard accelerator on Intel products. He is a Fellow of the IEEE and distinguished lecturer of IEEE solid-state circuits society. Krishnamurthy received his BE in electrical engineering from National Institute of Technology in India (1993), MS in electrical and computer engineering from State University of New York (1994), and PhD in electrical and computer engineering from Carnegie Mellon University (1997).

[Chapter 14]

Harish Krishnaswamy

Harish Krishnaswamy (S'03–M'09) received the B.Tech. degree in electrical engineering from IIT Madras, Chennai, India, in 2001, and the M.S. and Ph.D. degrees in electrical engineering from the University of Southern California (USC), Los Angeles, CA, USA, in 2003 and 2009, respectively. In 2009, he joined the Electrical Engineering Department, Columbia University, New York, NY, USA, where he is currently an Associate Professor and the Director of the Columbia High-Speed and Millimeter-Wave IC Laboratory (CoSMIC).

In 2017, he co-founded MixComm Inc., a venture-backed startup, to commercialize CoSMIC Laboratory's advanced wireless research. His current research interests include integrated devices, circuits, and systems for a variety of RF, mmWave, and sub-mmWave applications.

Dr. Krishnaswamy was a recipient of the IEEE ISSCC Lewis Winner Award for Outstanding Paper in 2007, the Best Thesis in Experimental Research Award from the USC Viterbi School of Engineering in 2009, the DARPA Young Faculty Award in 2011, a 2014 IBM Faculty Award, the Best Demo Award at the 2017 IEEE ISSCC, Best Student Paper Awards (First Place) at the 2015 and 2018 IEEE Radio Frequency Integrated Circuits Symposia, and the 2019 IEEE MTT-S Outstanding Young Engineer Award . He has been a member of the technical program committee of several conferences, including the IEEE International Solid-State Circuits Conference since 2015 and the IEEE Radio Frequency Integrated Circuits Symposium since 2013. He currently serves as a Distinguished Lecturer for the IEEE Solid-State Circuits Society and as a member of the DARPA Microelectronics Exploratory Council.

[Chapter 2]

Lukas Kull

Lukas Kull received the M.Sc. degree in electrical engineering from the Swiss Federal Institute of Technology, Zurich (ETH), Switzerland, in 2007 and the Ph.D. degree from the Swiss Federal Institute of Technology, Lausanne (EPFL), Switzerland, in 2014. He was with IBM Research - Zurich, Rüschlikon, Switzerland, from 2010 to 2018, where he was involved in analog circuit design for high-speed low-power ADCs. In 2018 he joined Cisco Systems. His research interests include analog circuit design, hardware for cognitive workloads, IR and THz imaging. In these areas he authored or co-authored more than 20 patents and 50 technical publications.

[Chapter 9]

Hanh-Phuc Le

Dr. Hanh-Phuc Le is an Assistant Professor of ECEE at the University of Colorado Boulder. He received the B.S. from Hanoi University of Science and Technology in Vietnam (2004), M.S. from KAIST, Korea (2006), and Ph.D. from UC Berkeley (2013), all in Electrical Engineering. In 2012, he co-founded, helped secure $14M in funding, and served as the CTO at Lion Semiconductor until October 2015. He also held R&D positions at Oracle, Intel, Rambus, JDA Tech in Korea and the Vietnam Academy of Science and Technology. His current research interests include miniaturized/on-die power conversions, large conversion ratios, smart power delivery for high performance IT systems, communication, mobile, wearable, and IoT applications.

Dr. Le received the 2012-2013 IEEE Solid-State Circuits Society Pre-doctoral Achievement Award, and the 2013 Sevin Rosen Funds Award for Innovation at UC Berkeley. He authored one book chapter, over forty journal and conference papers, and is an inventor with 13 U.S. patents (7 granted and 6 pending). He is an associate editor of the IEEE Journal of Emerging and Selected Topics in Power Electronics, and the Technical Program Committee (TPC) chair for the 2018 International Workshop on Power Supply On Chip (PwrSoC 2018).

[Chapter 13]

John R. Long

John R. Long received the B.Sc. in Electrical Engineering from the University of Calgary in 1984, and the M. Eng. and Ph.D. degrees in Electronics from Carleton University, Ottawa, Canada, in 1992 and 1996, respectively. He worked for 12 years in the Advanced Technology Laboratory at Bell-Northern Research in Ottawa, and then began his academic career at the University of Toronto (1996-2002). From 2002-2014 he was Chair of the Electronics Research Laboratory at TU Delft in the Netherlands. He is now a faculty member in the Department of Electrical and Computer Engineering at the University of Waterloo. His current research interests include low-power and broadband circuits for highly-integrated wireless transceivers, energy-efficient wireless sensors, and IC design for mm-wave and high-speed data communication.

[Chapter 3]

Danny Luu

Danny Luu received the B.Sc. and M.Sc. degrees in electrical engineering and information technology from the Swiss Federal Institute of Technology (ETH) Zurich, Switzerland, in 2013. He joined IBM Research – Zurich, Rueschlikon, Switzerland, in 2013, where he has been conducting research into analog circuit design for high-speed, high-resolution, low-power ADCs in collaboration with ETH Zurich towards his doctoral degree. He received his PhD degree from ETH Zurich in 2018 and is now with Cisco Systems, Switzerland.

[Chapter 9]

Sanu K. Mathew

Sanu K. Mathew (M'99–SM'15-F'18) received the B.Tech. degree in electronics and communications engineering from the College of Engineering, Trivandrum, India, in 1993, and the M.S. and Ph.D. degrees in electrical and computer engineering from The State University of New York at Buffalo, Buffalo, NY, USA, in 1996 and 1999, respectively. He joined Intel Corporation in 1999. He is currently a Senior Principal Engineer with Circuits Research Laboratories at Intel Hillsboro, OR, USA, where he leads the security arithmetic circuits research team, which is responsible for developing energy-efficient security circuit primitives for next-generation microprocessors. He also mentors Intel and SRC-funded research projects in leading universities. He holds 41 issued patents, has 63 patents pending, and has published over 77 conference/journal papers. His research interests are in the areas of high-speed/low-power computer arithmetic datapath circuits and special-purpose hardware accelerators for cryptography and security. Dr. Mathew was a recipient of the ISSCC Distinguished Technical Paper Award in 2012. He has served on the program committees of the ARITH, ISLPED, DAC, and SOCC conferences. He is a Fellow of IEEE.

[Chapter 14]

Bram Nauta

Bram Nauta was born in 1964 in Hengelo, The Netherlands. In 1987 he received the M.Sc degree (cum laude) in electrical engineering from the University of Twente, Enschede, The Netherlands. In 1991 he received the Ph.D. degree from the same university on the subject of analog CMOS filters for very high frequencies. In 1991 he joined the Mixed-Signal Circuits and Systems Department of Philips Research, Eindhoven the Netherlands. In 1998 he returned to the University of Twente, where he is currently a distinguished professor, heading the IC Design group. Since 2016 he also serves as chair of the EE department at this university. His current research interest is high-speed analog CMOS circuits, software defined radio, cognitive radio and beamforming.

He served as the Editor-in-Chief (2007-2010) of the IEEE Journal of Solid-State Circuits (JSSC), and was the 2013 program chair of the International Solid State Circuits Conference (ISSCC). He is currently the President of the IEEE Solid-State Circuits Society (2018-2019 term).

Also, he served as Associate Editor of IEEE Transactions on Circuits and Systems II (1997-1999), and of JSSC (2001-2006). He was in the Technical Program Committee of the Symposium on VLSI circuits (2009-2013) and is in the steering committee and programme committee of the European Solid State Circuit Conference (ESSCIRC). He served as distinguished lecturer of the IEEE, is co-recipient of the ISSCC 2002 and 2009 "Van Vessem Outstanding Paper Award" and in 2014 he received the 'Simon Stevin Meester' award (500.000€), the largest Dutch national prize for achievements in technical sciences. He is fellow of the IEEE and member of the Royal Netherlands Academy of Arts and Sciences (KNAW).

[Chapter 6]

Khiem Nguyen

Khiem Nguyen was born in Saigon, Vietnam, in 1970. He received his Bachelor and Master of Applied Sciences degrees in Electrical Engineering from the University of Toronto, Canada, in 1993 and 1995.

Since 1995, he has been with Analog Devices where he is currently the Design Engineering Manager of the High Performance Inertial product group. Khiem is a co-recipient of the 1998 ISSCC Outstanding Paper Award, and an author of several ISSCC and invited JSSC papers. Khiem is a senior member of the IEEE and had served in the Technical Program Committee for CICC in 2016 and 2017.

Khiem holds 25 issued patents in the area of mixed-signal and data converters. His technical interests include analog circuit design, data converters, fractional-N PLL-based clock synthesis, DSPs for audio, ASICs for MEMs accelerometer and gyro.

[Chapter 10]

Hui Pan

Hui Pan received the Ph.D. degree from UCLA in 1999 and joined Broadcom Corp upon graduation, where he developed four generations of multi-gigabit physical layer (PHY) products for enterprise, consumer, and automotive applications. Dr. Pan is currently a Technical Director and Distinguished Engineer with Broadcom Inc. and an Associate Editor of JSSC, developing high performance ADCs for 5G base station and satellite internet applications. He was the recipient of the 1998 Analog Devices Outstanding Student Designer Award and the Honorable Mention of DAC 2000 Student Design Contest. He served on the ISSCC TPC from 2006 to 2010 and the CSICS TPC from 2014 to 2017.

[Chapter 8]

Sandip Ray

Sandip Ray is an Endowed IoT Term Professor at the Department of Electrical and Computer Engineering, University of Florida. His research involves developing correct, dependable, secure, and trustworthy computing through cooperation of specification, synthesis, architecture and validation technologies. His research targets next-generation computing applications, including autonomous automotive systems, smart homes, intelligent implants, etc. Before joining University of Florida, Dr. Ray was a Senior Principal Engineer at NXP Semiconductors, where he led the R&D on security validation for automotive and Internet-of-Things applications. Prior to that, he was a Research Scientist at Intel Strategic CAD Labs, where he worked on pre-silicon and post-silicon validation of security and functional correctness of SoC designs, design-for-security and design-for-debug architectures, CAD tools, and specifications for SoC design requirements. Prior to joining industry, Dr. Ray was a Research Scientist at University of Texas at Austin, where he led several sponsored projects from DARPA, SRC, and National Science Foundation. Dr. Ray is the author of three books (one upcoming) and over 60 publications in peer-reviewed premier international journals and conferences. He served as guest editors for an IEEE Transactions on Multi-Scale Systems (TMSCS) special issue on Wearables, Implants, and Internet-of-Things, as well as special issues of ACM Transactions on Design Automation of Electronic Systems (TODAES) and Springer Journal on Electronic Testing Theory and Applications (JETTA). He has given number of invited, tutorial, and keynote presentations at several international forums on security, validation, and energy challenges in the IoT regime. During his tenure in industry, Dr. Ray served as Intel and NXP representative in Semiconductor Research Consortium (SRC) technical advisory board, and as semiconductor industry representative on trustworthy systems to the Semiconductor Industry Association (SIA). He has served as a program committee member for more than 50 international meetings and conferences, and as program chair for Formal Methods in Computer-Aided Design (FMCAD). He currently serves as an Associate Editor for IEEE TMSCS and Springer Journal on Hardware and Systems Security. He has a Ph.D. from University of Texas at Austin and is a Senior Member of IEEE.

[Chapter 15]

Sudhir Satpathy

Sudhir Satpathy received his Bachelor's degree from Indian Institute of Technology (IIT), Kanpur and doctorate degree from University of Michigan, Ann Arbor in electrical engineering. Since 2012, he has been a research scientist at Intel's Circuits Research Laboratory working on high performance energy-efficient hardware accelerator development for content processing in Intel's next generation microprocessors. His areas of expertise include symmetric and public key cryptography, secure key generation and storage, and data compression circuits. He has filed 65 patents and published 45 articles in these areas.

[Chapter 14]

Vikram Suresh

Vikram Suresh (M'10) received Bachelor's degree in engineering from Visvesvaraya Technological University, India, in 2007; M.S and Ph.D degrees from the University of Massachusetts, Amherst, MA, USA, in 2012, and 2014, respectively. He is currently a Staff Research Scientist at Intel Labs, Hillsboro, OR, USA. He has authored 32 publications at peer-reviewed conference and journals. He holds 10 issued patents with 29 pending applications. His research interests include variation-aware circuit design, impact of noise in nanometer CMOS, and hardware security.

[Chapter 14]

Hua Wang

Hua Wang received his M.S. and Ph.D. degrees in electrical engineering from the California Institute of Technology, Pasadena, in 2007 and 2009, respectively. He worked at Intel Corporation and Skyworks Solutions. He joined the School of Electrical and Computer Engineering (ECE) at Georgia Institute of Technology in 2012 and is now the Demetrius T. Paris associate professor.

Dr. Wang is interested in innovating mixed-signal, RF, and mm-Wave integrated circuits and hybrid systems for wireless communication, radar, imaging, and bioelectronics applications.

Dr. Wang received the DARPA Young Faculty Award in 2018, the National Science Foundation CAREER Award in 2015, the IEEE MTT-S Outstanding Young Engineer Award in 2017, the Georgia Tech Sigma Xi Young Faculty Award in 2016, the Georgia Tech ECE Outstanding Junior Faculty Member Award in 2015, and the Lockheed Dean's Excellence in Teaching Award in 2015. His research group Georgia Tech Electronics and Micro-System (GEMS) lab has won multiple best paper awards, including the IEEE RFIC Best Student Paper Awards (1st Place in 2014, 2nd Place in 2016, and 2nd Place in 2018), the IEEE CICC Outstanding Student Paper Awards (2nd Place in 2015 and 2nd Place in 2018), the IEEE CICC Best Conference Paper Award (2017), the 2016 IEEE Microwave Magazine Best Paper Award, the IEEE SENSORS Best Live Demo Award (2nd Place in 2016), as well as multiple best paper award finalists at IEEE conferences.

Dr. Wang is an Associate Editor of the IEEE Microwave and Wireless Components Letters (MWCL) and a guest Editor of the IEEE Journal of Solid-State Circuits (JSSC). He is a Technical Program Committee (TPC) Member for IEEE ISSCC, RFIC, CICC, and BCICTS conferences. He is a Steering Committee Member for IEEE RFIC and CICC. He serves as the Chair of the Atlanta's IEEE CAS/SSCS joint chapter that won the IEEE SSCS Outstanding Chapter Award in 2014.

[Chapter 7]

For Product Safety Concerns and Information please contact our EU
representative GPSR@taylorandfrancis.com
Taylor & Francis Verlag GmbH, Kaufingerstraße 24, 80331 München, Germany

www.ingramcontent.com/pod-product-compliance
Ingram Content Group UK Ltd.
Pitfield, Milton Keynes, MK11 3LW, UK
UKHW062047190926
475702UK00009B/1222

* 9 7 8 8 7 7 0 2 2 0 4 9 1 *